Aquatic Chemistry Concepts

Aquatic Chemistry Concepts

Second Edition

James F. Pankow

CRC Press
Taylor & Francis Group
Boca Raton London New York

CRC Press is an imprint of the
Taylor & Francis Group, an **informa** business

CRC Press
Taylor & Francis Group
6000 Broken Sound Parkway NW, Suite 300
Boca Raton, FL 33487-2742

First issued in paperback 2022

ISBN-13: 978-1-439-85440-2 (hbk)
ISBN-13: 978-1-03-233773-9 (pbk)
DOI: 10.1201/9780429198861

Publisher's Note

The publisher has gone to great lengths to ensure the quality of this reprint but points out that some imperfections in the original copies may be apparent.

Visit the Taylor & Francis Web site at
http://www.taylorandfrancis.com

and the CRC Press Web site at
http://www.crcpress.com

This book is dedicated to the three professors in my academic lineage:

Gilbert E. Janauer (dec.), James J. Morgan, and Werner Stumm (dec.), in appreciation for their contributions to their fields of science and engineering, and to me. And to Joachim Rønneberg (dec.), and his compatriots, who showed staggering bravery, cunning, and perseverance in their belief in a better world. Would that most will take up the current struggle against the many that in cancerous greed, apathy, and with a world population net growth rate of more than 200,000 per day chose to care little for our planetary home and all of its varied and astonishingly beautiful inhabitants.

James F. Pankow
July 2019
Portland, Oregon

Contents

PART II Acid/Base Chemistry

PART III Metal/Ligand Chemistry

PART IV Mineral Solubility

PART V Redox Chemistry

PART VI Effects of Electrical Charges on Solution Chemistry

Foreword

There's a lot to like about a book on water chemistry that lays it out simply. Einstein said that everything should be as simple as it can be, but not simpler. Wise advice. And that is what James F. Pankow has accomplished in the second edition of his textbook, *Aquatic Chemistry Concepts*. It covers the "waterfront" of essential inorganic chemistry topics, and it supplies enough examples to lead the student toward problem solving.

Pankow appropriately begins in the Introduction (Part I) with definitions and the theoretical basis for solving chemical equilibrium problems—thermodynamics. Part II of the book on Acid/Base Chemistry is a *tour de force*, from mass balances to the chemistry of dissolved carbon dioxide. Pankow's Aunt Fatima injects humor along the way, and flashbacks from his college chemistry instructors are especially pertinent. Acid/base examples from household vinegar to ammonia illuminate the path, while "Geek Optional" boxes lend additional insight. I especially enjoyed Chapter 9, which includes problems on acid rain, acidification of lakes, and "the most worrisome" example of all, ocean acidification.

When Professor Pankow published the first edition of *Aquatic Chemistry Concepts* in 1991, the global atmospheric concentration of carbon dioxide, a weak acid, stood at 354 parts per million (ppm). Today it is 415 ppm, a 17% increase in concentration in less than 30 years. Likewise, the pH of the surface of the ocean off Hawaii has declined from 8.12 to 8.05 – meaning a similar 17% increase in H^+ concentration (see Example 9.13). Imagine … we have increased $[H^+]$ in the ocean, the base of the food chain for all aquatic life, by 17% in one generation. Aquatic chemistry concepts don't lie.

We have known since 1862 and John Tyndall's famous experiments that CO_2 absorbs back radiation and heats the atmosphere. Our real-life planetary experiment verifies the physics every year, and the average global temperature is now 0.5 °C (0.9 °F) warmer than when the first edition was published. If we fail to rein in our fossil fuel emissions of CO_2 and other greenhouse gases, the signal will continue to grow ever hotter in coming decades. This textbook teaches an urgent lesson.

Today, all the "master variables" that control aquatic chemistry are changing—pH, temperature, salinity, and oxidation-reduction potential. Part III explains Metal/Ligand Chemistry, and Part IV discusses the topic of Mineral Solubility. Changing ocean salinity, temperature, and pH render all these reactions in a state of planetary flux. Even the quality of our drinking water is subject to change as the master variables change. Fortunately, foundational concepts in Pankow's book allow us to make the relevant calculations.

Part V on Redox Chemistry and Part VI on Effects of Electrical Charges on Solution Chemistry further elucidate systems that are strong functions of the changing state variables of pe and pH. Although some of the most difficult chemical concepts are contained in these final chapters, Pankow lays them bare—as simple as possible, but not simpler.

In the Preface to *Aquatic Chemical Concepts,* 1st Edition, Pankow warned, "The scope of local, regional, and global environmental problems seems to grow with each passing day. We are in a race which we do not wish to lose." That was certainly true in 1991, and it is even more cogent today. Our list of environmental nightmares is enduring and growing—climate change, toxic chemicals, eutrophication, harmful algal blooms, and unsafe drinking water. This book, the 2nd Edition, is an advance in stating the problems simply so we can analyze them quantitatively. Only then can we effect change.

Jerald L. Schnoor
July 2019
Iowa City, Iowa

Preface to the First Edition

In the last decade, the field of natural water chemistry has matured from an infancy born of L. G. Sillen to an acknowledged field of science as nurtured by W. Stumm, J. J. Morgan, J. D. Hem, R. Garrels, C. L. Christ, J. N. Butler, F. M. M. Morel, R. M. Pytkowicz, and many others. As a result, there now are several textbooks available which are very useful to the natural water chemist. Why then has this author spent so many hours creating another text? One answer to this question lies in the fact that I perceive that the field of aquatic chemistry needs a book that is very didactic in nature. That is, a book which spends considerable time going into the details behind some of the complicated equations and principles of aquatic chemistry. Thus, it is my hope that interested scientists and engineers will be able to use this text in a self-instructional manner. Of course, I also hope that the text sees some classroom use.

Anyone who is familiar with Stumm and Morgan's (1981) text Aquatic Chemistry will note certain similarities between that book and this one. The reason for this is two-fold. Firstly, given the natural order present in Aquatic Chemistry, it would be difficult to find a more logical sequence with which to present the various important principles. Secondly, so that these two books can reinforce one another, the specific chemical examples of some of those principles together with their accompanying figures have been intentionally structured in a similar manner.

One is finally brought to the question as to why an environmental researcher and teacher such as myself should put so much energy into collecting and structuring "old" science into a text. After all, conducting research and publishing "new" science sometimes seems to be so much more exciting. (Also, the simultaneous jobs of research and teaching seem to leave little time for anything else.) For me, the answer to this question is clear. Indeed, if one is lucky, perhaps ten or so articles written during a lifetime of research will have some significant impact; most of the articles written will be much less important. Thus, placing established science into a text that helps others to learn, to solve important practical environmental problems, as well as to do research just may provide a contribution that far outweighs the importance of one's own original research. The scope of local, regional, and global environmental problems seems to grow with each passing day. We are thus in a race which we do not wish to lose. My hope is that this text will help to move us all along in that race by at least a few steps.

James F. Pankow
Portland, Oregon
May 1991

Author

James F. Pankow earned a BA in chemistry at the State University of New York at Binghamton in 1973, training in the laboratory of Dr. Gilbert E. Janauer. He earned a PhD in Environmental Engineering Science at the California Institute of Technology in 1979, training in the laboratory of Dr. James J. Morgan (1966–1974, Editor-in-Chief, *Environmental Science and Technology*; 1999, Clarke Water Prize; 1999, Stockholm Water Prize). Dr. Pankow's awards include the John Wesley Powell (U.S. Geological Survey) National Citizen Achievement Award (1993), the American Chemical Society Award for Creative Advances in Environmental Science and Technology (1999), and the Haagen-Smit Prize (2005). He was elected to the National Academy of Engineering in 2009.

Part I

Introduction

1 Overview

1.1 GENERAL IMPORTANCE OF AQUATIC CHEMISTRY

A knowledge of water chemistry is critically important in understanding many processes in a wide variety of aquatic systems. Some important system types are summarized in Table 1.1. In the current age, no system is more important than the oceans, where the matter of falling pH values due to "ocean acidification" caused by rising atmospheric CO_2 levels is a serious issue for the many organisms that form shells and exoskeletons of solid calcium carbonate.

1.2 IMPORTANT TYPES OF CHEMICAL REACTIONS IN NATURAL WATERS

Most reactions of interest in aquatic systems can be classified into one of the following six categories. Example reactions are given for each. For all of the example reactions, the phase in which the chemical is present is indicated only when it is *not* water.

a. Acid/Base Reactions

Example 1. Dissociation of dissolved carbonic acid:

$$H_2CO_3 = H^+ + HCO_3^- \tag{1.1}$$

This reaction is important in virtually all natural waters, carbonated "mineral" waters, carbonated "soft" drinks, beer, and champagne.

Example 2. Protonation of ammonia by water in household ammonia:

$$NH_3 + H_2O = NH_4^+ + OH^- \tag{1.2}$$

This reaction is important in solutions of ammonia, such as may be purchased in stores selling cleaning supplies.

b. Complexation Reactions

Example 1. Combination of Fe^{3+} with OH^- from H_2O, as for Fe^{3+} found in acidic mine waters:

$$Fe^{3+} + H_2O = FeOH^{2+} + H^+ \tag{1.3}$$

Example 2. Chelation of Ca^{2+} with $EDTA^{4-}$:

$$Ca^{2+} + EDTA^{4-} = CaEDTA^{2-} \tag{1.4}$$

c. Dissolution/Precipitation Reactions

Example 1. Dissolution of calcite:

$$CaCO_{3(s)} = CO_3^{2-} + Ca^{2+} \tag{1.5}$$

TABLE 1.1
Important Environmental Systems in Which Water Chemistry Is of Interest

Surface Waters	Sub-Surface Waters	Atmospheric Waters
streams and rivers	sediments (in all surface water systems)	aerosol particles and droplets
ponds and lakes	"vadose zone" (unsaturated subsurface) water	cloud waters
estuaries	ground water	rain
oceans	deep geological water	
drinking water treatment systems		
wastewater treatment systems		

Example 2. Precipitation of the mineral struvite, as with kidney stones in dogs, cats, and humans, and in many domestic wastewater treatment systems (due to an abundance of ammonia and phosphate):

$$NH_4^+ + Mg^{2+} + PO_4^{3-} + 6H_2O = NH_4MgPO_4 \cdot (H_2O)_{6(s)} \tag{1.6}$$

d. Redox Reactions

Example 1. Role of oxygen as an oxidant (electron acceptor) (as in the oxidation of biologically related organic matter):

$$O_2 + 4e^- + 4H^+ = 2H_2O \tag{1.7}$$

Example 2. Oxidation of Fe^{2+} to form Fe^{3+} as when groundwater containing Fe^{2+} is exposed to oxygen:

$$Fe^{2+} = Fe^{3+} + e^- \tag{1.8}$$

e. Gas/Water Partitioning Reactions

Example 1. Dissolution of gaseous CO_2 into water with simultaneous hydration to form carbonic acid:

$$CO_{2(g)} + H_2O = H_2CO_3 \tag{1.9}$$

Example 2. Volatilization of H_2S (highly toxic to humans) from water to air:

$$H_2S = H_2S_{(g)} \tag{1.10}$$

f. Sorption Reactions

Example 1. Adsorption of the lead ion Pb^{2+} onto a mineral $SiO_{2(s)}$ surface:

$$SiO_{2(s)} + Pb^{2+} = SiO_{2(s)}{-}Pb^{2+} \tag{1.11}$$

Example 2. Sorption of cadmium ion Cd^{2+} onto organic material (OM) that may be present in dissolved form, or as part of particles that may be suspended in the water column or settled as sediments:

$$OM + Cd^{2+} = OM{-}Cd^{2+} \tag{1.12}$$

OM can be charged (usually negatively), and the nature of the interaction can be that of a complexation reaction.

1.3 CONCENTRATION SCALES

1.3.1 General

Brackets are often used to denote concentration. A variety of units can be chosen from for expressing the concentration of a given chemical species or material in a phase. For some generic monoprotic acid HA, then [HA] means "the concentration of HA". Likewise, $[H^+]$ means the concentration of H^+, $[OH^-]$ means the concentration of OH^-, and so on. The *units* for concentration are not specified just by writing "[HA]"; the units need to be specified.

Most concentration units in common use involve the "mole" concept. The unit "mole" is fundamentally no more complicated than is the unit "dozen". If you have one dozen of something, you have 12 items. If you have one mole of that item, you have 6.023×10^{23} of them. The number 6.023×10^{23} is very special in chemistry and is called *Avogadro's Number.* Except perhaps when used in the terms of a "mole fraction", the term "mole" is often abbreviated as "mol".

1.3.2 Molarity

When a solution contains 1.0 mol of a particular species per liter (L) of aqueous system, the solution is said to be 1.0 "molar", or simply 1.0 *M* in that species. "Molar" is thus abbreviated with an italicized *M*. Molarity can also be used when referring to a sum of molar concentrations. For example, when we add some acid HA to a solution, some of it will dissociate according to

$$HA = H^+ + A^-. \tag{1.13}$$

We can use molarity to refer to the individual molar concentrations of HA and A^-, as well as the *concentration sum* for $HA + A^-$. With the subscript T denoting "total", then

$$A_T\,(\text{in units of } M) = [HA] + [A^-]. \tag{1.14}$$

1.3.3 Molality

When 1 kilogram (kg) of a solution contains 1.0 mol of a particular species, the solution is said to be 1.0 "molal" (1.0 *m*) in that species. Eq.(1.14) can be re-written using units of molality for all of the terms. For aqueous solutions, because molality is defined as *mols per kg of solution* and not as mols per kg of water, in 1 kg of aqueous solution the actual mass of water molecules will always be less than 1 kg. Strictly speaking, this is true even in absolutely pure water because of the presence of trace amounts of H^+ and OH^-. At "ambient" temperatures and pressures (*e.g.*, at 25 °C and 1 atmosphere total pressure, abbreviated herein as 25 °C/1 atm), the densities of all "fresh" waters are close to 1 kg/L. Consequently, for "fresh" waters under ambient conditions, the concentration of a given species on the molal and molar scales are nearly equal, and one can use the two units essentially interchangeably.

However, as a practical matter, it is generally easier and more convenient to think of the *volume* of a solution than it is to think of its *weight.* As a result, in the field of natural water chemistry, units of molarity are used more frequently than are units of molality. It should be remembered though that when doing very careful work, molality is usually preferred over molarity. Indeed, physical chemists concerned with careful measurements and uses of equilibrium constants usually prefer the molal scale. The reason for this is that with the molal scale, one does not have to worry about changes in volume due to changes in temperature and pressure. For example, for a particular solution of glucose in water, while the molarity of the glucose will change with temperature and pressure, the molality will not.

Water undergoes the self-dissociation reaction

$$H_2O = H^+ + OH^- \qquad K_w = \frac{[H^+][OH^-]}{[H_2O]}. \tag{1.15}$$

The quotient for the equilibrium constant K_w in Eq.(1.15) involves concentrations unadorned by "activity coefficients". This is a very good approximation in dilute solutions, as is discussed further in Section 1.4.

The value of K_w is rather temperature dependent, but only weakly pressure dependent. If one is interested in measuring K_w over a range of temperatures (*e.g.*, Bandura and Lvov, 2006), by choosing to use the molal scale for $[H^+]$ and $[OH^-]$, one can just focus on the number of mols of H^+ and OH^- that are present in 1 kg of solution, and not worry about how the volume of the solution may be changing with temperature. It is for this reason that the most fundamental measurements of equilibrium constants for aqueous solutions utilize units of molality for the dissolved species. K_w is a partial exception because water itself is so abundant. In particular, as is discussed further below, in Eq.(1.15) $[H_2O]$ carries the units of *mole fraction*. In the types of aqueous solutions of interest in this text, the fraction of water on a mol per total mols basis is very close to 1. This is what leads us through $K_w = [H^+][OH^-]/1$ to the familiar $K_w = [H^+][OH^-]$.

Example 1.1 Calculating Molar and Molal Concentrations

Consider pure water at 25 °C/1 atm. The concentration of $[H^+]$ as well as $[OH^-]$ in such water is 1.005 $m \times 10^{-7}$ (mols/kg). The density of liquid water under these conditions is 0.997 kg/L. What are the molar concentrations of $[H^+]$ and $[OH^-]$? And, using the molality scale, what is the value of K_w?

Solution

$[H^+] = [OH^-] = (1.005 \times 10^{-7}$ mols/kg) $\times$ (0.997 kg/L) $= 1.002 \times 10^{-7}$ *M* (mols/L). Note the closeness of the numerical values of the two concentrations.

$K_w = (1.005 \times 10^{-7}\ m) \times (1.005 \times 10^{-7}\ m) = 1.01 \times 10^{-14}\ m^2$.

1.3.4 Formality

One formula weight (FW) of a given material corresponds to the weight of one formula unit of the material, whether or not the material actually exists as a *molecule* with that chemical formula. In the case of glucose, one FW of pure solid glucose does correspond to 1 mol of glucose molecules, and these do not dissociate when dissolved in water. Thus, adding 1 FW of glucose to exactly enough water to obtain 1.0 L of solution leads to a 1.0 *M* solution of glucose.

In contrast to the situation with glucose, one FW of the ionic salt $NaCl_{(s)}$ does not correspond to 1 mol of NaCl molecules: for our purposes, the NaCl molecule does not exist. One FW of $NaCl_{(s)}$ does correspond to 1 mol of Na^+ ions and 1 mol of Cl^- ions. Thus, in the salt $NaCl_{(s)}$, each Na^+ is not associated with a particular Cl^- but rather exists in a crystal lattice surrounded by Cl^- ions, and vice versa. Consequently, even though many chemists carelessly and incorrectly speak of "one mol of $NaCl_{(s)}$" (especially on the web), we here absolutely insist on "one FW of $NaCl_{(s)}$".

Consider now the chemistry of a solution obtained by adding 1.0 FW of $NaCl_{(s)}$ (22.990 g + 35.453 g = 58.443 g) to exactly enough water to obtain a 1.0 L solution. When the $NaCl_{(s)}$ dissolves, one gets

Na^+ and Cl^- ions in solution. The solution does contain 1 FW per liter of NaCl, so it is 1.0 formal (1.0 *F*) in NaCl. Last, Na^+ and Cl^- are each present at 1.0 *M*, but NaCl is not.

We go on and now consider the compound acetic acid. Pure liquid acetic acid is called "glacial acetic acid" (GAA). In GAA at 25 °C/1 atm, very nearly 100% of the acetic acid is in the molecular HA form: very little is present as H^+ and A^- ions. But when GAA is dissolved in water, a significant amount of the HA can dissociate to form H^+ and A^- (acetate) ions. Assume we now take 10^{-5} mol of GAA and add it to exactly enough water to obtain a 1.0 L solution. The resulting solution will be exactly 10^{-5} *F* acetic acid, but it is less than 10^{-5} *M* acetic acid.

Overall, the FW concept is absolutely indispensable when describing aqueous solutions of materials like: (1) NaCl that do not exist as molecules in their pure form; and (2) chemicals like acids that might exist nearly 100% as molecules when in pure form, but will dissociate to some extent when placed in water.

Example 1.2 Calculating Formal and Total Concentrations

Two different solutions are prepared. One contains 30.026 g/L of acetic acid and the other contains 41.017 g/L of sodium acetate. For each, what is the formality, and what is the total concentration $[HA]+[A^-]$? Here the acetic acid molecule is HA (the protonated form), and the acetate ion is A^- (the deprotonated form).

Solution

The FW of acetic acid is 60.052 g, and the FW of sodium acetate is 82.034 g.

$$\text{the acetic acid solution:} \quad \frac{30.026 \text{ g/L}}{60.052 \text{ g/FW}} = \mathbf{0.50}\ \boldsymbol{F}$$

$$\text{the sodium acetate solution:} \quad \frac{41.017 \text{ g/L}}{82.034 \text{ g/FW}} = \mathbf{0.50}\ \boldsymbol{F}$$

Both solutions are 0.50 *F* so that in both solutions $[HA]+[A^-]=0.50$ *M*. In the acetic acid solution, some A^- forms by the dissociation $HA=H^++A^-$, but the total A concentration remains 0.50 *M*, and is denoted as A_T. In the sodium acetate solution, after dissolution, some HA forms from the reaction $A^-+H_2O=HA+OH^-$, but A_T is again 0.50 *M*.

1.3.5 Mole Fraction

Represented with the symbol x, mole fraction is a fractional concentration. For some particular species i, the fraction is computed as

$$x_i = \frac{n_i}{\sum_j n_j} = \frac{n_i}{n_{\text{tot}}} \tag{1.16}$$

where for any some chosen volume (*e.g.*, 1 L) or weight (*e.g.*, 1 kg) of the solution, n_i is the number of mols of i, and $\sum_j n_j = n_{\text{tot}}$ is the sum of the numbers of mols of all species (including i). We are assuming the solution is homogeneous, so the size of the chosen portion does not matter; if the size were to double, the numbers of mols of each of the constituents would double, and each x_i value would remain the same.

In any pure material, the mole fraction of the chemical is exactly 1.0. But in no phase can any mole fraction value ever be greater than 1.0. And, the sum of all of the mole fraction values is always exactly 1.0:

$$\sum_j x_j = 1. \tag{1.17}$$

The mole fraction concentration scale can be used for gas mixtures, liquid solutions, and solid solutions. In most aqueous solutions of interest in this text, the mole fraction of liquid water x_w is sufficiently close to 1.0 to assume that $x_w = 1$. For gases, the reader has surely already made calculations based on the principle that in an ideal gas system, for each of the different gaseous constituents, the partial pressure of each constituent equals the mole fraction multiplied by the total pressure. In the case of $N_{2(g)}$ in the atmosphere, we have

$$p_{N_2} = x_{N_2} P_{tot} \tag{1.18}$$

where p_{N_2} is the partial pressure of $N_{2(g)}$, x_{N_2} is the mole fraction of $N_{2(g)}$ in air (about 0.78), and P_{tot} is the total air pressure. At the surface of the Earth, $P_{tot} \approx 1.0$ atm.

Last, in our lives we encounter examples of solid solutions all the time: these include metal alloys like 14 K gold, carbon steel, the silver-mercury amalgams used in dental fillings, brass, lead-tin solder, and thousands of other examples. Ionic solid solutions are also possible, as when $^{90}Sr^{2+}$ (a by product of nuclear fission in atomic bombs and nuclear reactors) substitutes for Ca^{2+} in the mineral hydroxyapatite $Ca_5(PO_4)_3OH_{(s)}$ which is an important phase in our bones. We can think of the resulting solution as a mixture of a small amount of $Sr_5(PO_4)_3OH_{(s)}$ dissolved in a relatively much larger amount of $Ca_5(PO_4)_3OH_{(s)}$.

Example 1.3 Calculating Mole Fraction Concentrations.1

An aqueous solution is prepared using 30.0 g of water and 0.010 FW of $NaCl_{(s)}$. When individually considering the ions Na^+ and Cl^-, but neglecting the formation of trace amounts of H^+ and some OH^- by the dissociation of a small amount of the water, what are the mole fractions of H_2O, Na^+, and Cl^-?

Solution

30.0 g of H_2O = 30.0 g/(18.015 g/mol) = 1.665 mols of H_2O. The $NaCl_{(s)}$ dissolves to form 0.01 mol of Na^+ and 0.01 mol of Cl^-. Representing the trace amounts of H^+ and OH^- that form as $trace_{H+}$ and $trace_{OH-}$, we have

$$x_{H_2O} = \frac{1.665}{1.665 + 0.010 + 0.010 + trace_{H^+} + trace_{OH^-}} = \mathbf{0.988} \tag{1.19}$$

$$x_{Na^+} = \frac{0.010}{1.665 + 0.010 + 0.010 + trace_{H^+} + trace_{OH^-}} = \mathbf{0.00594} \tag{1.20}$$

$$x_{Cl^-} = \frac{0.010}{1.665 + 0.010 + 0.010 + trace_{H^+} + trace_{OH^-}} = \mathbf{0.00594} \tag{1.21}$$

$$0.988 + 0.00594 + 0.00594 = 1.000$$

Example 1.4 Calculating Mole Fraction Concentrations.2

Brass used in plumbing and lighting fixtures, rivets, and screws is commonly about 70% by weight copper (Cu) and 30% by weight zinc (Zn). What are the mole fraction concentrations of copper and zinc in such brass?

Solution

Consider 100 g of the brass in question: then, 70 g is Cu, 30 g is Zn.

70 g Cu × 1 mol/63.546 g = 1.012 mols Cu
30 g Zn × 1 mol/65.380 g = 0.459 mols Zn

$$x_{Cu} = \frac{1.012}{1.012 + 0.459} = \mathbf{0.706} \qquad x_{Zn} = \frac{0.459}{1.012 + 0.459} = \mathbf{0.294} \tag{1.22}$$

$$0.706 + 0.294 = 1.000$$

In natural water chemistry, the most frequent application of the mole fraction scale occurs for water itself. Consider again the dissociation of water according to

$$H_2O = H^+ + OH^- \tag{1.23}$$

with

$$K_w = \frac{[H^+][OH^-]}{[H_2O]}. \tag{1.24}$$

As introduced above in Section 1.3.3, the reader is likely familiar with K_w written simply as

$$K_w = [H^+][OH^-] \tag{1.25}$$

for which, at 25 °C/1 atm, $K_w = 1.01 \times 10^{-14}\ m^2$. For Eq.(1.25) we have taken advantage of the fact that $[H_2O]$ is expressed on the mole fraction scale. Since on that scale $[H_2O] \approx 1$, there is no need to continue to represent it in the expression for K_w. Example 1.7 below shows why it is so useful to take this approach for K_w: it permits us to avoid multiplying or dividing by 55.5 every time we use K_w.

Example 1.5 Water Concentration in Molarity and Molality

At 25 °C/1 atm, a liter of fresh water weighs very close to 1 kg. Also, one kg of fresh water is very nearly all water when viewed on a molar basis. What is the concentration of H_2O in such a solution in units of molarity? In units of molality?

Solution

Since we are assuming that 1 liter of water contains very close to 1 kg of H_2O, the number of mols of water in the liter is 1000 g/(18.015 g/mol) = 55.5 mols. Therefore, the molar concentration of H_2O is **55.5** *M*. Similarly, since we are assuming that 1 kg of the solution is nearly all water, the molal concentration of H_2O is **55.5** *m*.

Example 1.6 Value of K_w Depends on the Concentration Units Chosen for Water

At 25 °C/1 atm, with units of mole fraction having been chosen for $[H_2O]$, the value of K_w as represented in Eq.(1.25) is 1.01×10^{-14} m^2. What would the value of K_w be if the **molal scale** had been selected for water rather than the mole fraction scale?

Solution

For the stated conditions, $[H^+][OH^-] = 1.01 \times 10^{-14}$ m^2. Using the molal scale for water, we would have

$$K_w = \frac{[H^+][OH^-]}{[H_2O]} = \frac{1.01 \times 10^{-14}\,m^2}{55.5\,m} = \mathbf{1.82 \times 10^{-16}\ m} \qquad (1.26)$$

Example 1.7 Using K_w if Water Concentration Was Expressed in Molalilty

At 25 °C/1 atm, the value of K_w as represented in Eq.(1.26) is 1.82×10^{-16} m. Starting with K_w in that form, then for pure water what are the values of $[H^+]$ and $[OH^-]$ at 25 °C/1 atm?

Solution

For pure water, every H_2O that dissociates gives one H^+ and one OH^- so that

$$[H^+] = [OH^-] = y \qquad (1.27)$$

We solve for y:

$$K_w = \frac{y^2}{[H_2O]} = 1.82 \times 10^{-16}\,m \qquad (1.28)$$

With this form of K_w, $[H_2O] = 55.5$ m so that

$$y^2 = 55.5\,m \times 1.82 \times 10^{-16}\,m = 1.01 \times 10^{-14}\,m^2 \qquad (1.29)$$

$$y = 1.00 \times 10^{-7}\,m = \left[H^+\right] = \left[OH^-\right] \qquad (1.30)$$

The point here is that for the many systems in which $[H_2O] \approx 55.5$ m, there is no reason to use the form of K_w such that $[H_2O]$ is expressed on the molal scale. If we did, for *every problem*, we would always have to multiply by 55.5 and get to $[H^+][OH^-] = 1.01 \times 10^{-14} m^2$, and then proceed from there. We might as well start with that, as in Eq.(1.25) version of K_w, and not waste time, right?

1.3.6 Weight Fractions: ppm, ppb, and ppt

When dealing with contaminants and other relatively minor species in water, it is popular to express concentrations in units of mass per volume, as for example in "milligrams per liter". Indeed, while a concentration in units of molarity does tell us "how much" of a chemical is present in water, it does not immediately give us the amount in terms of *mass per volume*, which is what government regulators and others doing mass-based calculations usually want to know. Because, for example, most considerations of the toxicology of some chemical i look at responses of organisms in units such "mg of chemical *i* per kg of body weight" and/or "mg of chemical *i* per kg of body weight per day".

Common mass per volume concentration units are milligrams (mg) per liter, micrograms (μg) per liter, and nanograms (ng) per liter. As noted above, in fresh waters at ambient temperatures and pressures, 1 L of aqueous solution weighs very close to 1 kg so that we have the correspondences given in Table 1.2.

Example 1.8 Converting from ppm to Molarity

Consider a sample of fresh water at 25 °C/1 atm for which the total dissolved concentration of inorganic Hg (mercury) is measured to be 0.004 mg/L (= 0.004 ppm). This is double the current U.S. EPA maximum concentration limit (MCL). The atomic weight of Hg is 200.59 g/mol. What is the concentration of Hg in the water in units of *M* (mols/L)?

Solution

200.59 g/mol corresponds to 200.59 × 1000 mg/mol, which corresponds to 1 mol/(200.59 × 1000 mg). Thus,

$(0.004 \text{ mg/L}) \times 1 \text{ mol}/(200.59 \times 1000 \text{ mg}) = \mathbf{1.99 \times 10^{-8}}\ M$ **(mols/liter)**

Example 1.9 Using the ppm Concentration Scale for Total Ammonia

In a particular solution of interest, the pH conditions are such that of all of the dissolved ammonia present, there are 15.0 mg/L of NH_3 as NH_3 *and* 5.0 mg/L of NH_4^+ as NH_4^+. What is the concentration of total ammonia species when *all* dissolved ammonia is expressed in terms of mg/L *as* NH_3?

Solution

The FW of NH_3 is 17.031 g/FW. The FW of NH_4^+ is 18.038 g/FW. Thus, with attention to units, we have:

concentration of total ammonia in mg NH_3 per liter =

$$\left(15.0 \text{ mg NH}_3/\text{L}\right) + 5.0 \text{ mg NH}_4^+/\text{L} \times \frac{17.031 \text{ g NH}_3/\text{FW} \times (1000 \text{ mg NH}_3/\text{g NH}_3)}{18.038 \text{ g NH}_4^+/\text{FW} \times (1000 \text{ mg NH}_4^+/\text{g NH}_4^+)}$$

Canceling units as possible and solving, the answer is **19.7 mg/L** as NH_3.

TABLE 1.2

Correspondences between Certain *Mass Per Volume* Concentrations and the *Mass Per Mass* Fractional Concentrations of ppm, ppb, and ppt, Assuming a Water Solution Density of 1 kg/L

mass per volume concentrations	weight fractions			
1 mg/L ≈	1 mg per 1 kg =	1 mg per 10^6 mg =	1 part in 10^6 parts =	**1 ppm (part per million)**
1 μg/L ≈	1 μg per 1 kg =	1 μg per 10^9 μg =	1 part in 10^9 parts =	**1 ppb (part per billion)**
1 ng/L ≈	1 ng per 1 kg =	1 ng per 10^{12} ng =	1 part in 10^{12} parts =	**1 ppt (part per trillion)**

Note: 1 kg = 10^6 mg = 10^9 μg = 10^{12} ng.

TABLE 1.3

Summary of Different Aqueous Phase Concentration Scales with Additional Examples

Molarity, *M*, mols/L = number of mols per liter of solution. Can be used to express the concentration of a single specific dissolved species, or the total concentration for some group of related dissolved species.

Example 1.10 In a particular solution, 1.7007 g of OH^- is present per liter. What is the molarity of OH^-?
Ans: The FW of OH^- is 17.007 g/FW. The concentration of OH^- is **0.10** *M*.

Example 1.11 The concentration of NH_3 in a particular solution is 1.0×10^{-4} *M*. In the same solution, the concentration of NH_4^+ is 1.5×10^{-4} *M*. What is the total concentration of all ammonia species?
Ans: total ammonia concentration in $M=[NH_3]+[NH_4^+]=1.0\times10^{-4}\,M+1.5\times10^{-4}\,M=\mathbf{2.5\times10^{-4}}\,M$.

Molality, *m*, mols/kg = number of mols per kg of solution rather than number of mols per liter of solution as in molarity. Molality can be used to express the concentration of a single specific dissolved species, or the total concentration for some group of related dissolved species. Molality is a particularly convenient concentration scale when one does not want to bother tracking changes in volume caused by changes in temperature and/or pressure.

Example 1.12 In a particular solution, 1.7007 g of OH^- is present per kg of solution. What is the molality of OH^-?
Ans: The FW of OH^- is 17.007 g/FW. The concentration of OH^- is therefore **0.10 *m***.

Formality, *F*, FW/L = number of formula weights (FW) per liter of solution.

Example 1.13 The FW of acetic acid is 60.052 g/FW. A solution is prepared such that $10^{-4}\times60.052$ g of acetic acid is added to 1.0 L of initially pure water. What is the formality of acetic acid in the solution? Assume that essentially all of the liquid volume is provided by the 1.0 L of water.
Ans: The number of FW of acetic acid added to the 1 L of water solution is 10^{-4}. Therefore, the formality of the acetic acid is $\mathbf{10^{-4}}\,\boldsymbol{F}$. The solution is **not** 10^{-4} *M* in acetic acid because some of the acetic acid will dissociate to form some acetate ion. However, the **sum** of the acetic acid and acetate ion molarities is 10^{-4} *M*.

Normality, *N*, equivalents/L = number of equivalents per liter of solution for an acid, base, or redox-active species. For a solution of an acid, the factor that inter-relates normality and formality is the number of protons on the acid.

Example 1.14 Consider two solutions. One is 0.01 *F* in HC1. The other is 0.01 *F* in H_2SO_4. What is the normality of each solution?
Ans: HCl is "monoprotic", so the normality of the HCl solution is 1 equivalent/FW × 0.01 *F* = **0.01 *N***.
H_2SO_4 is "diprotic", so the normality of the H_2SO_4 solution is 2 equivalents/FW × 0.01 *F* = **0.02 *N***.

(*Continued*)

TABLE 1.3 (CONTINUED)
Summary of Different Aqueous Phase Concentration Scales with Additional Examples

For a solution of a base, the factor that inter-relates normality and formality is the number of protonatable positions on the base.

Example 1.15 Consider two solutions. One is 0.001 *F* in sodium acetate. The other is 0.001 *F* in sodium oxalate. What is the normality of each solution?

Ans: Acetate is "mono-basic" (one protonatable position), so the normality of the sodium acetate solution is

1 equivalent/FW × 0.001 *F* = **0.001** *N*.

Sodium oxalate is di-basic (two protonatable positions), so the normality of the sodium oxalate solution is

2 equivalents/FW × 0.001 *F* = **0.002** *N*.

For a redox species, the factor that inter-relates normality and formality is the number of electrons that the chemical releases or picks up in its redox reaction.

Example 1.16 Consider a solution that contains 8.0 mg of dissolved oxygen (O_2) per liter of water. Eq.(1.7) shows that in its redox reaction to form H_2O, each O_2 molecule picks up 4 electrons. What is the redox normality of this solution of O_2?

Ans: The FW of O_2 is 31.999 g/FW. An 8.0 mg/L solution of O_2 is therefore

8.0 mg/L × 1 FW/31,999 mg = 2.50×10^{-4} *F*. The solution is also 2.50×10^{-4} *M*, since O_2 does not dissociate as an acid in water. The redox formality of O_2 in this solution is

4 equivalents/FW × 2.50×10^{-4} *F* = **0.0010** *N*.

Mole fraction, x = Fraction for the species of interest of all mols present. For the inorganic geochemistry that is the main focus of this text, the mole fraction concentration scale is most used for expressing the concentration of liquid water, and for the composition of mineral solids.

Example 1.17 Consider a solution that is 1.00×10^{-7} *m* in both H^+ and OH^- and in which water is the only other species present. What is the concentration of all of three species on the mole fraction scale?

Solution: For 1 kg of solution: number of mols of $H^+ = 1.00 \times 10^{-7}$ mols

number of mols of $OH^- = 1.00 \times 10^{-7}$ mols

weight of $H^+ = 1.00 \times 10^{-7}$ mols × 1.008 g/mol = 1.008×10^{-7} g

weight of $OH^- = 1.00 \times 10^{-7}$ mols × 17.007 g/mol = 17.007×10^{-7} g

weight of H_2O = 1 kg − 1.008×10^{-7} g − 17.007×10^{-7} g ≈ 1 kg.

number of mols of H_2O = 1000 g/(18.015 g/mol) = 55.5 mols

total number of mols = $2 \times 10^{-7} + 55.5 \approx 55.5$ mols

x for H^+ and OH^- = $1.00 \times 10^{-7}/55.5$ = $\mathbf{1.80 \times 10^{-9}}$

x for H_2O = 55.5/~55.5 ≈ **1.00**

(*Continued*)

TABLE 1.3 (CONTINUED)

Summary of Different Aqueous Phase Concentration Scales with Additional Examples

parts per million (ppm = mg/kg ≈ mg/L) = mass fraction (per million) of a particular constituent of interest in water. Often, the units "parts per million" are used synonymously with "mg/L" for fresh waters: 1.0 L of water weighs very close to 10^6 mg, so that 1 mg per liter is nearly exactly equivalent to 1 part in 10^6 parts.

Example 1.18 Consider a solution that contains all of the following elements in dissolved form, with all at 1.00×10^{-6} *M*: iron (Fe), lead (Pb), manganese (Mn) and sulfur.
Solution: See Column F.

A	B	C	D	E	F	G	H
				= D × B	**= 10^3 × E**	**= 10^6 × E**	**= 10^9 × E**
element	**atomic weight g/mol**	***M* mols/L**	***m* mols/kg**	**g/kg**	**ppm (mg/kg)**	**ppb (μg/kg)**	**ppt (ng/kg)**
Fe	55.845	1.00E-05	1.00E-05	0.000558	0.558	558	558,000
Pb	207.200	1.00E-05	1.00E-05	0.00207	2.07	2070	2,070,000
S	32.065	1.00E-05	1.00E-05	0.000321	0.321	321	321,000

parts per billion (ppb = *μ*g/kg ≈ *μ*g/L) = mass fraction (per billion) of a particular constituent of interest in water. Often, the units "parts per billion" are used synonymously with "μg/L" for fresh waters: 1.0 L of water weighs very close to 1 kg. 1 kg = 10^3 g, so 10^{-6} g per 10^3 g equals 1 part per 10^9 parts.

Example 1.19 Consider the solution from Example 1.18. What are the concentrations in units of ppb?
Solution. See Column G.

parts per trillion (ppt = ng/kg ≈ ng/L). Mass (in ng = 10^{-9} g) of a particular constituent of interest per kg of water. Often, the units "parts per trillion" are used synonymously with "ng/L" for fresh waters: 1.0 L of water weighs very close to 1 kg. 1 kg = 10^3 g. 10^{-9} g per 10^3 g equals 1 part per 10^{12} parts.

Example 1.20 Consider the solution from Example 1.18. What are the concentrations in units of ppt?
Solution. See Column H.

1.4 ACTIVITY AND ACTIVITY COEFFICIENTS

1.4.1 Basic Principles

As noted above, the concentration of a given species i is symbolized using brackets as $[i]$. However, depending on the nature of the phase in which it is present, a chemical may not *act* at the level one might expect based simply on its concentration. It might act *less concentrated*, or it might act *more concentrated*. How a chemical *acts* is called its *activity*. The activity of species i is usually symbolized using braces as $\{i\}$, or as a_i. Besides the verb *act*, there are many other words in English (all derived from the Latin noun *actum* = act or deed) that inform our understanding of the meaning of the chemical term *activity*. Examples are *actin*, *actor*, *react*, *interact*, and *transact*.

Whenever the activity of the species of interest in a particular solution does not equal its concentration, the species is said to be behaving **non-ideally**. In fresh waters, the degrees of non-ideality are usually small, but in waters that are not dilute (as with brines, seawater, etc.) they can be large enough that corrections are needed in the chemical calculations used to obtain the concentrations of all the species present.

When the **molal** concentration scale is being used for a dissolved aqueous species i, many practitioners write the relation between activity and concentration as

$$\{i\} = a_i = \gamma_i[i] \qquad \text{shortcut way to express } \{i\} \text{ for dissolved aqueous } i \text{ when } [i] \text{ is on the molal scale} \qquad (1.31)$$

so that a *dimensionless* **activity coefficient** γ_i provides the multiplicative factor that relates $[i]$ and $\{i\}$ (see Box 1.1). Because of the similarities of the molality and molarity scales in dilute water, γ_i generally adequately provides the activity coefficient for the molar scale. No activity coefficient can ever be zero or less than zero.

Since γ_i is a dimensionless correction factor, because $[i]$ carries concentration units (molality in the above discussion), then by Eq.(1.31) units of molality would be assigned to $\{i\}$. However, in many

BOX 1.1 Anthropomorphic Analogy for Activity Coefficients

Activity = (Activity Coefficient) × Concentration. Consider a room at a university that contains 100 persons.

Ideal behavior for the honors students is first defined as being how they act when surrounded only by "typical" students, as in an "infinitely dilute" solution of honor students.

Case A: There are 2 honors students and 98 typical students in the room; the *activity coefficient* for honors students is close to 1.

Case B: If we substitute 20 of the 98 typical students with 20 students that are interested only in partying and loud behavior, then the tendency of the honors students to leave the room will go up. Now, the honors student *activity coefficient* > 1. Indeed, standing outside the door to the room, one would get the feeling that the concentration of honors students in the room for Case B was greater than 2 per 100 people; the desire to escape from the room is greater than for Case A, even though the concentration is the same.

Case C: Now, Case A altered by substituting one of the 98 typical students with one Nobel Prize winner. The tendency of the honors students to leave the room will be lower than in Case A, because of their desire to hear what the Nobel Prize winner has to say; now, the honors student *activity coefficient* < 1. Standing outside the room, one would get the feeling that the concentration of honors students in the room was less than 2 per 100 people (given that the activity coefficient was defined as being 1 when each honors student is exclusively surrounded by typical students, which is the situation (very nearly) in Case A).

important equations in chemical thermodynamics, it is necessary to use the logarithms of chemical activities. Finicky mathematicians insist that one cannot take the logarithm of any parameter carrying units: "*You can only take the log of a pure number.*" To avoid this difficulty, the non-shortcut, precise way for strict physical chemists is to rewrite Eq.(1.31) as

$$\{i\} = a_i = \gamma_i \frac{[i]}{[i]^{\circ}}. \quad \text{thermodynamically precise way to express } \{i\} \text{ for dissolved aqueous } i \text{ ([}i\text{] is on the molal scale)} \tag{1.32}$$

As noted, the quantity γ_i is dimensionless. $[i]^{\circ}$ has a magnitude of 1.0 and carries the same dimensions as $[i]$. With this definition, $\{i\}$ has the same magnitude as $\gamma_i[i]$ but is dimensionless, as desired. Most people just use Eq.(1.31). When challenged, they either do not know about the exact way, or if they do, they respond with "yeah, yeah, I know that; it doesn't make any difference". The reader need not worry about this further. For us, Eq.(1.31) will be adequate, and we will go ahead and implicitly allow, by virtue of using Eq.(1.31), that activities can carry units (see Box 1.2).

Besides solving the above-described units problem for strict people, $[i]^{\circ}$ plays a special role in chemical thermodynamics because it refers to a **hypothetical standard state** in which i is not only at unit concentration but is also behaving ideally (*i.e.*, with $\gamma_i = 1.0$). For aqueous solutions and the molal scale, γ_i values are defined to be unity in very dilute solutions when the concentrations of i and all other dissolved species (besides water) are extremely low, so that i is in the **reference state** of "infinite dilution". The hypothetical standard state to which $[i]^{\circ}$ corresponds thus involves i at 1.0 molal concentration with each i behaving the same way it would in the reference state of "infinite dilution" in water, *i.e.*, as if: (a) all the surrounding entities were just water molecules and the very infrequently encountered H^+ and OH^-), and (b) there were no encounters or interactions with any

BOX 1.2 Chosen Units Affect *K* Values, but Not the Underlying Thermodynamics

As in everyday life, it seems that even the strictest people can choose to relax an important rule when the rule becomes too inconvenient. An example of this may be derived from Example 1.6 in which it is shown that careful chemists know that the numerical value of an equilibrium constant depends on the concentration scales chosen for the reactants and products, so for example, it is said that K_w usually is assigned units of m^2, as derived from the concentration scales used for the reactants and products. However, strictly speaking, using dimensionless activities in the expressions for equilibrium constants would make all K values unitless. Even very strict physical chemists ignore this. Instead, to avoid confusion on how to use a particular value of an equilibrium constant, they allow it to have units, because the units carry information on which concentration scales are associated with the activities. This is very useful, because otherwise one would have to give the *value* of the equilibrium constant (dimensionless), then *also* specify the concentration scales for all the reactants and products associated with that value.

To summarize, it would be most consistent and strict to always say:

"K_w has a value of 1.01×10^{-14} (dimensionless) at 25 °C/1 atm when molality is the concentration scale chosen for both H^+ and OH^-, and mole fraction is the concentration scale chosen for water".

That would obviously quickly become repetitive and annoying. So, what is said instead is:

"K_w has a value of 1.01×10^{-14} m^2 at 25 °C/1 atm". Unsaid but implied is: "We are allowing K_w to have units so that we all can infer what concentration scales are being used for H^+, OH^-, and H_2O). So, yes, we would like to be strict about things, but, well, everyone is busy doing important things and so we are taking some shortcuts".

TABLE 1.4
Summary of Possible Cases for Activity Coefficients for Constituents Dissolved in Water When Using the Molal Scale[a]

how species *i* is acting $\{i\} = a_i = \gamma_i\,[i]$	implications for activity coefficient	allowing shorthand of Eq.(1.31)
less concentrated than it is	$\gamma_i < 1$	$\{i\} < [i]$
more concentrated than it is	$\gamma_i > 1$	$\{i\} > [i]$
as concentrated as it is (ideal behavior)	$\gamma_i = 1$	$\{i\} = [i]$

[a] Also usually used for the molar scale in the case of dilute solutions, which means all fresh waters.

other *i*. This is a paradox: how can some species *i* ever be at unit molal concentration and also be completely and only surrounded by water (and the occasional H^+ and OH^-)? The way around the paradox involves admitting that while such a solution cannot actually exist, and *is* therefore only hypothetical, we can *imagine* exactly what the numerical magnitude of the activity would be in such a state: unity (see Box 1.3).

Each type of phase (aqueous, solid mineral, and gas) is assigned its own reference state for ideality, and the assignment depends on what is most convenient. And, a given species *i* will have an activity coefficient for each phase in which it is present (water is present in aqueous solutions (obviously), can be in the gas phase, and can be dissolved inside a solid mineral phase). For all species except water itself, in an aqueous phase, the concentration scale we use is molality (or molarity because usually it is very nearly the same), and the reference state is infinite dilution. For water in an aqueous phase, we use the mole fraction scale, and the reference state is pure liquid water. For a chemical *i* dissolved in a solid, the concentration scale we use is the mole fraction scale, and the reference state is pure solid *i* in some particular crystal structure. For the dimensionless mole fraction concentration scale, the analog of Eq.(1.31) is

$$\{i\} = a_i = \zeta_i x_i. \quad \text{shortcut way to express } \{i\} \text{ values in solid solutions, and in a liquid aqueous phase for } \{H_2O\} \qquad (1.33)$$

BOX 1.3 Anthropomorphic Analogy for the "Standard State"

Consider again some honors students in a room with a capacity of 100 persons. Let us say our unit of concentration is *one dozen per room (d)*, and that the concentration of honors students is now 1.0 *d*, which means 12 per room. The other 88 students are "typical students". In this case, the honors student *activity coefficient* is less than 1, because now the honors students encounter one another with some frequency, and since we will assume each feels more affinity with other honors students, each feels more comfortable than in Case A of the anthropomorphic analogy in Box 1.1. Thus, each feels less inclined to leave the room (though with 12 of them in the room, the aggregate leaving tendency is considerably higher than in Case A). The nature of this situation does not prevent us from imagining a room with a concentration of honors students equal to 1.0 *d* in which, some means – perhaps some Jedi mind trick – has caused each honors student to think that everyone in the room is a typical student. That is hypothetical, of course, but we can compute what the energy state would be.

where ζ_i is the activity coefficient on the mole fraction scale. In other texts and scientific and engineering writings, it is very common to see γ_i written as the activity coefficient on the mole fraction scale, not ζ_i. However, in this text, γ_i is strictly reserved for the activity coefficient on the molal scale and, when molality ≈ molarity, the molarity scale.

For gas phases, the scale for gas-phase activity is pressure, with units such as atm or Pascals. The gaseous pressure of *i* (namely p_i) is relatable to the gaseous volumetric concentration if the temperature *T* (Kelvin) is known: for an ideal gas, the needed relation is the Ideal Gas Law, *i.e.*, $p_iV = n_iRT$ which yields n_i/V (mols/L)$=p_i/RT$. *R* is the gas constant (= 0.082 L-atm/mol-K).

While this text does not consider any cases that require activity corrections for the gas phase, we now briefly address that subject so as to emphasize similarities across all phases. In the case of non-ideality in the gas phase, a special name is used for the gas phase activity: this is "fugacity" (*f*). This word can be traced back to the Latin verb *fugare*, which means "put to flight", which makes sense, because something having a high activity in the gas phase would certainly be flying around in abundance. As with liquid and solid solutions, strictly speaking, a fugacity is dimensionless. However, as with Eq.(1.31), for the sake of convenience, many practitioners allow fugacity to have units of pressure. We then write the gas-phase analog of Eq.(1.31) as:

$$\{i\} = f_i = \lambda_i p_i \qquad \text{shortcut way to express } f_i \text{ values for a gas phase} \qquad (1.34)$$

where λ_i is the gas-phase fugacity (activity) coefficient. This text thus recognizes three different symbols for activity coefficient: γ for the molal (≈ molarity) scale in aqueous solutions, ζ for the mole fraction scale as for solids and water in liquid water, and λ for the gas phase. Fugacity corrections are never needed when the gas phase pressures are low, *i.e.*, near the surface of the earth where pressures are less than a few tens of atmospheres. However, with increasing depth in the earth, gases under high pressures can exist wherein fugacity corrections will be required, as in deep geohydrology and petrochemical extraction. Since this text focuses on near-surface systems, we will always assume ideal behavior in the gas phase (all $\lambda_i = 1$), and we will not henceforth mention gas-phase fugacity coefficients.

1.4.2 γ_i (Molal Scale Activity Coefficient in Water) Depends on What *i* Is, and on the Nature of the Solution

Activity coefficients depend on the identity of the species and on the "matrix" in which the species is present. For example, in a 0.01 *F* solution of $MgCl_2$, for Mg^{2+}, it turns out that $\gamma=0.51$, while for Cl^-, we have $\gamma=0.85$. In the more concentrated solution of 0.10 *F* $MgCl_2$, the two values are again different: for Mg^{2+}, $\gamma=0.25$, and for Cl^-, $\gamma=0.71$.

Consider again the acid dissociation reaction introduced above:

$$HA = H^+ + A^-. \qquad (1.13)$$

HA represents a generic monoprotic acid, which might be acetic acid, or any other neutral monoprotic acid. If a given solution is 10^{-5} *F* in HA and also contains say 0.005 *F* dissolved NaCl, the H^+ ions

TABLE 1.5
Examples of the Species and Matrix Dependence of γ Values

	0.01 *F* $MgCl_2$	0.1 *F* $MgCl_2$
Mg^{2+}	$\gamma=0.51$	$\gamma=0.25$
Cl^-	$\gamma=0.85$	$\gamma=0.71$

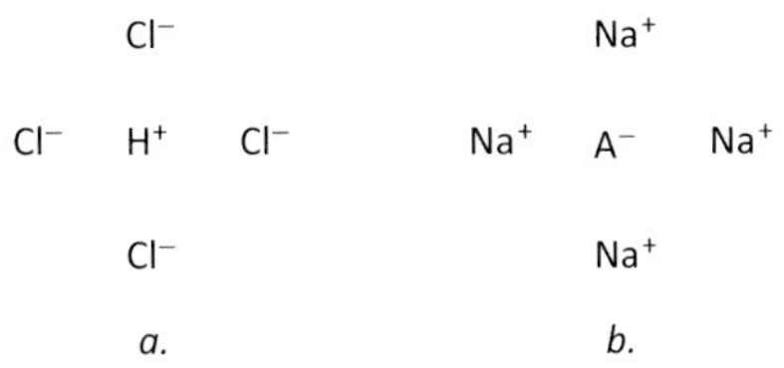

FIGURE 1.1 Schematic cross-sections of ion clouds surrounding H^+ and A^- ions in a solution in which most of the negative ions are Cl^-, and most of the positive ions are Na^+. *a*. Shielding of H^+ by Cl^- ions; *b*. Shielding of A^- by Na^+ ions.

that form by dissociation of the HA will tend to be surrounded somewhat by Cl^- ions, and the A^- ions will tend to be surrounded somewhat by Na^+ ions. This will lead to a partial electrical shielding of both H^+ and A^- so that neither H^+ nor A^- will be as fully visible (a.k.a. *active*) as they would be in the absence of the NaCl: the γ values for both H^+ and A^- will be about 0.93 (see Figure 1.1 and Box 1.4). For these reasons, γ values for ions tend to become increasingly less than 1.0 when moving from fresh water to brackish water to seawater.

In contrast to the situation for the H^+ and A^- ions in 0.005 F NaCl, the neutral molecule HA is only very weakly affected by ionic charges in solution: γ for HA is very close to 1.0 in 0.005 F NaCl. Because it carries no electrical charge, HA is much less affected by the presence of dissolved ions. Overall, in many cases for which the γ values of the ions present are noticeably different from 1.0, the γ values of the neutral species present are very close to 1.0.

For ions in solution, when the total ion concentration becomes large (*i.e.*, considerably greater than that for seawater), the γ values stop decreasing, then actually start to climb back towards 1.0. At high salt concentrations, γ values for ions in very salty solutions can be greater than 1.0. This happens because at high salt concentrations, the system does not have enough water to effectively solvate all the ions, and the result is higher γ values for all of the ions. In the infinite dilution reference state, there are many H_2O molecules available for solvating ions, and so the ions can acquire

BOX 1.4 Anthropomorphic Analogy for the Effects of Ions on the Activity Coefficients of Other Ions

For the reaction in Eq.(1.13), consider the following heterosexual relationship analogy (*NB*: This analogy is certainly not intended to be heteronormative, and a homosexual analogy could also be constructed). Let us say that: (1) a collection of HA represents some heterosexual married couples; (2) H^+ entities are single men (all of a same kind); and (3) A^- entities represent single women (all of a same kind). We first place a certain concentration (formality) of the married couples in the system and wait for equilibrium to be reached. At equilibrium, there will be a certain concentration of partnered couples HA, a certain concentration of single men H^+, and a certain concentration of single women A^-. These concentrations will depend on the inherent stability of the partnership between H^+ and A^-, and upon the formality (F) for the HA in the system (more about that in Chapter 5). Consider that we now add to the system some NaCl, which is made up of charming (charged), but non-committal Na^+, and similarly charming but non-committal Cl^-. Na^+ and Cl^- never partner with anyone, but do mingle with the others in solution. And, because they have some charm, there are always some Na^+: (1) attracted to the A^-; and (2) obscuring them from the H^+. The exact converse applies to the Cl^- ions. The overall result is that although adding the NaCl does not change the total formality of the HA solution, it will tend to cause the H^+ and A^- to be less active relative to combining to form HA: both γ_{H^+} and γ_{A^-} are less than 1.0.

whatever comfortable degree of solvation they desire. Thus, at infinite dilution, each ion is actually not a bare ion, but is well surrounded and partially shielded by water molecules. This is the reference character for the ion for $\gamma = 1$. However, as the total concentration of total ions becomes high, a competition for solvating water molecules begins, and there is no longer enough solvating water to go around: the ions become less and less well solvated, more bare and exposed, and thus more *active* than at infinite dilution: for some very salty solutions, for ions $\gamma > 1$ (See Chapter 2 for additional details).

1.5 EQUILIBRIUM *VS.* KINETIC MODELING

1.5.1 General

The goal of this text is to provide clear explanations of how to make equilibrium-based, predictive calculations of the "chemical speciation" in a wide range of natural water systems. "Chemical speciation" means the set of all concentration values for the species in solution, the levels of chemicals in the gas phase (if there is one), and which (if any) mineral solids are present.

A given chemical system is either at thermodynamic equilibrium or it is not. If not, some people will be interested in making "chemical kinetics" calculations to predict the time course of the speciation as the system moves towards either: (a) an equilibrium speciation; or (b) a permanent "*steady-state*" speciation (all $d[i]/dt = 0$), as for systems that are being continuously frustrated from reaching equilibrium by inputs of energy or mass. A gas-phase example of the latter is found in the stratosphere where daily incoming UV light energy leads to the continued creation of ozone molecules by a photochemical route. The formed ozone tends to be destroyed by sink reactions. When the formation rate equals the destruction rate, the result is a stable ozone concentration. When the steady-state ozone levels in the stratosphere are sufficiently high, the ozone protects us by acting as a "sunscreen", absorbing much dangerous incoming UV light. But when the sink reactions are favored, as can happen due to excessive ozone-depleting substances having been emitted to the atmosphere, the steady-state ozone levels are too low to provide the needed protection.

For a system that can, in adequate time, reach equilibrium for specified initial concentrations of chemicals A, B, C, D, etc., then for particular values of temperature T, pressure P, volume V, the **advantage of the equilibrium approach** is that we know with 100% certainty that we will be able to correctly determine the equilibrium speciation that will be attained if we: (1) know the correct values for all needed equilibrium constants at the T and P of interest; and (2) correctly execute any needed activity corrections. And, even if the system cannot reach equilibrium in our timeframe of interest, we will at least know that the system is headed towards that equilibrium state.

A **disadvantage of the equilibrium approach** is that except for simple acid/base problems, for which proton-transfer reactions proceed quickly, we may not know how much time will be required to reach equilibrium. Indeed, many important redox reactions are very slow, and attainment of equilibrium can require years and longer. By the way, let's all be thankful for that because the organic carbon in each of our bodies is highly unstable in the presence of ambient oxygen, as by conversion to CO_2. But thanks to some slow redox kinetics, I am writing this (without burning up), and you can read it (without burning up).

1.5.2 Kinetic Approaches

There are two major problems with applying kinetic models to natural water systems. First, relevant rate constants are usually not known very well at all, and trace species can sometimes act as catalysts that greatly affect rate constants. Second, rate expressions are often not as simple as in the example of Eq.(1.37) given below. For example, there is no guarantee that the rate of a given combination reaction for two species will be given simply by the product of a rate constant and the two concentrations. Overall, kinetic solutions to aquatic chemistry problems are usually sought only for highly specialized problems when the timeframe of interest is relatively short (minutes to days/weeks/months), and an equilibrium calculation would give a hopelessly unrealistic answer.

If every forward reaction has a corresponding back reaction, then the time-dependent speciation given by the kinetic model will eventually approach the same chemical speciation given by the corresponding equilibrium model. If every forward reaction does not have a corresponding back reaction, and if there are constant energy or mass inputs, then the time-dependent speciation given by the kinetic model will eventually approach the chemical speciation for a steady-state condition.

Examples of natural systems requiring kinetics-based treatments are:

1) Cloud droplets containing recently emitted SO_2 that is being oxidized to sulfuric acid;
2) A stream subjected to recent inputs of sewage/biological oxygen demand (BOD), the BOD being oxidized by and thus depleting the stream oxygen and possibly causing fish to die;
3) Pond water that is experiencing rapid (hours) algal photosynthesis such that CO_2 is being removed from the water and the pH is rising;
4) Groundwater that is seeping to the surface and bringing dissolved ferrous iron (Fe^{2+}) into exposure to oxygen in air; the resulting oxidation of the Fe^{2+} leads to ferric iron (Fe^{3+}) which then precipitates as a generally slimy-looking orange $Fe(OH)_{3(s)}$ (you may have seen this);
5) Contaminated groundwater that is flowing through an engineered subsurface trench that is filled with elemental iron (Fe°) chips/filings, for removal (by chemical reduction) of chlorinated solvents like trichloroethylene (TCE) and perchloroethylene (PCE) (Tratnyek et al., 2003).

For illustration, we consider two simple cases. For a simple one-phase (aqueous) system, consider the set of reactions (Pankow and Morgan 1981a, 1981b)

$$M + L \xrightarrow{k_f} ML \tag{1.35}$$

$$M + L \xleftarrow{k_b} ML. \tag{1.36}$$

M represents a generic dissolved metal ion (the charge is not shown), L represents a generic "ligand" that can bind to the metal ion, and ML is the metal-ligand complex. Reaction (1.35) describes the forward bimolecular combination of M with L to form ML; the rate constant for the combination is k_f. Reaction (1.36) describes the backward unimolecular decomposition of ML to release M and L; the rate constant for the decomposition is k_b. We now suppose that a system is created by establishing initial dissolved concentrations of M, L, and ML, and we are interested in the values of those concentrations at subsequent times.

From a speciation point of view, we have three time-dependent unknowns, namely [M], [L], and [ML]. We therefore need three independent equations to solve the problem. There are several possible sets of three independent equations that we could use. Each set contains the two governing mass balance equations and one of the possible differential equations. With t representing the variable time, neglecting activity corrections, one possible set of three independent reactions is:

$$d[\mathrm{M}]/dt = k_b[\mathrm{ML}] - k_f[\mathrm{M}][\mathrm{L}] \qquad \text{differential equation} \tag{1.37}$$

$$\mathrm{M_{tot}} = [\mathrm{M}]_o + [\mathrm{ML}]_o = [\mathrm{M}] + [\mathrm{ML}] \qquad \text{mass balance on total dissolved M} \tag{1.38}$$

$$\mathrm{L_{tot}} = [\mathrm{ML}]_o + [\mathrm{L}]_o = [\mathrm{ML}] + [\mathrm{L}]. \qquad \text{mass balance on total dissolved L} \tag{1.39}$$

Eq.(1.37) gives the time derivative of [M]. Additional differential equations besides Eq.(1.37) can be written, namely the time derivatives of [ML] and [L], but they are not linearly independent of Eqs.(1.37)–(1.39).

To solve the set Eq.(1.37)–(1.39), we substitute

$$[\mathrm{L}] = \mathrm{L}_{\mathrm{tot}} - [\mathrm{ML}] \tag{1.40}$$

and

$$[\mathrm{ML}] = \mathrm{M}_{\mathrm{tot}} - [\mathrm{M}] \tag{1.41}$$

into Eq.(1.37), thereby obtaining a nonlinear differential equation in the one unknown [M], namely

$$d[\mathrm{M}]/dt = k_b(\mathrm{M}_{\mathrm{tot}} - [\mathrm{M}]) - k_f[\mathrm{M}](\mathrm{L}_{\mathrm{tot}} - \mathrm{M}_{\mathrm{tot}} + [\mathrm{M}]). \tag{1.42}$$

Based on known values of k_b and k_f, and the initial conditions ($[\mathrm{ML}]_o$, $[\mathrm{M}]_o$, and $[\mathrm{L}]_o$), this equation can be integrated numerically to obtain [M] as a function of t. The mass balance equations then allow the calculation of [ML] and [L] as functions of t.

At equilibrium, $d[\mathrm{M}]/dt=0$. This occurs by definition when [M], [L], and [ML] take on their equilibrium values. Thus, at equilibrium, Eq.(1.42) gives

$$k_b(\mathrm{M}_{\mathrm{tot}} - [\mathrm{M}]) = k_f[\mathrm{M}](\mathrm{L}_{\mathrm{tot}} - \mathrm{M}_{\mathrm{tot}} + [\mathrm{M}]) \tag{1.43}$$

if $[\mathrm{ML}]_{\mathrm{eq}}$, $[\mathrm{M}]_{\mathrm{eq}}$, and $[\mathrm{L}]_{\mathrm{eq}}$ represent the equilibrium values of ML, M, and L, respectively,

$$k_b[\mathrm{ML}]_{\mathrm{eq}} = k_f[\mathrm{M}]_{\mathrm{eq}}[\mathrm{L}]_{\mathrm{eq}}]. \tag{1.44}$$

Thus, for this simple case, at long times, the kinetic model will yield concentration values that satisfy

$$k_f/k_b = \frac{[\mathrm{ML}]_{\mathrm{eq}}}{[\mathrm{M}]_{\mathrm{eq}}[\mathrm{L}]_{\mathrm{eq}}} = K. \tag{1.45}$$

and ratio k_f/k_b is identified with the equilibrium constant K for

$$\mathrm{M} + \mathrm{L} = \mathrm{ML}. \tag{1.46}$$

Thus, when a system of reactions allows that a state of equilibrium can in fact be reached (because all forward reactions have corresponding back reactions), we see how a correctly implemented kinetic model will predict the equilibrium distribution at t values large enough that equilibrium is attained.

An example of a system of reactions involving the molecule AB that can never reach an equilibrium distribution of concentrations is

$$\mathrm{AB} \xrightarrow{k_1} \mathrm{A} + \mathrm{B} \tag{1.47}$$

$$\mathrm{AB} \xleftarrow{k_2} \mathrm{A} + \mathrm{B} \tag{1.48}$$

$$\mathrm{A} \xrightarrow{k_3} \mathrm{C}. \tag{1.49}$$

There is no equilibrium solution because there is no reaction taking C back to A. No matter what the initial conditions are, eventually no AB or A will be present. Also, because there is no reaction taking C back to any of the reactants, no steady-state solution is possible: B and C just keep piling up in the system. There *would* be a steady-state solution (but still no true equilibrium solution) if we added the reaction

$$C + B \xrightarrow{k_4} AB. \tag{1.50}$$

1.5.3 Thermodynamic Equilibrium Approach

1.5.3.1 General

When the "thermodynamic equilibrium" is used here and elsewhere, it should be emphasized that full thermodynamic equilibrium is being considered, not simply some type of steady-state chemical stasis. However, since full thermodynamic equilibrium is the only form of "equilibrium" that will ever be considered in this text, for convenience's sake, here the adjective "thermodynamic" will usually be dropped.

As noted above: (1) equilibrium models tell you what the chemistry will be at equilibrium; they do not tell you about the rates according to which a system will move towards the equilibrium state; and (2) equilibrium models will provide good results when the governing reactions occur quickly relative to the timeframe of interest.

1.5.3.2 Systems at Constant *T* and *P* (the Emphasis Here) *vs.* Systems at Constant *T* and *V*

Equilibrium modeling can be carried out for constant T and P systems, and also for constant T and V systems. Most systems near the surface of the earth are of the first type, and happily this type is very much easier to deal with because the K values for the final system are knowable before the problem solving begins. As an example, say T=298.15 K and the problem is to solve for the aqueous chemistry that results after 1 mol of H_2 and ½ mol of O_2 are combined and come to equilibrium in a piston chamber that equalizes with P=1 atm: all K values are knowable since we know the final T and P.

In contrast, in a constant T and V system, the final equilibrium P value depends on the final equilibrium chemical distribution. For example, say T=298.15 K and the V of the system is 18.015 mL (which is the volume occupied by 1 mol of liquid water at T=298.15 K and P=1 atm). If we react 1 mol of H_2 with ½ mol of O_2 in this volume to form very nearly 1 mol of liquid H_2O, the pressure will be 1 atm and the pH of the water will be the same as that for the piston P=1 atm case. However, if in the same 18.015 mL system, we react 1.2 mols of H_2 with 0.6 mol of O_2 to form 1.2 mols of liquid H_2O, the pressure at equilibrium will be very much higher, and in fact we will have to consult the density of liquid water *vs.* P (at T=298.15), seeking P as required to yield a liquid water density of 1.2 mol/18.015 mL=1.2 g/mL. Only once we determine that P (which turns out to be about 9000 atm; Floriano and Nascimento, 2004) will we be able to look up the value of K_w at T=298.15 and that P so that we can compute the pH. It certainly would be difficult if all our modeling calculations were like this. Thankfully, as noted, most of the natural systems of interest for this text are constant T and P systems.

To summarize, in the equilibrium approach at constant T and P: (1) all of the reactions are considered to be reversible, and governed by the values of the equilibrium constants at the system values of T and P; (2) at equilibrium the system experiences no chemical changes with time. The exact nature of the equilibrium state is governed completely by the thermodynamic free energy (G) levels of all the chemical constituents involved at the system T and P, including all gaseous, dissolved, and solid species: the system seeks the equilibrium state by means of the reactions that adjust the amounts and concentrations of each species to find the lowest overall free energy state.

1.5.3.3 Equilibrium Constants – General

The K_w for water in Eq.(1.25) is an example of an equilibrium constant. As another, more general example, for the aqueous-phase reaction

$$aA + bB = cC + dD. \tag{1.51}$$

we have

$$K \equiv \frac{\{C\}^c\{D\}^d}{\{A\}^a\{B\}^b} = \frac{(\gamma_C[C])^c(\gamma_D[D])^d}{(\gamma_A[A])^a(\gamma_B[B])^b} \tag{1.52}$$

where: K is the T- and P-dependent equilibrium constant for reaction (1.51); {A}, {B}, {C}, and {D} are the chemical activities of species A, B, C, and D, respectively; and a, b, c, and d are the species' stoichiometric coefficients in the reaction. When activity corrections can be neglected in the aqueous phase, then all $\gamma_i = 1$ so that

$$K = \frac{[C]^c[D]^d}{[A]^a[B]^b}. \tag{1.53}$$

It was L.G. Sillén (1961) who popularized the application of equilibrium models in the solving of natural water chemistry problems. Prior to Sillén's work, natural waters had been viewed as too dynamic and subject to forcings away from equilibrium for applications of equilibrium models to be worthwhile.

The fact that equilibrium models of natural water systems are often useful means that

1) Equilibrium is often attained for species governing major aspects of natural water chemistry (*e.g.*, the reactions involving the acid/base chemistry of CO_2); and
2) When completely attained over all aspects of the system chemistry, then the changes that are occurring (as equilibrium is being sought by the system) are indicated by what equilibrium modeling says about where the system is headed, *e.g.*, oxygen levels will fall when BOD is added, and whether the system could become completely anoxic can be predicted by the starting oxygen and BOD levels.

Equilibrium models can only be as good as the equilibrium constants used in the calculations; if the value of some important equilibrium constant is in error due to problems in its original determination, then the predictions from using that value will be in error. And, of course, if an incorrect value for an equilibrium constant finds its way into a reference book, text, or website, then that value will remain incorrect no matter how universally that source is respected.

When modeling an actual natural water chemistry situation, many chemical equilibria are likely to be relevant. At equilibrium, all of the corresponding K expressions and the various pertinent mass balance conditions governing the system must be satisfied. Usually, many of the reactions are coupled (*i.e.*, many of the species are involved in multiple reactions). H^+ (for acid/base reactions) and the electron (for redox reactions) are two examples of species that are involved in many different natural water reactions. Over the last decades, considerable progress has been made in developing computer codes that can solve for the equilibrium conditions that will be attained in a complex geochemical system. The first of these was REDEQL (Morel et al., 1976). Much additional work has since been done. A recent descendant of REDEQL is MINTEQA2, currently available as freeware from a US EPA website (2006).

It is fortunate for equilibrium modeling that H^+ is an important variable in water chemistry because proton-transfer reactions between acids and bases are generally very fast. Similarly, many "complexation" reactions between metals and "ligands" are reasonably fast. However, as noted above, many redox reactions are not fast, even when mediated by bacteria and other organisms. A variety of dissolution/precipitation reactions can also be fairly slow, especially those involving very insoluble solids, *e.g.*, initial precipitates are often metastable and so tend to change into other more stable solids so that the solubility constants of the initial and final solids can be significantly different. For example, increasing the pH in a solution containing significant dissolved Fe^{3+} will initially lead to the precipitation of amorphous $Fe(OH)_{3(s)}$. In time, this solid will change into the mineral goethite (α-$FeOOH_{(s)}$), because goethite is more stable and less soluble.

Often it can be useful to ignore the fact that *full equilibrium* is not immediately attained. In that approach, one ignores the slow reaction(s) and computes the chemistry for the partial equilibrium that is attained in the intermediate term. A good example would be the Fe^{3+} chemistry in the above case: it can be assumed with good results for relatively short times (days to weeks), that $Fe(OH)_{3(s)}$ rather than α-$FeOOH_{(s)}$ governs the dissolved concentration of Fe^{3+}.

If the Earth was completely isolated from all energy sources, then it would not matter as much whether natural reactions were fast or slow, as the global system would eventually come to equilibrium. However, our planet is not isolated, but receives a constant stream of energy provided by the Sun. This energy stream continually disrupts the approach to equilibrium in aquatic and other systems. Solar energy is stored by photosynthetic organisms through their equilibrium-disrupting chemical reduction of CO_2 to form organic matter and O_2. The organic matter is then unstable in the presence of O_2 and other oxidants (*e.g.*, sulfate and nitrate) and so can be oxidized back to CO_2 by respiration in humans and other organisms. So, chemical thermodynamics is always trying to take the Earth system to full equilibrium, and energy inputs from the Sun are continually disrupting that movement towards equilibrium. Without the continual flow of solar energy, we would all be at equilibrium, *i.e.*, dead, and organisms do die when they become unable, for some reason, of maintaining the processes that frustrate equilibrium.

1.5.3.4 Concentration Scales and Equilibrium Constants

If the value of a given equilibrium constant K is to be determined, then the concentration/activity scales which are to be used for each of the reactants and products must first be selected. When the units chosen for an equilibrium constant are changed, then the numerical value may also change, as we saw to be the case for K_w depending on whether molality or mole fraction is used for water in K_w. As a further illustration, consider a reaction involving three aqueous species such that

$$A = B + C \qquad K = \frac{[B][C]}{[A]}. \tag{1.54}$$

As indicated by the expression given for K, we will be neglecting activity corrections. Assume now that the activity scale chosen is molality (m) for A, B, and C, and that with that choice, K happens to be measured to be 1.0 m (The units are m because $m^2/m = m$). Assume now that we create a system in which the initial concentrations are

$$[A] = 0.2\,m \qquad [B] = 0.0\,m \qquad [C] = 0.0\,m. \qquad \text{initial conditions} \tag{1.55}$$

Because [B] and [C] are initially zero, at equilibrium their concentrations will be equal, which we represent as y. At equilibrium $[A] = 0.2 - y$. We need to solve

$$\frac{y^2}{0.2 - y} = 1 \tag{1.56}$$

$$y^2 + y - 0.2 = 0. \tag{1.57}$$

Solving this quadratic equation gives $y = 0.171$ and so

$$[A] = 0.029\,m \qquad [B] = 0.171\,m \qquad [C] = 0.171\,m. \qquad \text{equilibrium} \tag{1.58}$$

To attain equilibrium, much of the initial A decomposes to form B and C at 0.171 m.

What would happen if we were to change the units in which we choose to express the concentration (activity)? Let us say that mole fraction rather than molal had been selected. Intuition tells us that the speciation at equilibrium cannot change because of this change in units. That is, the chemical species that participate in the reaction cannot know or care what units are chosen for expressing their concentration values. Nature does not respond to changes in units like mols or kg that Earthlings or Venusians or anybody else should happen to devise to numerically explain things. The physical height of a building 100 m tall from its base remains constant regardless of whether we chose to express the height in centimeters (10,000 cm tall), or feet (328.084 ft tall).

For the example at hand, for the aqueous equilibrium conditions described by Eq.(1.58) on the molal scale the concentration of water is ~55 *m* so for Eq.(1.16), n_{tot} in one liter will be ~55 (A, B, and C contribute negligibly to n_{tot}). The mole fraction values at equilibrium are then

$$x_A = \frac{0.029}{\sim 55} = 0.00053 \qquad x_B = \frac{0.171}{\sim 55} = 0.0031$$

$$x_C = \frac{0.171}{\sim 55} = 0.0031. \qquad \text{equilibrium} \tag{1.59}$$

With these units, the value of K would be $x_B x_C/x_A = (0.0031)^2/0.029 = 0.0018$. The numerical value and units have changed, but the composition at equilibrium has not changed. The lesson here is that a chemist can chose whatever units are desired for expressing a given equilibrium constant K, but the composition at equilibrium will independent of those units. However, whenever we solve a problem, we still have to be aware of the units that are associated with the needed K value. For example, assume again that we create an initial system like that in Eq.(1.55). And, assume that we find from a reference that $K = 1$. *But*, if we incorrectly come to believe that the concentration scales for A, B, and C for that K are mole fraction (rather than molal) so that we believed $x_B x_C/x_A = 1$ at equilibrium, then the solved values for the concentration results would be in great error.

REFERENCES

Bandura AV, Lvov SN (2006) The ionization constant of water over wide ranges of temperature and density. *J. Phys. Chem. Ref. Data*, **35**, 15–30.

Floriano WB, Nascimento MAC (2004) Dielectric constant and density of water as a function of pressure at constant temperature. *Brazilian J. Physics*, **34**, 38–41.

Morel FMM, McDuff RE, Morgan JJ (1976) Theory of interaction intensities, buffer capacities and pH stability in aqueous systems, with application to the pH of seawater and a heterogeneous model ocean system. *Marine Chemistry*, **4**, 1–26.

Pankow JF, Morgan JJ (1981a) Kinetics for the aquatic environment, I. *Environmental Science & Technology*, **15**, 1155–1164.

Pankow JF, Morgan JJ (1981b) Kinetics for the aquatic environment, II. *Environmental Science & Technology*, **15**, 1306–1313.

Sillén LG (1961) The physical chemistry of sea water. In: M Sears (Ed.), *Oceanography*, **67**, 549–581. American Association for the Advancement of Science, Washington DC.

Tratnyek PG, Scherer MM, Johnson TJ, Matheson LJ (2003) Permeable reactive barriers of iron and other zero-valent metals. In: MA Tarr (Ed.) *Chemical Degradation Methods for Wastes and Pollutants: Environmental and Industrial Applications*, Marcel Dekker, New York, pp. 371–421.

US EPA (2006) MINTEQA2 Equilibrium Speciation Model. http://www2.epa.gov/exposure-assessment-models/minteqa2.

2 Thermodynamic Principles

2.1 FREE ENERGY AND CHEMICAL CHANGE

2.1.1 Systems at Constant Temperature (*T*) and Pressure (*P*)

Determining the concentrations of all the relevant chemicals that are present at thermodynamic equilibrium in a given system can, in principle, be carried out by knowing and then solving, the *system* of equations governing all of the pertinent chemical species. At equilibrium, all of the governing equations must be satisfied. They are of three types:

a) Equilibrium constant equations;
b) Mass balance equations; and
c) Electroneutrality equations.

Equilibrium constants (*i.e.*, *K* values) are both *P*- and *T*-dependent. For example, the dissociation constant K_w for water is different for each of the following conditions:

$$\left.\begin{array}{l}\text{a) } T = 298.15\,\text{K}\,(25\,^{\circ}\text{C}),\ P = 1\,\text{atm},\ K_w = 1.01\ldots\times10^{-14}\\ \text{b) } T = 298.15\,\text{K}\,(25\,^{\circ}\text{C}),\ P = 2\,\text{atm},\ K_w = 1.01\ldots\times10^{-14}\end{array}\right\}\begin{array}{l}\text{though these two } K_w \text{ values are essentially equal}\\ \text{because the dependence on } P \text{ is very weak}\end{array}$$

$$\text{c) } T = 288.15\,\text{K}\,(15\,^{\circ}\text{C}),\ P = 1\,\text{atm},\ K_w = 0.45\ldots\times10^{-14}.$$

K values depend only weakly on *P*, so that near the surface of the Earth, where *P* ranges from less than 1 atmosphere to only a few tens of atmospheres, *K* values are virtually invariant. Thus, for conditions (a) and (b), K_w for water is almost exactly the same. However, many *K* values depend strongly on *T*, and so when $\Delta T = 10$ K (= 10 °C), there is often a significant change (which could be an increase or a decrease) in a *K* of interest that needs to be taken into consideration. In this chapter, for *K* values that pertain to some temperature (*e.g.*, 25 °C), the applicable value of *P* (usually 1 atm) will also be given.

In natural water systems, there are four main sources of pressure:

a) The weight of the overlying atmosphere (at sea level, this leads to $P = 1$ atm);
b) The weight of any overlying water;
c) The weight of any overlying confining rock and soil; and
d) Surface tension pressurization of small droplets and crystals.

For any aquatic system near sea level (collected rain, fresh surface waters (*e.g.*, rivers, shallow lakes, and reservoirs), shallow groundwaters, and near-surface seawater), $P \approx 1$ atm. When going above sea level (as for clouds or mountain streams), *P* decreases from 1 atm towards 0 atm. But, as noted, since *K* values change only weakly with changing *P*, even as *P* becomes a fraction of 1 atm, we can still use *K* values for $P = 1$ atm.

At significant water depths, large *P* values can be present. About 1 atm of pressure is added per ~10.7 m increase in depth. At the deepest point of the Pacific Ocean (the Mariana Trench, near The Philippines), the water is about 11 km deep, and $P \approx 1200$ atm. At such a pressure, the effect on *K* values must be considered.

As noted, another environment in which $P > 1$ atm is inside droplets and particles that are smaller than ~1 μm in diameter. The added pressure is due to the interfacial tension that acts over the aggregate interfacial skin of the droplet (or particle). (This is effect that causes the pressure inside a soap bubble (or balloon) to be greater than that outside.) The PV work (energy) that a material takes in during compression raises its Gibbs Free Energy G. If the per mol (molar) G of a substance in a phase is raised, there is an increase in its tendency to escape the phase. The result is that for a droplet of diameter d_p that is suspended in air, the phase making up the droplet is more volatile than when the same compound is volatilizing across a flat surface ($d_p = \infty$). Similarly, material making up a small mineral crystal of diameter d_p is more soluble in water than is a very large crystal for which dissolution is essentially taking place across a flat surface ($d_p = \infty$). This is the "Kelvin effect", and the magnitude of the effect increases as d_p decreases. For crystals surrounded by a solution, the relatively higher solubility of small crystals is what causes "Ostwald ripening": a collection of many small crystals surrounded by a liquid are slowly converted into fewer larger crystals. Many students of chemistry have observed this in a laboratory class: some small crystals made one day are left at the end of a lab class, then found a week or two later to have been transformed into a few bigger crystals.

In any system at constant T and P, the property that governs chemical equilibrium is G (kJ). For a particular reaction system, some of the G will be carried by the chemicals on the left of the = sign of the equilibrium, and the rest by the right of the = sign. The finding of an equality between

G for the remaining reactant chemicals vs. G for the produced chemicals

is what brings a particular chemical reaction into equilibrium. In a given overall system at constant T and P with many possible equilibria, each equilibrium must be in this type of balance, and moreover the overall equilibrium state is the one that gives the lowest total G.

At constant T and P, if any given chemical process lowers the total system G below the initial G, then that process:

a) Is thermodynamically favored; and
b) Will tend to occur.

The question of *how rapidly* a system will move towards the state giving the minimum G is not answered by thermodynamics. Thermodynamics only tells us that the system will, if given sufficient time, reach the predicted equilibrium state. The rate at which a chemical reaction moves at constant T and P towards a new chemical state is addressed by the field of chemical kinetics, which is largely outside the scope of this book.

2.1.2 Systems at Constant Temperature (T) and Volume (V)

Some systems of interest move to equilibrium at constant T and volume (V) rather than at constant T and P. An example would be some liquid in which a reaction is taking place, with the liquid occupying a compartment of fixed volume in rigid rock. P will change as the reaction proceeds because the produced amounts of the products will always occupy at least a slightly different volume than the corresponding reacted amounts of the reactants.

The thermodynamic function that must be minimized for equilibrium to be attained at constant T and V is not G, but rather the "Helmholtz Free Energy" A (kJ). The equilibrium state that is reached at constant T and V will be the one for the particular initial mix of reactants and products that minimizes A. This type of problem is as solvable as one with a constant T and P, but to *predict* the nature of the equilibrium state, we would find the final amounts of reactants and products that minimize A. The final state will, of course, be at some ultimately constant P. If that P can be measured, the values of the G-based chemical equilibrium constants for the T and P may then be used to compute the equilibrium speciation. This text focuses on G-based predictive calculations based on K values for specific T and P conditions; we will not trouble ourselves further with constant T and V systems.

2.2 A POTENTIAL ENERGY ANALOG FOR CHEMICAL ENERGY-DRIVEN CHEMICAL CHANGE

Consider a generic chemical reaction between a mols of A with b mols of B to give c mols of C and d mols of D:

$$a\mathrm{A} + b\mathrm{B} = c\mathrm{C} + d\mathrm{D} \qquad K = \frac{\{\mathrm{C}\}_{\mathrm{eq}}^{c}\{\mathrm{D}\}_{\mathrm{eq}}^{d}}{\{\mathrm{A}\}_{\mathrm{eq}}^{a}\{\mathrm{B}\}_{\mathrm{eq}}^{b}}. \tag{2.1}$$

Assume that the system is characterized by initial values of the activities, $\{\mathrm{A}\}_{\mathrm{init}}$, $\{\mathrm{B}\}_{\mathrm{init}}$, $\{\mathrm{C}\}_{\mathrm{init}}$, and $\{\mathrm{D}\}_{\mathrm{init}}$, and that these that do not satisfy K for reaction (2.1) at the T and P of interest. An analog for a chemical system governed by G is a mechanical system governed by potential energy (PE). In Figure 2.1.a, a ball is at height $h+H$. The lowest available energy state is at H. The initial state is

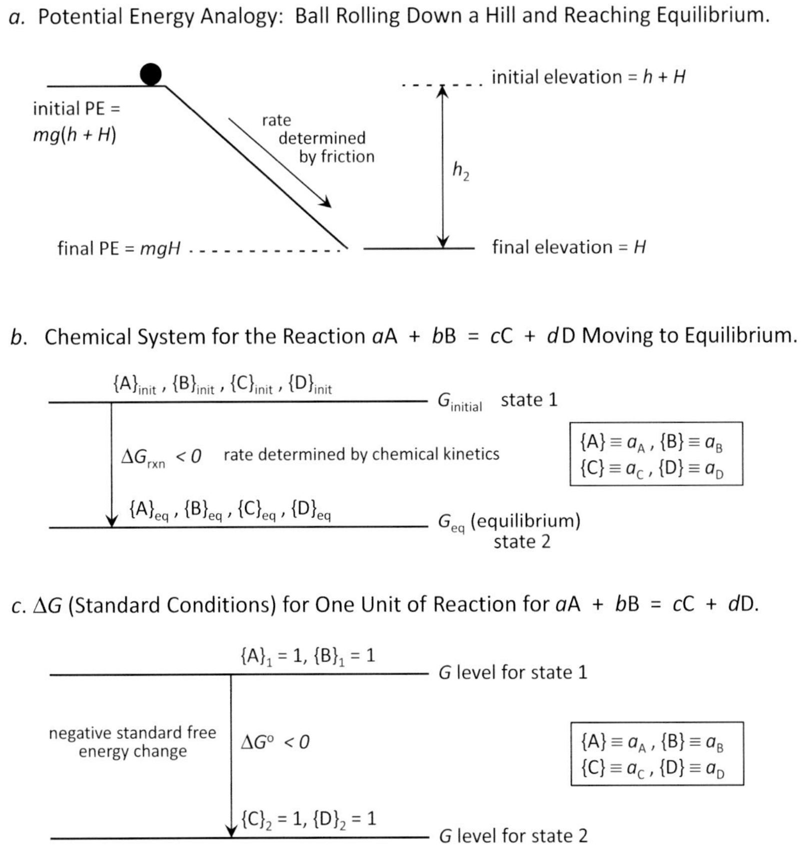

FIGURE 2.1 a. A potential energy (PE) system that can move from an initial PE state to a lower PE state. The PE change is controlled by the initial and final elevations, the mass of the ball m, and the gravitational acceleration g. Friction and drag will determine the rate at which the ball will roll down the incline. b. Because $\Delta G_{\mathrm{rxn}} < 0$, the initial, nonequilibrium chemical species distribution at some T and P will tend to move towards an equilibrium species distribution at the T and P of interest; chemical kinetics will determine the rate at which the system moves towards the equilibrium state. c. ΔG° is the change in G when the products (at standard conditions) are formed from the reactants at standard conditions.

not stable because a slight push will start the ball rolling down the slope towards *H*. In the chemical system, Figure 2.1.*b* shows that an initial, non-equilibrium chemical species distribution is unstable relative to the equilibrium species distribution, which gives the lowest possible *G*.

Thermodynamics can predict the nature of a final chemical equilibrium state just as Figure 2.1.*a* allows us to predict that the ball will no longer tend to change its level once it reaches *H*. However, our knowledge that the PE of the ball governs its equilibrium position in Figure 2.1.*a* does not tell us anything about how quickly the ball will roll down the slope; so too, thermodynamics does not tell us anything specific about the rate (*i.e.*, kinetics) according to which the chemical process in Figure 2.1.*b* will occur. *Friction* in Figure 2.1.*a* is the analog of the *chemical kinetics limitations* that impede thermodynamically possible chemical change in Figure 2.1.*b*.

At the *T* and *P* of the system, the height of the upper level for the total *G* in Figure 2.1.*b* is set by the initial values for the numbers of mols of the different chemicals and the per mol *G* values of the species, as set by their initial chemical activities. The total *G* level for the equilibrium state will be set by the equilibrium values of the numbers of mols and the per mol *G* values of the reactant and product species, as set by their final, equilibrium activities. At equilibrium, the concentration and activity values must satisfy: (a) all equilibrium constants for the system; (b) all pertinent mass balance requirements for the chemicals; and (c) electroneutrality.

2.3 CHEMICAL POTENTIAL AND ITS RELATIONSHIP TO FREE ENERGY *G*

In a given system of interest, the **chemical potential** of species *i* is the per mol amount of *G* of species *i* at *T*, *P*, and the activity of *i*. The chemical potential of *i* is represented as μ_i (kJ/mol). The activity of *i* we represent **interchangeably** as $\{i\}$ and as a_i. For a dissolved constituent, a_i is affected by the concentrations not only of *i*, but also of all other species in the solution. Consider two solutions, both at $T=298.15\ K$ and $P=1$ atm. One is pure water, and the other is some salty solution like seawater. In pure water, $[H^+]$ is $10^{-7.00}\ M$, and let us say we have picked a salty solution for which $[H^+]$ again happens to be $10^{-7.00}\ M$. In pure water, because there is nothing in the solution to affect $\{H^+\}$, then $\{H^+\}=[H^+]=10^{-7.00}\ M$. However, in the salty solution, attraction of H^+ to the negatively charged salt ions causes $\{H^+\}$ to be lower than $[H^+]$, *i.e.*, $\gamma_{H^+}<1$. In particular, the negative ions of the salty water tend to surround each H^+, which makes each H^+ more "comfortable" than in pure water: μ_{H^+} is lower in the salty solution than in pure water.

The mathematical definition of μ_i in a system of interest is

$$\mu_i = \left(\frac{\partial G}{\partial n_i}\right)_{T,P,n_j \neq n_i}. \tag{2.2}$$

For simplicity, we will assume here that the system is a single phase. At the given *T* and *P*, each μ_i value depends on the makeup of the phase. By definition, μ_i is the ratio of: (1) the infinitesimal change in the system *G* at constant *T* and *P*; to (2) the causal infinitesimal change in the number of mols of species *i*. Since μ_i is a partial derivative, the composition of the phase (as well as *T* and *P*) remains constant during the addition of ∂n_i. This means the numbers of mols of all species $j \neq i$ are held constant, and for *i* we already know that ∂n_i is very small.

Since μ_i represents the free energy content per mol of *i* under the specific conditions of the system, for a one-phase system, we can multiply μ_i by n_i (the number of mols of *i*) to obtain G_i, the contribution made by species *i* to the total system *G*:

$$G_i = n_i\mu_i. \tag{2.3}$$

For a species in a mixture (whether gaseous, solution, or solid), n_i will equal the product of the concentration of *i* (mol/L) and the volume *V* (L) of the phase so that:

$$G_i = [i]V\mu_i. \tag{2.4}$$

For a pure solid, n_i is given by the mass of the solid divided by the formula weight. Since μ_i values depend on T and P, so do G_i values. For the total G of a one-phase system, the contributions made by all chemical species present are summed:

$$G = \sum_i G_i = \sum_i n_i \mu_i. \tag{2.5}$$

For systems that have multiple phases, Eq.(2.3) needs to be summed over all phases to obtain the total G ascribable to i. And, the G values for the various phases, as obtained by Eq.(2.5), need to be summed to obtain the total G.

Example 2.1 G_i

The chemical potential of H^+ at 10^{-5} M is y kJ/mol in a certain one-phase aqueous system. The volume of that system is 5 L. Write the expression for G_i for H^+.

Solution

$G_i = 10^{-5}\ M \times 5\ \text{L} \times y$ kJ/mol.

Example 2.2 G_i and Multiple Phases

Assume that oxygen is present in a two-phase, air/water system. Write an expression for G_{O_2} that includes a term of the form given by Eq.(2.4) for each of the two phases.

Solution

$G_{O_2} = [O_2]_{air} V_{air} \mu_{O_2,air} + [O_2]_{water} V_{water} \mu_{O_2,water}$

Example 2.3 Total G over Multiple Phases

Assume that some initially pure water (IPW), some oxygen, and some nitrogen are combined to yield a two-phase, gas/water system. Write an expression for the total G that includes terms for H_2O, O_2, and N_2 in both the aqueous and gas phases, and terms for H^+ and OH^- in just the aqueous phase (*i.e.*, we are assuming that both H^+ and OH^- are not volatile, and so exist essentially only in the aqueous phase).

Solution

$$\begin{aligned} G &= G_{O_2} + G_{N_2} + G_{H_2O} + G_{H^+} + G_{OH^-} \\ &= [O_2]_{air} V_{air} \mu_{O_2,air} + [O_2]_{water} V_{water} \mu_{O_2,water} \\ &\quad + [N_2]_{air} V_{air} \mu_{N_2,air} + [N_2]_{water} V_{water} \mu_{N_2,water} \\ &\quad + [H_2O]_{air} V_{air} \mu_{H_2O,air} + [H_2O]_{water} V_{water} \mu_{H_2O,water} \\ &\quad + [H^+] V_{water} \mu_{H^+} + [OH^-] V_{water} \mu_{OH^-} \end{aligned}$$

2.4 PROPERTIES AND APPLICATIONS OF THE CHEMICAL POTENTIAL

2.4.1 General

A given μ_i can be positive, zero, or negative. By Eq.(2.5), this means that system G values can also be positive, zero, or negative. As with physical height, the magnitude and sign of a given μ_i value depends on the characteristics of what is being measured, and on the value that is chosen as the reference for the measurement. Consider the Empire State Building (ESB) in New York City which is 443 m from lobby to pinnacle. If the absolute reference zero height level is taken as current sea level, since the lobby level is 14 m above sea level, then the height of the building is 457 m. Other than inconvenience, there is no reason why one could not use the center of the Earth as the absolute zero for elevation measurements.

For matters related to the free energy, the reason we are interested in μ_i values is so that we can compute energy *differences* (*e.g.*, see Eq.(2.8) below). In the physical world, the analog of this type of focus would be having interest only in changes in elevation from one height to another. When moving up and down in the ESB, we can do this with either the lobby level or sea level serving as the reference level, as long as we are consistent (*i.e.*, remain with the same reference level within a particular difference calculation).

μ_i depends on the activity of i at a given T and P according to

$$\mu_i = \mu_i^{\text{o}} + RT \ln a_i \tag{2.6}$$

where μ_i^{o} is the reference level for μ_i, which is the chemical potential of i in the chosen *standard state* at T and P. When $a_i = 1$, then

$$\mu_i = \mu_i^{\text{o}} \qquad (a_i = 1). \tag{2.7}$$

2.4.2 Changes in Free Energy

Because of Eq.(2.5), the system G is composed of individual contributions from all of the species in the system. Therefore, when one or more chemical reactions occur which change the n_i and μ_i values, then

$$\Delta G = \sum_i \Delta G_i = \sum_i (n_i\mu_i)_{\text{final}} - \sum_i (n_i\mu_i)_{\text{initial}}. \tag{2.8}$$

Eq.(2.8) is used to compute ΔG for a chemical system that goes from one state to another. Eq.(2.8) can be expanded to give

$$\Delta G = \sum_i \Delta G_i = \sum_i n_{i,\text{final}}(\mu_i^{\text{o}} + RT \ln a_{i,\text{final}}) - \sum_i n_{i,\text{initial}}(\mu_i^{\text{o}} + RT \ln a_{i,\text{initial}}). \tag{2.9}$$

In Figure 2.1.*b*, the initial n_i and a_i values for A, B, C, and D are envisioned to be altered (by forward reaction or backward reaction) according to Eq.(2.1), to reach a final *equilibrium* set of n_i and a_i, for which the G level is lower than the initial level. During the reaction, the stoichiometry in Eq.(2.1) is followed until the lowest possible G level is attained. As noted, that level will be the one that satisfies the value of K, the relevant mass balance constraints, and electroneutrality if any species are charged.

For an example of a system subject to particular mass balance constraints, assume that: (1) the reaction in Eq.(2.1) takes place in an aqueous phase; (2) $a=b=c=d=1$; and (3) initially, [A] = 1 m, [B] = 1 m, [C] = 0 m, [D] = 0 m. Three independent mass balance constraints are then operational as the reaction progresses and at equilibrium:

$$[\text{A}] = [\text{B}] \qquad [\text{C}] = [\text{D}] \qquad [\text{A}] = 1 - [\text{C}].$$

It is also true that [B] = 1 − [C], but that is not an independent relation.

Understanding the meaning of chemical equilibrium involves looking at the ΔG difference between two states, *i.e.*, we are interested in knowing the G for the products under specific activity conditions relative to the G for the reactants under specific activity conditions. For reaction (2.1), we are interested in knowing what the ΔG difference is between

state 2: c mols of C at activity $\{C\}_{init}$ plus d mols of D at activity $\{D\}_{init}$ (no A or B)

and

state 1: a mols of A at activity $\{A\}_{init}$ and b mols of B at activity $\{B\}_{init}$ (no C or D)

$$\Delta G_{rxn} = G_{state\ 2} - G_{state\ 1} = c\mu_C + d\mu_D - (a\mu_A + b\mu_B). \tag{2.10}$$

At a given T and P, if $\Delta G_{rxn} < 0$, then the G level of *state 2* is lower than G level of *state 1*, and the reaction will tend to occur from left to right. If $\Delta G_{rxn} > 0$, then the G level of *state 2* is higher than the G level of *state 1*, and the reverse will tend to occur. But if $\Delta G_{rxn} = 0$, then the G levels of the two states are the same, so that a system with A at activity $\{A\}_{init}$, B at activity $\{B\}_{init}$, C at activity $\{C\}_{init}$ *and* with D at activity $\{D\}_{init}$ is already at equilibrium.

Usually, one does not bother including the subscript "rxn" in equations like Eq.(2.10); we rely on the context of the usage to convey that we are referring to a ΔG for some conversion of reactants under specific conditions to products under specific conditions. In general then, for the ΔG of a reaction,

$$\Delta G = \sum_i \nu_i \mu_i = \sum_i \nu_i (\mu_i^o + RT \ln a_i). \tag{2.11}$$

The "stoichiometric coefficients" ν_i have signs and are related to the n_i. For reaction (2.1),

$$\nu_C = c = n_C$$
$$\nu_D = d = n_D$$
$$\nu_A = -a = -n_A$$
$$\nu_B = -b = -n_B.$$

The ν_i are positive for products and negative for reactants. They are negative for reactants because of the minus sign in Eq.(2.8) in front of the summation $\sum_i (n_i \mu_i)_{initial}$.

Example 2.4 Stoichiometric Coefficients ν_i for a Reaction

What are the ν values for each of the following reactions?

a. Dissociation of hydrofluoric acid in water: $HF = H^+ + F^-$.
b. Dissolution of solid calcium carbonate in water to form dissolved calcium and carbonate ions: $CaCO_{3(s)} = Ca^{2+} + CO_3^{2-}$.
c. Dissolution in water of the sodium sulfate mineral *mirabilite* (contains 10 waters of hydration) to form dissolved sodium ions, sulfate ions, and water: $Na_2SO_4 \cdot 10H_2O_{(s)} = 2Na^+ + SO_4^{2-} + 10\,H_2O$.

Solution

a. $\nu_{H^+} = +1,\quad \nu_{F^-} = +1,\quad \nu_{HF} = -1.$
b. $\nu_{Ca^{2+}} = +1,\quad \nu_{CO_3^{2-}} = +1,\quad \nu_{CaCO_{3(s)}} = -1.$
c. $\nu_{Na^+} = +2,\quad \nu_{SO_4^{2-}} = +1,\quad \nu_{H_2O} = +10,\quad \nu_{Na_2SO_4 \cdot 10H_2O(s)} = -1.$

Specific values of T and P and of the activities for the reactants and products are needed when applying Eq.(2.11). Again, then, for reaction (2.1), Eq.(2.11) gives ΔG when a mols of A at activity a_A and b mols of B at activity a_B are converted at T and P to c mols of C at activity a_C and d mols of D at activity a_D.

Equation (2.11) can be expanded into two summations, *i.e.*,

$$\Delta G = \sum_i \nu_i \mu_i^o + \sum_i \nu_i RT \ln a_i. \tag{2.12}$$

Let

$$\boxed{\Delta G^o = \sum_i \nu_i \mu_i^o = \sum_i \nu_i \Delta G_f^o} \tag{2.13}$$

then

$$\boxed{\Delta G = \Delta G^o + RT \ln Q.} \tag{2.14}$$

Q is the **"reaction quotient"**, and can also be thought of as the **product** $\prod$ of the activities of the chemical products and reactants each raised to the power of its υ_i value. For reaction (2.1),

$$Q = \frac{a_C^c a_D^d}{a_A^a a_B^b} = \prod_i a_i^{\nu_i} \tag{2.15}$$

$$\Delta G^o = c\mu_C^o + d\mu_D^o - a\mu_A^o - b\mu_B^o. \tag{2.16}$$

$\Delta G°$ pertains only to the T and P for which the μ_i^o apply; most μ_i^o values are tabulated for 25 °C and 1 atm.

Equation (2.14) leads to two important conclusions. First, when all a_i = 1.0, then Q = 1 and

$$\Delta G = \Delta G^o \qquad (\text{all } a_i = 1). \tag{2.17}$$

As indicated in Figure 2.1.*c*, $\Delta G°$ is thus the difference in free energy at 25 °C and 1 atm for

state 2: c mols of C at a_C = 1 and d mols of D at a_D = 1 (no A or B)

vs.

state 1: a mols of A at a_A = 1 and b mols of B at a_B = 1 (no C or D)

Second, when a system at 25 °C and 1 atm is at equilibrium because ΔG=0, then Q=K at 25 °C and 1 atm:

$$0 = \Delta G^o + RT \ln K \qquad (\text{at 25 °C and 1 atm}) \tag{2.18}$$

and

$$\Delta G^o = -RT \ln K \qquad (\text{at 25 °C and 1 atm}) \tag{2.19}$$

$$\boxed{\Delta G^o = -2.303\ RT \log K.} \tag{2.20}$$

Combining Eqs.(2.14) and (2.20),

$$\Delta G = 2.303\ RT \log Q/K. \tag{2.21}$$

When $Q=1$, if the system is at 25 °C and 1 atm so that $\Delta G = \Delta G°$, then Eq.(2.21) reverts to Eq.(2.20). Any reaction for which $\Delta G < 0$ will tend to proceed as written because the system is moving from a higher G state to a lower G state: the system is seeking the minimum value of G.

2.4.3 The Role and Nature of μ_i^o

As discussed, μ_i (kJ/mol) is the G content *per mol* of species i at activity a_i. The PE analogy used in Figure 2.1.*a* can be adapted to better understand μ_i and Eq.(2.6). If h=height, then for some mass m, we have PE=mgh. If m has units of metric tons (1000 kg), g=9.8 m/s^2, and h has units of m, then mgh has units of kJ. Dividing mgh by m gives gh (kJ/ton) which is the potential energy per unit mass. Thus, gh (kJ/ton) is a PE analog of μ_i (kJ/mol). We can take this analogy further and write

$$gh = (gh)° + gh \tag{2.22}$$

so that $(gh)°$ is the standard reference value for gh. Although we don't have to, PE can be measured relative to $h=0$, and if we do then $(gh)°=0$. Eq.(2.22) has the same form as Eq.(2.6). It is no coincidence that for each pure element in its most stable form at 25 °C and 1 atm, it has been decided to set $\mu_i^o=0$. Thermodynamicists did not have to do that, they could have picked something else; zero for all elements in their most stable forms at 25 °C and 1 atm was convenient and easy to remember. For the element oxygen, at 25 °C and 1 atm, oxygen is gaseous O_2, so the standard state is pure O_2 gas at 25 °C/1 atm with $\mu_i^o=0$.

Usually, we are not dealing with pure elements, but rather elements as they may be dissolved in another chemical (as with O_2 dissolved in water), elements combined with other elements (as with the glucose molecule $C_6H_{12}O_6$), and myriad other cases. So, all subsequent μ_i values are usually measured using other non-zero μ_i^o values for chemicals in particular reference states, all measured relative to the special μ_i^o which are all 0 for the elements in their standard states. Consider what happens if you leave a boat at a dock (sea level) in New York City, walk to the Empire State Building (ESB), and go into the lobby which is at 14 m above current sea level. When in the boat, the PE value is 0 kJ/ton: this is the special zero reference, relative to sea level. The lobby level is then a non-zero reference state of great interest that we will want to measure against when going up and down in the elevator, so the lobby has a reference state value of $g \times 14$ kJ/ton. Table 2.1 summarizes this analogy, and analogizes the lobby with O_2 dissolved in water at 1 *m* (at 25 °C/1 atm), and analogizes being in the elevator with O_2 dissolved in water at other concentrations (at 25 °C/1 atm).

TABLE 2.1
Potential Energy and Chemical Potential Analogy: New York City and the Empire State Building *vs.* Oxygen

	potential energy		chemical potential	
	physical system	**PE (kJ/ton)**	**chemical system**	**μ_{O_2} (kJ/mol)**
special overall zero:	boat at NYC dock (sea level)	$g \times 0$	O_2, pure gas, 25 °C/1 atm	$\mu^o_{O_2}$(gaseous)=0.00 kJ/mol
other case of interest:	lobby, ESB	$g \times 14$	O_2, aqueous at 1 *m*, 25 °C/1 atm	$\mu^o_{O_2}$(aqueous)=16.32 kJ/mol
sub-cases measured relative to the other case of interest:	elevator, ESB; moving $\pm h$ from lobby	$g \times (14 \pm h)$	O_2, aqueous at x *m*, 25 °C/1 atm	μ_{O_2} = 16.32 kJ/mol + $RT \ln x$

Both gh and μ_i can take on an infinity of values. Figure 2.2 depicts gh and μ_i each on an energy *staircase*. Although Figure 2.2 shows finite heights for the steps in the staircases, for us, an infinitely fine gradation exists between the energy levels since both h and a_i are continuous. To summarize: (1) the reference level for the submarine is where $h=0$, and μ_i^o is the reference level for μ_i; (2) a species i may be a pure element in some form, the element dissolved in water, a compound of different elements, and/or an ion; (3) if i is an element, for only one of possible multiple forms will $\mu_i^o=0$; (4) μ_i^o does not have to be zero for a species for it to be useful as a reference – it just needs to be quantified and used according to consistent thermodynamic conventions.

The quantity $RT\ln a_i$ becomes increasingly negative as a_i decreases below unity, with $RT\ln a_i \to -\infty$ as $a_i \to 0$. (The μ_i staircase goes down forever.) No matter how large μ_i^o might be, one can always find a value of a_i that makes $\mu_i<0$, and thus when $a_i \to 0$, then μ_i becomes infinitely negative. This is the fundamental and immutable reason why it is: (1) so difficult to prepare materials that are nearly pure; and (2) impossible to prepare materials that are completely pure. Consider purifying some gold that also contains some silver. Some of the Ag can be dissolved out oxidatively using nitric acid (HNO_3), the result being a solution of silver nitrate, which can be removed by decantation. The remaining solid gold can be dissolved in aqua regia (~3:1 HCl:HNO_3) to form $AuCl_4^-$. A second attempt to remove Ag^+, this time by precipitation as $AgCl_{(s)}$, is thereby implemented. Then, the *nearly* Ag-free solution of $AuCl_4^-$ can be chemically reduced using a sulfur compound to yield a precipitate of *nearly* pure elemental gold. However, despite anyone's best efforts, there will *always* have been some Ag^+ in the solution from which the gold is precipitated, and some of

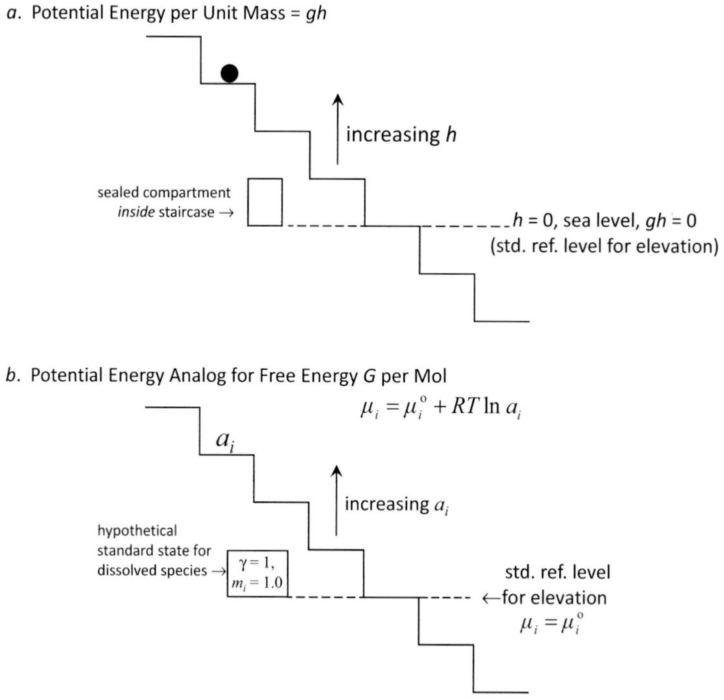

FIGURE 2.2 *a.* Potential energy (PE) staircase. Although one cannot access the sealed compartment in the staircase, the PE level that the ball would have if it were in that compartment is perfectly well defined. *b.* Chemical potential energy (μ) staircase. Although one cannot physically access the hypothetical standard state for which $\gamma_i=1$ and $[i]=1\ m$, the μ_i level that species i would have, if it were in that state, is perfectly well defined, and equal to μ_i^o. (Note that it *is* physically possible for $\mu_i = \mu_i^o$ without simultaneously having $\gamma_i=1$ and $[i]=1$: all that we need is $\gamma_i[i]=\{i\}=a_i=1$.)

that silver will end up in the precipitated gold. The solid gold will *always* be contaminated to some degree by the reaction

$$Ag^{\text{surroundings}} = Ag^{\text{solid gold}}. \quad (2.23)$$

When the a_{Ag} in the gold is very low, $\mu_{Ag}^{\text{solid gold}}$ becomes very negative, and ΔG for reaction (2.22) becomes highly negative:

$$\Delta G = \mu_{Ag}^{\text{solid gold}} - \mu_{Ag}^{\text{surroundings}} \ll 0. \quad (2.24)$$

Pure materials are therefore like *irresistible thermodynamic magnets* for all other species. Achieving and maintaining *near absolute purity* is difficult and expensive. As noted, achieving and maintaining *absolute purity* for any phase is impossible. The interested reader might examine the pricing from any supplier of chemicals for some element of interest at increasing purity levels, *e.g.*, 99.9% ("three nines") to 99.99% ("four nines") to 99.999% ("five nines").

Example 2.5 Chemical Potential Depends on Activity

For reaction (2.24), $\mu_{Ag}^{\text{gold}} = \mu_{Ag}^{\text{o,gold}} + RT\ln a_{Ag}^{\text{gold}}$

a. If the activity of silver in solid gold (namely a_{Ag}^{gold}) is measured on the mole fraction scale, what is the value of $RT\ln a_{Ag}^{\text{gold}}$ at 25 °C (=298.15 K) when $a_{Ag}^{\text{gold}} = 10^{-3}$, 10^{-13}, and 10^{-23}? Note: $R=0.008314$ kJ/mol-K.
b. Since $\mu_{Ag}^{\text{o,gold}}$ is a finite number, how does this trend affirm what has been said about the value of the chemical potential of silver dissolved in gold (μ_{Ag}^{gold}) as a_{Ag}^{gold} becomes very small?

Solution

a. The values of $RT\ln a_{Ag}^{\text{gold}}$ at $a_{Ag}^{\text{gold}} = 10^{-3}$, 10^{-13}, and 10^{-23} are −17.1, −74.2, and −131.3 kJ/mol, respectively.
b. The smaller that a_{Ag}^{gold} becomes, the smaller that μ_{Ag}^{gold} will become, because the value of $\mu_{Ag}^{\text{o,gold}}$ is a constant.

2.4.4 Standard Free Energies of Formation (ΔG_f^o) for Neutral Species

Continuing the discussion of the Figure 2.1.*a* analogy given above, the *change* in the PE of the ball does not depend on the specific value of the lower elevation *H*. For example, if $H=6$ cm and $h+H=10$ cm, the elevation change will be −4 cm. If *H* is assigned a value of 5 cm, the upper level will be at 9 cm, and the elevation change will still be −4 cm. Analogously, the difference in *G* between the lower and the upper levels in Figure 2.1.*b* does not depend on the absolute overall free energy value of the lower level. This is fortunate because *absolute values for the* μ_i are not known. This is why we can assign $\mu_i^o \equiv 0$ for the elements in their "standard states". So, as noted, for 25 °C and 1 atm, $\mu_i^o \equiv 0$ for pure O_2 gas at 1 atm. For pure gold at 25 °C and 1 atm, the most stable form is the solid in the *cubic close-packed* crystal structure gold, so for gold in that form, $\mu_i^o \equiv 0$. All μ_i^o values for elements in forms other than their standard states, and for chemical compounds, can be obtained on the $\mu_i^o \equiv 0$ values for the elements.

μ_i^o values (including those for the elements themselves) are referred to synonymously as "standard free energies of formation," with the symbol ΔG_f^o(kJ/mol). In words, we define that ΔG_f^o for a chemical is the free energy change when one mol of the chemical is formed in its standard state at 25 °C and 1 atm from its constituent elements in their most stable forms at 25 °C and 1 atm.

Since the standard state for any given chemical i calls for an activity of 1.0, then according to Eq.(2.6), the chemical potential μ_i (kJ/mol) of a species in its standard state will equal μ_i^o. Thus, at 25 °C and 1 atm, we have

$$\mu_i \text{ (under standard state conditions)} = \mu_i^o = \Delta G_f^o. \tag{2.25}$$

We consider some reactions that involve the formation of chemical molecules from their constituent elements:

$$H_{2(g)} + S_{(s)} = H_2S_{(g)} \tag{2.26}$$

$$H_{2(g)} + S_{(s)} = H_2S_{(aq)} \tag{2.27}$$

$$H_{2(g)} + ½O_{2(g)} = H_2O_{(g)} \tag{2.28}$$

$$H_{2(g)} + ½O_{2(g)} = H_2O_{(aq)} \tag{2.29}$$

$$S_{(s)} + O_{2(g)} = SO_{2(g)} \tag{2.30}$$

$$Hg_{(l)} + S_{(s)} = \text{(red) } HgS_{(s)} \tag{2.31}$$

$$O_{2(g)} = O_{2(aq)}. \tag{2.32}$$

Since $\Delta G_f^o = 0$ for all of the reactants, and since each reaction only gives one mol of one product, measuring $\Delta G°$ for each reaction provides $\mu_i^o = \Delta G_{f,i}^o$ for the product. For example, for reaction (2.26) at 25 °C and 1 atm,

$$\Delta G^o = -33.56 \text{ kJ} = (1 \text{ mol})\Delta G^o_{f,H_2S(g)} - 0 - 0 = (1 \text{ mol})\Delta G^o_{f,H_2S(g)}. \tag{2.33}$$

Table 2.2 gives the results for reactions (2.26)–(2.32). Stumm and Morgan (1996), Bard et al. (1985), and other sources provide tables of ΔG_f^o values.

The ΔG_f^o values measured when species are formed from their elements (*i.e.*, utilizing reactions such as (2.26)–(2.32)) are used to deduce $\Delta G°$ values at 25 °C/1 atm for other reactions of interest. For example, based on data in Table 2.2, we can conclude that at 25 °C and 1 atm that for

$$S_{(s)} + O_{2(aq)} = SO_{2(aq)} \tag{2.34}$$

$$\begin{aligned}\Delta G^o &= (1 \text{ mol})(-300.19 \text{ kJ/mol}) + (-1 \text{ mol})(0 \text{ kJ/mol}) \\ &\quad + (-1 \text{ mol})(16.52 \text{ kJ/mol}) = -316.71 \text{ kJ}.\end{aligned} \tag{2.35}$$

$\Delta G°$ has units of kJ rather than kJ/mol because the stoichiometric coefficients have the same units as n_i that is, "mols". Thus, at 25 °C and 1 atm, for

$$½S_{(s)} + ½O_{2(aq)} = ½SO_{2(g)}. \tag{2.36}$$

TABLE 2.2
Selected ΔG_f^o(kJ/mol)(= μ_i^o) Values for at 25 °C and 1 atm

species	phase	standard state: value & concentration scale	ΔG_f^o(kJ/mol)(= μ_i^o)
$H_2S_{(g)}$	gas	1 atm	–33.56
$H_2S_{(aq)}$	aqueous	1 *m* (molal)	–22.87
$H_2O_{(g)}$	gas	1 atm	–228.59
$H_2O_{(aq)}$	aqueous	$x = 1$ (mole fraction)	–237.18
$SO_{2(g)}$	gas	1 atm	–300.19
(red) $HgS_{(s)}$	solid	$x = 1$ (mole fraction)	–50.63
$O_{2(aq)}$	aqueous	1 *m* (molal)	16.52
acetic acid	aqueous	1 *m* (molal)	–396.4
H^+	aqueous	1 *m* (molal)	0.0
Cl^-	aqueous	1 *m* (molal)	–131.26
acetate ion	aqueous	1 *m* (molal)	369.2
Fe^{2+}	aqueous	1 *m* (molal)	–78.87

Note: For H^+, it is defined that $\Delta G_f^o \equiv 0.0$ kJ/mol.

$\Delta G°$ is one half that for reaction (2.34). In particular, we have that $\Delta G° = (½\text{ mol})(-300.19\text{ kJ/mol}) + (-½\text{ mol})(0\text{ kJ/mol}) + (-½\text{ mol})(16.52\text{ kJ/mol}) = -158.36\text{ kJ} = ½\,(-316.71)\text{ kJ}$.

Once a $\Delta G°$ has been computed at 25 °C/1 atm using tabulated ΔG_f^o values, then the value of the corresponding equilibrium constant at 25 °C and 1 atm can be calculated using Eq.(2.20). The value of the equilibrium constant at other T and P can then be determined using equations presented in Section 2.5.

Example 2.6: $\Delta G°$ for a Reaction

a. Compute the value of ΔG^o at 25 °C for the reaction by which gaseous O_2 dissolves in water, namely $O_{2(g)} = O_{2(aq)}$.
b. What are the units for the equilibrium constant $K = [O_{2(aq)}]/p_{O_2}$?
c. Use the value from part a to compute K.
d. When the total air pressure is 1 atm, the pressure of $O_{2(g)}$ is 0.21 atm. For that value, what is the concentration of $O_{2(aq)}$ in molality, and in mg of O_2 per liter.

Solution

a. From Table 2.2 and knowing that $\Delta G_f^o = 0$ for $O_{2(g)}$, then

$$\Delta G^o = (1\text{mol})(16.52\text{ kJ/mol}) - (1\text{mol})(0\text{ kJ/mol}) = 16.52\text{ kJ}.$$

b. Units of K are *m*/atm.
c. $\Delta G^o = -2.303\ RT \log K$.

$\log K = -\Delta G^o/2.303\ RT = -16.52\text{ kJ}/(2.303 \times 0.008314\text{ kJ/mol-K} \times 298.15\text{ K}) = -2.90$
$K = 1.27 \times 10^{-3}\ m/\text{atm}$

d. $K = [O_{2(aq)}]/p_{O_2}$

$$[O_{2(aq)}] = K\,p_{O_2} = 1.27 \times 10^{-3}(m/\text{atm}) \times 0.21\,\text{atm} = 2.67 \times 10^{-4}\ m \approx 2.67 \times 10^{-4}\ M.$$

The MW of O_2 is 31.999 g/mol = 31,999 mg/mol, so
$2.67 \times 10^{-4}\ M \times 31{,}999$ mg/mol = 8.55 mg/L.

2.4.5 Standard Free Energies of Formation (ΔG_f^o) Values for Ionic Species

The species formed by reactions (2.26–2.32) are all neutral, that is, not ionic. In aqueous systems, however, dissolved ions are very important. Indeed, besides water itself, ions are almost always the most abundant species. We need ΔG_f^o $(=\mu_i^o)$ values for dissolved ions.

Of all the ions present in water, H^+ is one of the most important, even when it is present at extremely low concentrations. Starting with the approach outlined above, consider a reaction by which H^+ is formed from $H_{2(g)}$, *e.g.*,

$$\tfrac{1}{2}H_{2(g)} + \tfrac{1}{2}Cl_{2(g)} = H^+ + Cl^-. \tag{2.37}$$

ΔG^o for this reaction can be measured by experiment to be −131.26 kJ. Since $\Delta G_f^o \equiv 0$ for both $H_{2(g)}$ and $Cl_{2(g)}$,

$$\Delta G^o = -131.26\text{ kJ} = (1\text{ mol})\Delta G^o_{f,H^+} + (1\text{ mol})\Delta G^o_{f,Cl^-} - 0 - 0. \tag{2.38}$$

However, there is no way to determine how much of the ΔG^o came from $\Delta G^o_{f,H^+}$ and how much from $\Delta G^o_{f,Cl^-}$. Because of that ambiguity, we are actually free to assign all of the ΔG^o to one and none to the other, as long the ΔG_f^o for every other ion is measured consistent with that same choice. Since H^+ is such a special species, it gets the honor that $\Delta G^o_{f,H^+} \equiv 0$, so that $\Delta G^o_{f,Cl^-} = -131.26$ kJ/mol.

Consider now

$$Cl_{2(g)} + Fe_{(s)} = Fe^{2+} + 2Cl^- \tag{2.39}$$

for which it is measured that

$$\Delta G^o = -\,341.39\text{ kJ} = \Delta G^o_{f,Fe^{2+}} + 2\Delta G^o_{f,Cl^-} - 0 - 0. \tag{2.40}$$

Thus,

$$\Delta G^o_{f,Fe^{2+}} = -341.39 - 2\Delta G^o_{f,Cl^-} = -341.39 - 2(-131.26) = -78.87\text{ kJ/mol}. \tag{2.41}$$

Physical chemists have used this approach to assign ΔG_f^o values for many ions of interest.

Example 2.7 ΔG^o and K

a. Compute the value of ΔG^o at 25 °C and 1 atm for the acid dissociation reaction for acetic acid to H^+ and acetate ion, represented as namely $HA = H^+ + A^-$.
b. What is the value of the acidity constant K_a for acetic acid?

Solution

a. Based on the values in Table 2.2, including that ΔG_f^o for H^+ is 0,

$$\begin{aligned}\Delta G^o &= (1\,\text{mol})(0\,\text{kJ/mol}) + (1\,\text{mol})(-369.2\,\text{kJ/mol}) \\ &\quad - (1\,\text{mol})(396.4\,\text{kJ/mol}) \\ &= 27.2\,\text{kJ}.\end{aligned}$$

b. $\Delta G^o = -2.303\, RT \log K_a$.
$\log K_a = -\Delta G^o/2.303\, RT = -27.2\,\text{kJ}/(2.303 \times 0.008314\,\text{kJ/mol-K} \times 298.15\,\text{K}) = -4.76$

$$K_a = 1.79 \times 10^{-5}\ m^2/m \qquad pK_a = 4.76$$

2.4.6 Concentration Scales and Standard States

As discussed in Chapter 1, a variety of scales can be used to express solution concentration and activity. In this text, both aqueous and solid solutions are of interest. For species i, the important scales are mole fraction (x_i), molality (m_i), and molarity (c_i or M_i). Thermodynamicists like the mole fraction scale for some problems because the degree to which a solution containing i exhibits the properties i is related directly to x_i with the solution approaching pure i as $x_i \to 1$. One application of this principle is Raoult's Law, which states that for an ideal liquid solution, the gaseous pressure of i above the solution equals the product of x_i with $p_{L,i}^o$, the vapor pressure of pure liquid i.

In the field of water chemistry, since liquid water and most mineral solids are generally found in the natural environment as nearly pure phases (*i.e.*, $x_i \approx 1$), the mole fraction scale is usually chosen to express their concentrations and activities. For dissolved species, however, molarity and molality are more convenient. For molarity, the results of analytical measurements are indeed most easily expressed as amount per unit volume of sample. As noted previously, given a choice between molarity and molality, thermodynamicists prefer molality over molarity because it is not affected by the volume changes that accompany changes in T and P. (A kg of solution remains a kg of solution independent of T and P.)

When we compute the ΔG^o and the K for a given reaction, if we use μ_i^o with multiple activity scales Eq.(2.13) for the different reactant and product species, we will be expressing the activities in the K obtained with the same set. For example, for reaction (2.35), the μ_i^o types used (as tabulated ΔG_f^o values) are $(\mu_i^o)_x$ for $S_{(s)}$, $(\mu_i^o)_m$ for $O_{2(aq)}$, and $(\mu_i^o)_{atm}$ for $SO_{2(g)}$. So, for the corresponding K, which has units of atm/m, we have

$$K = \frac{p_{SO_2}}{x_{S_{(s)}} m_{O_2}}. \tag{2.42}$$

K values involving dissolved species usually are measured and reported with the dissolved species expressed on the molality scale. It is usually more convenient, however, to measure and report aqueous concentrations in molarity. Fortunately, as noted above, except for extremely concentrated solutions, the differences between values reported on the molality and molarity scales are very small. The result is that concentrations expressed in units of molality and molarity are essentially equal and may usually be used interchangeably in equilibrium constants without any concern for significant error. This is certainly true for calculations made for fresh waters. Moreover, as Stumm and Morgan (1996, p.130) note, any differences between molarity and molality are usually very small in comparison to the uncertainties in: (1) the reported values of equilibrium constants; and (2) estimated activity coefficients.

2.4.7 The Standard State and the Activity Coefficient Reference Convention

2.4.7.1 General

From above,

$$\mu_i = \mu_i^{\text{o}} + RT \ln a_i \tag{2.6}$$

where μ_i^{o} is the chemical potential of species i, usually at 25 °C and 1 atm, and always in a particular *standard* state. We know that the standard state provides the energy level benchmark relative to which μ_i is measured, and that $\mu_i = \mu_i^{\text{o}}$ when $a_i = 1$ (on the concentration/activity scale for that μ_i^{o}). The standard state is defined as the state in which *both* the activity coefficient for species i and $[i]$ are equal to 1.0. For chemical species like water or gold (but not Na^+), this standard state does exist on the mole fraction scale because: (1) the pure material can be created (at least very nearly) so that $x_i = 1$; and (2) on the mole fraction scale $x_i = 1$ is in fact the reference state for ideal behavior, *i.e.*, it is the state in which it is defined that $\zeta_i = 1$. For the molality scale however, the reference state for $\gamma_i = 1$ is infinitely dilute water, not $m_i = 1$. So, when $m_i = 1$, then $\gamma_i \neq 1$. This means that on the molality scale,

$$m_i = 1, \ \gamma_i = 1 \qquad \text{standard state on molality scale} \tag{2.43}$$

is *hypothetical*. (This certainly does not mean that we can never have a real system with $a_i = 1.0$ on the molal scale: non-unity values of both m_i may γ_i can *combine* to give $a_i = m_i\gamma_i = 1$.)

2.4.7.2 Why Choose a Hypothetical Standard State for μ_i^{o} on the Molality Scale?

As noted, on the molality scale, in the hypothetical standard state $\gamma_i = 1.0$ and $m_i = 1.0$. Many students find the concept of a hypothetical standard state to be confusing. They ask: why not define a real standard state in which $a_i = 1$? From Eq.(2.6), on the molarity scale, if i is in its standard state, then we need $a_i = m_i\gamma_i = 1.0$. For a real state that satisfies that criterion, if there are no special requirements on γ_i and m_i individually, we do require that their *product* equals 1.0. The good news is that for any given species, there will be an infinite number of real solution compositions that give $m_i\gamma_i \approx 1.0$. The bad news is that: (1) we would have to do experiments in the laboratory to find a solution composition that gives that result for each every chemical species of interest (that's a lot of work); and (2) for every species we would end up with a different solution. Consider Na^+ in a solution of NaCl. Based on experimental studies, one composition that gives $m_i\gamma_i \approx 1.0$ is an NaCl solution at 1.3 *F* so that for Na^+ we have (1.3) (0.77) $\approx$ 1.0. But for Ca^{2+}, a 1.3 *F* solution of NaCl *does not* give $m_i\gamma_i \approx 1.0$ (nor for that matter, does a 1.3 *F* solution of $CaCl_2$).

2.4.7.3 Meaning of the Hypothetical $\gamma_i = 1.0$, $m_i = 1.0$ Standard State

Now that the attractiveness of the $\gamma_i = 1.0$, $[i] = 1.0$ hypothetical standard state has been established (well, at least any real alternative is unattractive), we need to convince ourselves that the hypothetical standard state represents something *meaningful*. Consider again Figure 2.2. The PE analog for the $\gamma_i = 1.0$, $m_i = 1.0$ hypothetical standard state is the *sealed* compartment *inside* the Figure 2.2.*a* staircase. This state is *not physically accessible* to the ball. Although the sealed compartment is not physically accessible, its energy level is perfectly well-defined, and we could place the ball in a real state at the same PE level. Similarly, although the particular chemical state of $\gamma_i = 1.0$, $m_i = 1.0$ in Figure 2.2.*b* is not physically accessible, the corresponding chemical energy level is well-defined, and it is possible to actually place a chemical species at the same energy level (recall Na^+ in 1.3 *F* NaCl).

2.5 EFFECTS OF *T* AND *P* ON EQUILIBRIUM CONSTANTS

K values based on Eq.(2.20) with Eq.(2.13) pertain to 25 °C and 1 atm, and will change if the *T* or *P* change; most of the world is not at 25 °C and 1 atm. From basic thermodynamics (Denbigh 1981, p.300),

$$R\, d\ln K = \frac{\Delta H^{\circ}}{T^2} dT - \frac{\Delta V^{\circ}}{T} dP. \tag{2.44}$$

With H_{f}° being the molar enthalpy of formation of reactant or product *i* (*i.e.*, the enthalpy analog of μ_i°),

$$\Delta H^{\circ} = \sum_i \upsilon_i H_{\mathrm{f}}^{\circ}. \tag{2.45}$$

ΔH° is the net change in enthalpy (*i.e.*, heat content) when the reaction is occurring under standard conditions (all $a_i = 1.0$). ΔH° is related to ΔG° through

$$\Delta G^{\circ} = \Delta H^{\circ} - T\Delta S^{\circ} \tag{2.46}$$

where ΔS° is the change in entropy when the reaction is proceeding under standard conditions.

Analogously with Eqs.(2.13) and (2.45),

$$\Delta V^{\circ} = \sum_i \upsilon_i \bar{V}_i^{\circ} \tag{2.47}$$

where $\bar{V}_i$ is the partial molar volume of reactant or product *i*. As usual, the stoichiometric coefficients υ_i have sign (positive for products, and negative for reactants). **With Eq.(2.47), all gaseous species in the reaction are left out of the summation**. This is because for a gaseous species *j*, just changing the total pressure does not change affect μ_j (which at constant *T* is only altered by changing the individual gas phase fugacity of *j*). (For a condensed species *i*, there is *PV* work done on *i* when the pressure goes up, and this raises μ_j°.) Both ΔH° and ΔV° are *T* and *P* dependent.

Two partial derivatives that may be obtained from (2.44) are

$$\left(\frac{\partial \ln K}{\partial T}\right)_P = \frac{\Delta H^{\circ}}{RT^2} \tag{2.48}$$

and

$$\left(\frac{\partial \ln K}{\partial P}\right)_T = -\frac{\Delta V^{\circ}}{RT}. \tag{2.49}$$

Convenient units for *R* for Eqs.(2.20) and (2.49) are 0.008314 kJ/mol-K; for Eq.(2.49), convenient units for *R* are 0.082 L-atm/mol-K.

When heat is taken up by the forward movement of a reaction, $\Delta H^{\circ} > 0$: the total enthalpy content of the chemical products under standard conditions is greater than the total enthalpy content of the chemical reactants under standard conditions. When $\Delta H^{\circ} > 0$, then heat may be thought of as one of the *reactants* of the reaction. When $\Delta H^{\circ} > 0$, Eq.(2.48) tells us that when *T* (which is connected to the total system *H*) increases, then *K* will increase. Le Chatelier's Principle tells us in general terms what Eq.(2.48) tells us in specific terms, namely that increasing *T* will tend to shift a reaction equilibrium towards the chemical products (increase *K*) when heat is a reactant ($\Delta H^{\circ} > 0$). The whole argument is reversed when $\Delta H^{\circ} < 0$ and heat is one of the *products* of the reaction, so that *K* decreases as *T* increases.

Because

$$d(1/T) = -dT/T^2 \tag{2.50}$$

then Eq.(2.48) can be converted into

$$\left(\frac{\partial \ln K}{\partial(1/T)}\right)_P = -\frac{\Delta H^\circ}{R}. \tag{2.51}$$

This can be integrated if ΔH° is known as a function of T. However, it is often a good assumption that ΔH° will stay reasonably constant over the T range of interest. If so, under conditions of constant P, integration gives

$$\log\frac{K_2}{K_1} = -\frac{\Delta H^\circ}{2.303\,R}\left(\frac{1}{T_2} - \frac{1}{T_1}\right). \tag{2.52}$$

K_1 can serve as the *known* value of K at temperature T_1, and K_2 is then the extrapolated value of K at T_2. Eq.(2.52) can be applied with confidence as long as the extrapolation temperature range (T_2–T_1) is not too large (*e.g.*, 15 degrees or less). As T_2 approaches T_1, Eq.(2.52) provides an increasingly good approximation of K as a function of T. More complex equations that do not make the assumption of a constant ΔH° are available in texts on physical chemistry.

When volume is increased by the forward movement of a reaction, $\Delta V^\circ > 0$: the total volume of the chemical products under standard conditions is greater than the total volume of the chemical reactants under standard conditions. When $\Delta V^\circ > 0$, then volume may be thought of as one of the *products* of the reaction. When $\Delta V^\circ > 0$, Eq.(2.49) tells us that when P increases, then K will decrease. Le Chatelier's Principle tells us in general terms what Eq.(2.49) tells us in specific terms, namely that increasing P will tend to shift a reaction equilibrium towards the chemical reactants (decrease K) when volume is a product ($\Delta V^\circ > 0$). The whole argument is reversed when $\Delta V^\circ < 0$ and volume is one of the *reactants*, so that K increases as P increases. Overall, increasing P tends to favor the side of the reaction with the least chemical volume. When $\Delta V^\circ = 0$, changing P has no effect on K.

At constant temperature, when ΔV° is assumed to remain constant with pressure, Eq.(2.49) may be integrated to yield

$$\log\frac{K_2}{K_1} = -\frac{\Delta V^\circ(P_2 - P_1)}{2.303\,RT}. \tag{2.53}$$

K_1 can serve as the *known* value of K at P_1, and K_2 is then the extrapolated value of K at P_2. In a manner that is analogous with Eq.(2.52), Eq.(2.53) will give an increasingly good estimate of K as a function of P as P_2 approaches P_1. However, even extrapolations over relatively large pressure ranges (*e.g.*, hundreds of atmospheres) may frequently be done with confidence since ΔV° often changes little with pressure. Indeed, the compressibilities of condensed phase species are very small. Thus, since the individual $\bar{V}_i$ change little with pressure, ΔV° will change little with pressure. In any case, for reactions and conditions of interest here, because ΔV° values are small, effects of changing P are small.

Example 2.8 Effect of *T* on a *K* Value

Consider the acid dissociation of water, namely $H_2O = H^+ + OH^-$. For these three species, H_f° has values of −285.83, 0, and −229.99 kJ/mol. If $K_w = 1.01 \times 10^{-14}$ m^2 at 25 °C and 1 atm, then what is the value of K_w at 37 °C and 1 atm?

Solution

37 C = 310.15 K

$$\Delta H^\circ = (1\,\text{mol})(0\,\text{kJ/mol}) + (1\,\text{mol})(-229.99\,\text{kJ/mol}) - (1\,\text{mol})(-285.83\,\text{kJ/mol}) = 55.84\,\text{kJ}.$$

$$\log\frac{K_w(310.15\,\text{K})}{1.01\times10^{-14}} = -\frac{55.84\,\text{kJ}}{2.303\times0.008314\,\text{kJ/mol-K}}\left(\frac{1}{310.15\,\text{K}}-\frac{1}{298.15\,\text{K}}\right) = 0.378$$

$$\frac{K_w(310.15\,\text{K})}{1.01\times10^{-14}} = 2.39,\ \text{so}\ K_w(310.15\,\text{K}) = 2.39\times1.01\times10^{-14}\,\text{m}^2 = 2.41\times10^{-14}\,\text{m}^2$$

Example 2.9 Effect of *P* on a *K* Value

Consider a water droplet with a radius $r = 0.01\ \mu\text{m} = 10^{-8}$ m that is suspended in air at 1 atm. Due to the "Kelvin Effect", the pressure inside the droplet is higher than the outside pressure (ΔP) by an amount equal to $2\sigma/r$ where σ is the surface tension of water. The vaporization of water takes place according to

$H_2O = H_2O_{(g)}$ with $K = p_{H_2O}/x_{H_2O}$, which for essentially pure liquid water gives $K = p_{H_2O}$. At 25 °C and 1 atm pressure, $K = p_{H_2O} = 0.0312$ atm. If $\sigma = 0.072$ N/m, what is the pressure inside the droplet, and what is the value of K for vaporization of water from the droplet? (N = Newton).

Solution

We will use Eq.(2.54) for this problem.
The density of water is 0.997 g/cm^3 = 997 g/L.
The MW of water is 18.015 g/mol
The molar volume of water is then (1 L/997 g) × (18.015 g/mol) = 0.0181 L/mol
ΔV° for the reaction, in which we do not include gaseous species (which in this case is $H_2O_{(g)}$) is then $\Delta V^\circ = -0.0181$ L/mol.
$\Delta P = (2\times0.072\ \text{N/m})/10^{-8}\ \text{m} = 1.44\times10^7\ \text{N/m}^2$.
1 atm (standard) = 101,325.01 N/m^2
$\Delta P = 1.44\times10^7\ \text{N/m}^2\times1\ \text{atm}/101{,}325.01\ \text{N/m}^2) = 142$ atm

$$\log\frac{p_{H_2O}(@142\,\text{atm})}{0.0312\,\text{atm}} = -\frac{\Delta V^\circ(P_2-P_1)}{2.303RT}$$

P_2 for this problem is the pressure inside the droplet, and $P_1 = 1$ atm, the ambient pressure, so $P_2 - P_1 = 142$ atm.

$$-\frac{\Delta V^\circ(P_2-P_1)}{2.303RT} = -\frac{-0.0181\,\text{L/mol}(142\,\text{atm})}{2.303\times(0.082\,\text{L-atm/mol-K})\times298.15K} = +0.0457$$

$$p_{H_2O}\,@142\,\text{atm} = 0.0312\times10^{0.0457} = 0.0347\,\text{atm}$$

The water in the droplet is therefore more volatile than water vaporizing from a flat surface. This is at the core of why formation of rain in the atmosphere needs to overcome a barrier when droplets initially form: at 25 °C when the relative humidity is 100% so that $p_{H_2O} = 0.0312$ atm, a small droplet with $r = 0.01\ \mu$m tends to create air around it for which the RH is higher than 100%, because $p_{H_2O} = 0.0347$ atm, and the droplet easily evaporates into the RH = 100% air because, effectively, right around the droplet, the air is at RH = 100% × 0.0347/0.0312 = 111%.

2.6 COMBINING EQUILIBRIUM EXPRESSIONS (TO GET NEW ONES)

Let us say that we know that at 25 °C/1 atm,

$$Fe^{3+} + OH^- = FeOH^{2+} \qquad K_{H1} = 10^{11.8}. \tag{2.54}$$

Also,

$$FeOH^{2+} + OH^- = Fe(OH)_2^+ \qquad K_{H2} = 10^{10.5}. \tag{2.55}$$

When these two equilibrium reactions are *added*, we obtain

$$\begin{array}{lcccl} & \text{summed reactant set} & & \text{summed product set} & \\ & \boxed{Fe^{3+} + OH^-} & = & \boxed{FeOH^{2+}} & K_{H1} = 10^{11.8} \\ + & \boxed{FeOH^{2+} + OH^-} & = & \boxed{Fe(OH)_2^+} & K_{H2} = 10^{10.5} \\ \hline \end{array} \tag{2.56}$$

$$Fe^{3+} + 2OH^- + FeOH^{2+} = FeOH^{2+} + Fe(OH)_2^+ \quad K = K_{H1}K_{H2} = 10^{22.3}.$$

Cancelling the two like $FeOH^{2+}$ terms, we have

$$Fe^{3+} + 2OH^- = Fe(OH)_2^+ \qquad K = K_{H1}K_{H2} = 10^{22.3}. \tag{2.57}$$

The equilibrium constant for the reaction in Eq.(2.57) is the *product* of the equilibrium constants for the individual reactions that were added together. This is an important general result. Another way of looking at this is

$$\frac{\{FeOH^{2+}\}}{\{Fe^{3+}\}\{OH^-\}} \times \frac{\{Fe(OH)_2^+\}}{\{FeOH^{2+}\}\{OH^-\}} = \frac{\{Fe(OH)_2^+\}}{\{Fe^{3+}\}\{OH^-\}^2}. \tag{2.58}$$

When one equilibrium reaction is subtracted from another, the equilibrium constant for the resulting reaction is given by the *quotient* of the corresponding individual equilibrium constants. To illustrate, we have

$$Fe^{3+} + 2OH^- = Fe(OH)_2^+ \qquad K = K_{H1}K_{H2} = 10^{22.3} \tag{2.57}$$

$$-\left(2H^+ + 2OH^- = 2H_2O \qquad K = K_w^{-2}\right) \tag{2.59}$$

$$Fe^{3+} + 2H_2O = Fe(OH)_2^+ + 2\,H^+ \qquad K = K_{H1}K_{H2}/K_w^{-2} = K_{H1}K_{H2}K_w^2. \tag{2.60}$$

Example 2.10 Combining Chemical Reactions and Corresponding *K* Values

Consider reactions 1–3 and their *K* values at 25 °C and 1 atm (we will discuss what the * on $H_2CO_3^*$ means in Chapter 9):

1. $CO_{2(g)} + H_2O = H_2CO_3^* \qquad K = 10^{-1.47}$
2. $H_2CO_3^* = H^+ + HCO_3^- \qquad K = 10^{-6.35}$
3. $H_2O = H^+ + OH^- \qquad K = 1.01 \times 10^{-14}$
   Obtain the *K* value for the reaction
4. $CO_{2(g)} + OH^- = HCO_3^- \qquad K = ?$

Solution

Reaction 4 can be obtained by adding reaction 1 with reaction 2 then subtracting reaction 3. Doing so gives the desired K as $10^{-1.47}\times 10^{-6.35}/1.01\times 10^{-14}=1.50\times 10^{6}$.

2.7 INFINITE DILUTION, CONSTANT CONCENTRATION, AND "MIXED" EQUILIBRIUM CONSTANTS

2.7.1 Infinite Dilution Constants

In its purest form, an equilibrium K value is the quotient of activities for the reactants and products. For any dissolved species, as has been noted, the concentration scale chosen is that of molality, with the reference state for $\gamma_i=1$ being highly (a.k.a. "infinitely") dilute water. Equilibrium constants obtained with Eqs.(2.14) and (2.20) are thus referred to as "infinite dilution" constants.

Consider the particular case of the acid dissociation reaction

$$HA = H^+ + A^-. \tag{2.61}$$

As usual,

$$K = \frac{\{H^+\}\{A^-\}}{\{HA\}} = \frac{[H^+]\gamma_{H^+}[A^-]\gamma_{A^-}}{[HA]\gamma_{HA}} \tag{2.62}$$

where the $[i]$ are molalities ($\approx$ molarities except for very concentrated solutions). For solutions that have total salt concentrations like 0.01 F, then for ions, the γ_i are sufficiently different from 1.0 that we need to estimate the γ_i when solving problems: for example, we cannot just assume $\{H^+\}\approx[H^+]$.

However, for fresh waters with hardly any dissolved salts, activity corrections are not very important, all $\gamma_i\approx 1$, and so the infinite dilution constants may be used with concentrations substituted for activities:

$$K = \frac{\{H^+\}\{A^-\}}{\{HA\}} \approx \frac{[H^+][A^-]}{[HA]}. \qquad \text{fresh waters} \tag{2.63}$$

Example 2.11 Using Infinite Dilution Constants

For the acid dissociation reaction in Eq.(2.61), consider that HA is acetic acid, A^- is acetate ion, and $K=10^{-4.76}$ at 25 °C and 1 atm. Note now that pH $\equiv -\log\{H^+\}$. In a solution that is 0.05 F KCl, say $\gamma_{H^+}=0.82$, $\gamma_{A^-}=0.82$, $\gamma_{HA}\approx 1$, and $[H^+]=1.19\times 10^{-5}\,M$.

If $[HA]=2\times 10^{-5}\,M$, what is $[A^-]$?

Solution

In 0.05 F KCl:

$$\frac{[H^+]\gamma_{H^+}[A^-]\gamma_{A^-}}{[HA]\gamma_{HA}} = 10^{-4.76}.$$

$$\frac{1.19\times 10^{-5}\;0.82[A^-]0.82}{2.0\times 10^{-5}(1)} = 10^{-4.76}.$$

$$[A^-] = 4.33\times 10^{-5}.$$

2.7.2 Constant Ionic Medium Constants

In aqueous solutions in which some ions are present at concentrations like ~0.01 *M* and higher, then the γ_i values for *all ions* will be sufficiently different from 1.0 that activity corrections are needed. In Example 2.11, K^+ and Cl^- are present at 0.05 *M*; while both $[H^+]$ and $[A^-]$ are only of order 10^{-5}, because of the overall presence of ionic charge in the solution (mostly from the K^+ and Cl^-), $\gamma_{H^+} = 0.82$ and $\gamma_{A^-} = 0.82$. There are many other examples like this of important systems in which low concentration species are reacting and equilibrating in the presence of major ionic species that: (1) provide the bulk of the system's ionic matrix; *and* (2) very nearly *constant* in concentration from one problem of interest to the next. Think seawater, biological fluids, brines. In seawater, both Na^+ and Cl^- are ~ 0.5 *m* and rather constant, but $[H^+]$ can easily vary from 10^{-9} to 5×10^{-8} *m*, and such minor variation can hardly affect the overall ionic nature of the solution.

Eq.(2.62) can be manipulated as follows:

$$K = \frac{\{H^+\}\{A^-\}}{\{HA\}} = \left(\frac{[H^+][A^-]}{[HA]}\right)\left(\frac{\gamma_{H^+}\gamma_{A^-}}{\gamma_{HA}}\right) \tag{2.64}$$

or equivalently,

$$\frac{[H^+][A^-]}{[HA]} = \left(\frac{\gamma_{HA}}{\gamma_{H^+}\gamma_{A^-}}\right) K \equiv {}^cK \tag{2.65}$$

So, if the ionic matrix of the medium remains essentially constant within a given set of problems, the term involving the activity coefficients is a constant, yielding ${}^cK$. The superscript "c" is intended to indicate that ${}^cK$ refers to some specific, **constant ionic medium**: ${}^cK$ is valid only for a *specific* solution matrix, that is, the matrix that gives a specific value of the term $(\gamma_{HA}/\gamma_{H^+}\gamma_{A^-})$. Once a given ${}^cK$ value is calculated or measured for a given matrix, *activity corrections are completely built into* ${}^cK$ *for that equilibrium*; one can then work directly with *concentrations* in all calculations.

If one needs a ${}^cK$ for some specific solution, but the corresponding infinite dilution *K* is not known, it is likely to be more attractive to measure ${}^cK$ rather than measure *K* plus then obtain the γ_i values. Indeed, in measuring ${}^cK$, all we need to do is obtain the *concentrations* of the species at equilibrium.

Example 2.12 Using Infinite Dilution Constants In a Similar Problem

As in Example 2.11, consider that HA is acetic acid, A^- is acetate ion, and $K = 10^{-4.76}$ at 25 °C and 1 atm. Assume that the solution is again 0.05 *F* KCl so that again $\gamma_{H^+} = 0.82$, $\gamma_{A^-} = 0.82$, $\gamma_{HA} \approx 1$. If we now have $[H^+] = 3.0 \times 10^{-5}$ *M* and $[HA] = 1.0 \times 10^{-5}$ *M*, what is $[A^-]$?

Solution

$$\frac{[H^+]\gamma_{H^+}[A^-]\gamma_{A^-}}{[HA]\gamma_{HA}} = 10^{-4.76}$$

$$\frac{3.0 \times 10^{-5}\ 0.82[A^-]0.82}{1.0 \times 10^{-5}(1)} = 10^{-4.76}$$

$$[A^-] = 8.61 \times 10^{-6}\ M$$

If we compare Examples 2.11 and 2.12, all the activity coefficient values are exactly the same in both problems. That is because 0.05 *F* KCl is dominating the ionic compositions of both solutions so that both solutions have the same set of γ_i values. Because of this, if one is going to be working repeatedly with this particular solution matrix, it makes sense to combine all the activity coefficient values into a new equilibrium constant, which we call ${}^cK$.

Example 2.13 Using ${}^cK$ Values

For problems like Example 2.11 and 2.12, what value for ${}^cK$ is obtained by combining all the activity coefficients into the K value?

Solution

In 0.05 F KCl:

$$\frac{[H^+]0.82[A^-]0.82}{[HA](1)} = 10^{-4.76}$$

$$\frac{[H^+][A^-]}{[HA]} = \frac{(1)}{0.82 \times 0.82} 10^{-4.76} = 10^{-4.59} = {}^cK$$

2.7.3 Mixed Constants

Some chemists use equilibrium constants in which some of the reactants/products appear as concentrations, while some appear as activities. This is especially common in the field of marine chemistry. These equilibrium constants are referred to as "mixed constants". As in the preceding section, this circumstance will again involve a constant ionic medium because such a medium allows us to use some concentrations in a K value rather than activities.

One type of equilibrium process for which mixed constants are particularly attractive are *acid dissociations* as in Eq.(2.61). It is convenient to use [HA] and $[A^-]$ in the expression for the mixed equilibrium constant because they are often **measurable** or **inferable**. However, it is convenient to retain $\{H^+\}$ because it is **measurable** by pH electrode. For the mixed constant for reaction (2.61) we obtain

$$K' = \frac{\{H^+\}[A^-]}{[HA]}. \tag{2.66}$$

Summarizing, we have

$$K = \frac{\{H^+\}\{A^-\}}{\{HA\}} = {}^cK \frac{\gamma_{H^+}\gamma_{A^-}}{\gamma_{HA}} = K' \frac{\gamma_{A^-}}{\gamma_{HA}}. \tag{2.67}$$

Using any of the three constants K, ${}^cK$, or K' is correct as long as each definition is clearly understood. For a given reaction, the three forms are merely different ways of quantitatively expressing the same equilibrium. However, if one takes a ${}^cK$ constant from the literature and uses it as if it were an infinite dilution constant, that will lead to errors. It can also be said that a given set K, ${}^cK$, and K' must satisfy Eq.(2.67). If not, they are *not consistent*, and at least one of the values is *incorrect*.

A discussion of how one pair of scientists dealt with the variety of acid/base equilibrium constants in the literature is found in the foreword for the Smith and Martell (1976) equilibrium data compilation:

Equilibria involving protons have been expressed as concentration constants [*i.e.*, ^{c}K values] in order to be more consistent with the [manner in which] metal ion stability constants … [are typically reported]. Concentration constants may be determined by calibrating the electrodes with solutions of known hydrogen ion concentrations or by conversion of [the directly measured] pH values using the appropriate hydrogen ion activity coefficient. When standard buffers are used, mixed constants [*i.e.*, K' values] (also known as Bronsted or practical constants) are obtained which include both activity and concentration terms. Literature values expressed as mixed constants have been converted to concentration constants by using the hydrogen ion activity coefficients determined in KCl solution before inclusion in the tables. In some cases, [some data from some] papers were … [not included in the database] because no indication was given as to the use of concentration or mixed constants.

Example 2.14 Using *K′* Values

In an example of "inferable", consider that we start with 0.01 *F* acetic acid in 0.1 *F* KCl, and add 0.0025 FW per L of NaOH. Because acetic acid is a rather weak acid, relatively little HA dissociates on its own to form A^-, so in the final solution, almost all of $[A^-]$ will be due to the reaction of the strong base with the HA.

a. Without activity corrections, infer the values of $[A^-]$ and [HA].
b. If a pH electrode was used to measure that the pH of the solution is 4.17, what is the value of K' for acetic acid in 0.1 *F* KCl?

Solution

a. $[A^-]=0.0025\ M$. $[HA]=0.01-0.0025=0.0075\ M$.
b. $\{H^+\}=10^{-pH}$.

In 0.1 *F* KCl:

$$K' = \frac{10^{-4.17}[A^-]}{[HA]} = \frac{10^{-4.17}\,0.0025}{0.0075} = 10^{-4.65}$$

2.8 ACTIVITY COEFFICIENT EQUATIONS

2.8.1 Activity Coefficient Equations for Single Ions

For both ionic as well as neutral species, the dominant property that affects γ_i values is the ionic strength I

$$I = ½\sum_i m_i z_i^2 \tag{2.68}$$

where m_i is the molal (mol/kg) concentration of *i*. (Molarity can also be used, with essentially no error.) z_i is the charge for species *i*. For example: $z_i=+1$ for Na^+; +2 for Ca^{2+}; −1 for Cl^-; and −2 for SO_4^{2-}. Because no type of ion can be present in a solution to any meaningful extent except when ions of opposite charge are also present, strictly speaking it is not possible to exactly determine (experimentally) the value of γ_i for a single type of ion such as H^+. However, because physical chemists have found some ways to minimize this problem, we will refer to and freely use single-ion γ_i values in this text.

Values of γ_i are most accurately predicted for aqueous solutions that are dilute with respect to total dissolved ions ($I<3\times10^{-3}$). Table 2.3 provides a summary of equations for predicting γ_i values for ions. When $I<3\times10^{-3}$, chemists generally agree that the best option is the Debye–Hückel (D–H) Equation. However: (a) in the development of the D–H Equation, it is assumed that *i* is a

TABLE 2.3
Equations for Activity Coefficient γ_i for Ionic Species

Name	Equation	Applicable Ionic Strength Range
Debye–Hückel	$\log \gamma_i = -Az_i^2\sqrt{I}$	$I < 10^{-2.3}$
Extended Debye–Hückel	$\log \gamma_i = -Az_i^2\left[\sqrt{I}/\left(1+Ba\sqrt{I}\right)\right]$ a is the "ion size parameter" for ion i; it should not be confused with the activity	$I < 10^{-1.0}$
Güntelberg	$\log \gamma_i = -Az_i^2\left[\sqrt{I}/\left(1+\sqrt{I}\right)\right]$ equivalent to Extended D–H with an average value of $a=3$	$I < 10^{-1.0}$
Davies	$\log \gamma_i = -Az_i^2\left[\left(\sqrt{I}/\left(1+\sqrt{I}\right)\right)-0.3I\right]$	$I < 0.5$

Note: The parameter A depends on T (K) according to $A=1.82\times 106(\varepsilon T)^{-3/2}$ where ε is the temperature-dependent dielectric constant of water. $B=50.3(\varepsilon T)^{-1/2}$. For water at 298.15 K (25 °C), $A=0.51$ and $B=0.33$. Applicable ionic strength ranges are from Stumm and Morgan (1996, p.135).

point charge (*i.e.*, has no volume); (b) as a solution becomes more concentrated ($3\times 10^{-3}< I<0.1$), the point charge assumption starts to fail (ions start to get closer to one another, and can begin to "sense" the size of other ions). The Extended D–H Equation recognizes the role of ion size and requires input of the radius a for i (Table 2.4). The Güntelberg Equation is an empirical generalization of the Extended D–H Equation that assumes that ions of interest have an average radius $a=3$ Å. For $I>0.1$, the Extended D–H Equation (and the Güntelberg Equation) begin to fail. For $0.1<I<0.5$, many geochemists use the Davies Equation, which can be viewed as being an extension

TABLE 2.4
Ion Size Parameters a for the Extended Debye–Hückel Equation (from Kielland 1937)

Ion	a	Ion	a	Ion	A
Ag^+	3	Fe^{2+}	6	Mn^{2+}	6
Al^{3+}	9	Fe^{3+}	9	Na^+	4
Ba^{2+}	5	H^+	9	NH_4^+	3
Be^{2+}	8	HCO_3^-	4	NO_3^-	3
Ca^{2+}	6	HPO_4^{2-}	4	OH^-	3
Ce^{3+}	9	$H_2PO_4^-$	4	Pb^{2+}	5
CH_3COO^-	4	HS^-	3	PO_4^{3-}	4
Cl^-	3	I^-	3	Sn^{2+}	6
ClO_4^-	3	K^+	3	SO_4^{2-}	4
CO_3^{2-}	5	La^{3+}	9	Sr^{2+}	5
Cu^{2+}	6	Mg^{2+}	8	Zn^{2+}	6

of the Davies Equation. For seawater, for which $I \approx 0.7$, chemical oceanographers have developed some complex empirical equations for predicting γ_i values. For this text, the version of the Davies Equation given in Table 2.3 will be adequate for most examples considered. For applications of the principles discussed in the text to solutions to solutions with $I > 0.5$, the reader can investigate more elaborate γ_i prediction equations.

For each equation in Table 2.3, the dependence of γ_i on z_i involves z_i^2. The result is that for a given value of I (and for a particular value of a in the case of the Extended D–H Equation), the same γ_i value is obtained for any ion with +1 or −1 charge, and again the same γ_i value for +2 and −2, etc.: at any given value of I, the γ_i value for $z_i = \pm 2$ is less than for $z_i = \pm 1$ (recall the dependence on z_i^2): the Coulombic interactions of a doubly charged ion with the surrounding ions in a solution of specific I are stronger than the interactions of a singly-charged ion. Figure 2.3 presents curves according to the Güntelberg Equation (a.k.a. the generalized Extended D–H Equation), and the Davies Equation for γ_i *vs.* log I for $z_i = \pm 1$, and for $z_i = \pm 2$.

Section 1.4 explains why $\gamma_i < 1$ for ions as I starts to rise above zero. With increasing I, the D–H, Extended D–H, and Güntelberg Equations all predict an *unending decrease* of predicted γ_i towards 0. This is not what happens. For the Davies Equation, there is a reversal around $I = 0.3$, and predicted γ_i values begin to increase. This reversal reproduces experimental observations for γ_i values as I approaches and passes 1. Indeed, at high I values, ions are starting to become starved for water molecules for their solvation shells (Bockris and Reddy 1970, p.238), and this starvation causes them become *more exposed* to one another, and thus *active*. In other words, because water is polar, ions in highly dilute solutions are partially shielded from one another by their solvation shells. At high I, ions are "fighting" with one another to get enough water for their solvation shells. There is no longer enough water to let every ion have as much solvation water as it "wants", and the charge character of each ion becomes more revealed relative to the case of infinite dilution (which is the reference state for all $\gamma_i = 1$ on the molal scale).

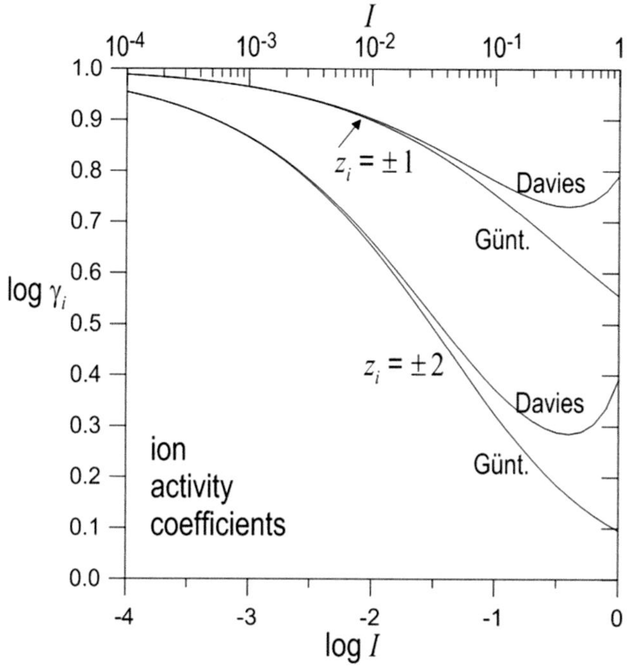

FIGURE 2.3 Activity coefficients *vs.* log I for $z = \pm 1$ and $z = \pm 2$ for the Davies Equation and the Güntelberg Equation.

Example 2.15 Using the Davies Equation

For 25 °C, use the Davies Equation to calculate the γ_i values for H^+, OH^-, *and* the salt ions present in the following two solutions: (a) 0.001 *F* NaCl, and (b) 0.001 *F* $CaSO_4$. Neglect the contribution of H^+ and OH^- to *I*.

Solution

a. The 0.001 *F* NaCl leads to $[Na^+] = 0.001$ *M*, and $[Cl^-] = 0.001$ *M*.

$$I = ½(0.001(+1)^2 + 0.001(-1)^2) = 0.001$$

For all the ions in the solution, $z_i = \pm 1$, and the Davies Equation will predict the same value, which is represented γ_1.

$$\log \gamma_1 = -0.51(\pm 1)^2[(\sqrt{0.001}/(1+\sqrt{0.001})) - 0.3(0.001)]$$

$$\gamma_1 - 0.96$$

b. We assume that the 0.001 *F* $CaSO_4$ leads to $[Ca^{2+}] = 0.001$ *M*, and $[SO_4^{2-}] = 0.001$ *M*.

$$I = ½(0.001(+2)^2 + 0.001(-2)^2) = 0.004$$

For H^+ and OH^-, $z_i = \pm 1$, the Davies Equation will predict the same value, γ_1:

$$\log \gamma_1 = -0.51(\pm 1)^2[(\sqrt{0.004}\ /\ (1+\sqrt{0.004})) - 0.3\ (0.004)]$$

$$\gamma_1 = 0.93$$

For Ca^{2+} and SO_4^{2-}, $z_i = \pm 2$, the Davies Equation will predict the same value, γ_2:

$$\log \gamma_2 = -0.51(\pm 2)^2\left[(\sqrt{0.004}/(1+\sqrt{0.004})) - 0.3\ (0.004)\right]$$

$$\gamma_2 = 0.76$$

2.8.2 Activity Coefficient Equations for Neutral Species

The activity coefficients for neutral species *i* in water containing a dissolved salt *j* at concentration c_j are usually found to depend on c_j according to

$$\log \gamma_i = k_{i,j} c_j \qquad \text{“Setschenow Equation”} \tag{2.69}$$

where $k_{i,j}$ is a constant that depends on the identity of both *i* and the salt *j*, and on *T* and *P*. Usually $k_{i,j} > 0$, and γ_i becomes increasingly greater than 1.0 as c_j increases. This behavior is sometimes referred to as "salting-out" because the salt ions are tending to drive *i* out of the solution by increasing a_i (through γ_i). When $k_{i,j} < 0$, then γ_i becomes increasingly less than 1.0 as c_j increases. This "salting in" behavior is due to a *favorable* interaction between *i* and the salt ions, which lowers a_i (through γ_i).

When $k_{i,j} > 0$, values usually range between 0.05 to 0.2. A typical value is ~0.15. For molecular O_2 in solutions of NaCl, $k_{i,j} = 0.13$ at 25 °C (Harned and Owen 1958, p.535). An example of salting-in involves the acetic acid (CH_3COOH) molecule in solutions of potassium nitrate (KNO_3), for which $k = -0.02$ (Harned and Owen 1958, p.536).

Example 2.16 Using the Setschenow Equation

For 25 °C, use the Setschenow Equation to calculate γ_i for O_2 in (a) 0.05 *F* NaCl; and (b) 0.7 *F* NaCl.

Solution

a. $\log \gamma_{O_2} = 0.13 \times 0.05, \; \gamma_{O_2} = 1.015$
b. $\log \gamma_{O_2} = 0.13 \times 0.7, \; \gamma_{O_2} = 1.23$

Fresh natural waters have less dissolved salts than what is present in 0.05 *F* NaCl. The results from the above example explain why: (1) activity corrections for neutral species in fresh waters can be neglected: the γ_i values are very close to 1.0; and (2) for sea water, γ_i corrections for neutral species are large enough to affect answers, but not hugely so (~20%). However, some atmospheric aerosol droplets have large ion concentrations, and it has been argued that these ions lead to large, required corrections when considering the dissolution (from the gas phase) of polar molecules such as glyoxal and methyl glyoxal (Waxman et al., 2015). Useful additional discussions on the topic of the Setschenow Equation can be found in Harned and Owen (1958), and Pytkowicz (1983).

REFERENCES

Bard AJ, Parsons R, Jordan J (1985) *Standard Potentials in Aqueous Solution*, Marcel Dekker, New York.
Bockris JO'M, Reddy AKN (1970) *Modern Electrochemistry, Vol. 1*, Plenum Press, New York.
Denbigh K (1981) *The Principles of Chemical Equilibrium*, 4th edn, Cambridge University Press, Cambridge, UK.
Harned HS, Owen BB (1958) *The Physical Chemistry of Electrolyte Solutions*, American Chemical Society Monograph Series, Reinhold Publishing, New York.
Kielland J (1937) Individual activity coefficients of ions in aqueous solutions. *Journal of the American Chemical Society*, **59**, 1675–1678.
Pytkowicz RM (1983) *Equilibria, Nonequilibria, and Natural Waters, Vol. 1*, Wiley-Interscience, New York.
Smith RM, Martell AE (1976) *Critical Stability Constants, Volume 4, Inorganic Complexes*, Plenum Press, New York.
Stumm W, Morgan JJ (1996) *Aquatic Chemistry: Chemical Equilibria and Rates in Natural Waters*, 3rd edn, Wiley Publishing Co, New York.
Waxman EM, Elm J, Kurtén T, Mikkelsen, KV, Ziemann PJ, Volkamer R (2015) Glyoxal and methylglyoxal Setschenow salting constants in sulfate, nitrate, and chloride solutions: measurements and Gibbs energies. *Environmental Science and Technology*, **49**, 11500–11508.

Part II

Acid/Base Chemistry

3 The Proton (H^+) in Aquatic Chemistry

3.1 GENERAL IMPORTANCE OF H^+ IN NATURAL WATERS

As discussed in Chapter 1, there are six main classes of reactions in natural water chemistry. These are:

- Acid/base
- Complexation
- Dissolution/precipitation
- Redox
- Gas/water partitioning
- Adsorption

In addition to playing its obvious role in acid/base reactions, the proton H^+ is a direct or indirect participant in reactions of the other types. $\{H^+\}$ therefore helps determine the equilibrium positions of many reactions: in aquatic chemistry, $pH = -\log\{H^+\}$ is often called a "master variable". See Table 3.1 for acidity constant values for different acids at 25 °C and 1 atm pressure; see Table 3.2 for K_w as a function of temperature.

Consider the following:

1. Acid/base reaction examples

$$H_2O = H^+ + OH^- \quad \text{water acting as an acid} \tag{3.1}$$

$$H_2CO_3^* = HCO_3^- + H^+ \quad \text{carbonic acid acting as an acid} \tag{3.2}$$

$$HCO_3^- = CO_3^{2-} + H^+ \quad \text{bicarbonate ion acting as an acid} \tag{3.3}$$

$$NH_4^+ = NH_3 + H^+ \quad \text{ammonium ion acting as an acid} \tag{3.4}$$

$$RCOOH = RCOO^- + H^+ \quad \text{organic carboxylic acid acting as an acid} \tag{3.5}$$

2. Complexation reaction examples

$$Fe^{3+} + H_2O = FeOH^{2+} + H^+ \quad \text{“hydrolysis” of } Fe^{3+} \text{ to form } FeOH^{2+} \tag{3.6}$$

$$Cu^{2+} + H_2O = CuOH^+ + H^+ \quad \text{“hydrolysis” of } Cu^{2+} \text{ to form } CuOH^+ \tag{3.7}$$

TABLE 3.1
Values of Acidity Constants for Acids of Interest in Natural Waters (with Temperature Dependence as ($dpK/d(1/T)$)

	Species	pK_a (pK_1)	$\Delta H°$	pK_2	$\Delta H°$	pK_3	$\Delta H°$
acetic acid	CH_3COOH/CH_3COO^-	4.76	−0.42				
ammonium ion	NH_4^+/NH_3	9.24	52.1				
arsenic acid	$H_3AsO_4/H_2AsO_4^-/HAsO_4^{2-}/AsO_4^{3-}$	2.24	−7.07	6.76	3.22	11.60	18.2
arsenious acid	$H_3AsO_3/H_2AsO_3^-/HAsO_3^{2-}/AsO_3^{3-}$	9.23	27.5	12.10	NA	13.41	NA
benzoic acid	benzoic acid/benzoate	4.20	−0.53				
boric acid	$H_3BO_4/B(OH)_4^-$	9.27	13.9				
carbonic acid	$H_2CO_3^*/HCO_3^-/CO_3^{2-}$	6.35	see Table 9.2	10.33	see Table 9.2		
citric acid	$H_3Cit/H_2Cit^-/HCit^{2-}/Cit^{3-}$	3.13	4.17	4.76	2.44	6.40	−3.36
hydrochloric acid	HCl/Cl^-	< −3	–				
hydrofluoric acid	HF/F^-	3.18	−13.2				
hydrogen cyanide	HCN/CN^-	9.21	42.9				
hydrogen sulfide	$H_2S/HS^-/S^{2-}$	7.02	22.2	12.9[a]	49.8		
hypochlorous acid	$HOCl/OCl^-$	7.4	15.1				
nitric acid	HNO_3/NO_3^-	< −1	–				
nitrous acid	HNO_2/NO_2^-	3.25	9.83				
oxalic acid	H_2Ox / HOx^{2-} / Ox^{2-}	1.25	−3.44	3.1	−6.63		
ortho silicic acid	$Si(OH)_4$ / $SiO(OH)_3^-$	9.86	NA				
perchloric acid	$HClO_4/ClO_4^-$	< −5	–				
phenol	phenol/phenolate	9.99	23.2				
phosphoric acid	$H_3PO_4/H_2PO_4^-/HPO_4^{2-}/PO_4^{3-}$	2.16	−7.61	7.21	3.49	12.32	14.4
sulfuric acid	$H_2SO_4/HSO_4^-/SO_4^{2-}$	< −3	–	1.99	−17.8		
sulfurous acid	$H_2SO_3/HSO_3^-/SO_3^{2-}$	1.85	−3.86	7.18	−11.7		

Note: Data are for 25 °C/1 atm pressure, and infinite dilution. Values for K_w for water are given in Table 3.2. For polyprotic acids, $pK_1 = pK_a$ as given. K data in part from Perrin (1982) and *CRC Handbook of Chemistry and Physics*, 84th Edition (2004). $\Delta H°$ (kJ/mol) data for temperature correction according to Eq.(2.52) are from Christensen et al. (1976).

[a] Value from Christensen et al. (1976). NA = not available.

TABLE 3.2
Acid Dissociation Constant K_w (m^2) for Water as a Function of Temperature at 1 atm Pressure; Infinite Dilution Scale

t (°C)	K_w	log K_w
0	0.12×10^{-14}	−14.93
5	0.18×10^{-14}	−14.73
10	0.29×10^{-14}	−14.53
15	0.45×10^{-14}	−14.35
20	0.68×10^{-14}	−14.17
25	1.01×10^{-14}	−14.00
30	1.47×10^{-14}	−13.83
35	2.10×10^{-14}	−13.68
40	2.94×10^{-14}	−13.53
50	5.48×10^{-14}	−13.26

Note: $\log K_w = -4470.99/T + 6.0875 - 0.01706\,T$, where T = temperature in degrees *K*. From Harned and Owen (1958). $T(\mathrm{K}) = t(°C) + 273.15$.

3. Dissolution/precipitation reaction examples

$CaCO_{3(s)} + H^+ = Ca^{2+} + HCO_3^-$ proton assisted dissolution of calcite (3.8)

$NaAlSi_3O_{8(s)} + H^+ + 9/2\,H_2O = 1/2\,Al_2Si_2O_5(OH)_{4(s)} + Na^+ + 2H_4Si$ conversion of albite to kaolinite (3.9)

4. Redox reaction examples

$O_2 + 4e^- + 4H^+ = 2H_2O$ reduction of oxygen to water (3.10)

$C_6H_{12}O_6 + 24e^- + 24H^+ = 6CH_{4(g)} + 6H_2O$ reduction of glucose to methane (3.11)

5. Gas/water partitioning reaction example

$CO_{2(g)} + H_2O = H_2CO_3^*$
$H_2CO_3^* = H_2CO_3^- + H^+$ dissolution of $CO_{2(g)}$ followed by $H_2CO_3^*$ dissociation (3.12)

6. Adsorption reaction example

$\equiv S{-}OH + Pb^{2+} = \equiv S{-}O{-}Pb^+ + H^+$ sorbing to a surface hydroxyl site (3.13)

3.2 SHOULD WE REFER TO THE PROTON IN WATER AS H^+ OR H_3O^+?

3.2.1 Definitions

For the dissociation of a simple monoprotic acid HA, we can write

$$HA = H^+ + A^-. \tag{3.14}$$

We will assume ideal behavior (*i.e.*, $\{i\} = [i]$) in the solution for this discussion, so that the acid dissociation equilibrium constant K_a for the particular acid HA of interest is given by

$$K_a = \frac{[H^+][A^-]}{[HA]}. \tag{3.15}$$

The literal interpretation of Eqs.(3.14) and (3.15) is that there is an equilibrium between HA and the bare ions H^+ and A^-. However, species in aqueous solutions (especially ions) do not exist all by themselves and are in fact closely surrounded by solvating water molecules (see discussion in Section 2.8.1 on the failure of the Extended Debye–Hückel Equation at high ionic strength). For example, when discussing the aqueous chemistry of transition metal ions, specialists sometimes specifically note the association with water molecules; *e.g.*, aqueous Cu^{2+} is sometimes represented as $Cu(H_2O)_6^{2+}$. Usually, this is not done, though, because such representations obscure most discussions of reactions and equilibria. In any case, whenever we consider any species in aqueous solution, it should be implicitly understood that each species is surrounded and solvated at least to some extent by water molecules. For some species, like H^+, Fe^{3+}, and Al^{3+}, that association with water is very close; for others, like I^-, the association is much weaker.

Many chemistry texts explicitly recognize the strong solvation of H^+ by water and argue that as soon as any H^+ is formed by a reaction such as Eq.(3.14), reaction with water proceeds according to

$$H^+ + H_2O = H_3O^+ \qquad K = \frac{[H_3O^+]}{[H^+][H_2O]} \tag{3.16}$$

with H_3O^+ named the "hydronium" ion. The net reaction for Eqs.(3.14) and (3.16) is then

$$HA + H_2O = H_3O^+ + A^- \tag{3.17}$$

$$K_a = \frac{[H_3O^+][A^-]}{[HA][H_2O]} = \frac{[H_3O^+][A^-]}{[HA]}. \tag{3.18}$$

Equation (3.18) follows the usual convention that the concentration scale chosen for water (as the solvent) is the mole fraction scale so that in most solutions $[H_2O] \approx 1$.

So, how can one text use the H^+ representation and another use the H_3O^+ representation, and both still come out with the same answer for pH and other parameters for a given solution of interest? The answer is based on the fact that neither representation is exactly correct, and in fact, they both refer symbolically to all forms of H^+ in aqueous solution, with every possible combination of interactions with H_2O. Since they both refer to the same quantity, with that quantity being physically meaningful and experimentally determinable, either one can be used.

3.2.2 The Meaning for Both $[H^+]$ and $[H_3O^+]$: Protons in Water Exist as a Set of Species H^+, H_3O^+, $H_5O_2^+$, ...

The beginning of the resolution of the H^+ *vs.* H_3O^+ quandary is found in the fact that while the equilibrium constant K for Eq.(3.16) may well be large, it cannot be infinitely large. If K is not infinitely

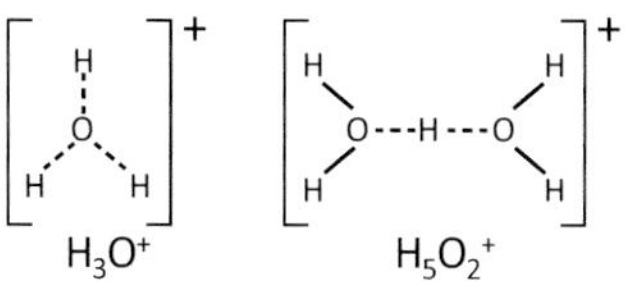

FIGURE 3.1 Simple schematic representations of the species H_3O^+ and $H_5O_2^+$.

large, the concentration of bare H^+ ions in any aqueous solution cannot be zero. So, even if the H_3O^+ view of things *were* more accurate (on average) than the H^+ view of things, it still would not be exactly right: the H^+ view and the H_3O^+ view are *both* approximations of reality. In fact, protons in solution exist as a whole *series* of species. Let us introduce some italics to help emphasize when we are referring to an actual particular species: a true bare proton is H^+; a proton truly tied to only one water is $\mathit{H_3O}^+$; a proton truly tied to two water molecules is $\mathit{H_5O_2}^+$, etc. (see Figure 3.1). The range of species can be abbreviated as $\mathit{H(H_2O)}_n^+$ where n starts at 0 and goes up. Spectral studies indicate that values as large as $n=6$ are important (Headrick et al., 2005); the same applies to other aqueous ions. Space limitations around any given ion coupled with decreasing strengths with distance for the interaction forces mean that species decrease in importance quickly as n rises above ~6.

We now have

$$\left[\mathrm{H}^+\right] \equiv \left[\mathrm{H_3O}^+\right] \equiv \left[H^+\right] + \left[H_3O^+\right] + \left[H_5O_2^+\right] + \left[H_7O_3^+\right] + \ldots \tag{3.19}$$

If the upper limit for importance is $n \approx 6$, one might conclude that Eq.(3.19) only has seven important terms on the right-hand side. However, each of the hydrated species actually has an infinite number of possibilities due to a continuum of interaction distances values, so even the summation in Eq.(3.19) is very much an abbreviation of reality Fortunately, the *distribution* among all the species is constant at given T and P, so that when $[H^+]$ ($\equiv [H_3O^+]$) is doubled, then the A^- really sees twice as much of the same group of H^+-containing species in solution, and the concept of "mass action" is retained; there is a doubling of the driving force to form HA.

Overall, we see that the equilibrium constant quotients in Eqs.(3.15) and (3.18) are representations of the same thing. This text will use the H^+ notation: the H_3O^+ notation is cumbersome, and besides, the same "problem" exists with all other species, and no-one bothers, for example, to suggest that any A^- species might be better represented as $A(H_2O)^-$.

REFERENCES

Christensen JJ, Hansen LD, Izatt RM (1976) *Handbook of Proton Ionization Heats and Related Thermodynamic Quantities*, J. Wiley and Sons, New York.

Harned HS, Owen BB (1958) *The Physical Chemistry of Electrolyte Solutions*, Reinhold, New York.

Headrick JM, Diken EG, Walters RS, Hammer NI, Christie RA, Cui J, Myshakin EM, Duncan MA, Johnson MA, Jordan KD (2005) Spectral signatures of hydrated proton vibrations in water clusters. *Science*, **308**, 1765–1769.

Perrin DD (1982) *Ionisation Constants of Inorganic Acids and Bases in Aqueous Solution*, 2nd edn, IUPAC Chemical Data Series, No. 29, Pergamon Press, Oxford, UK.

4 The Electroneutrality Equation, Mass Balance Equations, and the Proton Balance Equation

4.1 INTRODUCTION

Solving a chemical problem requires knowledge of the equations that govern the problem. For chemical speciation problems, equilibrium constant (K) expressions are among the equations that are needed. Other important equations include the electroneutrality equation (ENE), mass balance equations (MBEs), and the proton balance equation (PBE), if one can be written for the system. When any of these equations are written here using only terms like $[Na^+]$, $[A^-]$, $[Cl^-]$, etc., it is described as a "general" version of the equation; Eq.(4.4) below is a general ENE. A "specific" version of an equation contains one or more specific values, such as 10^{-3}. PBEs are usually left in general form because prior to solving a given problem, specific values of terms are usually not known.

4.2 THE ELECTRONEUTRALITY EQUATION (ENE)

Definition: The **ENE** is the mathematical statement that an aqueous solution must be electrically neutral. The ENE is also known as the charge balance equation. We can always write an ENE for an aqueous solution. Since the ENE involves multiple species in a solution, it is very useful in solving problems. Example cases follow. All expressions written for equilibrium constants in this section, as well as thereafter in this chapter, will use concentrations, not activities, and thus assume activity corrections can be neglected.

Case 1. Pure water. ENE. Water dissociates (weakly) according to

$$H_2O = H^+ + OH^- \qquad K_w = \frac{[H^+][OH^-]}{[H_2O]} = [H^+][OH^-]. \tag{4.1}$$

Both H^+ and OH^- are present in every aqueous solution, so both $[H^+]$ and $[OH^-]$ are present in every ENE. For pure water, H^+ and OH^- are the only ions. Electroneutrality means there is "charge balance" (total positive charge = total negative charge), so the ENE for pure water is

$$[H^+] = [OH^-]. \tag{4.2}$$

Because Eq.(4.2) holds, pure water is said to be "neutral" from an acid/base point of view. In other solutions discussed below, if $[H^+] > [OH^-]$, the solution is said to be "acidic"; if $[OH^-] > [H^+]$, the solution is said to be "basic".

Case 2. Pure water + HA. ENE. If we add some HA (a monoprotic acid, *e.g.*, acetic acid) to pure water, some of the HA will dissociate to form some H^+ and some A^-.

$$HA = H^+ + A^- \qquad K = \frac{[H^+][A^-]}{[HA]}. \tag{4.3}$$

Besides H^+ and OH^-, we now also have A^- in the solution. The ENE is

$$[H^+] = [OH^-] + [A^-]. \tag{4.4}$$

Case 3. Pure water + NaA. ENE. $NaA_{(s)}$ is a salt composed of the ions Na^+ and A^-, which enter the solution when some $NaA_{(s)}$ is dissolved. There is no chemical bond possible between Na^+ and A^-. All of the sodium goes into the solution as the "spectator ion" Na^+. Other than: (1) taking some water into its solvation shell; and (2) contributing to the ionic nature of the solution, a "spectator ion" just "watches" what goes on in the surrounding solution; it does not participate in any reactions. The ENE is

$$[Na^+] + [H^+] = [A^-] + [OH^-]. \tag{4.5}$$

Case 4. Pure water + H_2B. ENE. H_2B represents a diprotic acid (*e.g.*, carbonic acid $H_2CO_3^*$; the * will be explained in Chapter 9). Some of the H_2B will dissociate to form some H^+ and some HB^-:

$$H_2B = H^+ + HB^- \qquad K_1 = \frac{[H^+][HB^-]}{[H_2B]}. \tag{4.6}$$

Some of the HB^- will dissociate to form some more H^+, and some B^{2-}:

$$HB^- = H^+ + B^{2-} \qquad K_2 = \frac{[H^+][B^{2-}]}{[HB^-]}. \tag{4.7}$$

The ENE is

$$[H^+] = [HB^-] + 2[B^{2-}] + [OH^-]. \tag{4.8}$$

B^{2-} is doubly charged, hence the factor of 2 for its term.

Case 5. Pure water + NaHB. ENE. $NaHB_{(s)}$ is a salt composed of Na^+ and HB^- (*e.g.*, $NaHCO_3$). There is no chemical bond between Na^+ and HB^-. When some $NaHB_{(s)}$ is added to water, all of the sodium enters the solution as the spectator ion Na^+. Some of the HB^- can dissociate to form some H^+, and some B^{2-}. The ENE is

$$[H^+] + [Na^+] = [HB^-] + 2[B^{2-}] + [OH^-]. \tag{4.9}$$

Case 6. Pure water + Na_2B. ENE. $Na_2B_{(s)}$ is a salt composed of $2Na^+$ and B^{2-} (*e.g.*, $Na_2CO_{3(s)}$). There is no chemical bond between the Na^+ and B^{2-}. When Na_2B is added to water, all of the sodium goes into the solution as the spectator ion Na^+. Because K_1 and K_2 in Eqs.(4.6) and (4.7) are finite and positive-non-zero, at equilibrium, neither $[HB^-]$ or $[H_2B]$ can be zero: there must be some of all three B species.

One way to get some HB^- is by the reaction

$$B^{2-} + H_2O = HB^- + OH^-. \tag{4.10}$$

Another way is by reaction of B^{2-} with H^+.

One way to get some H_2B is by the reaction

$$HB^- + H_2O = H_2B + OH^-. \tag{4.11}$$

Another way is by reaction of HB^- with H^+. Overall, the ENE is the same as for Case 5, namely Eq.(4.9). Although the solution chemistry will be different, the list of ions is the same.

Example 4.1 The ENE for a Simple Salt Solution

Write the general ENE for a solution of some pure water plus some NaCl.

Solution

Dissolving some $NaCl_{(s)}$ in pure water adds Na^+ and Cl^-, which makes for a total of four ions in solution. The general ENE is

$$[Na^+]+[H^+]=[Cl^-]+[OH^-].$$

Example 4.2 The ENE for a Solution of Some Salt Plus a Diprotic Acid

An aqueous solution is prepared from pure water, some NaCl, some acetic acid (which can be represented as HA), and some oxalic acid (which can be represented as H_2B). Write the general ENE.

Solution

Two of the ions in this solution are H^+ and OH^- (as from dissociation of H_2O, though in many solution types there might be other sources for one or the other or both of these ions). From dissociation of some of the HA, there will also be some A^-; from dissociation of some of the H_2B, there will be some HB^- and some B^{2-}. The ENE is

$$[H^+]+[Na^+]=[A^-]+[HB^-]+2[B^{2-}]+[OH^-]+[Cl^-].$$

Example 4.3 The ENE for a Complex Solution

The ions in a solution are H^+, Na^+, K^+, Mg^{2+}, Ca^{2+}, Fe^{3+}, $FeOH^{2+}$, $Fe(OH)_2^+$, acetate ion, OH^-, HCO_3^-, CO_3^{2-}, Cl^-, HSO_4^-, SO_4^{2-}, and $Fe(OH)_4^-$. Acetate ion can be represented as A^-. Write the general ENE.

Solution

$$[H^+]+[Na^+]+[K^+]+2[Mg^{2+}]+2[Ca^{2+}]+3[Fe^{3+}]+2[FeOH^{2+}]+[Fe(OH)_2^+]$$
$$=[A^-]+[OH^-]+[HCO_3^-]+2[CO_3^{2-}]+[Cl^-]+[HSO_4^-]+2[SO_4^{2-}]+[Fe(OH)_4^-]$$

4.3 MASS BALANCE EQUATIONS (MBES)

Definition: An **MBE** is a statement of conservation of mass over a unit volume such as 1.0 liter, or over a unit solution mass (*e.g.*, 1.0 kg). Some example cases follow. In each, the general version(s) of the MBE(s) is(are) given.

Case 1. Pure water. MBE. When water dissociates according to $H_2O=H^++OH^-$, there is a 1:1 relationship between the amounts of H^+ and OH^- formed. Therefore, although it is not usually viewed from that perspective, for pure water Eq.(4.2) is an MBE as well as the ENE.

Case 2. Pure water + HA. MBE. Some of the HA will dissociate to form H^+ and A^-. Both the remaining HA and the formed A^- have the moiety A in them. If the total concentration of A moiety in the solution is denoted A_T (*M* or *m*), the MBE for this solution is

$$[HA]+[A^-]=A_T. \tag{4.12}$$

Case 3. Pure water + NaA. MBEs. When some $NaA_{(s)}$ is added to water, all of the sodium goes into the solution and remains as the spectator ion Na^+. Some of the A^- will pick up a proton and be converted to HA. Eq.(4.12) is an MBE for this solution, and a **second MBE** is

$$[Na^+]=A_T. \tag{4.13}$$

Case 4. MBE: pure water + H_2B. By analogy with Case 2, the MBE is

$$[H_2B]+[HB^-]+[B^{2-}]=B_T. \tag{4.14}$$

Case 5. MBEs: pure water + NaHB. All three B species must be present at equilibrium, so Eq.(4.14) again applies as an MBE. In addition, because of the 1:1 relationship between $[Na^+]$ and B_T, a **second MBE** is

$$[Na^+]=B_T. \tag{4.15}$$

Case 6. MBE: pure water + Na_2B. Eq.(4.14) again applies as an MBE. In addition, because of the 2:1 relationship between $[Na^+]$ and B_T, a **second MBE** is

$$[Na^+]=2B_T. \tag{4.16}$$

Example 4.4 MBE Example.1

Write the general MBEs for a solution of some initially pure water plus some NaCl.

Solution

If we dissolve some pure $NaCl_{(s)}$ in pure water, the salt carries no H^+ or OH^- that would alter the balance between H^+ or OH^- in pure water, so technically Eq.(4.1) is an MBE for this system. As in Case 1, we usually do not bother to write it as such, however, because it is derivable from the ENE and the MBE that we develop next. Cl^- is a spectator ion, like Na^+; because of the 1:1 relationship between the amounts of Na^+ and Cl^-, the general MBE for the solution is

$$[Na^+]=[Cl^-].$$

Example 4.5 MBE Example.2

10^{-3} FW of NaCl and 3×10^{-4} FW of acetic acid (which can be represented as HA) are placed in 2 L of initially pure water. For this solution, write the general MBEs and the corresponding specific MBEs.

Solution

General MBEs	Specific MBEs
$[Na^+] = [Cl^-]$	$[Na^+] = 10^{-3}/2\ M = [Cl^-] = 10^{-3}/2\ M\ (= 1.5 \times 10^{-3}\ M)$
$[HA] + [A^-] = A_T$	$[HA] + [A^-] = 3 \times 10^{-4}/2\ M\ (= 1.5 \times 10^{-4}\ M)$

Example 4.6 MBE Example.3

3×10^{-3} FW of Na_2CO_3 (sodium carbonate, which can be represented as Na_2B) is placed in 2 L of initially pure water. For this solution, what are the general MBEs, and what are the specific MBEs?

Solution

General MBEs	Specific MBEs
$[Na^+] = 2B_T$	$[Na^+] = (2 \times 3 \times 10^{-3})/2\ M = 3 \times 10^{-3}$
$[H_2B] + [HB^-] + 2[B^{2-}] = B_T$	$[H_2B] + [HB^-] + 2[B^{2-}] = 3 \times 10^{-3}/2\ M = 1.5 \times 10^{-3}\ M$

4.4 THE PROTON BALANCE EQUATION (PBE)

Definition: **If one can be written**, the **PBE** for a solution provides the balance sheet for the combined accounting for:

1) Which species present in the solution account for the concentration of H^+, and according to what factors?

   and

2) Which species present in the solution account for the concentration of OH^-, and according to what factors?

Because of the combined nature of the accounting, a more accurate name for the PBE would certainly be "the proton and hydroxide balance equation". However, PBE is the name in common use, and we will continue that tradition here.

It is always possible to write an ENE for an aqueous solution: one just has to know which ions are present in solution, and it can be written. And, if one knows how the solution came to be, governing MBE(s) for the solution can also always be written. However, the concept of "proton balance equation (PBE)" is different in that we only write a PBE for a given solution of acid or base (water is an acid all by itself) when the solution is at a **special compositional point (SCP)**. Except for the case of pure water, each of the SCPs amounts to an **"equivalence point" (EP)** in the titration curve for the acid or base of interest. This will become increasingly clear from the example cases below, and from material in Chapters 5 through 9. The steps for writing a PBE are given in Table 4.1. Each of the six cases below is at an SCP such that a PBE can be written.

Case 1. Pure water. PBE. Water dissociates (weakly) according to Eq.(4.1). The H^+ that is present is connected directly to the OH^- that is present. Vice versa, one could say that the OH^- that is present is connected directly to the H^+ that is present. The PBE is then

$$[H^+] = [OH^-]. \tag{4.17}$$

TABLE 4.1
Steps for Writing a Proton Balance Equation (PBE) for a Solution

1. Verify that the solution is either just pure water, or a solution of an acidic or basic species in **one** of its characteristic "equivalence point" forms (so that a PBE can be written).
2. Start with $[H^+]=[OH^-]$, because water is always present: $[H^+]$ and $[OH^-]$are present in every PBE (on opposite sides).
3. On the $[OH^-]$ side, write the concentration of every other species whose genesis led to H^+. Include the appropriate multiplying factor.
4. On the $[H^+]$ side, write the concentration of every other species whose genesis led to OH^-. Include the appropriate multiplying factor.
5. Check to be sure that what was added to make the solution is not in the PBE. *E.g.*, [HA] is not in the PBE for a solution of HA, $[A^-]$ is not in the PBE for a solution of NaA, etc.

We saw above that this equation is also the ENE for this solution. This type of equivalency occurs for other solutions as well. In fact, if a PBE can be written (because the solution is either pure water, or a solution of an acid or base system at one of its EPs), then the collection of equations composed of the ENE, the MBEs, and the PBE will not all be linearly independent. For Case 1, the PBE that is Eq.(4.17) is not independent of the ENE that is Eq.(4.1); they are obviously the same, though obtained by ostensibly different means.

Because all aqueous solutions contain water, and because water can always dissociate according to Eq.(4.1), **both $[H^+]$ and $[OH^-]$ are in every PBE that can be written**, $[H^+]$ is always on the left, and $[OH^-]$ is always on the right.

Case 2. Pure water + HA. PBE. Eq.(4.17) is the starting point for considering the PBE for this solution (and all others). The question is, what terms, if any, need to be added on each side? HA is an acid, not a base. So, the only source of OH^- in this solution is dissociation of water: nothing needs to be added to the $[H^+]$ side (to account for the OH^- present). But there are two sources of H^+ in the solution: dissociation of water (which is covered by the term $[OH^-]$) *and* dissociation of HA by Eq.(4.3). $[A^-]$ provides a measure of the production of H^+ from the dissociation of HA. The PBE is

$$[H^+] = [A^-] + [OH^-]. \tag{4.18}$$

As in Case 1, for this solution, the ENE and the PBE are the same. The solution is at least somewhat acidic because it must be true that $[H^+] > [OH^-]$ (the term $[OH^-]$ needs help on the right-hand side to equal the left-hand side).

Case 3. Pure water + NaA. PBE. Eq.(4.17) is again the starting point. NaA is not an acid. The only source of H^+ in this solution is, therefore, the dissociation of water, so nothing needs to be added to the $[OH^-]$ side (to account for the H^+ present). When a solution of NaA is prepared, the NaA that dissolves forms Na^+ and A^- ions. A^- will participate in an equilibrium with HA according to Eq.(4.2). K is not infinite, so [HA] cannot be zero: some of the A^- that is introduced must be converted to HA. Most of the protons in the solution are on H_2O, so one way that HA can form is by the reaction

$$A^- + H_2O = HA + OH^-. \tag{4.19}$$

A second way is by reaction of A^- with H^+. The amount of HA in the solution, therefore, accounts for how much OH^- was formed by Eq.(4.19) plus how much H^+ was depleted due to HA formation by reaction of A^- with H^+. The PBE is

$$[H^+] + [HA] = [OH^-]. \tag{4.20}$$

The solution must be at least somewhat basic because it must be true that $[OH^-] > [H^+]$ (the term $[H^+]$ needs help on the left-hand side to equal the right-hand side).

Case 4. Pure water + H_2B. PBE. Eq.(4.17) is the starting point. H_2B is an acid. As noted, an example chemical is $H_2CO_3^*$. The only source of OH^- in this solution is dissociation of water: nothing needs to be added to the $[H^+]$ side. Besides the dissociation of water as a source for H^+, there is dissociation of H_2B. If we add some H_2B to water, some will dissociate according to Eq.(4.6), so for each HB^- present, there is one H^+ present, and $[HB^-]$ needs to be on the right-hand side with $[OH^-]$. Also, some of the HB^- that forms will dissociate according to Eq.(4.7), and for each B^{2-} present, two protons be present. The PBE is

$$[H^+] = [OH^-] + [HB^-] + 2[B^{2-}]. \tag{4.21}$$

As in Cases 1 and 2, the PBE is the same as the ENE. The solution is at least somewhat acidic because it must be true that $[H^+] > [OH^-]$.

It is common for students to wonder why the factor of 2 is needed for $[B^{2-}]$. They think the first proton liberated in the formation of $[B^{2-}]$ is counted by the $[HB^-]$ term, and $1 \times [B^{2-}]$ would count the second one. However, any HB^- that goes on to form B^{2-} is no longer in the solution. So, $[HB^-]$ can only account for how many H^+ are in the solution due to the formation of the HB^- ions (that are **in the solution**). Thus, $2[B^{2-}]$ accounts for how many H_2B dissociated all the way to B^{2-}, giving two H^+. Figure 4.1 explains this schematically. The situation for pure water is represented in panel a. In panel b.1, some H_2B has been added, some of which has dissociated according to $H_2B = HB^- + H^+$. In panel b.2, some of the HB^- that was in panel b.1 has dissociated according to $HB^- = H^+ + B^{2-}$.

Case 5. Pure water + NaHB. PBE. For NaHB, an example chemical is sodium bicarbonate ($NaHCO_3$). We can view the first step for NaHB dissolution in water as 100% conversion to Na^+ and HB^- ions. A portion of the HB^- will dissociate according to Eq.(4.7), each HB^- giving

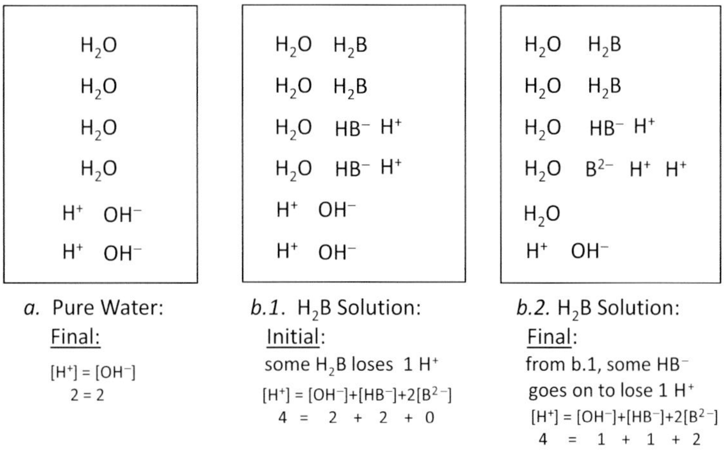

FIGURE 4.1 Schematic representation of two cases. **a.** PBE in pure water; **b.1** initial PBE in an H_2B solution after some dissociation of some H_2B; **b.2.** final PBE in an H_2B solution after some dissociation of HB^- and some re-formation of H_2O.

one H^+ and one B^{2-}. Every B^{2-} present in the solution is then responsible for the presence of one H^+. With Eq.(4.17) as the starting point for the PBE, we need to add the term $[B^{2-}]$ to the right-hand side. Since H_2O and HB^- are the only possible sources for H^+, we are done with the right-hand side.

Now, do we need to add anything to the left-hand side? Because K_1 in Eq.(4.6) is finite, $[H_2B]$ cannot be zero. The needed H_2B can form by Eq.(4.11) and by reaction of HB^- with H^+. For each H_2B formed, there is either one OH^- formed or one H^+ consumed; we need to add $[H_2B]$ to the left-hand side. The PBE is

$$[H^+] + [H_2B] = [OH^-] + [B^{2-}]. \tag{4.22}$$

In contrast to all the other cases considered so far, for this case, we cannot say for sure if the solution will be acidic, basic, or neutral. Depending on the values of $[H_2B]$ and $[B^{2-}]$, it could be that $[H^+] > [OH^-]$ or $[H^+] < [OH^-]$. To know which is true, we would need to know the identity of NaHB (so that we can look up the values of K_1 and K_2). Then, the final answer will depend on whether HB^- is stronger as an acid (by the equilibrium for Eq.(4.7)) or stronger as a base (by the equilibrium for Eq.(4.11)). If equally strong in both directions, the solution will be neutral because $[H^+] = [OH^-]$. Solutions of $NaHCO_3$ are basic because bicarbonate ion (HCO_3^-) is stronger as a base than as an acid (see Chapter 9).

Case 6. Pure water + Na_2B. PBE. This solution is a "reverse" analogy of Case 4. Here, instead of adding the fully acid form of B to the solution, namely H_2B, we are adding the fully base form, namely B^{2-}. With Eq.(4.17) as the starting point for the PBE, because of Eq.(4.10), we need to add $[HB^-]$ to the left-hand side. Some of the HB^- formed will go on to form some H_2B, according to Eq.(4.11), so that for each H_2B formed, two HB^- are formed. The PBE is

$$[H^+] + [HB^-] + 2[H_2B] = [OH^-]. \tag{4.23}$$

The equations developed in this chapter are summarized in Table 4.2.

TABLE 4.2
Summary of General PBEs, ENEs, and MBEs for Examples of Aqueous Solutions

Solution Case	ENE	MBE(s)	PBE	not present in PBE
1. Pure Water	$[H^+]=[OH^-]$	$[H^+]=[OH^-]$	$[H^+]=[OH^-]$	$[H_2O]$
2. HA	$[H^+]=[A^-]+[OH^-]$	$A_T=[HA]+[A^-]$	$[H^+]=[A^-]+[OH^-]$	$[HA]$
3. NaA	$[Na^+]+[H^+]=[OH^-]+[A^-]$	$A_T=[HA]+[A^-]$ $[Na^+]=A_T$	$[H^+]+[HA]=[OH^-]$	$[A^-]$
4. H_2B	$[H^+]=[OH^-]+[HB^-]+2[B^{2-}]$	$B_T=[H_2B]+[HB^-]+[B^{2-}]$	$[H^+]=[OH^-]+[HB^-]+2[B^{2-}]$	$[H_2B]$
5. NaHB	$[Na^+]+[H^+]=[OH^-]+[HB^-]+2[B^{2-}]$	$B_T=[H_2B]+[HB^-]+[B^{2-}]$ $[Na^+]=B_T$	$[H^+]+[H_2B]=[OH^-]+[B^{2-}]$	$[HB^-]$
6. Na_2B	$[Na^+]+[H^+]=[OH^-]+[HB^-]+2[B^{2-}]$	$B_T=[H_2B]+[HB^-]+[B^{2-}]$ $[Na^+]=2B_T$	$[H^+]+[HB^-]+2[H_2B]=[OH^-]$	$[B^{2-}]$

Example 4.7 Writing PBEs

Consider the following specific identities.

1. HA is CH_3COOH (acetic acid);
2. NaA is sodium acetate;
3. H_2B is $H_2CO_3^*$ (for HB^- use HCO_3^-, for B^{2-} use CO_3^{2-});
4. NaHB is $NaHCO_3$ (for H_2B use $H_2CO_3^*$, for B^{2-} use CO_3^{2-});
5. Na_2B is Na_2CO_3 (for H_2B use $H_2CO_3^*$, for HB^- use HCO_3^-).

Write the PBE for each of the following solutions	Solution
a. Solution of acetic acid:	$[H^+]=[OH^-]+[CH_3COO^-]$
b. Solution of sodium acetate:	$[H^+]+[CH_3COOH]=[OH^-]$
c. Solution of $H_2CO_3^*$:	$[H^+]=[OH^-]+[HCO_3^-]+2[CO_3^{2-}]$
d. Solution of $NaHCO_3$	$[H^+]+[H_2CO_3^*]=[OH^-]+[CO_3^{2-}]$
e. Solution of Na_2CO_3	$[H^+]+[HCO_3^-]+2[H_2CO_3^*]=[OH^-]$

Example 4.8 Using Linearly Independent Equation Sets

It is discussed above that when a PBE can be written, the PBE is usually not independent of the ENE and MBE(s). This is obvious for Cases 1, 2, and 4; in each, the ENE and the PBE are the same. For a solution of the NaHB type:

a. Write the ENE, MBEs, and PBE; and
b. Show that the PBE can be obtained from the ENE and the MBEs.

Solution

a. ENE: $[H^+] + [Na^+] = [OH^-] + [HB^-] + 2[B^{2-}]$ (1)

MBEs: $B_T=[H_2B] + [HB^-] + [B^{2-}]$ (2)

$[Na^+] = B_T$ (3)

PBE: $[H^+] + [H_2B] = [OH^-] + [B^{2-}]$ (4)

b. Subtract Eq.(2) from Eq.(1). From the result, subtract Eq.(3). The result is the PBE. The PBE is not independent of the other three because it can be generated using the other three.

5 Quantitative Acid/Base Calculations for Any Solution of Acids and Bases

5.1 INTRODUCTION

As discussed in Chapter 2, the final equilibrium position that a given aqueous system will take is determined by: (1) the T- and P-dependent value(s) of the pertinent equilibrium constant(s); (2) the mass balance and other equation(s) governing the system; and (3) how the activity coefficients γ_i depend on chemical composition. The γ_i will be determined by the *final* equilibrium composition of the system, and so will not be exactly knowable *a priori* for use in the calculations. However, for dilute solutions, or when the reaction medium contains relatively large amount(s) of background dissolved salt(s) not participating in the reactions, then the γ_i in the final equilibrium system may be estimated accurately.

5.2 SOLUTION OF THE GENERIC ACID HA, ALL $\gamma_i = 1$

5.2.1 INTRODUCTION

There are four species in a solution of HA (in addition to H_2O):

$$H^+ \quad A^- \quad OH^- \quad HA.$$

This means there are four unknowns $[H^+]$, $[A^-]$, $[OH^-]$, and [HA]. We will assume all $\gamma_i = 1$ so that concentrations may be used in the equilibrium K expressions rather than activities. How we can address cases when $\gamma_i \neq 1$ will be considered in Section 5.7.

With four unknowns, four independent equations are required to solve the problem:

$$K_w = [H^+][OH] \qquad \text{first chemical equilibrium equation} \tag{5.1}$$

$$K_a = \frac{[H^+][A^-]}{[HA]} \qquad \text{second chemical equilibrium equation} \tag{5.2}$$

$$[HA] + [A^-] = A_T = C \qquad \text{mass balance equation (MBE) on total A} \tag{5.3}$$

$$[H^+] = [A^-] + [OH^-] \qquad \text{electroneutrality equation (ENE).} \tag{5.4}$$

For our initial discussions in this chapter, the subscript "a" has been included in the K for the *acid* dissociation constant of HA. Later, we will drop this subscript. The variable C has been introduced as **synonymous** with A_T. This is because C is commonly used in treatments of this problem by others. We will use C when we wish to emphasize prior treatments that the reader may have seen, and A_T when the emphasis is a more general.

Regarding the MBE, if HA is a very weak acid (small K_a), most of the A_T will tend to be present as HA. For this pair, HA is the "conjugate acid", and A^- is the "conjugate base". If HA is an acid of intermediate strength (intermediate K_a), then a more even balance of A^- and HA will tend to be present. If HA is a rather strong acid (large K_a), most of the A_T will tend to be present as A^-. The value of A_T also affects the extent dissociation of HA, as will be seen in Figure 5.2: for a solution of HA, for a given value K_a, at high A_T, most of the A_T can be present as HA, while at lower A_T most can be present as A^-.

Before proceeding, for any solution that contains HA and A^-, for convenience we define

$$\alpha_0 = \frac{[\mathrm{HA}]}{[\mathrm{HA}]+[\mathrm{A}^-]} \quad \text{fraction of } \mathrm{A_T} \text{ present as HA} \tag{5.5}$$

$$\alpha_1 = \frac{[\mathrm{A}^-]}{[\mathrm{HA}]+[\mathrm{A}^-]} \quad \text{fraction of } \mathrm{A_T} \text{ present as } \mathrm{A}^-. \tag{5.6}$$

In this text, the symbol α always means "fraction". Here, α_0 means the fraction of HA because HA has lost 0 protons relative to HA. And α_1 means the fraction of A^- because A^- has lost 1 proton relative to HA. By definition,

$$[\mathrm{HA}] \equiv \alpha_0 \mathrm{A_T} \tag{5.7}$$

$$[\mathrm{A}^-] \equiv \alpha_1 \mathrm{A_T} \tag{5.8}$$

$$\alpha_0 + \alpha_1 = 1. \tag{5.9}$$

Dividing the numerator and denominator of Eq.(5.5) by [HA], and numerator and denominator of Eq.(5.6) by $[A^-]$, expressions for α_0 and α_1 are

$$\alpha_0 = \frac{1}{1+\dfrac{[\mathrm{A}^-]}{[\mathrm{HA}]}} \tag{5.10}$$

$$\alpha_1 = \frac{1}{\dfrac{[\mathrm{HA}]}{[\mathrm{A}^-]}+1}. \tag{5.11}$$

Now using K_a,

$$\frac{[\mathrm{A}^-]}{[\mathrm{HA}]} = \frac{K_a}{[\mathrm{H}^+]} \tag{5.12}$$

so

$$\alpha_0 = \boxed{\frac{1}{1+\dfrac{K_a}{[\mathrm{H}^+]}}} = \boxed{\frac{[\mathrm{H}^+]}{[\mathrm{H}^+]+K_a}} \tag{5.13}$$

$$\alpha_1 = \frac{1}{\dfrac{[\mathrm{H}^+]}{K_a}+1} = \boxed{\frac{K_a}{[\mathrm{H}^+]+K_a}}. \tag{5.14}$$

To help remember the expressions for α_0 and α_1, it is useful to focus on the singly boxed versions. Think first that only the two things that affect each of these are $[H^+]$ and K_a. Put the sum $[H^+]+K_a$ on the bottom for both. All that is left is to remember is which α has $[H^+]$ on top. Since α_0 is for the acid form, α_0 has $[H^+]$ on top; that leaves K_a on top for α_1. The doubly boxed expression for α_0 will be very useful for inferring the expressions for α values in diprotic systems, triprotic systems, etc. so we especially emphasize it here now: it is the base equation for the recursive expression for the α_0 of diprotic, triprotic, etc. acids as summarized below in Box 5.7.

5.2.2 Solving for the Speciation for a Solution of HA – All $\gamma_i = 1$, Otherwise No Simplifying Assumptions

For this problem with four equations in four unknowns, we need to select one starting equation from among Eqs.(5.1–5.4), then substitute into that one using the other equations. The goal will be one equation in one unknown. The best choice for the starting equation is the ENE because it has all **linear** terms, *i.e.*, no products or quotients. (If we were to select the K_a expression, there would be a quotient at the beginning of the process, and substituting into that would get messy quickly.) We substitute for $[A^-]$ into the ENE using Eq.(5.8):

$$[H^+] = \alpha_1 A_T + [OH^-]. \tag{5.15}$$

This has utilized K_a (because of Eq.(5.14)), *and* the MBE on A_T. Using the last equation, K_w, we then obtain

$$[H^+] = \alpha_1 A_T + \frac{K_w}{[H^+]} \tag{5.16}$$

or taking (LHS – RHS) = 0,

$$\boxed{\begin{gathered}(\text{LHS} - \text{RHS}) = 0 \\ [H^+] - \alpha_1 A_T - \frac{K_w}{[H^+]} = 0 \\ \text{or} \\ [H^+] - \frac{K_a}{[H^+] + K_a} A_T - \frac{K_w}{[H^+]} = 0.\end{gathered}} \tag{5.17}$$

For given values of K_a, K_w, and A_T, we solve Eq.(5.17) by finding the "root" value of $[H^+]$ that satisfies the equation. This is directly comparable to the process of solving a polynomial of the form $ax^3 + bx^2 + cx^3 + d = 0$. A polynomial of that type has three roots, and actually so does Eq.(5.17). In the case of Eq.(5.17), for meaningful values of K_a, K_w, and A_T (*e.g.*, no negative values allowed), there is only one positive real root value of $[H^+]$: that is the root we want. The other roots are negative or imaginary, and are not relevant for us.

Solving for the root of Eq.(5.17) (with specific K_a, K_w, and A_T) can be done "by hand" on a calculator (the author did it that way in the 1970s with an HP35), making an initial "guess" for $[H^+]$, evaluating (LHS – RHS), and seeing how close the result is to 0. **One keeps changing $[H^+]$ ("iterating") in a "trial and error" search for (LHS – RHS) = 0 until the successive $[H^+]$ values stop changing to any meaningful extent, because one has reached (LHS – RHS) $\approx$ 0.** Computers of course are faster at this than humans, and the spreadsheet Excel includes an Add-In called "Solver" that will do this very quickly. The MATLAB® software is another possibility. This text will provide examples using Excel.

To develop an understanding of what "finding the root" means, we can consider a specific problem. First, let

$$f_a([H^+]) = (LHS - RHS) = [H^+] - \alpha_1 A_T - K_w/[H^+]. \tag{5.18}$$

Finding the root means finding the value of $[H^+]$ that makes $f_a([H^+])=0$. Consider the case when $A_T = 10^{-4}\ M$, $K_a = 10^{-5}$, and $K_w = 1.01 \times 10^{-14}$. Values of $f_a([H^+])$ for a range of pH are given in Table 5.1. And $f_a([H^+])$ is plotted *vs.* pH in Figure 5.1. (The *x*-axis for the plot is pH rather than $[H^+]$ because if we used the $[H^+]$, the huge variation in $[H^+]$ over the common pH range of interest would compress one side of the scale enormously, and make it difficult to see what is going on.) As seen in Figure 5.1, $f_a([H^+])$ passes through zero at $[H^+] = 10^{-4.57} = 2.69 \times 10^{-5}$ (pH = 4.57). Table 5.1 illustrates that, as one gets close to the root value, then increasingly small changes in the guessed value of $[H^+]$ are needed to "home in" on the root value. Figure 5.1 and Table 5.1 also illustrate that the shape of $f_a([H^+])$ allows that an initial guess on *either* side of the root value will successfully lead to the root.

With the equilibrium (root) value of $[H^+]$ in hand, the concentrations of the remaining species can be calculated. With $K_a = 10^{-5}$, $A_T = 10^{-4}\ M$, and $K_w = 1.01 \times 10^{-14}$,

$$[OH^-] = K_w/[H^+] = 1.01 \times 10^{-14}/(2.69 \times 10^{-5}) = 3.72 \times 10^{-10}\ M \tag{5.19}$$

$$[HA] = \alpha_0 A_T = \frac{[H^+]}{[H^+] + K_a} A_T = \frac{10^{-4.57}}{10^{-4.57} + 10^{-5}} 10^{-4} = 7.29 \times 10^{-5}\ M \tag{5.20}$$

$$[A^-] = \alpha_1 A_T = \frac{K_a}{[H^+] + K_a} A_T = \frac{10^{-5}}{10^{-4.57} + 10^{-5}} 10^{-4} = 2.71 \times 10^{-5}\ M. \tag{5.21}$$

We can repeat the solution process for any combination of K_a and A_T. Figure 5.2 shows how the equilibrium pH decreases as either K_a or A_T increases. When either K_a or A_T is very small, pH ≈ 7.00.

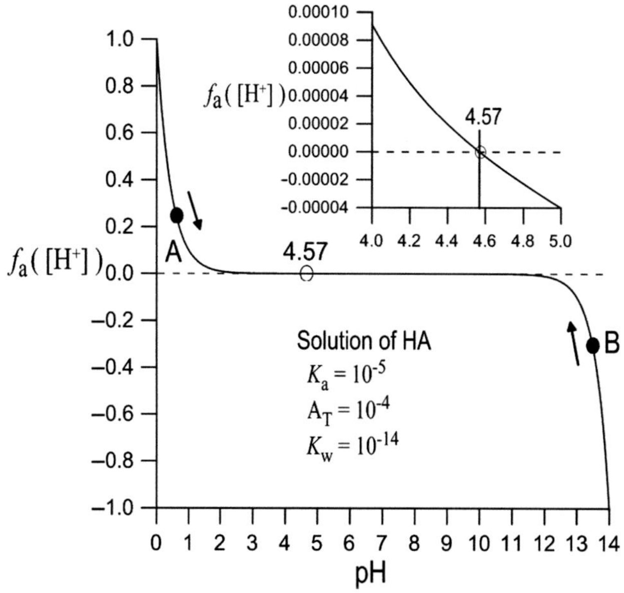

FIGURE 5.1 $f_a([H^+]) = (LHS - RHS)$ *vs.* pH for a solution of a monoprotic acid HA with $K_a = 10^{-5}$ and $C = A_T = 10^{-4}\ M$ (with $K_w = 1.01 \times 10^{-14}$, as for 25 °C and 1 atm pressure).

TABLE 5.1
Values of $f_a([H^+])$ for pH between 1.0 and 13.0 for $K_a = 10^{-5}$ m, $C = 10^{-4}$ M, and $K_w = 10^{-14}$ m^2

pH	$f_a([H^+])$	pH	$f_a([H^+])$
1.0	1.00E-01	4.8	−2.28E-05
2.0	1.00E-02	4.9	−3.17E-05
3.0	9.99E-04	5.0	−4.00E-05
3.5	3.13E-04	5.5	−7.28E-05
4.0	9.09E-05	6.0	−8.99E-05
4.1	6.83E-05	7.0	−9.90E-05
4.2	4.94E-05	8.0	−1.01E-04
4.3	3.35E-05	9.0	−1.10E-04
4.4	1.97E-05	10.0	−2.01E-04
4.5	7.60E-06	11.0	−1.11E-03
4.6	−3.36E-06	12.0	−1.02E-02
4.7	−1.34E-05	13.0	−1.01E-01

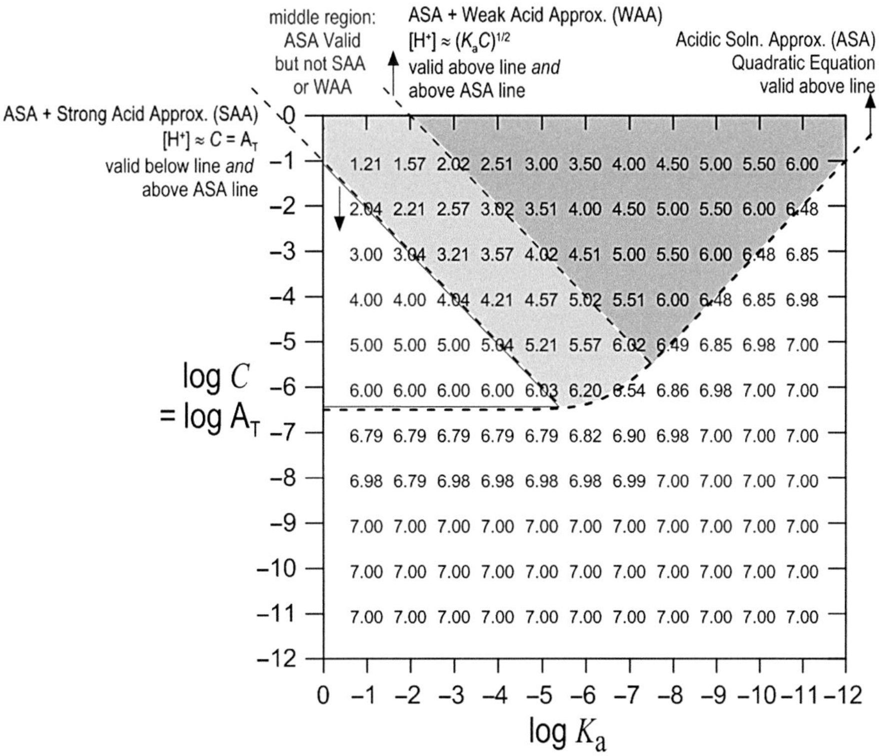

FIGURE 5.2 (See color insert.) Grid of pH values for solutions of HA as a function of log C and log K_a (with $K_w = 1.01 \times 10^{-14}$, as for 25 °C and 1 atm pressure). The region where the acidic solution approximation (ASA) is valid is shown, along with the individual sub-regions for *additional* applicability of the strong acid approximation (SAA), and the weak acid approximation (WAA). A color version of this figure is found in the center pages.

Important Practical Point 1 – Amplify the Function: Referring to the text in bold above, for (LHS – RHS) ≈ 0, how close to 0 will be good enough? In any numerical code (including that for Solver), a "convergence criterion" ε instruction is set so that the code "knows" to stop when the function has been brought to within the small number ε of 0. However, the functions we are interested in within this text can have **very small values** to begin with, even far (in pH) from the root. This means one can be quite far from the root and yet have (LHS – RHS) within the default ε. If so, Solver, MATLAB, whatever, will stop the search too early. One can deal with this by adjusting the convergence criterion ε to *a really small value*. But, rather than having to figure out how to do that, it is easier to amplify the function and solve for when (LHS – RHS) × $10^6 = 0$.

Important Practical Point 2 – Drive the Search by Varying pH, not [H⁺]: At the root for the case in Figure 5.1, $[H^+] = 2.69 \times 10^{-5}$ (pH = 4.57). Earlier, we said that the function $f_a([H^+])$ has an attractive shape such that the root can be deduced even starting with a horrendous initial guess of $[H^+] = 10^{-14}$. Say we did start with an initial guess of $[H^+] = 10^{-14}$. Not being at the root, in an algorithm like Solver another try might be 10% bigger than the initial guess, so the second guess in that case would be 1.1×10^{-14}. At that kind of rate, there is a real danger that a "maximum number of iterations" setting in the algorithm will be reached before reaching the root. A simple solution to this problem is to have the scheme vary pH in the search for the root, not $[H^+]$. With pH being varied, we can move very quickly through a logarithmically wide range of $[H^+]$.

Example 5.1 Building an Excel Spreadsheet to Solve for pH of a Solution of HA

The solution will be based on Eq.(5.17). Set up the spreadsheet so that Solver varies pH, and not $[H^+]$ directly, to find the root value of $[H^+]$. Use the spreadsheet to find the pH and speciation at 25 °C and 1 atm for a 10^{-4} *F* solution of HA with. $K_a = 10^{-5}$

Solution

The formulas in column E are the ones that underlie the adjacent cells in column D.

	A	B	C	D	E
1	Pankow Water Chemistry Spreadsheet #5.1				
2	***Solution of HA***				
3	ENE: [H+] = [A-] + [OH-]				
4	[A-] = α_1*A_T				
5	[OH-] = *K*_w/[H+]			results:	formulas:
6					
7		inputs:			
8	*K*_a =	1.00E-05	[H+] =	*2.70E-05*	= 10^-B18
9	*K*_w =	1.01E-14	[OH-] =	*3.74E-10*	= B9/D8
10	HA Conc. =	1.00E-04	A_T =	1.00E-04	= B10
11			for HA: α_0 =	*7.30E-01*	= D8/(D8+B8)
12			for A-: α_1 =	*2.70E-01*	= B8/(D8+B8)
13			α_0 + α_1 =	*1.00000*	= D11+D12
14	Set Solver Target Cell: D18		[HA]=α_0 *A_T =	*7.30E-05*	= D11*D10
15	Equal to Value of 0		[A-]=α_1*A_T =	*2.70E-05*	= D12*D10
16	By Changing Cell: B18		LHS=[H+] =	*2.70E-05*	= D8
17			RHS=α_1*A_T+[OH-] =	*2.70E-05*	= D15+D9
18	**pH guess for Solver=**	***4.57***	(LHS-RHS)*1E6 =	*-7.67E-07*	= (D16-D17)*1e6
19					

5.2.3 The [H+] Polynomial Version in the HA Solution Problem

We now investigate what the polynomial in $[H^+]$ version of the problem can teach us. We do this only for instructional purposes, because while most math applications offer polynomial solvers, there is no payoff for us from converting Eq.(5.17) (or the (LHS – RHS) expressions of other problems) into polynomials in $[H^+]$: (1) it requires work; (2) an algebra mistake might be made; (3) the mathematical properties of the polynomial may make it harder to solve than the original (LHS – RHS) = 0 version.

Multiplying Eq.(5.17) by $[H^+]([H^+]+K_a)$ gives

$$[H^+]^2([H^+]+K_a) - K_a A_T[H^+] - K_w([H^+]+K_a) = 0 \quad (5.22)$$

$$\boxed{[H^+]^3 + K_a[H^+]^2 - (K_a A_T + K_w)[H^+] - K_w K_a = 0 \quad \text{cubic equation.}} \quad (5.23)$$

The common way to express a cubic equation is

$$a\left[H^+\right]^3 + b\left[H^+\right]^2 + c\left[H^+\right] + d = 0. \quad (5.24)$$

For Eq.(5.23), then $a=1$, $b=K_a$, $c=-(A_T K_a + K_w)$, and $d=-K_w K_a$. These values can be inputted to a polynomial solver, which will proceed by a process of successive guesses and return three roots. For the same conditions as Figure 5.1, the positive real root for Eq.(5.23) is (and must be) $[H^+]=10^{-4.57}$.

Let

$$g_a([H^+]) = [H^+]^3 + K_a[H^+]^2 - (A_T K_a + K_w)[H^+] - K_w K_a. \quad (5.25)$$

Figure 5.3 plots $g_a([H^+])$ *vs.* pH with $K_a = 10^{-5}$, $A_T = 10^{-4}$ *M*, and $K_w = 1.01 \times 10^{-14}$. $g_a([H^+])$ crosses the pH axis at 4.57 and goes negative, has a minimum at about pH = 4.8, then asymptotically approaches 0 as pH increases, never crossing the pH axis again. This shape is **unattractive** from a root-solving point of view. Recall that the root-solving process involves making a guess for $[H^+]$ and calculating how close to zero that guess makes the function. For $g_a([H^+])$, if the first guess is to the left of the

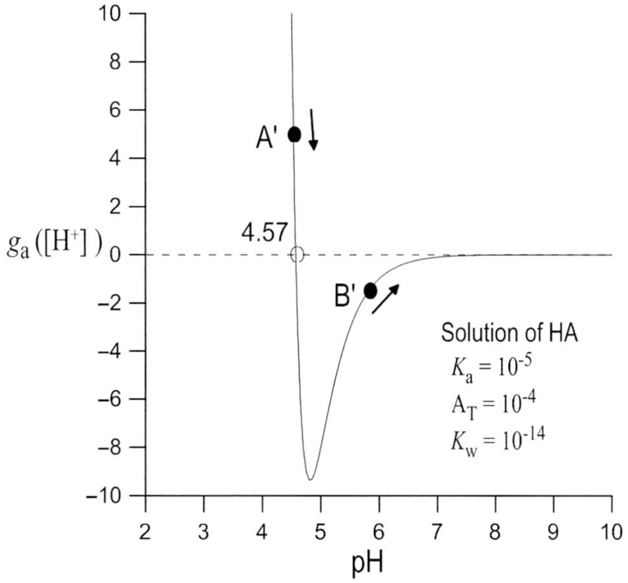

FIGURE 5.3 $g_a([H^+])$ for the cubic equation *vs.* pH for a solution of a monoprotic acid HA with $K_a = 10^{-5}$ and $C = A_T = 10^{-4}$ *M* (with $K_w = 1.01 \times 10^{-14}$, as for 25 °C and 1 atm pressure).

minimum in pH space, as with point A′, the process will take you to the root. But, if the first guess is to the right of the minimum in pH space, as with point B′, then the process will not take you to the root, but to +∞. This is what happens with Solver in Excel with $g_a([H^+])$ if you have a value like 6 in the cell for pH (the "guess" cell"), and then ask Solver to start the search for the pH such that $g_a([H^+])=0$. Solver tries, but gives up, and says it cannot find a solution. However, Solver *will* succeed if you start with a guess of pH=4.

Happily, $f_a([H^+])$ has the absolutely charming feature that no matter how horrendous your first guess is for pH, the process will lead to the root. Even if you make the ridiculous guess that pH=14 for the *acidic* Figure 5.1 case, the root-finding process will make its way all the way back to acidic pH values and succeed. As an occasion for even more happiness, this turns out to be the case for any version of $f_a([H^+])$ obtainable by for different combinations of K_a, A_T, and K_w: the $f([H^+])$ function **always has the same general shape**, going from large-positive at low pH to large-negative at high pH, with the root lying in between.

5.2.4 Simplifying Using the Acidic Solution Approximation (ASA) for a Solution of HA

5.2.4.1 ASA Alone

While Eq.(5.17) gives us a means to solve for the pH of any solution of HA without any simplifying assumptions (except that all $\gamma_i=1$), it is very useful to be able to solve problems with a piece of scratch paper, or in your head, and not have to rely on a computer or calculator root solver. And, indeed, the introduction to the chemistry of a solution of acid given to beginning chemistry students is always aided with one or more simplifying assumptions.

A very logical first assumption to make for a solution of HA is that the solution is, well, acidic. Going back to the starting point for the solution process, we have

$$[H^+] = [A^-] + [OH^-] \qquad \text{electroneutrality equation (ENE).} \tag{5.4}$$

If the solution is even modestly acidic, then $[OH^-] << [H^+]$, which by the ENE means that $[OH^-] << [A^-]$:

$$[H^+] = [A^-] + \underset{\text{assume small}}{\mathbf{[OH^-]}} \tag{5.26}$$

$$[H^+] \approx [A^-] \qquad \text{acidic solution approximation.} \tag{5.27}$$

By this assumption, the $K_w/[H^+]$ term in Eq.(5.17) is neglected and we obtain

$$[H^+] - \frac{K_a}{[H^+] + K_a} A_T \approx 0 \tag{5.28}$$

which gives

$$\boxed{[H^+]^2 + K_a[H^+] - K_a A_T \approx 0. \quad \text{second order equation.}} \tag{5.29}$$

Considering the cubic equation representation of the problem that is Eq.(5.23), if we were to discard all terms with K_w, which means neglecting $[OH^-]$ (because it is the $[OH^-]$ term in the ENE that got K_w got into the cubic equation), then we would get Eq.(5.29) by that route. By the way, discarding the terms with K_w is a result of the fact that water is an exceedingly weak acid compared to HA, so $[OH^-]$ is small, viz., the ASA applies.

Eq.(5.29) can be solved using the solution for "quadratic equations", which gives

$$[H^+] \approx \frac{-K_a + \sqrt{K_a^2 + 4K_a A_T}}{2} \tag{5.30}$$

where the + sign has been selected in front of the square root, because that is what is needed to give a positive value for $[H^+]$. For the case when $K_a = 10^{-5}$, $A_T = 10^{-4}$ *M*, and $K_w = 10^{-14}$, Eq.(5.30) yields $[H^+] = 2.70 \times 10^{-5}$ *M* (pH = 4.57). For this case, there is essentially no error in using the ASA. If this seems familiar, let's look at the problem like we did in freshman college chemistry.

1st Flashback: College chemistry instructor lecturing:
"... We want to calculate the pH of a solution of an acid HA of concentration C. As an acid, HA will dissociate according to

$$HA = H^+ + A^- \tag{5.31}$$

with

$$K_a = \frac{[H^+][A^-]}{[HA]}. \tag{5.32}$$

There is 1:1 correspondence between the amount of H^+ formed and the amount of A^- formed. Let x equal the concentration of both:

$$x = [H^+] = [A^-]. \tag{5.33}$$

Then,

$$[HA] = C - x. \tag{5.34}$$

This gives

$$K_a = \frac{x^2}{C - x} \tag{5.35}$$

or

$$K_a = \frac{[H^+]^2}{C - [H^+]} \tag{5.36}$$

$$x^2 + K_a x - K_a C = 0 \tag{5.37}$$

which can be solved by the quadratic equation according to

$$x = [H^+] = [A^-] = \frac{-K_a \pm \sqrt{K_a^2 + 4K_a C}}{2}. \tag{5.38}$$

~end of first flashback~

Eq.(5.38) is obviously the same as Eq.(5.30), except that Eq.(5.30) uses an ≈ sign and the symbol A_T, and Eq.(5.38) uses an = sign and the symbol *C*. They are the same because Eq.(5.33) is, in fact, the ASA, though no one said at the time it was an approximation.

BOX 5.1 (Geek Optional) Specific Criterion for Applying the ASA

We define that the ASA is adequately accurate when

$$[OH^-] \le 0.1[H^+].$$

The above relation with $K_w = [H^+][OH^-] = 10^{-14}$ (25 °C/1 atm, and neglecting γ_i corrections) gives

$$0.1[H^+]^2 \ge 10^{-14} \qquad [H^+] \ge 10^{-6.5}\,M.$$

Combining that with Eq.(5.36) yields

$$K_aC - 10^{-6.5} \qquad K_a \ge 10^{-13.0}$$

as the criterion that the ASA applies with less than ~10% error in $[H^+]$. Using an equals sign lets us put a boundary line in Figure 5.2. The ASA fails when C (=A_T) gets very low (<$10^{-6.5}$ M, so not enough HA has been added to the system to make the solution sufficiently acidic, even if K_a is large). The ASA can also fail when the value of K_a becomes too small (pK_a too large) so that the acid HA is too weak to make the solution adequately acidic. Plots showing regions where different approximations hold in acid/base systems are also discussed by Narasaki (1987).

5.2.4.2 ASA + WAA: Acidic Solution Approximation *Plus* Weak Acid Approximation (WAA) for a Solution of HA

If in addition to being able to use the ASA we are also able to assume that the acid is behaving as a "weak" acid (*i.e.*, mostly does not dissociate), then we are assuming Eq.(5.27) *and*

$$[HA] \approx C = A_T. \tag{5.39}$$

From Eq.(5.13), this means $\alpha_0 \approx 1$, and therefore $K_a << [H^+]$. Then, because

$$[H^+] - \frac{K_a}{[H^+] + K_a} A_T \approx 0 \tag{5.28}$$

by neglecting K_a in the denominator for α_1 we obtain

$$\boxed{[H^+] \approx \sqrt{K_a A_T} \quad \text{first order equation.}} \tag{5.40}$$

The ASA reduced the order of the problem from a cubic equation to a second order equation. Adding the WAA reduced the order again, from second order to first order.

2nd Flashback: College chemistry instructor:
"... Now, if HA is a weak acid so that [HA] ≈ C, then instead of Eq.(5.36), we have

$$K_a = \frac{[H^+]^2}{C} \tag{5.41}$$

$$[H^+] = \sqrt{KC}. \tag{5.42}$$

~end of second flashback~

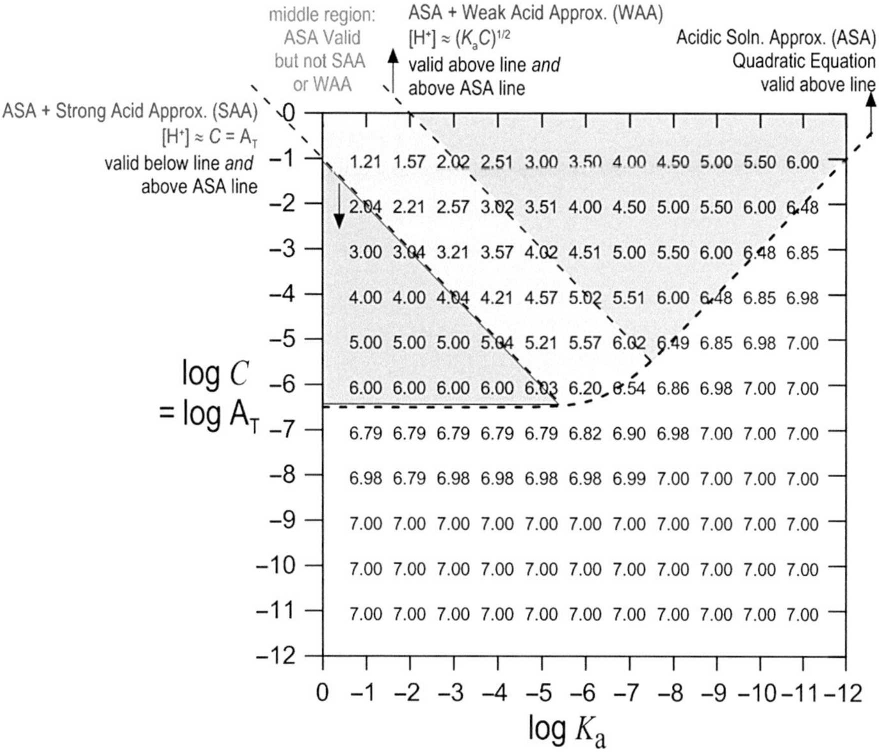

FIGURE 5.2 Grid of pH values for solutions of HA as a function of log C and log K_a (with $K_w = 1.01 \times 10^{-14}$, as for 25 °C and 1 atm pressure). The region where the acidic solution approximation (ASA) is valid is shown, along with the individual sub-regions for *additional* applicability of the strong acid approximation (SAA), and the weak acid approximation (WAA).

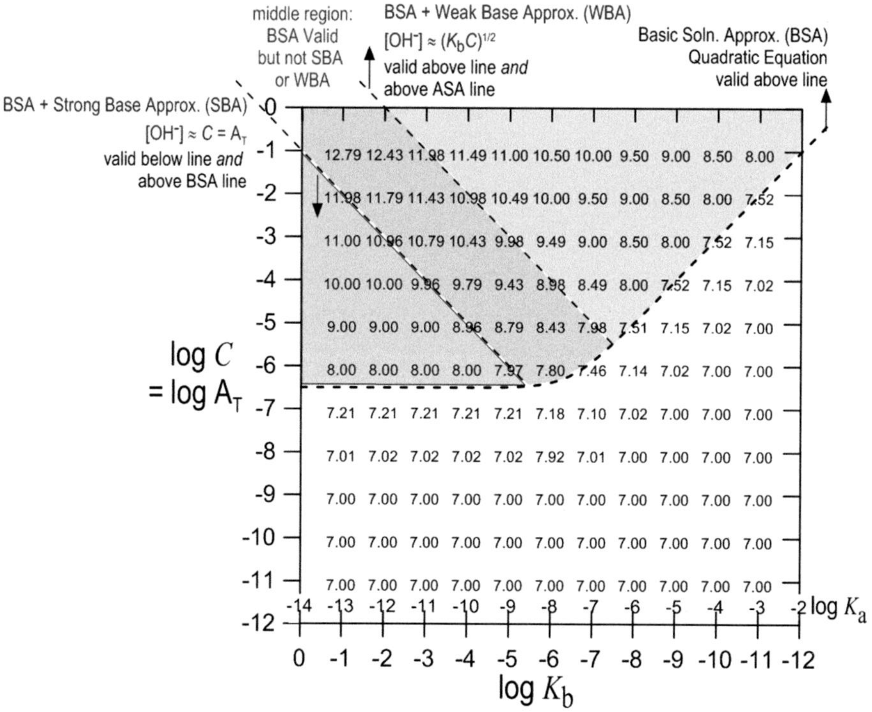

FIGURE 5.5 Grid of pH values for solutions of NaA as a function of log C and log K_b (with $K_w = 1.01 \times 10^{-14}$, as for 25 °C and 1 atm pressure). The region where the basic solution approximation (BSA) is valid is shown, along with the individual sub-regions for *additional* applicability of the strong base approximation (SBA), and the weak base approximation (WBA).

BOX 5.2 (Geek Optional) Specific Criterion for Applying the ASA and Then Also the WAA

We define that the WAA is adequately accurate when $[H^+] \approx [A^-] \leq 0.1C$. The WAA says $[H^+] = \sqrt{K_a C}$. Thus,

$$\sqrt{K_a C} \leq 0.1C$$

or

$$C \geq 100\,K_a.$$

Using an equals sign lets us plot a boundary line in Figure 5.2. The line demarcates the portion of the ASA region in which the ASA+WAA holds. This is when the combination of the pK_a and a not-dilute C combine to maintain the WAA (a not-dilute C value reduces dissociation).

5.2.4.3 ASA + SAA: Acidic Solution Approximation *Plus* Strong Acid Approximation (SAA) for a Solution of HA

If in addition to being able to use the ASA we are also able to assume that the acid is behaving as a "strong" acid (*i.e.*, nearly completely dissociated), then we are assuming Eq.(5.27) *and*

$$\left[A^-\right] \approx C = A_T. \tag{5.43}$$

Because of Eq.(5.27), we obtain

$$\boxed{[H^+] \approx C = A_T \quad \text{first order equation.}} \tag{5.44}$$

3rd Flashback: College chemistry instructor:
"... Now, if HA is a strong acid and is completely dissociated, as with HCl, then we have

$$[H^+] = C \tag{5.45}$$

For example, the pH of 10^{-3} F HCl is 3.0, with $[H^+] = 10^{-3}$ M."

~end of third flashback~

Example 5.2 For a Particular Solution of HA, Comparing the Exact Solution with the Solutions with ASA, ASA + WAA, and ASA + SAA

Consider a solution of HA for which $C = 10^{-3}$ F and $K_a = 10^{-5}$. Compute the pH of the solution exactly (though assuming $\gamma_i = 1$), and compare the result with what is obtained with the ASA, ASA + WAA, and ASA + SAA.

Solution

Eq.(5.17) gives pH = 4.07. The ASA gives pH = 4.07, matching the exact solution: we are well within the zone of validity for the ASA. But ASA + WAA gives pH = 4.00, and ASA + SAA gives pH = 3.50; neither of those more aggressive approximation approaches works well; as indicated by Figure 5.2, we in the transition zone between applicability for SAA and WAA.

BOX 5.3 (Geek Optional) Specific Criterion for Applying the ASA and Then Also the SAA

We define that the SAA is adequately accurate when $[HA] \leq 0.1C$. The SAA says $[H^+]=[A^-]=C$. The K_a expression is

$$K_a = \frac{[H^+][A^-]}{[HA]}.$$

Making the substitutions noting the direction of the inequality,

$$K_a \geq \frac{C^2}{0.1C}$$

$$K_a \geq 10C.$$

Using an equals sign lets us put a boundary line in Figure 5.2. The line demarcates the portion of the ASA region in which the ASA+SAA holds. This is when the combination of the pK_a and an adequately dilute value of C combine to maintain the SAA (a dilute C value enhances dissociation).

Example 5.3 pH of Vinegar

"Distilled" vinegar in the grocery store is usually labeled as "5% acidity", meaning 5% by weight acetic acid, *i.e.*, 50 g of acetic acid per kg of solution. Since 1 kg of a mostly aqueous solution occupies close to 1 L, this means the concentration is close to 50 g/L. **a.** What is the formality of acetic acid in 5% vinegar? **b.** If the $pK_a=4.76$ for acetic acid at 25 °C and 1 atm pressure, if we can ignore activity corrections and the effects of minor components (such as CO_2 from the atmosphere), what is the pH at at 25 °C and 1 atm pressure?

Solution

a. The formula weight of acetic acid (CH_3COOH) is 60.052 g/FW.

$$50 \text{ g/L} \times 1\text{FW}/60.052\text{g} = 0.833\ F.$$

b. For this solution $A_T=C=0.833\ M$. This is a rather concentrated solution, and the acid is fairly weak, which makes for excellent conditions for applying ASA+WAA.

$$[H^+] = \sqrt{K_aC} = \sqrt{10^{-4.76} \times 0.833} = 0.00380$$

$$pH = 2.42.$$

Solving "exactly" using Eq.(5.17) gives the same result. This pH is rather low, which accounts for the strongly sour taste of vinegar, not to mention the fact that chronic dietary consumption of vinegar, as in "oil and vinegar" salad dressing, can be hard on one's teeth.

5.3 SOLUTION OF THE GENERIC BASE NaA, ALL $\gamma_i = 1$

5.3.1 Introduction

NaA is the "sodium salt of the acid HA". NaA can be made by neutralizing a portion of the acid HA with an equivalent amount of the strong base sodium hydroxide (NaOH). For example, we make sodium acetate by neutralizing acetic acid with NaOH, and we make NaCl by neutralizing HCl with NaOH. The discussion that follows utilizes cases involving solutions of NaA, but the chemistry is equivalent for solutions of potassium (K) salts, of the formula KA: Na^+ and K^+ behave the same way in water, as spectator ions.

For a solution of NaA, there are five species and therefore five unknowns, $[H^+]$, $[OH^-]$, [HA], $[A^-]$, and $[Na^+]$. Five equations are required to solve the problem. Assuming all $\gamma_i = 1$ so that concentrations may be used in equilibrium K expressions rather than activities, the five equations are:

$$K_w = [H^+][OH] \tag{5.1}$$

$$K_a = \frac{[H^+][A^-]}{[HA]} \tag{5.2}$$

$$[HA] + [A^-] = A_T = C \qquad \text{MBE} \tag{5.3}$$

$$[Na^+] = A_T \qquad \text{MBE} \tag{5.46}$$

$$[H^+] + [Na^+] = [A^-] + [OH^-] \qquad \text{ENE.} \tag{5.47}$$

NaA is a base: when it dissolves in water, some of the A^- must react with water to form some HA; [HA] cannot be zero because K_a is finite. The reaction of A^- with water is

$$A^- + H_2O = HA + OH^-. \tag{5.48}$$

We obtain the equilibrium constant for Eq.(5.48) as follows. The *reverse* of the equilibrium for K_a involves the conversion of A^- to HA:

$$H^+ + A^- = HA \qquad (K_a)^{-1}. \tag{5.49}$$

The dissociation reaction for water is

$$H_2O = H^+ + OH^- \qquad K_w. \tag{5.50}$$

Adding these two reactions gives

$$A^- + H_2O = HA + OH^- \qquad K = \frac{K_w}{K_a} \equiv K_b = \frac{[HA][OH^-]}{[A^-]} \tag{5.51}$$

in which A^- is acting as a **base** and the equilibrium constant is given the symbol K_b. The relation between any K_a and the *corresponding* K_b is therefore

$$K_a \times K_b = K_w \tag{5.52}$$

$$\log K_a + \log K_b = \log K_w \tag{5.53}$$

$$pK_a + pK_b = pK_w. \tag{5.54}$$

At 25 °C and 1 atm,

$$pK_a + pK_b = 14.0. \tag{5.55}$$

As an example use of "corresponding", when HA represents acetic acid, then A^- is the acetate ion.

Since we know the value of K_w, *there is no information contained in the value of K_b that is not in K_a*. K_a tells us how much A^- "likes" a proton from the perspective of the willingness of HA to give up an H^+ (think of a dollar) and become A^-. So, K_a is analogous to a measure of *generosity*. Alternatively, K_b tells us how much A^- "likes" a proton from the point of view of the tendency of A^- to take a proton (dollar) from water (the reference person with a dollar), and become HA, while forming OH^- in the process. K_b is then analogous to a measure of *miserliness*. Carrying this anthropomorphic analogy one step further, if you happen to know that your Aunt Fatima is a quite generous person, you automatically know that after she has readily given up her one dollar to the community, she will not be prone to take the dollar back. Fatima with a dollar is like HF (rather large K_a, about 10^{-3}); Fatima without a dollar is like F^- (rather small K_b, about 10^{-11}). The given-up dollar is like the solvated proton in solution.

There is much symmetry between the problem of a solution of HA and that for a solution of NaA. With a solution of HA, we are focused on acidic solutions, and determining the value of $[H^+]$, with the help of the value of K_a. With a solution of NaA, we are focused on basic solutions and determining the value of $[OH^-]$, and we can frame the problem using K_b. The results for the two problems (Figures 5.2 and 5.5) and the simplifying assumptions available are therefore highly symmetrical. Also, we will see that if the pH of a solution of some specific HA at a particular concentration C is x, then the pH of a solution of the corresponding NaA equals ($pK_w - x$). The symmetries results and in the equations are summarized in Table 5.2.

5.3.2 Solving for the Speciation for a Solution of NaA – All $\gamma_i = 1$, Otherwise No Simplifying Assumptions

As with the problem of a solution of HA, of the five equations for the problem we chose to build the solution on the ENE,

$$[H^+] + [Na^+] = [A^-] + [OH^-]. \tag{5.47}$$

Using what we have learned so far, it is now very easy to develop our needed "one equation in one unknown" by the following steps:

1) Substitute the value of A_T for $[Na^+]$ (based on the MBE Eq.(5.46));
2) Substitute $\alpha_1 A_T$ for $[A^-]$ (based on K_a and the MBE that is Eq.(5.3)); and
3) Substitute $K_w/[H^+]$ for $[OH^-]$.

The result is

$$[H^+] + [Na^+] = \alpha_1 A_T + K_w / [H^+] \tag{5.56}$$

so that

$$\boxed{\begin{array}{c}(\text{LHS} - \text{RHS}) = 0 \\ [H^+] + [Na^+] - \alpha_1 A_T - K_w/[H^+] = 0.\end{array}} \tag{5.57}$$

TABLE 5.2
Symmetry between the Chemistry of the Problem of a Solution of HA, and That of a Solution of NaA

Solution of HA with $A_T = C$.	Solution of NaA with $A_T = C$.
pH Result	**pH Result**
Say the pH $= x$ for specific C and K_a.	…for same C and K_a, pH $= pK_w - x$
PBE	**PBE**
$[H^+] = [A^-] + [OH^-]$ solution must be acidic	$[H^+] + [HA] = [A^-]$ solution must be basic
ENE	**ENE**
$[H^+] = [A^-] + [OH^-]$ same as PBE	$[H^+] + [Na^+] = [A^-] + [OH^-]$. (Substituting $[Na^+] = A_T$ gives PBE.)
solving exactly amounts to finding the positive real root of a cubic equation	solving exactly amounts to finding the positive real root of a cubic equation
Possible Approximations	**Possible Approximations**
ASA (acidic solution approximation) may be applicable so $[H^+] \approx [A^-]$ and $[H^+] \approx \dfrac{-K_a + \sqrt{K_a^2 + 4K_aC}}{2}$	BSA (basic solution approximation) may be applicable so $[HA] \approx [OH^-]$ so that $[OH^-] \approx \dfrac{-K_b + \sqrt{K_b^2 + 4K_aC}}{2}$
possibly with:	**possibly with:**
WAA (weak acid approximation) $[HA] \approx A_T = C$ so that $[H^+] \approx \sqrt{K_aC}$	WBA (weak base approximation) $[A^-] \approx A_T = C$ so that $[OH^-] \approx \sqrt{K_bC}$
or	*or*
SAA (strong acid approximation) $[A^-] \approx A_T = C$ so that $[H^+] \approx C$	SBA (strong base approximation) $[HA] \approx A_T = C$ so that $[OH^-] \approx C$

We are done with the setup. (Remember, we know what $[Na^+]$ is, so this really is one equation in one unknown.) We search for the solution (root value of $[H^+]$) by setting

$$f_b([H^+]) = [H^+] + [Na^+] - \alpha_1 A_T - K_w/[H^+]. \tag{5.58}$$

This equation is very similar to Eq.(5.18). The only difference is the term $[Na^+]$. This term moves the root *from* the acid side (for Eq.(5.18)) *to* the base side (for Eq.(5.58)); the bigger it is, the farther the move, and the more basic the result.

Consider the example case, the subject of Figure 5.4, for which $A_T = 10^{-4}\,M$, $K_a = 10^{-5}$ ($K_b = 10^{-9}$), and $K_w = 1.01 \times 10^{-14}$). This is the NaA analog of the HA case of Figure 5.1. Finding the root gives $[H^+] = 3.02 \times 10^{-8}\,M$, or pH $= 7.52$. This pH is close to being neutral because K_b is very small (and A_T is not particularly large). The concentrations of the other species can now be calculated:

$$[OH^-] = K_w/[H^+] = 10^{-14}/(3.02 \times 10^{-8}) = 3.31 \times 10^{-7}\,M \tag{5.59}$$

$$[A^-] = \alpha_1 A_T = \frac{K_a}{[H^+] + K_a} A_T = \frac{10^{-5}}{10^{-7.52} + 10^{-5}} 10^{-4} = 9.97 \times 10^{-5}\,M \tag{5.60}$$

$$[HA] = \alpha_0 A_T = \frac{[H^+]}{[H^+] + K_a} A_T = \frac{10^{-7.52}}{10^{-7.52} + 10^{-5}} 10^{-4} = 3.01 \times 10^{-7}\,M. \tag{5.61}$$

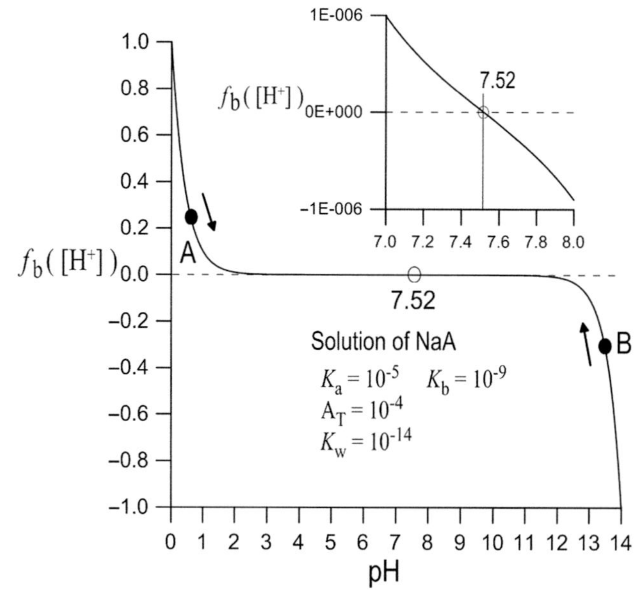

FIGURE 5.4 $f_b([H^+]) = (LHS - RHS)$ *vs.* pH for a solution of a NaA with $K_a = 10^{-5}$ ($K_a = 10^{-9}$) and $C = A_T = 10^{-4}$ M (with $K_w = 1.01 \times 10^{-14}$, as for 25 °C and 1 atm pressure).

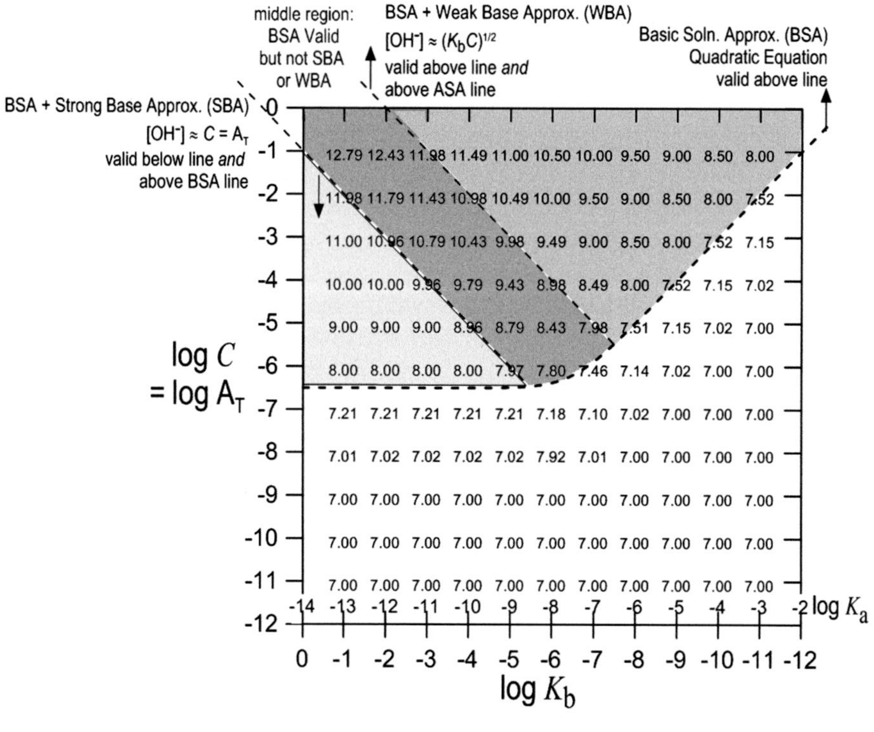

FIGURE 5.5 (See color insert.) Grid of pH values for solutions of NaA as a function of log C and log K_b (with $K_w = 1.01 \times 10^{-14}$, as for 25 °C and 1 atm pressure). The region where the basic solution approximation (BSA) is valid is shown, along with the individual sub-regions for *additional* applicability of the strong base approximation (SBA), and the weak base approximation (WBA). A color version of this figure is found in the center pages.

Figure 5.5 illustrates how the equilibrium pH of a solution of NaA varies with K_b and C at 25 °C and 1 atm. Because of the symmetry in the chemistry, each pH value in Figure 5.5 is given by pK_w (=14.00) minus the corresponding Figure 5.2 value.

Example 5.4 Building an Excel Spreadsheet to Solve for pH of a Solution of NaA

The solution will be based on Eq.(5.57). Set up the spreadsheet so that Solver varies pH, and not [H$^+$] directly, to find the root value of [H$^+$]. Use the spreadsheet to find the pH and speciation at 25 °C and 1 atm for a 10^{-4} *F* solution of NaAcetate. $K_a = 1.74 \times 10^{-5}$.

Solution

The formulas in column E are the ones that underlie the adjacent cells in column D.

The only differences between this spreadsheet and the one for Example 5.1 are: (a) the definitions in cells A2 and A10; (b) the cells for Na$^+$, namely C6, D6, and E6; and (c) the revised equations for LHS in cells C16, D16, and E16.

	A	B	C	D	E
1	Pankow Water Chemistry Spreadsheet #5.2				
2	***Solution of NaA***				
3	ENE: [H+] [Na+] = [A-] + [OH-]				
4	[A-] = α_1*A_T				
5	[OH-] = *K*_w/[H+]			results:	formulas:
6			[Na+] =	1.00E-04	= B10
7		inputs:			
8	*K*_a =	1.74E-05	[H+] =	*3.87E-08*	= 10^-B18
9	*K*_w =	1.01E-14	[OH-] =	*2.61E-07*	= B9/D8
10	NaA Conc. =	1.00E-04	A_T =	*1.00E-04*	= B10
11			for HA: α_0 =	*2.22E-03*	= D8/(D8+B8)
12			for A-: α_1 =	*9.98E-01*	= B8/(D8+B8)
13			α_0 + α_1 =	*1.0000*	= D11+D12
14	Set Solver Target Cell: D18		[HA]=α_0 *A_T =	*2.22E-07*	= D11*D10
15	Equal to Value of 0		[A-]=α_1*A_T =	*9.98E-05*	= D12*D10
16	By Changing Cell: B18		LHS=[H+]+[Na+] =	*1.00E-04*	= D8+D6
17			RHS=α_1*A_T+[OH-] =	*1.00E-04*	= D15+D9
18	pH guess for Solver=	*7.41*	(LHS-RHS)*1E6 =	*-1.75E-07*	=(D16-D17)*1000000
19					
20					

5.3.3 Simplifying Using the Basic Solution Approximation (BSA) for a Solution of NaA

5.3.3.1 BSA Alone

The ENE for a solution of NaA is

$$[H^+] + [Na^+] = [A^-] + [OH^-]. \tag{5.47}$$

Substituting for [Na$^+$] using Eqs.(5.3) and (5.47) leads to the proton balance equation for this solution

$$[H^+] + [HA] = [OH^-] \tag{5.62}$$

which we can use to start consideration of simplifying assumptions. By analogy with the acidic solution approximation (ASA) for a solution of HA, a logical assumption to consider for a solution of NaA is that the solution is basic. If adequately basic, then $[H^+]$ is small relative to $[OH^-]$, which means $[H^+] << [HA]$. That is,

assume small ↘

$$[H^+] + [HA] = [OH^-] \tag{5.63}$$

$$[HA] = [OH^-] \quad \text{basic solution approximation.} \tag{5.64}$$

Then, by Eq.(5.51),

$$\frac{[OH^-]^2}{[A^-]} = K_b. \tag{5.65}$$

Since $[A^-] = A_T - [HA]$,

$$\frac{[OH^-]^2}{A_T - [OH^-]} \approx K_b \tag{5.66}$$

which gives the K_b analog of Eq.(5.29), namely

$$\boxed{[OH^-]^2 + K_b[OH^-] - K_bA_T \approx 0. \quad \text{2nd order equation}} \tag{5.67}$$

$$[OH^-] \approx \frac{-K_b + \sqrt{K_b^2 + 4K_bA_T}}{2}. \tag{5.68}$$

For the case when $K_a = 10^{-5}$ (so $K_b = 10^{-9}$), $A_T = 10^{-4}\,M$, and $K_w = 10^{-14}$, Eq.(5.68) gives $[OH^-] = 3.16 \times 10^{-7}\,M$, so $[H^+] = 3.16 \times 10^{-8}\,M$ (pH = 7.49). The exact solution as given in Figure 5.4 is pH = 7.52, so there is a noticeable (though small) error. If either A_T or K_b were smaller, the error would be larger: we are right on the boundary for adequacy of the BSA.

BOX 5.4 (Geek Optional) Specific Criterion for Applying the BSA

We define that the BSA is adequately accurate when

$$[H^+] \le 0.1[OH^-].$$

Procedures analogous to those discussed in Box 5.1 can be used to develop a criterion for the a priori prediction of when Eq.(5.68) can be applied with little error. The result is

$$K_bC - 10^{-6.5}K_b \ge 10^{-13.0}$$

as the criterion that the BSA applies with less than ~10% error in $[OH^-]$. Using an equals sign lets us plot the line in Figure 5.5. The BSA fails when C ($=A_T$) gets very low ($<10^{-6.5}\,M$, so not enough A^- has been added to the system to make the solution sufficiently basic, even if K_a is large); or at higher C, if the value of K_b becomes too small (pK_b too large), the base A^- is too weak to make the solution adequately basic.

5.3.3.2 BSA + WBA: Basic Solution Approximation *Plus* Weak Base Approximation (WBA) for a Solution of NaA

From above, by the BSA,

$$\frac{[\mathrm{OH}^-]^2}{[\mathrm{A}^-]} \approx K_\mathrm{b}. \tag{5.65}$$

If in addition to the BSA we are assuming A^- is behaving as a "weak" base (*i.e.*, mostly does not become protonated), then

$$[\mathrm{A}^-] \approx C = \mathrm{A_T} \tag{5.69}$$

so that

$$\boxed{[\mathrm{OH}^-] \approx \sqrt{K_\mathrm{b}C} \quad \text{1st order equation}} \tag{5.70}$$

which is the NaA solution analog of Eq.(5.42) for a solution of HA.

BOX 5.5 (Geek Optional) Specific Criterion for Applying the BSA and Then Also the WBA

We define that the WBA is adequately accurate when $[\mathrm{OH}^-] \approx [\mathrm{HA}] \leq 0.1C$. Assuming the BSA is valid, procedures analogous to those discussed in Box 5.2 can be used to develop a criterion for the a priori prediction of when Eq.(5.70) can be applied with little error. The result is

$$\sqrt{K_\mathrm{b}C} \leq 0.1C$$

or

$$C \geq 100\,K_b.$$

Using an equals sign lets us plot the line in Figure 5.5. The line demarcates the portion of the BSA region in which the BSA + WBA holds.

BOX 5.6 (Geek Optional) Specific Criterion for Applying the BSA and Then Also the SBA

We define that the SBA is adequately accurate when $[\mathrm{A}^-] \leq 0.1C$. Assuming the BSA is valid, procedures analogous to those discussed in Box 5.3 can be used to develop a criterion for the a priori prediction of when Eq.(5.72) can be applied with little error. The result is

$$K_\mathrm{b} \geq 10\,C.$$

Using an equals sign lets us plot the line in Figure 5.5. The line demarcates the portion of the BSA region in which the BSA + SBA holds.

5.3.3.3 BSA + SBA: Basic Solution Approximation *Plus* Strong Base Approximation (SBA) for a Solution of NaA

If in addition to being able to use the BSA we are also able to assume that the base is behaving as a "strong" base (*i.e.*, becomes mostly protonated), then

$$[\mathrm{HA}] \approx C \tag{5.71}$$

with the BSA, namely Eq.(5.65), we have

$$\boxed{[\mathrm{OH}^-] \approx C \quad \text{first order equation}} \tag{5.72}$$

which is the NaA solution analog of Eq.(5.45) for a solution of HA.

5.4 SOLUTIONS OF H_2B, NaHB, AND Na_2B, ALL $\gamma_i = 1$

5.4.1 INTRODUCTION – α VALUES FOR THE SPECIES RELATED TO A DIPROTIC ACID

As seen above, the use of α values greatly simplifies the pursuit of the mathematical solution representing the chemistry of acid/base problems involving HA and A^-. The same is true for solutions involving some diprotic acid H_2B and the related species HB^- and B^{2-} (and indeed any solution that is a mixture of any acids and bases). Viewed one at a time, the two sets H_2B/HB^- and HB^-/B^{2-} each comprise a conjugate acid/conjugate base pair. In natural water chemistry, the most important acid of the H_2B type is $H_2CO_3^*$ which forms when CO_2 dissolves in water and then is hydrated by a molecule of water: $CO_2 + H_2O = H_2CO_3^*$. The purpose of the * will be explained in Chapter 9. As a diprotic acid, $H_2CO_3^*$ leads to both bicarbonate (HCO_3^-), and carbonate (CO_3^{2-}).

If we just have some H_2B, HB^-, and B^{2-} in the system and no HA and A^-, then we can use α_0 to refer unambiguously to the fraction of B_T that is H_2B. If it was possible that there is an HA of some kind in the solution, then we will need a different α_0 for the fraction of A_T present as HA. (We will get to problems of that type in Section 5.6.) For the subscripts on the α values, we count the number of protons that have been lost relative to H_2B:

$$\alpha_0 \equiv \frac{[\mathrm{H_2B}]}{\mathrm{B_T}} = \frac{[\mathrm{H_2B}]}{[\mathrm{H_2B}] + [\mathrm{HB^-}] + [\mathrm{B^{2-}}]} \tag{5.73}$$

$$\alpha_1 \equiv \frac{[\mathrm{HB^-}]}{\mathrm{B_T}} = \frac{[\mathrm{HB^-}]}{[\mathrm{H_2B}] + [\mathrm{HB^-}] + [\mathrm{B^{2-}}]} \tag{5.74}$$

$$\alpha_2 \equiv \frac{[\mathrm{B^{2-}}]}{\mathrm{B_T}} = \frac{[\mathrm{B^{2-}}]}{[\mathrm{H_2B}] + [\mathrm{HB^-}] + [\mathrm{B^{2-}}]} \tag{5.75}$$

$$[\mathrm{H_2B}] \equiv \alpha_0 \mathrm{B_T} \qquad [\mathrm{HB^-}] = \alpha_1 \mathrm{B_T} \qquad [\mathrm{B^{2-}}] = \alpha_2 \mathrm{B_T}. \tag{5.76}$$

Definitions of fractions do not depend on whether the solution has non-unity γ_i values: the values of the α *values* do depend on activity corrections.

Taking Eq.(5.73) and dividing the top and bottom by $[H_2B]$ gives

$$\alpha_0 = \frac{1}{1 + \dfrac{[\mathrm{HB^-}]}{[\mathrm{H_2B}]} + \dfrac{[\mathrm{B^{2-}}]}{[\mathrm{H_2B}]}}. \tag{5.77}$$

Dividing Eq.(5.74) by Eq.(5.73) gives

$$\frac{\alpha_1}{\alpha_0} = \frac{[HB^-]}{[H_2B]} \qquad \alpha_1 = \alpha_0 \frac{[HB^-]}{[H_2B]}. \tag{5.78}$$

Dividing Eq.(5.75) by Eq.(5.74) gives

$$\frac{\alpha_2}{\alpha_1} = \frac{[B^{2-}]}{[HB^-]} \qquad \alpha_2 = \alpha_1 \frac{[B^{2-}]}{[HB^-]}. \tag{5.79}$$

So that we can also use just concentration values in the expressions for the acidity constants, we assume for the moment that all $\gamma_i = 1$ so that

$$H_2B = H^+ + HB^- \qquad K_1 = \frac{[H^+][HB^-]}{[H_2B]} \qquad \frac{[HB^-]}{[H_2B]} = \frac{K_1}{[H^+]} \tag{5.80}$$

$$HB^- = H^+ + B^{2-} \qquad K_2 = \frac{[H^+][B^{2-}]}{[HB^-]} \qquad \frac{[B^{2-}]}{[HB^-]} = \frac{K_2}{[H^+]} \tag{5.81}$$

$$H_2B = 2H^+ + B^{2-} \qquad K_1K_2 = \frac{[H^+]^2[B^{2-}]}{[H_2B]} \qquad \frac{[B^{2-}]}{[H_2B]} = \frac{K_1K_2}{[H^+]^2}. \tag{5.82}$$

Substituting Eqs.(5.80) and (5.82) into Eq.(5.77) gives

$$\boxed{\alpha_0 = \frac{1}{1 + K_1/[H^+] + K_1K_2/[H^+]^2}.} \tag{5.83}$$

We could get equations like this for α_1 and for α_2 but it is easier to get the higher α values by means of ratios. Substituting Eq.(5.80) into Eq.(5.78) gives

$$\frac{\alpha_1}{\alpha_0} = \frac{[HB^-]/B_T}{[H_2B]/B_T} = \frac{[HB^-]}{[H_2B]} = \frac{K_1}{[H^+]} \qquad \boxed{\alpha_1 = \alpha_0 \frac{K_1}{[H^+]}.} \tag{5.84}$$

Substituting Eq.(5.81) into Eq.(5.79) gives

$$\frac{\alpha_2}{\alpha_1} = \frac{[B^{2-}]/B_T}{[HB^-]/B_T} = \frac{[B^{2-}]}{[HB^-]} = \frac{K_2}{[H^+]} \qquad \boxed{\alpha_2 = \alpha_1 \frac{K_2}{[H^+]}.} \tag{5.85}$$

The patterns among these α values are summarized in Box 5.7.

5.4.2 Solution of H_2B of Concentration C (F)

The species present in a solution of H_2B are

$$H_2B \quad HB^- \quad B^{2-} \quad H^+ \quad OH^-.$$

so we have five concentration unknowns. Neglecting activity corrections, the equations needed to solve for the five unknowns in such systems are K_1, K_2, K_w, the MBE on B_T, and the ENE. The latter two are

$$[H_2B] + [HB^-] + [B^{2-}] = C = B_T. \qquad \text{MBE on total B} \tag{5.86}$$

$$[H^+] = [HB^-] + 2[B^{2-}] + [OH^-] \qquad \text{ENE.} \tag{5.87}$$

Choosing the ENE as usual for the "backbone" for the solution, we use all four of the other equations to immediately obtain

$$\underline{[H^+] = \alpha_1 B_T + 2\alpha_2 B_T + K_w/[H^+]} \tag{5.88}$$

which can be solved by seeking the root value of $[H^+]$ that gives (LHS – RHS) = 0.

BOX 5.7 The Powerful Recursive Patterns in the Equations for α Values in Acid/Base Systems (All $\gamma_i = 1$)

For any <u>monoprotic acid</u> HA, neglecting activity corrections, we first recall that

$$\alpha_0 = \frac{[H^+]}{[H^+] + K_a}$$

which gives

$$\alpha_0 = \frac{1}{1 + \dfrac{K_a}{[H^+]}}.$$

For a monoprotic acid, K_a is K_1 (there are no K_2, etc.). Thus, the equation for α_0 for a diprotic acid looks just like α_0 for a monoprotic acid, except that there is the second term $K_1K_2/[H^+]^2$. Namely,

$$\alpha_0 = \frac{1}{1 + \dfrac{K_1}{[H^+]} + \dfrac{K_1K_2}{[H^+]^2}}.$$

This suggests (*and it's true*) that for any <u>triprotic acid</u> H_3C:

$$\alpha_0 = \frac{1}{1 + \dfrac{K_1}{[H^+]} + \dfrac{K_1K_2}{[H^+]^2} + \dfrac{K_1K_2K_3}{[H^+]^3}}.$$

The equation for α_0 for a polyprotic acid can be inferred from this pattern. Expressions for the α_1 etc. can be deduced as given in the table below.

Acid Type	Equations		
HA (monoprotic)	$\alpha_1 = \alpha_0 \dfrac{K_a}{[H^+]}$ for A^-		
H_2B (diprotic)	$\alpha_1 = \alpha_0 \dfrac{K_1}{[H^+]}$ for HB^-	$\alpha_2 = \alpha_1 \dfrac{K_2}{[H^+]}$ for B^{2-}	
H_3C (triprotic)	$\alpha_1 = \alpha_0 \dfrac{K_1}{[H^+]}$ for H_2C^-	$\alpha_2 = \alpha_1 \dfrac{K_2}{[H^+]}$ for HC^{2-}	$\alpha_3 = \alpha_2 \dfrac{K_3}{[H^+]}$ for C^{3-}
etc.	in general, for $j \geq 1$, use $\alpha_j = \alpha_{j-1} \dfrac{K_j}{[H^+]}$		

5.4.3 Solution of NaHB of Concentration *C* (*F*)

The species in a solution of NaHB are the same as those in a solution of H_2B, plus Na^+. Thus, there are six unknown concentrations in systems of this type. For the equations, we again have K_1, K_2, K_w, the MBE on B_T, an MBE for $[Na^+]$, and the ENE. The latter two are

$$[Na^+] = C = B_T \qquad \text{MBE on } Na^+ \tag{5.89}$$

$$[H^+] + [Na^+] = [HB^-] + 2[B^{2-}] + [OH^-] \qquad \text{ENE.} \tag{5.90}$$

With the ENE as the "backbone" for the solution, we use all five of the other equations to obtain

$$\underline{[H^+] + [Na^+] = \alpha_1 B_T + 2\alpha_2 B_T + K_w/[H^+].} \tag{5.91}$$

We know $[Na^+] = B_T$, so Eq.(5.91) is one equation in one unknown, and it is solved by finding root value of $[H^+]$ that gives (LHS – RHS) = 0. This equation is similar to Eq.(5.88). The only difference is the term $[Na^+]$, which moves the root *from* the acid side (for Eq.(5.88)) *towards* the base side (for Eq.(5.91)).

5.4.4 Solution of Na_2B of Concentration *C* (*F*)

The ENE to be used for solving the problem (with (LHS – RHS) = 0) is again

$$\underline{[H^+] + [Na^+] = \alpha_1 B_T + 2\alpha_2 B_T + K_w/[H^+]} \tag{5.91}$$

but now

$$[Na^+] = 2C = 2B_T \qquad \text{MBE on } Na^+. \tag{5.92}$$

For the same value of B_T, a solution of Na_2B will have a higher pH than a solution of NaHB: for the solution of Na_2B, the $[Na^+]$ term in Eq.(5.91) is larger, and that moves the root to a higher pH.

5.4.5 A Solution of H_2B, or NaHB, or Na_2B of Concentration *C* (*F*)

Although Eq.(5.88) for the solution of H_2B seems different from Eq.(5.91) (which applies for solutions of NaHB and Na_2B), actually, Eq.(5.88) is a special case of Eq.(5.91), namely that with $[Na^+] = 0$. Because of this, we can build *one spreadsheet* to handle all three problems, and wherein we input the values of *C* and $[Na^+]$ for the specific problem of interest.

Example 5.5 Building an Excel Spreadsheet to Solve for pH for Any Solution of H_2B, NaHB, and Na_2B

Every solution can be based on Eq.(5.91). Set up the spreadsheet so that Solver varies pH, and not $[H^+]$ directly, to find the root value of $[H^+]$. Use the spreadsheet to find the pH and speciation at 25 °C and 1 atm for 10^{-3} *F* $H_2CO_3^*$, 10^{-3} F $NaHCO_3$, and 10^{-3} F Na_2CO_3. $K_1 = 10^{-6.35}$, $K_2 = 10^{-10.33}$.

Solution

$B_T = 10^{-3}$ *M* for each of these solutions. For the $H_2CO_3^*$ solution, $[Na^+] = 0$ and pH = 4.68. For the $NaHCO_3$ solution (see example spreadsheet), $[Na^+] = 10^{-3}$ *M* and pH = 8.30. For the Na_2CO_3 solution, $[Na^+] = 2 \times 10^{-3}$ *M*, and pH = 10.56. The screenshot of the spreadsheet given applies to the $NaHCO_3$ solution.

	A	B	C	D	E	F
1	Pankow Water Chemistry Spreadsheet #5.3					
2	*Solution of H2B, NaHB, or Na2B*					
3	ENE: [H+] + [Na+] = [HB-] + 2[B2-] + [OH-]					
4	[HB-] = α_1*B_T					
5	[B2-] = α_2*B_T			results:	formulas:	
6	[OH-] = *K*_w/[H+]		[Na+] =	1.00E-03	= B12	
7		inputs:				
8	*K*_1=	4.47E-07	[H+] =	*5.05E-09*	= 10^-B18	
9	*K*_2=	4.68E-11	[OH-] =	*2.00E-06*	= B10/D8	
10	*K*_w =	1.01E-14	B_T=	*1.00E-03*	= B11	
11	B_T =	1.00E-03	for H2B: α_0 =	*1.11E-02*	= 1/(1+B8/D8+B8*B9/D8^2)	
12	[Na+] =	1.00E-03	for HB-: α_1 =	*9.80E-01*	= D11*B8/D8	
13			for B2-: α_2 =	*9.08E-03*	= D12*B9/D8	
14	Set Solver Target Cell: D18		[H2B]=α_0*B_T=	*1.11E-05*	= D11*D10	
15	Equal to Value of 0		[HB-]=α_1*B_T=	*9.80E-04*	= D12*D10	
16	By Changing Cell: B18		[B2-]=α_2*B_T=	*9.08E-06*	= D13*D10	
17			LHS=[H+]+[Na+] =	*1.00E-03*	= D8+D6	
18	**pH guess for Solver=**	*8.30*	RHS=[HB-]+2[B2-]+[OH-] =	*1.00E-03*	=D15+2*D16+D9	
19			(LHS-RHS)*1E6 =	*-9.30E-08*	= (D17-D18)*1E6	
20						

5.5 SOLUTIONS OF AMMONIA AND AMMONIUM SALTS

5.5.1 The Monoprotic Acid Form (NH_4^+) Is Ionic, the Conjugate Base Form (NH_3) Carries No Charge

The ammonium ion (NH_4^+) and ammonia (NH_3) are such important chemical species in natural systems and in waste waters that they deserve some special treatment, especially since, unlike the generic monoprotic acid "HA", obviously NH_4^+ is charged. The acid dissociation reaction is

$$NH_4^+ = NH_3 + H^+ \qquad K_a = \frac{[H^+][NH_3]}{[NH_4^+]} \tag{5.93}$$

where for the expression for K_a, we are taking all $\gamma_i = 1$ either as an approximation, or because the solution is dilute. By the usual counting system for the subscript for acid/base α values, in this system α_0 refers to the fraction of N_T present as NH_4^+, and α_1 refers to the fraction present as NH_3:

$$\alpha_0 = \frac{[NH_4^+]}{N_T} \qquad \alpha_1 = \frac{[NH_3]}{N_T} \qquad N_T = [NH_4^+] + [NH_3]. \tag{5.94}$$

5.5.2 Solution of NH_4Cl of Concentration C (F)

If we want to prepare a solution of NH_4^+, because of the ionic charge, NH_4^+ has to come into the solution with a "counterion": we cannot have a bottle of just NH_4^+ as it would not be electrically neutral. Let us say the counterion is Cl^-, which under most circumstances can be considered to be a spectator ion. For a solution of the monoprotic acid NH_4Cl, the ENE involves four of the five unknowns and is

$$[NH_4^+] + [H^+] = [OH^-] + [Cl^-]. \tag{5.95}$$

The fifth unknown is $[NH_3]$. Eq.(5.95) can be set up for the solution by (LHS – RHS) = 0 by writing it as

$$\alpha_0 N_T + [H^+] = K_w/[H^+] + [Cl^-]. \tag{5.96}$$

Eq.(5.96) is based on the ENE plus all of the four other equations, namely the expressions for K_a and K_w, the MBE that $N_T = C$, and the MBE that $[Cl^-] = C = N_T$. That a solution of NH_4Cl *must be acidic* follows from the substitution of the expression in Eq.(5.94) for N_T into $[Cl^-]$ in Eq.(5.95); $[OH^-]$ must be less than $[H^+]$ because

$$[H^+] = [OH^-] + [NH_3]. \tag{5.97}$$

Eq.(5.97) can be compared with Eq.(5.4).

5.5.3 Solution of NH_3 of Concentration *C* (*F*)

The unknowns in this problem are $[NH_3]$, $[NH_4^+]$, [H+], and $[OH^-]$. In addition to the ENE, the equations we have are the expressions for K_a and K_w, and the MBE that $N_T = C$. By analogy for $[Na^+]$ in the discussion for the solutions of H_2B, NaHB, and Na_2B, just because $[Cl^-] = 0$ does not mean we cannot use Eq.(5.96), because it does qualify as the ENE for any solution of NH_4Cl or NH_3. So, again

$$\alpha_0 N_T + [H^+] = K_w/[H^+] + [Cl^-]. \tag{5.96}$$

That a solution of NH_3 *must be basic* follows from $[Cl^-] = 0$ so that

$$[NH_4^+] + [H^+] = [OH^-]. \tag{5.98}$$

Eq.(5.98) can be compared in this context with the PBE for a solution of NaA, namely $[HA] + [H^+] = [A^-]$.

Example 5.6 pH of Household Ammonia

As sold in grocery stores, "household ammonia" is often 5% by weight NH_3, or 50 g of NH_3 per kg of water. Since 1 kg of a mostly aqueous solution occupies close to 1 L, this means the concentration is close to 50 g/L. **a.** What is the formality of NH_3 in 5% ammonia? **b.** If the $pK_a = 9.24$ for NH_4^+ at 25 °C and 1 atm pressure, if we can ignore activity corrections and assume that no CO_2 from the atmosphere has entered the solution, what is the pH at at 25 °C and 1 atm pressure?

Solution

a. The formula weight of NH_3 is 17.031 g/FW.

$$50\ \text{g/L} \times 1\,\text{FW}/17.031\,\text{g} = 2.94\ F.$$

b. For this solution $N_T = C = 2.94$ *M*. This is a very concentrated solution. $pK_a = 9.24$ which means $pK_b = 4.76$: ammonia is as weak as a base as acetic acid is as an acid. This makes for perfect conditions for applying the BSA + WBA.

$$[OH^-] = \sqrt{K_b C} = \sqrt{10^{-4.76} \times 2.94} = 0.00715$$

$$\text{pH} = 11.85.$$

Solving "exactly" using Eq.(5.96) with $[Cl^-] = 0$ gives the same result. This pH is rather high, which accounts for why household ammonia is good for cleaning: the high pH helps saponify grease and dissolve it, making soap in the process. Note also that $\alpha_1 \approx 1$ in this solution ($\text{pH} >> pK_a$), which accounts for the fact that a bottle of ammonia smells very strongly of ammonia: N_T is high, and most of the N_T is present as NH_3, the volatile form.

5.6 SETTING UP THE ENE TO SOLVE FOR THE SPECIATION OF ANY ACID/BASE PROBLEM

5.6.1 Foundational Principles

This text nearly always uses the ENE as the algebraic backbone when solving for the speciation in a chemical problem. The great power of this approach for aqueous solutions of acids and bases stems from the fact that for all conjugate acid/base pairs since there is loss of H^+ between each pair, one or the other (and maybe both) of the species is charged, and so will appear in the ENE. This appearance brings in the acidity constant value(s) through the α value(s), and the associated MBE for each acid/base family present. (For example, acetic acid and acetate ion make a family; $H_2CO_3^*$, HCO_3^-, and CO_3^{2-} make another family; NH_4^+ and NH_3 make yet another family.) We account for the effects of chemicals like NaOH and HCl on the system through $[Na^+]$ and $[Cl^-]$ in the ENE.

5.6.2 A Set of α Values for Each Acid/Base Family

We can use a superscript on an acid/base α value to denote the family to which the α applies, and therefore which acidity constant values are to be used to evaluate each α. For example, say we have an aqueous solution in which the species present include species from three families: Family A) acetic acid and acetate ion; Family C) $H_2CO_3^*$, HCO_3^-, and CO_3^{2-}; and Family N) NH_4^+ and NH_3. The ENE for any such solution plus possibly Na^+ and/or Cl^- is then

$$\alpha_0^{\mathrm{N}}\mathrm{N_T} + [\mathrm{H^+}] + [\mathrm{Na^+}] = K_\mathrm{w}/[\mathrm{H^+}] + \alpha_1^{\mathrm{C}}\mathrm{C_T} + 2\alpha_2^{\mathrm{C}}\mathrm{C_T} + \alpha_1^{\mathrm{A}}\mathrm{A_T} + [\mathrm{Cl^-}] \tag{5.99}$$

which can be solved by (LHS − RHS) = 0. Very happily, for Eq.(5.99) **and for every other such equation for a mixture of acids and bases**, the function (LHS − RHS) always has the general shape of $f_a([H^+])$ in Figure 5.1, which means we will never have any trouble finding the root, no matter how horrendous the initial guess. How cool is that?

Example 5.7 The pH of a Complex Mixture of Acids and Bases

Consider 1 L of aqueous solution with the FW/L values of the following added chemicals.

chemical	FW/L	chemical	FW/L
NH_3	0.001	$NaHCO_3$	0.01
NH_4Cl	0.0005	Na_2CO_3	0.001
acetic acid	0.002	HCl	0.0001
Na acetate	0.001	NaOH	0.001

a. For use in Eq.(5.99), what are the values of N_T, C_T, A_T, $[Na^+]$, and $[Cl^-]$?
b. Using the same values of the acidity constant values for each family as introduced above for 25 °C and 1 atm, what is the pH?

Solution

a.

$$\begin{aligned}
N_T &= 0.001\,M + 0.0005\,M = 0.0015\,M \\
C_T &= 0.01\,M + 0.001\,M = 0.011\,M \\
A_T &= 0.002\,M + 0.001\,M = 0.003\,M \\
[Na^+] &= 0.001\,M + 0.01\,M + 2(0.001)\,M + 0.001\,M = 0.014\,M \\
[Cl^-] &= 0.0005\,M + 0.0001\,M = 0.0006\,M
\end{aligned}$$

b. The problem is set up for solution in the Excel spreadsheet indicated below. Neglecting activity corrections, solving by (LHS – RHS) = 0 gives pH = 8.93 Raising the value of $[Na^+]$ without changing C_T, as by adding more NaOH, or as by increasing the proportion of Na_2CO_3 to $NaHCO_3$, will move the root to a higher pH. Raising the value of $[Cl^-]$ without changing N_T, as by adding more HCl, or as by increasing the proportion of NH_4Cl to NH_3, will move the root to a lower pH. Raising the value of C_T without changing $[Na^+]$ will move the root to lower pH because this has to occur by adding $H_2CO_3^*$, which is an acid. Raising the value of N_T without changing $[Cl^-]$ will move the root to higher pH because this has to occur by adding NH_3, which is a base.

	A	B	C	D	E	F	G	H
1	Pankow Water Chemistry Spreadsheet #5.4							
2	*Complex Mixture Example*							
3								
4	ENE							
5	[NH4+] + [H+] + [Na+] = [OH-]+ [HCO3-] + 2[CO32-] + [A-] + [OH-]							
6								
7	ammonia family			using [H+]	formulas	Inputs		
8	K=	5.75E-10	αN_0 =	6.69E-01	= G15/(G15+B8)	NT=	0.0015	
9			αN_1 =	3.31E-01	= D8*B8/G15	CT=	0.011	
10	acetic family					AT=	0.003	
11	K=	1.74E-05	αA_0 =	6.71E-05	= G15/(G15+B11)	[Na+]=	0.014	
12			αA_1 =	1.00E+00	= D11*B11/G15	[Cl-] =	0.0006	
13	CO2 family					Solver		
14	K1=	4.47E-07	αC_0 =	2.50E-03	=(1+B14/G15+B14*B15/G15^2)^-1	guess pH=	8.93	formulas
15	K2=	4.68E-11	αC_1 =	9.59E-01	= D14*B14/G15	[H+]=	1.17E-09	= 10^-F14
16			αC_2=	3.85E-02	= D15*B15/G15	LHS=	1.50E-02	= D8*F8+F15+F11
17	water					RHS=	1.50E-02	= B18/F15+D15*F9+2*D16*F9+D12*F10+F12
18	Kw=	1.01E-14				(LHS-RHS)*1e6=	-7.22E-08	= (F16-F17)*1000000

5.7 GENERAL APPROACH FOR SOLVING FOR THE SPECIATION INCLUDING ACTIVITY CORRECTIONS

5.7.1 USING ^cK VALUES WHEN MAKING ACTIVITY CORRECTIONS

All MBE and ENE expressions use concentrations. But, equilibrium constants K of the "infinite dilution type" require usage of activities, not concentrations. We dealt with this problem in the above examples by assuming that the solutions were sufficiently dilute that all $\gamma_i = 1$. We could then express all K using concentrations, and move forward with various combinations of equations. If all $\gamma_i \neq 1$, we need to modify the approach. For ions, to calculate γ_i values we use the Davies Equation with the value of the ionic strength I:

$$I = \frac{1}{2}\sum_i c_i z_i^2. \tag{5.100}$$

Recall now that Section 2.7.2 introduced the concept of ^cK values. Namely, for the acid dissociation constant for some monoprotic acid HA,

$$K_a = \frac{\{H^+\}\{A^-\}}{\{HA\}} = \left(\frac{[H^+][A^-]}{[HA]}\right)\left(\frac{\gamma_{H^+}\gamma_{A^-}}{\gamma_{HA}}\right) \tag{5.101}$$

$$ {}^{c}K_{a} = \frac{[H^{+}][A^{-}]}{[HA]} = \left(\frac{\gamma_{HA}}{\gamma_{H^{+}}\gamma_{A^{-}}}\right)K_{a} = \frac{1}{\gamma_{1}^{2}}K_{a} \tag{5.102} $$

where the third equality makes use of: (1) the common assumption that $\gamma_i = 1$ for a neutral species (*e.g.*, CH_3COOH (molecular acetic acid), $H_2CO_3^*$, NH_3, etc.) even in rather salty solutions which, for ions, causes $\gamma_i \neq 1$; and (2) the assumption that to an adequate approximation, any singly charged species will be characterized by the same γ_i, denoted γ_1. By extension, for any doubly charged ion (Ca^{2+}, Mg^{2+}, CO_3^{2-}, SO_4^{2-}, etc.), we can use γ_2.

Since ${}^{c}K$ values use concentrations and not activities, they can be used directly with the ENE and any MBE expressions. So, for any problem of the type already discussed, if we had a case in which $\gamma_i \neq 1$, we can still use all the equations developed after merely substituting the ${}^{c}K$ value for each K. Expressions for ${}^{c}K$ for some other equilibria besides $HA = H^+ + A^-$ are given below.

$$ {}^{c}K_{1} = \frac{[H^{+}][HB^{-}]}{[H_2B]} = \left(\frac{\gamma_{H_2B}}{\gamma_{H^{+}}\gamma_{HB^{-}}}\right)K_{1} = \frac{1}{\gamma_{1}^{2}}K_{1} \tag{5.103} $$

$$ {}^{c}K_{2} = \frac{[H^{+}][B^{2-}]}{[HB^{-}]} = \left(\frac{\gamma_{HB^{-}}}{\gamma_{H^{+}}\gamma_{B^{2-}}}\right)K_{2} = \frac{1}{\gamma_{2}}K_{2} \tag{5.104} $$

$$ {}^{c}K_{w} = [H^{+}][OH^{-}] = \left(\frac{1}{\gamma_{H^{+}}\gamma_{OH^{-}}}\right)K_{w} = \frac{1}{\gamma_{1}^{2}}K_{w} \tag{5.105} $$

$$ {}^{c}K_{a} = \frac{[H^{+}][NH_3]}{[NH_4^{+}]} = \left(\frac{\gamma_{NH_4^{+}}}{\gamma_{H^{+}}\gamma_{NH_3}}\right)K_{a} = K_{a}. \tag{5.106} $$

For Eq.(5.106), the γ_i values for the ions cancel out, and it is assumed that $\gamma_{NH_3} \approx 1$.

5.7.2 Activity Corrections When the Final Ionic Strength *I* Is Known *A Priori*

There are many cases when the final ionic strength I in the solution is known before the problem is solved, to a good approximation. An example would be a slightly briny stream in which one wanted to make acid/base calculations pH etc.: the value of I is being set by ions that are not participating in the reactions of interest, and I calculated after the problem (when all concentrations are known) will be close to what can be guessed at the beginning. This is the easiest type of activity correction situation. The steps are:

1) Set up the problem using the same approach used in Sections 5.1 through 5.6.
2) Calculate all needed γ_i values for ions (*i.e.*, γ_1 and γ_2, rarely γ_3) using the known I and the Davies Equation;
3) Calculate all needed ${}^{c}K$ values based on the calculated γ_i values and tabulated K values;
4) Apply ${}^{c}K$ values in the problem and solve for $[H^+]$ – note here that if we are driving the solution by searching for the value of $-\log [H^+]$, we will call this p_cH not pH, the c referring to *concentration*;

5) Use α values to obtain other concentrations, as desired;
6) Compute $pH = -\log\{\gamma_1[H^+]\}$.

Example 5.8 pH Calculations when *I* Is Known *A Priori*

Some salty water known to be about 0.2 *F* NaCl is amended with 0.001 *F* Na_2CO_3. For 25 °C and 1 atm, assuming that NaCl and the Na_2CO_3 are the only chemicals dissolved in the water, compute the pH (recall that $pH \equiv -\log\{H^+\}$).

Solution

This problem has Eq.(5.91) as its governing equation.

Neglecting the contribution of the water and the Na_2CO_3 to the ionic strength,

$$I = \frac{1}{2}\left((+1)^2(0.2) + (-1)^2(0.2)\right) = 0.2.$$

The Davies Equation is $\log \gamma_i = -Az_i^2\left[\sqrt{I}/(1+\sqrt{I}) - 0.3I\right]$. This gives $\gamma_1 = 0.74$ and $\gamma_2 = 0.31$. The ${}^cK$ values are given the spreadsheet below. The result is $p_cH = 10.20$, with $pH = 10.33$.

	A	B	C	D	E	F	G	H
1	Pankow Water Chemistry Spreadsheet #5.5							
2	***Activity Correction Problem 1. Vaue of I is known* a priori.**							
3	*I* =	0.2						
4	ENE							
5	[H+] + [Na+] = [HCO3-] + 2[CO32-] + [OH-]				Inputs			
6					γ1 =	0.75		
7	CO2 family		formulas		γ2 =	0.31		
8	K1=	4.47E-07			CT=	0.001		
9	K2=	4.68E-11			[Na+]=	0.002		
10	cK1=	7.94E-07	=B8/F6^2					
11	cK2=	1.51E-10	=B9/F7		Solver		formulas	
12	αC_0 =	2.32E-05	=1/(1+B10/F13+B10*B11/F13^2)		guess pcH=	10.20		
13	αC_1 =	2.94E-01	=B12*B10/F13		[H+]=	6.28E-11	=10^-F12	
14	αC_2=	7.06E-01	=B13*B11/F13		LHS=	2.00E-03	=F13+F9	
15	water				RHS=	1.99E-03	=B13*F8+2*B14*F8+B17/F13	
16	Kw=	1.01E-14			(LHS-RHS)*1e6=	7.78E+00	=(F14-F15)*1000000	
17	cKw=	1.80E-14	=B16/F6^2		pH =	10.33	=-LOG(F13*F6)	

5.7.3 Activity Corrections When *A Priori* the Final Ionic Strength *I* Is Not Known

When the value of *I* in the solution is not known to some reasonable approximation at the beginning of the problem, one cannot calculate the γ_i values needed to solve the problem as in Example 5.8. Now we can use an iterative approach, and guess what *I* might be, use that *I* to calculate the needed γ_i values, and solve the problem (see Table 5.3). The results for the speciation can be used to calculate a new and hopefully better estimate of *I*, which can be used to calculate new and better γ_i values, and we solve the problem again. Hopefully, after repeating this over and over again, the solution results stop changing and "converge" on the actual true *I* and speciation. In Example 5.9, it is discussed how this iteration process can be nested within the iteration sequence that Solver uses in Excel to find the root, that nesting achievable by means of circular references in the spreadsheet.

TABLE 5.3
Summary of Process for Applying the Iterative Procedure for Calculating a Given Chemical Speciation, Including the Effects of Activity Corrections

Iteration	Steps
first	a. Guess a value for the ionic strength *I*. Calculate corresponding values for γ_1 and γ_2 (and γ_3 if needed). b. Solve and compute first estimate of speciation.
second and subsequent	a. Compute the *I* based on speciation from previous iteration. b. Compute next estimate γ_i values based on the new estimate of *I*. c. Solve and compute next estimate of speciation; and d. Check to see if speciation has changed significantly from previous iteration. If yes, do another iteration. If no, stop and accept calculated speciation.

Note: This is either done "by hand", or built into a spreadsheet solution allowing circular references.

Example 5.9 pH Calculations When *I* Is Not Known *A Priori*

A solution that is 0.1 *F* Na_2CO_3 is prepared at for 25 °C and 1 atm. The value of *I* in the solution at equilibrium is not known *a priori*. Compute the pH. (Recall that pH ≡ −log{H+}.)

Solution

Results for the problem after having run Solver for the first iteration are contained in the spreadsheet below. To execute the second iteration, the *values* for γ_1 and γ_2 in cells I23 and I24 are cut and pasted into cells F6 and F7, respectively, and Solver is run a second time. The process converges quite quickly, and really we just need the two iterations to get to convergence, as seen in the table. (A common error among students learning this approach is that rather than update the values of γ_i used to convert the *K* values to ${}^cK$ values in each iteration, the updated γ_i values are applied to the preceding ${}^cK$ values. This double-, then triple-, then quadruple-(etc.) corrects the *K* values, and the solution results keep changing, not to mention keep getting worse.) To avoid the need to cut and paste each updated set of γ_i values, Excel allows use of the new γ_i in the formulas for the ${}^cK$ values. In the spreadsheet below, we can change the formulas for three cells as follows: B10 becomes=B8/I23^2; B11 becomes=B9/I24; and B17 becomes=B16/I23^2. Unless the toggle is already on, upon such changes, Excel alerts the user that these changes introduce circular references. "Enable iterative calculation" can then be turned on in Excel Options/Formulas, while also applying a suitably small "Maximum change" convergence criterion (0.0001 worked in this case). Solver can then be invoked with some initial guesses for p_cH and the γ_i, and the entire solution process is one step (for the user, not for Solver!).

	A	B	C	D	E	F	G	H	I	J
1	Pankow Water Chemistry Spreadsheet #5.6									
2	*Activity Correction Problem 2. Value of I is not known a priori.*									
3										
4	ENE									
5	[H+] + [Na+] = [HCO3-] + 2[CO32-] + [OH-]				Inputs					
6					guess γ1 =	1.00				
7	CO2 family		formulas		guess γ2 =	1.00				
8	K1=	4.47E-07			CT=	0.1				
9	K2=	4.68E-11			[Na+]=	0.1				
10	cK1=	4.47E-07	=B8/F6^2							
11	cK2=	4.68E-11	=B9/F7		Solver		formulas			
12	αC_0 =	1.00E-02	=1/(1+B10/F13+B10*B11/F13^2)		guess pcH=	8.34				
13	αC_1 =	9.80E-01	=B12*B10/F13		[H+]=	4.58E-09	=10^-F12			
14	αC_2=	1.00E-02	=B13*B11/F13		LHS=	1.00E-01	=F13+F9			
15	water				RHS=	1.00E-01	=B13*F8+2*B14*F8+B17/F13			
16	Kw=	1.01E-14			(LHS-RHS)*1e6=	-1.18E-07	=(F14-F15)*1000000			
17	cKw=	1.01E-14	=B16/F6^2		pH =	8.34	=-LOG(F13*F6)			
18										
19					Speciation for *I* for next iteration		formulas			
20					[H+]=	4.58E-09	=F13			
21					[Na+]=	0.1	=F9		*I* =	9.90E-02
22					[HCO3-]=	9.80E-02	=B13*F8		New γ	
23					[CO32-]=	1.00E-03	=B14*F8		γ1 =	0.78178
24					[OH-]=	2.21E-06	=B17/F13		γ2 =	0.37355
25										

ITERATION RESULTS FOR EXAMPLE 5.9

iteration	I	γ_1	γ_2		p_cH	pH
1	0	1.00000	1.00000	with γ_1 and γ_2 →	8.34	8.34
	use above p_cH to calculate:					
2	9.90E-02	0.78178	0.37355	with new γ_1 and γ_2 →	8.02	8.13
	use above p_cH to calculate:					
3	9.87E-02	0.78194	0.37384	with new γ_1 and γ_2 →	8.02	8.13
	use above p_cH to calculate:					
4	9.87E-02	0.78194	0.37384	with new γ_1 and γ_2 →	8.02	8.13

REFERENCE

Narasaki H (1987) The range of application of the approximation formulae in acidbase equilibria. *Fresenius' Zeitschrift für Analytische Chemie*, **328**, 633–638.

6 Dependence of α Values on pH, and the Role of Net Strong Base

6.1 INTRODUCTION

Chapter 5 focused on problems in which the task is to determine the pH, and thereby the general speciation in particular solutions, including: (1) solutions of an acid HA with some specific C and K_a values; (2) solutions of some NaA with specific C and K_b (= K_w/K_a) values; (3) solutions of some H_2B, or NaHB, or Na_2B, with specific C, K_1, and K_2 values; and then (4) complicated mixtures of acids and bases. Now we wish to broaden the vision from the idea of *particular* solutions, to how one moves *within a system*, from one solution chemistry to another. For example, if one started with a solution of acetic acid as HA with $C = A_T = 10^{-3}$ *M*, how much NaOH needs to be added to move to some different pH, and what is the new speciation there? Or, for a more complicated natural water chemistry situation that we are not ready yet to handle, if some water in a stream has a pH of 8.1, and if a certain amount of some HCl is spilled into the stream, will the pH go low enough to kill fish? This topic of how water solutions respond to acid/base changes is the subject of Chapters 6 through 9.

6.2 Logα AND Log[ ] *VS.* pH PLOTS FOR MONOPROTIC ACID SYSTEMS

6.2.1 Log α *vs.* pH Plots

As introduced in Chapter 5, for a monoprotic acid/base pair (which could be HA and A^-, or NH_4^+ and NH_3) we can define fractional α values. For HA and A^-,

$$\alpha_0 = \frac{[\text{HA}]}{\text{A}_\text{T}} \qquad \alpha_1 = \frac{\left[\text{A}^-\right]}{\text{A}_\text{T}} \tag{6.1}$$

$$\log[\text{HA}] = \log\alpha_0 + \log\text{A}_\text{T} \qquad \log\left[\text{A}^-\right] = \log\alpha_1 + \log\text{A}_\text{T}. \tag{6.2}$$

Neglecting activity corrections (or substituting ${}^cK$ values to include them), we have

$$\alpha_0 = \frac{[\text{H}^+]}{[\text{H}^+] + K_a} = \frac{1}{1 + \dfrac{K_a}{[\text{H}^+]}} \tag{6.2}$$

$$\alpha_1 = \frac{K_a}{[\text{H}^+] + K_a} = \frac{1}{\dfrac{[\text{H}^+]}{K_a} + 1}. \tag{6.4}$$

For solutions involving HA and A^-, for contextual relevance to high school and undergraduate chemistry courses, Chapter 5 discussed solutions of HA and solutions of NaA of concentration C, with $C = A_T$. In this chapter, A_T is preferred, as it is a more specific use of nomenclature.

Figure 6.1 is a plot of log α_0 and log α_1 for pK_a=5.0. Each of the two log α lines has two nearly straight asymptotic legs, one on each side of a **transition region** defined by about p$K_a \pm 1$. Each log α

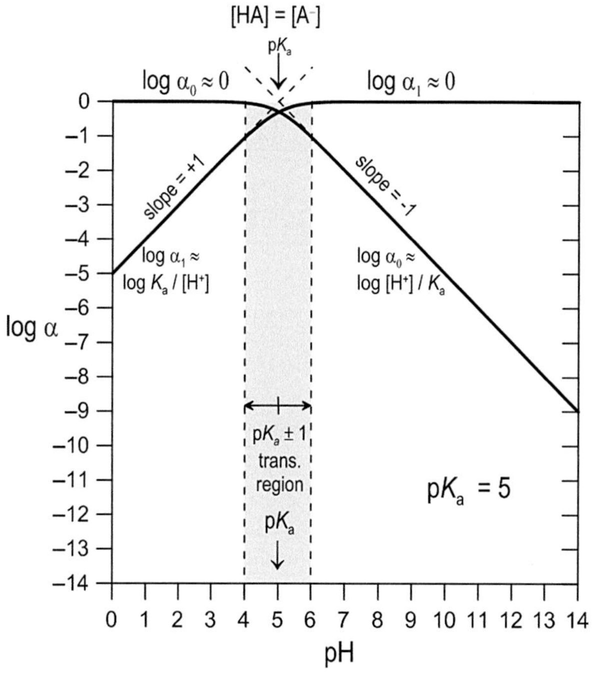

FIGURE 6.1 Log α_0 and log α_1 *vs.* pH for an HA/A⁻ system with $K_a = 10^{-5}$. Behaviors in asymptotic regions are given. Activity corrections neglected (not valid at very low or very high pH).

line has one leg such that log $\alpha_0 \approx \log 1 = 0$. For the other two legs, the asymptotic slope is either +1 or −1, and when extrapolated, the two asymptotic legs cross at pH = pK_a, $y = 0$. Table 6.1 summarizes the important characteristics of the Figure 6.1 plot. Figure 6.2 is a plot without any labeling for the axes; it can be used to make a transparency that can be overlayed on the log [ ] (*i.e.*, log concentration) *vs.* pH grid discussed below (Figure 6.4).

6.2.2 pH ≪ pK_a

When

$$[H^+] > 10K_a \tag{6.5}$$

$$pH < pK_a - 1 \tag{6.6}$$

which is commonly abbreviated as

$$pH \ll pK_a. \tag{6.7}$$

The ≪ sign in Eq.(6.7) seems an overstatement of Eq.(6.6), but a pH value that is one unit removed from the pK will create an HA/A⁻ distribution (or NH_4^+/NH_3 distribution) that is very different from when pH = pK. To the left of the transition region in Figure 6.1, where Eq.(6.5) is satisfied, then

$$\alpha_0 \approx \frac{[H^+]}{[H^+]} = 1 \qquad \alpha_1 \approx \frac{K_a}{[H^+]} \tag{6.8}$$

$$\frac{d \log \alpha_0}{d\,pH} \approx 0 \qquad \frac{d \log \alpha_1}{d\,pH} \approx +1. \tag{6.9}$$

TABLE 6.1
Comments about Important Characteristics of the Log α Lines in a Monoprotic Acid System

1	The curves for log α_0 and log α_0 cross at $pH = pK_a$ because then $\alpha_0 = \alpha_1 = ½$.
2	$\log ½ = -\log 2 = -0.301$: the two log α lines cross at the (x,y) point on the grid that is (pK_a, −0.301).
3	Each of the log α lines has two nearly straight asymptotic legs, one on each side of a transition region defined by about $pK_a \pm 1$. Each log α line has one leg with slope ≈ 0, and one with log $\alpha \approx \log 1 = 0$.
4	For the log α_0 line, the second leg has a slope ≈ −1; for the log α_1 line the second leg has a slope ≈ +1. *Extrapolation* of the two legs through the transition zone leads to intersection at the point (pK_a, 0).

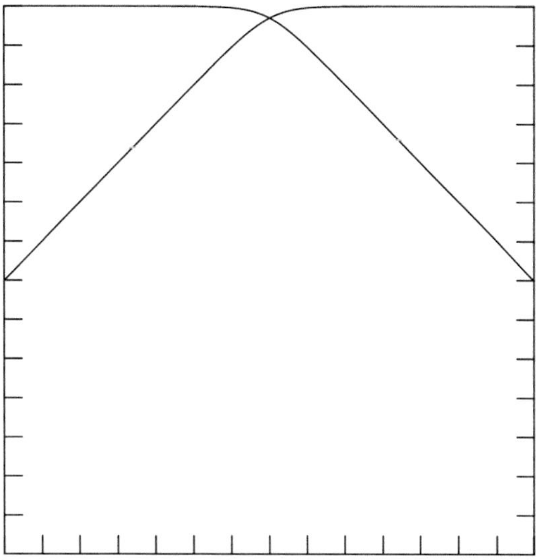

FIGURE 6.2 Generic log α_0 and log α_1 *vs.* pH plot for an HA/A^- system; applicable as a transparency with Figure 6.4, to create a log [ ] *vs.* pH diagram for any pK_a and A_T.

Because of Eq.(6.8), if we are more acidic than $pH = pK_a$ by at least 1 pH unit (Eq.(6.6)), it is very useful to remember that for every pH unit we move further to lower pH below pK_a, the value of α_1 goes down by a factor of 10. For example, if $pK_a = 5$ and $pH = 4$, then $5 - 4 = 1$, and $\alpha_1 \approx 10^{-1}$ (and $\alpha_0 \approx 0.9$); if $pK_a = 5$ and $pH = 3$, then $\alpha_1 \approx 10^{-2}$ (and $\alpha_0 \approx 0.99$), etc. This is because the sloped portion of the log α_1 line extrapolates through the point (pK_a, 0) in Figure 6.1. (See comment 4 in Table 6.1.)

6.2.3 $pH \gg pK_a$

When

$$[H^+] < \frac{1}{10} K_a \tag{6.10}$$

$$pH > pK_a + 1 \tag{6.11}$$

which is commonly abbreviated as

$$pH \gg pK_a. \tag{6.12}$$

To the right of the transition region in Figure 6.1, where Eq.(6.10) is satisfied, then

$$\alpha_0 \approx \frac{[\mathrm{H}^+]}{K_\mathrm{a}} \qquad \alpha_1 \approx \frac{K_\mathrm{a}}{K_\mathrm{a}} = 1 \tag{6.13}$$

For the slopes,

$$\frac{d \log \alpha_0}{d\,\mathrm{pH}} \approx -1 \qquad \frac{d \log \alpha_1}{d\,\mathrm{pH}} \approx 0. \tag{6.14}$$

Because of Eq.(6.13), if we are more basic than $\mathrm{pH} = \mathrm{p}K_\mathrm{a}$ by at least 1 pH unit (Eq.(6.11)), it is very useful to remember for every pH unit we move to further raise the pH above $\mathrm{p}K_\mathrm{a}$ the value of α_0 goes down by a factor of 10. If $\mathrm{p}K_\mathrm{a} = 5$ and $\mathrm{pH} = 6$, then $6 - 5 = 1$, $\alpha_0 \approx 10^{-1}$ (and $\alpha_1 \approx 0.9$); if $\mathrm{p}K_\mathrm{a} = 5$ and $\mathrm{pH} = 7$, then $\alpha_0 \approx 10^{-2}$ (and $\alpha_1 \approx 0.99$), etc. This is because the sloped portion of the log α_0 line extrapolates through the point ($\mathrm{p}K_\mathrm{a}$, 0) in Figure 6.1. (See comment 4 in Table 6.1.)

Example 6.1 Quickly Estimating α Values Based on pH and $\mathrm{p}K_\mathrm{a}$

The temperature and pressure are 25 °C and 1 atm.

a. A solution containing some ammonia has a pH of 8.10. Without any calculations, give estimates of α_0 and α_1. $\mathrm{p}K_\mathrm{a} = 9.24$ for the ammonium ion.
b. A solution containing some acetic acid has a pH of 6.60. Without any calculations, give estimates of α_0 and α_1. $\mathrm{p}K_\mathrm{a} = 4.76$ for acetic acid.

Solution

a. $9.24 - 8.10 = 1.14$. $\alpha_1 \approx 10^{-1.14}$, $\alpha_0 \approx 1 - 10^{-1.14}$
b. $6.60 - 4.76 = 1.34$. $\alpha_0 \approx 10^{-1.84}$, $\alpha_1 \approx 1 - 10^{-1.84}$

6.2.4 Log[ ] *vs.* pH Plots

Because of Eq.(6.2), when $\mathrm{A_T}$ is constant, a log concentration *vs.* pH plot for [HA] and [A^-] looks very much like the log α plot, except that the log[HA] and log[A^-] lines are vertically displaced by the value of log $\mathrm{A_T}$. With $\mathrm{A_T} = 10^{-3}$ *M*, the displacement is −3, and for $K_\mathrm{a} = 10^{-5}$, the result is Figure 6.3a. The horizontal legs of the log α lines lie on log[ ] = −3 (the vertical displacement), and the transition region in which extrapolations of the other the two legs cross is centered on $\mathrm{pH} = \mathrm{p}K_\mathrm{a}$. For the same K_a, but decreasing $\mathrm{A_T}$ by a factor of 10, the displacement is −4 (Figure 6.3b). For that $\mathrm{A_T}$ and decreasing K_a by a factor of 10, the [HA] and [A^-] lines move to the right by one unit (Figure 6.3c). The grid in Figure 6.4 can be used with a transparency of Figure 6.2 to create all of the Figure 6.3 plots. Last, to generate plots like in Figure 6.3, it is easy to enter the equations for [HA], [A^-], [H^+], and [OH^-] in a spreadsheet, and have a computer draw the lines, but there is definitely intellectual value in being able to do it quickly by hand. The sequence of steps for that process is given in Box 6.1.

Example 6.2 Hand-Drawing a log[ ] *vs.* pH Diagram for a Monoprotic Acid HA

Follow the steps in Box 6.1 to hand-draw the log[ ] *vs.* pH diagram for an acid HA for which $\mathrm{p}K_\mathrm{a} = 6.0$ and $\mathrm{A_T} = 10^{-4}$ *M*.

Solution

See the accompanying figure. Dotted points connect the two legs for each of the bold log [ ] lines lines, which cross at log(½) = −0.301 below log A_T.

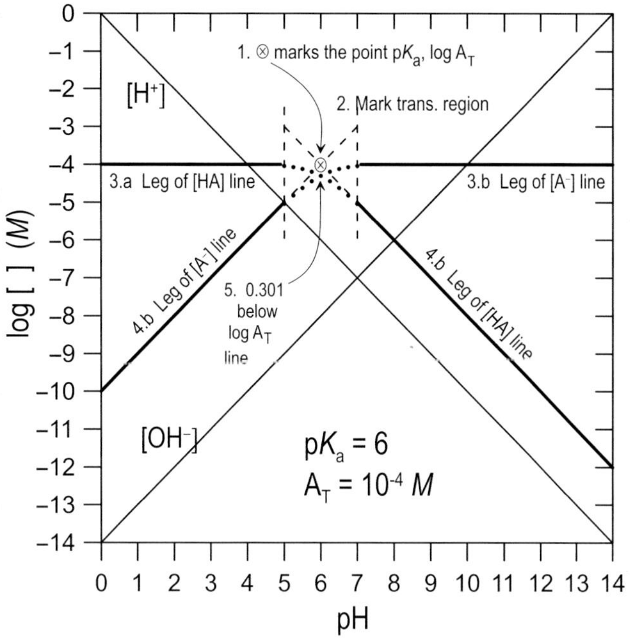

6.3 Log [ ] *VS.* pH PLOTS AND THE RANGE OF CHEMISTRIES OF SOLUTIONS IN A MONOPROTIC ACID SYSTEM

6.3.1 Solution of HA – "The HA Equivalence Point"

Over the range of pH values covered in Figure 6.3a, there is an infinity of different possible solutions, each of which is characterized by $A_T = 10^{-3}$ *M*, $K_a = 10^{-5}$, and $K_w = 1.01 \times 10^{-14}$. At each pH value, we can envision a vertical line on which we can locate the values of the concentrations of the different species.

At each pH, the values satisfy $A_T = 10^{-3}$ *M*, $K_a = 10^{-5}$, and $K_w = 1.01 \times 10^{-14}$. One of these pH values is for the solution of HA with $A_T = 10^{-3}$ *M*, $K_a = 10^{-5}$. A solution of HA is said to be at the HA "equivalence point" (EP), obtainable by starting with a solution of HA, or starting with a solution of NaA and adding an **equivalent** amount of strong acid like HCl. For example, both of the two following solutions are at the HA EP for the acetic acid system: (1) 10^{-3} FW of acetic acid in 1 liter of water; and (2) 10^{-3} FW of sodium acetate plus 10^{-3} FW of HCl in 1 liter of water.

Recall now that the ENE (and PBE) for a solution of HA is

$$[H^+] = [OH^-] + [A^-] \tag{6.15}$$

If the acidic solution approximation (ASA) can be made, we can neglect $[OH^-]$ so that

$$[H^+] \approx [A^-] \qquad \text{ASA} \tag{6.16}$$

This indicates that the process of finding the pH for the solution of HA might proceed by preparing the plot and graphically locating where the log $[H^+]$ line crosses the log $[A^-]$ line. In Figure 6.3a,

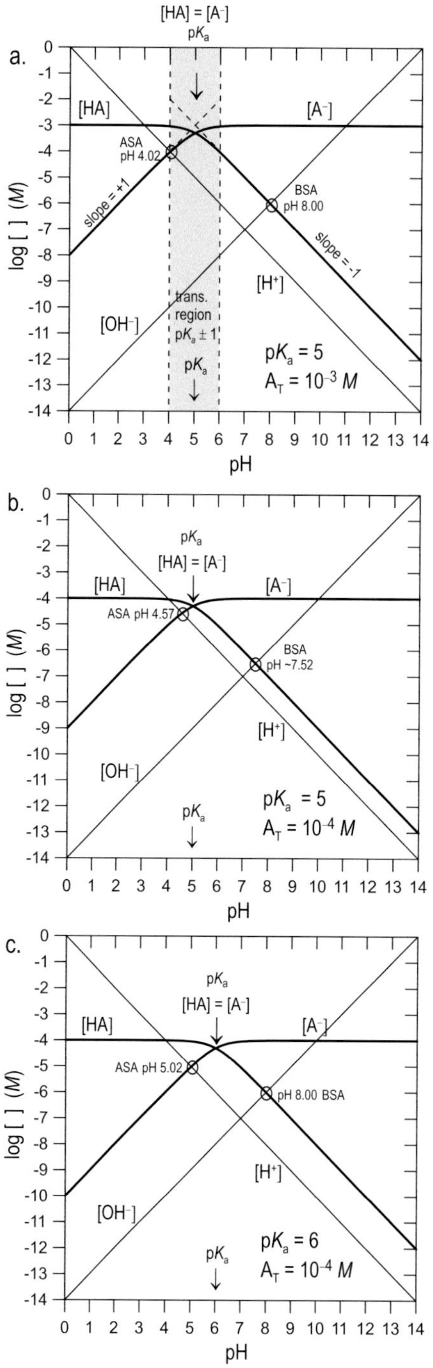

FIGURE 6.3 Log[ ] *vs.* pH diagrams for three different combinations of pK_a and A_T for HA/A^- systems. 25 °C and 1 atm pressure, with $K_w = 1.01 \times 10^{-14}$. The transition region is marked for panel a, but not for panels b and c.

that intersection occurs at pH = 4.02, as corroborated by Figure 5.2. Figure 6.3a, moreover, clearly shows that neglecting $[OH^-]$ in the ENE is an excellent approximation, given how far the log$[OH^-]$ line lies below the log$[H^+]$ = log$[A^-]$ intersection. For each of the cases in Figure 6.3b and 6.3c, at the HA EP, the ASA is again a good approximation, the pH values at the intersections being 4.57 and 5.02 respectively.

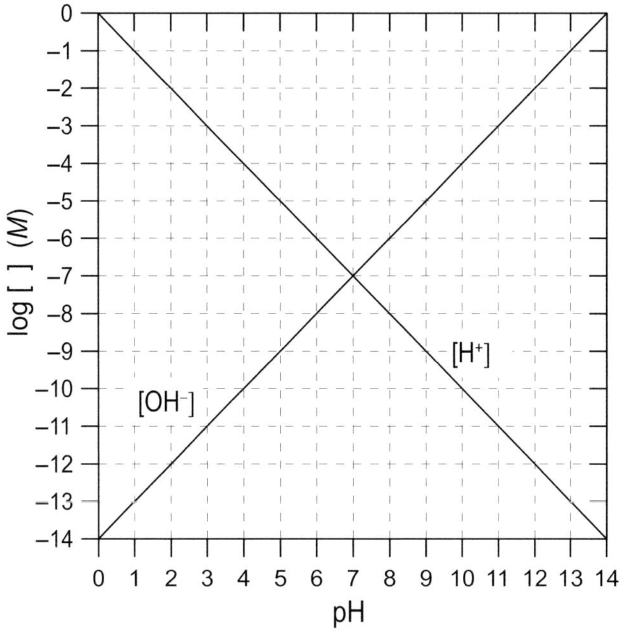

FIGURE 6.4 Generic grid giving log $[H^+]$ and log $[OH^-]$ *vs.* pH. 25 °C and 1 atm pressure, with $K_w = 1.01 \times 10^{-14}$.

> **BOX 6.1 Hand Drawing a Log [ ] *vs.* pH Plot in an HA/A⁻ System**
>
> **Step 1.** On a grid with log $[H^+]$ and log $[OH^-]$ plotted *vs.* pH, locate and mark the (x,y) point given by pK_a and log A_T.
>
> **Step 2.** Note or lightly mark that the "transition region" pH regime is $pK_a \pm 1$.
>
> **Step 3.** Draw two horizontal lines at the value of log A_T: (a) for HA, from low pH almost up to $pK_a - 1$; and (b) for A^- from a bit past $pK_a + 1$ to high pH.
>
> **Step 4.** (a) for HA, draw a line with slope −1 from high pH to lower pH that **passes through** the point marked in Step 2; for $pH < pK_a + 1$, draw it as a dashed line. (b) for A^-, draw a line with slope +1 from low pH to higher pH that **passes through** the point marked in Step 2; for $pH > pK_a - 1$, draw it as a dashed line.
>
> **Step 5.** As the point for where log [HA] and log $[A^-]$ cross, mark the (x,y) point given by $pH = pK_a$ and log [ ] = log A_T −0.301.
>
> **Step 6.** Use the point from Step 6 as a means to smoothly connect the two asymptotic legs for the log [HA] line and similarly for the log $[A^-]$ line.

Graphical means were commonly used to solve acid/base problems up to about the early 1970s, at which point hand calculators and computers became readily available. Historical examples of graphical solution methods in the U.S. are found in the work of Henry Freiser, as in Freiser and Fernando (1963), and Freiser (1967, 1970). While we are not going to get out pieces of graph paper and carefully draw plots like those in Figure 6.3 in order to solve problems, it is certainly very useful to be able to understand acid base chemistry in these terms. For example, at the HA EP, we can see that the weak acid approximation (WAA) is reasonable for the cases in Figures 6.3a and c, but is not very good for the case in Figure 6.3b.

6.3.2 Solution of NaA – "The NaA Equivalence Point"

For the range of pH values covered in Figure 6.3a, one of these is a solution of NaA with $A_T = 10^{-3}$ *M*, $K_a = 10^{-5}$. A solution of NaA is said to be at the NaA "equivalence point", obtainable by starting with a solution of NaA, or starting with a solution of HA and adding an **equivalent** amount of a strong base like NaOH. For example, both of the two following solutions are at the NaA EP for acetic acid system: (1) 10^{-3} FW of sodium acetate in 1 liter of water; and (2) 10^{-3} FW of acetic acid plus 10^{-3} FW of NaOH in 1 liter of water.

The ENE for a solution of NaA is

$$[H^+] + [Na^+] = [OH^-] + [A^-]. \tag{6.17}$$

The two MBEs for such a system give $[Na^+] = A_T = [HA] + [A^-]$ so

$$[H^+] + [HA] = [OH^-] \tag{6.18}$$

which is the PBE for a solution of NaA. If the basic solution approximation (BSA) can be made, we can neglect $[H^+]$ so that

$$[HA] \approx [OH^-] \qquad \text{BSA.} \tag{6.19}$$

This indicates that the process of finding the pH for the solution of NaA can proceed by preparing the plot and graphically locating where the log[HA] line crosses the log$[OH^-]$ line. That intersection is noted in Figure 6.3a as occurring at pH = 8.00, which is indeed corroborated by Figure 5.5. Figure 6.3a, moreover, clearly shows that neglecting $[H^+]$ in the ENE is a very good approximation (though not as good as neglecting $[OH^-]$ for the corresponding solution of HA). For each of the cases in Figures 6.3b and 6.3c, at the NaA EP the BSA is again an adequate approximation, the pH values at the intersections being ~7.5 and 8.0, respectively.

6.3.3 Solutions Other than Simply "Solution of HA" or "Solution of NaA"

For a monoprotic acid system with a given value of A_T, the solution at the HA EP and the solution at the NaA EP are only two points on the pH continuum for the log[ ] *vs.* pH diagram. If we want to access other pH values without changing A_T, we need to use some acid other than HA, or some base other than NaA. For maximum acid effect, we will assume that the acid chosen for lowering the pH is a strong acid, like HCl. HCl has such a large K_a that essentially no HCl species is present in water: H^+ is completely liberated to the solution, and Cl^- is relegated to the role of "spectator ion". For maximum base effect, we will assume that the base chosen for raising the pH is a strong base, like NaOH (KOH would also work). NaOH dissolves completely as ions in water: OH^- is completely liberated to the solution, and Na^+ is relegated to the role of "spectator ion".

We know now that the ENE for any solution of HA and/or NaA plus some HCl and/or NaOH is

$$[H^+] + [Na^+] = [A^-] + [OH^-] + [Cl^-]. \tag{6.20}$$

The quantity $[Na^+]$ is an exact tracer for concentration of strong base (C_B) that has been added to the solution. This is true whether the Na^+ came in as NaOH, or as NaA. (We view the latter as an equivalent combination of HA with NaOH.) Similarly, the value of $[Cl^-]$ is an exact tracer for the concentration of strong acid (C_A) that has been added to the solution. These definitions, plus substituting $[A^-] = \alpha_1 A_T$ give

$$[H^+] + C_B = \alpha_1 A_T + [OH^-] + C_A \tag{6.21}$$

which can, of course, be solved by finding the root value of $[H^+]$ that gives (LHS − RHS) = 0.

If the goal was merely determining the pH of some arbitrary solution, our development of Eq.(6.21) is no different from the method described in Section 5.6, except that here we add the terms for C_B and C_A. Here, however, we are interested not only in "what is the pH?" of some particular solution, but also "how does the pH change?" and "how do we move in a plot like Figure 6.3 as we change C_B and/or C_A?" These questions can be considered with the assistance of Figure 6.5.

6.3.4 $C_B - C_A$: The Units of Net Strong Base Are Equivalents of Charge per Liter

Eq.(6.21) can be rearranged to obtain

$$C_B - C_A = \alpha_1 A_T + [OH^-] - [H^+]. \tag{6.22}$$

The quantity $C_B - C_A$ is called the "net strong base" (NSB) because it is the net excess of strong base over strong acid. Like a bank deposit account, which can have a positive balance, a zero balance, or a negative (overdrawn) balance, NSB can be positive, zero, or negative. The solution at the NaA EP has a positive NSB that is numerically equal to A_T. The solution at the HA EP has a zero NSB. A solution with a pH that is lower than the pH for the solution of HA has a negative NSB, which means it has positive net strong acid (NSA):

$$C_B - C_A = \text{net strong base} \qquad C_A - C_B = \text{net strong acid} \tag{6.23}$$

$$\begin{aligned} C_B - C_A &= -(C_B - C_A) \\ \text{NSB} &= -\text{NSA}. \end{aligned} \tag{6.24}$$

Anticipating material to be discussed in Chapter 9, the definitions in Eqs.(6.23) and Eq.(6.24) provide explanations of how a lake that has some positive NSB, and is fish-friendly with a pH of say

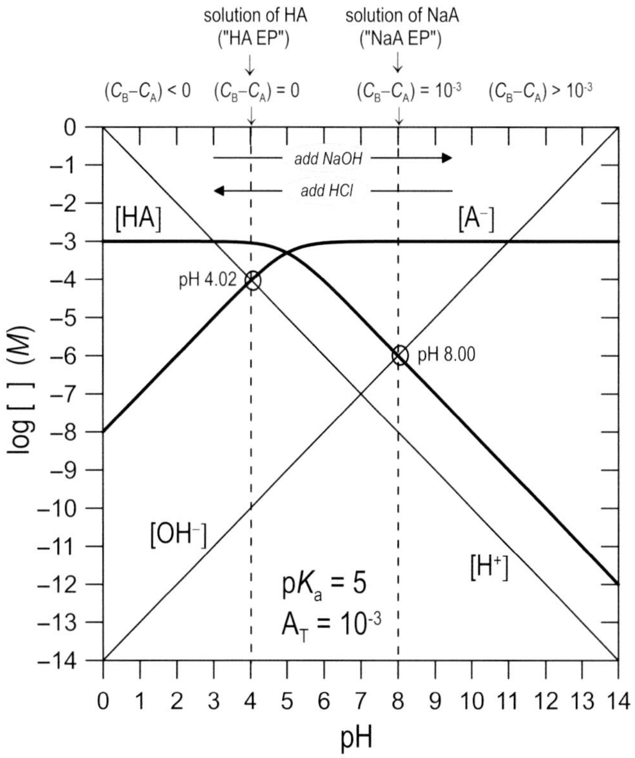

FIGURE 6.5 Annotated log[ ] *vs*. pH plot for an HA/A⁻ system with $pK_a = 5.0$ and $A_T = 10^{-3}$ *M*. 25 °C and 1 atm pressure, with $K_w = 1.01 \times 10^{-14}$.

~8, can be adversely affected by the input of rain with positive NSA, namely "acid rain". If over a number of years, more and more strong acid is added to the lake, less and less NSB will remain in the lake, and the pH may be lowered to not-fish-friendly values like 6 and 5, and worse. There are many lakes in the Adirondack Mountains of New York State, Canada, Scandinavia, and elsewhere, that have such low pH values that all the fish have died. Oh well, that's the price of "progress", right?

The units for $C_B - C_A$ are "equivalents of base per liter", which appear to be inconsistent with the RHS of Eq.(6.22), given its seeming use of units of molarity (mols/L). However, in the ENE (which led to Eq.(6.21)), there are implicit factors in front of each concentration term that involve equivalents of charge. Explicitly written, these are

$$\frac{1\text{ eq. pos. charge}}{\text{mol H}^+}[\text{H}^+] + \frac{1\text{ eq. pos. charge}}{\text{mol Na}^+}[\text{Na}^+]$$
$$= \frac{1\text{ eq. neg. charge}}{\text{mol A}^-}[\text{A}^-] + \frac{1\text{ eq. neg. charge}}{\text{mol OH}^-}[\text{OH}^-] \tag{6.25}$$
$$+ \frac{1\text{ eq. neg. charge}}{\text{mol Cl}^-}[\text{Cl}^-]$$

so that each term has overall units of "equivalents of charge (positive or negative) per liter": the RHS of Eq.(6.22) then gives the "equivalents of net negative charge per liter" as being carried by all the ions that are not spectator ions.

We are getting closer to a resolution of the matter of units, since now for the LHS $= (C_B - C_A)$ for Eq.(6.22) we have units of *equivalents of net strong base per liter*, and for the RHS we have units of *equivalents of net negative charge per liter* carried by non-spectator ions. But what does "non-spectator ion" mean in this context? It means a species that can accept 1 or more protons because it is a base. The units are resolved because we can now view the units of the RHS as being the same as *equivalents of basic species per liter*. The RHS then tells us how the base added to the solution is being carried among the basic species in solution, with H^+ carrying negative strong base.

Consider some groundwater that, upon analysis, reveals the presence of some Na^+, Cl^-, Ca^{2+}, carbonate species from the CO_2 system. The ENE is

$$[\text{H}^+] + [\text{Na}^+] + 2[\text{Ca}^{2+}] = [\text{HCO}_3^-] + 2[\text{CO}_3^{2-}] + [\text{OH}^-] + [\text{Cl}^-] \tag{6.26}$$

which can be rearranged to

$$[\text{Na}^+] + 2[\text{Ca}^{2+}] - [\text{Cl}^-] = [\text{HCO}_3^-] + 2[\text{CO}_3^{2-}] + [\text{OH}^-] - [\text{H}^+]. \tag{6.27}$$

On the RHS, we have all the non-spectator ion basic species. $[CO_3^{2-}]$ takes a factor of 2 because it is doubly charged, but we can also say it takes a factor of 2 because it is dibasic: two protons are required to bring it back to $H_2CO_3^*$. $[H^+]$ again takes a minus sign because it is, well, $[H^+]$. On the LHS, we have $[Na^+] - [Cl^-]$ as before, *and* $2[Ca^{2+}]$. Like Na^+, Ca^{2+} can come into solution basic species like OH^-, or HCO_3^-, or CO_3^{2-}, so it (like Na^+) is a tracer for strong base. Eq.(6.27) can then be viewed as

$$C_B - C_A = [\text{HCO}_3^-] + 2[\text{CO}_3^{2-}] + [\text{OH}^-] - [\text{H}^+] \tag{6.28}$$

with

$$[\text{Na}^+] + 2[\text{Ca}^{2+}] - [\text{Cl}^-] = C_B - C_A. \tag{6.29}$$

If any increment of x FW/L of NaCl was part of the original mix of chemicals for the solution (which may also have included some NaOH, HCl, $NaHCO_3$, $CaCO_3$, etc.), then doing so would have added zero net increment to $C_B - C_A$: the $+1x$ for Na^+ in Eq.(6.27) would be canceled by the $-1x$ for the Cl^-. If an increment of y FW/L of $CaCl_2$ was part of the mix, the $+2y$ for Ca^{2+} would be canceled by $-2y$ for the Cl^-.

The NSB concept is so important in water chemistry that use of the ENE in solving problems is frequently placed in the NSB format (as in Eq.(6.26)) rather than left as the straightforward ENE (as in Eq.(6.25)). From the mathematical process of obtaining the root value of $[H^+]$, there is no difference after picking the format, because the expression that results from (LHS – RHS) is the same for both. Some example cases for computing $C_B - C_A$ are given in Table 6.2.

Example 6.3 Calculating pH and C_B–C_A from an Acid/Base α Value

For the system in Figure 6.3a ($K_a = 10^{-5}$, $A_T = 10^{-3}$), some strong base is added to a solution at the HA equivalence point. The value of α_0 is lowered to 0.3. Assume 25 °C and 1 atm pressure and that activity corrections can be neglected. a. What is the pH? b. How many equivalents of strong base per liter need to be added to accomplish this change?

Solution

a. $\alpha_0 = [H^+]/([H^+] + K_a) = 0.3$. $K_a = 10^{-5}$
   $[H^+] = 0.3\,[H^+] + 0.3\,K_a$. $[H^+] = 0.3\,K_a/0.7 = (3/7)\times10^{-5} = 4.28\times10^{-6}$. pH = 5.37.

b. $C_B - C_A = \alpha_1 A_T + [OH^-] - [H^+] = 0.7\times10^{-3} + 1.01\times10^{-14}/10^{-5.47} - 10^{-5.47} = 7.00\times10^{-4}$.
   For the starting solution, $C_B - C_A = 0$, so the amount of strong base that has to be added to reach pH 5.37 is 7.00×10^{-4} equivalents per liter.

TABLE 6.2
Determination of Value of $C_B - C_A$ for Various Cases Each of Which May Be Solved Using $C_B - C_A = \alpha_1^A A_T + \alpha_1^C C_T + 2\alpha_2^C C_T + K_w/[H^+] - [H^+]$

	Formality Values (*F*)					
	Case 1	Case 2	Case 3	Case 4	Case 5	Case 6
Contributions to the Dissolved Mix						
HA (acetic acid)	1.0×10^{-4}	5.0×10^{-5}	1.0×10^{-3}	5.0×10^{-4}		5.0×10^{-3}
NaA (sodium acetate)	5.0×10^{-5}	5.0×10^{-5}		5.0×10^{-4}		5.0×10^{-4}
NaOH	1.0×10^{-5}			1.0×10^{-4}		1.0×10^{-4}
HCl				6.0×10^{-4}		6.0×10^{-3}
NaCl			1.0×10^{-4}		2.0×10^{-4}	5.0×10^{-4}
$NaHCO_3$						5.0×10^{-4}
$CaCO_3$					2.0×10^{-5}	1.0×10^{-4}
Results						
A_T (*M*)	1.5×10^{-4}	1.0×10^{-4}	1.0×10^{-3}	1.0×10^{-3}	0	5.5×10^{-3}
C_T (*M*) (*total carbonate species*)	0	0	0	0	2.0×10^{-5}	6.0×10^{-4}
$C_B - C_A$ (eq/L)	6.0×10^{-5}	5.0×10^{-5}	0	0	4.0×10^{-5}	-4.7×10^{-3}

Example 6.4 Calculating C_B–C_A Values from pH and A_T

For the system in Figure 6.3a ($K_a = 10^{-5}$, $A_T = 10^{-3}$), consider a solution that is at pH = 10.0. a. What is the initial $C_B - C_A$? b. How much net strong base needs to be added to the solution per liter to *lower* the pH to *3.0*? Assume 25 °C and 1 atm pressure and that activity corrections can be neglected. (Hint: Part b is a "trick question" because adding actual base never lowers the pH of a solution; it must be that the amount of net strong base to be added is *negative, i.e.*, that positive net strong acid needs to be added.)

Solution

a. pH = 10. We can see from the figure that this pH is **higher** than the pH at the NaA EP, so the initial value of $C_B - C_A$ must be greater than 10^{-3} eq/L.
$C_B - C_A = \alpha_1 A_T + [OH^-] - [H^+] = [10^{-5}/(10^{-10} + 10^{-5})] \times 10^{-3} + 1.01 \times 10^{-14}/10^{-10} - 10^{-10}$
$= 1.10 \times 10^{-3}$ eq/L
b. pH = 3.0
$C_B - C_A = \alpha_1 A_T + [OH^-] - [H^+] = [10^{-5}/(10^{-3} + 10^{-5})] \times 10^{-3} + 1.01 \times 10^{-14}/10^{-3} - 10^{-3}$
$= -9.90 \times 10^{-4}$ eq/L
$\Delta(C_B - C_A) = [(-9.90 \times 10^{-4}) - (1.10 \times 10^{-3})]$ eq/L $= -2.09 \times 10^{-3}$ eq/L
So, the amount of net strong base that needs to be added is -2.09×10^{-3} eq/L, which means that positive net strong acid needs to be added, specifically 2.09×10^{-3} eq/L.

REFERENCES

Freiser H, Fernando Q (1963) *Ionic Equilibria in Analytical Chemistry*, John Wiley and Sons, New York. Reprinted by Robert E. Krieger Publishing Company, Malabar, FL.

Freiser H (1967) Simplifying and strengthening the teaching of pH concepts and calculations using log-chart transparencies. *School Science and Mathematics*, **67**, 227–242.

Freiser H (1970) Acid-base reaction parameters. *Journal of Chemical Education*, **47**, 809–811.

7 Titrations of Acids and Bases

7.1 INTRODUCTION

In Chapter 5, we discussed how to determine the ***one*** equilibrium pH and other concentration values in a particular solution of a chemical like HA, NaA, H_2B, NaHB, or Na_2B. Then in Chapter 6, we discussed how, if we add strong acid or strong base, a whole ***range*** of pH values becomes accessible in a solution with a specific value of A_T (or B_T), not just the one pH value for the solution of HA, or NaA, etc. In this chapter, we consider how the incremental addition of a strong base or acid causes incremental changes in pH, focusing on monoprotic (A_T) systems. Chapter 9 will consider diprotic systems, in the context of CO_2 chemistry.

Many readers will recognize the concept of "incremental addition of strong base or acid" as relating to *titration*. In an **alkalimetric** titration, successive increments of strong base are added to a solution while recording the resulting successively higher equilibrium pH values. In an **acidimetric** titration, successive increments of strong acid are added to a solution while recording the resulting successively lower equilibrium pH values. Alkalimetric titration and acidimetric titration are negative analogs of one another, just like going west is a negative analog of going east, and *going west negatively* amounts to going east. So, if we add some strong base (*e.g.*, NaOH), we can negate that by adding the equivalent amount of strong acid (*e.g.*, HCl): we end up where we started, plus some dissolved salt (*e.g.*, NaCl).

A titration of a given volume of a sample solution is usually carried out by adding *small* volume increments of a *concentrated* (*e.g.*, 0.1 *N*) strong base or acid solution (the titrant). For alkalimetric and acidimetric titrations, the **titration curve** is usually plotted as pH *vs.* amount (*e.g.*, **volume**) of strong base or strong acid added. Here, however, since we want to discuss things in a general context, for the *x*-axis rather than volume, we will mostly use the value of $C_B - C_A$ that is produced in the solution (or the related parameter *f*, defined below). If the concentration in the titrant is sufficiently large relative to the concentration of what is being titrated (say 20×), then near-constancy in the value of A_T (or B_T, etc.) in the solution can be assumed over the course of the titration: the volume of titrant added does not significantly increase the volume of solution being titrated so that A_T (or B_T, etc.) is not significantly lowered during the titration. This assumption simplifies computing titration curves.

While the mathematics of titrations might be viewed as, well, tedious and confusing, it is fundamentally important for understanding how aqueous solutions behave in response to changes in acid or base content. First, most of the salt in the oceans is the result of the grand back-and-forth chemical neutralization that has occurred over geologic time of metal oxide bases from terrestrial solids reacting with added HCl from volcanism and vice versa (Schilling et al., 1978). Second, natural waters undergo titration changes whenever there are spills of strong acid or base, and when acid rain falls on a lake watershed system. Third, laboratory titrations are routinely used in analytical determinations of "alkalinity" in samples of: (1) natural water; and (2) water flowing through waste and drinking water treatment plants, to provide input information for calculations regarding pH control. Last, if a given sample is titrated with a strong base or strong acid, the local slope of a titration curve reveals the ability of that solution to resist changes in pH: at each pH, the **shallower the slope** of the curve for pH *vs.* the amount of titrant added, the **greater the ability of the solution to resist pH changes, at that pH** (*i.e.*, the greater the capacity of the solution to absorb acid or base, with only a small change in pH, at that pH). So, the entire concept of **buffer solutions** comes directly from titration curve shape. Overall, for most water chemists nowadays, there is usually more need to understand why titration curves look the way they do than to have to actually do them in the lab or compute them in spreadsheets.

7.2 TITRATIONS IN A MONOPROTIC ACID SYSTEM

7.2.1 General Considerations and Two Instructional Limiting Cases

For a solution of HA having some value of A_T, the ENE during the titration can be written as

$$C_B - C_A = [A^-] + [OH^-] - [H^+] \tag{7.1}$$

$$= \alpha_1 A_T + [OH^-] - [H^+]. \tag{7.2}$$

We begin by considering two limiting cases:

Case 1. HA is an exceedingly strong acid (K_a is exceedingly large) so we can take $\alpha_1 = 1$ throughout the titration: in a solution of HA, the HA is completely dissociated; adding more H^+ to the solution using strong acid cannot drive H^+ onto A^- and decrease α_1; and, adding strong base to the solution does not increase α_1.

Case 2. HA is an exceedingly weak acid (K_a is exceedingly small) so that we can take $\alpha_1 = 0$ throughout the titration: in a solution of HA, none of the HA has dissociated; adding more H^+ to the solution using strong acid cannot increase the amount of HA (already $\alpha_1 = 1$), so α_1 remains 0; and, if we add strong base to the solution to attain high $[OH^-]$, we cannot deprotonate HA.

$$C_B - C_A = A_T + [OH^-] - [H^+] \quad \text{Case 1. exceedingly strong acid (ESA)} \tag{7.3}$$

$$C_B - C_A = 0 + [OH^-] - [H^+] \quad \text{Case 2. exceedingly weak acid (EWA).} \tag{7.4}$$

If one is starting with a solution of HA, the titration direction of most interest is the alkalimetric direction, in which case C_A remains zero because a strong base is being added. Nevertheless, we retain C_A and the general difference term $C_B - C_A$. In this way, later we can consider how, in a given titration curve we can go not only in an alkalimetric direction towards higher pH by adding strong base with rising C_B (*e.g.*, a strong base is spilled in a river), but also in an acidimetric direction towards lower pH by adding strong acid with rising C_A (*e.g.*, strong acid is spilled in a river).

Recall now that from Chapter 5, neglecting activity corrections (or substituting ${}^cK$ values to include them), that

$$HA = H^+ + A^- \qquad K_a = \frac{[H^+][A^-]}{[HA]} \tag{7.5}$$

$$A^- + H_2O = HA + OH^- \qquad K_b = \frac{[HA][OH^-]}{[A^-]} \tag{7.6}$$

$$K_a \times K_b = K_W \qquad pK_a + pK_b = pK_W. \tag{7.7}$$

As the acid dissociation constant for HA, K_a is the measure of the strength of HA as an acid. K_b is the basicity constant for A^-, and is the measure of the strength of A^- as a base.

If HA is an exceedingly strong acid (ESA, very large K_a, let's say 10^4 as an example, so $pK_a = -4$), then A^- is an exceedingly weak base (very small K_b). On the other hand, if HA is an exceedingly weak acid (EWA, very small K_a, let's say 10^{-18} as an example so $pK_a = 18$), then A^- is an exceedingly strong base (very large K_b).

To calculate the titration curves for Cases 1 and 2, set up a spreadsheet set up with rows spanning a wide range of pH (*e.g.*, 0.00, 0.05, 0.10, … 13.90, 13.95, 14.00). Then, after specifying A_T, $C_B - C_A$ can be calculated for each row using Eq.(7.3) for Case 1, and Eq.(7.4) for Case 2. One can then pick the rows of interest, and plot pH *vs.* $C_B - C_A$. Figure 7.1 provides a plot of the curves for the two cases, with $A_T = 10^{-3}$ *M*. Some notes about considering dilution in a specific titration are given in Box 7.1.

For Case 1 (HA = ESA), at point *A*, with no strong base added, pH = 3 because the acid is essentially 100% dissociated, so $[H^+] = [A^-] = A_T = 10^{-3}$ *M*. At point *B* (equivalent to a solution of NaA), because A^- has essentially no basicity (exceedingly weak base), again $[A^-] = A_T$ because [HA] is essentially zero, and pH = 7 (the only source of OH^- is dissociation of water). Viewed alternatively, if we had started at point A and gone to point *B*, the added OH^- exactly has neutralized the initial H^+, so pH = 7.

For Case 2 (HA = EWA), at point *R*, with no strong base added, pH = 7 (because the acid is essentially 0% dissociated). If we move to point *S* by adding strong base, essentially none of the HA releases protons to neutralize the added strong base, so $[OH^-] = 10^{-3}$ *M* and pH = 11. Viewed alternatively, point *S* is equivalent to a solution of NaA where A^- is an exceedingly strong base (because HA = ESA; see Eq.7.7). So, if we created the solution at point *S* by dissolving 10^{-3} FW of NaA in one liter of water, essentially all of the A^- will have been converted to HA so that $[HA] = A_T$ and $[OH^-] = A_T = 10^{-3}$ *M*.

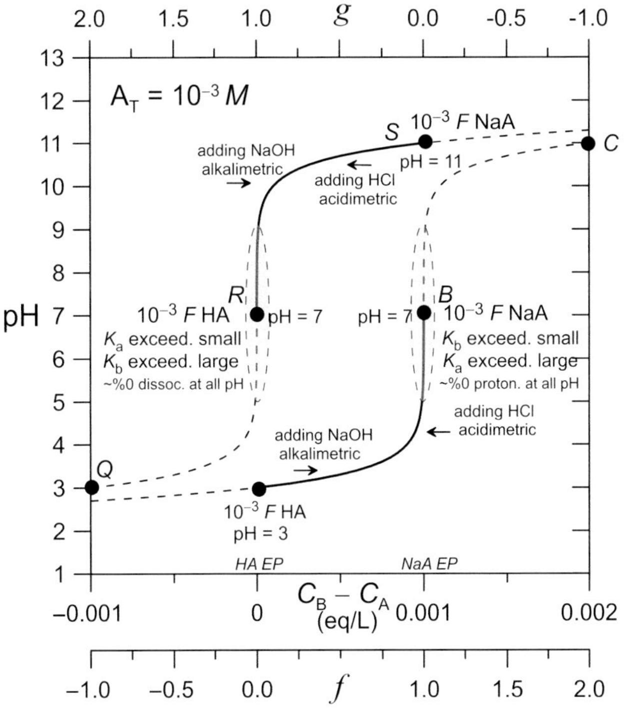

FIGURE 7.1 Titration curves calculated for an HA system with $A_T = 10^{-3}$ *M* when HA is an exceeding strong acid, and when HA is an exceedingly weak acid. For the exceedingly strong acid, at $f = 0$, pH = 3.0 (point *A*); and at $f = 1$, pH = 7.0). For the exceedingly weak acid, at $f = 1$, pH = 7.0 (point *R*); and at $f = 1$, pH = 11.0). The two shaded ellipses mark the abrupt pH transition jumps for the two titration curves.

BOX 7.1 Considering Dilution during an Actual Titration

When actually carrying out a titration, depending on the circumstances, A_T may be reduced somewhat over the course of the titration because titrant is being added. This dilution effect is frequently assumed to be of negligible effect. If the effect is not negligible, let V_o (L) = sample volume of solution selected for titration, $A_{T,o}$ (M) = initial value of A_T, and V_t (L) = volume of titrant that has been added (so far). Because the number of mols of total A does not change during the titration, if A_T is the concentration in the solution during the titration, then

$$A_{T,o} V_o = A_T(V_o + V_t)$$

$$A_T = A_{T,o}\frac{V_o}{V_o + V_t}.$$

Therefore, multiplying Eq.(7.1) by ($V_o + V_t$) and dividing by $A_{T,o}V_o$, which is the total equivalent of chemical under titration, gives

$$\frac{(C_B - C_A)(V_o + V_t)}{A_{T,o} V_o} \equiv f = \frac{[A^-]}{A_T} + \frac{[OH^-] - [H^+]}{\dfrac{A_{T,o} V_o}{V_o + V_t}} \qquad f = \alpha_1 + \frac{[OH^-] - [H^+]}{A_{T,o}\left(\dfrac{V_o}{V_o + V_t}\right)}.$$

7.2.2 *f* AND *g* – THE MATH AND THE MEANINGS

When on a trip, it is natural to refer to how far we have gotten by expressing the progress as a percentage, or fraction, of the goal. For example, if on a titration "trip" starting with a solution of the acid HA with $A_T = 10^{-3}\,M$, and we have "progressed" as far as adding $C_B - C_A = 0.4 \times 10^{-3}$ eq/L, then we are 0.4 of the way to reaching the NaA equivalence point (EP). This fraction is called f:

$$f \equiv \frac{C_B - C_A}{A_T} = \frac{\alpha_1 A_T + [OH^-] - [H^+]}{A_T} \tag{7.8}$$

$$= \alpha_1 + \frac{[OH^-] - [H^+]}{A_T} \tag{7.9}$$

$f = 0$ and $f = 1$ are of special importance. When $f = 0$, $C_B - C_A = 0$, and we are at the HA EP. When $f = 1$, $C_B - C_A = A_T$, and we are at the NaA EP. The fraction g is defined in Box 7.2.

While $f = 0$ and $f = 1$ are important mileposts, f is not limited to that range. We can make $C_B - C_A > A_T$ so that $f > 1$. And, $C_B - C_A$ can be negative, so it is possible that $f < 0$. In Figure 7.1, for Case 1 and Case 2, these other portions of the titration curves are shown with dashed lines. The segment A to B to C is the alkalimetric titration curve for a solution of a strong acid starting at $f = 0$ and going to $f = 1$ (NaA EP), then past the NaA EP to $f = 2$. The segment S to R to Q is the acidimetric titration curve for a solution of a strong base starting at $g = 0$ through $g = 1$ (the HA EP), then past the HA EP to $g = 2$.

In a lab class, the EP at point B in the alkalimetric titration curve for an ESA might have been located by the reader by watching for the sharp rise in pH at the EP either on a pH meter, or by having added a small amount of an indicator dye like phenolphthalein which changes from colorless to pink as the pH jumps from low values to values > 8.2. The EP at point R in an acidimetric titration curve might have been located by watching for the sharp drop in pH at the EP on a pH meter, or by having added a small amount of an indicator dye like methyl red which changes from yellow to red as the pH drops sharply from values > 6.2 to values < 4.4.

For an HA/A$^-$ system, both the HA and NaA EPs are mileposts that are equally useful for knowing position. If we know $f = 0.4$ so that we are 40% of the way from the HA EP towards the NaA EP,

BOX 7.2 *g*: Viewing Things from the NaA Equivalence Point as in an Acidimetric Titration

We can define another fraction, g, as being the fraction of the distance from the NaEP towards the HA EP. Since $f + g = 1$,

$$g = 1 - \left[\alpha_1 + \frac{[\mathrm{OH}^-] - [\mathrm{H}^+]}{\mathrm{A_T}}\right]$$

Since $\alpha_0 + \alpha_1 = 1$,

$$g = \alpha_0 + \frac{[\mathrm{H}^+] - [\mathrm{OH}^-]}{\mathrm{A_T}}$$

which is a symmetrical analog of Eq.(7.9) for f. If dilution during the titration needs to be considered, substitute the expression for $\mathrm{A_T}$ as given in Box 7.1.

we also know that we are 60% of the distance from the NaA towards the HA EP. (On a west–east, border-to-border transect of a country, if we are 40% of the way to the east of the western border, we know that we are 60% of the way going west from the eastern border, and of course this is regardless of at which side of the country we may have started.)

As noted, in a titration of a monoprotic acid, f gives the **ratio of the eq/L of net strong base added, relative to $\mathrm{A_T}$**. It is tempting to express this simply as "f is the fraction of the acid that has been titrated". That is ok, but dangerous, because it will lead to conceptual errors if it is confused with the idea that f equals the fraction of the HA that exists as A^-, which is α_1. By Eq.(7.9), f does not exactly equal α_1, though there certainly are non-ESA and non-EWA cases **when for $0 < f < 1$** that $f \approx \alpha_1$ because the value $([\mathrm{OH}^-]-[\mathrm{H}^+])/\mathrm{A_T}$ is small. But, for the ESA titration curve in Figure 7.1, $\alpha_1 \approx 1$ everywhere along the curve so $f \neq \alpha_1$. And, for the EWA titration case $\alpha_1 \approx 0$ everywhere along the curve so again $f \neq \alpha_1$.

Example 7.1 Calculating a Titration Curve

Calculate the detailed titration curve for f = 0 to 2, as pH *vs.* f, with HA = acetic acid ($\mathrm{p}K_\mathrm{a} = 4.76$), $V_\mathrm{o} = 0.025$ L, $A_{\mathrm{T,o}} = 10^{-3}$ M, and using basic titrant concentration $c_{\mathrm{t,b}} = 0.1$ N NaOH. Neglect dilution effects caused by volume changes due to the addition of the titrant volume v_t (*i.e.*, assume $\mathrm{A_T} = \mathrm{A_{T,o}}$ during the titration). Neglect activity corrections. Assume 25 °C. Include three versions of the *x*-axis: v_t (μL), $C_\mathrm{B} - C_\mathrm{A}$ (eq/L), and f.

Solution

From Eq.(7.9),

$$f = \alpha_1 + \frac{[\mathrm{OH}^-] - [\mathrm{H}^+]}{\mathrm{A_{T,o}}}$$

For $f = 0$, this gives pH = 3.91. Then, from Box 7.3, neglecting dilution,

$$\frac{c_{\mathrm{t,b}} \dfrac{v_\mathrm{t}}{10^6}}{\mathrm{A_{T,o}} V_\mathrm{o}} = \alpha_1 + \frac{[\mathrm{OH}^-] - [\mathrm{H}^+]}{\mathrm{A_{T,o}}}$$

$$c_{\mathrm{t,b}} \frac{v_\mathrm{t}}{10^6} = \alpha_1 \mathrm{A_{T,o}} V_\mathrm{o} + V_\mathrm{o}([\mathrm{OH}^-] - [\mathrm{H}^+])$$

$$v_t = 10^6 \frac{\alpha_1 A_{T,o} V_o + V_o([OH^-] - ([H^+])}{c_{t,b}}$$

$$= 10^6 \frac{[K_a/([H^+] + K_a)]A_{T,o} V_o + V_o(K_w/[H^+] - [H^+])}{c_{t,b}}.$$

For known values of K_a, $A_{T,o}$, V_o, K_w, and $c_{t,b}$, then v_t can be calculated for a range of input pH values (pH ≥ 13 is not accessible because we cannot make the solution more basic than the titrant). We then select the results from the spreadsheet for $v_t \geq 0$. With values for $v_t \geq 0$ in hand as a function of pH, the corresponding values of $C_B - C_A = c_{t,b} v_t/(10^6 V_o)$ are calculated, along with $f = c_{t,b} v_t/(10^6 A_{T,o} V_o)$.

Example 7.1 Values for selected points

pH	v_t (μL)	f	$C_B - C_A$
3.91	0	0.0	0
4.0	12.0	0.0480	4.80E-05
4.2	38.2	0.1528	1.53E-04
4.4	66.0	0.26	2.63E-04
4.6	95.9	0.38	3.82E-04
4.8	127	0.51	5.05E-04
5.0	156	0.62	6.21E-04
5.2	182	0.73	7.22E-04
5.4	202	0.81	8.03E-04
5.6	218	0.87	8.64E-04
5.8	229	0.91	9.07E-04
6.0	236	0.94	9.36E-04
7.0	249	0.99	9.84E-04
8.0	250	1.00	9.91E-04
9.0	253	1.01	1.00E-03
10.0	276	1.10	1.09E-03
11.0	508	2.03	1.97E-03

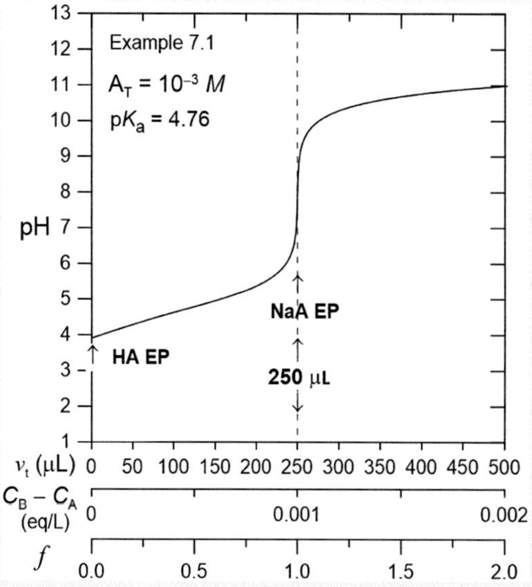

BOX 7.3 $C_B - C_A$ (and f or g) in Relation to the Volume of Titrant Added v_t (μL)

When physically doing a titration, there will be values for the following:

V_o (L) = initial volume of solution selected for titration;
$A_{T,o}$ (M) = initial value of A_T;
v_t (μL) = volume of titrant that has been added (so far);
$c_{t,b}$ (eq/L) = concentration of strong base in titrant solution (value is negative if the titrant is a strong acid solution).

Using the factor of 10^6 to account for the fact that v_t has units of μL,

$$C_B - C_A = \frac{c_{t,b}\dfrac{v_t}{10^6}}{V_o + \dfrac{v_t}{10^6}}.$$

The $v_t/10^6$ term in the denominator accounts for the extent that the volume increases, causing A_T to drop during the titration. (For the titration curves given in Figure 7.1, it was assumed that A_T did not change significantly.) Considering the possibility of dilution, from Box 7.1, during the titration

$$A_T = \frac{A_{T,o}\,V_o}{V_o + \dfrac{v_t}{10^6}}.$$

<u>If we are starting the titration at $f=0$</u>: then the number equivalents of strong base added by the addition of volume v_t equals $c_{t,b}v_t/10^6$. The number of equivalents of HA under titration is given by $A_{T,o}V_o$. So,

$$f = \frac{c_{t,b}\dfrac{v_t}{10^6}}{A_{T,o}\,V_o} \quad ✓$$

which can also obtained as usual by $f \equiv (C_B - C_A)/A_T$. Similarly, during the titration,

$$C_B - C_A = \frac{A_T c_{t,b}\dfrac{v_t}{10^6}}{A_{T,o}\,V_o} = \frac{c_{t,b}\dfrac{v_t}{10^6}}{V_o + \dfrac{v_t}{10^6}}.$$

7.2.3 Titrations with a Range of pK_a Values, and Inflection Points (IPs)

In a monoprotic acid system, there can be three (inflection point) IPs, or one. These can be at a "low" (L) pH that occurs very close to $f=0$, a "medium" pH (M) that occurs close to $f=0.5$, and a "high" pH (H) that occurs very close to $f=1$. In general, there is not an *ultra-exact* equivalency between the locations of the IPs and the associated f titration landmarks. Nevertheless, the positions of the IPs are usually so close to those landmarks that essentially no error is introduced by assuming equivalency in location: *e.g.*, if you can find the IP_H for a titration curve for a solution of HA, you have found $f=1$; and, going the other way, if you can find the IP_L for a titration curve for a solution of NaA, you have found $f=0$. Chapter 8 will discuss the connection between the presence and pH locations of IPs and the concept of buffers.

Figure 7.2 provides pH *vs.* f curves for systems with $A_T = 10^{-3}\,M$ for a range of pK_a values. The curve for p$K_a = 3$ with its single strong IP (an IP_H) lies slightly above the curve for the ESA found in

Figure 7.1: for example, the pH at $f=0$ for p$K_a=3$ is slightly higher than the pH at $f=0$ for the ESA case because, while $K_a=10^{-3}$ makes for a decent acid, is not enormous. Continuing the trend, the curve for p$K_a=5$ lies above the curve for p$K_a=3$, and so on. At $f=0$ and ~0.5, a slight IP_L and a slight IP_M are visible in the curve for p$K_a=5$, but not for p$K_a=3$. With p$K_a=7$ etc., the IPs at $f=0$ and ~0.5 are more pronounced. With p$K_a=9$, the IP at $f=0$ is even more pronounced, but now the IPs at $f=$~0.5 and 1 are slight (*cf.* the curve for p$K_a=5$ at $f=$~0.5 and 0). Finally, for p$K_a=11$, the curve lies slightly below the EWA curve in Figure 7.1, and the single strong IP at $f=0$ (IP_L) mirrors the IP at $f=1$ (IP_H) for p$K_a=3$.

Figure 7.3 provides an enlarged view of the region $f=0$ to 0.5 from Figure 7.2. As we know, at $f=0$, as pK_a rises, the solution must become less acidic (pH rises). At $f=0.1$, there is the same trend, but we also note that as pK_a increases, the pH *increase* when going from $f=0$ to $f=0.1$ increases. The p$K_a=3$ "army" is very willing to release its H^+ into solution even before any battle with added strong base starts at f = 0, and so it gives the highest $[H^+]$ (lowest pH) at $f=0$. The p$K_a=5$ HA army is less willing and gives a lower $[H^+]$ (higher pH) at $f=0$. And, regarding the increase in pH, at $f=0.1$ the alkalimetric invasion of an OH^- army has gotten to the point that 0.1 OH^- has been added for every one HA, the factor by which $[H^+]$ has gone down when the p$K_a=5$ HA army is fighting is greater than when the p$K_a=3$ HA army is fighting: even though H^+ is being depleted in the solution, the p$K_a=5$ HA army is not as willing to fight (replace protons on the battlefield; the pK_a value is the measure of how high the pH has to get for the acid to really start to let go of protons). The p$K_a=7$ HA army is even less enthusiastic about fighting than the p$K_a=5$ HA army, and so on for p$K_a=9$ and 11. Figure 7.1 shows what an EWA (pK_a >> 11) will do ... which is essentially not fight at all: for $f=0$, pH=7.0; and if $A_T=10^{-3}$ *M*, then at $f=0.1$, we have $C_B-C_A=10^{-4}=[OH^-]$ giving pH=10: the EWA does essentially nothing against an OH^- invasion.

As noted, there is no IP_L in the titration curve for p$K_a=3$, but there is in the titration curve for p$K_a=5$. To understand why an IP appears at $f=0$ as pK_a increases, consider Figure 7.4 which provides α_1 (and α_0) *vs.* pH curves on a linear scale for p$K_a=3$, 5, and 7. Moving from low pH to higher pH, for p$K_a=5$, the value of α_1 does not start to rise appreciably above zero until we are within about 1 unit of p$K_a=5$, namely as pH approaches 4. This means that upon adding strong base at $f=0$ with p$K_a=5$, the HA is not going to start fighting in a significant way by releasing H^+ until pH starts to

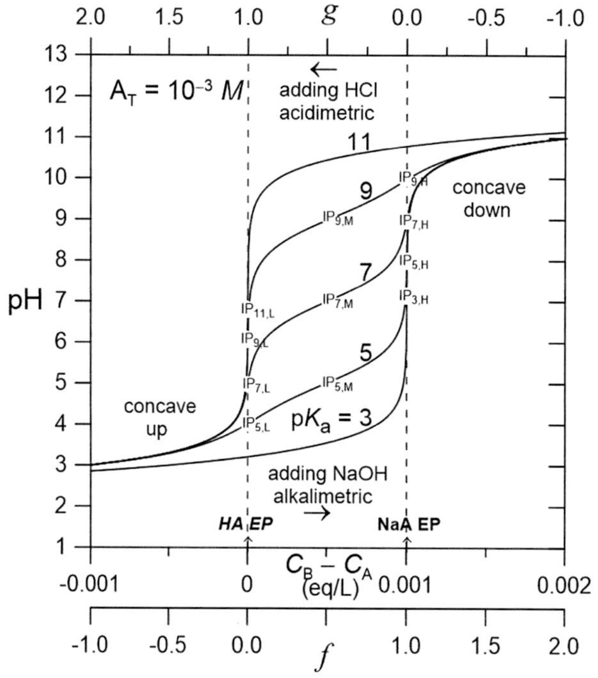

FIGURE 7.2 Titration curves calculated for HA systems with $A_T=10^{-3}$ *M* for a range of pK_a values. All inflection point (IPs) are indicated.

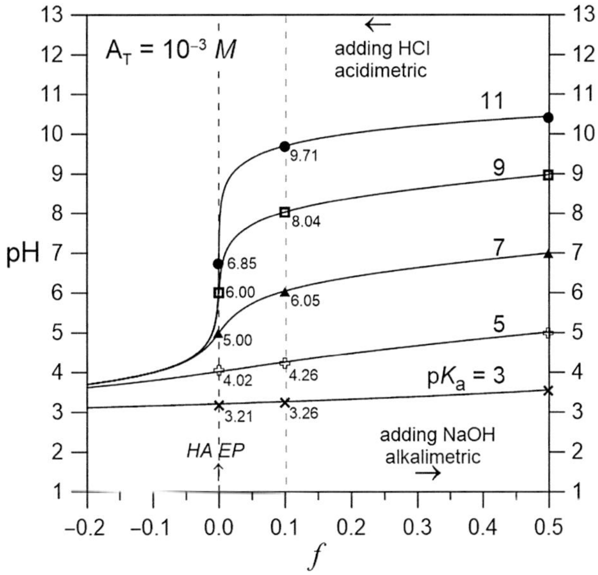

FIGURE 7.3 Titration curves calculated for HA systems with $A_T = 10^{-3}\,M$ for a range of pK_a values, focusing on the behavior in the region from $f=0$ to $f=0.5$.

get near 4 (one unit less the pK_a=5). So, [OH⁻] and pH rise until the acid engages ("oh crap, I have to start fighting"). There is no need to "convince" a strong acid to engage, it throws all its protons on the battlefield, and then it is just a matter of the pH rising as they are used up. But, as the pH rises alkalimetrically with a more reluctant acid, when the acid does start to engage, the rate of the rise in pH temporarily slows. This is the cause of the IP that is slight at $f=0$ for pK_a=5, very visible for pK_a=7 and 9. For pK_a=11, somewhat past $f=0$, the alkalimetric rise in pH is not being slowed because the acid is engaging, but rather because [OH^-] is jumping so high so fast that very soon it is not easily made much larger by further addition of OH^- (if an invading army comes in, once the

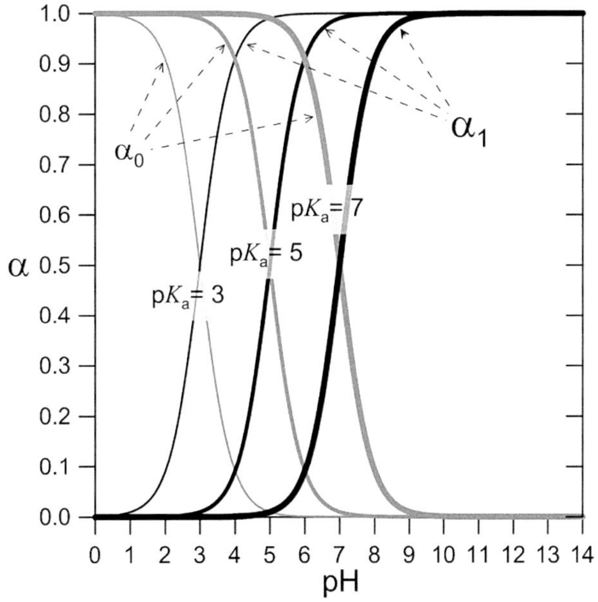

FIGURE 7.4 α_0 and α_1 *vs.* pH curves with a linear y-axis for HA systems with pK_a=3, 5 and 7.

invading troops have reached a high concentration, it is hard to raise the concentration by another factor of 2. (Note: log 2 = 0.301, so at high [OH⁻], raising pH by 0.301 becomes difficult.)

All monoprotic system titration curves are concave up for $f<0$ and concave down for $f>1$. This is because pH is a log scale, and the titration x-axis is a linear scale. If $f<0$, there is net strong acid in the solution. So, as we are linearly moving towards to the left of $f=0$, we are linearly adding strong acid carrying free H^+, but pH (because it is on a log scale) drops less rapidly than linearly, and the curve must be concave up. And, on the other side, as we linearly add strong base carrying OH^- while moving alkalimetrically to the right of $f=1$, pH rises less rapidly than linearly, and the curve must be concave down. To go from concave down to concave up requires an odd number of IPs. Both of the titration curves in Figure 7.1 have one IP, as do curves for $pK_a=3$ and $pK_a=11$ in Figure 7.2. But for $pK_a=5$, 7, 9 in Figure 7.2, there are three IPs. For $pK_a=5$ there are the slight ones discussed above at $f=0$ and ~0.5, and a very noticeable one at $f=1$.

So, what causes an IP moving alkalimetrically through $f=0.5$? Continuing the battle analogy, the answer is a tipping point between the HA being more and more willing to fight (α_1 rising) and give protons as pH rises, and a running out of ammunition (the level of undissociated HA) as evidenced by α_0 falling significantly towards 0. As α_0 falls to close to 0, the HA army has been defeated and the pH starts to rise more and more rapidly. But since the titration must eventually become concave down again, the titration curve goes through a third IP near at (essentially) $f=1$. That third IP has the same cause as the single IP at $f=0$ (essentially) in the $pK_a=11$ curve and in the EWA curve in Figure 7.1: continuing significant rise in $[OH^-]$ is hard to maintain once $[OH^-]$ becomes large.

For an acidimetric titration starting at $g=0$ ($f=1$), the above story can be retold as a description of armies of type A^- with different degrees of willingness to fight an invading army of strong acid H^+. For $pK_b=3$ ($pK_a=11$), there is good enthusiasm for a fight: much of the A^- reacts with water at $g=0$ to make OH^- to react with the incoming H^+ even before the H^+ arrives: the invading strong acid is thereby resisted until we run out of ammunition, and there is one IP. For $pK_b=5$ ($pK_a=9$), there is somewhat less enthusiasm for a fight; there is an IP at (essentially) $g=1$, along with two more IPs further on, acidimetrically. For a base with little enthusiasm to fight, *e.g.*, $pK_b=11$ ($pK_a=3$), the invading acid H^+ immediately overruns the situation, the pH drops immediately, and there is only one IP.

Example 7.2 Hand Drawing Some Titration Curves

Without looking at Figure 7.2, reproduce it using a pencil by drawing on the accompanying plot.

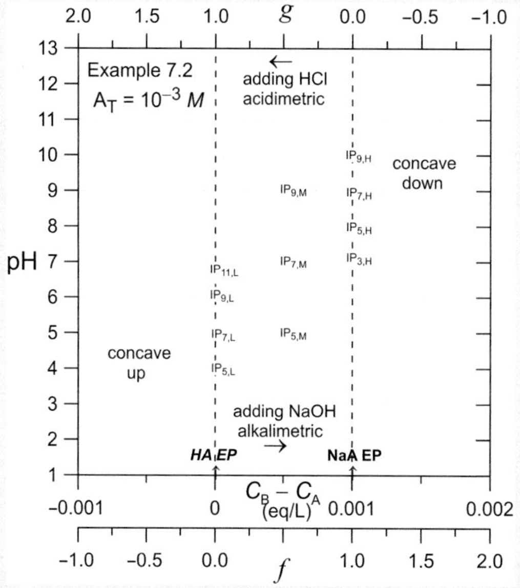

7.3 USING THE 1ST DERIVATIVE TO FIND TITRATION CURVE EQUIVALENCE POINTS (EPS)

In Figures 7.1 through 7.3, because we know the exact nature of each of the solutions, we know where the EPs are to be found on the *x*-axis for $(C_B - C_A)$. But for titrations of unknown solutions, accurately finding an EP by visually identifying the associated IP (if it exists) can be difficult in some titration curves, as when the concentration of the species under titration is low, and/or the pK_a values are such that a sought for IP is not prominent. An example would be the slight IP at $f=1$ for p$K_a=9$ in Figure 7.1, and the curve for the unknown solution in Example 7.3. Happily, however, mathematical methods exist for processing titration data to help locate EPs.

For a curve *y vs. x*, by elementary calculus, if there is an IP at some *x*, then at that *x* there is a maximum or minimum in the first derivative dy/dx. In titration curves plotted with increasing $(C_B - C_A)$ to the right (alkalimetrically), IPs at EPs occur when the titration goes from concave up to concave down; this change gives a maximum in the first derivative $d\text{pH}/d(C_B - C_A)$, the location of which then locates an alkalimetric EP (usually highly exactly, see the discussion above regarding EP location ≈ IP location). Acidimetrically, EPs are located at minima in the first derivative $d\text{pH}/d(C_A - C_B)$. Figure 7.5 plots the log of the first derivative for the curves in Figure 7.2. A log scale would not generally be selected for processing experimental data in the region of an IP, but is used here to allow all of the curves to be plotted on the same vertical scale. The strength of each IP in Figure 7.2 is directly related to the magnitude of its associated $d\text{pH}/d(C_A - C_B)$ peak in Figure 7.5. With experimental data, usually titration curves are processed as pH *vs.* volume of titrant v_t, so the version of the first derivative used is $d\text{pH}/dv_t$.

Example 7.3 Using the First Derivative to Find an Equivalence Point (EP)

Here is a plot for experimental data obtained when titrating 25 mL of a solution of some acid HA using 0.1 *N* NaOH as the titrant ($c_{t,b}=0.1$ *N*). It is known that the titration started at $f=0$, but neither A_T or K_a are known. An IP can be seen to be present in the titration curve near the titrant volume v_t value of ~20 to 24 μL, but it is hard to be more specific on just a visual basis. pH *vs.* v_t data are given for the marked portion (heavy dashes) of the curve, which includes the IP/EP. For each value of v_t given (except the first and last), calculate the slope $d\text{pH}/dv_t$ using the succeeding and preceding values for v_t and for pH. For example, for point 2, $d\text{pH}/dv_t=(7.19-7.00)/(19.6-18.0)=0.11875$. Plot the results, and locate the IP and thus the EP, and estimate the concentration A_T. Then, if the initial pH for the sample ($f=0$, pH = 5.77 at) was caused only by the HA (*e.g.*, the sample was not contaminated with CO_2 from the atmosphere), neglecting activity corrections, estimate pK_a using the determined value of A_T.

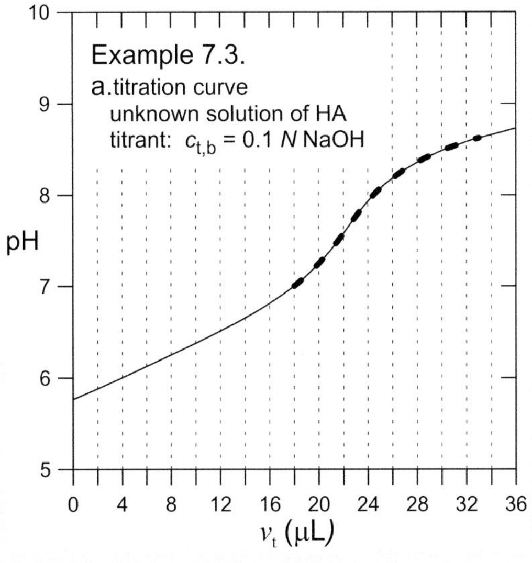

Example 7.3 Values for selected points

	Given		Calculated
point	v_t (μL)	**pH**	dpH/dv_t
1	18.0	7.00	–
2	18.9	7.10	0.11875
3	19.6	7.19	0.13636
4	20.0	7.25	0.15000
5	21.0	7.40	0.15625
6	21.6	7.50	0.18000
7	22.0	7.58	0.19091
8	22.7	7.71	0.18333
9	23.2	7.80	0.17273
10	23.8	7.90	0.16154
11	24.5	8.01	0.14167
12	25.0	8.07	0.12222
13	25.4	8.12	0.10556
14	26.8	8.26	0.08750
15	28.6	8.40	0.06857
16	30.3	8.50	0.05263
17	32.4	8.60	–

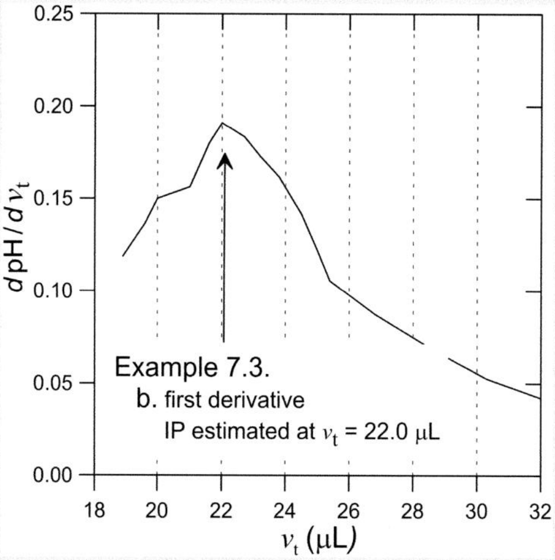

Solution

Plotting the results indicates that the IP and thus the EP is at $v_t \approx 22.0\ \mu$L. The location of the peak in the discrete, point-by-point is an estimate of the location of the IP. The estimate might be improved by fitting the points to a polynomial curve and finding the IP based on the fitted curve. (See Example 7.6 for application of that approach in a different problem.) Modern titration instruments perform this process automatically. The amount of base added to reach the EP is [22.0 μL/(10^6 L/μL)] $\times$ 0.1 $N = 2.2 \times 10^{-6}$ eq. This is the amount of HA in the 25 mL sample. HA is monoprotic, so the number of mols of HA in the 25 mL is 2.2×10^{-6} mol. So, we estimate: $A_T = 2.2 \times 10^{-6}$ mol/(0.025 L) $= 8.8 \times 10^{-6}$ M. If the initial pH of the solution is 5.77, we can use the ENE for a solution of HA to solve for K_a. Namely, [H$^+$] = [A$^-$]+[OH$^-$]. [H$^+$] $= \alpha_1 A_T +$ [OH$^-$]. Neglecting activity corrections, $10^{-5.77} = A_T K_a/(10^{-5.77} + K_a) + K_w/10^{-5.77}$. p$K_a = 6.30$.

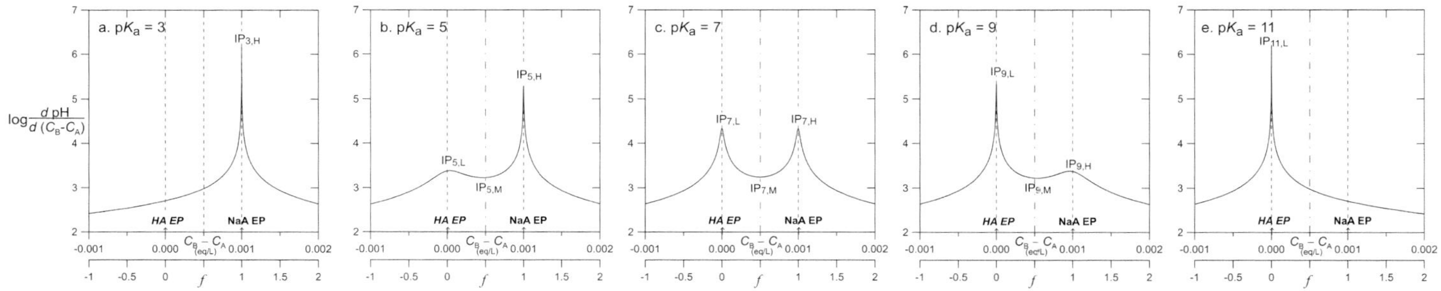

FIGURE 7.5 Titration first derivative curves with $A_T = 10^{-3}\,M$ and five pK_a values. The number of maxima and minima in each log dpH/$d(C_B - C_A)$ curve corresponds to the number of IPs that are present in the underlying titration curves in Figure 7.2 (one for p$K_a = 3$ and 11, and three for p$K_a = 5$, 7, and 9). An IP_H corresponds to a visible IP at the $f = 1$ EP; an IP_L corresponds to a visible IP at the $f = 0$ EP. The use of a log scale for the y-axis allows all panels to be presented on the same vertical scale. For experimentally determined titration curves, a log y-axis scale would not generally be used in the region of an IP, and the x-axis would be the volume of titrant v_t added to the sample, not $(C_B - C_A)$ or f in the sample (because they would not be known *a priori*). An experimentally determined first derivative curve would thus generally be plotted as dpH/dv_t *vs.* v_t.

7.4 GRAN TITRATION FUNCTIONS

7.4.1 General Considerations

As discussed above, for a titration curve of interest, it is not uncommon that an IP is either very weak or non-existent at an EP of interest; locating the EP by means of computing and plotting the first derivative is then difficult or impossible. As examples, Figure 7.6 provides titration curves from $f < 0$ to $f > 1$ for an acid HA with $pK_a = 5$ and two values of A_T. In both cases, the IP at $f = 1$ is distinct and easily located. But at $f = 0$ the IP is weak when $A_T = 10^{-3}$ *M*, and at $A_T = 3 \times 10^{-4}$ *M*, the IP is barely present.

Sørensen (1951) noted for the very high pH side of alkalimetric titrations of monoprotic acids, that beyond $f = 1$, $[OH^-]$ increases linearly, so a plot of $[OH^-]$ *vs.* v_t can be used for finding that EP. This approach was modified and extended by Gran (1952, 1988) for a range of different types of EPs (including $f = 0$), and has since come to be known as the "Gran titration method". To be clear, this method is not a type of titration *per se*. Rather, it involves a group of "Gran functions" F for processing titration data that allows EPs of interest to be located because each function tends to change proportionally with distance moving away from its EP: thus, looking back on a given F line towards the EP leads to an intersection ($F = 0$) with the titration v_t axis at the EP. Second, for Gran functions in the interior of a titration curve (in a monoprotic case, that means between $f = 0$ and $f = 1$), the functions allow deduction of the K_a value governing that region of the titration. This second feature makes the Gran method useful even when it is not required to locate an EP.

The Gran method has been widely applied to alkalimetric titrations in monoprotic acid HA systems, and also in diprotic acid H_2B systems such as with CO_2-related solutions in which initially $f > 0$. Each Gran function uses data on one side or the other side of an EP. Since there are two data sides for each EP, there can be two Gran functions for locating each EP. For a monoprotic acid system, there are two EPs ($f = 0$ and $f = 1$) so one can define four Gran functions. In the CO_2 system, there are three EPs ($f = 0$, $f = 1$, and $f = 2$), so one can define six Gran

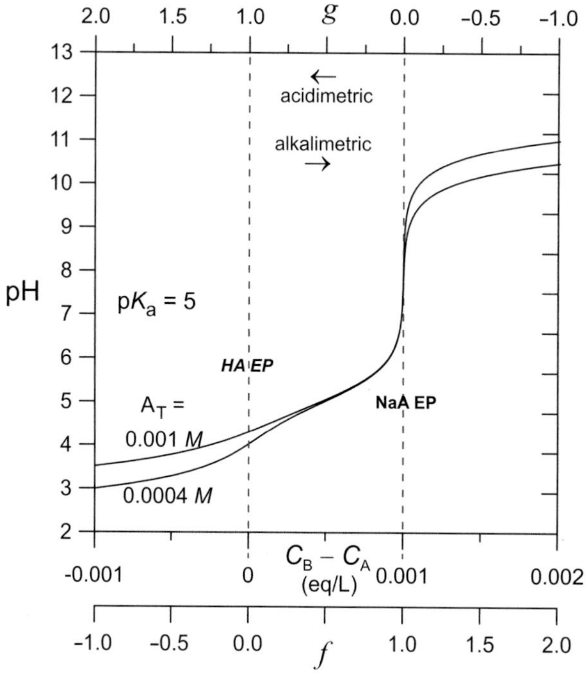

FIGURE 7.6 Titration curves calculated for two HA systems with $pK_a = 5.0$: $A_T = 10^{-3}$ *M* and 4×10^{-4} *M*.

functions (*e.g.*, see Rounds, 2012). Here we will discuss the four Gran functions in a monoprotic (HA) system with alkalimetric titration, and one in the CO_2 system with acidimetric titration.

A new Gran function nomenclature has been developed for this text that indicates the type of acid, the EP side, and whether the titration is alkalimetric or acidimetric. Using that nomenclature, in order of presentation, the Gran functions discussed here are: $F_{f<0}^{HA}$, $F_{f>1}^{HA}$, $F_{f<1}^{HA}$, $F_{f>0}^{HA}$, and $F_{f<0}^{CO_3^{2-}}$. In this nomenclature, if the superscript is the most protonated acid form possible in the system (HA in an HA system, H_2B in an H_2B system, etc.), then the titration is alkalimetric; if the superscript is the completely deprotonated form (A^- in an HA system, B^{2-} in an H_2B system, etc.) the titration is acidimetric. For example, $F_{f<0}^{HA}$ as defined below applies in an alkalimetric titration in a solution involving some strong acid plus some weak acid HA, with the EP being sought being that for $f=0$ (to determine much strong acid is present), using data on the $f<0$ side of the $f=0$ EP. And, for the CO_2 system, $F_{f<0}^{CO_3^{2-}}$ is applied in an acidimetric titration in a solution that involves dissolved CO_2, with the EP sought being that for $f=0$ (to determine the value of "alkalinity", "Alk", see Chapter 9), using data on the $f<0$ side of the $f=0$ EP.

7.4.2 The "Outer" Gran Functions in an HA System: Quantifying Strong Acid and Weak Acid Levels with $F_{f<0}^{HA}$ and $F_{f>1}^{HA}$

7.4.2.1 Background

An important case of when an $f=0$ EP is of interest pertains to the study of rain that is contaminated with some strong acid (like H_2SO_4 and/or HNO_3), plus some amount of weak monoprotic organic acid HA (*e.g.*, $pK_a \approx 5$). The common term for such rain is "acid rain". At the time of collection of the rain, $f<0$. It is generally easy to alkalimetrically titrate such a sample from $f<0$ to $f=1$ and find a distinct IP at $f=1$. The amount of base so added will be equivalent to the initial total amount of strong *plus* weak acid. But what if, as with Keene et al. (1983) and Molvaersmyr and Lund (1983), we want to know the individual amounts of strong and weak acid, *i.e.*, what if we want to know both: (1) the volume of titrant v needed just to reach $f=0$; *and* (2) the volume of titrant then needed to cover the distance going from $f=0$ to $f=1$? To answer both questions, we need to be able to find $f=0$ (where $v_t=v_{t,f=0}$) and $f=1$ (where $v_t=v_{t,f=1}$).

Earlier in this chapter, we considered why it is that monoprotic titration curves plotted as pH *vs.* f (and pH *vs.* (C_B-C_A), and pH *vs.* volume of titrant v_t) are always *concave up* when $f<0$, and always *concave down* when $f>1$. The explanation given (which is closely related to the findings of Sørensen (1951)) is the root reason why the function we will define as $F_{f<0}^{HA}$ can be used to find $f=0$, and why the function we will define as $F_{f>1}^{HA}$ can be used to find $f=1$. Namely, it was pointed out that pH is a log scale, whereas all the various versions of the x-axis (f, (C_B-C_A), and v_t) are linear scales. So, while $[H^+]$ tends to be increasing linearly going from $f=0$ to $f<0$, pH which is a log scale does not decrease linearly: the observed y *vs.* x functionality is flattened: $y=-\log[H^+]$, not $-[H^+]$). And, when moving from $f=1$ to $f>1$, $[OH^-]$ tends to be increasing linearly (on an equivalents basis, essentially no more HA is left to fight the incoming alkalimetric OH^-); but, being on a log scale, pH does not increase linearly: the observed functionality in the pH *vs.* f titration curve is again flattened: $y=-\log[H^+]=-\log K_w+\log[OH^-]$, not some linear function of $[OH^-]$. The Gran function $F_{f<0}^{HA}$ makes use of the near linearity in $[H^+]$ when $f<0$ to find $f=0$. And the Gran function $F_{f>1}^{HA}$ makes use of the near linearity in $[OH^-]$ when $f>1$ to find $f=1$.

7.4.2.2 $F_{f<0}^{HA}$

The ENE during a titration in a monoprotic system is

$$C_B - C_A = [A^-] + [OH^-] - [H^+] \tag{7.1}$$

or

$$C_A - C_B = [H^+] - [A^-] - [OH^-] \quad (7.10)$$

$$f \equiv \frac{(C_B - C_A)}{A_T}. \quad (7.8)$$

When $f<0$, then in the solution we have $(C_B - C_A)<0$, so $(C_A - C_B)>0$: net positive strong acid is present. At the low pH values generally encountered when $f<0$, the values of $[A^-]$ and $[OH^-]$ are small, so

$$C_A - C_B \text{ (eq/L)} \approx [H^+]. \quad (7.11)$$

During a titration, the total solution volume V_T (L) is given by the sum of the initial sample volume V_o (L) plus the volume of titrant v_t (μL) added to that point:

$$V_T = V_o + \frac{v_t}{10^6}. \quad (7.12)$$

We could use liters as the units for v_t, and that would avoid the annoying factor of 10^6 μL/L that will now pester us from here forward, but in a laboratory setting titration volumes are rarely expressed in liters. The number of equivalents of net strong acid in the solution is always exactly given by the product of $(C_A - C_B)$ with V_T. Thus, as we are moving from $f<0$ towards $f=0$, by Eqs.(7.11) and (7.12), to obtain the definition of $F_{f<0}^{HA}$ we write

$$\begin{matrix}\text{eq of net strong acid remaining}\\ \text{in solution when } f<0\end{matrix} \approx [H^+]\left(V_o + \frac{v_t}{10^6}\right). \quad (7.13)$$

By consideration of equivalent amounts of acids and bases, if $c_{t,b}$ (eq/L, N) is the concentration of strong base in the titrant solution, then $c_{t,b}(v_{t,f=0}/10^6)$ gives the total eq of strong acid initially in V_o, so $c_{t,b}(v_{t,f=0} - v_t)/10^6$ gives the remaining eq of strong acid at any point in the titration. Thus, neglecting activity corrections between [H⁺] and pH, for the region $f<0$,

$$\underline{F_{f<0}^{HA}\text{(eq)} \equiv 10^{-pH}\left(V_o + \frac{v_t}{10^6}\right)} \approx c_{t,b}\frac{(v_{t,f=0} - v_t)}{10^6}. \quad (7.14)$$

Plotting $F_{f<0}^{HA}$ (as underlined) *vs.* v_t in the region $f<0$ then gives a line that decreases about linearly towards $F_{f<0}^{HA}=0$ at $f=0$, where $v_t=v_{t,f=0}$. The slope is $-c_{t,b}/10^6$. The linearity deteriorates as $f \rightarrow 0$ because the approximation in Eq.(7.11) is deteriorating (a significant portion of the $[H^+]$ is starting to come from the HA). Taking the slope at any point then means that $v_{t,f=0}$ is always overestimated *to some degree* using $F_{f<0}^{HA}$. If activity corrections cannot be neglected because we do not have $\gamma_{H^+} \approx 1$, then the definition of $F_{f<0}^{HA}$ as underlined in Eq.(7.14) can still be used to find $v_t=v_{t,f=0}$; the only difference is that the fitted line $F_{f<0}^{HA}$ *vs.* v_t will have a slope of $-\gamma_{H^+}c_{t,b}/10^6$.

7.4.2.3 $F_{f>1}^{HA}$

At $f=1$, the equivalent amount of strong base needed to titrate the strong acid plus the weak acid has been added: $v_t=v_{t,f=1}$. For $f>1$, $[H^+]$ will be small so by Eq.(7.1)

$$C_B - C_A \approx [A^-] + [OH^-]. \quad (7.15)$$

Unless HA is an exceedingly weak acid, which is not of interest at this moment, for $f > 1$ then $[A^-] \approx A_T$, so

$$C_B - C_A \approx A_T + [OH^-] \tag{7.16}$$

$$(C_B - C_A) - A_T \approx [OH^-]. \tag{7.17}$$

Eq.(7.17) gives the eq/L of strong base that is in the solution *beyond* what is required to reach $f = 1$. (Note for that titration of a sample of "acid rain" containing both some strong acid and some weak acid, the value of C_A in Eq.(7.17) is due to the initial strong acid, while the value of C_B is due to the addition of the strong base titrant.) For $f > 1$,

$$\begin{matrix}\text{eq of net strong base in solution} \\ \text{beyond what is required to reach } f = 1\end{matrix} \approx [OH^-]\left(V_o + \frac{v_t}{10^6}\right) = \frac{K_w}{10^{-pH}}\left(V_o + \frac{v_t}{10^6}\right) \tag{7.18}$$

where the substitution of $K_w/10^{-pH}$ for $[OH^-]$ assumes activity corrections can be neglected. By consideration of equivalent amounts of acids and bases, $c_{t,b}(v_{t,f=1}/10^6)$ gives the total initial eq of strong plus weak acid in V_o, so $c_{t,b}(v_t - v_{t,f=1})/10^6$ gives the excess strong base beyond $f = 1$. Thus,

$$\underline{F_{f>1}^{HA}\,(\text{eq}) \equiv 10^{pH} K_w\left(V_o + \frac{v_t}{10^6}\right)} \approx c_{t,b}\frac{(v_t - v_{t,f=1})}{10^6}. \tag{7.19}$$

Plotting $F_{f>1}^{HA}$ (as underlined) *vs.* v_t in the region $f > 1$ then gives a line that rises increasingly linearly, with slope $+c_{t,b}/10^6$. The linear portion extrapolates to $F_{f>1}^{HA} = 0$ at $f = 1$, where $v_t = v_{t,f=1}$. The linearity improves as f increases beyond 1 because the approximation in Eq.(7.17) is improving, though with not-too-dilute solutions of acids of even moderate pK_a strength, the points become linear very quickly. Taking the slope at any point then means that $v_{t,f=1}$ is always underestimated *to some degree* using $F_{f>1}^{HA}$. If activity corrections cannot be neglected because we do not have $\gamma_{OH^-} \approx 1$, then the definition of $F_{f>1}^{HA}$ as underlined in Eq.(7.19) can still be used to find $f = 1$; the only difference is that the slope of the fitted line *vs.* v_t will be $+\gamma_{OH^-} c_{t,b}/10^6$. To simplify the definition of the Gran function for this region, some chemists move the K_w over to the $c_{t,b}$ side of Eq.(7.19). Indeed, that was the approach used in the first edition of this text. But doing that would obscure the analogous nature of $F_{f>1}^{HA}$ in relation to $F_{f<0}^{HA}$.

Example 7.4 "Acid Rain": Determining the Strong Acid and Weak Acid Using the Outer Gran Functions

A 25.0 mL sample of "acid rain" containing some strong acid and some weak acid HA was collected and subjected to alkalimetric titration with strong base with $c_{t,b} = 0.1$ *N*. The pH *vs.* v_t data are given. Plot the titration curve, and use the data to estimate $v_{t,f=0}$ and $v_{t,f=1}$. Then calculate the concentrations of strong acid and weak acid HA in the original sample. It can be assumed that dissolved CO_2 is negligible and does not affect the titration curve. (Little dissolved CO_2 was present initially, and the solution under titration was protected with a flow of N_2 gas to prevent dissolution of atmospheric CO_2 in the increasingly basic titration solution.)

v_t (μL)	pH	v_t (μL)	pH	v_t (μL)	pH	v_t (μL)	pH
0	3.39						
5	3.41	55	3.72	105	4.45	155	5.89
10	3.44	60	3.76	110	4.57	160	6.40
15	3.46	65	3.81	115	4.68	165	8.99
20	3.49	70	3.86	120	4.80	170	9.47
25	3.51	75	3.92	125	4.91	175	9.69
30	3.54	80	3.99	130	5.03	180	9.84
35	3.57	85	4.07	135	5.15	185	9.95
40	3.60	90	4.15	140	5.29	190	10.03
45	3.64	95	4.24	145	5.44	195	10.11
50	3.68	100	4.35	150	5.63	200	10.17

Example 7.4 Values for selected points

v_t (μL)	pH	$F^{HA}_{f<0}$	$F^{HA}_{f>1}$
0	3.39	**1.02 E-05**	6.22 E-13
20	3.49	**8.19 E-06**	7.72 E-13
40	3.60	**6.24 E-06**	1.01 E-12
60	3.76	4.34 E-06	1.46 E-12
80	3.99	2.56 E-06	2.48 E-12
100	4.35	1.13 E-06	5.61 E-12
120	4.80	4.02 E-07	1.58 E-11
140	5.29	1.29 E-07	4.94 E-11
160	6.40	1.01 E-08	6.35 E-10
180	9.84	3.66 E-12	**1.75 E-06**
200	10.17	1.71E-12	**3.75 E-06**
based on bolded values:			
slope =		-9.77×10^{-8}	1.00×10^{-7}
implied			
x-intercept =		$v_{t,f=0} = 103.9$ μL	$v_{t,f=1} = 162.5$ μL

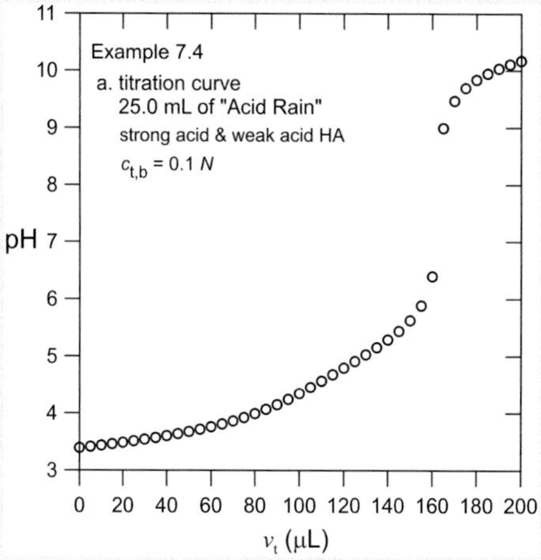

Solution

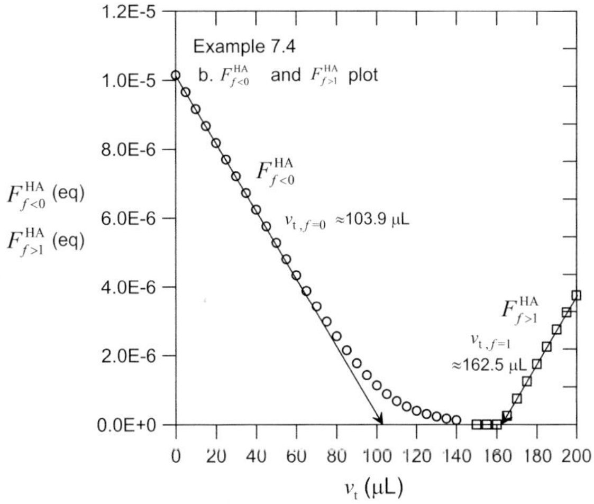

The values for $F^{HA}_{f<0}$ indicate $v_{t,f=0} \approx$ <u>103.9 μL</u>; the values for $F^{HA}_{f<0}$ indicate $v_{t,f=1} \approx$ <u>162.5 μL</u>. The concentration of strong acid in the original 25.0 mL sample ($V_o = 0.025$ L) is calculated as $c_{t,b}v_{t,f=0}/(V_o 10^6) = 0.1 \times 103.9/(0.025 \times 10^6) =$ <u>4.15×10^{-4} *N*</u>. The concentration of weak acid HA is calculated as $c_{t,b}(v_{t,f=1} - v_{t,f=0})/(V_o 10^6) = 0.1 \times 58.6/(0.025 \times 10^6) =$ <u>2.34×10^{-4} *F*</u>. We can use formality for the HA concentration since we are assuming the acid is monoprotic.

7.4.3 The Inner Gran Functions ($0 < f < 1$) in an HA System: $F^{HA}_{f<1}$ and $F^{HA}_{f>0}$

7.4.3.1 $F^{HA}_{f<1}$

First, in a titration of a solution of HA, when $f < 1$ and $f = 1$ is being approached (alkalimetrically), the concentration of HA is continually decreasing, about linearly. As a weak acid, there is not a strong tendency for HA to dissociate, so the great majority of the A^- present is due to the fact that there has been a reaction of HA with strong base, since we are past $f = 0$. (When $f < 0$, the base titrant is being consumed to titrate the strong acid.) This line of thinking gives

$$[A^-]\left(V_o + \frac{v_t}{10^6}\right) \approx c_{t,b}\frac{(v_t - v_{t,f=0})}{10^6}. \tag{7.20}$$

This equation will be totally invalid when $f > 1$ because once the HA is essentially exhausted, $[A^-]$ can no longer measurably increase.

Second, since we are past $f = 0$ but have not yet reached $f = 1$; however, much HA is left is mostly due to the fact that we have not yet reached the NaA EP ($f = 1$). This line of thinking gives

$$[HA]\left(V_o + \frac{v_t}{10^6}\right) \approx c_{t,b}\frac{(v_{t,f=1} - v_t)}{10^6}. \tag{7.21}$$

This equation will be totally invalid when $f < 0$ because there, [HA] is essentially unchanging.

Dividing Eq.(7.21) by Eq.(7.20) gives

$$\frac{[HA]}{[A^-]} \approx \frac{(v_{t,f=1} - v_t)/10^6}{(v_t - v_{t,f=0})/10^6}. \tag{7.22}$$

We have retained each of the 10^6 factors (which are actually 10^6 μl/L) to retain units of L, as possible. Eq.(7.22) considers the $f=0$ to $f=1$ distance to be composed of two lengths, an HA length, and an A^- length, so that the concentration ratio is the ratio of the lengths. This approximation is most valid when $0<f<1$. For example, $[A^-]$ is not zero at $f=0$, and [HA] is not zero at $f=1$.

Neglecting activity corrections, $[HA]/[A^-]=[H^+]/K_a$ so

$$\frac{[H^+]}{K_a} \approx \frac{(v_{t,f=1} - v_t)\ /\ 10^6}{(v_t - v_{t,f=0})\ /\ 10^6}. \tag{7.23}$$

Neglecting activity corrections, we can then define

$$\underline{F_{f<1}^{HA}\ (\text{eq}) \equiv 10^{-pH}\frac{(v_t - v_{t,f=0})}{10^6}} \approx K_a\frac{(v_{t,f=1} - v_t)}{10^6}. \tag{7.24}$$

Using the value of $v_{t,f=0}$ as evaluated with $F_{f<0}^{HA}$, then we can calculate and plot $F_{f<1}^{HA}$ (as underlined) *vs.* v_t in the region $f<1$ to obtain a line that decreases about linearly (with slope $-K_a/10^6$, to within activity corrections) towards 0 at $f=1$, to provide an estimate of $v_{t,f=1}$. The linearity breaks down close to $f=1$, and when $f>1$ because Eq.(7.20) is not valid there. Overall, $F_{f<1}^{HA}$ provides: (1) an estimate of $v_{t,f=1}$ (which, because more assumptions are involved, is probably less reliable than that obtained with $F_{f>1}^{HA}$); and (2) an estimate of K_a via the slope.

7.4.3.2 $F_{f>0}^{HA}$

Eq.(7.24) can be rearranged to use the other difference in titrant volume and obtain a fourth Gran function (which has unusual units), *i.e.*,

$$\underline{F_{f>0}^{HA}\ (\text{L}^2/\text{eq}) \equiv 10^{pH}\frac{(v_{t,f=1} - v_t)}{10^6}} \approx \frac{(v_t - v_{t,f=0})}{K_a 10^6}. \tag{7.25}$$

Using the value of $v_{t,f=1}$ as evaluated with $F_{f>1}^{HA}$ or $F_{f<1}^{HA}$, we can calculate and plot $F_{f>0}^{HA}$ (as underlined) *vs.* v_t in the region $f>0$ to obtain a line that decreases about linearly (with slope $+1/(K_a 10^6)$, to within activity corrections) towards 0 at $f=0$ to provide an estimate of $v_{t,f=0}$. The linearity breaks down close to $f=0$, and when $f<0$ because Eq.(7.21) is not valid there. Overall, $F_{f>0}^{HA}$ provides: (1) an estimate of $v_{t,f=0}$ (which, because more assumptions are involved, is probably less reliable than that obtained with $F_{f<0}^{HA}$); and (2) an estimate of K_a via the slope.

Example 7.5 "Acid Rain": Determining the Strong Acid and Weak Acid and the Weak Acid K_a Value Using the Inner Gran Functions

For the titration data in Example 7.4, use $v_{t,f=1}=162.5$ μL (from Example 7.4) to compute and plot $F_{f>0}^{HA}$ for the range $v_t=70$ to 180 μL ($F_{f>0}^{HA}$ goes strongly negative at $v_t=165$ μL). And, use $v_{t,f=0}=103.9$ μL (from Example 7.4) to compute and plot $F_{f<1}^{HA}$ for the same range. Choose linear regions to compute the slopes and use the values to obtain estimates of $v_{t,f=1}$ and $v_{t,f=0}$, and two estimates of K_a (and the average).

Solution

For $F_{f<1}^{HA}$, using the bolded values as a linear range, the slope is (1.09 E-10 − 2.43 E-10)/(150 − 130) = −6.74 × 10^{-12} = $-K_a/10^6$. The implied *x*-intercept is $v_{t,f=1}$ = <u>166.1 μL</u>, and K_a is estimated as 6.74 × 10^{-6}. For $F_{f>0}^{HA}$, using the bolded values as a linear range, the slope is (4.38 − 2.65)/(140 − 120) = 0.0863 = $+1/(K_a 10^6)$. The implied *x*-intercept is $v_{t,f=0}$ = <u>89.3 μL</u>, and K_a is estimated as 1.15 × 10^{-5}. The average of the two K_a estimates is <u>0.916 × 10^{-5}</u>.

Example 7.5 Values for selected points

v_t (μL)	pH	$F^{HA}_{f>0}$	$F^{HA}_{f<1}$
70	3.86	0.677	−4.63 E-09
80	3.99	0.809	−2.44 E-09
90	4.15	1.03	−9.83 E-10
100	4.35	1.38	−1.76 E-10
110	4.57	1.93	1.66 E-10
120	4.80	**2.65**	2.58 E-10
130	5.03	**3.49**	**2.43 E-10**
140	5.29	**4.38**	**1.85 E-10**
150	5.63	5.31	**1.09 E-10**
160	6.40	6.25	2.25 E-11
170	9.47	−22,000	2.24 E-14
180	9.84	−120,000	1.11 E-14
Based on bolded values:			
slope =		-6.74×10^{-12}	0.0863
implied K_a =		6.74×10^{-6}	1.16×10^{-5}
average K_a =		$\underline{9.16\times10^{-6}}$	
implied x-intercept =		$v_{t,f=1}=\underline{166.1\ \mu L}$	$v_{t,f=0}=\underline{89.3\ \mu L}$

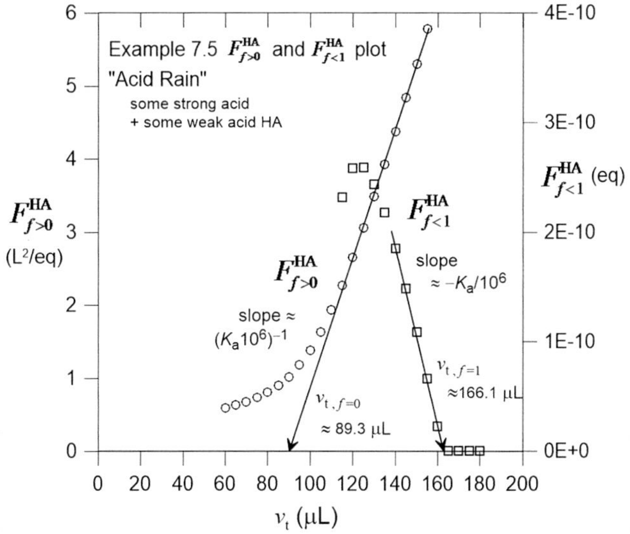

7.4.4 Determining Alkalinity (Alk) in the CO_2 System: $F^{CO_3^{2-}}_{f<0}$

With no other acid or base present, the ENE in a CO_2 system is

$$C_B - C_A = [HCO_3^-] + 2[CO_3^{2-}] + [OH^-] - [H^+] \tag{7.26}$$

$$C_B - C_A = \alpha_1^C C_T + 2\alpha_2^C C_T + [OH^-] - [H^+] \tag{7.27}$$

where C_T is the total dissolved CO_2-related carbon, and the α values are the fractional acid/base α values in the CO_2 system (see Chapter 9). Here,

$$f \equiv \frac{(C_B - C_A)}{C_T}. \tag{7.28}$$

Most waters in streams, rivers, lakes, and aquifers have positive net strong base so that $f > 0$. As will be discussed in detail in Chapter 9, the concentration of net strong base in such systems is referred to as the system "alkalinity", or "Alk" for short. The Alk value is a key piece of information for such a solution because if we know its value, and pH, by Eq.(7.27) we can calculate C_T. Or, if we know Alk and C_T, we can calculate the system pH. For its particular value of $(C_B - C_A)$, a particular water can have a positive value of C_A and a positive value of C_B, giving the net value $(C_B - C_A)$.

Positive Alk values ($f > 0$) are determined by measuring how much strong acid is needed to move a given sample volume from the starting point to $f=0$, where Alk$=0$. By definition this is done by adding increments ΔC_A. Once we go past $f=0$ into $f<0$, we enter a region where the solution has positive net strong acid, $(C_A - C_B) > 0$. *Wherever* we are in the titration, the net strong acid value is always the negative of the net strong base value. By Eq.(7.26)

$$C_A - C_B = [H^+] - [HCO_3^-] - 2[CO_3^{2-}] - [OH^-]. \tag{7.29}$$

So, in an acidimetric titration to determine Alk, we start with $(C_B - C_A) > 0$, and add enough strong acid to reach $f=0$ with $(C_B - C_A) = 0$ (which we want to find). Then, continued addition of strong acid leads to $(C_A - C_B) > 0$ and $f<0$. As with $F^{HA}_{f<0}$, we can use the linearly changing net strong acid character in the region $f<0$ to find $v_{t,f=0}$. The only difference is that with $F^{HA}_{f<0}$ the net strong acid is decreasing with v_t (the titration is alkalimetric), and here the net strong acid in solution will be increasing with v_t (the titration is acidimetric).

When $f<0$, then in the solution $(C_A - C_B) > 0$. At the low pH values generally encountered when $f<0$, the values of $[HCO_3^-]$, $2[CO_3^{2-}]$, and $[OH^-]$ are all small, so

$$C_A - C_B \text{ (eq/L)} \approx [H^+]. \tag{7.30}$$

During the titration, the total solution volume V_T (L) is again given by Eq.(7.12). The number of equivalents of net strong acid in the solution is always exactly given by the product of $(C_A - C_B)$ with V_T. Thus, as we are moving from $f=0$ into $f<0$, by Eqs.(7.12) and (7.30), to obtain the definition of $F^{CO_3^{2-}}_{f<0}$, we write

$$\begin{matrix}\text{eq of net strong acid in}\\ \text{solution when } f < 0\end{matrix} \approx [H^+]\left(V_o + \frac{v_t}{10^6}\right). \tag{7.31}$$

By consideration of equivalent amounts of acids and bases, if $c_{t,a}$ (eq/L, N) is the concentration of strong acid in the titrant solution, then $c_{t,a}(v_{t,f=0}/10^6)$ gives the total eq of net strong base (Alk) initially in V_o, so $c_{t,a}(v_t - v_{t,f=0})/10^6$ gives the eq of strong acid at any point in the titration. Thus, neglecting activity corrections between $[H^+]$ and pH, for the region $f<0$,

$$F^{CO_3^{2-}}_{f<0}\text{ (eq)} \equiv 10^{-pH}\left(V_o + \frac{v_t}{10^6}\right) \approx c_{t,a}\frac{(v_t - v_{t,f=0})}{10^6}. \tag{7.32}$$

Plotting $F_{f<0}^{CO_3^{2-}}$ (as underlined) *vs.* v_t in the region $f<0$ then gives a line that increases about linearly upwards from $F_{f<0}^{CO_3^{2-}}=0$ at $f=0$, where $v_t=v_{t,f=0}$. The slope is $+c_{t,a}/10^6$. The linearity deteriorates as $f \to 0$ because the approximation in Eq.(7.30) is deteriorating (a significant portion of the $[H^+]$ is starting to come from the carbonic acid). If activity corrections cannot be neglected because we do not have $\gamma_{H^+} \approx 1$, then the definition of $F_{f<0}^{CO_3^{2-}}$ as underlined in Eq.(7.32) can still be used to find $v_t=v_{t,f=0}$; the only difference is that the fitted line $F_{f<0}^{CO_3^{2-}}$ *vs.* v_t will have a slope of $+\gamma_{H^+}c_{t,a}/10^6$.

Example 7.6 Estimating Alk by Titration Using the First Derivative and the $F_{f<0}^{CO_3^{2-}}$ Gran Function

A 25.0 mL sample of stream water with a relatively low Alk value was collected and subjected to acidimetric titration with strong acid with $c_{t,a}=0.05$ *N*. The pH *vs.* v_t data are given. Plot the titration curve as pH *vs.* v_t. Calculate and plot the derivative dpH/dv_t *vs.* v_t to locate the $v_{t,f=0}$ at the IP for $f=0$. Calculate and plot $F_{f<0}^{CO_3^{2-}}$ as a function of v_t to estimate $v_{t,f=0}$. Use both estimates of $v_{t,f=0}$ to estimate Alk for the sample.

v_t (μL)	pH	v_t (μL)	pH	v_t (μL)	pH	v_t (μL)	pH
0	7.54						
5	7.29	55	6.26	105	5.01	155	4.00
10	7.11	60	6.19	110	4.80	160	3.96
15	6.97	65	6.10	115	4.62	165	3.92
20	6.86	70	6.02	120	4.48	170	3.89
25	6.76	75	5.93	125	4.38	175	3.85
30	6.67	80	5.83	130	4.29	180	3.83
35	6.58	85	5.72	135	4.21	185	3.80
40	6.50	90	5.58	140	4.15	190	3.77
45	6.42	95	5.43	145	4.09	195	3.75
50	6.34	100	5.23	150	4.04	200	3.72

Solution

Visual examination of the titration curve indicates an IP around $v_t \approx 105$ μL. The presence of this IP is indicated by the minimum in the calculated dpH/dv_t plot. (In acidimetric titrations, IPs at EPs are located at *minima* in dpH/dv_t *vs.* v_t curves; in alkalimetric titrations, they are at *maxima*, cf., Figure 7.5 and Example 7.3.) Finding the exact location of an IP requires curve-fitting, either curve-fitting pH *vs.* v_t then taking 2nd derivative of the fit to set $d^2pH/dv_t^2=0$ to find the IP, or equivalently computing dpH/dv_t *vs.* v_t, curve-fitting, then taking the 1st derivative of the fit to again reach $d^2pH/dv_t^2=0$. This example uses the version of the IP approach. For the bolded values in the results table, a fourth order polynomial fit gives $dpH/dv_t=-8.467E\text{-}08\ v_t^4+3.533E\text{-}05\ v_t^3-5.438E\text{-}03\ v_t^2+3.658E\text{-}01\ v_t-9.103$. So, $d^2pH/dv_t^2=-3.387E\text{-}07\ v_t^3+1.060E\text{-}04\ v_t^2-1.088E\text{-}02\ v_t+3.658E\text{-}01$, which is 0 at $v_t=103.7$ μL $(=v_t,f_{=0})$. This gives Alk $=103.7\times0.05/(10^6\times0.025)=2.07\times10^{-4}$ eq/L.

For the Gran approach, $F_{f<0}^{CO_3^{2-}}$ values are computed as part of the solution, and indicate $v_t,f_{=0}=$ <u>103.9 μL</u>, which gives Alk $=103.9\times0.05/(10^6\times0.025)=$ <u>2.08×10^{-4} eq/L.</u>

Example 7.6 Values for selected points

v_t (μL)	pH	$\frac{dpH}{dv_t}$	$F_{f<0}^{CO_3^{2-}}$
50	6.34	−1.55 E-02	1.14 E-08
60	6.19	−1.60 E-02	1.63 E-08
70	6.02	−1.77 E-02	2.40 E-08
80	5.83	**−2.13 E-02**	3.73 E-08
90	5.58	**−2.88 E-02**	6.54 E-08
95	5.43	**−3.50 E-02**	9.39 E-08
100	5.23	**−4.15 E-02**	1.46 E-07
105	5.01	**−4.38 E-02**	2.44 E-07
110	4.80	**−3.91 E-02**	4.01 E-07
120	4.48	**−2.45 E-02**	8.25 E-07
130	4.29	−1.63 E-02	1.30 E-06
140	4.15	−1.20 E-02	**1.78 E-06**
150	4.04	−9.33 E-03	**2.28 E-06**
160	3.96	−7.77E-03	**2.77 E-06**

Based on bolded values:

IP approach	Gran approach
dpH/dv_t minimizes at 103.7 μL	slope = 4.94×10^{-8}, implied x-intercept $v_{t,f=0}$ = 103.9 μL

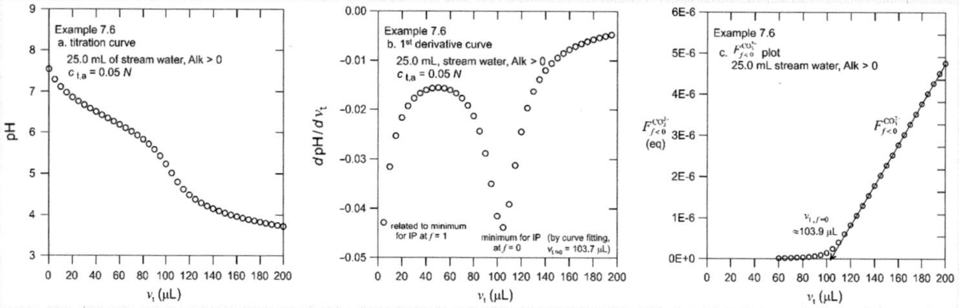

REFERENCES

Gran G (1952) Determination of the equivalence point in potentiometric titrations. Part II. *Analyst (London)*, **77**, 661–671.

Gran G (1988) Equivalence volumes in potentiometric titrations. *Analytica Chimica Acta*, **206,** 111–123.

Keene WC, Galloway JN, Holden JD (1983) Measurement of weak organic acidity in precipitation from remote areas of the world. *J. Geophysical Research: Oceans*, **88**, 5122–5130.

Molvaersmyr K, Lund W (1983) Acids and bases in fresh-waters. Interpretation of results from Gran plots. *Water Research*, **17**, 303–307.

Rounds SA (2012) Section 6.6. Alkalinity and Acid Neutralizing Capacity, Version 4, 9/2012). Section 6.6.4.B Inflection Point Titration Method and Section 6.6.4.C Gran-Function Plot Method. In: U.S. Geological Survey (2012) *National Field Manual for the Collection of Water-Quality Data (NFM). Techniques and Methods*, Book 9, Chapter A6, Field Measurements. Available at: https://water.usgs.gov/owq/FieldManual/Chapter6/section6.6/pdf/6.6.pdf (Accessed 8 September 2017).

Schilling J-G, Unni CK, Bender ML (1978) Origin of chlorine and bromine in the oceans. *Nature*, **273**, 631–636.

Sorensen P (1951) Transformation af Titreringskurver til Rette Linier. *KEMISK Maanedsblad og Nordisk. Handelsblad for Kemisk Industri*, **32**, 73–76.

8 Buffer Intensity β

8.1 INTRODUCTION

If any amount of strong base is added to any aqueous solution, the pH will go up. Conversely, if any amount of strong acid is added, the pH will go down. Chemists are often interested in knowing how pH-sensitive a system is to incremental addition of strong base or acid; the parameter that quantifies this concept is the buffer intensity β. In the environment, many aquatic organisms depend on the pH of the water to be in some particular range, and deviations from that range cause stress, or in extreme cases, mass die-offs. In the human body, a great many important biochemical reactions are very pH sensitive, so buffering is again extremely important. For humans, nothing makes this point more clearly than noting that human blood needs to be in a very narrow pH range so that pHdependent reactions can proceed at needed rates. For arterial blood, the normal range is 7.35 to 7.45; for venous blood, 7.32 to 7.42. Having recently passed through the lungs, arterial blood is slightly more alkaline because of loss of some metabolically generated CO_2 to exhaled air. "Acidosis" is the condition when blood pH is too low, and "alkalosis" is the condition when blood pH is too high. (Acidosis in humans often leads to tachypnea/hyperventilation as a means to off-gas CO_2 and raise blood pH.) Most natural water organisms benefit when their aquatic environment is characterized by an adequately large β so as to prevent significant swings in pH due to acid/base additions or losses.

If adding a relatively large amount of strong base or acid causes only a small pH change, then β is large. If adding a small amount of strong base or acid causes a large pH change, then β is small. With $(C_B - C_A)$ being the total net strong base in the solution that is subject to change when there is an incremental addition of strong base or strong acid, then

$$\beta \equiv \frac{d(C_B - C_A)}{d\text{pH}}. \tag{8.1}$$

The units of β are eq/L-pH unit (eq/L per pH unit), so β is a measure of the ability of a solution to resist a change in pH due to a change in $(C_B - C_A)$.

Biologically, the addition of acid or base will often not be the result of an addition or removal of some *strong acid*, or from an addition or removal of some *strong base*, but rather from an addition or removal of some *weak acid*, or an addition or removal of some *weak base*. Examples would be the loss by off-gassing of metabolic CO_2 (loss of a weak acid), and the hydrolysis of acetylcholine to release choline and acetic acid (addition of a weak acid). Removing or adding some amount of any weak acid or any weak base will cause less of a pH change than removal or addition of strong acid or strong base.

As discussed in Chapter 7, an acid/base titration curve shows how a system responds to the addition of strong acid or strong base. The curve can be drawn as pH *vs.* v_t (volume of titrant), pH *vs.* f, or pH *vs.* $(C_B - C_A)$. For the latter, the slope $S = d\text{pH}/d(C_B - C_A)$. So, $\beta = 1/S$. When S is large, then β is small; when S is small, then β is large.

When we are just adding strong base, $d(C_B - C_A) = dC_B$ and

$$\beta = \frac{dC_B}{d\text{pH}} \quad \text{(just strong base added).} \tag{8.2}$$

When we are just adding strong acid, $d(C_B - C_A) = -dC_A$ and

$$\beta = -\frac{dC_A}{d\text{pH}} \quad \text{(just strong acid added).} \tag{8.3}$$

When $dC_B > 0$, then $d\text{pH} > 0$, so by Eq.(8.2) we have $\beta > 0$. When $dC_A > 0$, then $d\text{pH} < 0$, so by Eq.(8.3) we again have $\beta > 0$. β is always **positive and nonzero**. When finite amounts of strong base (or strong acid) are added to some initial solution, then the value of β for the initial solution can be estimated as

$$\beta \approx \frac{\Delta C_B}{\Delta \text{pH}} \tag{8.4}$$

$$\beta \approx -\frac{\Delta C_A}{\Delta \text{pH}}. \tag{8.5}$$

As with any mathematical curve, here for any given pH, there is only one value for the slope S, regardless of the direction (left or right) in which the tangent line is viewed. Looking towards larger pH corresponds to determining the pH change when $dC_B > 0$. Looking towards lower pH corresponds to determining the pH change when $dC_A > 0$.

β does not provide a quantity we could not have already figured out how to obtain. Indeed, if we are interested in the pH sensitivity of a particular system, we could always compute what a specific increment of strong base ΔC_B (or strong acid ΔC_A) will do to the system in terms of ΔpH: we would just compute the pH for the initial value of $(C_B - C_A)$, then hypothesize the addition of some finite increment of strong base ΔC_B (or strong acid ΔC_A), and re-compute the pH for the new value of $(C_B - C_A)$. We would then compute the ratio $\Delta C_B/\Delta\text{pH}$ (or $-\Delta C_A/\Delta\text{pH}$), and obtain a finite difference version of β. Nevertheless, if for a given system we can develop an expression for β that depends on the system parameters, then the pH sensitivity can be obtained in one calculation instead of two, and moreover we have an expression that provides a functional understanding of how β depends on the nature of the system.

Example 8.1 β in a Dilute HCl Solution

Consider 1 L of a 10^{-3} *F* HCl solution. 0.05×10^{-3} equivalents of NaOH are added. What is the initial pH, and what is the pH after the NaOH addition? Estimate $\beta = d(C_B - C_A)/d\text{pH}$ using $\beta \approx \Delta C_B/\Delta\text{pH}$. The $\approx$ is used because the addition is finite, not differential. Activity corrections can be neglected.

Solution

HCl is a very strong acid, so initially $[H^+] = 10^{-3}$ *M*, pH = 3.00. After the addition of the strong base, $[H^+] = (10^{-3} - 0.05 \times 10^{-3}) = 0.95 \times 10^{-3}$ *M*, pH = 3.02228 (wherein we retain a number of significant figures for the benefit of Example 8.2). $\Delta C_B = +0.05 \times 10^{-3}$. $\Delta\text{pH} = +0.02228$. $\beta \approx (0.05 \times 10^{-3}/0.02228) = 0.00224$ eq/L-pH unit.

Example 8.2 β Is a Slope

Consider 1 L of a 10^{-3} *F* HCl solution. 0.05×10^{-3} equivalents of HCl are added. What is the initial pH, and what is the pH after the HCl addition? Estimate $\beta \approx -\Delta C_A/\Delta\text{pH}$. Activity corrections can be neglected.

Solution

Initially $[H^+] = 10^{-3}$ *M*, pH = 3.0. After the addition of the strong acid, $[H^+] = (10^{-3} + 0.05 \times 10^{-3}) = 1.05 \times 10^{-3}$ *M*, pH = 2.97881. $\Delta C_A = +0.05 \times 10^{-3}$. $\Delta\text{pH} = -0.02119$ $\beta \approx -0.05 \times 10^{-3}/(-0.02119) = 0.00236$ eq/L-pH unit. The estimated value of β is similar to the value in Example 8.1 because β is the *slope* of a curve, and curve slopes are independent of the direction of evaluation. In the limit of increasingly small additions of HCl and NaOH, the β estimates will be identical.

8.2 β IN MONOPROTIC ACID SYSTEMS

In the titration curve in a monoprotic acid system, depending on the values of K_a and A_T, the slope at the equivalence points (EPs) in a titration curve can be large, and if so β there is small. Second, in a monoprotic system, around $f \approx 0.5$, the slope can be shallow, as discussed in Chapter 7; if so, β there will be appreciable. Last, at both $f << 0$ and at $f >> 1$, the slope is low and β is large. These observations can be considered by means of Figure 8.1, which is the titration curve for a monoprotic system in which $A_T = 10^{-3}$ M and $pK_a = 7.0 = ½pK_w$ (25 °C). Note that when $pK_a = pK_b = ½pK_w$, the overall titration curve is exactly symmetrical. Figure 8.2 gives log β *vs.* pH for the same conditions, and plots the log of each of the three terms shown below to comprise β.

In the Figure 8.1 curve, there are two regions in which β is small, Regions II and IV. There are three regions of the curve in which the slope of the curve is relatively small (so β is large), Regions I, III, and V. For the following discussions of Regions I through V in Figure 8.1, note that a concentration of 10^{-3} M for A_T is not "low", and that with $pK_a = pK_b = ½pK_w$, HA is not a strong acid, and A^- is not a strong base. A low concentration would be something like 10^{-6} M, which is low enough that the acid/base properties of water itself are starting will be important in affecting the pH (*cf.* Figures 5.2 and 5.5).

Region I. β is Large: Low pH with $f < 0$. In this region, there is considerable free H^+ in the solution. Adding some strong base (dC_B) will not decrease $[H^+]$ very much. Adding some more strong acid (dC_A) will not increase $[H^+]$ very much. In both cases, the change in pH is small, so β is large. As an example, when $f = -1$, the solution is equivalent to HCl at 10^{-3} F plus HA at 10^{-3} F; this makes a good "buffer" (large β), but pH < 3. So, just because a solution has a high β does not mean the solution is buffered at a pH that is appropriate for planned experiments. A lab supervisor who issues the order "prepare a solution with a high β" should receive the question "at what pH?"

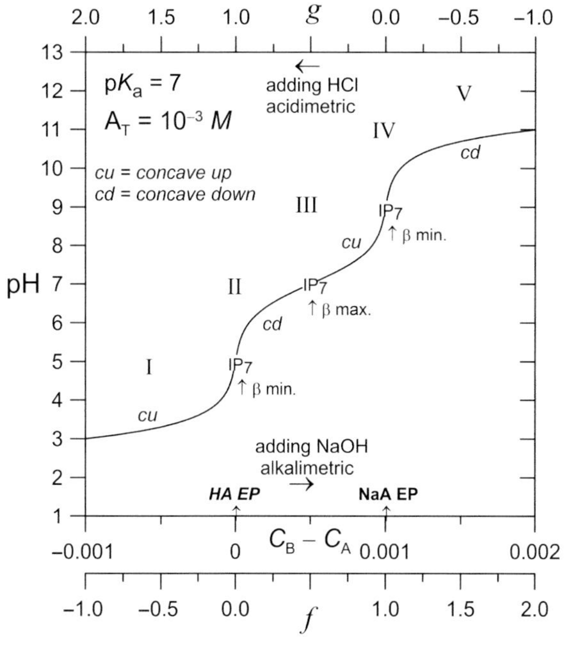

FIGURE 8.1 Titration curve for a monoprotic acid system with $pK_a = 7$ and $A_T = 10^{-3}$ M. The locations of the three inflection points (IPs) are marked. The IPs are where the curve goes from *concave up (cu)* to *concave down (cd)*, then *cd* to *cu*, then *cu* to *cd*. The value of the buffer intensity $\beta = [d(C_B - C_A)/d\text{pH}]$ is large in Region I where the pH is low. At the three successive IPs, β goes through a minimum (Region II, $f = 0$), a maximum (Region III, $f = 0.5$ and pH = pK_a), then another minimum (Region IV, at $f = 1$). β becomes large again in Region V where the pH is high. The x positions of the IPs are generally quite close to the corresponding f values.

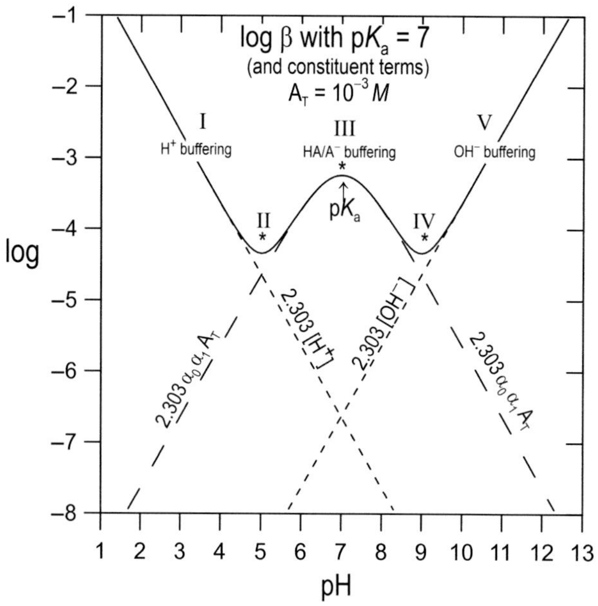

FIGURE 8.2 Log β for a monoprotic acid system with pK_a = 7 and $A_T = 10^{-3}$ *M*. Dashed lines are for the three terms comprising β. Regions I to V are marked. Each of the three inflection points (IPs) is marked with an *. β is large in Region I where the pH is low. β goes through a minimum in Region II (at $f = 0$), a maximum in Region III (at $f = 0.5$ and pH = pK_a), then another minimum in Region IV (at $f = 1$). β becomes large again in Region V where the pH is high.

Region II. β is Small: Acidic pH with $f \approx 0$, HA Not a Strong Acid. Because HA is not a strong acid, with $f = 0$, pH is not particularly low. Adding a small amount of strong base (dC_B) causes the increase in pH to be large because pH is rising from the initial acidic pH towards a pH much closer to pK_a, because protons are starting to come off (by reaction of HA with OH^-), and for that to happen it must be true that α_1 is being raised above ~0. Looking in the more acidic direction in the titration curve, adding a small amount of strong acid (dC_A) leads to a significant increase in [H^+] and a significant drop in pH because HA in the starting solution is not a strong acid and at $f = 0$, the pH is not particularly low. In either direction, the change in pH is large, so β is small.

Region III. β is Large: pH ≈ pK_a with $f \approx 0.5$, HA Not a Strong Acid and A^- Not a Strong Base. We have reached [A^-] ≈ $0.5A_T$. Note that in general, as can be derived from the expression for K_a,

$$\mathrm{pH} = \mathrm{p}K_a + \log\frac{[\mathrm{A}^-]}{[\mathrm{HA}]} \tag{8.6}$$

Eq.(8.6) is the frequently cited "Henderson–Hasselbalch Equation". (This goes to show that if you can take the log of a simple mathematical expression – in this case, the K_a concentration quotient – and do one line of rearranging, you might be able to get the new result named after yourself.) Viewing the titration curve alkalimetrically, at $f \approx 0.5$, if A_T is not small, then [HA] is still relatively large and so there is enough HA left to react with most of an increment dC_B: there will not be a large increase in the [A^-]/[HA] ratio, so the increase in pH will be small. Viewed acidimetrically, when adding a small amount of strong acid dC_A, [A^-] is relatively large, and there is enough to react with most of dC_A, so there will not be a large decrease in the [A^-]/[HA] ratio, so the drop in pH will be small.

Region IV. β is Small: Basic pH with $f \approx 1$, A^- Not a Strong Base. Because A^- is not a strong acid, with $f = 1$, pH is not particularly high. Adding a small amount of strong base (dC_B) leads to a significant increase in [OH^-], and the rise is pH is large. Adding a small amount of strong acid (dC_A) causes a large drop in pH because pH is moving from an alkaline pH to a pH much closer to pK_a, because protons are starting to move onto A^- (by reaction of H^+ with A^-) and for that to happen it must be that α_o is being raised significantly above ~0. In either case, the change in pH is large, so β is small.

Region V. High pH, $f > 1$. In this region, there is considerable free OH^- present. Viewed alkalimetrically, adding some additional strong base (dC_B) does not increase [OH^-] very much, and pH does not rise significantly. Viewed acidimetrically, adding some strong acid (dC_A) will not decrease [OH^-] very much. In either case, the change in pH is small, so β is large.

8.3 EQUATION FOR β IN A MONOPROTIC ACID SYSTEM

During the titration of a monoprotic acid system, the ENE is

$$C_B - C_A = [A^-] + [OH^-] - [H^+]. \tag{8.7}$$

$$\beta \equiv \frac{d(C_B - C_A)}{dpH} = \frac{d[A^-]}{dpH} + \frac{d[OH^-]}{dpH} - \frac{d[H^+]}{dpH}. \tag{8.8}$$

For the second and third terms on the RHS of Eq.(8.8), assuming that activity corrections can be neglected (or that a constant ionic medium constant ${}^cK_w$ is used),

$$[H^+][OH^-] = K_w \tag{8.9}$$

$$\log[H^+] + \log[OH^-] = \log K_w \tag{8.10}$$

$$d(\log[H^+]) + d(\log[OH^-]) = 0 \tag{8.11}$$

$$d(\log[H^+]) = -dpH = -d(\log[OH^-]). \tag{8.12}$$

From the study of logarithms and calculus,

$$d\log x = \frac{1}{2.303}\frac{dx}{x}. \tag{8.13}$$

Based on Eqs.(8.12–8.13),

$$dpH = -d(\log[H^+]) = -\frac{1}{2.303}\frac{d[H^+]}{[H^+]}. \tag{8.14}$$

By Eqs.(8.12–8.13), we also have

$$dpH = \frac{1}{2.303}\frac{d[OH^-]}{[OH^-]}. \tag{8.15}$$

Thus, for use in Eq.(8.8), we obtain

$$-\frac{d[H^+]}{dpH} = \frac{-d[H^+]}{-(1/2.303)d[H^+]/[H^+]} = 2.303[H^+] \tag{8.16}$$

and

$$\frac{d[\mathrm{OH^-}]}{d\mathrm{pH}} = \frac{d[\mathrm{OH^-}]}{(1/2.303)d[\mathrm{OH^-}]/[\mathrm{OH^-}]} = 2.303[\mathrm{OH^-}]. \tag{8.17}$$

Thus,

$$\beta = \frac{d[\mathrm{A^-}]}{d\mathrm{pH}} + 2.303[\mathrm{OH^-}] + 2.303[\mathrm{H^+}]. \tag{8.18}$$

Taking $\mathrm{A_T}$ to be constant during the titration, since

$$[\mathrm{A^-}] = \alpha_1 \mathrm{A_T} \tag{8.19}$$

then

$$\beta = \mathrm{A_T}\frac{d\alpha_1}{d\mathrm{pH}} + 2.303[\mathrm{OH^-}] + 2.303[\mathrm{H^+}]. \tag{8.20}$$

The chain rule says

$$\frac{du}{dz} = \frac{du}{dx}\frac{dx}{dz}. \tag{8.21}$$

Applied here,

$$\frac{d\alpha_1}{d\mathrm{pH}} = \frac{d\alpha_1}{d[\mathrm{H^+}]}\frac{d[\mathrm{H^+}]}{d\mathrm{pH}}. \tag{8.22}$$

The second factor is given by Eq.(8.16). For the first factor,

$$\alpha_1 = \frac{K_\mathrm{a}}{[\mathrm{H^+}] + K_\mathrm{a}}. \tag{8.23}$$

For the derivative of a quotient,

$$d\frac{u}{v} = \frac{vdu - udv}{v^2} \tag{8.24}$$

$$\frac{d\alpha_1}{d[\mathrm{H^+}]} = -\frac{K_\mathrm{a}}{([\mathrm{H^+}] + K_\mathrm{a})^2}. \tag{8.25}$$

So by Eqs.(8.16), (8.22), and (8.25)

$$\frac{d\alpha_1}{d\mathrm{pH}} = -\frac{K_\mathrm{a}}{([\mathrm{H^+}] + K_\mathrm{a})^2}(-2.303[\mathrm{H^+}]) = 2.303\frac{[\mathrm{H^+}]}{[\mathrm{H^+}] + K_\mathrm{a}} \times \frac{K_\mathrm{a}}{[\mathrm{H^+}] + K_\mathrm{a}} \tag{8.26}$$

so

$$\frac{d\alpha_1}{d\mathrm{pH}} = 2.303\alpha_0\alpha_1. \tag{8.27}$$

By Eqs.(8.20) and (8.27), for a monoprotic acid system, we obtain the important equation

$$\beta = \underbrace{2.303[\mathrm{H^+}]}_{\substack{\text{major term in}\\ \text{very acidic solutions}\\ (f<0)}} + \underbrace{2.303[\mathrm{OH}]}_{\substack{\text{major term in}\\ \text{very basic solutions}\\ (f>1)}} + \underbrace{2.303\alpha_0\alpha_1\mathrm{A_T}}_{\substack{\text{major term when } \mathrm{A_T} \text{ is}\\ \text{not small and } \mathrm{pH}\approx\mathrm{p}K_\mathrm{a}\\ (f\approx 0.5)}}. \tag{8.28}$$

β can never be zero because $[H^+]$ and $[OH^-]$ are never zero. β can be large just by having low pH conditions ($[H^+]$ is large), or rather high pH conditions solution ($[OH^-]$ is large). But for β to be large at some **intermediate** pH (say, 4 to 10), we need some significant values of both [HA] and $[A^-]$, and a pK_a near the pH value of interest.

The three terms in Eq.(8.28) are plotted individually in Figure 8.2 ($pK_a = 7$ and $A_T = 10^{-3}$ *M*). In Region I, the term $2.303[H^+]$ dominates. In Region V, the term $2.303[OH^-]$ dominates. In Region III, the term $2.303\alpha_0\alpha_1 A_T$ dominates: at intermediate pH and when A_T is not too low, most of the buffering is provided by the HA/A^- term:

$$\beta \approx 2.303\alpha_0\alpha_1 A_T. \tag{8.29}$$

When most chemists think of "buffers", it is in the context of Eq.(8.29) and most of the buffering being provided by some pair HA/A^- (or possibly H_2B/HB^-, or HB^-/B^{2-}, or H_3C/H_2C^-, etc.). It is often not mentioned that there are in fact terms in β involving $[H^+]$ and $[OH^-]$.

With Eq.(8.29), for β to be appreciable, neither α_0 nor α_1 can be small. So, we can have neither $pH \ll pK_a$ because then α_1 is small, nor $pH \gg pK_a$ because then α_0 is small. The product $\alpha_0\alpha_1$ maximizes at $pH = pK_a$ (see Table 8.1 and Figure 8.3).

Example 8.3 β Values in a Monoprotic System at Different *f* Values

Consider a solution in the acetic acid system with $A_T = 10^{-2}$ *M*. $pK_a = 4.76$ at 25 °C. Calculate pH, $\alpha_0\alpha_1$, $[HA]/[A^-]$, and β at $f = 0$, 0.5, and 1.0. Neglect activity corrections.

Solution

$f = \alpha_1 + ([OH^-] - [H^+])/A_T$. We substitute $\alpha_1 = K_a/(K_a + [H^+])$. For each value of *f*, we solve for pH by LHS − RHS = 0. The different terms can then be evaluated based on pH.

f	pH	$\alpha_0\alpha_1$	$[HA]/[A^-]$	β
0	3.39	0.039	23.4	1.76E-03
0.5	4.76	0.25	0.99	5.55E-03
1.0	8.38	0.00024	0.00024	1.06E-05

Example 8.4 Designing a Buffer Solution

Consider a solution that is to be buffered at pH = 7.40 in an HA system with $pK_a = 7.20$ at 25 °C. The buffer solution is to be used in some experiments in which up to 10^{-4} eq of strong base will be added per liter of water. If the maximum rise in pH that can be tolerated is 0.05, what is the required value of β and what is the needed value of A_T? Neglect activity corrections.

Solution

The needed $\beta = 10^{-4}/0.05 = 2 \times 10^{-3}$ eq/L-pH unit.

$\beta = 2.303([H^+] + [OH^-] + \alpha_0\alpha_1 A_T)$. At pH = 7.4, if $\beta = 2 \times 10^{-3}$ eq/L-pH unit, the terms involving $[H^+]$ and $[OH^-]$ are negligible, so $\beta \approx 2.303\alpha_0\alpha_1 A_T$. At pH = 7.4, $\alpha_0 = 0.39$, $\alpha_1 = 0.61$, $2.303\alpha_0\alpha_1 = 0.55$, so the needed $A_T = 3.66 \times 10^{-3}$ *M*.

TABLE 8.1
Values of α_0 and α_1, the Product $\alpha_0\alpha_1$, and the Ratio [HA]/[A⁻] in a Monoprotic Acid System as Functions of pH in the Range pH = p$K_a \pm 1$

pH=	α_0	α_1	$\alpha_0\alpha_1$	[HA]/[A⁻]
pK_a − 1	0.91	0.09	0.08	10.0
pK_a − 0.8	0.86	0.14	0.12	6.3
pK_a − 0.6	0.80	0.20	0.16	4.0
pK_a − 0.4	0.72	0.28	0.20	2.5
pK_a − 0.2	0.61	0.39	0.24	1.6
pK_a	0.50	0.50	0.25	1.0
pK_a+0.2	0.39	0.61	0.24	0.63
pK_a+0.4	0.28	0.72	0.20	0.40
pK_a+0.6	0.20	0.80	0.16	0.25
pK_a+0.8	0.14	0.86	0.12	0.16
pK_a+1	0.09	0.91	0.08	10.0

Note: Values are valid for any pK_a.

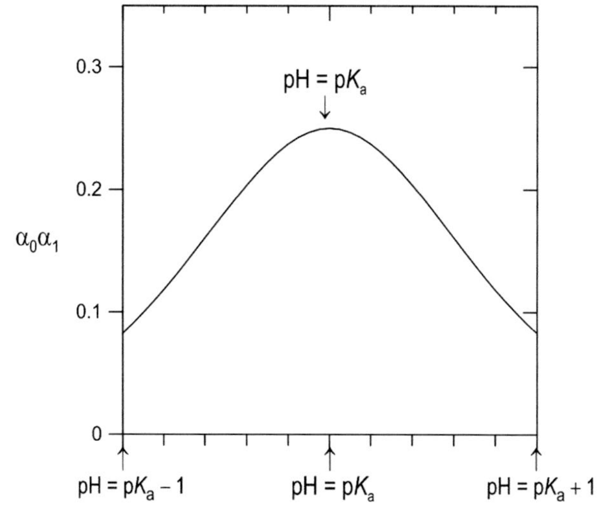

FIGURE 8.3 The product $\alpha_0\alpha_1$ *vs.* pH for values in the range pH = p$K_a \pm 1$.

8.4 β AS A FUNCTION OF K_A, A_T, AND pH

8.4.1 General

In the Figure 8.1 alkalimetric titration curve, there is a sharp rise in pH at both $f = 0$ (HA EP) and $f = 1$ (NaA EP). At each of these EPs, the titration curve changes from being *concave up* (*cu*) to *concave down* (*cd*). Any point at which a curve changes from *cu* to *cd* (or vice versa) is an *inflection point* (IP). At an IP in any function $y(x)$, the second derivative of the function is zero: $d^2y(x)/dx^2 = 0$. So, there will be an IP when the first derivative of the first derivative is zero. That will happen when the curve for the first derivative of the curve, namely the curve for $dy(x)/dx$, is at a minimum or maximum.

Although Figure 8.1 plots pH *vs.* $(C_B - C_A)$ (or f), for β we are interested in $(C_B - C_A)$ *vs.* pH (because $\beta = d(C_B - C_A)/d\text{pH}$). But an IP in a curve for pH *vs.* $(C_B - C_A)$ will also be an IP in a curve for $(C_B - C_A)$ *vs.* pH. So, an IP in a curve for pH *vs.* $(C_B - C_A)$ indicates that $(d/d\text{pH})\big(d(C_B - C_A)/d\text{pH}\big) = 0$. Namely, $d\beta/d\text{pH} = 0$. But for any curve, wherever the first derivative = 0, there is a maximum or minimum. Thus wherever there is an IP in a titration curve, β is going through an either maximum or minimum. For Figure 8.1, as shown in Figure 8.2, the IP at each EP ($f = 0$ and $f = 1$) is due to a minimum in β; the IP at $f \approx 0.5$ is due to a local maximum in β. (Taking the locations for the IPs to be $f \approx 0$, ≈ 0.5, and ≈ 1 is not exact to many decimal points, but for any practical purpose, this assumption is excellent unless A_T is very low.)

As discussed in Chapter 7, when $f = 1$ for some HA system at some A_T can be located in an alkalimetric titration by finding the IP (or by using an $f > 1$ Gran function), one can determine how much HA is present in an HA solution based on how much strong base is required to reach $f = 1$. Conversely, when $f = 0$ for some HA system at some A_T can be located in an acidimetric titration by finding the IP (or by using an $f < 0$ Gran function), one can determine how much NaA is present in an NaA solution based on how much strong acid is required to reach $f = 0$.

8.4.2 Comparing the $2.303\alpha_0\alpha_1 A_T$ Curve to the Lines for $2.303[H^+]$ and $2.303[OH^-]$

The number of IPs in a titration curve is always odd. There will always be at least one IP in every titration curve that corresponds to a minimum in log β *vs.* pH. For a monoprotic system, in order for there to be three IPs, the term $2.303\alpha_0\alpha_1 A_T$ has to be large enough to create a "hump" in β so that there is a local maximum from the hump and plus a second minimum, giving, overall, one maximum and two minima. In Figure 8.2, how high the hump is for $2.303\alpha_0\alpha_1 A_T$ determines how large β will be at pH = pK_a, The height is determined by the magnitude of A_T. Figure 8.4 plots log β *vs.* pH for pK_a = 5, 7, and 9 when $A_T = 10^{-3}$ *M*. Going from pK_a = 7 to 5, the hump moves towards the $2.303[H^+]$ portion of β. Lowering pK_a further yet would eventually cause the low-pH minimum and the adjacent maximum to disappear (simultaneously), and with them, two IPs. See Figure 8.5 for pK_a = 3 (cf. Figure 7.2). Similarly, going from pK_a = 7 to 9, the hump moves towards the $2.303[OH^-]$ portion of β, and raising pK_a further yet would eventually cause the high-pH minimum and the

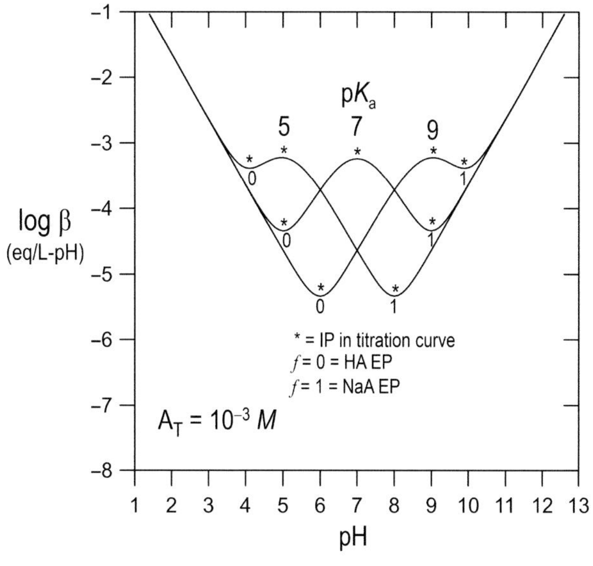

FIGURE 8.4 Log β *vs.* pH at $A_T = 10^{-3}$ *M* for pK_a = 5, 7, and 9.

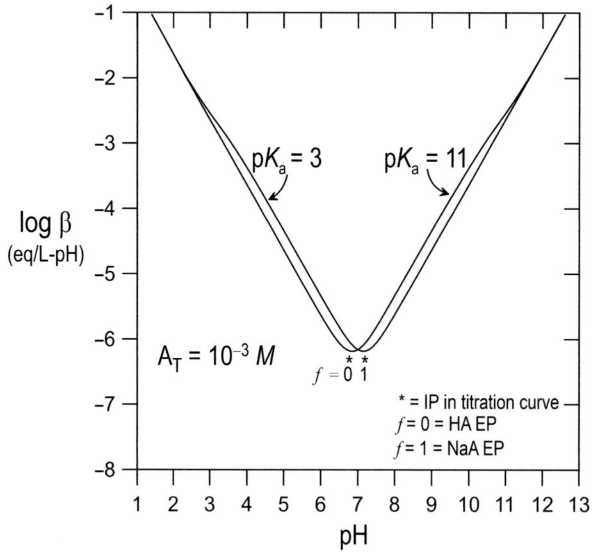

FIGURE 8.5 Log β *vs.* pH when $A_T = 10^{-3}$ M for pK_a = 3 and pK_a = 11.

adjacent maximum to disappear (simultaneously), and with them two IPs. See Figure 8.5 for pK_a = 9 (*cf.* Figure 7.2).

If three IPs are present for some pair of pK_a and A_T values, lowering A_T far enough will eventually cause two IPs to disappear (simultaneously) as the pH for one of the β minima moves closer and closer to the pH for the β maximum. The disappearance happens at a higher A_T for pK_a = 5 (or 9) than for pK_a = 7, because the pK_a = 7 hump is more distant from the line for 2.303[H^+] (or the line for 2.303[OH^-]). This helps us understand why you cannot measure the level of Cl^- in a solution by acidimetric titration: there is no drop IP-related drop in pH at the f = 0 EP. Cl^- is an exceedingly weak base, and does not want to take protons, and only will when {H^+} is raised to exceedingly high values (pH <<< 0): for HCl, pK_a is −4 or less) so the $2.303\alpha_0\alpha_1A_T$ curve is *way way* over to the left (and thus far far under) the line for 2.303[H^+] line. This means that as the pH would be lowered in a titration effort to put H^+ on Cl^-, even though {H^+} rising and this is creating tiny amounts of molecular HCl, there is still near-zero-protonation success by the time f = 0 has been reached, even as [H^+] buffering has become large. Looking the other way, you can of course measure the level of HCl (a strong acid) in a solution by alkalimetric titration with strong base, by observation of the sharp rise in pH moving through f = 1 EP (if A_T is large enough): the IP present is the one for f = 1. (See the Figure 7.1 curve for titration of an exceedingly strong acid/weak base.)

By analogy with the impossible acidimetric determination of Cl^-, you cannot determine the level of an exceedingly weak acid by alkalimetric titration, by observation of a noticeable jump in pH at the f = 1 EP: an exceedingly weak acid does not want to give up protons, and only will when {OH^-} is raised to exceedingly high values (pH >>> 14). With a very large pK_a, the $2.303\alpha_0\alpha_1A_T$ curve is *way way* over to the right (and thus far far under) the line for 2.303[OH^-]. This means that as the pH rises in the effort to pull off H^+, [OH^-] rises and OH^- buffering becomes large even as near-zero-deprotonation success is accomplished by the time f = 1 is reached. An alcohol like ethanol is an example of an exceedingly weak acid (p$K_a \approx 16$). You cannot measure the concentration of ethanol by alkalimetric titration, but you can measure the level of the sodium salt $CH_3CH_3O^-Na^+$ by acidimetric titration, by observation of the sharp drop in pH moving through the f = 0 EP (if A_T is large enough): the IP present is the one for f = 0. (See the Figure 7.1 curve for titration of an extremely weak acid/strong base.)

8.5 β IN SYSTEMS CONTAINING MORE THAN ONE ACID/BASE PAIR, OR A POLYPROTIC ACID

There are many different monoprotic acids. We can distinguish different possible acid–base pairs using one ′ as with HA′/A′⁻, two ′ as with HA″/A″⁻, three ′ as with HA‴/A‴⁻, etc. If more than one of these pairs is present in a solution, then there will be contributions to β from each so that

$$\beta = 2.303([H^+] + [OH^-] + \alpha_0'\alpha_1'A_T' + \alpha_0''\alpha_1''A_T'' + \alpha_0'''\alpha_1'''A_T''' + \ldots). \quad (8.30)$$

For a solution of single polyprotic acid, the value of β can be *approximated* as that for a solution of the individual acid/base pairs within the polyprotic acid, all of the same total concentration. The approximation improves as the pK values for the sequential protons are increasingly distant from one another. Dissolved CO_2 gives a diprotic system: CO_2 reacts with water according to $CO_2 + H_2O = H_2CO_3$. As discussed further in Chapter 9, both dissolved CO_2 and actual H_2CO_3 are capable of leading to the release of up to two protons, though the former, like an army "reserve", must first be activated ("called up") by reaction with H_2O. So, we define $[H_2CO_3^*] \equiv [CO_2] + [H_2CO_3]$. At 25 °C, with p$K_1$ = 6.35 (for $H_2CO_3^*$) and pK_2 = 10.33, the pK values are well separated.

$$C_T = [H_2CO_3^*] + [HCO_3^-] + [CO_3^{2-}]. \quad (8.31)$$

For a CO_2 system that is "closed" (C_T cannot change due to CO_2 coming in from, or going out to, a gas phase), by the above argument we have the approximation

$$\beta \approx 2.303[H^+] + [OH^-] + \alpha_0^C\alpha_1^C C_T + \alpha_1^C\alpha_2^C C_T). \quad (8.32)$$

As will be shown in Chapter 9, the exact expression is

$$\beta = 2.303[H^+] + [OH^-] + \alpha_0^C\alpha_1^C C_T + \alpha_1^C\alpha_2^C C_T + 4\alpha_0^C\alpha_2^C C_T \quad (8.33)$$

Eq.(8.32) provides a slight underestimate of β: the term $4\alpha_0^C\alpha_2^C C_T$ is never significant relative to the other terms because with pK_1 and pK_2 well separated, the product $\alpha_0^C\alpha_2^C$ is always small. When the pH is near pK_1, then $\alpha_0^C\alpha_1^C C_T >> \alpha_1^C\alpha_2^C C_T$, and if $[H^+]$ and $[OH^-]$ can be neglected (C_T is not small), then

$$\beta \approx 2.303\alpha_0^C\alpha_1^C C_T \qquad pH = pK_1 \pm 1. \quad (8.34)$$

When the pH is near pK_2, then $\alpha_1^C\alpha_2^C C_T >> \alpha_0^C\alpha_1^C C_T$, and if $[H^+]$ and $[OH^-]$ can be neglected (C_T is not small), then

$$\beta \approx 2.303\alpha_1^C\alpha_2^C C_T \qquad pH = pK_2 \pm 1. \quad (8.35)$$

Analogously, in a system containing species related to phosphoric acid (and no CO_2), with

$$P_T = [H_3PO_4] + [H_2PO_4^-] + [HPO_4^{2-}] + [PO_4^{3-}] \quad (8.36)$$

then, and so to a first approximation,

$$\beta \approx 2.303\left([H^+] + [OH^-] + \alpha_0^P\alpha_1^P P_T + \alpha_1^P\alpha_2^P P_T + \alpha_2^P\alpha_3^P P_T\right). \quad (8.37)$$

Eq.(8.37) always underestimates β because minor terms have been neglected (as with Eq.(8.32)). The error, however, is always very small because the pK values in the phosphate system (pK_1 = 2.16, pK_2 = 7.20, and pK_1 = 12.35 at 25 °C) are even more separated than in the CO_2 system. When P_T is not small,

$$\beta \approx 2.303\alpha_0^P\alpha_1^P P_T \qquad pH = pK_1 \pm 1 \tag{8.38}$$

$$\beta \approx 2.303\alpha_1^P\alpha_2^P P_T \qquad pH = pK_2 \pm 1 \tag{8.39}$$

$$\beta \approx 2.303\alpha_2^P\alpha_3^P P_T \qquad pH = pK_3 \pm 1. \tag{8.40}$$

Example 8.5 Calculating pH in a Phosphate Buffer Solution with Activity Corrections

A laboratory supply company states that a phosphate buffer with pH = 7.2 at 25 °C can be prepared by mixing 14 mL of 0.2 F NaH_2PO_4 with 14 mL of 0.2 F Na_2HPO_4, then diluting to a total volume of 100 mL. The pH is what a pH meter would read, *i.e.*, pH = $-\log\{H^+\}$. At 25 °C, pK_1 = 2.16, pK_2 = 7.20 and pK_1 = 12.35. (Note: Strictly speaking, if atmospheric CO_2 is not excluded, we should consider the effects of dissolved CO_2 on the pH; we will neglect that for now and consider it in Chapter 9.)

a. What are $[Na^+]$ and P_T in the 100 mL solution?
b. If activity corrections can be neglected, what is the pH at 25 °C and what is β?
c. Taking activity corrections into account, what is pH at 25 °C? (We will not worry about activity effects on β: as long as we know about what it is, that is all that usually matters).

Solution

a. $$[Na^+] = \frac{0.014\text{ L} \times 0.2\,M + 0.036\text{ L} \times 0.4\,M}{0.10\text{ L}} = 0.172\,M$$

$$P_T = \frac{0.014\text{ L} \times 0.2\,M + 0.036\text{ L} \times 0.2\,M}{0.10\text{ L}} = 0.100\,M$$

b. The ENE is

$$[Na^+] + [H^+] = \alpha_1 P_T + 2\alpha_2 P_T + 3\alpha_3 P_T + [OH^-]$$

Neglecting activity corrections, solving gives pH = $-\log[H^+]$ = 7.61, and Eq.(8.37) gives β = 0.0464 eq/L-pH. Based on computed concentrations of all the ions, the ionic strength I = 0.244, which is too high to be neglecting activity corrections.

c. $$K_w = \{H^+\}\{OH^-\} = \gamma_1[H^+]\gamma_1[OH^-] \qquad \text{so } {}^cK_w = K_w / \gamma_1^2$$

$$K_1 = \frac{\{H^+\}\{H_2PO_4^-\}}{\{H_3PO_4\}} = \frac{\gamma_1[H^+]\gamma_1[H_2PO_4^-]}{\gamma_0[H_3PO_4]} \qquad \text{so } {}^cK_1 = K_1/\gamma_1^2 \ (\gamma_0 = 1)$$

$$K_2 = \frac{\{H^+\}\{HPO_4^{2-}\}}{\{H_2PO_4^-\}} = \frac{\gamma_1[H^+]\gamma_2[HPO_4^{2-}]}{\gamma_1[H_2PO_4^-]} \qquad \text{so } {}^cK_2 = K_2/\gamma_2$$

$$K_3 = \frac{\{H^+\}\{PO_4^{3-}\}}{\{HPO_4^{2-}\}} = \frac{\gamma_1[H^+]\gamma_3[PO_4^{3-}]}{\gamma_2[HPO_4^{2-}]} \qquad \text{so } {}^cK_3 = K_3\gamma_2/(\gamma_1\gamma_3)$$

With the estimate that $I = 0.24399\ldots$, by the Davies Equation, $\gamma_1 = 0.739$, $\gamma_2 = 0.298$, $\gamma_3 = 0.066$.
Conversion of the K values to ${}^cK$ values gives

$pK_w = 14.00$	$\rightarrow$	$p^cK_w = 13.733$
$pK_1 = 2.16$	$\rightarrow$	$p^cK_1 = 1.897$
$pK_2 = 7.20$	$\rightarrow$	$p^cK_2 = 6.675$
$pK_3 = 12.35$	$\rightarrow$	$p^cK_3 = 11.562$

Re-solving gives $p_cH = 7.08$, so $pH = -\log(\gamma_1 10^{-7.08}) = 7.22$, with I again very close to 0.244. Since the I value is very similar to that computed initially, only one iteration was needed. pH = 7.22 agrees well with the quoted value for the buffer.

Example 8.6 β in the Citrate System

The pK_a values for citric acid at at 25 °C are pK_1 = 3.16, pK_2 = 4.78, and pK_3 = 6.38. Consider a system in which total citrate = Cit_T = 0.05 M. For pH = 1 to 13 in steps of 0.05, compute f values, then plot the titration curve as pH *vs.* f from $f = -1$ to 4. Convert the f values to $(C_B - C_A)$ values. Use the results to estimate the titration-curve slope $dpH/d(C_B - C_A)$. Invert those slopes to obtain $\beta = d(C_B - C_A)/dpH$. Also compute the estimate using citrate analog of Eq.(8.37). Plot $\log\beta$ *vs.* pH using the values obtained nearly exactly from the slope estimates, and the values obtained from the approximation $\beta \approx 2.303\left([H^+] + [OH^-] + \alpha_0^{Cit}\alpha_1^{Cit}Cit_T + \alpha_1^{Cit}\alpha_2^{Cit}Cit_T + \alpha_2^{Cit}\alpha_3^{Cit}Cit_T\right)$. Neglect activity corrections.

Solution

In this triprotic system, $f = \alpha_1^{Cit} + 2\alpha_2^{Cit} + 3\alpha_3^{Cit} + ([OH^-] - [H^+])/Cit_T$, which is easily calculated over a range of pH values. Then, $f = (C_B - C_A)/Cit_T$, so $(C_B - C_A) = f \times Cit_T$. For the slope at some pH = x, we use the values at pH = x+0.05 and pH = x−0.05, *e.g.*, at pH = 1.05, $dpH/d(C_B - C_A) = (1.10 - 1.00)/(-0.0790 - (-0.0997)) = 4.841$. Then, $\beta = 1/4.841 = 0.207$, $\log\beta = -0.685$. By the β approximation equation, $\log\beta = -0.686$. The results are plotted. The β approximation underestimates the actual value in the pH region ~2 to ~7 because of the missing terms. −25% is the worst that the error is, which occurs at pH = 5.5. The author has never bothered to derive the exact equation for β in a triprotic system, so as to identify the missing terms; it would be a good calculus exercise. In any case, an interesting feature of the citric acid system is that, because of the closeness of the pK_a values, β remains largely constant over quite a wide pH range. There are three humps contributing to $\log\beta$ between pH = 2 to pH = 7. The first hump is due to maximization of $2.303\alpha_0^{Cit}\alpha_1^{Cit}Cit_T$. The second hump is due to maximization of $2.303\alpha_1^{Cit}\alpha_2^{Cit}Cit_T$. The third hump is due to maximization of $2.303\alpha_2^{Cit}\alpha_3^{Cit}Cit_T$.

Example 8.6 Values for selected points

pH	α_0^{Cit}	α_1^{Cit}	α_2^{Cit}	α_3^{Cit}	f	$(C_B - C_A)$	$\frac{dpH}{d(C_B - C_A)}$	$\beta = \frac{d(C_B - C_A)}{dpH}$	$\log\beta$	$\log\beta$ (approx.)
1.00	9.93E-01	6.87E-03	1.14E-06	4.75E-12	−1.993	−0.0997				−0.636
1.05	9.92E-01	7.70E-03	1.43E-06	6.71E-12	−1.775	−0.0887	4.841	0.207	−0.685	−0.686
1.10	9.91E-01	8.63E-03	1.80E-06	9.47E-12	−1.580	−0.0790	5.426	0.184	−0.734	−0.735
1.15	9.90E-01	9.68E-03	2.27E-06	1.34E-11	−1.406	−0.0703	6.080	0.164	−0.784	−0.785
1.20	9.89E-01	1.08E-02	2.85E-06	1.88E-11	−1.251	−0.0626	6.810	0.147	−0.833	−0.834
1.25	9.88E-01	1.22E-02	3.59E-06	2.66E-11	−1.113	−0.0556	7.624	0.131	−0.882	−0.883
1.30	9.86E-01	1.36E-02	4.51E-06	3.75E-11	−0.989	−0.0494	8.532	0.117	−0.931	−0.932
etc.										

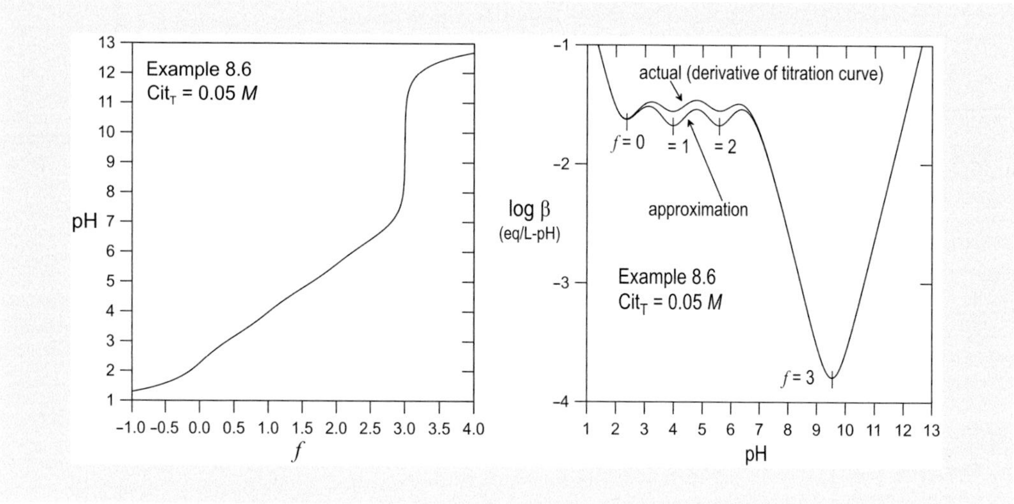

8.6 EFFECTS OF DILUTION ON pH IN BUFFER SOLUTIONS

8.6.1 General

Dilution of an aqueous solution with pure water will move the pH towards a neutral value, namely ½pK_w, or 7.00 at 25 °C. If the solution is initially acidic, the pH will become less acidic; if the solution is initially basic, the pH will become less basic. pH = ½pK_w will always be reached at very high dilution. In addition to being able resist changes in pH due to addition of acids or bases, a well-buffered solution will effectively resist changes in pH due to modest dilution, say by up to a factor of 2. A situation where minor dilution comes up involves calibration of a pH electrode with a series of buffer solutions, with rinsing of the electrode with deionized water between buffer solutions; the rinse water remaining on the electrode dilutes the next buffer solution. Minor electrode rinsing will not be a problem, though it would not be good laboratory practice to do this every day for a week while not using fresh buffer solutions.

8.6.2 Specifics

As has been discussed above, except for solutions at very high [H$^+$] or at very high [OH$^-$], a buffer solution owes much of its β value to some term of the form $2.303\alpha_j^Q\alpha_{j+1}^Q Q_T$. The system may be monoprotic ($j = 0$), diprotic ($j = 0$ or 1), triprotic ($j = 0$ or 1 or 2), etc. We can consider the matter of buffer solution dilution using the case of a monoprotic system for which

$$f \equiv \frac{C_B - C_A}{A_T} = \frac{\alpha_1 A_T + [OH^-] - [H^+]}{A_T} = \alpha_1 + \frac{[OH^-] - [H^+]}{A_T}. \tag{8.41}$$

Because $\alpha_0\alpha_1$ maximizes at $f = 0.5$, if β is significant, the solution will be at a pH for which f is inside the range $0.2 < f < 0.8$. If A_T is large (say $\geq 10^{-3}\ M$) relative to both [OH$^-$] and [H$^+$] (which are both less than 10^{-5} if the pH is between $5 < \text{pH} < 9$), then

$$\left| \frac{[OH^-] - [H^+]}{A_T} \right| \leq 0.01 \tag{8.42}$$

$$f \approx \alpha_1. \tag{8.43}$$

Dilution with pure water cannot change f: both $(C_B - C_A)$ and A_T will change by the same factor. Thus, the only way that the pH could change significantly on dilution is if the approximation in Eq.(8.43) deteriorates significantly. Upon dilution, the difference between $[OH^-]$ and $[H^+]$ will decrease. Therefore, the *most* that the absolute value function in Eq.(8.42) could increase by when A_T is diluted to $A_T/2$ is a factor of 2, giving 0.02 (smaller for smaller dilutions), which will preserve the approximation in Eq.(8.43) (if $A_T \geq 10^{-3}$ M) and the 2× dilution does not significantly affect the pH.

Example 8.7 Dilution of an Acetate Buffer

In Example 8.3, in the acetic acid system, it was determined that at 25 °C, pH = 4.76 when $A_T = 10^{-2}$ M and $f = 0.5$. Calculate the pH to more decimal points, and then compare the value with the predicted pH upon dilution by a factors of 2, 10, and 100 with pure water. Neglect activity corrections.

Solution

The value of f does not change on dilution, so for every value of A_T, we solve $f = \alpha_1 + ([OH^-] - [H^+])/A_T$. With dilution, all the values move towards pH = 7.00. For dilution of 2, the pH change is only observable in the third decimal place.

Example 8.7 Results

Dilution Factor	A_T (M)	pH
1	10^{-2}	4.7630
2	0.5×10^{-2}	4.7660
10	10^{-3}	4.7883
100	10^{-4}	4.9556

Example 8.8 Dilution of a Concentrated Phosphate Buffer, Considering Changing Activity Coefficient Effects

For the solution in Example 8.5, calculate the pH to more decimal points, and then calculate the pH upon dilution by factors of 2, 10, and 100 with pure water. Make the calculations without activity corrections, then repeat with activity corrections.

Solution

Example 8.8 Results

			No γ Corrections	With γ Corrections		
Dilution Factor	$[Na^+]$ (M)	P_T (M)	pH	I	$p_cH = -\log[H^+]$	$pH = -\log\{H^+\}$
1	0.172	0.1	7.6101	0.244	7.0849	7.2162
2	0.086	0.05	7.6101	0.122	7.1567	7.2700
10	0.0172	0.01	7.6101	0.0244	7.3494	7.4146
100	0.00172	0.001	7.6093	0.00244	7.5149	7.5385

Dilution causes almost no change in pH when activity corrections are not considered, which is exactly analogous to the result in Example 8.7. The pH change with dilution is larger when activity corrections are included (because the ${}^{c}K_a$ values are changing), but still not large for a factor of 2 dilution. Many pH calibration buffers are phosphate solutions; dilution by 5% (factor of 1.05 dilution) due to electrode rinsing will barely change the pH.

9 Chemistry of Dissolved CO_2

9.1 INTRODUCTION

Like $O_{2(g)}$, $N_{2(g)}$, and all other gases, $CO_{2(g)}$ will dissolve in water. The dissolution reaction is

$$CO_{2(g)} = CO_2 \tag{9.1}$$

where the species on the RHS represents dissolved aqueous CO_2. As the CO_2 molecule carries no protons, it is not an acid in and of itself. However, CO_2 can undergo hydration to form the diprotic species H_2CO_3:

$$\underset{\leftarrow\ \textit{dehydration}}{\overset{\textit{hydration}\ \rightarrow}{CO_2 + H_2O = H_2CO_3}} \qquad K_{\text{hydration}}^{CO_2} = \frac{\{H_2CO_3\}}{\{CO_2\}}. \tag{9.2}$$

We choose to not write $\{H_2O\}$ in the denominator for $K_{\text{hydration}}^{CO_2}$ because, for the aqueous solutions of interest in this text, $\{H_2O\}$ is expressed on the mole fraction (x) scale, and we are always near $x_{H_2O} = 1$.

At 25 °C, $K_{\text{hydration}}^{CO_2} = 1/650$, so $[CO_2] >> [H_2CO_3]$. In living cells, there are many circumstances when catalysis of hydration/dehydration is needed (Berg et al., 2015). The enzymes that perform this catalysis are *carbonic anhydrases*. At a normal human blood pH value of ~7.4, α_1 in the CO_2 system is ~0.9 = $[HCO_3^-]/C_T$, so ~90% of the total dissolved CO_2 is present as HCO_3^- (bicarbonate). This means that if blood C_T from metabolic CO_2 is to be rapidly exhaled via the lungs, bicarbonate needs to be protonated, then speedily dehydrated to dissolved CO_2 for subsequent outgassing to lung air:

$$HCO_3^- + H^+ \rightarrow H_2CO_3 \xrightarrow{\text{carbonic anydrase}} CO_2 + H_2O \xrightarrow{\text{outgassing}} CO_{2(g)} + H_2O. \tag{9.3}$$

To adjust the outgassing rate and control blood pH, animals with blood and lungs can adjust the "respiratory rate" (*i.e.*, breaths per minute). If the blood pH is too low, this is "acidosis", and the classic symptom is rapid breathing ("tachypnea"): the body is trying to raise pH values by increasing the CO_2 loss rate. And, if someone breathes rapidly when it is *not* needed, *i.e.*, "hyperventilates" as during a panic attack, the result can be "respiratory alkalosis": the CO_2 loss rate by respiration has been higher than optimal, and the blood pH becomes too high; faintness and tingling in the extremities are symptoms.

H_2CO_3 in a solution has protons to give. In addition, whatever CO_2 is present can potentially react with water according to Eq.(9.2) to form more H_2CO_3. Thus, when counting how many acidic protons are present (because of the H_2CO_3) *as well as* potentially present (because of the CO_2) we need to count both those on the H_2CO_3 molecules *and* those that can be made if the dissolved CO_2 molecules were to undergo hydration. For an analogy, consider that H_2CO_3 represents the active army of a nation, and CO_2 represents the army reserves. If there is a war (attack by addition of OH^-), the active army will fight immediately, and the reserves will, as necessary, be equipped with their gear (H_2O in Eq.(9.2)) and then react, as necessary, with OH^-. In effect, the total concentration of diprotic acid in a CO_2 system is $[H_2CO_3] + [CO_2]$. To represent this chemistry, the concept of $H_2CO_3^*$ (active army + the reserves) has been developed:

$$[H_2CO_3^*] = [H_2CO_3] + [CO_2]. \tag{9.4}$$

$H_2CO_3^*$ is not a chemical species *per se*, but it can be treated as such in chemical equilibria. Actually, in exactly the same way, when we write $[H^+]$, there is no single H^+ species either, but rather a range of species with varying degrees of interaction with the solvating water molecules (see Chapter 3), so we have been doing this type of thing already.

Overall, when CO_2 dissolves in water,

$$CO_{2(g)} + H_2O = H_2CO_3^* \qquad K_H\ (M/\text{atm}) = \frac{\{H_2CO_3^*\}}{p_{CO_2}} = \frac{[H_2CO_3^*]\gamma_{H_2CO_3^*}}{p_{CO_2}} \tag{9.5}$$

where K_H is the "Henry's Law" constant for dissolution of CO_2 in water. K_H has units of M/atm: it is the constant of proportionality between $[H_2CO_3^*]$ and the gaseous pressure p_{CO_2}. As with Eq.(9.2), in this text we do not write $\{H_2O\}$ in the denominator for K_H because, for most solutions of interest here, $\{H_2O\} \approx 1$ (mole fraction scale). Because $H_2CO_3^*$ carries no ionic charge with which to interact with dissolved ions (the major cause of non-idealities in aqueous solutions), in fresh waters $\gamma_{H_2CO_3^*} \approx 1$ so that

$$[H_2CO_3^*] = K_H p_{CO_2}. \tag{9.6}$$

(For seawater, we need to correct for the fact that $\gamma_{H_2CO_3^*} \neq 1$, as illustrated in Example 9.13 below.)

Example 9.1 $[H_2CO_3^*]$ and p_{CO_2}

Assume equilibrium and that activity corrections can be neglected.

a. At 25 °C, when $p_{CO_2} = 10^{-3.40}$ atm, what is $[H_2CO_3^*]$ at pH = 5, 7, and 9?
b. At 25 °C, what is $[H_2CO_3^*]$ when $p_{CO_2} = 10^{-4.40}$, $10^{-3.40}$, $10^{-2.40}$, and 2 atm? (p_{CO_2} = 2 atm is typical for bottles of "carbonated" beverages.)

Solution

a. At 25 °C, $K_H = 10^{-1.47}$ M/atm. Eq.(9.6) does not depend on pH, so at all three pH values, $[H_2CO_3^*] = (10^{-1.47}\ M/\text{atm}) \times 10^{-3.40}\ \text{atm} = 10^{-4.87}\ M$.
b. $p_{CO_2} = 10^{-4.40}$ atm, $[H_2CO_3^*] = (10^{-1.47}\ M/\text{atm}) \times 10^{-4.40}\ \text{atm} = 10^{-5.87}\ M$.

   $p_{CO_2} = 10^{-3.40}$ atm, $[H_2CO_3^*] = (10^{-1.47}\ M/\text{atm}) \times 10^{-3.40}\ \text{atm} = 10^{-4.87}\ M$.

   $p_{CO_2} = 10^{-2.40}$ atm, $[H_2CO_3^*] = (10^{-1.47}\ M/\text{atm}) \times 10^{-2.40}\ \text{atm} = 10^{-3.87}\ M$.

   $p_{CO_2} = 2$ atm, $[H_2CO_3^*] = (10^{-1.47}\ M/\text{atm}) \times 2\ \text{atm} = 10^{-1.17}\ M$.

There are significant amounts of carbon dioxide in the atmosphere, and also in most natural water systems near the surface of the Earth. The fact that $H_2CO_3^*$ is acidic is therefore of enormous importance for natural water chemistry: the prevalence of dissolved CO_2 means that CO_2 plays some role in determining the pH of virtually all natural waters. Moreover, many subsurface geological systems can contain very large amounts of carbon dioxide: in areas where CO_2 is venting from the mantle, dissolved CO_2 levels can become enormous. Consider Lake Nyos, which sits in the crater of an inactive volcano in Cameroon. Magma below the crater continually leaks large amounts of CO_2 into the lake water. With a maximum depth of 208 m, the maximum hydraulic pressure is 20 atm, so that the maximum total pressure (adding in the 1 atm from the atmosphere) that could be applied to a dissolving gas is 21 atm. At that p_{CO_2}, the value of $[H_2CO_3^*]$ is very large: 0.7 *M*. On August 21, 1986, some kind of disturbance in the lake caused a "turnover" event so that waters at the bottom were brought towards the surface. The loss of system pressure for the water caused a violent bubbling of CO_2, as when a bottle of strongly carbonated beverage is suddenly uncapped. The bubbling introduced much further upward momentum in the lakewater, which caused further turnover and violent bubbling. The end result was that a large amount CO_2 was outgassed. Being denser than air, the $CO_{2(g)}$ did not rise in the atmosphere, but rather displaced the surrounding air near the surface; more than 1700 people perished by suffocation.

As a diprotic acid, $H_2CO_3^*$ can lose two protons:

$$H_2CO_3^* = H^+ + HCO_3^- \qquad K_1 = \frac{\{H^+\}\{HCO_3^-\}}{\{H_2CO_3^*\}} \tag{9.7}$$

$$HCO_3^- = H^+ + CO_3^{2-} \qquad K_2 = \frac{\{H^+\}\{CO_3^{2-}\}}{\{HCO_3^-\}}. \tag{9.8}$$

Since

$$C_T = [H_2CO_3^*] + [HCO_3^-] + [CO_3^{2-}], \tag{9.9}$$

$$[H_2CO_3^*] = \alpha_0 C_T \qquad [HCO_3^-] = \alpha_1 C_T \qquad [CO_3^{2-}] = \alpha_2 C_T \tag{9.10}$$

where the values of α_0, α_1, and α_2 are obtained according to equations given in Chapter 5. This text has used H_2B to refer to a generic diprotic acid in a system of some B_T concentration, and H_2C has been used to refer to a generic triprotic acid in a system of some C_T concentration. However, when discussing dissolved CO_2, we here redirect C_T as indicated in Eq.(9.9). No ambiguities will be introduced because from here forward, specific nomenclature will always be used as necessary for any triprotic acids under consideration (*e.g.*, P_T for total phosphate species, Cit_T for total citrate species, etc.). Some of the more important of the natural processes involving CO_2 are listed in Table 9.1, and summarized schematically in Figure 9.1. Table 9.2 gives values for K_w, K_1, K_2, and K_H for CO_2 as a function of temperature. Table 9.3 compares K_H values for CO_2 with those of other gases. A table in Example 9.13 provides specialized *K* values for seawater.

Example 9.2 α_0, α_1, and α_2 in the CO_2 System

Consider an aqueous solution at 25 °C with pH = 8.33 and $C_T = 10^{-3}$ *M*. What are the values of $[H_2CO_3^*]$, $[HCO_3^-]$, and $[CO_3^{2-}]$?

Solution

At 25 °C, $K_1 = 10^{-6.35}$ and $K_1 = 10^{-10.33}$. At pH = 8.33, then $\alpha_0 = 0.0103$; $\alpha_1 = 0.980$; $\alpha_2 = 0.00980$. $[H_2CO_3^*] = 1.03 \times 10^{-5}$ *M*; $[HCO_3^-] = 0.98 \times 10^{-3}$ *M*; and $[CO_3^{2-}] = 0.98 \times 10^{-5}$ *M*.

9.2 TERMS DESCRIBING THE TITRATION POSITION OF A CO_2 SYSTEM

9.2.1 General

Extending the definition for *f* given in Chapter 7, in a CO_2 system of a particular C_T

$$f = \frac{(C_B - C_A)}{C_T}. \tag{9.11}$$

TABLE 9.1
Important Processes Involving CO_2 Species in the Environment

Human Processes	→ Reaction	← Reaction
$C_{(s)} + O_{2(g)} \rightarrow CO_{2(g)}$	burning coal for energy	carbon sequestration as biochar
$C_nH_{2n+2} + (1.5n + 0.5)O_{2(g)} \rightarrow nCO_{2(g)} + (n+1)H_2O$	burning hydrocarbons (natural gas, gasoline, oil, etc.) for energy	
$CO_{2(g)} \rightarrow CO_2$(geological)	carbon sequestration as CO_2	
$CaCO_{3(s)} \rightarrow CaO_{(s)} + CO_{2(g)}$	calcination of limestone to obtain $CaO_{(s)}$ (lime) for cement	
Selected Natural Processes		
CO_2(geological) $\rightleftarrows CO_{2(g)}$	emission by the mantle	
$CO_{2(g)} \rightleftarrows CO_2$	dissolution of atmospheric CO_2	outgassing of dissolved CO_2
$6CO_{2(g)} + 6H_2O \rightleftarrows C_6H_{12}O_6 + 6O_{2(g)}$	photosynthesis	respiration
$C_6H_{12}O_6 \rightarrow 3CO_{2(g)} + 3CH_{4(g)}$	anaerobic fermentation	
$Ca^{2+} + CO_3^{2-} \rightleftarrows CaCO_{3(s)}$	precipitation of metal carbonates	dissolution of metal carbonates
$CaCO_{3(s)} + H_2O + CO_2 \rightleftarrows Ca^{2+} + 2HCO_3^-$	dissolution of limestone rock	precipitation of stalagtites/stalagmites

In a titration of a monoprotic acid HA, $f = 0$ and $f = 1$ are the only equivalence point (EP) conditions. For the diprotic acid $H_2CO_3^*$ and a titration with some C_T, there will be three EP conditions: $f = 0$, $f = 1$, and $f = 2$. At the EP for $f = 0$ the solution chemistry will be that of a C_T F solution of $H_2CO_3^*$. At the EP for $f = 1$, the number of equivalents of strong base present equals the number of equivalents of protons for one of the $H_2CO_3^*$ protons (the eq/L of NaOH added to reach that point equals C_T); if the strong base used in the titration is NaOH, the $f = 1$ EP corresponds to a C_T F solution of $NaHCO_3$. The $f = 2$ EP corresponds to a C_T F solution of Na_2CO_3 (the eq/L of NaOH added to reach that point equals $2C_T$).

For any system of dissolved CO_2 plus strong base and/or plus strong acid, the ENE is

$$(C_B - C_A) = [HCO_3^-] + 2[CO_3^{2-}] + [OH^-] - [H^+] \tag{9.12}$$

$$= \alpha_1 C_T + 2\alpha_2 C_T + [OH^-] - [H^+]. \tag{9.13}$$

By Eqs.(9.11) and (9.13),

$$f = \alpha_1 + 2\alpha_2 + \frac{[OH^-] - [H^+]}{C_T}. \tag{9.14}$$

Eqs.(9.12–9.13) are valid whether C_T is fixed (closed system), or C_T is not fixed because exchange of CO_2 may occur between the solution and a gas phase (open system). The same is true for Eqs. (9.11) and (9.14), though these two equations are mostly used to discuss titration in fixed C_T systems.

Figure 9.2 provides curves of pH *vs.* f for systems with $C_T = 10^{-3.0}$, $10^{-2.5}$, $10^{-2.0}$ and $10^{-1.5}M$. For the three lowest C_T values, as may be understood based on Chapter 8, a distinct upturn in pH is

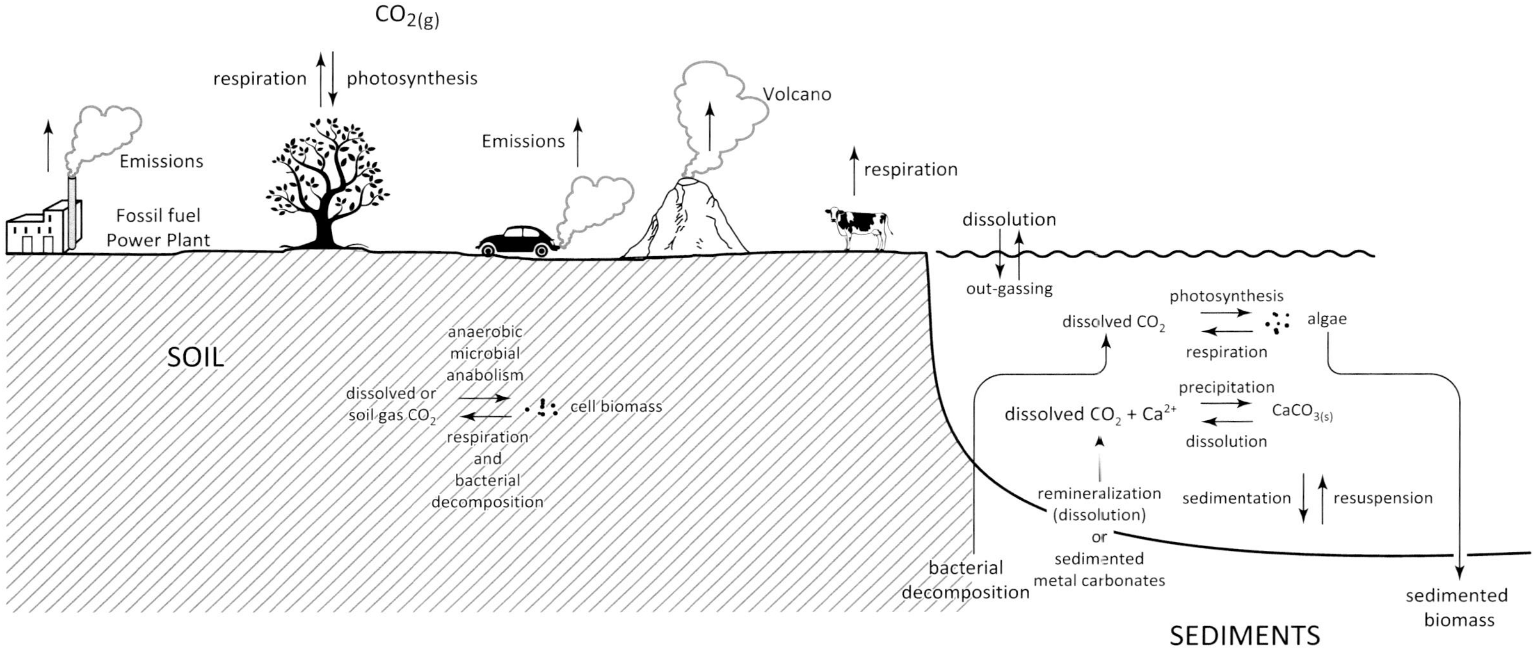

FIGURE 9.1 Processes involving CO_2.

TABLE 9.2
Equilibrium Constants for the Carbon Dioxide System as a Function of Temperature (Infinite Dilution)[a]

Temperature (°C)	pK_1	pK_2	$\log K_H$ (M/atm)	$\log K_w$
0	6.58	10.63	−1.11	−14.93
5	6.53	10.56	−1.19	−14.73
10	6.46	10.49	−1.27	−14.53
15	6.42	10.43	−1.32	−14.35
20	6.38	10.38	−1.41	−14.17
25	6.35	10.33	−1.47	−14.00
30	6.33	10.29	−1.53	−13.83
35	6.31	10.25	−1.58[b]	−13.68
40	6.30	10.22	−1.64	−13.53
50	6.29	10.17	−1.72	−13.26

[a] Values from for CO_2 constants from: Harned and Scholes (1941), Harned and Davies (1943), Ellis (1959), and Buch (1960). Values for K_w are based on $\log K_w = -4470.99/T + 6.0875 - 0.01706\,T$, where T (K) = temperature; from Harned and Owen (1958).

[b] Interpolated.

TABLE 9.3
Values of Log K_H (M/atm) for Five Gases

Temperature (°C)	CO_2	O_2	N_2	CH_4	H_2
10	−1.27	−2.77	−3.09	−2.73	−3.06
15	−1.34	−2.82	−3.13	−2.78	−3.08
20	−1.41	−2.86	−3.17	−2.83	−3.09
25	−1.47	−2.90	−3.20	−2.87	−3.11
30	−1.53	−2.93	−3.23	−2.91	−3.12
35	−1.58	−2.96	−3.26	−2.94	−3.13
40	−1.64	−2.98	−3.28	−2.97	−3.13

absent at $f = 2$ (Na_2CO_3 EP). The reason is that the second proton does not start to come off until the pH is approaching pK_2 (= 10.33), and $[OH^-]$ is by then approaching 10^{-4} M, and the OH^- buffering term ($2.303[OH^-]$) has become at least comparable to if not larger than the $[HCO_3^-]/[CO_3^{2-}]$ buffering term ($2.303\alpha_1\alpha_2C_T$). So, at $f = 2$ an inflection point (IP) is: (a) not present for $C_T = 10^{-3.0}$ M or $C_T = 10^{-2.5}$ M; and (b) only faintly present for $C_T = 10^{-2.0}M$.

In a CO_2 titration, the amount of strong acid or strong base that needs to be added to move from one point to another depends on C_T. Consider the travel analog in Figure 9.3 on the hypothetical Gaia Island. Mid City is located exactly between West City and East City. First, if the overall width of Gaia Island is 200 km, then the distance from East City to Mid City is 100 km, as is the distance from Mid City to West City. (If the width of Gaia Island is 300 km, each distance is 150 km.) Second, once the scale of the island is established, knowing how far you are east or west of any of the landmarks tells you how far you are from the other landmarks. For an island width of 200 km, if we are 50 km east of West City, we know we are 50 km west of Mid City, and 150 km west of East City.

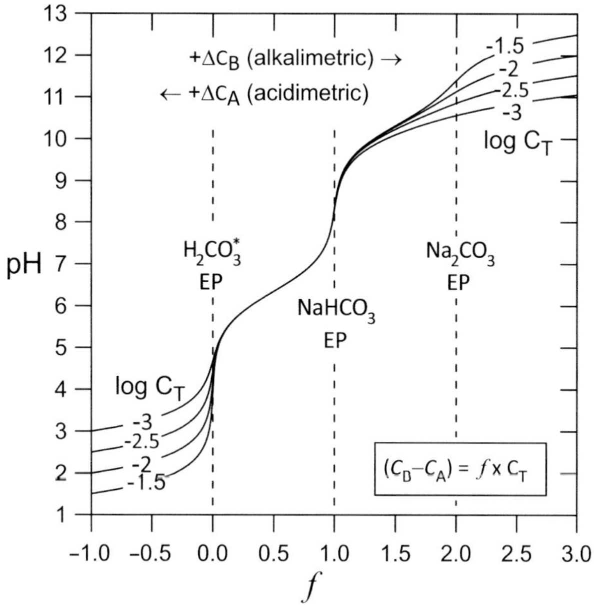

FIGURE 9.2 CO_2 system titration curves at 25 °C for $C_T = 10^{-3}$, $10^{-2.5}$, 10^{-2}, and $10^{-1.5}$ M. Activity corrections neglected.

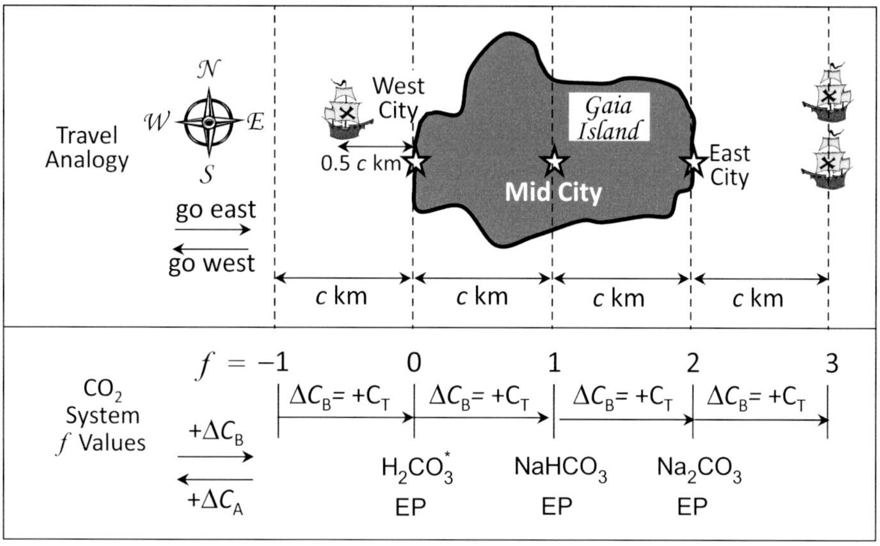

FIGURE 9.3 Travel on a hypothetical island as an analog for using $f = 0$, or 1, or 2 as reference points for location. The width of the island is $2c$ km. Mid City is in the middle of the island.

Thus, for a CO_2 system, if we know that we are at $f = 0.5$ and that $C_T = 10^{-3}$ M, we know that ΔC_A to take us back to $f = 0$ is 0.5×10^{-3} eq/L, and that ΔC_B to take us to $f = 2$ is 1.5×10^{-3} eq/L. Third, as in physical travel in the area of an island, it is possible to be out in the water west or east of the island: we could be 50 km west of West City, which by the way, is the same as being –50 km east of West City (cf. Eq.(9.17) below).

Example 9.3 C_T, Alk, and *f*

1 L of water has chemicals added to it in a series of 11 sequential steps. The water volume does not change. In step 1, 0.001 FW of $H_2CO_3^*$ is added. In step 2, 0.001 FW of $NaHCO_3$ is added, etc. Take the information in columns A through E to develop the results in columns F through I for the chemistry at the end of each step.

Solution

The values are provided; each situation where the resulting mixture to that point is at an "equivalence point" (EP) for the CO_2 system is identified.

		given			to be determined			
A	B	C	D	E	F	G	H	I
Step	chemical added in step	FW added in step	C_T factor for the step	strong base factor for the step	C_T (*M*)	Alk C_B-C_A (eq/L)	*f*	EP?
1	$H_2CO_3^*$	0.001	1	0	0.001	0.000	0.0	yes
2	$NaHCO_3$	0.001	1	1	0.002	0.001	0.5	
3	NaOH	0.001	0	1	0.002	0.002	1.0	yes
4	Na_2CO_3	0.001	1	2	0.003	0.004	1.33	
5	NaOH	0.002	0	1	0.003	0.006	2.0	yes
6	KOH	0.003	0	1	0.003	0.009	3.0	
7	HCl	0.003	0	−1	0.003	0.006	2.0	yes
8	HCl	0.003	0	−1	0.003	0.003	1.0	yes
9	HCl	0.003	0	−1	0.003	0.000	0.0	yes
10	HCl	0.003	0	−1	0.003	−0.003	−1.0	
11	NaOH	0.003	0	1	0.003	0.000	0.0	yes

9.2.2 Alk and H^+-Acy: The $f = 0$ ($H_2CO_3^*$) Equivalence Point

Alk (alkalinity) and H^+-Acy (a.k.a. H^+-acidity or "mineral acidity") are measures of how far a CO_2 system is from the $f = 0$ EP. Alk corresponds to the value of $(C_B - C_A)$, so Alk is synonymous with the concentration of net strong base, namely, how far are we to the "right" of $f = 0$. H^+-Acy corresponds to the value of $(C_A - C_B)$, so H^+-Acy is synonymous with the concentration of net strong acid, namely, how far are we to the "left" of $f = 0$:

$$\text{Alk} \equiv (C_B - C_A) \tag{9.15}$$

$$\text{H}^+\text{-Acy} \equiv (C_A - C_B) \tag{9.16}$$

$$\text{Alk} = -\text{H}^+\text{-Acy}. \tag{9.17}$$

If we know Alk, we know H^+-Acy. (If we know how far east we are of West City, we know how far west we are of West City.) The nature of the solution of interest will determine which is most convenient to use. If the solutions of interest generally have positive Alk values, it will be most natural to describe them all in terms of positive Alk values, not negative H^+-Acy values, *e.g.*:

You: "Hi Mom, I'm just calling to tell you I am now negative 50 km west of West City."
Mom: "What? You mean you are 50 km east of West City?"

Most natural surface and ground waters have positive Alk values, so chemical analysis reports for such waters (and the drinking waters derived from them) typically contain values as measured for Alk. On the other hand, "acid rain" samples have positive H^+-Acy values, so reports about them typically contain values as measured for H^+-Acy.

Alk and H^+-Acy are related in a fundamental way to the PBE at $f = 0$ (solution of $H_2CO_3^*$) which is

$$[H^+] = [OH^-] + [HCO_3^-] + 2[CO_3^{2-}] \qquad \text{PBE at } f = 0,\ H_2CO_3^* \text{ EP.} \tag{9.18}$$

If $f = 0$, by Eq.(9.11) then $(C_B - C_A) = 0$ and the above PBE is obviously satisfied.

But for $f > 0$, then relative to Eq.(9.18) we have RHS > LHS. Alk ≡ RHS − LHS, and as required by the ENE,

$$\text{Alk} \equiv (C_B - C_A) = [HCO_3^-] + 2[CO_3^{2-}] + [OH^-] - [H^+]. \tag{9.19}$$

Eq.(9.19) makes sense because it is a *sum of terms* for basic species in the solution, with $[CO_3^{2-}]$ needing a factor of 2 because it is dibasic, and $[H^+]$ needing the factor −1 because the proton is the anti-base. When $f > 0$ so that Alk > 0, then getting back to $f = 0$ requires addition of strong acid. For water samples with Alk > 0, Alk is typically measured in a laboratory by determining the value of ΔC_A required for that trip by acidimetric titration (see Figure 9.4), with the titration location of $f = 0$ identified by one of the methods in Table 9.4.

For $f < 0$, then relative to Eq.(9.18) we have LHS > RHS; so H^+-Acy ≡ LHS − RHS, and as required by the ENE,

$$H^+\text{-Acy} \equiv (C_A - C_B) = [H^+] - [HCO_3^-] - 2[CO_3^{2-}] - [OH^-]. \tag{9.20}$$

If $f < 0$ so that H^+-Acy > 0, then getting back to $f = 0$ requires addition of strong base. Indeed, for water samples with H^+-Acy > 0, like samples of "acid rain", then H^+-Acy is typically measured in a laboratory by determining the value of ΔC_B required for that trip by alkalimetric titration. Arrival at $f = 0$ is identified by when the pH jumps up as it goes through the inflection point at $f = 0$ (or by a Gran function) H^+-Acy is often referred to as the "mineral acidity" since strong acids are usually inorganic acids that are related to *minerals* (*e.g.*, HCl is relatable to $NaCl_{(s)}$ (*halite*), or $KCl_{(s)}$ (*sylvite*) etc.; HNO_3 is relatable to $NaNO_{3(s)}$ (*nitratine*) or $KNO_{3(s)}$ (*niter*), etc.; and H_2SO_4 is relatable to minerals of sodium sulfate and calcium sulfate.

Some natural waters contain basic species other than HCO_3^-, CO_3^{2-}, and OH^- that will take protons as we add strong acid to get back to $f = 0$ for the CO_2 system. The borate ion $B(OH)_4^-$, which is present in seawater at significant levels, is an important example. Ammonia is another. Considering these bases yields

$$\text{Alk} = [HCO_3^-] + 2[CO_3^{2-}] + [B(OH)_4^-] + [OH^-] + NH_3 - [H^+]. \tag{9.21}$$

The criterion for whether to include a particular base in the expression for Alk is that the K_b for the base is equal or greater than the K_b for HCO_3^-, as given by $K_w/K_1 = 10^{-14.0}/10^{-6.35} = 10^{-7.65}$ at 25 °C. Thus, the criterion for inclusion in Alk is that the base must take protons at least as readily as HCO_3^-. For boric acid ($B(OH)_3$) the infinite dilution value for K_a is $10^{-9.30}$:

$$B(OH)_3 + H_2O = B(OH)_4^- + H^+ \qquad K_a = 10^{-9.30}. \tag{9.22}$$

K_b for $B(OH)_4^-$ is then $10^{-14.0}/10^{-9.30} = 10^{-4.70}$ at 25 °C, which is indeed greater than K_b for HCO_3^-, hence Eq.(9.21).

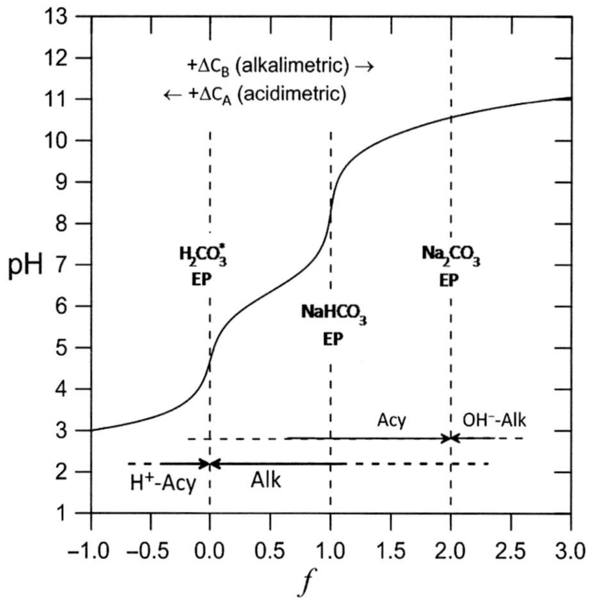

FIGURE 9.4 Representations of Alk and H^+-Acy given as distances from the $f = 0$ EP. Alk corresponds to the number of equivalents per liter of strong acid that need to be added to get back to $f = 0$. H^+-Acy corresponds to the number of equivalents of strong base that need to be added per liter of water to get back to $f = 0$. Representations of quantities called OH^--Alk and Acy are given in terms of distances from the $f = 2$ EP. While the titration curve shown is for $C_T = 10^{-3}$ *M*, the schematic representations of Alk, H^+-Acy, OH^--Alk, and Acy apply to any value of C_T.

TABLE 9.4
Methods for Locating the $f = 0$ Endpoint in Determination of Alk by Acidimetric Titration

Method 1. Make the best possible guess for C_T in the solution, then calculate the pH for a solution of $H_2CO_3^*$ of that C_T. Measure the value of v_t required to reach that pH. (See Example 9.6.) Because the pH drops steeply near $f = 0$, some error in the best guess for pH for $f = 0$ will not translate into a problematic error unless C_T or Alk is small.

Method 2. Record the pH *vs.* v_t titration curve past $f = 0$ and into the region where $f < 0$, and plot dpH/dv_t, the derivative of the titration curve. The sought for IP in the titration curve (*i.e.*, $d^2\text{pH}/dv_t^2 = 0$) occurs essentially at $f = 0$ where $d\text{pH}/dv_t$ has a minimum (because then $(d/dv_t)(d\text{pH}/dv_t) = 0 = d^2\text{pH}/dv_t^2 = 0$).

Method 3. Record the pH *vs.* v_t titration curve as in Method 2. Then, for sample volume V_o (L), plot the Gran function $F_{f<0}^{CO_3^{2-}} = (V_o(\text{L}) + v_t(\mu\text{L})/(10^6\mu\text{L/L})\times 10^{-\text{pH}}$ *vs.* v_t (see Chapter 7). In many cases adding v_t to the sample volume hardly changes $(V_o + v_t/(10^6)$ during the titration, so one can simply plot $10^{-\text{pH}} = [H^+]$ *vs.* v_t. Method 3 works because Alk is the acid-neutralizing capacity of the solution. That capacity is brought to zero once $f = 0$ is reached. As the solution moves beyond $f = 0$, H^+-Acy $\approx [H^+]$ increases linearly with further increases in v_t; extrapolating the line for $F_{f<0}^{CO_3^{2-}}$ (or simply $10^{-\text{pH}} = [H^+]$) down to the x-axis ($[H^+] \approx 0$) gives the value for v_t at $f = 0$.

Note: Methods 1 and 3 are included in the section for Alk in *Standard Methods for Examination of Water and Wastewater*, APHA et al., 2012. Method 3 is recommended for cases when C_T or Alk is small.

Example 9.4 Alk from C_T and pH

Some fresh water at 25 °C was measured to have pH = 8.20 and $C_T = 1.1 \times 10^{-3}$ *M*. What is Alk? Neglect activity corrections.

Solution

At pH 8.20, $\alpha_1 = 9.79 \times 10^{-1}$; $\alpha_2 = 7.26 \times 10^{-3}$; $[HCO_3^-] = \alpha_1 C_T = 1.08 \times 10^{-3}$ *M*; $[CO_3^{2-}] = \alpha_2 C_T = 7.98 \times 10^{-6}$ *M*; $[OH^-] = 10^{-5.80}$ *M*; and $[H^+] = 10^{-8.20}$ *M*.

$\text{Alk} = [HCO_3^-] + 2[CO_3^{2-}] + 10^{-5.80} - 10^{-8.20} = 1.09 \times 10^{-3}$ eq/L.

For fresh waters, if Alk > 0, often a significant portion of the Alk is a consequence of the dissolution of some type of carbonate rock. The most important of these are forms is $CaCO_{3(s)}$. Alk is therefore often expressed in units of mg of dissolved $CaCO_{3(s)}$ per liter of water. The FW of $CaCO_{3(s)}$ is 100.087 g/FW, which very conveniently rounds closely to 100 g/FW, or 100,000 mg/FW. Each FW of $CaCO_{3(s)}$ carries 2 eq of net strong base, giving 50,000 mg of $CaCO_{3(s)}$ per eq. To convert between the two sets of units for Alk we have

$$\text{Alk (mg CaCO}_3\text{/L)} = \text{Alk (eq/L)} \times \frac{50{,}000 \text{ mg CaCO}_3}{\text{eq}}. \tag{9.23}$$

Example 9.5 Units for Alk

a. For Alk = 30, 150, and 500 mg $CaCO_3$/L, convert to units of eq/L.
b. For Alk = 0.25×10^{-3}, 0.5×10^{-3}, and 1.0×10^{-3} eq/L, convert to units of mg $CaCO_3$/L.

Solution

Alk (mg $CaCO_3$/L) = Alk (eq/L) × 50,000 mg $CaCO_3$/L

	a.			b.		
mg $CaCO_3$/L	30	150	500	**12.5**	**25**	**50**
eq/L	**6.0×10^{-4}**	**3.0×10^{-3}**	**1.0×10^{-2}**	2.5×10^{-4}	5.0×10^{-4}	1.0×10^{-3}

As noted above, Alk > 0 values are measured most commonly by acidimetric titration, recording the pH during incremental additions in the volume v_t (μL) of a strong acid titrant solution with concentration $c_{t,a}$ (*N*). If Alk > 0, then $f > 0$ at the start of the titration; the titration is usually carried out through $f = 0$ and into the region where $f < 0$. The goal is to determine the value of v_t that brought the solution to $f = 0$ as the endpoint. Identifying when we have arrived at $f = 0$ can be determined by the three different methods in Table 9.4; the second and third are the most accurate.

Example 9.6 Using a Predicted Endpoint pH (Method 1) to Find the Endpoint in a Titration for Alk

The following is an excerpt from *Standard Methods for Examination of Water and Wastewater* (APHA et al., 2012):

> When alkalinity is due entirely to carbonate or bicarbonate content, the pH at the equivalence point of the titration is determined by the concentration of carbon dioxide (CO_2) at that stage. CO_2 concentration depends, in turn, on the total carbonate species originally present and any losses that may have occurred during titration. The [following] pH values ... are suggested as the equivalence points for the corresponding alkalinity concentrations as milligrams $CaCO_3$ per liter ... 30 mg/L, pH = 4.9; 150 mg/L, pH = 4.6; 500 mg/L, pH = 4.3.

When pH is inside the range ~6 < pH < 10, Alk is borne mostly by $[HCO_3^-]$, and for ambient atmospheric p_{CO_2} values, C_T (*M*) ≈ Alk (eq/L). Example 9.5 converted Alk values of 30, 150, and 500 mg/L to units of eq/L. Determine the pH at 20 °C (a typical laboratory temperature) at $f = 0$ for the corresponding C_T values: 6.0×10^{-4}, 3.0×10^{-3}, and 1.0×10^{-2} *M*.

Solution

We use the *K* values for 20 °C in Table 9.2 and solve the ENE for $f = 0$ for each C_T. This gives pH = 4.81, 4.45, and 4.19, respectively. These values are all about 0.1 pH unit lower than the values cited above from *Standard Methods*. This difference is consistent with an apparent assumption in *Standard Methods* of some "losses" of CO_2 during titration. (As strong acid titrant is being added, $[H_2CO_3^*]$ rises in the solution under titration, causing some CO_2 to leave, which makes the pH not as low at $f = 0$ than if no losses had occurred.)

Example 9.7 Using the First Derivative (Method 2) and 10^{-pH} (Method 3) to Find the Endpoint in a Titration for Alk

A 100 mL (0.1 L) sample of water is titrated at 20 °C with 0.1 *N* HCl ($c_{t,b} = 0.1$ eq/L). pH is measured *vs.* the volume v_t (μL) of titrant added. Data are provided, though somewhat coarsely from $v_t = 0$ to 950 μL, and for $v_t \geq 975$ μL. Finer resolution will usually be available over a whole titration curve in most laboratory analyses, but here such a table would be too long. Plot the titration curve as pH *vs.* v_t. Calculate and plot the first derivative $d\text{pH}/dv_t$ *vs.* v_t, and 10^{-pH} *vs.* v_t, and use each to determine Alk for the water sample. (Example 7.3 suggests a way to calculate $d\text{pH}/dv_t$ with data from adjacent points.)

Solution

Figures 9.5.a-c give the titration, $d\text{pH}/dv_t$, and 10^{-pH} plots. The application of the $d\text{pH}/dv_t$ method indicates that v_t for the endpoint is ~970 μL. The number of equivalents of strong acid in that volume is $0.1\ \text{eq/L}\times970\times10^{-6}\ \text{L} = 9.70\times10^{-5}$ eq. This number of equivalents of strong acid neutralized the same number of eq of Alk in the original 0.1 L sample, so for the original sample, Alk = 9.70×10^{-5} eq/(0.1 L) = 9.70×10^{-4} eq/L. The application of the Gran-based 10^{-pH} method by indicates that v_t for the endpoint is ~969 μL. The number of equivalents of strong acid in that volume is then 9.69×10^{-5} eq, so Alk = 9.69×10^{-4} eq/L.

Example 9.7 Values for titration curve

Given		Calculated	
v_t (μL)	pH	dpH/dv_t	10^{-pH}
0	7.90	–	1.26E-08
50	7.46	–6.80E-03	3.47E-08
100	7.22	–4.09E-03	6.02E-08
150	7.05	–3.05E-03	8.90E-08
200	6.92	–2.51E-03	1.22E-07
250	6.80	–2.19E-03	1.59E-07
300	6.70	–1.99E-03	2.01E-07
350	6.60	–1.86E-03	2.51E-07
400	6.51	–1.78E-03	3.09E-07
450	6.42	1.74E-03	3.77E-07
500	6.34	–1.74E-03	4.61E-07
550	6.25	–1.78E-03	5.64E-07
600	6.16	–1.85E-03	6.94E-07
650	6.06	–1.98E-03	8.64E-07
700	5.96	–2.18E-03	1.10E-06
750	5.85	–2.49E-03	1.43E-06
800	5.71	–2.83E-03	1.95E-06
825	5.63	–3.33E-03	2.33E-06
850	5.54	–3.84E-03	2.86E-06
875	5.44	–4.58E-03	3.63E-06
900	5.32	–5.67E-03	4.84E-06
925	5.16	–7.33E-03	6.97E-06
950	4.95	–9.36E-03	1.12E-05
965	4.80	–1.05E-02	1.60E-05
975	4.69	–1.03E-02	2.05E-05
980	4.64	–9.00E-03	2.32E-05
1000	4.44	–7.74E-03	3.67E-05
1025	4.24	–6.83E-03	5.77E-05
1050	4.09	–5.16E-03	8.06E-05
1075	3.98	–4.06E-03	1.04E-04
1100	3.89	–3.33E-03	1.29E-04
1125	3.81	–2.81E-03	1.53E-04
1150	3.75	–2.42E-03	1.78E-04
1175	3.69	–2.13E-03	2.03E-04
1200	3.64	–	2.27E-04

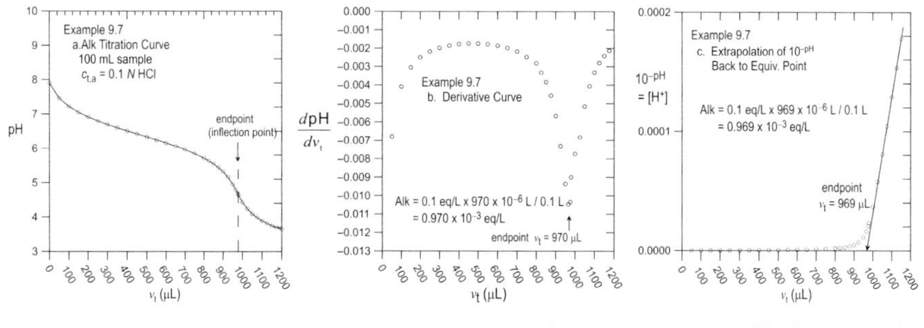

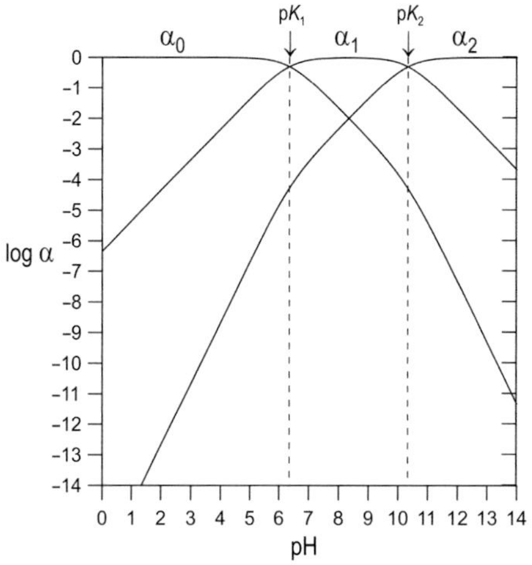

FIGURE 9.5 Log α *vs.* pH curves for the CO_2 system at 25 °C. $\alpha_0 = \alpha_1$ at pH = pK_1; and $\alpha_1 = \alpha_2$ at pH = pK_2.

9.2.3 The Base-Side Analogs of H^+-Acy and Alk: OH^--Alk and Acy and the $f = 2$ (Na_2CO_3) Equivalence Point

Just as CO_2 systems are commonly described using Alk and H^+-Acy to indicate how far each is from $f = 0$, we can describe a CO_2 system by how far it is from $f = 1$ or $f = 2$. While $f = 1$ is rarely used, $f = 2$ as a reference is occasionally used; we discuss it briefly here.

The PBE at the Na_2CO_3 EP where $f = 2$ is

$$[H^+] + 2[H_2CO_3^*] + [HCO_3^-] = [OH^-] \qquad \text{PBE at } f = 2,\ Na_2CO_3 \text{ EP.} \tag{9.24}$$

Proceeding in a manner analogous to that in the previous section, for $f > 2$, we have RHS > LHS, so OH^--Alk ≡ RHS − LHS:

$$OH^-\text{-Alk} \equiv [OH^-] - [H^+] - 2[H_2CO_3^*] - [HCO_3^-]. \tag{9.25}$$

We note that OH^--Alk is sometimes called the "caustic" alkalinity.

$$C_T = [H_2CO_3^*] + [HCO_3^-] + [CO_3^{2-}] \tag{9.9}$$

$$2C_T = 2[H_2CO_3^*] + 2[HCO_3^-] + 2[CO_3^{2-}] \tag{9.26}$$

$$2[H_2CO_3^*] + [HCO_3^-] = 2C_T - [HCO_3^-] - 2[CO_3^{2-}]. \tag{9.27}$$

Combining Eqs.(9.25) and (9.27) give

$$OH^-\text{-Alk} = [OH^-] - [H^+] - 2C_T + [HCO_3^-] + 2[CO_3^{2-}] \tag{9.28}$$

$$= [HCO_3^-] + 2[CO_3^{2-}] + [OH^-] - [H^+] - 2C_T \tag{9.29}$$

$$= (C_B - C_A) - 2C_T. \tag{9.30}$$

If we are at $f = 2$, then $(C_B - C_A) = 2C_T$ (as with a solution of Na_2CO_3), and OH^--Alk = 0. Thus, OH^--Alk represents however much net strong base is in the solution in *excess* of that required to reach $f = 2$.

For $f < 2$, we have LHS > RHS of Eq.(9.24); so Acy ≡ LHS − RHS:

$$\text{Acy} \equiv [H^+] + 2[H_2CO_3^*] + [HCO_3^-] - [OH^-] \quad (9.31)$$

$$OH^-\text{-Alk} = -\text{Acy}. \quad (9.32)$$

Like Alk and H^+-Acy, the quantities OH^--Alk and Acy have units of eq/L. OH^--Alk gives the total eq/L of all bases in the solution that are at least as basic as OH^-. Acy gives the total eq/L of all protons in the solution that are at least as acidic as HCO_3^-. In water, in fact, there cannot be bases stronger than OH^- because if a base stronger than OH^- were placed in water, call it Z^-, then Z^- would react essentially stoichiometrically with water to yield OH^-. The reaction would be $Z^- + H_2O = HZ + OH^-$. This is the base version of the "acid/base leveling effect" of water. (For the acid version of that effect, there is no acid in water stronger than water-solvated H^+ because if an acid stronger than water-solvated H^+ were placed in water, it would virtually completely give up its protons to water to form water-solvated H^+; that is what happens when molecular HCl dissolves in water.)

9.2.4 Alk = ANC, Acy = BNC

The terms Acid Neutralizing Capacity (ANC) and Base Neutralizing Capacity (BNC) are sometimes used to refer to how much **strong acid** or **strong base** a system is capable of neutralizing, respectively. Since Alk is a measure of how much strong acid would be neutralized if a system with $f > 0$ was moved to the $f = 0$ EP (to form a solution of $H_2CO_3^*$),

$$\text{Alk} = \text{ANC}. \quad (9.33)$$

Since Acy is a measure of how much strong base would be neutralized if a system with $f < 2$ was moved to the $f = 2$ EP (to form a solution Na_2CO_3),

$$\text{Acy} = \text{BNC}. \quad (9.34)$$

9.3 CONSERVATION OF ALKALINITY

9.3.1 General

There are many situations in which the *number of equivalents* of net strong base do not change as a system moves from one condition to another. In such circumstances, constancy in the equivalents of net strong base becomes a governing constraint on the problem.

9.3.2 Conservation of Alk during Addition or Removal of CO_2

If CO_2 is **added** to a solution, although C_T will increase, by itself the addition will not cause a change in either C_B or C_A, so while the pH will go down, Alk = $(C_B - C_A)$ stays constant. Examples:

1) Some CO_2 from the atmosphere dissolves into some water for any reason;
2) Dissolved CO_2 is added to water by organisms carrying out respiration according to $C_6H_{12}O_6 + 6O_2 = 6CO_2 + 6H_2O$.

Similarly, if CO_2 is **removed** from a solution, although C_T will decrease, by itself the removal will not cause a change in either C_B or C_A, so while the pH will go up, Alk stays constant. Examples:

1) Lakewater or some carbonated beverage is bubbling and losing CO_2;
2) Lakewater or a carbonate beverage is not releasing bubbles of CO_2, but is still off-gassing CO_2; and
3) Dissolved CO_2 is removed from lakewater by algae carrying out photosynthesis according to $6CO_2 + 6H_2O = C_6H_{12}O_6 + 6O_2$. This can occur over large portions of lakes and sections of the oceans during algae blooms (see Section 9.3.5). The loss of CO_2 from solution can subsequently lead to a "whiting event" because of precipitation of $CaCO_3$ (which *then* will reduce Alk), visible from planes and satellites (such images are easily found on the internet).

9.3.3 Conservation of Alk during Changes in *T* or *P*

A change in T (or P) does not by itself add any equivalents of strong base or strong acid. So, when a change in T (or P) does not cause a significant change in the volume of a sample, there is no significant change in Alk (eq/L). When a change in T (or P) does cause a significant change in the volume, there are two options: (1) because the number of equivalents of net strong base for some specific initial volume does not change, just correct for the volume change to obtain the new Alk in units of eq/L; or (2) as is commonly used for seawater, utilize Alk in units of eq/kg of solution, which will not change with T (or P).

9.3.4 Conservation of Total Equivalents of Net Strong Base during Mixing of Solutions

Consider water sample 1 with value Alk_1 and volume V_1 (L) that is being mixed with water sample 2 with value Alk_2 and volume V_2 (L). The number of equivalents of net strong base in sample 1 is Alk_1V_1. The number of equivalents of net strong base in sample 2 is Alk_2V_2. The total overall amount number of equivalents of net strong base is $\text{Alk}_1V_1 + \text{Alk}_2V_2$. The new volume is $V_1 + V_2$. For the mixture,

$$\text{Alk}_{\text{mix}} = \frac{\text{Alk}_1V_1 + \text{Alk}_2V_2}{V_1 + V_2}. \tag{9.35}$$

Example 9.8 Conservation of Alk with Mixing

100 mL of 10^{-3} F $NaHCO_3$ is mixed with 500 mL of 10^{-4} F Na_2CO_3 in a closed system at 25 °C. Neglect activity corrections.

a. Calculate Alk for the mixture.
b. Calculate C_T for the mixture.
c. Based on Alk and C_T, what is the pH of the closed-system mixture?

Solution

a. $\text{Alk} = \dfrac{0.1\,\text{L} \times 10^{-3}\ \text{eq/L} + 0.5\,\text{L} \times 2 \times 10^{-4}\ \text{eq/L}}{0.1\,\text{L} + 0.5\,\text{L}} = 3.33 \times 10^{-4}\ \text{eq/L}$

b. $C_T = \dfrac{0.1\,\text{L} \times 10^{-3}\ M + 0.5\,\text{L} \times 10^{-4}\ M}{0.1\,\text{L} + 0.5\,\text{L}} = 2.50 \times 10^{-4}\ M$

c. $\text{Alk} = \alpha_1C_T + 2\alpha_2C_T + K_w/[H^+] - [H^+]$. Solving by LHS – RHS = 0 for 25 °C gives pH = 9.62.

9.3.5 Lack of Conservation of Alk during Precipitation or Dissolution of Carbonate Minerals

Alk is not conserved in a water sample whenever a mineral with acid or base properties either dissolves into the solution or precipitates from the solution as in a "whiting event" in a lake or the oceans. Carbonate minerals like calcite, aragonite, and dolomite are basic. In order, these minerals are $CaCO_{3(s)}$ in the "hexagonal" crystal form, $CaCO_{3(s)}$ in the "orthorhombic" crystal form, and $CaMgCO_{3(s)}$ in the "hexagonal" crystal form.

When $CaCO_{3(s)}$ or $CaMgCO_{3(s)}$ dissolves, Alk increases. Each mol of dissolving metal carbonate brings in two eq of net strong base (manifested in the CO_3^{2-} and traced by Ca^{2+} and/or Mg^{2+}). Ca^{2+} and Mg^{2+} behave as spectator ions, except at very high pH. For dissolution of $CaCO_{3(s)}$, then

$$CaCO_{3(s)} \rightarrow Ca^{2+} + CO_3^{2-} \qquad \Delta\text{Alk} = 2\Delta\left[Ca^{2+}\right] = 2\Delta C_T > 0. \tag{9.36}$$

When $CaCO_{3(s)}$ precipitates, Alk decreases, and again $\Delta\text{Alk} = 2\Delta[Ca^{2+}]$ (because $\Delta[Ca^{2+}]$ is now negative):

$$Ca^{2+} + CO_3^{2-} \rightarrow CaCO_{3(s)} \qquad \Delta\text{Alk} = 2\Delta\left[Ca^{2+}\right] = 2\Delta C_T < 0. \tag{9.37}$$

The rise in pH discussed in Section 9.3.2 during an algae bloom can be sufficiently large that even though C_T goes down, the product $\alpha_2 C_T = [CO_3^{2-}]$ goes up enough that aragonite $CaCO_{3(s)}$ becomes supersaturated: the ion product $[Ca^{2+}][CO_3^{2-}]$ becomes greater than the solubility product for aragonite, and the result is a "whiting event" (particles of white $CaCO_{3(s)}$ in the water). Alk in the solution is thereby reduced because precipitation of $CaCO_{3(s)}$ removes CO_3^{2-} from the solution, along with the corresponding amount of the strong-base-tracer Ca^{2+}.

9.4 Log CONCENTRATION DIAGRAMS IN CLOSED (FIXED C_T) CO_2 SYSTEMS

9.4.1 General

For a system with a specific C_T, the log[ ] *vs.* pH diagram is obtained based on the relations in Eq.(9.10) so that

$$\log[H_2CO_3^*] = \log\alpha_0 + \log C_T \tag{9.38}$$

$$\log[HCO_3^-] = \log\alpha_1 + \log C_T \tag{9.39}$$

$$\log[CO_3^{2-}] = \log\alpha_2 + \log C_T. \tag{9.40}$$

Each log[ ] expression contains a $\log\alpha$ term and the offset term $\log C_T$. Since $H_2CO_3^*$ is a diprotic acid, from Chapter 6, neglecting activity corrections

$$\alpha_0 = \frac{1}{1 + K_1/[H^+] + K_1K_2/[H^+]^2} \qquad \alpha_1 = \alpha_0\frac{K_1}{[H^+]} \qquad \alpha_2 = \alpha_1\frac{K_2}{[H^+]}. \tag{9.41}$$

9.4.2 Log Concentration *vs.* pH Diagrams for Systems with Fixed C_T

Figure 9.5 is a plot with $\log\alpha_0$, $\log\alpha_1$, and $\log\alpha_2$, *vs.* pH. Figure 9.6 is a $\log\alpha$ plot on the same scale for use as a transparency with Figure 6.4, to create diagrams like Figure 9.7.

In a log[ ] *vs.* pH plot for a monoprotic acid HA, such as Figure 6.3, at pH << pK_a, for the log[HA] line, the asymptotic slope is 0 and for the log[A^-] line the asymptotic slope is +1. Conversely, for pH >> pK_a, for the log[HA] line, the asymptotic slope is −1, and for the log[A^-] line the asymptotic

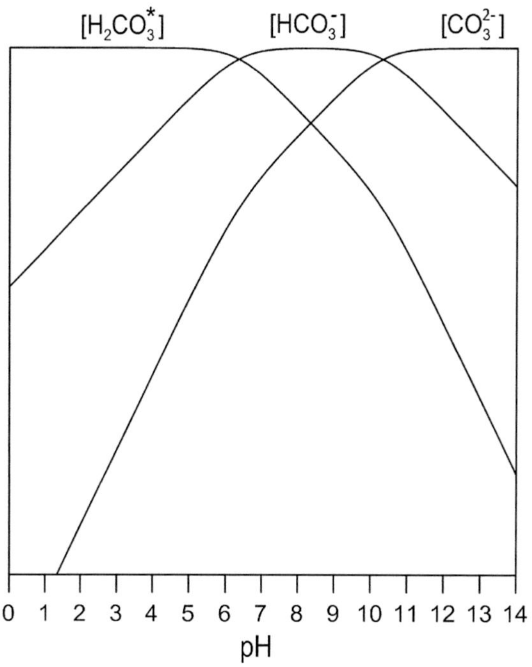

FIGURE 9.6 Generic log[ ] *vs.* pH diagram for CO_2 systems with fixed C_T at 25 °C; for use as a transparency with Figure 6.4.

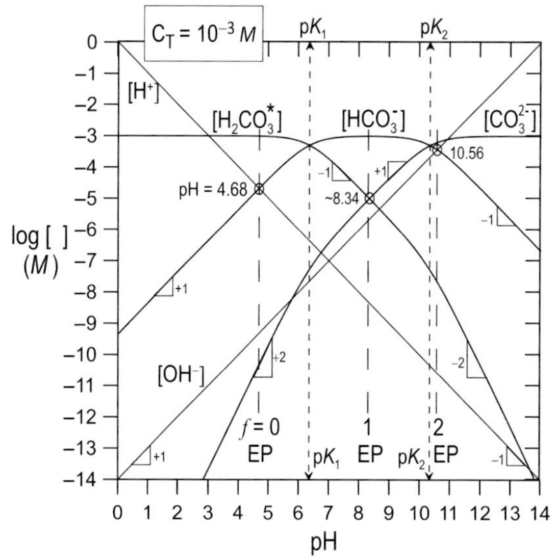

FIGURE 9.7 Log[ ] *vs.* pH diagram with $C_T = 10^{-3}$ *M* at 25 °C. At pH = pK_1, then $[H_2CO_3^*] = [HCO_3^-]$; the linearized and extrapolated legs for log[ ] (not shown) cross at the point log C_T, pK_1. At pH = pK_2, then $[HCO_3^-] = [CO_3^{2-}]$; the linearized and extrapolated legs for log[ ] (not shown) cross at the point log C_T, pK_2. The pH for the $f = 0$ EP is marked, as located by the approximation (very good at this C_T) for the PBE at $f = 0$ that $[H^+] \approx [HCO_3^-]$. The pH for the $f = 1$ EP is marked, as located by the approximation (fairly good at this C_T) for the PBE at $f = 1$ that $[HCO_3^-] \approx [CO_3^{2-}]$. The pH for the $f = 2$ EP is marked, as located by the approximation (very good at this C_T) for the PBE at $f = 2$ that $[HCO_3^-] \approx [OH^-]$.

slope is 0. This behavior can be seen in Figure 9.7 with $H_2CO_3^*$ and HCO_3^- behaving like an HA/A^- pair around pK_1, and HCO_3^- and CO_3^{2-} behaving like another HA/A^- pair around pK_2. However, at pH << pK_1, the $\log[CO_3^{2-}]$ line has a steepened asymptotic slope of +2, and at pH >> pK_2, the $\log[H_2CO_3^*]$ line has a steepened asymptotic slope of −2.

9.4.3 The f = 0, 1, and 2 Equivalence Points (EPs) in a Log Concentration *vs.* pH Diagram with Fixed C_T

The pH range in Figure 9.7 extends from low pH values to high pH values so the plot extends from $f < 0$ to $f > 2$. In this range, the pH value for each of the three EPs ($f = 0, f = 1$, and $f = 2$) will be found. As in Chapter 6 where solutions for a monoprotic acid system are considered, we can locate the approximate pH position for each EP by consideration of the PBE for the EP.

$H_2CO_3^*$ (f = 0) Equivalence Point. We have:

$$[H^+] = [OH^-] + [HCO_3^-] + 2[CO_3^{2-}] \qquad \text{PBE for } H_2CO_3^* \text{ EP } (f = 0). \tag{9.42}$$

In Figure 9.7, for which $C_T = 10^{-3}$ M, we seek the pH value where Eq.(9.42) is satisfied. Since C_T is not small and since $H_2CO_3^*$ is acidic, it is reasonable to try and assume that both $[OH^-]$ and $[CO_3^{2-}]$ are negligible for the RHS of Eq.(9.42). Thus, we seek the point where

$$[H^+] \approx [HCO_3^-] \qquad \text{approximate PBE for } H_2CO_3^* \text{ EP } (f = 0). \tag{9.43}$$

In Figure 9.7, the lines $\log[H^+]$ and $\log[HCO_3^-]$ cross at pH = 4.68. At this pH, both $\log[OH^-]$ and $[CO_3^{2-}]$ are far below the intersection, so the speciation for a 10^{-3} F solution of $H_2CO_3^*$ is indeed given to a very good approximation by the pH at the intersection $\log[H^+] = \log[HCO_3^-]$.

Further, we note that since the first proton on $H_2CO_3^*$ is weakly acidic, and since the second proton is *very* weakly acidic, the pH of a not-too dilute solution of $H_2CO_3^*$ may be considered to behave as a solution of a weak *monoprotic* weak acid, and $[H^+]$ may be estimated using Eq.(5.40) with K_1 substituted for K_a, and C_T substituted for A_T. This amounts to adopting the acidic solution approximation (ASA) *plus* the weak acid approximation (WAA). Thus,

$$[H^+] \approx \sqrt{K_1 C_T}. \tag{9.44}$$

At 25 °C, where $K_1 = 10^{-6.35}$, for $C_T = 10^{-3}$, this gives pH = 4.68. For ever-decreasing values of C_T, the abilities of Eq.(9.43) and Eq.(9.44) to combine to give a good estimate of the pH will ultimately fail as the pH approaches 7.00. This can demonstrated using a transparency of Figure 9.6 with Figure 6.4 to find the C_T-dependent pH that satisfies Eq.(9.42).

$NaHCO_3$ (f = 1) Equivalence Point. We have:

$$[H^+] + [H_2CO_3^*] = [OH^-] + [CO_3^{2-}] \qquad \text{PBE for } NaHCO_3 \text{ EP } (f = 1). \tag{9.45}$$

In Figure 9.7, for which $C_T = 10^{-3}$ M, we seek the pH value where Eq.(9.45) is satisfied. Since HCO_3^- is very weakly acidic but also weakly basic, whenever C_T is not too small, it will be reasonable to try and assume that both $[H^+]$ and $[OH^-]$ are negligible in Eq.(9.45):

$$[H_2CO_3^*] \approx [CO_3^{2-}] \qquad \text{approximate PBE for } NaHCO_3 \text{ EP } (f = 1). \tag{9.46}$$

In Figure 9.7, the lines $\log[H_2CO_3^*]$and $\log[CO_3^{2-}]$ cross at pH = 8.34. At this pH, $\log[H^+]$ is far below the intersection, but $\log[OH^-]$ is only ~0.7 log units below the intersection. So, at this C_T, neglecting $[H^+]$ in the PBE is an excellent approximation, but neglecting $[OH^-]$, while not an excellent assumption, is not a horrendous assumption either: the exact pH is 8.30 ≈ 8.34.

The pH of the intersection $\log[H_2CO_3^*] = \log[CO_3^{2-}]$ does not change with C_T. This means that as long as C_T is not low (say $\geq \sim 10^{-3}$ M) all solutions of $NaHCO_3$ will have a pH that is midway between the two pK values. We see this in Figure 9.2 in that the titration curves for the different C_T values very nearly lie directly on top of one another at $f=1$, with pH ≈ 8.34.

The above conclusion about the pH at $f=1$ can also be obtained by examining the product K_1K_2. Neglecting activity corrections,

$$K_1K_2 = \frac{[CO_3^{2-}][H^+]^2}{[H_2CO_3^*]}. \tag{9.47}$$

When Eq.(9.46) is a reasonable assumption, then

$$K_1K_2 \approx [H^+]^2, \qquad \text{pH} = \tfrac{1}{2}(\text{p}K_1 + \text{p}K_2). \tag{9.48}$$

The fact that a solution of $NaHCO_3$ is slightly **basic** is a consequence of the fact that HCO_3^- is a stronger base than it is an acid. At 25 °C, K_b for HCO_3^- is $1.01\times10^{-14}/10^{-6.35} = 2.26\times10^{-8}$, while K_a for HCO_3^- is $10^{-10.33} = 4.68\times10^{-11}$. In contrast, solutions of $NaHSO_4$ are acidic because HSO_4^- is a much stronger acid than it is a base (K_2 is $\sim10^{-2}$ as compared to $1.01\times10^{-14}/K_1 = 1.01\times10^{-14}/10^4 = 10^{-18}$; K_1 for H_2SO_4 in water is very poorly known, and is here only roughly estimated to be 10^4);

Na_2CO_3 ($f = 2$) Equivalence Point. We have

$$[H^+] + [HCO_3^-] + 2[H_2CO_3^*] = [OH^-] \qquad \text{PBE for } Na_2CO_3 \text{ EP } (f = 2). \tag{9.49}$$

In Figure 9.7, for which $C_T = 10^{-3}$ M, we seek the pH value where Eq.(9.49) is satisfied. The solution will be basic. Since C_T is not small, it will be reasonable to try and assume that both $[H^+]$ and $[H_2CO_3^*]$ are negligible. Thus, we seek the point where

$$[HCO_3^-] \approx [OH^-] \qquad \text{approximate PBE for } Na_2CO_3 \text{ EP } (f = 2). \tag{9.50}$$

In Figure 9.7, the lines for $\log[HCO_3^-]$ and $\log[OH^-]$ cross at pH = 10.56. At this pH, both $\log[H^+]$ and $\log[H_2CO_3^*]$ are far below the intersection, so Eq.(9.50) is an excellent approximation. The manner in which the pH will approach 7.00 (at 25 °C) as C_T is lowered can be examined using a transparency of Figure 9.6 with Figure 6.4.

Since CO_3^{2-} is not a strong base, and since HCO_3^- is an even weaker base, the pH of a not-too-dilute solution of Na_2CO_3 may be considered to behave like a solution of NaA where A^- is basic, and $[OH^-]$ may be estimated with K_w/K_2 (which is K_b for CO_3^{2-}) substituted for K_b in Eq.(5.70). This amounts to making the basic solution approximation (BSA) *plus* the weak base approximation (WBA). Thus, we have

$$[OH^-] \approx \sqrt{K_w C_T / K_2} \tag{9.51}$$

At 25 °C where $K_1 = 10^{-6.35}$, for $C_T = 10^{-3}$ M, Eq.(9.51) gives pH = 10.68, which is higher than the actual value. As can be seen from Figure 9.8, the reason is that CO_3^{2-} is too strong a base for the WBA, *i.e.*, there is some error in assuming $[CO_3^{2-}] \approx C_T$.

9.4.4 Buffer Intensity for Closed (Fixed C_T) CO_2 Systems

The buffer intensity β is defined in Chapter 8 as

$$\beta \equiv \frac{d(C_B - C_A)}{d\text{pH}}. \tag{9.52}$$

When C_T is constant as in a closed system, by Eq.(9.12),

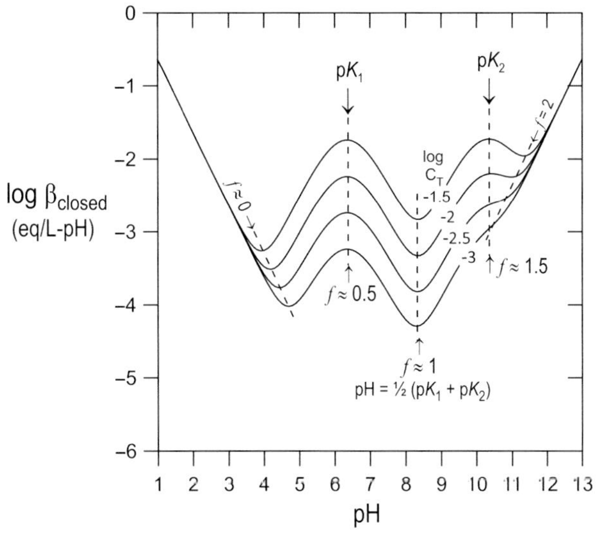

FIGURE 9.8 Log β_{closed} *vs.* pH diagram for closed CO_2 systems at 25 °C and four values of log C_T.

$$\beta_{\text{closed}} = \left(\frac{\partial(C_B - C_A)}{\partial \text{pH}}\right)_{C_T} = C_T\frac{d\alpha_1}{d\text{pH}} + 2C_T\frac{d\alpha_2}{d\text{pH}} + \frac{d[\text{OH}^-]}{d\text{pH}} - \frac{d[\text{H}^+]}{d\text{pH}}. \tag{9.53}$$

Expressions for the last two terms in Eq.(9.53) are given by Eqs.(8.16) and (8.17); including signs, they are 2.303[OH^-] and 2.303[H^+], respectively. The derivatives $d\alpha_1/d$pH and $d\alpha_2/d$pH are obtained in Box 9.1. The overall result is that for a closed system so that C_T remains constant, then

$$\beta_{\text{closed}} = 2.303\,(\alpha_0\alpha_1 C_T + \alpha_1\alpha_2 C_T + 4\alpha_0\alpha_2 C_T + [\text{H}^+] + [\text{OH}^-]). \tag{9.54}$$

Since pK_1 and pK_2 are separated by four, of the three terms with C_T, the $4\alpha_0\alpha_2 C_T$ term is always minor (2% or less). Eq.(9.54) thus simplifies with little error to

$$\beta_{\text{closed}} \approx 2.303\,(\alpha_0\alpha_1 C_T + \alpha_1\alpha_2 C_T + [\text{H}^+] + [\text{OH}^-]). \tag{9.55}$$

Figure 9.8 provides log β_{closed} as a function of pH for the four C_T values used for Figure 9.2. For each curve, the minimum seen at low pH corresponds to the $H_2CO_3^*$($f = 0$) EP. For $C_T = 10^{-3}$ *M*, Figure 9.9 aligns the curves for *f*, log β_{closed}, and log [ ] *vs.* pH.

9.5 OPEN CO_2 SYSTEMS WITH FIXED p_{CO_2}

9.5.1 General

When equilibrium is established between a gas phase containing $CO_{2(g)}$ and an aqueous phase, the CO_2 partition equilibrium is described by Eq.(9.5). This Henry's Law relationship describes the proportionality between the solution phase concentration of $H_2CO_3^*$ and the gas phase pressure of $CO_{2(g)}$. If an aqueous solution is "open" to a gas phase, CO_2 will exchange between the two phases until an equilibrium is reached. If the conditions in the solution change, CO_2 will enter or leave, and C_T will change. The value of K_H for CO_2 is $10^{-1.47}$ *M*/atm at 25 °C. This is about 1.5 orders or magnitude larger than the K_H values for $O_{2(g)}$ and $N_{2(g)}$, which are $10^{-2.89}$ and $10^{-3.19}$ *M*/atm, respectively, at 25 °C (Sander, 2015); CO_2 is much more soluble in water than O_2 and N_2.

The simplest type of open system is one in which p_{CO_2} is fixed. Examples include systems in which the volume of the gas phase is effectively infinitely large relative to the volume of the aqueous

BOX 9.1 $d\alpha_1/d$pH and $d\alpha_2/d$pH for β

For $d\alpha_1/d$pH,

$$\alpha_1 = \frac{1}{\frac{[H^+]}{K_1} + 1 + \frac{K_2}{[H^+]}}.$$

By the chain rule,

$$\frac{d\alpha_1}{dpH} = \frac{d\alpha_1}{d[H^+]}\frac{d[H^+]}{dpH}.$$

Because $d(u/v) = (vdu - udv)/v^2$

$$d\alpha_1/d[H^+] = \frac{-\frac{1}{K_1} + \frac{K_2}{[H^+]^2}}{\left[\frac{[H^+]}{K_1} + 1 + \frac{K_2}{[H^+]}\right]^2} = \left[-\frac{1}{K_1} + \frac{K_2}{[H^+]^2}\right]\alpha_1^2.$$

Multiplying by $d[H^+]/d$pH = –2.303[H$^+$] gives

$$d\alpha_1/dpH = 2.303\left[\frac{[H^+]}{K_1} - \frac{K_2}{[H^+]}\right]\alpha_1^2.$$

Since $([H^+]/K_1)\alpha_1 = \alpha_0$ and $(K_2/[H^+])\alpha_1 = \alpha_2$,

$$d\alpha_1/dpH = 2.303\,(\alpha_0\alpha_1 - \alpha_2\alpha_1).$$

For $d\alpha_2/d$pH,

$$\alpha_2 = \frac{1}{\frac{[H^+]^2}{K_1K_2} + \frac{[H^+]}{K_2} + 1}.$$

By the chain rule,

$$\frac{d\alpha_2}{dpH} = \frac{d\alpha_2}{d[H^+]}\frac{d[H^+]}{dpH}.$$

For use in the chain rule,

$$d\alpha_2/d[H^+] = \frac{-\frac{2[H^+]}{K_1K_2} + \frac{1}{K_2}}{\left[\frac{[H^+]^2}{K_1K_2} + \frac{[H^+]}{K_2} + 1\right]^2} = \left[-\frac{2[H^+]}{K_1K_2} - \frac{1}{K_2}\right]\alpha_2^2.$$

Multiplying by $d[H^+]/d$pH = –2.303[H$^+$] gives

$$d\alpha_2/d[H^+] = 2.303\left[\frac{2[H^+]^2}{K_1K_2} + \frac{[H^+]}{K_2}\right]\alpha_2^2.$$

Since $([H^+]^2/K_1K_2)\alpha_2 = \alpha_0$ and $([H^+]/K_2)\alpha_2 = \alpha_1$,

$$d\alpha_2/dpH = 2.303\,(2\alpha_0\alpha_2 + \alpha_1\alpha_2).$$

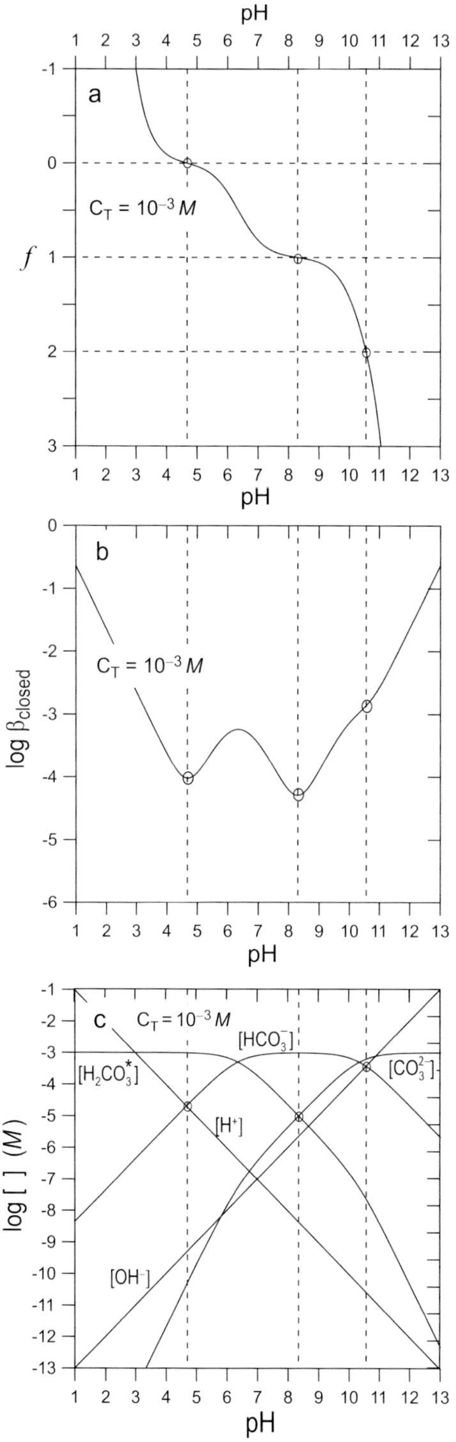

FIGURE 9.9 For $C_T = 10^{-3}$ *M* at 25 °C: *f*, log β_{closed}, and log[ ] *vs.* pH. The minima in the log β_{closed} *vs.* pH curve at the *f* = 0 and *f* = 1 equivalence points (EPs) give the inflection points (IPs) in the titration curve at those two EPs. At $C_T = 10^{-3}$ *M*, the absence of a minimum at *f* = 2 in the log β_{closed} *vs.* pH curve means there is no IP at *f* = 2 in the titration curve.

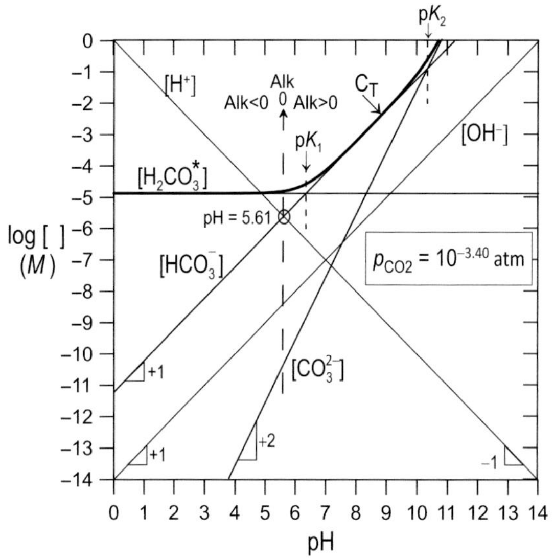

FIGURE 9.10 Log[ ] *vs*. pH diagram for an open CO_2 system with $p_{CO_2} = 10^{-3.40}$ atm at 25 °C. To increase pH, net strong base (+ΔAlk) must be added; to decrease pH, net strong acid (−ΔAlk) must be added.

solution, or when a gas with fixed p_{CO_2} is being bubbled or otherwise continually equilibrated with the solution. Systems at the surface of the Earth are commonly treated by assuming that the atmospheric p_{CO_2} value is fixed, and this is true for timescales of a few years. For longer timescales, it is regrettably true that one must take into account that the atmospheric p_{CO_2} level is slowly rising.

Because dissolved CO_2 can form an acid, if some strong base is added to an open-system solution, additional $CO_{2(g)}$ will dissolve and C_T will increase as the new equilibrium position is attained. Or, if some strong acid is added to an open-system solution, some of the added H^+ will react with HCO_3^- and CO_3^{2-}. This will temporarily increase the concentration of $H_2CO_3^*$, some of which will then outgas as CO_2, and the net result is that that C_T will decrease as the new equilibrium position is attained.

9.5.2 Log Concentration *vs.* pH Diagrams for Open Systems with Fixed p_{CO_2}

Neglecting activity corrections, *regardless* of whether p_{CO_2} is fixed, from Eq.(9.5), at equilibrium,

$$[H_2CO_3^*] = K_H p_{CO_2}. \tag{9.56}$$

But we also have

$$[H_2CO_3^*] = \alpha_0 C_T. \tag{9.57}$$

Combining Eqs.(9.56) and (9.57) gives the important relation that

$$\boxed{\alpha_0 C_T = K_H p_{CO_2}} \tag{9.58}$$

and two variations thereof,

$$\boxed{C_T = \frac{K_H p_{CO_2}}{\alpha_0}} \tag{9.59}$$

$$\boxed{p_{CO_2} = \frac{\alpha_0 C_T}{K_H}.} \tag{9.60}$$

There are three variables in Eqs.(9.58–9.60): p_{CO_2}, C_T, and pH (because α_0 depends on $[H^+]$). Knowledge of any two allows calculation of the third. All three equations apply not only to open systems, but also to closed systems by invoking the concept of the *in-situ* p_{CO_2}, which is discussed below.

For the three species comprising C_T, neglecting activity corrections, we have

$$[H_2CO_3^*] = \alpha_0 C_T \boxed{= K_H p_{CO_2}} \tag{9.61}$$

$$[HCO_3^-] = \alpha_1 C_T = \frac{\alpha_1}{\alpha_0} K_H p_{CO_2} \boxed{= \frac{K_1}{[H^+]} K_H p_{CO_2}} \tag{9.62}$$

$$[CO_3^{2-}] = \alpha_2 C_T = \frac{\alpha_2}{\alpha_0} K_H p_{CO_2} \boxed{= \frac{K_1 K_2}{[H^+]^2} K_H p_{CO_2}.} \tag{9.63}$$

All three concentrations are proportional to $K_H p_{CO_2}$.

In an open system with fixed p_{CO_2}, when activity corrections are neglected, all log[ ] *vs.* pH lines for individual CO_2 species are *straight lines*:

$$\log[H_2CO_3^*] = \log K_H p_{CO_2} \qquad d\log[H_2CO_3^*]/d\text{pH} = 0 \tag{9.64}$$

$$\log[HCO_3^-] = \log K_1 K_H p_{CO_2} + \text{pH} \qquad d\log[HCO_3^-]/d\text{pH} = +1 \tag{9.65}$$

$$\log[CO_3^{2-}] = \log K_1 K_2 K_H p_{CO_2} + 2\text{pH} \qquad d\log[CO_3^{2-}]/d\text{pH} = +2. \tag{9.66}$$

In 1750, at the surface of the Earth, $p_{CO_2} = 280\times10^{-6}$ atm. By 2016, $p_{CO_2} = 400\times10^{-6}$ atm (= $10^{-3.40}$ atm). A latter day Cassandra, Broecker (1975) issued a prescient, stark, and largely unheeded (by governments) warning about the great dangers of rising atmospheric CO_2 levels; sadly, the world has been very slow to action. Figure 9.10 uses $p_{CO_2} = 10^{-3.40}$ atm with K_H, K_1 and K_2 at 25 °C. Some observations based on Figure 9.10 are:

a. For pH << pK_1, $C_T \approx [H_2CO_3^*]$;
b. At pH = pK_1, as usual, $[H_2CO_3^*] = [HCO_3^-]$;
c. As pH moves beyond pK_1 towards pK_2, $C_T \approx [HCO_3^-]$ increases rapidly with pH;
d. At pH = pK_2, as usual, $[HCO_3^-] = [CO_3^{2-}]$;
e. For pH >> pK_2, $C_T \approx [CO_3^{2-}]$, and increases even more rapidly with pH.

In natural surface waters, one rarely finds pH values much higher than 10. This is because very high Alk values are required to achieve pH > 10: the acidic CO_2 that can enter from the gas phase "fights" the pH-raising effects of added strong base. On the other hand, if CO_2 is absent ($p_{CO_2} = 0$), getting to pH = 11 is easy: make a solution of 10^{-3} *F* NaOH (Alk = 10^{-3} eq/L). But when $p_{CO_2} = 10^{-3.40}$ atm, if Alk = 10^{-3} eq/L, pH is only 8.21 at 25 °C (see Example 9.9).

In Figure 9.10, both Alk (=$C_B - C_A$) and C_T are changing with pH, so it would be complicated to talk about how f is changing with pH. That is not really a problem because Figure 9.10 is not directly related to a titration process. Rather, it shows how the chemistry varies when $p_{CO_2} = 10^{-3.40}$ atm, and the pH is varied by the addition of strong acid or base. We can, nevertheless, identify the special point at which Alk = 0, with $f = 0$ for the C_T given by the system pH. This point corresponds to the chemistry of initially pure water that has been equilibrated with $p_{CO_2} = 10^{-3.40}$ atm at 25 °C. When Alk = 0, the PBE for a solution of $H_2CO_3^*$ applies, namely Eq.(9.42). When p_{CO_2} is not low, so that the solution is somewhat acidic, then the approximation Eq.(9.43) will apply: $[H^+] \approx [HCO_3^-]$. In Figure 9.10, this occurs at pH = 5.61, and the terms $2[CO_3^{2-}]$ and $[OH^-]$ are indeed negligible in the PBE. So, "pure" rainwater equilibrated at 25 °C with 2016 levels of atmospheric CO_2 (p_{CO_2} = $10^{-3.40}$ atm) will have a pH of 5.61; although acidic, this is not "acid rain". For $p_{CO_2} = 10^{-3.40}$ atm and 25 °C, if some rain is characterized by pH < 5.61, the rain contains net additional acidic species, likely strong acid(s) like H_2SO_4 and HNO_3, as with "acid rain". Conversely, if some rain is characterized by pH > 5.61, the rain contains some basic species, possibly suspended crustal minerals, or ammonia from facilities engaging in large-scale so-called "animal husbandry" (a term which, in most cases, is certainly a cruel oxymoron if there ever was one). Figure 9.11 is a generic open system plot that can be used as a transparency with Figure 6.4 to obtain a log[ ] *vs.* pH plot at 25 °C for any desired value of log p_{CO_2}: the transparency is placed such that the horizontal line for $\log[H_2CO_3^*]$ lies on the value for $\log K_H + \log p_{CO_2} = -1.47 + \log p_{CO_2}$. For Figure 9.10, this is −4.87.

9.5.3 Alkalinity Expressed as a Function of C_T and pH, and as a Function of p_{CO_2} and pH

When bases not related to CO_2 (*e.g.*, borate in seawater) can be ignored, then

$$\text{Alk} = [HCO_3^-] + 2[CO_3^{2-}] + [OH^-] - [H^+]. \qquad (9.18)$$

$$\boxed{\text{Alk} = \alpha_1 C_T + 2\alpha_2 C_T + [OH^-] - [H^+].} \qquad (9.67)$$

By Eq.(9.62) and (9.63),

$$\boxed{\text{Alk} = \frac{K_1 K_H p_{CO_2}}{[H^+]} + 2\frac{K_1 K_2 K_H p_{CO_2}}{[H^+]^2} + \frac{K_w}{[H^+]} - [H^+].} \qquad (9.68)$$

Along with Eq.(9.19), both Eqs.(9.67) and Eq.(9.68) are valid for closed *as well as* open systems, p_{CO_2} fixed or not. Eq.(9.67) tends to be more convenient in closed systems when C_T is fixed, and Eq.(9.68) is generally more convenient in open systems when p_{CO_2} is fixed. For Eq.(9.67), there are three variables, Alk, C_T, and pH; specifying any two allows calculation of the third. For Eq.(9.68), there are three variables, Alk, p_{CO_2}, and pH; specifying any two allows calculation of the third. For an open system with fixed p_{CO_2}, it does not matter how a solution has acquired its Alk value: neglecting activity corrections, if two solutions have the same Alk the same p_{CO_2}, the pH and carbonate chemistry will be the same.

Example 9.9 Computing pH in Open *vs.* Closed Systems when Alk Is Known

Consider the following solutions:

a. 10^{-3} *F* NaOH;
b. 10^{-3} *F* $NaHCO_3$;
c. 10^{-3} *F* $NaHCO_3$ and 10^{-3} *F* $H_2CO_3^*$;
d. 0.5×10^{-3} *F* Na_2CO_3; and
e. 10^{-3} *F* KOH;

Compute the pH of each solution for two cases at 25 °C: **I)** assuming closed systems and no externally applied fixed p_{CO_2}; then **II)** after equilibration in an open system with $p_{CO_2} = 10^{-3.40}$ atm. Neglect activity corrections.

Solution

For all parts in both cases **I** and **II**, Alk = 10^{-3} eq/L. The C_T and pH values will be different between the two cases for each part, but Alk remains at 10^{-3} eq/L because in or outgassing of CO_2 does not change Alk.

I) We know C_T for each solution. We find the pH that gives LHS−RHS = 0 for Eq.(9.67).
   a. $C_T = 0$. $[H^+]$ will be negligible, so $[OH^-] = 10^{-3}$ *M*. pH = 11.0.
   b. $C_T = 10^{-3}$ *M*. pH = 8.30.
   c. $C_T = 2\times10^{-3}$ *M*. pH = 6.35.
   d. $C_T = 0.5\times10^{-3}$ *M*. pH = 10.37.
   e. This is the same as for solution **a.** The only difference is that KOH is the source of the Alk rather than NaOH. $C_T = 0$. $[OH^-] = 10^{-3}$ *M*. pH = 11.0.

II) We know $p_{CO_2} = 10^{-3.40}$ atm for each part of the problem. We seek the pH that gives LHS−RHS = 0 for Eq.(9.68). With $p_{CO_2} = 10^{-3.40}$ atm and Alk = 10^{-3} eq/L for all parts, solutions **a** through **d** will be exactly identical. The only exceedingly small difference characterized by solution **e** (which has essentially no effect on pH at this low ionic strength) is that the tracer for the strong base is K^+ rather than Na^+. Neglecting activity corrections, Eq.(9.68) gives pH = 8.21 for all parts.

Example 9.10 Alk *vs.* pH in an Open System with Fixed p_{CO_2}

Plot log of the absolute value of Alk *vs.* pH for $p_{CO_2} = 10^{-3.40}$ atm and 25 °C. Consider the nature of the curve, and the behavior around pH 5.61 which is the pH for initially pure water (Alk = 0).

Solution

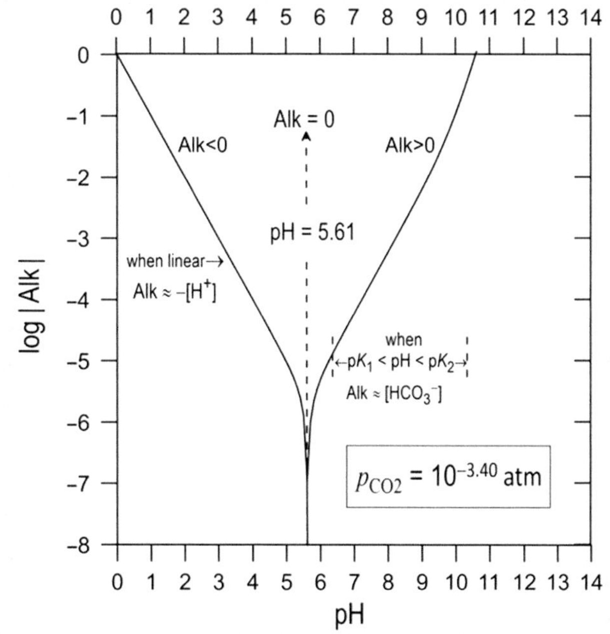

EXAMPLE 9.10 Log |Alk| *vs.* pH for an open system at 25 °C with $p_{CO_2} = 10^{-3.40}$ atm. When pH = 5.61, Alk = 0; when pH < 5.61, Alk < 0; when pH > 5.61, Alk > 0.

As discussed for Figure 9.11, when Alk = 0, pH = 5.61. If pH > 5.61, Alk > 0. If pH < 5.61, Alk < 0. Because all incoming CO_2 can act as an acid, a large Alk is required to attain a pH value such as 9. At pH << 5.61, most of the H^+ in the solution is due to the added strong acid, with Alk ≈ $-[H^+]$, so that |Alk| ≈ $[H^+]$; log |Alk| increases linearly as pH decreases. At pH = 0 where, neglecting activity corrections, $[H^+] = 1$ and log |Alk| = 0.

Example 9.11 Composition of "Acid Rain"

"Acid rain" is rain that has a pH below the value that results when initially pure water is equilibrated with atmospheric levels of CO_2. At 25 °C and $p_{CO_2} = 10^{-3.40}$ atm, this is pH = 5.61. The data below were obtained as part of the National Atmospheric Deposition Program at Hubbard Brook, New Hampshire for the week January 13, 2015 to January 20, 2015. The water is affected by combustion emissions of: (1) nitrogen oxides which are converted to nitric acid in the atmosphere; and (2) sulfur oxides which are converted to sulfuric acid in the atmosphere. The pH measured by pH electrode was 4.58. Assuming the analysis is complete, check this value by using the composition data to solve the ENE for $[H^+]$. Consider that the sulfate would have been present as a combination of HSO_4^- and SO_4^{2-}. However, for NH_4^+, because of its high pK_a value and the low pH, one can assume that total ammonia = $[NH_4^+] + [NH_3] \approx [NH_4^+]$. Solve two ways: **a.1.** Include the terms $[HCO_3^-]$, $2[CO_3^{2-}]$, and $[OH^-]$; **a.2.** Neglect those terms and comment on the validity of that approach. Neglect activity corrections.

	Ca^{2+}	Mg^{2+}	K^+	Na^+	NH_4^+	NO_3^-	Cl^-	SO_4^{2-}
mg/L:	0.062	0.015	0.006	0.103	0.064	2.028	0.183	0.14
important sources:	flyash, crustal particles	flyash, crustal, particles	flyash, biomass fires	sea salt, flyash	livestock emissions	combustion emissions	sea salt	coal burning

Solution

	Ca^{2+}	Mg^{2+}	K^+	Na^+	NH_4^+	NO_3^-	Cl^-	SO_4^{2-}
mg/FW:	40078	24305	39098	22990	18038	62005	35453	96063
M:	1.55×10^{-6}	6.17×10^{-7}	1.53×10^{-7}	4.48×10^{-6}	3.55×10^{-6}	3.27×10^{-5}	5.16×10^{-6}	1.46×10^{-6}
eq/L:	3.09×10^{-6}	1.23×10^{-6}	1.53×10^{-7}	4.48×10^{-6}	3.55×10^{-6}	3.27×10^{-5}	5.16×10^{-6}	–
	above cations: 1.25×10^{-5} eq/L					above anions excluding sulfate: 3.79×10^{-5} eq/L		

a. ENE including the terms $[HCO_3^-]$, $2[CO_3^{2-}]$, and $[OH^-]$:

$$[H^+] + 2[Ca^{2+}] + 2[Mg^{2+}] + [K^+] + [Na^+] + [NH_4^+] = [NO_3^-] + [Cl^-] + [HSO_4^-] + 2[SO_4^{2-}]$$
$$+ [HCO_3^-] + 2[CO_3^{2-}] + [OH^-]$$

$$[H^+]+1.25\times10^{-5}\text{ eq/L} = 3.79\times10^{-5}\text{ eq/L} + (\alpha_1^S + 2\alpha_2^S)S_T$$
$$+[HCO_3^-] + 2[CO_3^{2-}] + [OH^-]$$

a.1. Assume that the HSO_4^-/SO_4^{2-} pair are behaving as a monoprotic system with acidity constant K_a (=K_2 = $10^{-1.99}$ at 25 °C). $(\alpha_1^S + 2\alpha_2^S) = \frac{[H^+]}{[H^+] + K_2} + 2\frac{K_2}{[H^+] + K_2}$

$$[H^+] + 1.25 \times 10^{-5}\ \text{eq/L} = 3.79 \times 10^{-5}\ \text{eq/L} + (\alpha_1^S + 2\alpha_2^S)S_T + K_1K_Hp_{CO_2}/[H^+]$$

$$+ 2K_1K_2K_Hp_{CO_2}/[H^+]^2 + K_w/[H^+]$$

Solving, **pH = 4.55**. (At this pH, $\alpha_2^S \approx 1$, *i.e.*, $[HSO_4^-] << 2[SO_4^{2-}]$ so we could have neglected $[HSO_4^-]$ (*i.e.*, $\alpha_1^S \approx 0$).

a.2. The ENE neglecting the terms $[HCO_3^-]$, $2[CO_3^{2-}]$, and $[OH^-]$ is

$$[H^+] + 1.25 \times 10^{-5}\ \text{eq/L} = 3.79 \times 10^{-5}\ \text{eq/L} + 2S_T$$

Solving, **pH = 4.55**.

The dissolution of CO_2 has essentially no effect on the pH of this rain: $H_2CO_3^*$ is a weak acid, and $p_{CO_2} = 10^{-3.40}$ atm is too low to have an effect given the amount of net strong acid present. And, at pH ≈ 4.5, $[OH^-]$ is certainly negligible. Overall, the value for the total eq/L of anionic tracers for strong acid (NO_3^-, SO_4^{2-}, Cl^-) is greater than the total eq/L of cationic tracers for strong base (the metal cations) plus ammonium (as a tracer for the weak base ammonia). Most of the strong acid is nitric acid. The good agreement between the measured pH and the pH calculated using the ENE with the analytical data is a necessary, though not sufficient, condition that the analytical data is accounting for all major acid/base species.

Example 9.12 Lake Acidification due to "Acid Rain"

Many lakes in Scandinavia and parts of New York, California, and elsewhere are highly vulnerable to being acidified by "acid rain". This is due to an absence in the local geology of carbonate rocks, so that lake pH values in such geologies can naturally be <7. Renberg et al. (1993) reported reconstructed pH values for Lake Gaffeln in south-west Sweden for 1930 to 1980. In a later study, Anderson (1994) describes the lake as being fishless at pH = 4.4. The data are given here. The large drop in pH has been the consequence of deposition of strong acids on the watershed. For each year, calculate Alk assuming a temperature of 15 °C and the global p_{CO_2} value provided. Calculate ΔAlk over each decade, and over the total period 1930 to 1980.

Given				Solution	
Year	pH	p_{CO_2} (ppm)†	p_{CO_2} (sea level) (atm)	Alk (eq/L)	ΔAlk/y (eq/L-y)
1930	6.15	307.5	307.5×10^{-6}	$\mathbf{0.072 \times 10^{-4}}$	–
1940	5.9	311.3	311.3×10^{-6}	$\mathbf{0.032 \times 10^{-4}}$	$\mathbf{-4.0 \times 10^{-7}}$
1950	5.85	311.3	311.3×10^{-6}	$\mathbf{0.026 \times 10^{-4}}$	$\mathbf{-0.6 \times 10^{-7}}$
1960	5.64	317.1	317.1×10^{-6}	$\mathbf{0.002 \times 10^{-4}}$	$\mathbf{-2.4 \times 10^{-7}}$
1970	5.35	325.5	325.5×10^{-6}	$\mathbf{-0.031 \times 10^{-4}}$	$\mathbf{-3.4 \times 10^{-7}}$
1980	4.7	339	339.0×10^{-6}	$\mathbf{-0.196 \times 10^{-4}}$	$\mathbf{-16.5 \times 10^{-7}}$
1994	4.4	358.6	358.6×10^{-6}	$\mathbf{-0.396 \times 10^{-4}}$	$\mathbf{-14.3 \times 10^{-7}}$
			ΔAlk for 64 y =	$\mathbf{-0.468 \times 10^{-4}}$	

† Data from NASA for global mean CO_2 "mixing ratios" over time https://data.giss.nasa.gov/modelforce/ghgases/Fig1A.ext.txt.

Solution

When activity corrections can be neglected, for fresh waters

$$\text{Alk} = \frac{K_1 K_H p_{CO_2}}{[H^+]} + 2\frac{K_1 K_2 K_H p_{CO_2}}{[H^+]^2} + \frac{K_w}{[H^+]} - [H^+]. \tag{9.68}$$

In 1930, this lake had virtually no Alk, which is why it was so susceptible to the ill effects of "acid rain". A plot of pH *vs.* Alk (not shown) does have the appearance of a titration curve, with an IP near Alk = 0 (*i.e.*, $f = 0$; $(C_B - C_A) = 0$), though unlike the titration in Example 9.7, here we assume maintenance of equilibrium, during the titration, with the atmospheric p_{CO_2} values. The smallest drop in Alk was for the decade 1940 to 1950, assumedly due to the disruption of normal economic activity in Europe during World War II. (Note also that p_{CO_2} did not rise from 1940 to 1950.) By the 1970s, the problem of "acid rain" had become of great concern in Scandinavia and Europe, and indeed for this lake the steepest drop in Alk occurred during that decade. (Recognition of the seriousness of "acid rain eventually facilitated broad new regulations on sulfur emissions from coal-fired power plants and vehicles. In the U.S., "acid rain" was addressed under Title IV of the 1990 Clean Air Act.) The calculated ΔAlk values may be underestimates of the eq/L of strong acid added because: (1) dissolution of some alkaline geological materials would have neutralized some added strong acid; and (2) some biomediated removal of nitrate and sulfate likely occurred, along with the corresponding strong acid protons, by redox reaction with organic matter. Assuming carbohydrate organic matter, the reactions are

$$H^+ + NO_3^- + \tfrac{5}{24}C_6H_{12}O_6 \rightarrow \tfrac{1}{2}N_2 + \tfrac{5}{4}CO_2 + \tfrac{7}{4}H_2O \quad \text{loss of strong acid by reduction of nitrate}$$

$$2H^+ + SO_4^{2-} + \tfrac{1}{3}C_6H_{12}O_6 \rightarrow H_2S + 2CO_2 + 2H_2O \quad \text{loss of strong acid by reduction of sulfate.}$$

Example 9.13 Ocean Acidification due to Rising p_{CO_2} (The Most-Worrisome Conservation-of-Alk Example): Lowering of Oceanic pH Values due to Rising p_{CO_2}

Over geological time, in the oceans, a value of Alk ≈ 2.5×10^{-3} eq/kg (seawater concentration values are typically expressed on a per kg basis) has developed, and can be expected to remain at that value for any foreseeable future, though it seems that in the "Anthropocene", we cannot be particularly sure about anything. "Acid rain" has not greatly affected ocean Alk values: the oceans are enormous, and moreover biological processes can remove HNO_3 and H_2SO_4 by redox reduction reactions as mentioned in Example 9.12. *But*, steadily rising p_{CO_2} values due to burning of coal and petroleum are causing decreasing ocean pH ($\equiv -\log\{H^+\} = -\log[H^+]\gamma_{H^+}$). This is CO_2-driven ocean acidification. It is causing some ocean pH values to drop by amounts on the scale of 0.1 pH unit. (That type of change would not be a problem in freshwater systems, but ocean ecosystems are very sensitive because of the easily disrupted chemistry of the precipitation of calcium carbonate minerals: all $CaCO_{3(s)}$-forming life forms such as corals, shell-forming organisms, and calcifying foraminifera require seawater to be supersaturated with $CaCO_{3(s)}$ for the precipitation to proceed.) We start with Eq.(9.21); for seawater, a term for borate must be included in Alk. To account for necessary activity corrections in seawater, for the acidity constants we can use K' values which utilize $\{H^+\}$ for H^+, but concentrations for other species (see Section 2.7.3). For solubility of $CO_{2(g)}$ in seawater, we use ${}^cK_H = [H_2CO_3^*]/p_{CO_2}$ for seawater. Eq.(9.68) then becomes

$$\text{Alk} = \frac{K_1'\,{}^cK_H p_{CO_2}}{\{H^+\}} + 2\frac{K_1'K_2'\,{}^cK_H p_{CO_2}}{\{H^+\}^2} + \frac{K_w'}{\{H^+\}} + [B(OH)_4^-] - \frac{\{H^+\}}{\gamma_{H^+}} \tag{9.69}$$

with $[B(OH)_4^-] = \alpha_1^B B_T = \dfrac{K_a^{\prime B}}{K_a^{\prime B} + \{H^+\}} B_T$.

Values for the equilibrium constants are given in the accompanying table for a range of temperatures. (pH-dependent equilibrium constants used for seawater are usually taken to be of the "mixed" type; see Eq.(2.66) et seq.) As with Alk, values of B_T in seawater vary a bit, $0.42 \times 10^{-3}\ m$ is typical (Zeebe et al., 2001). 280 ppm was the pre-industrial level for CO_2; we flew past 400 ppm in 2016. If we continue on a "business as usual" track, we will cross the 500 ppm mark before the year 2040. Calculate pH at 15 °C for p_{CO_2} values corresponding to 280, 300, 400, 500, and 600 ppm (p_{CO_2}(atm) = ppm/10^6).

Values of Equilibrium Constants in Seawater at Salinity (S) Values of 35 Parts Per Thousand (g/kg).

source	a	a	b	a	c,d
t (°C)	$\log K_1'$	$\log K_2'$	$\log {}^cK_H$	$\log K_a'^B$	$\log K_w'$
5	−6.11	−9.33	−1.28	−8.89	−13.29
10	−6.07	−9.27	−1.36	−8.84	−13.06
15	−6.04	−9.21	1.43	−8.79	−13.85
20	−6.01	−9.15	−1.49	−8.74	−13.64
25	−5.98	−9.09	−1.55	−8.69	−13.44

Equilibrium Constant Definitions

$$K_1' = \frac{\{H^+\}[HCO_3^-]}{[H_2CO_3^*]} \quad \left(= K_1 \frac{\gamma_{H_2CO_3^*}}{\gamma_1}\right) \qquad \log {}^cK_H = \frac{[H_2CO_3^*]}{p_{CO_2}} \quad \left(= K_H \frac{1}{\gamma_{H_2CO_3^*}}\right)$$

$$K_2' = \frac{\{H^+\}[CO_3^{2-}]}{[HCO_3^-]} \quad \left(= K_2 \frac{\gamma_1}{\gamma_2}\right) \qquad K_a'^B = \frac{\{H^+\}[B(OH)_4^-]}{[B(OH)_3]} \quad \left(= K_a^B \frac{\gamma_{B(OH)_3}}{\gamma_1}\right)$$

$$K_w' = \{H^+\}[OH^-] \quad \left(= K_w \frac{1}{\gamma_1}\right)$$

a. Data of Lyman (1956) for $S \approx 35$, as discussed in Culberson and Pytkowicz (1968), with fitting here for the dependence on T.

b. Weiss (1974).

c. As follows: (1) Equation 63 of Millero (1995) gives an expression for a parameter termed K_w in terms of T (K) and S; this is actually ${}^cK_w \equiv [H^+][OH^-]$; (2) this expression gives ${}^cK_w = [H^+][OH^-] = 10^{-13.21}$ for $S = 35$ and $T = 298.15$ K (25 °C); (3) Zeebe and Wolf-Gladrow, 2001 (Table 1.1.3) give $\gamma_{H^+}' \approx 0.590$ for $S = 35$ and $T = 298.15$ K (25 °C); so (4) $K_w' \equiv \{H^+\}[OH^-] = \gamma_{H^+}' [H^+][OH^-] = 0.590 \times 10^{-13.21} = 10^{-13.44}$.

d. It is assumed here that $\gamma_{H^+}' = 0.590$ for all temperatures.

Solution

For CO_2 at 280, 300, 400, 500, and 600 ppm, 15 °C we obtain pH = 8.31, 8.28, 8.18, 8.09, and 8.03. This trend is greatly threatening and probably has already damaged ocean health. It would not be so troubling if this was just going to cause humans some real difficulties: "we" created this mess, no? The problem is we are taking down many unspeakably beautiful ecosystems and species on this planet. For what? So that, instead of a more reasonable number, there can be many billions of us with our gigatons of concrete and other all-important stuff?

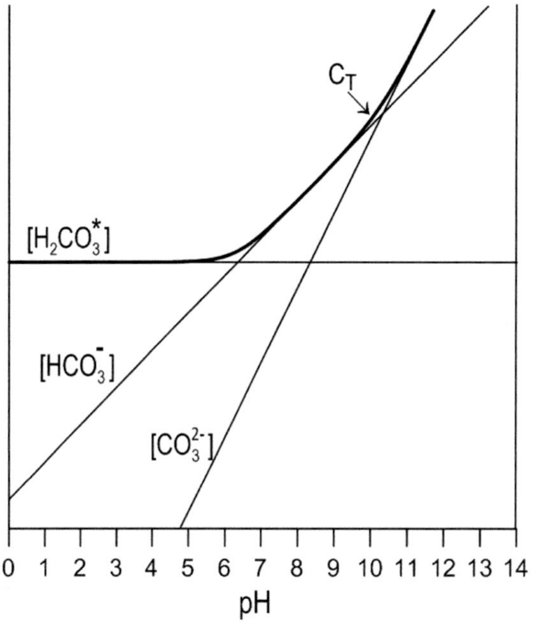

FIGURE 9.11 Generic log[ ] *vs.* pH diagram for an open CO_2 system at 25 °C with unspecified value of p_{CO_2}, for use as a transparency with Figure 6.4. The vertical position of the transparency is set such that $\log[H_2CO_3^*] = \log K_H + \log p_{CO_2} = -1.47 + \log p_{CO_2}$.

9.5.4 The Universal Acidification Plot (UAP)

Examples 9.11–9.13 consider how open system natural waters can become acidified by two mechanisms: (1) reduction of Alk as due to "acid rain"; and (2) elevation of p_{CO_2} by emissions of CO_2. For both mechanisms, the equation utilized in Example 9.13 applies, namely

$$\text{Alk} = \frac{K_1'\,{}^cK_H p_{CO_2}}{\{H^+\}} + 2\frac{K_1'K_2'\,{}^cK_H p_{CO_2}}{\{H^+\}^2} + \frac{K_w'}{\{H^+\}} + \frac{K_a'^B}{K_a'^B + \{H^+\}}B_T - \frac{\{H^+\}}{\gamma_{H^+}}. \tag{9.69}$$

For fresh waters, $B_T \approx 0$, and the K' values and ${}^cK_H$ value can usually be taken as equal to the infinite dilution values, along with $\{H^+\}/\gamma_{H^+} = [H^+]$. For higher ionic strength values, the corresponding K' values and ${}^cK_H$ value can be calculated using the equilibrium constant definitions provided in Example 9.13, with $\gamma_{B(OH)_3} \approx \gamma_{H_2CO_3} \approx 1$) and γ_{H^+} estimated from the Extended Debye–Hückel Equation as γ_1.

When "acid rain" affects fresh waters, by Eq.(9.69) the decreasing Alk will cause $\{H^+\}$ to rise. For rising p_{CO_2} values, with ocean acidification the result is that a constant Alk value again requires $\{H^+\}$ to increase (along with C_T). This text then introduces the concept of the **universal acidification plot (UAP)** in which the changes due to "acid rain" as well as ocean acidification can be represented on a single diagram. With straight isopleth lines of constant pH drawn on a plot of p_{CO_2} *vs.* Alk, both "acid rain" and ocean acidification lead to movement across isopleths from higher to lower pH. To calculate the isopleths, rearrangement of Eq.(9.69) gives

$$p_{CO_2} = \frac{1}{\dfrac{K_1'\,{}^cK_H}{\{H^+\}} + 2\dfrac{K_1'K_2'\,{}^cK_H}{\{H^+\}^2}}\left[\text{Alk} + \left(\frac{\{H^+\}}{\gamma_{H^+}} - \frac{K_w'}{\{H^+\}} - \frac{K_a'^B}{K_a'^B + \{H^+\}}B_T\right)\right] \tag{9.70}$$

or

$$p_{CO_2} = \frac{1}{c}\text{Alk} + \frac{1}{c}\left(\frac{\{H^+\}}{\gamma_{H^+}} - \frac{K'_w}{\{H^+\}} - \frac{K_a^{'B}}{K_a^{'B} + \{H^+\}}B_T\right) \tag{9.71}$$

for which

$$\begin{matrix} x \text{ intercept} \\ \text{(at pH for isopleth)} \end{matrix} : \ \text{Alk} = -\left(\frac{\{H^+\}}{\gamma_{H^+}} - \frac{K'_w}{\{H^+\}} - \frac{K_a^{'B}}{K_a^{'B} + \{H^+\}}B_T\right) \tag{9.72}$$

and if p_{CO_2} is plotted with Alk decreasing towards the right,

$$\text{slope} = -\frac{1}{c}. \tag{9.73}$$

When $B_T \approx 0$ (fresh waters) and all $\gamma = 1$, then

$$\begin{matrix} x \text{ intercept} \\ \text{(at pH for isopleth)} \end{matrix} : \ \text{Alk} = -\left([H^+] - [OH^-]\right). \tag{9.73}$$

As pH becomes lower, the slope $-1/c$ becomes increasingly steep; at pH < 6.5, at atmospherically environmentally relevant p_{CO_2} values, the isopleths are essentially vertical. Figure 9.12 provides a UAP plot for environmentally relevant values of p_{CO_2} and Alk.

For "acid rain effects", the reduction in pH is usually viewed as due to reduction in Alk, with little effect from rising p_{CO_2} values. Some "acid rain" effects have accrued over many decades

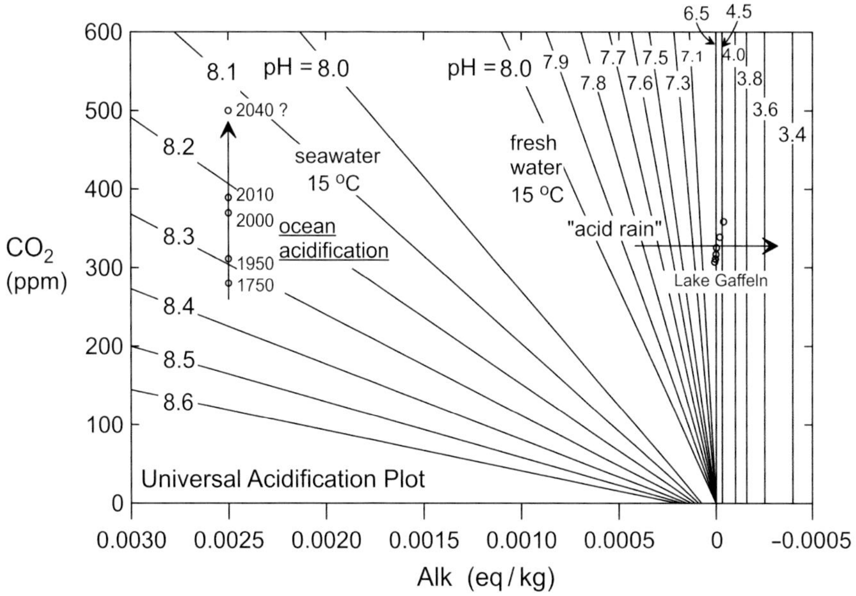

FIGURE 9.12 Universal acidification plot (UAP) based on the open system Alk equation. The y-axis has units of ppm, but in all equations p_{CO_2}(atm) = ppm/10^6. Ocean acidification involves rising p_{CO_2} values with Alk remaining constant. The seawater isopleths are based on the equilibrium constant values for seawater for 15 °C/1 atm and total borate = 0.42×10^{-3} m as in Example 9.13 (concentration units for seawater are expressed on a per kg basis). The fresh water isopleths are based on infinite dilution equilibrium constant values for 15 °C/1 atm and zero dissolved borate. The effects of "acid rain" involve addition of the strong acids H_2SO_4 and HNO_3 to surface waters. If "acid rain" effects occur over a decade or less, the acidification occurs with p_{CO_2} remaining approximately constant. The 1930 to 1994 data for Lake Gaffeln (Example 9.12) are shown.

during which p_{CO_2} has risen significantly, as with Lake Gaffeln (see Example 9.12 and Figure 9.12). However, surface waters that are most subject to the effects of "acid rain" have low initial Alk values and approximately neutral initial pH values. For initial pH < 7, the subsequent UAP isopleths are essentially vertical, and nearly all of the accrued "acid-rain" induced reduction in pH has been due to reduction in Alk, with only minor effects from the rising p_{CO_2} values.

As regards ocean acidification, as discussed in Example 9.13, falling pH values in ocean waters hinders the precipitation of $CaCO_{3(s)}$, generally aragonite, by a host of calcifying organisms that include corals, shell-forming organisms, and many species of foraminifera. The dissolution of aragonite occurs according to

$$CaCO_{3(s)}(\text{aragonite}) = Ca^{2+} + CO_3^{2-}. \quad (9.74)$$

The "apparent solubility product" equilibrium constant for aragonite in seawater, which uses the concentration scale (and so builds in activity corrections) is defined as

$$K'_{s0,\,\text{aragonite}} = \frac{[Ca^{2+}][CO_3^{2-}]}{\{CaCO_{3(s)}\}} \quad (9.75)$$

where $\{CaCO_{3(s)}\}$, the activity of the solid aragonite (mole fraction scale) is generally taken to unity (pure aragonite) so that

$$K'_{s0,\,\text{aragonite}} = [Ca^{2+}][CO_3^{2-}]. \quad (9.76)$$

The saturation state Ω of a given water with respect to aragonite is thus given by the ratio of the actual ion product $[Ca^{2+}][CO_3^{2-}]$ in the solution to $K'_{s0,\,\text{aragonite}}$:

$$\Omega = \frac{[Ca^{2+}][CO_3^{2-}]}{K'_{s0,\,\text{aragonite}}}. \quad (9.77)$$

When $\Omega > 1$, a solution is supersaturated and the solid will tend to precipitate. When $\Omega = 1$, a solution is exactly saturated and the solid will neither tend to precipitate, or dissolve (if any solid is present). When $\Omega < 1$, a solution is undersaturated, and the solid will tend to dissolve (if any is present).

With Alk and $[Ca^{2+}]$ roughly constant, rising p_{CO_2} values are causing a lowering in $[CO_3^{2-}]$ so that the product $[Ca^{2+}][CO_3^{2-}]$ is decreasing: $[Ca^{2+}]$ is roughly constant and the product $[CO_3^{2-}]$ is decreasing. For seawater, Mucci (1983) gives

$$\begin{aligned}\log K'_{s0,\,\text{aragonite}} = &\left(-171.945 - 0.077993T + \frac{2903.293}{T} + 71.595\log T\right)\\ &+\left(-0.068393 + 0.0017276T + \frac{88.135}{T}\right)S^{\frac{1}{2}}\\ &- 0.10018S + 0.0059415S^{\frac{3}{2}}\end{aligned} \quad (9.78)$$

where T (K) is temperature, and S is salinity in parts per thousand. At 15 °C ($T = 288.15$) and $S = 35‰$, $\log K'_{s0,\,\text{aragonite}} = -6.17$. For seawater at $S = 35‰$, $[Ca^{2+}] = 0.01\ m$. Table 9.5 and Figure 9.13 give values of $[CO_3^{2-}]$ calculated for seven of the pH isopleth values in Figure 9.12, along with corresponding $[Ca^{2+}][CO_3^{2-}]$ and Ω values. Even CO_2 levels of ~300 ppm, ocean waters have only been slightly supersaturated with aragonite, which speaks to the considerable vulnerability of marine ecosystems to ocean acidification. Chemical conditions in the oceans vary with location and time so that pH, temperature, salinity, and $[Ca^{2+}]$ do vary; for detailed maps of aragonite saturation state

TABLE 9.5
Values of pH, $[CO_3^{2-}]$, $[Ca^{2+}][CO_3^{2-}]$, and Ω for Aragonite for Seawater at 15 °C at Varying ppm Values for CO_2 (p_{CO_2} (atm) = ppm/10^6) and Alk = 2.5 × 10^{-3} eq/kg

pH (15 °C)	ppm CO_2	$[CO_3^{2-}]$ (15 °C)	$[Ca^{2+}][CO_3^{2-}]$ (15 °C)	Ω (aragonite) (15 °C)
8.6	119	4.51E-04	4.51E-06	6.70
8.5	164	3.94E-04	3.94E-06	5.86
8.4	225	3.40E-04	3.40E-06	5.06
8.3	304	2.90E-04	2.90E-06	4.31
8.2	407	2.45E-04	2.45E-06	3.64
8.1	538	2.05E-04	2.05E-06	3.04
8.0	707	1.70E-04	1.70E-06	2.52

Note: $[Ca^{2+}]$ is taken to be 0.01 *m* and $\log K'_{s0,\text{aragonite}} = -6.17$.

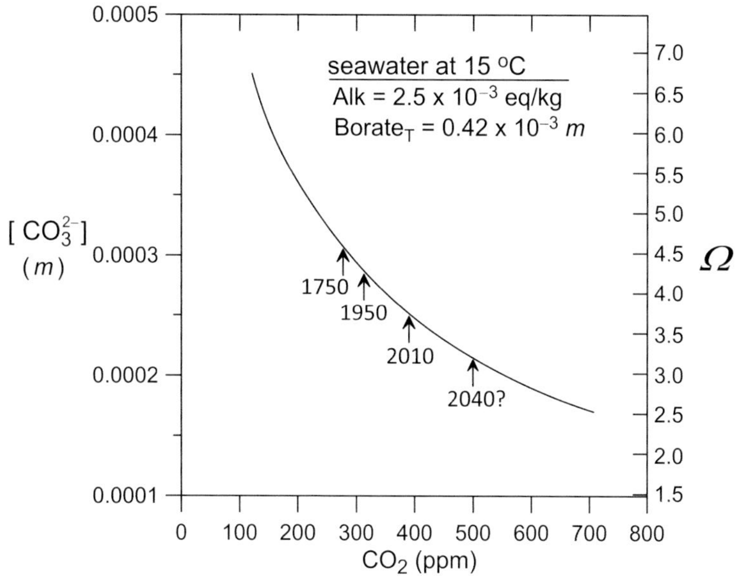

FIGURE 9.13 Values of $[CO_3^{2-}]$ (*m*) and the saturation state for aragonite as a function of CO_2 level (ppm) in the atmosphere (p_{CO_2}(atm) = ppm/10^6).

in the oceans, see Jiang et al. (2015). Because it will not alter atmospheric CO_2 levels, using atmospheric geoengineering with stratospheric particles (*e.g.*, see Irvine et al., 2019) to reduce incoming sunlight for mitigation of the temperature effects of climate change will do **nothing** to address the anthropogenic effects on calcium carbonate saturation state in the oceans.

9.5.5 Buffer Intensity in Open CO_2 Systems with Fixed p_{CO_2}

For any system,

$$\beta \equiv \frac{d(C_B - C_A)}{d\text{pH}}. \tag{9.52}$$

For an open system with p_{CO_2} fixed, applying Eq.(9.52) to Eq.(9.68) and using the chain rule as $dy/d\text{pH} = (dy/d[H^+])(d[H^+]/d\text{pH})$, we obtain

$$\beta_{\text{open}} = \left(\frac{\partial(C_B - C_A)}{\partial \text{pH}}\right)_{p_{CO_2}} = -\frac{K_1 K_H p_{CO_2}}{[H^+]^2}\frac{d[H^+]}{d\text{pH}} - 4\frac{K_1 K_2 K_H p_{CO_2}}{[H^+]^3}\frac{d[H^+]}{d\text{pH}} + \frac{d[OH^-]}{d\text{pH}} - \frac{d[H^+]}{d\text{pH}}. \quad (9.79)$$

By Eq.(8.16), $d[OH^-]/d\text{pH} = 2.303[OH^-]$; by Eq.(8.17), $d[H^+]/d\text{pH} = -2.303[OH^-]$, so

$$\beta_{\text{open}} = \left(\frac{\partial(C_B - C_A)}{\partial \text{pH}}\right)_{p_{CO_2}} = 2.303\left(\frac{K_1 K_H p_{CO_2}}{[H^+]} + 4\frac{K_1 K_2 K_H p_{CO_2}}{[H^+]^2} + [OH^-] + [H^+]\right) \quad (9.80)$$

$$= 2.303([HCO_3^-] + 4[CO_3^{2-}] + [OH^-] + [H^+]). \quad (9.81)$$

Figure 9.14 gives log β_{open} *vs.* pH for $p_{CO_2} = 10^{-3.40}$ atm, and K_H, K_1, and K_2 for 25 °C.

It is instructive to compare β_{open} for a given p_{CO_2} and pH with the value for β that would result for the same water composition (C_T and pH) if the water was closed off from any gas phase directly prior to addition of $d(C_B - C_A)$, *i.e.*, all other things being the same, how does β_{open} compare with the β if the system became closed and C_T fixed before the incremental change $d(C_B - C_A)$? This β_{closed} is obtained by computing C_T for each pH using Eq.(9.59) with the p_{CO_2} of interest, then using that C_T with Eq.(9.54) which gives log β_{closed} as a function of pH as plotted in Figure 9.14. Also included in

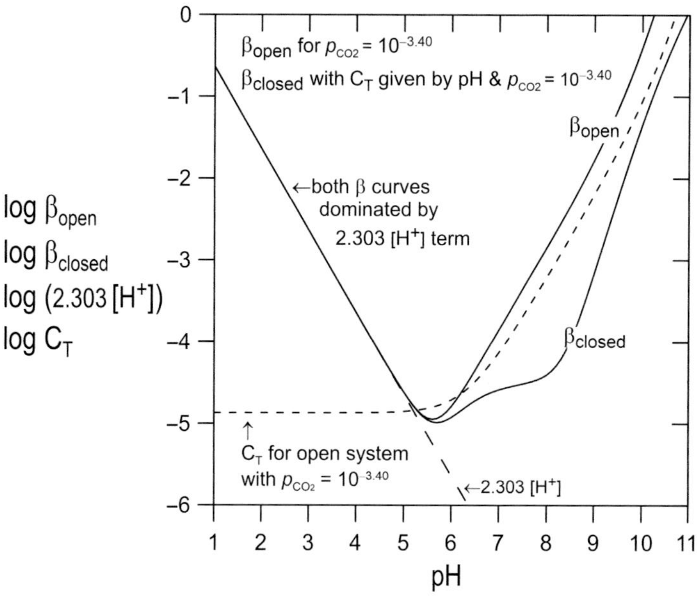

FIGURE 9.14 Log $\beta_{\text{open}} = \log\left(\partial(C_B - C_A)/\partial\text{pH}\right)_{p_{CO_2}}$ for an open system with $p_{CO_2} = 10^{-3.4}$ atm at 25 °C, *vs.* $\beta_{\text{closed}} = \left(\partial(C_B - C_A)/\partial\text{pH}\right)_{C_T}$ as for a closed system for the same value of C_T for the open system. The two curves allow comparison of β for the open system if it remains in equilibrium with the atmosphere and C_T can change, *vs.* β for a system at the same initial pH and C_T, but with system closed off from a gas phase immediately before the addition of dC_B or dC_A, so that C_T cannot change (but p_{CO_2} will).

Figure 9.14 is the line for the open system log C_T for $p_{CO_2} = 10^{-3.4}$ atm, based on Eq.(9.59). At pH greater than about ~6 so that HCO_3^- is starting to become important in C_T, then $\beta_{open} >> \beta_{closed}$. This is because in an open system, added acid can cause outgassing of CO_2 (which means losing weak acid), and added strong base can cause some ingassing of CO_2 (which means gaining weak acid). Last, as we know, with $p_{CO_2} = 10^{-3.40}$ atm, the pH for Alk = 0 is 5.61 (see Figure 9.8). At pH < 5, there is significant net strong acid in the solution, and the pH is largely determined by the net strong acid, not the dissolved CO_2 from $p_{CO_2} = 10^{-3.40}$ atm. So, at low pH, for both curves in Figure 9.14, then $\log \beta \approx \log 2.303[H^+]$.

Example 9.14 Buffer Intensity in Closed System *vs.* An Open Systems with Fixed p_{CO_2}

A river flowing at 1 mgd (million gallons per day) flows past an industrial facility. The river is in equilibrium with the atmosphere at 15 °C, with Alk = 2.75 × 10^{-3} eq/L. Use *K* values from Table 9.2. The facility wishes to discharge strong acid waste effluent to the river.

a. What is the initial pH and C_T in the river?
b. If the river does not equilibrate quickly with the atmosphere, but the discharged acid does mix quickly within the flow, then use β_{closed} for the initial conditions to estimate how many equivalents of strong acid could be discharged to the river per day without lowering the pH by more than 0.1 pH units.
c. If the river does equilibrate quickly with the atmosphere during release of the strong acid, use β_{open}, to estimate how many equivalents of strong acid could be discharged to the river per day without lowering the pH by more than 0.1 pH units.

Solution

a. Alk = 2.75 × 10^{-3} eq/L= $K_1K_Hp_{CO_2}/[H^+] + K_1K_2K_Hp_{CO_2}/[H^+]^2 + K_w/[H^+] - [H^+]$. With $p_{CO_2} = 10^{-3.40}$ atm, using the *K* values for 15 °C, then the initial pH = 8.57, and $C_T = 2.73 \times 10^{-3}$ *M*.
b. From Eq.(9.54), $\beta_{closed} = 1.32 \times 10^{-4}$ eq/L-pH.
   Releasable eq/day = (1.32 × 10^{-4} eq/L-pH) × 0.1 pH × 1 mgd × (3,785,000 L/day)/mgd = 50 eq/day
c. From Eq.(9.80), $\beta_{open} = 6.50 \times 10^{-3}$ eq/L-pH.

Releasable eq/day = (6.50 × 10^{-3} eq/L-pH) × 0.1 pH × 1 mgd × (3,785,000 L/day)/mgd = 2460 eq/day.

9.6 OPEN CO_2 SYSTEMS WITH p_{CO_2} A VARIABLE

In this text, when a gas phase is considered to be present and in equilibrium with an aqueous phase, the great majority of the cases consider p_{CO_2} to be some constant value. However, innumerable circumstances exist in which both aqueous and gas phases are present and p_{CO_2} is not fixed. When solving a problem of this type, since we no longer have an equation specifying p_{CO_2}, a replacement equation is required. That equation is provided by a mass balance equation (MBE) which constrains the total number of mols M_{CO_2} of CO_2 related species, as summed over the phases present. For a two phase, aqueous+gas system,

$$M_{CO_2} = V_w C_T + V_g c_{g,CO_2} \tag{9.82}$$

where V_w (L) and V_g (L) are the volumes of the aqueous and gas phases respectively, and c_{g,CO_2} (mols/L) is the gas phase CO_2 concentration. If n_{g,CO_2} is the number of mols of CO_2 in the gas phase, the ideal gas law gives $c_{g,CO_2} = n_{g,CO_2}/V_g = p_{CO_2}/RT$. Thus,

$$M_{CO_2} = V_w C_T + V_g p_{CO_2}/RT \tag{9.83}$$

so

$$M_{CO_2} = V_w \frac{K_H p_{CO_2}}{\alpha_0} + V_g \frac{p_{CO_2}}{RT} \quad (9.84)$$

$$p_{CO_2} = \frac{M_{CO_2}}{V_w \frac{K_H}{\alpha_0} + V_g \frac{1}{RT}}. \quad (9.85)$$

The functionality of Eq.(9.85) may be rationalized as follows: p_{CO_2} increases with: (1) increasing M_{CO_2} because there is more total CO_2 in the overall system; (2) decreasing V_w because then there is less water in which $CO_{2(g)}$ can dissolve; (3) decreasing V_g because then the volume of the gas phase is smaller, making the $CO_{2(g)}$ more concentrated; (4) decreasing K_H because then the $CO_{2(g)}$ is less soluble in the water; and (5) increasing α_0 because then more of C_T in the water is $H_2CO_3^*$ (*cf.* Eq.(9.57).

Example 9.15 Calculating the Composition of a Water/Gas System in Which p_{CO_2} Is Not Initially Known

Consider the case in Example 9.9.c (initial solution composition is 10^{-3} *F* $NaHCO_3$ with 10^{-3} *F* $H_2CO_3^*$). **Part I.** V_w = 1 L is equilibrated at 25 °C with V_g = 1 L of air that is initially free of CO_2. Assume that outgassing of CO_2 from the solution does not change either the 1 L gas volume, or the 1 L water volume. As usual, Alk remains constant in the solution as CO_2 outgasses. Eq.(9.68) applies with Alk = 0.001 eq/L, though we do not know p_{CO_2} until we have solved the problem. **a.** Use Eq.(9.85) for p_{CO_2} in Eq.(9.68) to compute the final equilibrium pH, C_T, and p_{CO_2}. **b.** What is the change in the number of dissolved mols of CO_2 caused by volatilization? **Part II.** Repeat the problem for V_g = 10,000 L.

Solution

Alk = 0.001 eq/L. M_{CO_2} = 0.002 mols. We solve Eq.(9.68) for pH using the method LHS–RHS = 0, with Eq.(9.85) for p_{CO_2} as it depends on pH. Then, $C_T = K_H p_{CO_2}/\alpha_0$. Results: **I.a.** pH = 6.69, p_{CO_2} = 1.34 × 10^{-2} atm, C_T = 1.45 × 10^{-3} *M*. **I.b.** The change in number of mols of dissolved CO_2 is given by V_w multiplied by the change in concentration = $V_w \Delta C_T$ = 1 L × (0.00145 – 0.002) *M* = –5.47 × 10^{-4} mols. **II.a.** pH = 9.99, p_{CO_2} = 3.21 × 10^{-6} atm, C_T = 6.87 × 10^{-4} *M*. **II.b.** The change in number of mols of dissolved CO_2 = $V_w \Delta C_T$ = 1 L × (6.87 × 10^{-4} – 0.002) *M* = –1.31 × 10^{-3} mols. Much more CO_2 outgasses when V_g = 10,000 L as compared to when V_g = 1 L. The much greater outgassing leads to a much higher pH. As $V_g \rightarrow \infty$, all of the CO_2 can be pulled into the gas phase, and the solution will become a 10^{-3} *F* solution of NaOH (under the assumption that none of the water evaporates!).

9.7 *IN-SITU* p_{CO_2}

In this text, when a system is described as being "closed", the meaning is that no gas phase is present so there is no actual *gas-phase* pressure of $CO_{2(g)}$ with which the solution is in equilibrium. *However*, in any solution containing CO_2, open or closed, the solution can be considered to be characterized by a particular *in-situ* p_{CO_2} value. Indeed, for any solution, by Eq.(9.60), $p_{CO_2} = \alpha_0 C_T / K_H$: (a) if the solution is in equilibrium with a gas phase, the result equals the gaseous p_{CO_2} value; (b) if the solution is in a closed system and there is no gas phase, then the result equals the value for the gaseous p_{CO_2} with which the solution *would* be in equilibrium if there were a gas phase (Figure 9.15). The in-situ p_{CO_2} measures the contribution that the dissolved CO_2 could make (along with all other dissolved gases plus water itself) to form a bubble against the ambient pressure. If the sum of

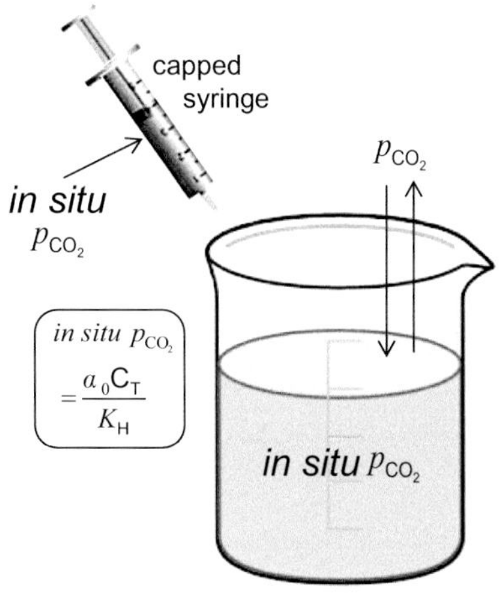

FIGURE 9.15 Beaker holding water that has reached equilibrium with some gas phase p_{CO_2} value at 25 °C. The solution has attained an "*in-situ*" p_{CO_2} value that is an inherent property of the solution. The water in the capped syringe was drawn from the beaker. The p_{CO_2} value for the gas phase equals the *in-situ* p_{CO_2} in the water in the beaker and in the water in the capped syringe: *in situ* $p_{CO_2} = \alpha_0 C_T / K_H$.

those contributions becomes larger than the system, a bubble can form. For example, for a closed container of some carbonated beverage, for *within the liquid* Eq.(9.53) might give $p_{CO_2} = 2$ atm. If the bottle is opened to a total atmospheric pressure of 1 atm, bubbles of gas (mostly CO_2) will form. Cavitation behind propeller blades is another example of such bubble formation, though with cavitation, the gas in the bubbles is not mostly CO_2, but rather mostly a combination of N_2, O_2, and H_2O. Another example of this process relates to life-threatening cases of "the bends" for scuba divers (breathing compressed air causes the *in-situ* value of $(p_{N_2} + p_{O_2})$ to become greater than 1, so as a diver comes back to the surface, bubbles can form in the blood). We also note that Eq.(9.85) is consistent with the view that we can calculate p_{CO_2} even when $V_g = 0$.

Example 9.16 First Example for Using the *In-Situ* p_{CO_2} of a Solution

Consider again the closed-system solution mixture discussed in Example 9.8.

a. What is the value of p_{CO_2} for the solution mixture?
b. Based on that value, if the solution was opened to the ambient atmosphere with $p_{CO_2} = 10^{-3.40}$ atm, would CO_2 enter or leave?
c. The solution is opened to the atmosphere with $p_{CO_2} = 10^{-3.40}$ atm. Calculate the new pH and the new C_T.

Solution

a. pH = 9.62. $K_H p_{CO_2} = \alpha_0 C_T$.
b. $p_{CO_2} = \alpha_0 C_T / K_H = 4.45 \times 10^{-4} \times 2.50 \times 10^{-4}\ M/(10^{-1.47}\ M/\text{atm}) = 3.29 \times 10^{-6}$ atm.
c. The value of p_{CO_2} is less than the value in the ambient atmosphere, so CO_2 will enter the solution.
d. Alk stays constant as CO_2 enters or leaves (if CO_2 entering or leaving is all that happens):

$$\text{Alk} = 3.33 \times 10^{-4}\ \text{eq/L} = \frac{K_1 K_H p_{CO_2}}{[H^+]} + 2\frac{K_1 K_2 K_H p_{CO_2}}{[H^+]^2} + [OH^-] - [H^+]$$

With $p_{CO_2} = 10^{-3.40}$ atm, then pH = 7.74.

Example 9.17 Second Example for Using the *In-Situ* p_{CO_2} of a Solution

Consider the closed-system solutions discussed in part I of Example 9.9. What is p_{CO_2} for each of the solutions? And, if each solution is opened to the ambient atmosphere with $p_{CO_2} = 10^{-3.40}$ atm, will CO_2 enter or leave, and will C_T rise or fall?

Solution

For all parts, $p_{CO_2} = \alpha_0 C_T / K_H$.

a. 10^{-3} *F* NaOH. Alk = 0.001 eq/L. $C_T = 0$, pH = 11.0. $p_{CO_2} = 0$. CO_2 will enter, so C_T will rise.
b. 10^{-3} *F* $NaHCO_3$. Alk = 0.001 eq/L. $C_T = 10^{-3}$ *M*. pH = 8.30. $\alpha_0 = 1.11 \times 10^{-2}$. $p_{CO_2} = 10^{-3.49}$ atm. CO_2 will enter, so C_T will rise, though not a great deal.
c. 10^{-3} *F* $NaHCO_3$ and 10^{-3} *F* $H_2CO_3^*$. Alk = 0.001 eq/L. $C_T = 2 \times 10^{-3}$ *M*. pH = 6.35. $\alpha_0 = 5.00 \times 10^{-1}$. $p_{CO_2} = 10^{-1.53}$ atm. CO_2 will leave, so C_T will fall.
d. 0.5×10^{-3} *F* Na_2CO_3. Alk = 0.001 eq/L. $C_T = 0.5 \times 10^{-3}$ *M*. pH = 10.37. $\alpha_0 = 4.52 \times 10^{-5}$. $p_{CO_2} = 10^{-6.18}$ atm. CO_2 will enter, so C_T will rise.
e. 10^{-3} *F* KOH. Alk = 0.001 eq/L. $C_T = 0$, pH = 11.0. $p_{CO_2} = 0$. CO_2 will enter, so C_T will rise.

Example 9.18 Third Example for Using the *In-Situ* p_{CO_2} of a Solution

What value of p_{CO_2} will result for part b of Example 9.14, and what will happen to pH and C_T as the river flows and seeks equilibrium with the atmosphere at 15 °C?

Solution

After the addition of the acid, pH = 8.57 − 0.10 = 8.47; $C_T = 2.73 \times 10^{-3}$ *M*.

In the stream, $p_{CO_2} = \alpha_0 C_T / K_H = 8.74 \times 10^{-3} \times 2.73 \times 10^{-3} / 10^{-1.32} = 10^{-3.30}$ atm. This is greater than the assumed ambient value of $10^{-3.40}$ atm, which means the stream will outgas CO_2, causing the pH to move back towards the initial value of 8.57.

REFERENCES

American Public Health Association (APHA), American Water Works Association (AWWA), Water Environment Federation (WEF) (2012) *Standard Methods for the Examination of Water and Wastewater*, 22nd edn, Rice EW, Baird RB, Eaton AD, Clesceri LS (Eds.), pp. 2–34.

Anderson NJ (1994) Comparative planktonic diatom biomass responses to lake and catchment disturbance. *Journal of Plankton Research*, **16**, 133–150.

Berg JM, Tymoszko JL, Gatto GJ, Stryer LB (2015). *Biochemistry*, 8th edn, Freeman Macmillan Publishing Co., New York.

Buch K (1960) Dissoziation der Kohlensäure, Gleichgewichte und Puffersystem. *Handbuch der Pflanzenphysiologie/Encyclopedia of Plant Physiology*, **5**, 1–11.

Broecker WS (1975) Climatic Change – Are we on the brink of a pronounced global warming? *Science*, **189**, 460–463.

Culberson CH, Pytkowicz RM (1968) Effect of pressure on carbonic acid, boric acid, and the pH in seawater. *Limnology and Oceanography*, **13**, 403–417.

Ellis AJ (1959) The solubility of carbon dioxide in water at high temperatures. *American Journal of Science*, **257**, 217–234.

Harned HS, Owen BB (1958) *The Physical Chemistry of Electrolyte Solutions*. Reinhold Publishing Co., New York.

Harned HS, Davies Jr R (1943) The ionization constant of carbonic acid in water and the solubility of carbon dioxide in water and aqueous salt solutions from 0 to 50°. *Journal of the American Chemical Society*, **65**, 2030–2037.

Harned HS, Scholes SR (1941) The ionization constant of HCO_3^- from 0 to 50°. *Journal of the American Chemical Society*, **63**, 1706–1709.

Irvine P, Emanuel K, He J, Horowitz LW, Vecchi G, Keith D (2019) Halving warming with idealized solar geoengineering moderates key climate hazards. *Nature Climate Change*, **9**, 295–299.

Jiang L-Q, Feely RA, Carter BR, Greeley DJ, Gledhill DK, Arzayus KM (2015) Climatological distribution of aragonite saturation state in the global oceans. *Global Biogeochemical Cycles*, **29**, 1656–1673.

Lyman J (1956) *Buffer Mechanism of Seawater*, PhD Thesis, Univ. Calif., Los Angeles, CA.

Millero FJ (1995) Thermodynamics of the carbon dioxide system in the oceans. *Geochimica et Cosmochimica Acta*, **59**, 661–677.

Mucci A (1983) The solubility of calcite and aragonite in seawater at various salinities, temperatures, and one atmosphere total pressure. *American Journal of Science*, **283**, 780–799.

Renberg I, Korsman T, Anderson NJ (1993). A temporal perspective of lake acidification in Sweden. *Ambio*, **22**, 264–271.

Sander R (2015). Compilation of Henry's Law constants (version 4.0) for water as solvent. *Atmospheric Chemistry and Physics*, **15**, 4399–4981.

Weiss R (1974). Carbon dioxide in water and seawater. The solubility of a non-ideal gas. *Marine Chemistry*, **2**, 203–215.

Zeebe RE, Sanyal A, Ortiz JD, Wolf-Gladrow DA (2001). A theoretical study of the kinetics of the boric acid–borate equilibrium in seawater. *Marine Chemistry*, **73**, 113–124.

Zeebe RE, Wolf-Gladrow D (2001) *CO_2 in Seawater: Equilibrium, Kinetics, Isotopes*, 1st edn, Volume 65, *Elsevier Oceanography Series*. Elsevier Science, Amsterdam, Netherlands.

Part III

Metal/Ligand Chemistry

10 Complexation of Metal Ions by Ligands

10.1 INTRODUCTION

"Coordination chemistry" is the subdiscipline in chemistry concerned with the association of metal ions with species like OH^-, Cl^-, SO_4^{2-}, NH_3, EDTA, and many others. In this context, species like OH^- are referred to as "ligands". As with "ligament", the word "ligand" derives from the Latin verb *ligare*, "to bind". Ligands are often negatively charged, but this is not always the case (cf., NH_3). The combined metal–ligand species is a "complex".

$$\begin{array}{c} Cd^{2+} + Cl^- = CdCl^+ \\ \text{metal ion} + \text{ligand} = \text{complex} \end{array} \qquad K_{Cl1} = \frac{\{CdCl^+\}}{\{Cd^{2+}\}\{Cl^-\}}. \tag{10.1}$$

The subscript Cl followed by 1 (one) denotes addition of a first Cl^- to the metal ion. A second Cl^- be added, leading to a higher complex:

$$CdCl^+ + Cl^- = CdCl_2^o \qquad K_{Cl2} = \frac{\{CdCl_2^o\}}{\{CdCl^+\}\{Cl^-\}}. \tag{10.2}$$

Before proceeding, we make two points. First, the specific value of a complexation constant such as K_{Cl1} will be different for Cd^{2+} as compared to Fe^{2+}, and for Fe^{2+} as compared to Fe^{3+}, etc. In complicated problems, metal-specific notation will be needed, as with $K_{Cl1}^{Cd(II)}$, $K_{Cl1}^{Fe(II)}$, and $K_{Cl1}^{Fe(III)}$, etc.; within this chapter we will consider one metal at a time. Second, in preceding chapters, we have considered Cl^- to be a spectator ion. Here too, most of the chloride present in a solution remains as a spectator ion, even if trace metals like Cd and Au as in sea waters and brines can be mostly complexed with Cl^- (Byrne, 2002); not much Cl^- can be drawn into trace metals.

H_2O itself is an important ligand in aqueous solutions because it has two pairs of outer shell non-bonding electrons, either of which can be used to bind to a metal. In Chapter 3, it was discussed that aqueous H^+ is actually present in water as a mix of solvated species with the general formula $H(H_2O)_n^+$, with $n = 6$ being an important example. Likewise, the reaction in Eq.(10.1) is more accurately written as

$$Cd(H_2O)_6^{2+} + Cl^- = Cd(H_2O)_5Cl^+ + H_2O \qquad K_{Cl1} = \frac{\{Cd(H_2O)_5Cl^+\}\{H_2O\}}{\{Cd(H_2O)_6^{2+}\}\{Cl^-\}} \tag{10.3}$$

in which we here include $\{H_2O\}$ though it will generally be assumed to be unity. On average, six ligand H_2O molecules are coordinated with aqueous Cd^{2+}, and other aqueous metal ions M^{z+} (Figure 10.1), so complexation of an aqueous metal ion with a ligand is actually a ligand exchange reaction. Fortunately, as in Chapter 3, we do not need to bother with explicitly acknowledging the underlying presence of H_2O in metal species. For K_{Cl1} for Cd^{2+}, when we write Cd^{2+} for the aqueous ion, we know that it is actually $Cd(H_2O)_6^{2+}$, and we know that $CdCl^+$ is actually $Cd(H_2O)_5Cl^+$. And, in most aqueous solutions, $\{H_2O\} = 1$. So overall, the Eq.(10.1) and Eq.(10.3) representations are mathematically equivalent when the water is relatively dilute ($\{H_2O\} = 1$).

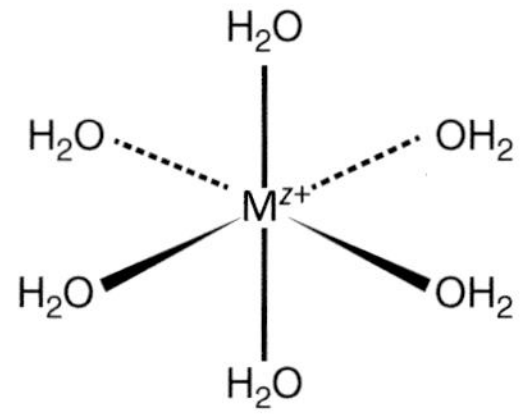

FIGURE 10.1 Typical "octahedral" orientation of six water molecules around am aqueous metal ion M^{z+}.

Taking H^+ from H_2O to create OH^- makes for three pairs of outer shell non-bonding electrons on OH^-; usually OH^- uses only one of these electron pairs for binding to one metal ion. For example, with the subscript H referring to hydroxide, for ferric iron,

$$Fe^{3+} + OH^- = FeOH^{2+} \qquad K_{H1} = \frac{\{FeOH^{2+}\}}{\{Fe^{3+}\}\{OH^-\}} \tag{10.4}$$

$$FeOH^{2+} + OH^- = Fe(OH)_2^+ \qquad K_{H2} = \frac{\{Fe(OH)_2^+\}}{\{FeOH^{2+}\}\{OH^-\}} \tag{10.5}$$

$$Fe(OH)_2^+ + OH^- = Fe(OH)_3^o \qquad K_{H3} = \frac{\{Fe(OH)_3^o\}}{\{Fe(OH)_2^+\}\{OH^-\}} \tag{10.6}$$

$$Fe(OH)_3^o + OH^- = Fe(OH)_4^- \qquad K_{H4} = \frac{\{Fe(OH)_4^-\}}{\{Fe(OH)_3^o\}\{OH^-\}}. \tag{10.7}$$

When OH^- uses two pairs of electrons in ligand complexation, the result is an hydroxide bridge, as with the ferric species $Fe_2(OH)_2^{4+}$ (Figure 10.2) which can be important in Fe(III) solutions that are relatively concentrated and of a low pH. An example is an $FeCl_3$ "coagulant feed" solution used for metering into "raw" drinking water; most of the diluted Fe(III) precipitates soon as $Fe(OH)_{3(s)}$, and the settled particles of $Fe(OH)_{3(s)}$ remove suspended particles (see also Chapter 11).

$$2Fe^{3+} + 2OH^- = Fe_2(OH)_2^{4+} \qquad \beta_{H22} = \frac{\{Fe_2(OH)_2^{4+}\}}{\{Fe^{3+}\}^2\{OH^-\}^2}. \tag{10.8}$$

The first 2 in the subscript 22 for the equilibrium constant denotes 2 OH^- having been added; the second 2 refers to the presence of 2 metal ions in the complex. β is used rather than K, because the constant pertains to the aggregated addition of multiple ligands, in this case 2.

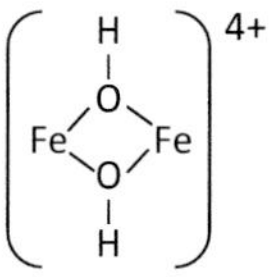

FIGURE 10.2 Structure of $Fe_2(OH)_2^{4+}$ with bridging OH. Complexes with multiple Fe(III) such as this are not important at equilibrium unless dissolved $Fe(III)_T$ values are ~10^{-2} or higher, *i.e.*, much higher than in natural waters. (Molecules of H_2O coordinating with the Fe centers are not shown.)

A chemical bond is formed when a pair of electrons is shared between two atoms. With carbon-carbon bonds, carbon-oxygen bonds, and many other types of bonds, the atoms participating in the bond bring one electron to the bond. For metal–ligand complexes, the ligand brings both electrons. The electron pair used for interaction with the metal is not involved in any bonding within the ligand itself. Once the metal–ligand complex is formed, the ligand and the metal ion share the electron pair in a manner that is usually unequal in character, with the ligand retaining a majority of the "control" on the electron pair.

Acid/base chemistry and coordination chemistry are fundamentally very similar. A conjugate base has an electron pair that can make a bond with H^+: the conjugate base plays the role of a ligand, and H^+ plays the role of a metal ion. When the conjugate base is negatively charged,

$$\underset{\text{proton ("metal ion")}}{H^+} + \underset{\text{conjugate base (ligand)}}{A^-} = \underset{\text{conjugate acid (complex).}}{HA} \tag{10.9}$$

All conjugate bases have an electron pair available for sharing, so all conjugate bases can also act as ligands with true metal ions. Thus, when a metal ion like Fe^{2+} is present in a solution, Fe^{2+} and H^+ are competing for OH^-. So, H^+ acts just like a metal, and as a result, some compilations of equilibrium constant values list H^+ among the metal ions considered (*e.g.*, Smith and Martell 1976).

With the expressions for acid/base α values, products of K values appear (Chapters 5 to 9), and the same is true with governing expressions for metal–ligand complexes. To facilitate the algebra, it is common to use the symbol β for a product of metal–ligand K values. For Eqs.(10.4–10.7),

$$Fe^{3+} + OH^- = FeOH^{2+} \qquad K_{H1} \equiv \beta_{H1} = \frac{\{FeOH^{2+}\}}{\{Fe^{3+}\}\{OH^-\}} \tag{10.10}$$

$$Fe^{3+} + 2OH^- = Fe(OH)_2^+ \qquad K_{H1}K_{H2} \equiv \beta_{H2} = \frac{\{Fe(OH)_2^+\}}{\{Fe^{3+}\}\{OH^-\}^2} \tag{10.11}$$

$$Fe^{3+} + 3OH^- = Fe(OH)_3^o \qquad K_{H1}K_{H2}K_{H3} \equiv \beta_{H3} = \frac{\{Fe(OH)_3^o\}}{\{Fe^{3+}\}\{OH^-\}^3} \tag{10.12}$$

$$Fe^{3+} + 4OH^- = Fe(OH)_4^- \qquad K_{H1}K_{H2}K_{H3}K_{H4} \equiv \beta_{H4} = \frac{\{Fe(OH)_4^-\}}{\{Fe^{3+}\}\{OH^-\}^4}. \tag{10.13}$$

10.2 FORMATION CONSTANTS *VS.* DISSOCIATION CONSTANTS (STABILITY CONSTANTS *VS.* INSTABILITY CONSTANTS)

When two species combine to form a composite species, one can write the equilibrium either as a formation reaction with an associated *stability* constant,

$$A + B = C \qquad K^{\text{stab}} = \frac{\{C\}}{\{A\}\{B\}} \tag{10.14}$$

or as a dissociation reaction with an associated *instability* constant,

$$C = A + B \qquad K^{\text{instab}} = \frac{1}{K^{\text{stab}}} = \frac{\{A\}\{B\}}{\{C\}}. \tag{10.15}$$

For metal–ligand complexation reactions, the uniformly accepted approach is to write the reaction in a "stability" format: the larger the equilibrium constant, the more stable is the complex. Indeed, when coordination chemists first began studying metal–ligand complexes, they were viewing things from the stability point of view: "how strongly do a metal and ligand tend to combine?" Early chemists interested in acid/base reactions, on the other hand, were viewing things from the instability point of view: "how strongly does an acid dissociate to give protons?" In both cases, A and B combine to form C, or alternatively C can break up to give A and B. Early chemists looking at the *solubility* of ionic solids adopted the instability point of view: "how strongly does a solid dissociate to yield ions in solution?", and hence the K_{s0} a.k.a. K_{sp} ("solubility product") format.

The reader, having by now considered the chemistry of acids with a range of dissociation constants, will understand that acids with dissociation constants of $\gtrsim 10^{-2}$, ~10^{-5}, and $\lesssim 10^{-8}$ may be characterized as strong, weak, and very weak, respectively. We can make this understanding useful in considering what types of metal–ligand formation (*i.e.*, stability) constant values characterize very weak, weak, and strong complexes. Consider:

$$Ca^{+2} + OH^{-} = CaOH^{+} \qquad K_{H1} = 20. \tag{10.16}$$

For the value expressed in the same way as an acidity constant, that is, as an instability constant:

$$CaOH^{+} = Ca^{2+} + OH^{-} \qquad \frac{1}{K_{H1}} = \frac{1}{20} = 5 \times 10^{-2}. \tag{10.17}$$

An acid with a dissociation constant of $5 \times 10^{-2} = 10^{-1.30}$ is fairly strong, that is, unstable. Thus, a metal–ligand formation constant of 20 corresponds to a weak complex. Overall, for K values of $\lesssim 10^{2}$, ~10^{5}, and $\gtrsim 10^{8}$, formation constants may be characterized as being weak, strong, and very strong respectively (Table 10.1).

Metal–ligand formation constants span a wide range of values. Complexes between metal ions from the first column of the periodic table (*e.g.*, Na^{+}, K^{+}, etc.) with ligands binding using one pair of electrons ("monodentate ligands") are generally very weak. For example, the complexes $NaOH^{\circ}$ and $NaCl^{\circ}$ are extremely weak and can essentially always be neglected (Na^{+} and Cl^{-} have been labeled spectator ions). An example of an extremely strong complex is $Co(NH_3)_6^{3+}$, the formation of which involves six constants denoted $K_{NH_3 1}$, $K_{NH_3 1}$, ...$K_{NH_3 5}$, and $K_{NH_3 6}$. This species is so stable that the core Co^{3+} ion can hold onto the six NH_3 in a hot solution of HCl.

TABLE 10.1
General Values of Metal–Ligand Equilibrium Constants as Both Formation and Dissociation Constants

	Strength of Complex		
Type of Constant	**Weak**	**Strong**	**Very Strong**
Metal–ligand *formation* constant (accepted format)	$<10^{2}$	~10^{5}	$>10^{8}$
Metal–ligand *dissociation* constant (format used for acids)	$>10^{-2}$ (strong acid)	~10^{-5} (weak acid)	$<10^{-8}$ (very weak acid)

10.3 HYDROLYSIS OF METAL IONS

10.3.1 Metal Ions Act as Acids

In Eqs.(10.4–10.8) and (10.10–10.13), the equilibrium constants entail $\{OH^-\}$, while for aqueous solutions, it is usually more convenient to deal with $\{H^+\}$ (via pH). For Eq.(10.4), we can make the switch by adding the water dissociation reaction:

$$Fe^{3+} + OH^- = FeOH^{2+} \qquad K_{H1} = \frac{\{FeOH^{2+}\}}{\{Fe^{3+}\}\{OH^-\}} \tag{10.4}$$

$$H_2O = H^+ + OH^- \qquad K_w = \{H^+\}\{OH^-\} \tag{10.18}$$

+

$$Fe^{3+} + H_2O = FeOH^{2+} + H^+ \qquad \log {}^*K_{H1} = \log(K_{H1}K_w) = 10^{-2.20}. \tag{10.19}$$

In Eq.(10.19), the metal ion reacts with water to form the complex with hydroxide, and is said to thereby be "hydrolyzed". (Note that this usage of "hydrolysis" is distinctly different from "hydrolysis of an ester, or amide", in which when water is said to lyse (cut) a molecule into two parts.) The asterisk in Eq.(10.19) indicates that the metal species is reacting a protonated form of the ligand, in this case H_2O. Any other complexation reaction with OH^- can be converted to this form, as with

$$FeOH^{2+} + H_2O = Fe(OH)_2^+ + H^+ \qquad \log {}^*K_{H2} = \log(K_{H2}K_w) = 10^{-3.50} \tag{10.20}$$

$$Fe(OH)_2^+ + H_2O = Fe(OH)_3^o + H^+ \qquad \log {}^*K_{H3} = \log(K_{H3}K_w) = 10^{-7.33} \tag{10.21}$$

$$Fe(OH)_3^o + H_2O = Fe(OH)_4^- + H^+ \qquad \log {}^*K_{H4} = \log(K_{H4}K_w) = 10^{-8.57}. \tag{10.22}$$

By this analysis, Fe^{3+} and indeed all metal ions M^{z+} are seen as being capable of acting as acids, so that here Fe^{3+} (which is actually $Fe(H_2O)_6^{3+}$) can generate a total of four H^+ by conversion to $Fe(OH)_4^-$ (which is actually $Fe(H_2O)_2(OH)_4^-$). ${}^*K_{H1}$, ${}^*K_{H2}$, ${}^*K_{H3}$, and ${}^*K_{H4}$ thus act exactly like K_1, K_2, K_3, and K_4 of a tetraprotic acid, and acid/base α values can be defined accordingly. When "polynuclear" Fe species like $Fe_2(OH)_2^{4+}$ can be neglected (which is essentially always for natural water conditions), the total dissolved $Fe(III)_T$ in solution is given by

$$Fe(III)_T = [Fe^{3+}] + [FeOH^{2+}] + [Fe(OH)_2^+] + [Fe(OH)_3^o] + [Fe(OH)_4^-] \tag{10.23}$$

with

$$\alpha_0 \equiv \frac{[Fe^{3+}]}{Fe(III)_T}, \quad \alpha_1 \equiv \frac{[FeOH^{2+}]}{Fe(III)_T}, \quad \alpha_2 \equiv \frac{[Fe(OH)_2^+]}{Fe(III)_T}, \quad \alpha_3 \equiv \frac{[Fe(OH)_3^o]}{Fe(III)_T}, \quad \alpha_4 \equiv \frac{[Fe(OH)_4^-]}{Fe(III)_T}. \tag{10.24}$$

Neglecting activity corrections, and since Fe^{3+} acts as a tetraprotic acid,

$$\alpha_0 = \frac{1}{1 + \frac{{}^*K_{H1}}{[H^+]} + \frac{{}^*K_{H1}\,{}^*K_{H2}}{[H^+]^2} + \frac{{}^*K_{H1}\,{}^*K_{H2}\,{}^*K_{H3}}{[H^+]^3} + \frac{{}^*K_{H1}\,{}^*K_{H2}\,{}^*K_{H3}\,{}^*K_{H4}}{[H^+]^4}} \tag{10.25}$$

with

$$\alpha_1 = \alpha_0 \frac{{}^*K_{H1}}{[H^+]}, \qquad \alpha_2 = \alpha_1 \frac{{}^*K_{H2}}{[H^+]}, \qquad \alpha_3 = \alpha_2 \frac{{}^*K_{H3}}{[H^+]}, \qquad \alpha_4 = \alpha_3 \frac{{}^*K_{H4}}{[H^+]}. \tag{10.26}$$

One could rewrite Eq.(10.25) in the β format using ${}^*\beta_{H1} = {}^*K_{H1}$, ${}^*\beta_{H2} = {}^*K_{H1}{}^*K_{H2}$, etc., but Eq.(10.25) is more readily compared with expressions used in Chapters 5 to 9 for acid/base α values. Values of $\log {}^*K_{H1}$ to $\log {}^*K_{H4}$ are given in Table 10.2 for Fe^{3+}. A $\log \alpha$ *vs.* pH plot or the five Fe(III) species is given in Figure 10.3.

Example 10.1 Computing α Values for Metal Ions

Compute the values of the five α values for Fe(III) at pH = 3.5 and 25 °C/1 atm.

Solution

$\alpha_0 = 2.44 \times 10^{-2}$, $\alpha_1 = 4.88 \times 10^{-1}$, $\alpha_2 = 4.88 \times 10^{-1}$, $\alpha_3 = 7.21 \times 10^{-5}$, and $\alpha_4 = 6.14 \times 10^{-10}$.

10.3.2 Polynuclear Hydroxo Complexes at High Total Metal Concentrations

A polynuclear complex is a complex containing more than one metal center. These can be important with high-oxidation state metals like Fe(III), Cr(III), and Al(III) which have a high affinity for OH^- as a bridging ligand. For Fe(III), both $Fe_2(OH)_2^{4+}$ and $Fe_3(OH)_4^{5+}$ are known:

$$2Fe^{3+} + 2H_2O = Fe_2(OH)_2^{4+} + 2H^+ \qquad \log {}^*\beta_{H22} = -2.89 \tag{10.27}$$

$$3Fe^{3+} + 4H_2O = Fe_3(OH)_4^{5+} + 4H^+ \qquad \log {}^*\beta_{H43} = -6.28. \tag{10.28}$$

TABLE 10.2
Infinite Dilution Constants for Hydrolysis of Fe(III) at 25 °C/1 atm

$\log K_{H1} = 11.80$[a]	$\log {}^*K_{H1} = -2.20$
$\log K_{H2} = 10.50$[a]	$\log {}^*K_{H2} = -3.50$
$\log K_{H3} = 6.67$	$\log {}^*K_{H3} = -7.33$[b]
$\log K_{H4} = 5.43$[a]	$\log {}^*K_{H4} = -8.57$
	$\log {}^*\beta_{22} = 10^{-2.89}$
	$\log {}^*\beta_{43} = 10^{-6.28}$

Note: $\beta_{H4} \equiv K_{H1}K_{H2}K_{H3}K_{H4}$ and ${}^*\beta_{H4} \equiv {}^*K_{H1}{}^*K_{H2}{}^*K_{H3}{}^*K_{H4}$.

[a] Smith and Martell (1976).

[b] Millero and Pierrot (2007) give an equation for ${}^*\beta = {}^*K_{H1}{}^*K_{H2}{}^*K_{H3}$, so ${}^*K_{H3}$ here is obtained using that ${}^*\beta$ with ${}^*K_{H1}$ and ${}^*K_{H2}$ as derived from K_{H1} and K_{H2} of Smith and Martell (1976).

[c] Obtained using $\beta_{H4} = \{Fe(OH)_4^-\}/(\{Fe^{3+}\}\{OH^-\}^4)$ and K_{s0} for $Fe(OH)_{3(s)}$ as reported by Smith and Martell (1976).

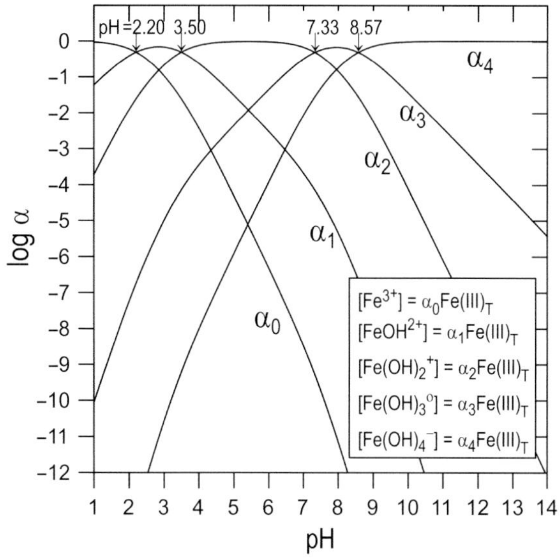

FIGURE 10.3 Log α *vs.* pH plot at 25 °C/1 atm for the five acid/base α values for dissolved Fe(III). Polynuclear Fe(III) species such as $Fe_2(OH)_2^{4+}$ are neglected under the assumption that the dissolved $Fe(III)_T$ is low.

As usual, the specific equilibrium constant values are for 25 °C/1 atm. Polynuclear complexes of this type are only important in relatively concentrated solutions of metal and so they are often neglected. They are important for when Fe(III) and Al(III) solutions of 0.01 F and higher are metered as coagulant feed solutions into both raw drinking water and treated wastewater prior to discharge.

For coagulant feed solutions of Fe(III), we then write

$$\begin{aligned} Fe_T &= [Fe^{3+}] + [FeOH^{2+}] + [Fe(OH)_2^+] + [Fe(OH)_3^o] \\ &\quad + [Fe(OH)_4^-] + 2[Fe_2(OH)_2^{4+}] + 3[Fe_3(OH)_4^{5+}]. \end{aligned} \tag{10.29}$$

From Eqs.(10.24) and (10.29),

$$\alpha_0 = \frac{[Fe^{3+}]}{Fe(III)_T} = \frac{1}{\left(1 + \dfrac{[FeOH^{2+}]}{[Fe^{3+}]} + \dfrac{[Fe(OH)_2^+]}{[Fe^{3+}]} + \dfrac{[Fe(OH)_3^o]}{[Fe^{3+}]} + \dfrac{[Fe(OH)_4^-]}{[Fe^{3+}]} + \dfrac{2[Fe_2(OH)_2^{4+}]}{[Fe^{3+}]} + \dfrac{3[Fe_3(OH)_4^{5+}]}{[Fe^{3+}]} \right)}. \tag{10.30}$$

Neglecting activity corrections,

$$\alpha_0 = \frac{[Fe^{3+}]}{Fe(III)_T} = \frac{1}{\left(1 + \dfrac{{}^*K_{H1}}{[H^+]} + \dfrac{{}^*K_{H1}\,{}^*K_{H2}}{[H^+]^2} + \dfrac{{}^*K_{H1}\,{}^*K_{H2}\,{}^*K_{H3}}{[H^+]^3} + \dfrac{{}^*K_{H1}\,{}^*K_{H2}\,{}^*K_{H3}\,{}^*K_{H4}}{[H^+]^4} + \dfrac{2[Fe_2(OH)_2^{4+}]}{[Fe^{3+}]} + \dfrac{3[Fe_3(OH)_4^{5+}]}{[Fe^{3+}]} \right)}. \tag{10.31}$$

From Eq.(10.27) and (10.28),

$$\alpha_0 = \frac{[Fe^{3+}]}{Fe(III)_T} = \frac{1}{\left(1+\frac{{}^*K_{H1}}{[H^+]}+\frac{{}^*K_{H1}{}^*K_{H2}}{[H^+]^2}+\frac{{}^*K_{H1}{}^*K_{H2}{}^*K_{H3}}{[H^+]^3}+\frac{{}^*K_{H1}{}^*K_{H2}{}^*K_{H3}{}^*K_{H4}}{[H^+]^4}+\frac{2[Fe^{3+}]{}^*\beta_{H22}}{[H^+]^2}+\frac{3[Fe^{3+}]^2{}^*\beta_{H43}}{[H^+]^4}\right)} \quad (10.32)$$

or

$$Fe(III)_T = [Fe^{3+}]\left(1+\frac{{}^*K_{H1}}{[H^+]}+\frac{{}^*K_{H1}{}^*K_{H2}}{[H^+]^2}+\frac{{}^*K_{H1}{}^*K_{H2}{}^*K_{H3}}{[H^+]^3}+\frac{{}^*K_{H1}{}^*K_{H2}{}^*K_{H3}{}^*K_{H4}}{[H^+]^4}+\frac{2[Fe^{3+}]{}^*\beta_{H22}}{[H^+]^2}+\frac{3[Fe^{3+}]^2{}^*\beta_{H43}}{[H^+]^4}\right) \quad (10.33)$$

which, given $Fe(III)_T$, can be solved for $[Fe^{3+}]$ as a function of pH. α_0 is then obtained by use of $Fe(III)_T$. For α_1 to α_4, the relations in Eq.(10.26) apply. For the two polynuclear complexes, using $[Fe^{3+}] = \alpha_0 Fe(III)_T$,

$$\alpha_{22} = \frac{2[Fe_2(OH)_2^{4+}]}{Fe(III)_T} = \frac{2[Fe^{3+}]^2\,{}^*\beta_{H22}}{Fe(III)_T[H^+]^2} = \frac{2\alpha_0^2 Fe(III)_T\,{}^*\beta_{H22}}{[H^+]^2} \quad (10.34)$$

$$\alpha_{43} = \frac{3[Fe_3(OH)_4^{5+}]}{Fe(III)_T} = \frac{3[Fe^{3+}]^3\,{}^*\beta_{H43}}{Fe(III)_T[H^+]^4} = \frac{3\alpha_0^3 Fe(III)_T^2\,{}^*\beta_{H43}}{[H^+]^4}. \quad (10.35)$$

The factors of 2 and 3 in Eqs.(10.34) and (10.35), respectively, account for the fact that the α system concerns the fractional distribution of mols of dissolved Fe(III) among the different species, and each of the polynuclear species contains more than one mol of Fe(III) per mol of the species.

When $Fe(III)_T$ are not low, polynuclear complexes become important, and all the α values become noticeably dependent on $Fe(III)_T$ in addition to pH. With activity corrections neglected, Figure 10.4a and b provide $\log\alpha$ *vs.* pH plots for two values of dissolved $Fe(III)_T$, *e.g.*, 0.005 *F* $FeCl_3$ and 0.025 *F* $FeCl_3$, respectively; $(C'_B - C'_A)$ (defined in Chapter 11) would be used to vary pH. Polynuclear complexes are more important in Figure 10.4b as compared to Figure 10.4a. Each figure indicates the region where $(am)Fe(OH)_{3(s)}$ (am = amorphous) becomes supersaturated because $[Fe^{3+}][OH^-]^3 > K_{s0}^{Fe(OH)3(s)} = 10^{-38.79}$ (solubility product). 0.025 *F* $FeCl_3$ is sufficiently high that including activity corrections would lower the Figure 10.4b pH = 1.65 estimate for precipitation of $(am)Fe(OH)_{3(s)}$.

Example 10.2 Calculating the pH of an Fe(III) Solution, Including Polynuclear Complexes

Including the polynuclear complexes, compute the pH of a 0.025 *F* solution of $FeCl_3$ at 25 °C/1 atm. Check for superstation with respect to $(am)Fe(OH)_{3(s)}$, for which $K_{s0}^{Fe(OH)3(s)} = 10^{-38.79}$. Neglect activity corrections.

Solution

a. The ENE is

$$[H^+] + 3[Fe^{3+}] + 2[FeOH^{2+}] + [Fe(OH)_2^+] + 4[Fe_2(OH)_2^{4+}] + 5[Fe_3(OH)_4^{5+}]$$

$$= [Fe(OH)_4^-] + [OH^-] + [Cl^-]$$

for which $[Fe_2(OH)_2^{4+}] = (\frac{1}{2}\alpha_{22})Fe(III)_T$, and $[Fe_3(OH)_4^{5+}] = (\frac{1}{3}\alpha_{43})Fe(III)_T$.

$$[H^+] + 3\alpha_0 Fe(III)_T + 2\alpha_1 Fe(III)_T + \alpha_2 Fe(III)_T + 4(\tfrac{1}{2}\alpha_{22})Fe(III)_T + 5(\tfrac{1}{3}\alpha_{43})Fe(III)_T$$

$$= \alpha_4 Fe(III)_T + \frac{K_w}{[H^+]} + [Cl^-].$$

Using Eqs.(10.34) and (10.35), in expanded, LHS–RHS = 0 format and neglecting all activity corrections,

$$[H^+] + Fe(III)_T\left(3\alpha_0 + 2\alpha_0\frac{{}^*K_{H1}}{[H^+]} + \alpha_0\frac{{}^*K_{H1}\,{}^*K_{H2}}{[H^+]^2} + \frac{4\alpha_0^2 Fe(III)_T\,{}^*\beta_{H22}}{[H^+]^2}\right.$$

$$\left. + \frac{5\alpha_0^3 Fe(III)_T^2\,{}^*\beta_{H43}}{[H^+]^4} - \alpha_0\frac{{}^*K_{H1}\,{}^*K_{H2}\,{}^*K_{H3}\,{}^*K_{H4}}{[H^+]^4}\right) - \frac{K_w}{[H^+]} - [Cl^-] = 0.$$

Eq.(10.32) gives α_0 as a function of pH and $[Fe^{3+}]$. A spreadsheet can be built that calculates α_0 based on a guess cell for pH, and a second cell that calculates $[Fe^{3+}] = \alpha_0 Fe(III)_T$. This leads to a "circular reference", because α_0 depends on pH and $[Fe^{3+}]$, and $[Fe^{3+}]$ depends on α_0. The circular reference can be allowed in Excel, under File> Options>Formulas, "enable interative calculation". Then, invoking Solver to vary pH such that the expanded, LHS–RHS = 0 version of the ENE is satisfied, the solution is pH = 1.93, $[OH^-] = 10^{-12.06}$, $\alpha_0 = 0.546$, $[Fe^{3+}] = 0.546 \times 0.025\ M = 0.0136\ M$. $[Fe^{3+}][OH^-]^3 = 8.93 \times 10^{-39}$, which is about $5 \times K_{s0}^{Fe(OH)3(s)}$. The solution will be supersaturated with respect to this solid if activity corrections could indeed be neglected. However, the ionic strength (*I*) of this solution is larger than 0.025; activity corrections will reduce if not eliminate the state of supersaturation.

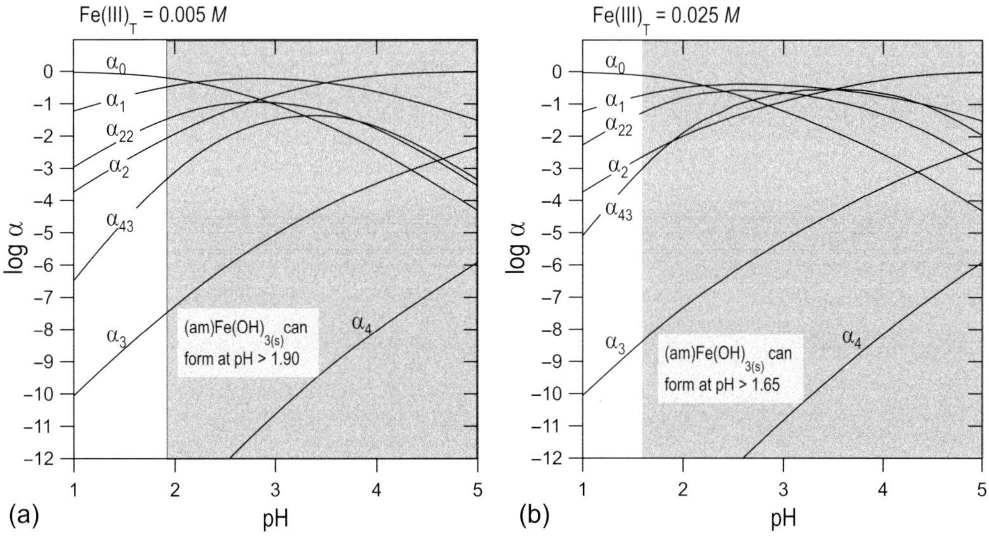

FIGURE 10.4 **a.** Dissolved $Fe(III)_T = 0.005\ M$, polynuclear Fe(III) complexes included among lines of $\log\alpha$ *vs.* pH at 25 °C/1 atm. Activity corrections neglected. **b.** Dissolved $Fe(III)_T = 0.025\ M$, polynuclear Fe(III) complexes included among lines of $\log\alpha$ *vs.* pH at 25 °C/1 atm. Activity corrections neglected.

10.4 COMPLEXES WITH Cl^- IN BRINES AND SEAWATER

As has been discussed, most aquo-metal ions can accommodate six solvating, ligand water molecules. One or more of these ligand water molecules can be exchanged for other ligands like OH^-, Cl^-, NH_3, etc. Other than water itself, in natural waters, OH^- is the most generally important ligand, especially for metal ions with a core charge like +3 (see Eqs.(10.4–10.7, Byrne, 2002). However, in brines and seawater where Cl^- levels are high, certain trace metals such as Cd and Au can be complexed to a significant degree by Cl^-. This nevertheless does not change our general view that Cl^- behaves mostly as a spectator ion, because in such circumstances, the vast majority of the Cl^- itself remains as Cl^-. In brines and seawater, ${}^cK$ values are more directly useful than K values; at an ionic strength I of 1.0, we have

$$Cd^{2+} + Cl^- = CdCl^+ \qquad \log {}^cK_1 = 1.35 \qquad (\log {}^c\beta_1 = 1.35) \tag{10.36}$$

$$CdCl^+ + Cl^- = CdCl_2^o \qquad \log {}^cK_2 = 0.35 \qquad (\log {}^c\beta_2 = 1.70) \tag{10.37}$$

$$CdCl_2^o + Cl^- = CdCl_3^- \qquad \log {}^cK_3 = -0.20 \qquad (\log {}^c\beta_3 = 1.50). \tag{10.38}$$

Complexation of Cd^{2+} by OH^- also occurs to a limited extent, according to

$$Cd^{2+} + H_2O = CdOH^+ + H^+ \qquad \log {}^cK_{H1} = -10.30 \qquad (\log {}^{c*}\beta_{H3} = -10.30) \tag{10.39}$$

$$CdOH^+ + H_2O = CdOH_2^o + H^+ \qquad \log {}^{c*}K_{H2} = -10.20 \qquad (\log {}^{c*}\beta_{H2} = -20.50). \tag{10.40}$$

The concentrations of mixed chloro-hydroxo species (*e.g.*, $CdCl_2^o$) will be non-zero, but their K values are not well-known. Neglecting such species, we then have

$$Cd_T = [Cd^{2+}] + [CdCl^+] + [CdCl_2^o] + [CdCl_3^-] + [CdOH^+] + [CdOH_2^o] \tag{10.41}$$

$$\alpha_0 \equiv \frac{[Cd^{2+}]}{Cd_T} = \frac{[Cd^{2+}]}{[Cd^{2+}] + [CdCl^+] + [CdCl_2^o] + [CdCl_3^-] + [CdOH^+] + [CdOH_2^o]}. \tag{10.42}$$

Using the β notation for products of the K values,

$$\alpha_0 = \frac{[Cd^{2+}]}{Cd_T} = \frac{1}{1 + {}^c\beta_1[Cl^-] + {}^c\beta_2[Cl^-]^2 + {}^c\beta_3[Cl^-]^3 + {}^{c*}\beta_{H1}/[H^+] + {}^{c*}\beta_{H2}/[H^+]^2} \tag{10.43}$$

$$\alpha_{1Cl} = \frac{[CdCl^+]}{Cd_T} = \alpha_0\, {}^c\beta_1[Cl^-]$$

$$\alpha_{2Cl} = \frac{[CdCl_2^o]}{Cd_T} = \alpha_0\, {}^c\beta_2[Cl^-]^2 \tag{10.44}$$

$$\alpha_{3Cl} = \frac{[CdCl_3^-]}{Cd_T} = \alpha_0\, {}^c\beta_3[Cl^-]^3$$

$$\alpha_{1OH} = \frac{[CdOH^+]}{Cd_T} = \alpha_0\, {}^{c*}\beta_{H1}/[H^+] \qquad \alpha_{2OH} = \frac{[CdOH_2^o]}{Cd_T} = \alpha_0\, {}^{c*}\beta_{H2}/[H^+]^2. \tag{10.45}$$

Example 10.3 Complexation of Cd^{2+} by Chloride and Hydroxide

Consider a brine derived from seawater at 25 °C/1 atm in which $[Cl^-] = 1.0$ *M*, and $I = 1.0$, and $pH \equiv \log\{H^+\} = \log(\gamma_{H^+}[H^+]) = 7.90$. The total dissolved Cd(II) concentration is 80 ng/L. Determine the six α values and $[Cd^{2+}]$, and identify the major species.

Solution

For Cd, 80 ng/L corresponds to 7.12×10^{-11} mol/L. We use ${}^c\beta$ throughout the problem to include activity corrections. The Davies Equation gives $\gamma_{H^+} = 0.79$. $[H^+] = 10^{-7.90}/0.79 = 1.59 \times 10^{-8}$. The Cd concentration is very low; there is no chance complexation of Cd by chloride will lower [Cl–] below 1.0 *M* to any relevant degree. For Eq.(10.43), $[Cl^-] = [Cl^-]^2 = [Cl^-]^3 = 1$, and so

$$\alpha_0 = \frac{[Cd^{2+}]}{Cd_T} = \frac{1}{1 + {}^c\beta_1 + {}^c\beta_2 + {}^c\beta_3 + {}^{c*}\beta_{H1}/(1.59 \times 10^{-8}) + {}^{c*}\beta_{H2}/(1.59 \times 10^{-8})^2}$$

$$= 9.51 \times 10^{-3}.$$

Using the relations in Eq.(10.44) and (10.45): $\alpha_{1Cl} = 0.213$; $\alpha_{2Cl} = 0.476$; $\alpha_{3Cl} = 0.300$; $\alpha_{1OH} = 0.0030$; $\alpha_{2OH} = 1.20 \times 10^{-7}$. $[Cd^{2+}] = (9.51 \times 10^{-3})(7.12 \times 10^{-11}) = 6.77 \times 10^{-12}$ *M*. The major species is the dichloride complex $CdCl_2^o$. The total chloride concentration bound to Cd is given by $[CdCl^+] + 2[CdCl_2^o] + 3[CdCl_3^-]$, which here equals 1.47×10^{-9} *M*. This is a very tiny fraction of the total chloride in solution, preserving $[Cl^-] \approx 1.0$ *M*.

10.5 CHELATES

Deprotonation of a carboxylic acid group –R–COOH gives the carboxylate group $-R-COO^-$, an effective ligand. It is true that even the carboxylic acid group itself has pairs of non-bonding electrons on the oxygen atoms that can become involved in ligand interactions with metal ions, but the carboxylate group is an order of magnitude stronger ligand.

Numerous organic acids are polyprotic. The fully deprotonated forms of such acids have more than one carboxylate group, hence more than one possible point of ligand attachment to a metal ion, and so may bind to a metal ion at more than one point: they are "multidentate". Malonic acid is diprotic, and so is of the H_2B type. When fully protonated as B^{2-}, the result is a bidentate ligand $(L-R-L)^{2-}$ where R is the carbon chain linking the two carboxylate ligand groups. So, just as two Cl^- can bind to Cd^{2+} yielding $CdCl_2^o$, the malonate ion (2–) can bind to Cd^{2+} giving $[R(L)_2Cd]^{o}$.

Multidentate ligands are often called chelates. The word "chelate" originates from the Greek "chele" (χελε) meaning "crab's claw": a crab's claw is bidentate. Two well-known, man-made chelates used for manipulating metal ion chemistry are nitrilotriacetic acid (NTA) and ethylenediaminetetraacetic acid (EDTA). The structures of the protonated and fully deprotonated forms of the two compounds are given in Figures 10.5 and 10.6. When fully deprotonated, the amine nitrogen on NTA offers one of the points of attachment to a metal ion. For deprotonated EDTA, two of the ligand points are amine nitrogens.

NTA^{3-} is characterized by large formation constants with many metal ions. At 25 °C/1 atm and $I = 0.1$, binding of NTA^{4-} with Mg^{2+} and Ca^{2+} is characterized by *K* values of $10^{6.5}$ and $10^{7.6}$, respectively. $EDTA^{4-}$ is an even more powerful chelate because its functionalities can occupy all *six* of the coordination locations of a typical metal ion. The four identical arms of EDTA and the $-CH_2-CH_2-$ segment between the two nitrogens are good lengths to allow the ligand functionalities to interact at six ligand points with most metal ions (Figure 10.7). *K* values with $EDTA^{4-}$ are quite large for many metal ions (10^6 or greater). Both EDTA and NTA have been used extensively to bind Ca^{2+} and Mg^{2+}

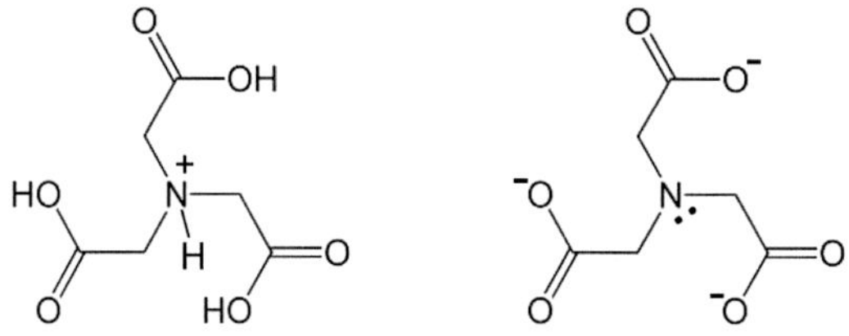

FIGURE 10.5 Nitrilotriacetic acid (NTA) in fully protonated and fully deprotonated forms. When fully deprotonated, NTA is a tetradentate ligand. The unshared pair of electrons on the nitrogen in the deprotonated form provides one of the ligand points.

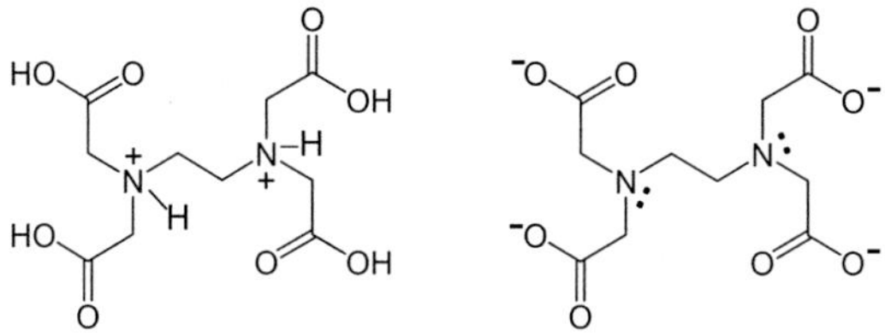

FIGURE 10.6 Ethylenediaminetetraacetic acid (EDTA) in fully protonated and fully deprotonated forms. When fully deprotonated, EDTA is a hexadentate ligand. Unshared pairs of electrons on the two nitrogens in the deprotonated form provide two of the ligand points.

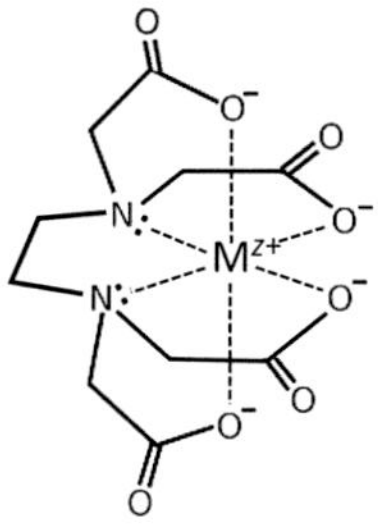

FIGURE 10.7 Chelate-metal ion complex between fully deprotonated EDTA and a metal ion M^{z+}. Two of the ligand points of attachment on EDTA are the amine nitrogens, each with one pair of non-bonding electrons.

and so prevent them from interfering in the performance of soaps and detergents. EDTA has been widely used in "chelation therapy" to help people with mercury and lead poisoning by solubilizing mercury and lead for excretion. In nature, the natural organic matter materials "humic acid" and "fulvic acid" act as metal chelates. Within the structures of these heterogenous materials, multiple carboxylate groups can exist in close proximity, and so bind metal ions in a multidentate fashion.

For EDTA, using Ca^{2+} as the metal ion, the metal-chelate formation constant is defined according to

$$Ca^{2+} + EDTA^{4-} = CaEDTA^{2-} \qquad K_{EDTA}^{Ca} \equiv \frac{\{CaEDTA^{2-}\}}{\{Ca^{2+}\}\{EDTA^{4-}\}} = 10^{10.7}. \tag{10.46}$$

The value for K_{EDTA}^{Ca} pertains to 25 °C/1 atm. The total EDTA in solution is the sum of the total "free" EDTA and the "metal-bound" EDTA:

$$EDTA_T = EDTA_{free} + EDTA_{bound}. \tag{10.47}$$

The total free EDTA is given by

$$EDTA_{free} = [H_6EDTA^{2+}] + [H_5EDTA^{+}] + [H_4EDTA^{o}] + [H_3EDTA^{-}] + [H_2EDTA^{2-}] + [HEDTA^{3-}] + [EDTA^{4-}]. \tag{10.48}$$

The bound EDTA and bound Ca are given by

$$EDTA_{bound} = Ca_{bound} = [CaEDTA^{2-}]. \tag{10.49}$$

Very few studies have reported infinite dilution constants for equilibrium constants involving EDTA, ionic concentration conditions usually pertaining to ${}^{c}K$ values measured at an ionic strength I of 0.1 or higher. Using ${}^{c}K$ values, the fraction of the free EDTA that is in the fully-protonated, H_6EDTA^{2+} form is given by

$$\alpha_0^{EDTA,free} = \frac{1}{\left(1+\frac{{}^cK_1}{[H^+]}+\frac{{}^cK_1\,{}^cK_2}{[H^+]^2}+\frac{{}^cK_1\,{}^cK_2\,{}^cK_3}{[H^+]^3}+\frac{{}^cK_1\,{}^cK_2\,{}^cK_3\,{}^cK_4}{[H^+]^4} + \frac{{}^cK_1\,{}^cK_2\,{}^cK_3\,{}^cK_4\,{}^cK_5}{[H^+]^5}+\frac{{}^cK_1\,{}^cK_2\,{}^cK_3\,{}^cK_4\,{}^cK_5\,{}^cK_6}{[H^+]^6}\right)} \tag{10.50}$$

so that

$$\alpha_6^{EDTA,free} = \alpha_0^{EDTA,free}\frac{{}^cK_1\,{}^cK_2\,{}^cK_3\,{}^cK_4\,{}^cK_5\,{}^cK_6}{[H^+]^6} \tag{10.51}$$

$$[EDTA^{4-}] = \alpha_6^{EDTA,free}\,EDTA_{free}. \tag{10.52}$$

At 25 °C/1 atm and ionic strength $I = 0.1$, the following values pertain: ${}^cK_1 = 10^{0.0}$; ${}^cK_2 = 10^{-1.5}$; ${}^cK_3 = 10^{-2.0}$; ${}^cK_4 = 10^{-2.66}$; ${}^cK_5 = 10^{-6.16}$; and ${}^cK_6 = 10^{-10.24}$.

The total Ca in solution is the sum of the total "free" Ca and the "EDTA-bound" Ca:

$$Ca_T = Ca_{free} + Ca_{bound}. \tag{10.53}$$

Ca^{2+} is complexed only very weakly by OH^-, so that only ${}^*K_{H1}$ ($=10^{-12.7}$ at 25 °C/1 atm) is usually relevant:

$$Ca_{free} = [Ca^{2+}] + [CaOH^+]. \tag{10.54}$$

The fraction of the free (non-EDTA-bound) Ca that is in the Ca^{2+} form is then given by

$$\alpha_0^{Ca,free} = \frac{1}{1+\frac{{}^*K_{H1}}{[H^+]}} \tag{10.55}$$

$$[Ca^{2+}] = \alpha_0^{Ca,free}\,Ca_{free}. \tag{10.56}$$

Combining Eqs.(10.46), (10.52), and (10.56),

$$K_{EDTA}^{Ca} = 10^{10.7} = \frac{Ca_{bound}}{\left(\alpha_0^{Ca,free}\,Ca_{free}\right)\left(\alpha_6^{EDTA,free}\,EDTA_{free}\right)} \tag{10.57}$$

where the value given pertains to 25 °C/1 atm and $I = 0.1$. With Eqs.(10.47), (10.49), and (10.53),

$$ {}^{c}K_{\mathrm{EDTA}}^{\mathrm{Ca}} = \frac{\left(\mathrm{Ca_T} - \mathrm{Ca_{free}}\right)}{\left(\alpha_0^{\mathrm{Ca,free}}\mathrm{Ca_{free}}\right)\left[\alpha_6^{\mathrm{EDTA,free}}\left(\mathrm{EDTA_T} - \left(\mathrm{Ca_T} - \mathrm{Ca_{free}}\right)\right)\right]}. \quad (10.58) $$

Because the two free α values depend only on pH, then given $\mathrm{Ca_T}$ and $\mathrm{EDTA_T}$, the extent of complexation (and then full speciation) can be determined by solving Eq.(10.58) to obtain $\mathrm{Ca_{free}}$, as by varying $\log \mathrm{Ca_{free}}$. For the numerical solution, two comments are pertinent: (1) Eq.(10.57) could, alternatively, have been modified so that $\mathrm{Ca_{bound}}$ was the remaining unknown, but when the formation of the complex is nearly 100%, the mathematical solution would need to be sought using very small fractional changes in $\mathrm{Ca_{bound}}$; the Eq.(10.58) form provides a more robust basis for the solution process. (2) Using Excel and Solver, preceding chapters have emphasized numerical solutions that seek LHS − RHS = 0. For Eq.(10.58) the LHS and RHS values during a numerical solution are $\sim 10^{10.7}$. Finding the condition that gives $\sim 10^{10.7} - \sim 10^{10.7} = 0$ to within some small convergence criterion makes less sense than finding when LHS/RHS = 1 to within the same small convergence criterion.

Example 10.4 Complexation of Ca^{2+} by EDTA

A water sample contains calcium at 40 mg/L (=0.0010 *M*). EDTA as the disodium salt Na_2H_2EDTA is added to give $\mathrm{EDTA_T} = 0.001$ *M*, and the pH is adjusted. Consider two pH values: **a.** pH = 6.5; **b.** pH = 7.5. For each pH, compute $\alpha_0^{\mathrm{Ca,free}}$ and $\alpha_6^{\mathrm{EDTA,free}}$. Then vary $\log\mathrm{Ca_{free}}$ to find the value that gives LHS/RHS = 1 for Eq.(10.58). Also find $\mathrm{ETDA_{free}}$ and $[\mathrm{CaEDTA^{2-}}] = \mathrm{Ca_{bound}} = \mathrm{EDTA_{bound}}$.

Solution

a. At pH = 6.5, $\alpha_0^{\mathrm{Ca,free}} \approx 1$, $\alpha_6^{\mathrm{EDTA,free}} = 1.25 \times 10^{-4}$. $\mathrm{Ca_{free}} = 10^{-4.90} = 1.26 \times 10^{-5}$ *M*. $\mathrm{ETDA_{free}} = 1.26 \times 10^{-5}$ *M*. $[\mathrm{CaEDTA^{2-}}] = \mathrm{Ca_{bound}} = \mathrm{EDTA_{bound}} = 9.87 \times 10^{-4}$ *M*. 98.7% of the Ca is bound as the complex.

b. At pH = 7.5, $\alpha_0^{\mathrm{Ca,free}} \approx 1$, $\alpha_6^{\mathrm{EDTA,free}} = 1.74 \times 10^{-3}$. $\mathrm{Ca_{free}} = 10^{-5.47} = 3.38 \times 10^{-6}$ *M*. $\mathrm{ETDA_{free}} = 3.38 \times 10^{-6}$ *M*. $[\mathrm{CaEDTA^{2-}}] = \mathrm{Ca_{bound}} = \mathrm{EDTA_{bound}} = 9.97 \times 10^{-4}$ *M*. 99.7% of the Ca is bound as the complex. More Ca is bound at the higher pH because the higher pH has increased $\alpha_6^{\mathrm{EDTA,free}}$ but has not significantly changed $\alpha_0^{\mathrm{Ca,free}}$.

REFERENCES

Byrne RH (2002) Inorganic speciation of dissolved elements in seawater: the influence of pH on concentration ratios. *Geochemical Transactions*, **3**, 11–16.

Millero FJ, Pierrot D (2007) The activity coefficients of Fe(III) hydroxide complexes in NaCl and $NaClO_4$ solutions. *Geochimica et Cosmochimica Acta*, **71**, 4825–4833.

Smith RM, Martell AE (1976) *Critical Stability Constants, Volume 4, Inorganic Complexes*. Plenum Press, New York.

Part IV

Mineral Solubility

11 Simple Salts and Metal Oxides/Hydroxides/Oxyhydroxides

11.1 INTRODUCTION

The dissolution of some solids produces only neutral aqueous species. An organic compound example is *para*-dichlorobenzene (MW = 147.00 g/mol) as used in "moth balls" and other applications. The dissolution reaction for this solid can be written:

$$C_6H_4Cl_{2(s)} = C_6H_4Cl_2 \qquad K = \frac{\{C_6H_4Cl_2\}}{\{C_6H_4Cl_{2(s)}\}} = \{C_6H_4Cl_2\}. \tag{11.1}$$

We abbreviate *para*-dichlorobenzene as *p*-dichlorobenzene. As with all K values, the numerical value depends on the units chosen for it. Here the units may be molality or molarity (in which case the numerical values are nearly the same), but solubility data are elsewhere often found in units such as mg/liter.

The simplification in Eq.(11.1) that $K = \{C_6H_4Cl_2\}$ pertains to when: (1) the concentration scale chosen for $\{C_6H_4Cl_{2(s)}\}$ is the mole fraction scale, *and* (2) we are interested in cases involving the dissolution of the pure or nearly-pure solid so that $\{C_6H_4Cl_{2(s)}\} = 1$, or at least ~1. As with all K values, literature values vary somewhat (though for specific conditions there is always only one true value). For pure solid *p*-dichlorobenzene in water at 25 °C/1 atm, the solubility is ~91 mg/L. For non-charged species, aqueous phase activity corrections on the infinite-dilution referenced scale are negligible, except for very salty waters. For our purposes then, using MW = 147.00 g/mol, we obtain for 25 °C/1 atm that $K = \{C_6H_4Cl_2\} \approx [C_6H_4Cl_2] = [(91 \text{ mg/L})/(147{,}000 \text{ mg/mol})] \times (1 \text{ kg/L}) = 6.2 \times 10^{-4}$ m. This is essentially $6.2 \times 10^{-4}\,M$.

A natural inorganic mineral that predominantly forms a non-charged species when it dissolves is $SiO_{2(s)}$ (quartz):

$$SiO_{2(s)} + 2H_2O = Si(OH)_4 \qquad K = \frac{\{Si(OH)_4\}}{\{SiO_{2(s)}\}} = \{Si(OH)_4\}. \tag{11.2}$$

The simplification in Eq.(11.2) that $K = \{Si(OH)_4\}$ is again allowed if we take the dissolving solid to be pure ($x = 1$) or nearly pure ($x \approx 1$).

For an example of a dissolution that leads to ions, consider the dissolution of $NaCl_{(s)}$ (halite):

$$NaCl_{(s)} = Na^+ + Cl^- \qquad K_{s0} = \frac{\{Na^+\}\{Cl^-\}}{\{NaCl_{(s)}\}} = \{Na^+\}\{Cl^-\}. \tag{11.3}$$

The simplification in Eq.(11.3) that $K = \{Na^+\}\{Cl^-\}$ is again allowed if we take the dissolving solid to be pure ($x = 1$) or nearly pure ($x \approx 1$). Making that assumption will be the general case in this chapter; the implications of non-pure mineral solids are discussed in Chapter 14.

At a given T and P, at solubility equilibrium, there will be only one value for $\{C_6H_4Cl_2\}$ in Eq.(11.1), and one value for $\{Si(OH)_4\}$ in Eq.(11.2). However, for $NaCl_{(s)}$, which *dissociates* into ions when it dissolves, any combination of ion activities that satisfies the equilibrium solubility expression is valid: now, a *product* of activities must equal the equilibrium constant. Beginning chemistry texts refer to equilibrium constants for dissolution reactions that produce dissolved ions as "solubility products", *e.g.*, as K_{sp} values, or as "ion activity product" (IAP) values. As introduced in Eq.(11.3), here such K values are given the symbol K_{s0}, with s for solubility, and 0 indicating that the identities of species moving into solution are the same (zero change) relative to the solid: for Eq.(11.3), the dissolved ions are Na^+ and Cl^-, the same as written in the formula for the solid.

11.2 DEFINING WHERE A SOLUTION IS RELATIVE TO EXACT SATURATION

11.2.1 Undersaturation, Saturation, and Supersaturation in Terms of K_{s0}

For the acid dissociation reaction $HA = H^+ + A^-$ with equilibrium constant K, the reaction quotient $Q = \{H^+\}\{A^-\}/\{HA\}$. At equilibrium, $Q = K$. If $Q < K$, then some HA will tend to dissociate to form more H^+ and A^- until $Q = K$. If $Q > K$, then some H^+ will tend to combine with an equal amount of A^- to form more HA until $Q = K$. The situation for dissolution/precipitation chemistry is analogous. In general, for dissolution of a pure or nearly pure solid,

$$\text{saturation state} \equiv \Omega \equiv \frac{Q}{K_{s0}}. \tag{11.4}$$

For $NaCl_{(s)}$, then $Q = \{Na^+\}\{Cl^-\}$,

$$\Omega = \frac{Q}{K_{s0}} = \frac{\{Na^+\}\{Cl^-\}}{K_{s0}}. \tag{11.5}$$

$\Omega < 1$ indicates undersaturation; $\Omega = 1$ indicates saturation equilibrium; $\Omega > 1$ indicates supersaturation. Some geochemists use

$$\text{saturation index SI} = \log \Omega = \log \frac{Q}{K_{s0}}. \tag{11.6}$$

$SI < 0$ indicates undersaturation; $SI = 0$ indicates equilibrium; $SI > 0$ indicates supersaturation (Table 11.1).

TABLE 11.1
Implications Concerning the Relative Values of Q and K for a Dissolution Reaction

Condition			State	For $NaCl_{(s)}$	Behavior
$Q<K_{s0}$	$\Omega<1$	$SI<0$	undersaturated	$\{Na^+\}\{Cl^-\}<K_{s0}$	a. if solid is present, not at equilibrium; some solid will tend to dissolve; b. if solid is not present, it will not form; the solution can be at equilibrium, just not at saturation equilibrium.
$Q=K_{s0}$	$\Omega=1$	$SI=0$	saturated	$\{Na^+\}\{Cl^-\}=K_{s0}$	a. if solid is present, no more solid will dissolve; at saturation equilibrium. b. if solid is not present, it will not form; the solution can be saturated without solid being present.
$Q>K_{s0}$	$\Omega>1$	$SI>0$	supersaturated	$\{Na^+\}\{Cl^-\}>K_{s0}$	a. if solid is present, not at equilibrium; more solid will tend to form; b. if solid is not present, not at equilibrium; some solid will tend to form.

Assumptions: if present, the solid is pure (mole fraction $x = 1$).

11.2.2 The Solid POV: The Solid Is Present; How Much Will Dissolve to Reach Equilibrium? *vs.* The Solution POV: The Solution Has a Specific Chemistry; Will Precipitation Occur, and If Yes, How Much?

Usually, in a chemical question involving solubility, built into the problem is one or the other of two of points of view (POV). *Solid POV* example questions are: (a) if the solid is present, based on the equilibrium constant(s) governing the solubility process, how much of some solid-related element of interest (like Fe) is dissolved at equilibrium for some particular conditions in the solution (like pH)? (b) if the solid is present, and the solution has some particular conditions, how might the system have gotten there? *Solution POV* example questions are: (a) for a solution with some particular characteristics (like dissolved Fe level and pH), based on the equilibrium constant(s) governing the solubility process, what is the saturation state Ω for a particular solid (like $Fe(OH)_{3(s)}$)?; (b) for a solution with some particular characteristics: if undersaturated ($\Omega < 1$), what could be done to move the solution to a point of saturation ($\Omega = 1$)? If it is established in a solution POV problem that the solid must be present (at equilibrium), then calculations can move forward taking the solid POV and using the available K values.

11.3 SIMPLE SALTS INVOLVE ONLY "SPECTATOR IONS"

Halite. Taking the solid POV, at equilibrium with $NaCl_{(s)}$, from Eq.(11.3),

$$\log\{Na^+\} = \log K_{s0} - \log\{Cl^-\} \qquad \frac{d\log\{Na^+\}}{d\log\{Cl^-\}} = -1. \tag{11.7}$$

Our iconic examples of spectator ions in aqueous solution are Na^+ and Cl^-. $NaCl_{(s)}$ (halite) is then an iconic example of a simple mineral salt. K_{s0} values for some other materials that can dissolve as simple salts (under some conditions) are given in Table 11.2. In problems involving dissolution of $NaCl_{(s)}$ in water, the ions in the solid are simply transferred to the dissolved phase, hydrated (*i.e.*, solvated) by water molecules, then henceforth behave as spectator ions. For saturation equilibrium with halite, the total amount of sodium in solution is governed only by Eq.(11.3), which tells the whole story for the dissolution of $NaCl_{(s)}$.

Very few minerals always dissolve as simple salts. Those that do, like $NaCl_{(s)}$, are highly soluble, and so only precipitate under very concentrated conditions, as when water evaporates in an arid basin, creating a "salt flat" such as Salar de Uyuni in Bolivia. Three examples of other natural minerals that behave as simple salts under some conditions are gypsum, cotunnite, and mirabilite.

Gypsum. For the dissolution of $CaSO_4{\cdot}2H_2O_{(s)}$ (gypsum)

$$CaSO_4 \cdot 2H_2O_{(s)} = Ca^{2+} + SO_4^{2-} + 2H_2O \qquad K_{s0} = \{Ca^{2+}\}\{SO_4^{2-}\}. \tag{11.8}$$

There are two molecules of water per $CaSO_4$ in the gypsum crystal. Regardless of whether the dissolution is simple, for saturation equilibrium with gypsum and neglecting solution phase activity corrections (or substituting ${}^cK$ values to consider them), with the solid POV,

$$\log[Ca^{2+}] = \log K_{s0} - \log[SO_4^{2-}] \qquad \frac{d\log[Ca^{2+}]}{d\log[SO_4^{2-}]} = -1. \tag{11.9}$$

When Ca^{2+} and SO_4^{2-} are behaving mostly as spectator ions, so that $[Ca^{2+}] \approx Ca_T$, and $[SO_4^{2-}] \approx S_T$, then gypsum dissolves as a simple salt and Eq.(11.8) is all we need to describe the dissolution. Ca^{2+} and SO_4^{2-} behave mostly as spectator ions when: (1) $[OH^-]$ is not very high (pH <11.7) so that formation of $CaOH^+$ is small; (2) $[H^+]$ is not very high (pH > 3) so that formation of HSO_4^- is small.

TABLE 11.2
Log K_{s0} Values at 25 °C/1 atm for Dissolution of the Solids in Figure 11.1

	log K_{s0}
$NaCl_{(s)}$ (halite) = $Na^+ + Cl^-$	1.58
$\log[Na^+] = 1.58 - \log[Cl^-]$	
$Na_2SO_4 \cdot 10H_2O_{(s)}$ (mirabillite) = $2Na^+ + SO_4^{2-} + 10H_2O$	−1.6
$\log[Na^+] = -0.8 - \frac{1}{2}\log[SO_4^{2-}]$	
$CaSO_4 \cdot 2H_2O_{(s)}$ (gypsum) = $Ca^{2+} + SO_4^{2-} + 2H_2O$	−4.58
$\log[Ca^{2+}] = -4.58 - \log[SO_4^{2-}]$	
$PbCl_{2(s)}$ (cotunnite) = $Pb^{2+} + 2Cl^-$	−4.8
$\log[Pb^{2+}] = -4.8 - 2\log[Cl^-]$	
$SrSO_{4(s)}$ (celestine) = $Sr^{2+} + SO_4^{2-}$	−6.50
$\log[Sr^{2+}] = -6.50 - \log[SO_4^{2-}]$	
$PbSO_{4(s)}$ (anglesite) = $Pb^{2+} + SO_4^{2-}$	−7.79
$\log[Pb^{2+}] = -7.79 - \log[SO_4^{2-}]$	
$MgF_{2(s)}$ (sellaite) = $Mg^{2+} + 2F^-$	−8.1
$\log[Mg^{2+}] = -8.1 - 2\log[F^-]$	
$AgCl_{(s)}$ (chlorargyrite) = $Ag^+ + Cl^-$	−9.74
$\log[Ag^+] = -9.74 - \log[Cl^-]$	
$BaSO_{4(s)}$ (barite) = $Ba^{2+} + SO_4^{2-}$	−9.97
$\log[Ba^{2+}] = -9.97 - \log[SO_4^{2-}]$	
$CaF_{2(s)}$ (fluorite) = $Ca^{2+} + 2F^-$	−10.3
$\log[Ca^{2+}] = -10.3 - 2\log[F^-]$	
$AgI_{(s)} = Ag^+ + I^-$	−16.0
$\log[Ag^+] = -16.0 - \log[I^-]$	

Note: All K_{s0} values pertain to infinite dilution. The expressions for concentrations as functions of anion concentrations take the solid point of view (POV), with the assumption that activity corrections can be neglected. Values from Smith and Martell (1976), Ball and Nordstrom (1991), and Aqion (2017).

But, if $[SO_4^{2-}]$ is high so that $[Ca^{2+}]$ based on the K_{s0} is low, then the ion pair $CaSO_4^o$ can contribute significantly to Ca_T (see Box 11.1 and Figure 11.1).

Cotunnite. For $PbCl_{2(s)}$ (cotunnite):

$$PbCl_{2(s)} = Pb^{2+} + 2Cl^- \qquad K_{s0} = \{Pb^{2+}\}\{Cl^-\}^2. \tag{11.10}$$

For saturation equilibrium, neglecting solution phase activity corrections (or substituting ${}^cK$ values to consider them), with the solid POV,

$$\log[Pb^{2+}] = \log K_{s0} - 2\log[Cl^-] \qquad \frac{d\log[Pb^{2+}]}{d\log[Cl^-]} = -2. \tag{11.11}$$

BOX 11.1 Effect of Formation of $CaSO_4^o$, $SrSO_4^o$, $PbSO_4^o$, and $BaSO_4^o$ "Ion Pairs" on Solubility of Divalent Metal Sulfates

Dissolution of gypsum produces a $z=2+$ metal ion, and a −2 anion. The electrostatic force between two ions is proportional to the product of the charges, so at a given distance, the attraction force between a +2/−2 pair is four times as large as for a +1/−1 pair like Na^+/Cl^-. The tendency for Ca^{2+} and SO_4^{2-} to combine to form an "ion pair" is large enough that the neutral species $CaSO_4^o$ can be important under some conditions; when $CaSO_4^o$ is important, gypsum will not dissolve as a simple salt. The species $SrSO_4^o$, $PbSO_4^o$, and $BaSO_4^o$ can be important in their respective systems.

For the dissolution of gypsum, we have

$$CaSO_4 \cdot 2H_2O_{(s)} = Ca^{2+} + SO_4^{2-} + 2H_2O \qquad K_{s0} = \{Ca^{2+}\}\{SO_4^{2-}\}. \qquad (11.8)$$

The formation of $CaSO_4^o$ in solution occurs according to

$$Ca^{2+} + SO_4^{2-} = CaSO_4^o \qquad K_{CaSO_4^o} = \frac{\{CaSO_4^o\}}{\{Ca^{2+}\}\{SO_4^{2-}\}}. \qquad (11.14)$$

At 25 °C, $K_{CaSO_4^o} = 199$. For the other +2 metal sulfates, $K_{SrSO_4^o} = 355$; $K_{PbSO_4^o} = 55$; and $K_{BaSO_4^o} = 501$.

Combining the equilibria in Eqs.(11.8) and (11.14) gives

$$CaSO_4 \cdot 2H_2O_{(s)} = CaSO_4^o + 2H_2O \qquad K_{s0}K_{CaSO_4^o} \equiv K_{s,CaSO_4^o} = \{CaSO_4^o\}. \qquad (11.15)$$

so that the solid will be in direct solubility equilibrium with $CaSO_4^o$. Neglecting activity corrections, $\log Ca_T = \log(K_{s0}/[SO_4^{2-}] + K_{s,CaSO_4^o})$. Figure 11.1 provides a dashed line for $\log Ca_T$ for gypsum by this equation, and also corresponding lines for $\log Sr_T$, $\log Pb_T$, and $\log Ba_T$ for the other sulfate solids. For each, by definition, the dissolution remains simple until $[MSO_4^o]$ becomes significant in M_T.

When Pb^{2+} is behaving mostly as a spectator ion so that $[Pb^{2+}] \approx Pb_T$, then cotunnite dissolves as a simple salt and Eq.(11.10) is all we need to describe the dissolution. Pb^{2+} behaves mostly as a spectator ion when the pH is not very high, so that formation of $PbOH^+$ is small.

Mirabillite. For $Na_2SO_4 \cdot 10H_2O_{(s)}$ (mirabillite):

$$Na_2SO_4{\cdot}10H_2O_{(s)} = 2Na^+ + SO_4^{2-} + 10H_2O \qquad K_{s0} = \{Na^+\}^2\{SO_4^{2-}\}. \qquad (11.12)$$

With the solid POV, neglecting solution phase activity corrections (or substituting ${}^cK$ values to consider them),

$$\log[Na^+] = \tfrac{1}{2}\log K_{s0} - \tfrac{1}{2}\log[SO_4^{2-}] \qquad \frac{d\log[Na^+]}{d\log[SO_4^{2-}]} = -\tfrac{1}{2}. \qquad (11.13)$$

When SO_4^{2-} is behaving mostly as a spectator ion so that $[SO_4^{2-}] \approx S_T$, then mirabillite dissolves as a simple salt and Eq.(11.12) is all we need to describe the dissolution. As with gypsum, SO_4^{2-} behaves mostly as a spectator ion when $[H^+]$ is not very high (pH >3) so that formation of HSO_4^- is small.

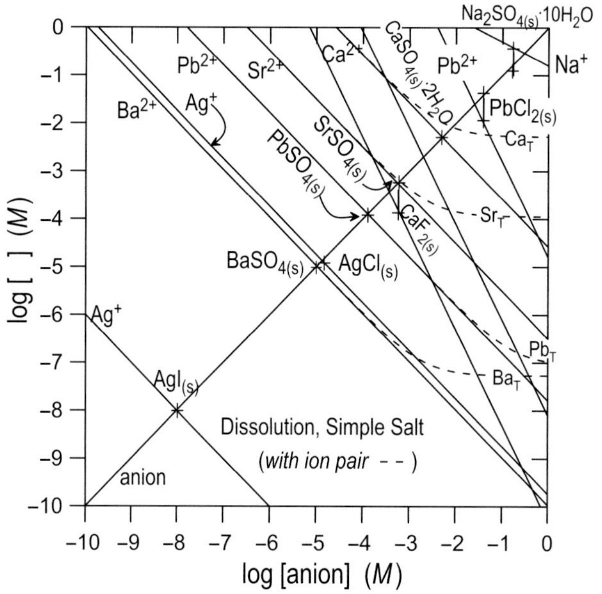

FIGURE 11.1 Solubility of various salt compounds taking the solid point of view (POV) as a function of the anion concentration at 25 °C/1 atm. Crosses and vertical lines with crosses mark the composition of the system satisfying the mass balance equation (MBE) governing the metal and the anion when the solid is equilibrated with initially-pure water. Dashed lines show total dissolved metal concentration lines for metal sulfates as affected by formation of the neutral complex MSO_4^o, *e.g.*,$CaSO_4^o$ for dissolution of $CaSO_4 \cdot 2H_2O_{(s)}$. Activity corrections neglected; adapted from Figure 7.1 of Stumm and Morgan (1996).

Following the above approach, Figure 11.1 is a solid-POV plot showing the metal ion concentration for dissolution of a variety of solid salts *vs.* the solution concentration of the anion of the salt. Of all the solids in Figure 11.1, perhaps the most environmentally relevant is gypsum, being found in many natural systems, including important Florida aquifers.

Simple Salt Plus Initially Pure Water. For any 1:1 solid (one cation and one anion), when the dissolution is simple, by mass balance the speciation that results when the dissolution occurs into initially pure water corresponds to equal concentrations of the cation and the anion. For the specific case of $BaSO_{4(s)}$ dissolving into initially pure water, with the solid POV, then at equilibrium,

$$[Ba^{2+}] = [SO_4^{2-}] \qquad \text{MBE} \tag{11.16}$$

$$\log[Ba^{2+}] = \log[SO_4^{2-}]. \tag{11.17}$$

So, neglecting solution phase activity corrections (or substituting ${}^cK$ values to consider them), at equilibrium, the speciation is given by the intersection $\log[M^{z+}] = \log[\text{anion}]$. Points on the $\log[M^{z+}]$ line to the right of the intersection are accessed by adding some of the anion, but none of the particular metal ion for the line. Points to the left of the intersection are accessed by adding some of the particular metal of interest, but none of the anion.

For the simple dissolution of any 1:2 solid (one metal ion per two anions) into initially pure water, the equilibrium saturation speciation that results will be characterized by an anion concentration that is twice the dissolved metal ion concentration. For $PbCl_{2(s)}$,

$$[Cl^-] = 2[Pb^{2+}] \qquad \text{MBE} \tag{11.18}$$

$$\log[Pb^{2+}] = \log[Cl^-] - \log 2 = \log[Cl^-] - 0.301. \quad (11.19)$$

Neglecting solution phase activity corrections (or substituting ${}^cK$ values to consider them), the equilibrium saturation speciation for dissolution into initially pure water will be that for which the $\log[M^{z+}]$ line is 0.301 log units below the log [anion] line. On the other hand, for a 2:1 solid, say $Na_2SO_4{\cdot}10H_2O_{(s)}$, the equilibrium saturation speciation for dissolution into initially pure water will be that for which the $\log[M^{z+}]$ line is 0.301 log units above the log [anion] line.

Example 11.1 Solubility of $Na_2SO_4 \cdot 10H_2O$ (mirabillite)

In a solution, $[SO_4^{2-}]=0.01$ *M*. Assuming saturation equilibrium with mirabillite, what is $[Na^+]$? This example has the solid POV. Include activity corrections. Comments: (a) Mirabillite is very soluble, so the ionic strength *I* will be very high. And, while the given value $[SO_4^{2-}]$ is very large compared to what is typical for natural waters, the calculated $[Na^+]$ will be much larger, requiring that some counterion like Cl^- is also present. Most of ionic strength would then be from the Na^+ and Cl^-, as in some salty brine. (b) At very high *I* values, the Davies Equation predicts an upturn in γ towards and past unity; (c) While the Davies Equation may not be accurate at $I \approx 1$ and higher, we will assume it is adequate for this problem.

Solution

$Na_2SO_4{\cdot}10H_2O$ (mirabillite) $= 2Na^+ + SO_4^{2-} + 10H_2O$. $[SO_4^{2-}]=0.01$ *M*. $\log K_{s0}=-1.6$, so $10^{-1.6}= [Na^+]^2(\gamma_1)^2[SO_4^{2-}]\gamma_2$. *First iteration*: guess $I=0$ so $\gamma_1=\gamma_2=1$. $[Na^+]^2=10^{-1.6}/0.01$, $[Na^+]=1.58$ *M*. Since $[SO_4^{2-}]$ is only 0.01 *M*, there must be another "counter ion" to the Na^+, probably something singly charged, like Cl^-. *Second iteration*: by the ENE, $[Cl^-] \approx [Na^+] - 2[SO_4^{2-}]=1.56$ *M* and we estimate, $I=½((+1)^2 1.58+(-1)^2 1.56 +(-2)^2 0.01)=1.59$. This is too high for accurate use of the Davies Equation, but we proceed anyway, obtaining $\gamma_1=0.91$, $\gamma_2=0.69$. With those γ values: ${}^cK_{s0}=10^{-1.6}/(\gamma_1^2\gamma_2)=10^{-1.36}=[Na^+]^2\,[SO_4^{2-}]$, so $[Na^+]=2.10$ *M*, and $[Cl^-] \approx [Na^+]-2[SO_4^{2-}]=2.08$ *M*. *Third iteration*: $I=2.11$ giving $\gamma_1=1.05$, $\gamma_2=1.22$, so ${}^cK_{s0}=10^{-1.6}/(\gamma_1^2\gamma_2)=10^{-1.72}=[Na^+]^2\,[SO_4^{2-}]$, so $[Na^+]=1.37$ *M*, and $[Cl^-] \approx [Na^+]-2[SO_4^{2-}]=1.35$ *M*. This is not converging. The $[Na^+]$ values for the three iterations are 1.58, 2.08, and 1.37 *M*; the solution is oscillating and with increasing amplitude. This is because we are in the region for *I* where the Davies Equation gives γ values can be above or below unity. There are two options. A manual option that succeeds is to not directly use the next estimate of *I*, but rather take the average using the prior estimate, and then proceed iterating. *E.g.*, for a fourth iteration, use $I=(1.35\ M+2.11\ M)/2=1.73\ M$, which is already close to the final converged result (read on). A Solver-based option is to prepare a spreadsheet to carry out the concentration calculations for a particular iteration in one block of cells based on a guess value for *I* that is located in, say, cell B12. The new value for the *I* inferred by that iteration is placed in another cell, say B20. If the difference (B12–B20) is not close to zero, then the solution has not converged. We can invoke Solver to vary cell B12 to find the value of *I* that gives (B12–B20) = 0. In other words, the guess value for *I* yields a speciation that has the guess value of *I*. In this problem, an initial guess of 0 placed in B12 does not work well because the functionality is too steep in that region. Something like 0.5 works fine. The final converged result is $I=1.80$ giving ${}^cK_{s0}=10^{-1.50}$, $[Na^+]=1.79$ *M*, and $[Cl^-]=[Na^+]-2[SO_4^{2-}]=1.77$ *M*. The fact that the Davies Equation is not accurate for γ values at such high *I* values means that a more exact solution to this problem will require more advanced γ estimation tools such as are provided by the Pitzer equations (*e.g.*, see Plummer et al., 1988).

Example 11.2 Saturation State for $CaSO_4 \cdot 2H_2O$ (Gypsum) in an Aquifer

In a study of groundwater in a Florida aquifer where gypsum is present, Sacks (1996) reported for "well #18" that the groundwater concentrations for total calcium and total sulfate were 330 mg/L and 900 mg/L, respectively, with pH=7.31. At 24.6 °C, the system was close to 25 °C/1 atm. What was the saturation state of this water relative to gypsum at 25 °C/1 atm, for which $K_{s0}=10^{-4.58}$? Consider the role of $CaSO_4^o$. For $Ca^{2+}+SO_4^{2-}=CaSO_4^o$, at 25 °C/1 atm the formation constant $K_{CaSO_4^o}= 10^{2.30}$. This example has the solution POV. For the solution, assume: (1) Ca^{2+} and $CaSO_4^o$ are the only important Ca-containing species; and (2) SO_4^{2-} and $CaSO_4^o$ are the only important SO_4-containing species. Calculate the solution speciation and the value of Ω. Neglect activity corrections.

Solution

	total (mg/L)	FW (g/mol)	total mol/L
Ca	330	40.01	8.25×10^{-3}
SO_4	900	96.06	9.37×10^{-3}

$$\frac{[CaSO_4^o]}{[Ca^{2+}][SO_4^{2-}]} = 10^{2.30}.$$

Let x = $[CaSO_4^o]$.

$$\frac{x}{(8.25\times10^{-3}-x)(9.37\times10^{-3}-x)} = 10^{2.30}.$$

Solving, $x=[CaSO_4^o]=4.19\times10^{-3}\,M$. $[Ca^{2+}]=8.25\times10^{-3}-4.19\times10^{-3}=4.04\times10^{-3}\,M$. $[SO_4^{2-}]= 9.37\times10^{-3}-4.19\times10^{-3}=5.18\times10^{-3}\,M$.

$$\Omega = \frac{(4.04\times10^{-3})(5.18\times10^{-3})}{10^{-4.58}} = 0.80$$

This value is close to 1.0. Given that there are uncertainties in the analytical data and in the equilibrium K values, water in the aquifer for that well is almost certainly saturated with gypsum.

11.4 METAL HYDROXIDES, OXIDES, AND OXYHYDROXIDES

11.4.1 THE 1ST (K_{s0}) PART OF THE STORY – IF THE DISSOLUTION WERE SIMPLE

Ferric hydroxide $Fe(OH)_{3(s)}$ is known as ferrihydrite. It involves Fe in the III oxidation state. When it forms, ferrihydrite is often essentially amorphous. To denote that, the prefix (am) is often tacked onto the front of the formula, as in (am)$Fe(OH)_{3(s)}$. Molecules of water can be removed from the solid, *e.g.*, for (am)$Fe(OH)_{3(s)}$ by converting $2OH^-$ to one O^{2-} and one H_2O, leaving $FeOOH_{(s)}$, or for $2Fe(OH)_{3(s)}$ by converting $6OH^-$ to $3O^{2-}$ and $3H_2O$, leaving $Fe_2O_{3(s)}$. Ferrihydrite is sometimes called hydrated ferric iron oxide. It is generally true that when a metal hydroxide or oxyhydroxide (like (am)$Fe(OH)_{3(s)}$) is dehydrated, the result is a metal solid that is less soluble than the hydrated form.

(am)$Fe(OH)_{3(s)}$ dissolves according to

$$s0: \quad (am)Fe(OH)_{3(s)} = Fe^{3+} + 3OH^- \qquad K_{s0} = \{Fe^{3+}\}\{OH^-\}^3. \tag{11.20}$$

Calculating $\{Fe^{3+}\}$ by means of the K_{s0} value thus involves $\{OH^-\}$. It is much more convenient, though, to use $\{H^+\}$. So, noting

$$3H_2O = 3H^+ + 3OH^- \qquad K_w^3 = \{H^+\}^3\{OH^-\}^3 \tag{11.21}$$

we subtract Eq.(11.21) from Eq.(11.20) to obtain

$$*s0 \quad (am)Fe(OH)_{3(s)} + 3H^+ = Fe^{3+} + 3H_2O \qquad {}^*K_{s0} = K_{s0} / K_w^3 = \{Fe^{3+}\}/\{H^+\}^3. \tag{11.22}$$

Table 11.3 contains ${}^*K_{s0}$ values for a range of oxides and hydroxides, and Figure 11.2 provides lines for the corresponding metal ion concentrations for solubility equilibrium (neglecting activity corrections). Figure 11.2 has the same general character as Figure 11.1, but here for all the lines the underlying anion is OH^-, and for the x axis we use pH (= $pK_w + \log\{OH^-\}$).

TABLE 11.3
${}^*K_{s0}$ Values at 25 °C/1 atm for the Hydroxides and Oxide Solids in Figure 11.2

	$\log {}^*K_{s0}$
$z = +3$	
$(am)Fe(OH)_{3(s)}$ (ferrihydrite) $+ 3H^+ = Fe^{3+} + 3H_2O$	3.2
$\log[Fe^{3+}] = 3.2 - 3pH$	
$(am)Al(OH)_{3(s)} + 3H^+ = Al^{3+} + 3H_2O$	10.8
$\log[Al^{3+}] = 10.8 - 3pH$	
$Cr(OH)_{3(s)} + 3H^+ = Cr^{3+} + 3H_2O$	12.2
$\log[Cr^{3+}] = 12.2 - 3pH$	
$z = +2$	
$HgO_{(s)} + 2H^+ = Hg^{2+} + H_2O$	2.6
$\log[Hg^{2+}] = 2.6 - 2pH$	
$Cu(OH)_{2(s)} + 2H^+ = Cu^{2+} + 2H_2O$	8.7
$\log[Cu^{2+}] = 8.7 - 2pH$	
$Zn(OH)_{2(s)} + 2H^+ = Zn^{2+} + 2H_2O$	11.5
$\log[Zn^{2+}] = 11.5 - 2pH$	
α-$PbO_{(s)}$ (litharge, "red") $+ 2H^+ = Pb^{2+} + H_2O$	12.60
$\log[Pb^{2+}] = 12.7 - 2pH$	
β-$PbO_{(s)}$ (massicot, "yellow") $+ 2H^+ = Pb^{2+} + 2H_2O$	12.78
$\log[Pb^{2+}] = 12.9 - 2pH$	
$Fe(OH)_{2(s)} + 2H^+ = Fe^{2+} + 2H_2O$	12.9
$\log[Fe^{2+}] = 12.9 - 2pH$	
$Cd(OH)_{2(s)} + 2H^+ = Cd^{2+} + 2H_2O$	13.6
$\log[Cd^{2+}] = 13.6 - 2pH$	
$Mn(OH)_{2(s)} + 2H^+ = Mn^{2+} + 2H_2O$	15.2
$\log[Mn^{2+}] = 15.2 - 2pH$	
$Mg(OH)_{2(s)} + 2H^+ = Mg^{2+} + 2H_2O$	16.9
$\log[Mg^{2+}] = 16.9 - 2pH$	
$Ca(OH)_{2(s)} + 2H^+ = Ca^{2+} + 2H_2O$	22.8
$\log[Ca^{2+}] = 22.8 - 2pH$	

Note: All ${}^*K_{s0}$ values pertain to infinite dilution, except for $Cr(OH)_{3(s)}$, for which the value is for ${}^{*c}K_{s0}$ at ionic strength = 0.1 *M*. The expressions for concentrations as functions of pH take the solid point of view (solid POV), with the assumption that activity corrections for the metal ions can be neglected.

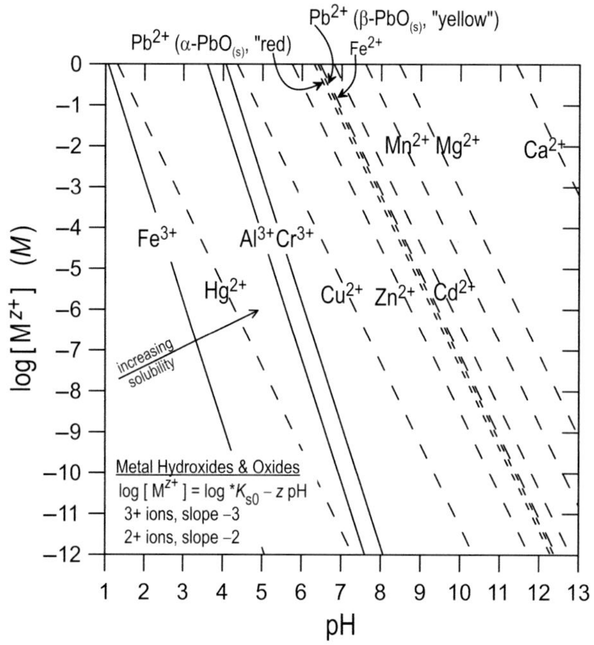

FIGURE 11.2 Concentration of various metal ions as controlled by their hydroxides or oxides taking the solid point of view (POV) as a function of pH at 25 °C/1 atm. Activity corrections for the metal ions are neglected (not valid at very low or very high pH where the ionic strength will be relatively large, *e.g.*, if pH = 1, then the ionic strength is at least ~0.1). Adapted from Figure 7.3 of Stumm and Morgan, 1996.

Lines in Figure 11.2 for three solids involve metal ions with +3 charge (thus in the +III oxidation state), namely (am)$Fe(OH)_{3(s)}$, (am)$Al(OH)_{3(s)}$, and $Cr(OH)_{3(s)}$. For all three, the solubility equilibrium is of the type

$$M(OH)_{3(s)} + 3H^+ = M^{3+} + 3H_2O \qquad {}^*K_{s0} = K_{s0}/K_w^3 = \{M^{3+}\}/\{H^+\}^3. \tag{11.23}$$

Neglecting activity corrections for M^{3+}, with the solid POV,

$$\log[M^{3+}] = \log {}^*K_{s0} - 3pH \tag{11.24}$$

$$\frac{d\log[M^{3+}]}{dpH} = -3. \tag{11.25}$$

For solids of metal ions with +2 charge (and in the +II oxidation state), with the solid POV, we have either

$$M(OH)_{2(s)} + 2H^+ = M^{2+} + 2H_2O \qquad {}^*K_{s0} = \{M^{2+}\}/\{H^+\}^2 \tag{11.26}$$

or

$$MO_{(s)} + 2H^+ = M^{2+} + H_2O \qquad {}^*K_{s0} = \{M^{2+}\}/\{H^+\}^2. \tag{11.27}$$

In both cases (though the two $^*K_{s0}$ will differ), neglecting activity corrections for M^{2+},

$$\log[M^{2+}] = \log {}^*K_{s0} - 2pH \tag{11.28}$$

$$\frac{d\log[M^{2+}]}{dpH} = -2. \tag{11.29}$$

For metal ions with $z=+1$ (and in the +I oxidation state), hydroxide, oxide, and oxyhydroxide solids are certainly known in chemistry, but they are all extremely water soluble, and so not relevant here. At the other extreme, for $z=+4$, solids like $TiO_{2(s)}$ and solid $PbO_{2(s)}$ are so *insoluble* that they are usually not considered to contribute significant dissolved metal to a solution. Figure 11.2 shows that at a given pH, neglecting activity corrections, the different solids prescribe widely varying values of $\log[M^{z+}]$. The particularly low solubility of (am)$Fe(OH)_{3(s)}$ is highly significant for environmental chemistry.

11.4.2 The Rest of the Story – Complexation by OH^-: K_{s1}, K_{s2}, etc. (or $^*K_{s1}$, $^*K_{s2}$, etc.) and the Not-Simple Solubility of Metal Hydroxides, Oxides, and Oxyhyroxides

The Fe^{3+} ion has a very high charge-to-size ratio, as does H^+. So, like H^+, the Fe^{3+} ion has considerable affinity for OH^- ions. As discussed in Chapter 10, an important equilibrium reaction is then

$$Fe^{3+} + OH^- = FeOH^{2+} \qquad K_{H1} = \frac{\{FeOH^{2+}\}}{\{Fe^{3+}\}\{OH^-\}}. \tag{11.30}$$

In this reaction, OH^- is a "ligand" that "complexes" the metal ion Fe^{3+}. As a result, (am)$Fe(OH)_{3(s)}$ does not dissolve as a simple salt. Since Eq.(11.30) involves OH^-, the extent to which $FeOH^{2+}$ forms depends on pH. In fact, Fe^{3+} is sometimes referred to as an "acid metal ion" (with accompanying α values, see Chapter 10) which makes sense since we can convert the above reaction to

$$Fe^{3+} + H_2O = FeOH^{2+} + H^+ \qquad {}^*K_{H1} = \frac{\{FeOH^{2+}\}\{H^+\}}{\{Fe^{3+}\}}. \tag{11.31}$$

The * nomenclature denotes the switch from an equilibrium constant involving OH^- to one involving H^+ (Chapter 10). With $^*K_{H1}=10^{-2.20}$ at 25 °C/1 atm Fe^{3+} is a modestly strong acid, not quite as strong as HSO_4^- ($K_a=10^{-1.99}$). The $\log\alpha$ *vs.* pH plot for the five α values for dissolved Fe(III) at 25 °C/1 atm is provided in Figure 10.3.

For Eqs.(11.30) and (11.31), the subscript H1 refers to "hydroxide one" added to the core metal ion (Fe^{3+}), to yield the metal-ligand complex $FeOH^{2+}$. One way to consider the effects of the formation of $FeOH^{2+}$ on the solubility of (am)$Fe(OH)_{3(s)}$ would be to first use $^*K_{s0}$ to compute how much Fe^{3+} is produced in solution, then consider how the $^*K_{H1}$ reaction promotes additional dissolution by equilibrium with Fe^{3+}. It is more direct to factor in the effects of $^*K_{H1}$ on the dissolution by combining equilibrium constants to obtain $^*K_{s1}$:

$$*s0 \quad (am)Fe(OH)_{3(s)} + 3H^+ = Fe^{3+} \qquad {}^*K_{s0} = \frac{\{Fe^{3+}\}}{\{H^+\}^3} \tag{11.22}$$

$$+\left(Fe^{3+} + H_2O = FeOH^{2+} + H^+ \qquad {}^*K_{H1} = \frac{\{FeOH^{2+}\}\{H^+\}}{\{Fe^{3+}\}}\right) \tag{11.31}$$

$$*s1 \quad (am)Fe(OH)_{3(s)} + 2H^+ = FeOH^{2+} + 2H_2O \qquad {}^*K_{s1} = {}^*K_{s0}\,{}^*K_{H1} = {}^*K_{s0}\,{}^*\beta_{H1} = \frac{\{FeOH^{2+}\}}{\{H^+\}^2}. \tag{11.32}$$

Because the high charge on Fe^{3+} gives it significant affinity for OH^-, addition of more than one OH^- is possible:

$$FeOH^{2+} + H_2O = Fe(OH)_2^+ + H^+ \qquad {}^*K_{H2} = \frac{\{Fe(OH)_2^+\}\{H^+\}}{\{FeOH^{2+}\}} \tag{11.33}$$

$$Fe(OH)_2^+ + H_2O = Fe(OH)_3^o \qquad {}^*K_{H3} = \frac{\{Fe(OH)_3^o\}\{H^+\}}{\{Fe(OH)_2^+\}} \tag{11.34}$$

$$Fe(OH)_3^o + H_2O = Fe(OH)_4^- \qquad {}^*K_{H4} = \frac{\{Fe(OH)_4^-\}\{H^+\}}{\{Fe(OH)_3^o\}}. \tag{11.35}$$

At high applied Fe(III) levels, as may be quickly created under water treatment situations at pH = ~6 to 8 involving coagulation with precipitated (am)$Fe(OH)_{3(s)}$, positively charged “polymeric” species involving more than one metal can be temporarily important, *e.g.*, $Fe_2(OH)_2^{4+}$ (see Section 10.3.3). The same applies for when Al^{3+} is added to precipitate (am)$Al(OH)_{3(s)}$.

The goal of Fe^{3+} or Al^{3+} addition in water treatment is to remove suspended colloidal particles, which are generally negatively charged (*e.g.*, organic matter is often negatively charged because of ionization of carboxyl groups, and clay particles are negatively charged by loss of lattice cations). Because the negatively charged particles repel one another, they do not readily coagulate on their own, even if they might very well stick together if they did manage to actually touch solid-to-solid (two people might not be attracted to one another at a distance, but if they did manage to actually touch, then they might like how that feels and stay together). So, small negatively charged particles tend to stay suspended in solution. Upon the initial dosing of Fe^{3+} or Al^{3+} in water treatment, the solutions become grossly supersaturated with the hydroxide solid (see Figures 11.3 and 11.4 below), so species with two and more metal ions can be important for short periods, as they are “on their way” to forming the solid with many metal ion locked together. Positively charged iron and aluminum ions can adsorb to negative particles and neutralize the surface charge. Stumm and Morgan (1996, p.852) comment as follows:

> Often, hydrolyzing metal ions, Al(III) or Fe(III), are used as coagulants … [An] Al(III) salt (or Fe(III) salt) added undergoes hydrolysis and forms polymeric or oligomeric charged hydroxo complexes of various structure … Such species adsorb specifically [to the surfaces of negatively-charged suspended particles] and modify the surface charge of the colloids.

With the surface charge reduced, the particles can more easily approach one another and agglomerate, then settle from the solution, as assisted by a sweep of settling flocs of (am)$Fe(OH)_{3(s}$ or (am) $Al(OH)_{3(s)}$.

For the additional needed solubility equilibria for (am)$Fe(OH)_{3(s)}$, Eqs.(11.33–11.35) may be combined with Eqs.(11.22) and (11.30) to obtain Eq.(11.36) to (11.38). (For $Fe(OH)_3^o$, there is no need for an *s3 constant, as there is no switch from using $\{OH^-\}$ to $\{H^+\}$.)

$$\text{*s2} \quad (am)Fe(OH)_{3(s)} + H^+ = Fe(OH)_2^+ + H_2O \qquad {}^*K_{s2} = {}^*K_{s0}\,{}^*K_{H1}\,{}^*K_{H2} = {}^*K_{s0}\,{}^*\beta_{H2} = \frac{\{Fe(OH)_2^+\}}{\{H^+\}} \tag{11.36}$$

$$\text{s3} \quad (am)Fe(OH)_{3(s)} = Fe(OH)_3^o \qquad K_{s3} = {}^*K_{s0}\,{}^*K_{H1}\,{}^*K_{H2}\,{}^*K_{H3} = {}^*K_{s0}\,{}^*\beta_{H3} = \{Fe(OH)_3^o\} \tag{11.37}$$

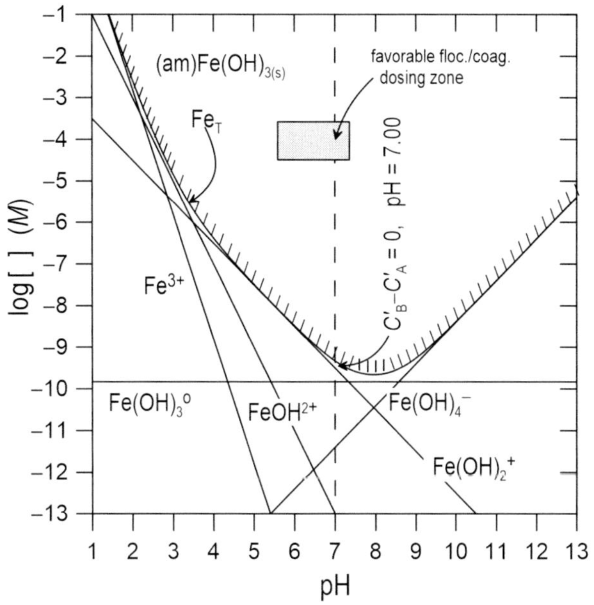

FIGURE 11.3 Log concentration (M) *vs.* pH for various Fe(III) species and Fe(III) in equilibrium with (am)$Fe(OH)_{3(s)}$ at 25 °C/1 atm (solid POV). Dashed vertical line gives speciation for dissolution into initially pure water (*i.e.*, for $(C'_B - C'_A) = 0$). Species with more than one Fe (*e.g.*, $Fe_2(OH)_2^{4+}$) have been neglected. Activity corrections neglected (not valid at very low or very high pH where the ionic strength will be relatively large, *e.g.*, if pH = 1, then the ionic strength is likely ~0.1). According to Edzwald (2018), the favorable coagulation/flocculation dosing zone for use of Fe(III) in drinking water treatment is pH = 5.5 to 7.2, and 2 to 20 mg/L iron as Fe. Increasing Fe(III) doses are required for increasing total organic carbon (TOC) in the raw drinking water. Quoting Edzwald (2018), "Dose range of ~2 mg/L (low TOC of ~ 2 mg/L or less and alkalinity of 50 mg/L or less) to 20 mg/L as Fe for waters with greater TOC and/or alkalinity." Treatment of raw drinking water by use of Fe(III) may well take place at temperatures different from 25 °C; the treatment zone given is far inside the region for formation of (am)$Fe(OH)_{3(s)}$ so at doses in the "favorable" region, some shifting of the Fe_T line with temperature will not significantly affect how much (am)$Fe(OH)_{3(s)}$ will form.

$$\text{*s4} \quad \text{(am)Fe(OH)}_{3(s)} + H_2O = Fe(OH)_4^- + H^+ \qquad {}^*K_{s4} = {}^*K_{s0}\,{}^*K_{H1}\,{}^*K_{H2}\,{}^*K_{H3}\,{}^*K_{H4} = {}^*K_{s0}\,{}^*\beta_{H4} = \{Fe(OH)_4^-\}\{H^+\} \tag{11.38}$$

For saturation equilibrium with (am)$Fe(OH)_{3(s)}$, neglecting species like $Fe_2(OH)_2^{4+}$ that are important only at high Fe_T with low pH (Chapter 10), the total dissolved Fe(III) is

$$Fe_T = [Fe^{3+}] + [FeOH^{2+}] + [Fe(OH)_2^+] + [Fe(OH)_3^o] + [Fe(OH)_4^-]. \tag{11.39}$$

Neglecting activity corrections, we have (solid POV)

$$Fe_T = {}^*K_{s0}[H^+]^3 + {}^*K_{s1}[H^+]^2 + {}^*K_{s2}[H^+] + K_{s3} + \frac{{}^*K_{s4}}{[H^+]}. \tag{11.40}$$

Values for the constants at 25 °C/1 atm for (am)$Fe(OH)_{3(s)}$ are given in Table 11.4.

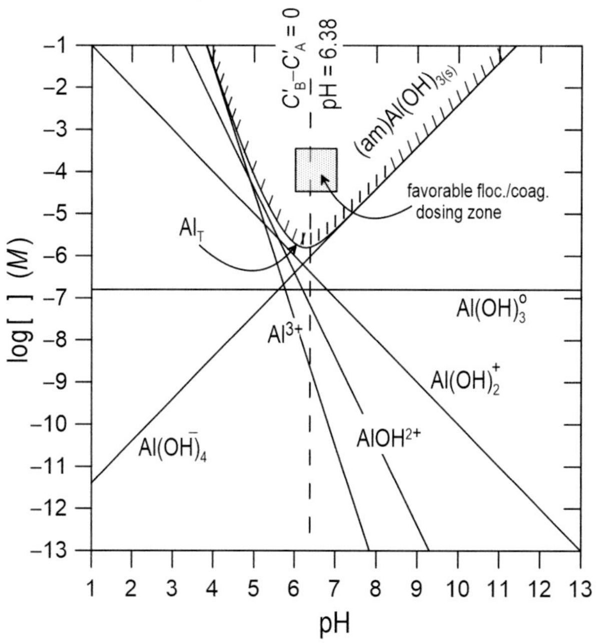

FIGURE 11.4 Log concentration (*M*) *vs.* pH for various Al(III) species and Al_T in equilibrium with $(am)Al(OH)_{3(s)}$ at 25 °C/1 atm (solid POV). Dashed vertical line gives speciation for dissolution into initially pure water (*i.e.*, for $(C'_B - C'_A) = 0$). Species with more than one Al (*e.g.*, $Al_2(OH)_2^{4+}$) have been neglected. Activity corrections neglected (not valid at very low or very high pH where the ionic strength will be relatively large, *e.g.*, if pH = 1, then the ionic strength is likely ~0.1). According to Edzwald (2018), the favorable coagulation/flocculation dosing zone for use of Al(III) in drinking water treatment is pH = 6.0 to 7.0, and ~1.3 to 10 mg/L aluminum as Al. Increasing Al(III) doses are required for increasing total organic carbon (TOC) in the raw drinking water. Quoting Edzwald (2018), "For low TOC (~2 mg/L and less) and alkalinity ~ 50 mg/L or less, doses are low at about 1.3 to 2.0 mg/L as Al. For TOC of 2–8 mg/L and alkalinity of 50 mg/L or less, doses run 1.3 to 5 mg/L as Al. For higher TOC and or alkalinity ([with] use of alum as an acid to lower pH), doses can be 8 to 10 mg/L as Al." Treatment of raw drinking water by use of Al(III) may well take place at temperatures different from 25 °C; the treatment zone given is far inside the region for formation of $(am)Al(OH)_{3(s)}$ so at doses in the "favorable" region, some shifting of the Al_T line with temperature will not significantly affect how much $(am)Al(OH)_{3(s)}$ will form.

11.4.3 Total Solubility of a Particular Metal Hydroxide, Oxide, or Oxyhydroxide *vs.* pH

For $(am)Fe(OH)_{3(s)}$, based on the *K* values in Table 11.4, Figure 11.3 provides the solid-POV line for $Fe(III)_T$ as per Eq.(11.40), and lines for the individual dissolved Fe(III) species. The hashmarks on the interior of each of the $\log Fe(III)_T$ curve infer the presence of a solid "wall", by analogy with construction engineering diagrams, as will be discussed below for Figure 11.7. The slopes of the lines for $\log[Fe^{3+}]$, $\log[FeOH^{2+}]$, $\log[Fe(OH)_2^+]$, $\log[Fe(OH)_3^o]$ and $\log[Fe(OH)_4^-]$ are –3, –2, –1, 0, and +1. Each slope equals the number of OH^- groups that the corresponding species is "away from the solid", the solid having 3 OH^- groups. $(am)Fe(OH)_{3(s)}$ becomes increasingly soluble as the pH decreases below ~8, and also as the pH increases above ~8. The presence of a solubility minimum is exhibited by most metal oxides, hydroxides, and oxyhydroxides. An important conclusion from Figure 11.3 is that at typical natural water pH values (~6 to ~8), the solubility of $(am)Fe(OH)_{3(s)}$ is exceedingly low; it is even lower for both α-$FeOOH_{(s)}$ (goethite) and α-$Fe_2O_{3(s)}$ (hematite) (see Eqs. (11.45–11.46)). This has significant, observable biological consequences: under aerobic conditions,

TABLE 11.4
Metal Hydrolysis and Metal Hydroxide Solubility Equilibrium Constant Values at 25 °C/1 atm for Fe(III), Al(III), and Pb(II)

Fe(III) equilibria

Hydrolysis	log *K*	log *β*
$Fe^{3+}+H_2O=FeOH^{2+}$	$\log {}^*K_{H1}=-2.2$	$\log {}^*\beta_{H1}=-2.2$
$FeOH^{2+}+H_2O=Fe(OH)_2^++H^+$	$\log {}^*K_{H2}=-3.5$	$\log {}^*\beta_{H2}=-5.7$
$Fe(OH)_2^++H_2O=Fe(OH)_3^o+H^+$	$\log {}^*K_{H3}=-7.33$	$\log {}^*\beta_{H3}=-13.03$
$Fe(OH)_3^o+H_2O=Fe(OH)_4^-+H^+$	$\log {}^*K_{H4}=-8.57$	$\log {}^*\beta_{H4}=-21.6$
Solubility		solubility line
$(am)Fe(OH)_{3(s)}+3H^+=Fe^{3+}+3H_2O$	$\log {}^*K_{s0}=3.2$	$\log[Fe^{3+}]=3.2-3pH$
$(am)Fe(OH)_{3(s)}+2H^+=FeOH^{2+}+2H_2O$	$\log {}^*K_{s1}=1.0$	$\log[FeOH^{2+}]=1.0-2pH$
$(am)Fe(OH)_{3(s)}+H^+=Fe(OH)_2^++H_2O$	$\log {}^*K_{s2}=-2.5$	$\log[Fe(OH)_2^+]=-2.5-pH$
$(am)Fe(OH)_{3(s)}=Fe(OH)_3^o$	$\log K_{s3}=-9.83$	$\log[Fe(OH)_3^o]=-9.83$
$(am)Fe(OH)_{3(s)}+H_2O=Fe(OH)_4^-+H^+$	$\log {}^*K_{s4}=-18.4$	$\log[Fe(OH)_4^-]=-18.4+pH$

Al(III) equilibria

Hydrolysis	log *K*	log *β*
$Al^{3+}+H_2O=AlOH^{2+}$	$\log {}^*K_{H1}=-4.95$	$\log {}^*\beta_{H1}=-4.95$
$AlOH^{2+}+H_2O=Al(OH)_2^++H^+$	$\log {}^*K_{H2}=-5.6$	$\log {}^*\beta_{H2}=-10.55$
$Al(OH)_2^++H_2O=Al(OH)_3^o+H^+$	$\log {}^*K_{H3}=-6.7$	$\log {}^*\beta_{H3}=-17.25$
$Al(OH)_3^o+H_2O=Al(OH)_4^-+H^+$	$\log {}^*K_{H4}=-5.6$	$\log {}^*\beta_{H4}=-22.85$
Solubility		solubility line
$(am)Al(OH)_{3(s)}+3H^+=Al^{3+}+3H_2O$	$\log {}^*K_{s0}=10.5$	$\log[Al^{3+}]=10.5-3pH$
$(am)Al(OH)_{3(s)}+2H^+=AlOH^{2+}+2H_2O$	$\log {}^*K_{s1}=5.6$	$\log[AlOH^{2+}]=5.6-2pH$
$(am)Al(OH)_{3(s)}+H^+=Al(OH)_2^++H_2O$	$\log {}^*K_{s2}=0.0$	$\log[Al(OH)_2^+]=0.0-pH$
$(am)Al(OH)_{3(s)}=Al(OH)_3^o$	$\log K_{s3}=-6.8$	$\log[Al(OH)_3^o]=-6.8$
$(am)Al(OH)_{3(s)}+H_2O=Al(OH)_4^-+H^+$	$\log {}^*K_{s4}=-12.4$	$\log[Al(OH)_4^-]=-12.4+pH$

Pb(II) equilibria

Hydrolysis	log *K*	log *β*
$Pb^{2+}+H_2O=PbOH^+$	$\log {}^*K_{H1}=-7.22$	$\log {}^*\beta_{H1}=-7.22$
$PbOH^++H_2O=Pb(OH)_2^o+H^+$	$\log {}^*K_{H2}=-9.69$	$\log {}^*\beta_{H2}=-16.91$
$Pb(OH)_2^o+H_2O=Pb(OH)_3^-+H^+$	$\log {}^*K_{H3}=-11.17$	$\log {}^*\beta_{H3}=-28.08$
$Pb(OH)_3^-+H_2O=Pb(OH)_4^{2-}+H^+$	$\log {}^*K_{H4}=-11.64$	$\log {}^*\beta_{H4}=-39.72$
Solubility		solubility line
α-$PbO_{(s)}$ (litharge) $+2H^+=Pb^{2+}+H_2O$	$\log {}^*K_{s0}=12.6$	$\log[Pb^{2+}]=12.6-2pH$
α-$PbO_{(s)}$ (litharge) $+H^+=PbOH^++H$	$\log {}^*K_{s1}=5.38$	$\log[PbOH^+]=5.38-pH$
α-$PbO_{(s)}$(litharge) $+H_2O=Pb(OH)_2^o$	$\log K_{s2}=-4.31$	$\log[Pb(OH)_2^o]=-4.31$
α-$PbO_{(s)}$(litharge) $+2H_2O=Pb(OH)_3^-+H^+$	$\log {}^*K_{s3}=-15.48$	$\log[Pb(OH)_3^-]=-15.48+pH$
α-$PbO_{(s)}$(litharge) $+3H_2O=Pb(OH)_4^{2-}+2H^+$	$\log {}^*K_{s4}=-27.12$	$\log[Pb(OH)_4^{2-}]=-21.72+2pH$

Note: All constants pertain to infinite dilution. The expressions for concentrations as functions of pH take the solid point of view (POV), with the assumption that activity corrections for the metal-related ions can be neglected (pH itself does not require activity corrections). *K* and *β* values derived from examination of multiple sources, including Ball and Nordstrom (1991), Pernitsky and Edzwald (2003), and Wesolowski and Palmer (1994).

where Fe(III) is by far the dominant oxidation state, many aquatic organisms have difficulty acquiring adequate systemic Fe from their surrounding solutions. Likewise, for humans, little dietary Fe is provided by drinking water, so the considerable Fe we need must come from our food, and/or cast iron cookware, and/or supplement pills (Cheng and Brittin, 1991). Another consequence of the low solubility of Fe(III) solids relates to the iron stains that can develop on sinks, tubs, toilets, concrete, and masonry: it is essentially impossible to remove such stains by washing with water at ~neutral pH where the solubility of Fe(III) solids is exceedingly low. The only alternative is to expose the stains to a solution of low pH, or of high pH, far away from the solubility minimum; low pH is usually used because the solubility increases more rapidly for pH decreasing below 7 than for pH increasing above 7.

Example 11.3 Solubility of (am)Fe(OH)$_{3(s)}$ as a Function of pH

Based on Eq.(11.42), what is the total solubility of (am)Fe(OH)$_{3(s)}$ at pH values of 6.00, 8.00, and 10.00? at 25 °C/1 atm? (This example has the solid POV.)

Solution

Using the K values in Table 11.4, at the three pH values, Fe(III)$_T = 3.32 \times 10^{-9}$, 2.19×10^{-10} (near the minimum), and 4.13×10^{-9} M, respectively. These are very low numbers.

For (am)Al(OH)$_{3(s)}$, which, like (am)Fe(OH)$_{3(s)}$ has commonly been used as a coagulant/flocculant in water treatment, neglecting activity corrections, the total solubility equation takes the same form as Eq.(11.40):

$$\boxed{\mathrm{Al_T} = {}^*K_{s0}[\mathrm{H^+}]^3 + {}^*K_{s1}[\mathrm{H^+}]^2 + {}^*K_{s2}[\mathrm{H^+}] + K_{s3} + \frac{{}^*K_{s4}}{[\mathrm{H^+}]}.} \tag{11.41}$$

Unlike iron, aluminum does not have a stable (II) oxidation state. So, when talking about dissolved Al, we are always speaking of Al(III). Figure 11.4 provides the solid-POV line for Al_T, and lines for the individual dissolved Al(III) species. A comparison of Figure 11.4 with Figure 11.3 shows that (am)Al(OH)$_{3(s)}$ is several orders of magnitude more soluble than (am)Fe(OH)$_{3(s)}$. The significant solubility of (am)Al(OH)$_{3(s)}$ is one of the reasons why acidified lakes are toxic to fish: Al(III) is very abundant in rocks, soils, and sediments, so when the pH starts to get below 6, there is plenty to dissolve. Example 9.12 discusses pH values as low as ~4.4 in a lake in Sweden.

For PbO$_{(s)}$, neglecting activity corrections, the total solubility equation takes a form that is related to Eqs.(11.40) and (11.41). namely

$$\mathrm{Pb_T} = [\mathrm{Pb^{2+}}] + [\mathrm{PbOH^+}] + [\mathrm{Pb(OH)_2^o}] + [\mathrm{Pb(OH)_3^-}] \tag{11.42}$$

$$= {}^*K_{s0}[\mathrm{H^+}]^2 + {}^*K_{s1}[\mathrm{H^+}] + K_{s2} + \frac{{}^*K_{s3}}{[\mathrm{H^+}]}. \tag{11.43}$$

Dissolved inorganic lead is generally all Pb(II); there is a stable (IV) oxidation state, but it is essentially insoluble except at very high pH (see Chapter 19). So, when talking about dissolved Pb, we are generally speaking of Pb(II). Figure 11.5 provides the solid-POV line for Pb_T for α-PbO$_{(s)}$ (a.k.a. litharge or "red PbO$_{(s)}$") and lines for the individual dissolved Pb(II) species. An application of this type of analysis in the study of how dissolved Pb(II) dissolves from lead-contaminated soils is found in Stanforth and Cui (2001), see especially their Figure 4. Here, in Figure 11.6, the line for log Pb_T for α-PbO$_{(s)}$ is compared with the lines for log Fe(III)$_T$ for (am)Fe(OH)$_{3(s)}$ and log Al_T for

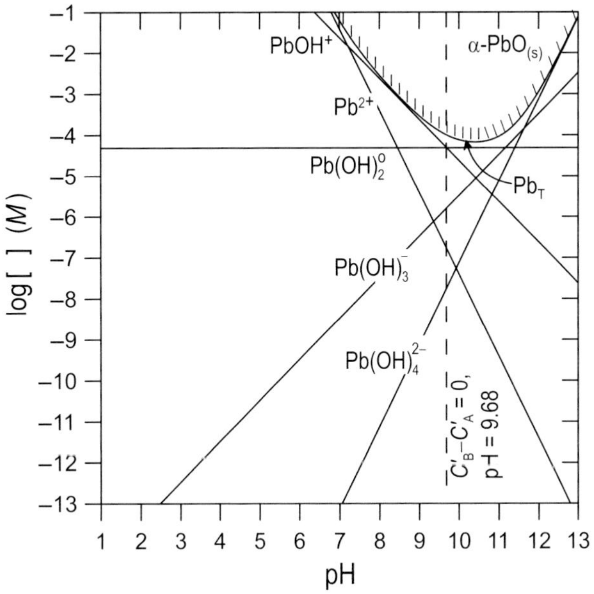

FIGURE 11.5 Log concentration (M) *vs.* pH for various Pb(II) species and Pb_T in equilibrium with α-$PbO_{(s)}$ at 25 °C/1 atm (solid POV). Dashed vertical line gives speciation for dissolution into initially-pure water (*i.e.*, for $(C'_B - C'_A) = 0$). Activity corrections neglected (not valid at very low or very high pH where the ionic strength will be relatively large, *e.g.*, if pH = 1, then the ionic strength is likely ~0.1).

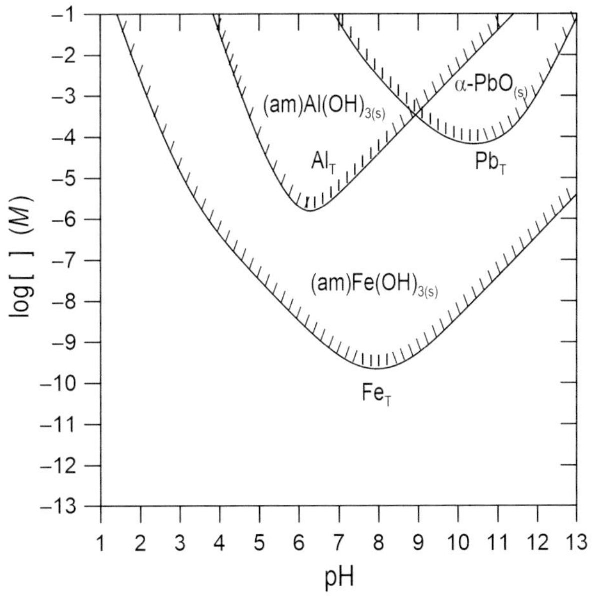

FIGURE 11.6 Log concentration (M) *vs.* pH for Fe_T (= $Fe(III)_T$) in equilibrium with (am)$Fe(OH)_{3(s)}$, Al_T in equilibrium with (am)$Al(OH)_{3(s)}$, and Pb_T in equilibrium with α-$PbO_{(s)}$ at 25 °C/1 atm (all curves take the solid POV). Activity corrections neglected (not valid at very low or very high pH where the ionic strength will be relatively large, *e.g.*, if pH = 1, then the ionic strength is likely ~0.1).

(am)$Al(OH)_{3(s)}$. For all the solids, solubility rises as pH becomes low, and as pH becomes high. For each of the solids, the rise at high pH is due to a negatively charged species for which the number of OH^- in the ion is the oxidation state plus 1; these are $Fe(OH)_4^-$, $Al(OH)_4^-$, and $Pb(OH)_3^-$.

Plots like Figures 11.3 through 11.6 illustrate the pH dependence of the solid-POV equilibrium solution chemistry of a system containing a metal hydroxide, oxide, or oxyhydroxide mineral. Now, let's move to a solution POV circumstance. Consider a system at 25 °C/1 atm at pH = 1 that contains a total of 10^{-6} mols of Fe(III) per liter of system water, so that $Fe_{T,sys} = Fe(III)_{T,sys} = 10^{-6}$ *M*. We are here extending the applicability of "molarity" to **mols per liter of system water**, whether all dissolved or not. This important concept is used routinely in geochemistry calculations. It allows MBEs to be written governing a sum of concentrations for dissolved *and* precipitated species.

For saturation at pH = 1, (am)$Fe(OH)_{3(s)}$ would give $Fe(III)_T \approx 10^{-2}$ *M*, so if $Fe(III)_{T,sys} = 10^{-6}$ *M*, then at equilibrium all the Fe(III) must be in solution. If we start to raise the pH upwards from 1.0, then the value of the dissolved $Fe(III)_T$ moves along the bold line in Figure 11.7. The Fe(III) speciation is shifting in favor of the hydroxide-containing complexes, but all the Fe(III) remains in solution. At pH = 3.71, the bold line "hits" the hash-marked "solid" line because Eq.(11.40) and the solid POV gives log $Fe(III)_T = -6$ at that pH. The solution is now saturated with the solid, but having just become saturated, all the Fe(III) is still in solution. It is only when we push beyond pH = 3.71 that some solid becomes present, and $Fe(III)_T$ in solution becomes less than $Fe(III)_{T,sys}$. At any pH, the difference between 10^{-6} *M* and the value given by the bold line is the number of FW/L of precipitated (am)$Fe(OH)_{3(s)}$. As the pH moves above 3.71, the dissolved $Fe(III)_T$ decreases and the amount of solid present per liter of system water increases until the minimum in the log $Fe(III)_T$ curve is reached at pH ≈ 8.0. At that pH, the great majority of the system Fe(III) is precipitated. In the region pH = 5.5 to about 7.5, $Fe(OH)_2^+$ is the dominant dissolved Fe(III) species. Beyond pH ≈ 8.0, the dissolved $Fe(III)_T$ starts to increase again because the solubility of the solid is increasing again

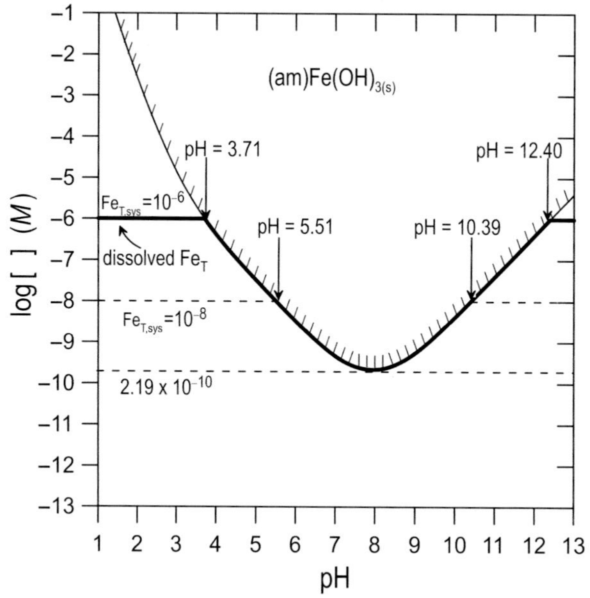

FIGURE 11.7 For $Fe_{T,sys}$ (= $Fe(III)_{T,sys}$) = 10^{-6} *M* at 25 °C/1 atm, the bold line gives dissolved log Fe_T as the pH is varied. All Fe_T is dissolved for low pH to pH = 3.71, and for pH = 12.40 and above. Inside that pH range (3.71 < pH < 12.40), some (am)$Fe(OH)_{3(s)}$ is present. If $Fe_{T,sys} = 10^{-8}$ *M*, all Fe(III) is dissolved for pH ≤ 5.51, and for pH ≥ 10.39; inside that pH range (5.51 < pH < 10.39), some (am)$Fe(OH)_{3(s)}$ is present. No solid can form at any pH if Fe(III) is at or below 2.19×10^{-10} *M*. Activity corrections neglected (not valid at very low or very high pH where the ionic strength will be relatively large, *e.g.*, if pH = 1, then the ionic strength is likely ~0.1).

due to the formation of an increasing amount of $Fe(OH)_4^-$. For pH ≥ ~ 9, $Fe(OH)_4^-$ is the dominant dissolved species. Once pH = 12.40, the solid POV gives $[Fe(OH)_4^-] \approx 10^{-6}$ M, so all the solid has become re-dissolved.

Whenever we have a situation that there is no solid present but the solution is exactly saturated, then Eq.(11.40) gives $Fe(III)_{T,sys}$:

$$^*K_{s0}[H^+]^3 + {^*K_{s1}}[H^+]^2 + {^*K_{s2}}[H^+] + K_{s3} + \frac{^*K_{s4}}{[H^+]} = Fe_{T,sys}. \tag{11.44}$$

We can solve Eq.(11.44) as usual by the LHS – RHS = 0 method. With $Fe(III)_{T,sys} = 10^{-6}$ M, the polynomial (fourth order) has two positive real roots for $[H^+]$. We know that one of the roots is pH = 3.71, and the other is pH = 12.40. Initiating Solver in Excel with a pH guess on the acid side leads to the low pH root. Initiating Solver with an adequately basic pH guess leads to the high pH root. For $Fe(III)_{T,sys} = 10^{-8}$ M, the two roots are closer together, pH = 5.51 and pH = 10.39. At the minimum of the $Fe(III)_T$ curve, pH = 7.95 and $Fe(III)_T = 2.19 \times 10^{-10}$ M. For $Fe(III)_{T,sys} = 2.19 \times 10^{-10}$ M, the two roots are equal (a.k.a. "degenerate"): the solution is saturated at pH = 7.95, but no solid can form. If $Fe(III)_{T,sys}$ is below 2.19×10^{-10} M, the solution cannot become saturated, and there are no positive, real roots. To summarize, as $Fe(III)_{T,sys}$ drops, the two positive-real roots move closer and closer together, become degenerate at pH = 7.95, then go imaginary. (*N.B.*: For (am)$Fe(OH)_{3(s)}$, the LHS – RHS function for Eq.(11.44) has little slope at pH ≈ 7.95, so in a slope-based numerical solution, a guess near that pH may shoot the algorithm exceedingly far away from the nearest root, and the algorithm may fail. A revised guess away from the minimum, will, however, work fine.)

Example 11.4 Saturation pH for (am)$Fe(OH)_{3(s)}$ when $Fe_{T,sys} = 10^{-9}$ M

$Fe(III)_{T,sys} = 10^{-9}$ M, initially pH = 6.20; the system is at 25 °C/1 atm. **a.** Considering (am)$Fe(OH)_{3(s)}$ and assuming equilibrium, show why some solid is not present. **b.** If the pH is raised, at what pH does the solution become saturated with (am)$Fe(OH)_{3(s)}$? **c.** Once the pH is raised to 7.95, in units of FW/L, how much solid is present? **d.** If the pH continues to be raised, at what pH does all the (am)$Fe(OH)_{3(s)}$ dissolve? Neglect activity corrections.

Solution

a. At pH = 6.20, Eq.(11.40) and the solid POV gives $Fe(III)_T = 2.15 \times 10^{-9}$ M. This is greater than $Fe(III)_{T,sys}$, so no solid is present. **b.** For $Fe(III)_{T,sys} = 10^{-9}$ M, the low pH root of Eq.(11.44) is 6.57: the solution becomes saturated at pH = 6.57. **c.** At pH = 7.95, Eq.(11.40) and the solid POV gives $Fe(III)_T = 2.19 \times 10^{-10}$ M. The amount of solid present is $Fe(III)_{T,sys} - Fe(III)_T = 10^{-9}$ $M - 2.19 \times 10^{-10}$ $M = 7.81 \times 10^{-10}$ FW/L. **d.** For $Fe(III)_{T,sys} = 10^{-9}$ M, the high pH root of Eq.(11.44) is 9.33: all the solid is gone at pH = 9.33.

Example 11.5 Calculating pH-Dependent Concentrations of Dissolved Fe(III) Species

In Example 11.4, it is shown that no solid is initially present if $Fe(III)_{T,sys} = 10^{-9}$ M, pH = 6.20, and the system is at 25 °C/1 atm. If the solid is not present, and the solution is not saturated with the solid, we cannot use the solid POV to solve for the concentrations of the various Fe(III) species. Use the fact that Fe^{3+} acts as a multiprotic acid (see Section 10.3.1) to compute the acid/base α values for these species at pH = 6.20, then calculate the corresponding concentrations. Neglect activity corrections.

Solution

The values of α_0 through α_4 for pH = 6.20 can be calculated using the values of $^*K_{H1}$ through $^*K_{H4}$ in Table 10.1. The results are:

$$\alpha_0 = 1.85 \times 10^{-7}. \qquad [Fe^{3+}] = \alpha_0 \, Fe(III)_{T,sys} = 1.85 \times 10^{-16} M.$$

$$\alpha_1 = 1.85 \times 10^{-3}. \qquad [FeOH^{2+}] = \alpha_1 \, Fe(III)_{T,sys} = 1.82 \times 10^{-12} M.$$

$$\alpha_2 = 9.29 \times 10^{-1}. \qquad [Fe(OH)_2^+] = \alpha_2 \, Fe(III)_{T,sys} = 9.29 \times 10^{-10} M.$$

$$\alpha_3 = 6.89 \times 10^{-2}. \qquad [Fe(OH_3^\circ)] = \alpha_3 \, Fe(III)_{T,sys} = 6.89 \times 10^{-11} M.$$

$$\alpha_4 = 2.94 \times 10^{-4}. \qquad [Fe(OH)_4^-] = \alpha_4 \, Fe(III)_{T,sys} = 2.94 \times 10^{-13} M.$$

11.4.4 Solubility Differences among the Different Fe(III) Metal Hydroxides, Oxyhydroxides, and Oxides

A given metal ion may be able to form multiple, chemically similar mineral solids. Fe^{3+}, for example, can precipitate as multiple hydroxide/oxide/oxyhydroxide solids. **The different solids will have different solubilities**. For Fe^{3+}, when the hydroxide is found, it is usually the (am)$Fe(OH)_{3(s)}$ form. The reader may have seen this red/brown material in a water treatment plant, or on a hillside where groundwater containing dissolved Fe(II) (mostly as Fe^{2+}) is seeping, becomes exposed to atmospheric oxygen, and is then oxidized to form a slimy mass of (am)$Fe(OH)_{3(s)}$.

As discussed above, we have

$$(am)Fe(OH)_{3(s)} + 3H^+ = Fe^{3+} + 3H_2O \qquad ^*K_{s0} = K_{s0} / K_w^3 = \{Fe^{3+}\}/\{H^+\}^3. \qquad (11.22)$$

Another solid of Fe^{3+} is the α form of $FeOOH_{(s)}$. This is the much-studied mineral goethite, named after Wolfgang von Goethe (1749–1832), a natural scientist and German literature icon (*e.g., Faust*). Goethite dissolves according to

$$\alpha\text{-}FeOOH_{(s)} + 3H^+ = Fe^{3+} + 2H_2O \qquad ^*K_{s0} = \{Fe^{3+}\}/\{H^+\}^3. \qquad (11.45)$$

Another solid of Fe^{3+} is α-$Fe_2O_{3(s)}$, which is hematite; it dissolves according to

$$\tfrac{1}{2}\alpha\text{-}Fe_2O_{3(s)} + 3H^+ = Fe^{3+} + \tfrac{3}{2}H_2O \qquad ^*K_{s0} = \{Fe^{3+}\}/\{H^+\}^3. \qquad (11.46)$$

The assumptions of unity for $\{H_2O\}$ and for {solid} in each of the above three cases lead to the same expression for saturation equilibrium, namely $\{Fe^{3+}\} = {}^*K_{s0}/\{H^+\}^3$. The $^*K_{s0}$ values are different at 25 °C/1 atm. The order of decreasing $^*K_{s0}$ gives the order of increasing stability. At any pH, increasing stability means lower $\{Fe^{3+}\}$, lower $\{FeOH^{2+}\}$, lower $\{Fe(OH)_2^+\}$, lower $\{Fe(OH)_3^\circ\}$, lower $\{Fe(OH)_4^-\}$, and lower Fe_T at any given pH, because with increasing stability of the solid the Fe^{3+} has increasing tendency to stay in the solid, as opposed to dissolving.

At 25 °C/1 atm, the order of decreasing $^*K_{s0}$ is ferrihydrite > goethite > hematite, with $^*K_{s0}$ values of $10^{3.2} > 10^{0.49} > 10^{-0.71}$ (Jang et al., 2007). So at 25 °C/1 atm, (am)$Fe(OH)_{3(s)}$ is about 10,000 times more soluble than hematite. More Fe(III) hydroxide/oxide/oxyhydroxide minerals are known. Why so many? Because when a metal ion is very insoluble, there tends to be more than one solid that can form that *at least* gets the species out of solution. Analogy: I hate working in the yard on a hot day. Being in the shade right next to my house will be a big improvement. Just inside the door of my house will be better. All the way inside and sitting on a soft couch will be even better, and is where I end up, but I can always morph towards that "phase" with the passage of time once I get in the shade next to the house. Likewise, if (am)$Fe(OH)_{3(s)}$ precipitates first, then in time it can convert to one of the less soluble forms. Liquid water will facilitate the conversion. Consider some solid form #1 with $^*K_{s0}^{\#1}$, and some solid form #2 with $^*K_{s0}^{\#2}$. For the two forms, $^*K_{s0}^{\#1} > {}^*K_{s0}^{\#2}$. If form #1 is at saturation equilibrium with a water phase, then form #2 is supersaturated and will (eventually) start to precipitate. This will lower the dissolved concentration around form #1, and cause it to dissolve, causing more precipitation of form #2. The process continues until all form #1 is converted into form #2.

If the applicable equilibration times are very long for a given problem, then calculations regarding the solubility of Fe(III) should be carried out in terms of one of the more stable forms. But, for short equilibration times, like hours to weeks, consideration of a less stable form is always advised, and certainly if it is known that some particular less stable form is the one present. For Fe(III), this is often (am)$Fe(OH)_{3(s)}$.

As noted at the start of this section, the circumstance that chemically similar solids of a given metal ion have different solubilities is not specific to Fe^{3+}. For Pb^{2+}, six different $Pb(OH)_{2(s)}$ solids, and two different $PbO_{(s)}$ solids (yellow and red) have been identified, giving eight different values for $^*K_{s0} = \{Pb^{2+}\}/\{H^+\}^2$. And for Ca^{2+}, there are three carbonate solids (vaterite, aragonite, and calcite), with three different values of $K_{s0} = \{Ca^{2+}\}\{CO_3^{2-}\}$. Calcite is the least soluble at surface temperatures and pressures, but aragonite is frequently the one that is initially formed. Overall, in this text the emphasis is more on shorter time frame systems, and less on longer, geologic time-scale systems. So, when a choice of solid is made for examples in this text, there is some bias towards shorter term, transient minerals; a text written by a full-fledged geochemist might have the opposite bias.

A word of caution before proceeding: a direct relation between K_{s0} values and solubilities applies only when the solids being compared all lead to the same form of the solubility expression in as Eqs.(11.22), (11.45) and (11.46) for Fe(III) hydroxide/oxide/oxyhydroxide minerals. The same can be done among the K_{s0} values for the three carbonates of Ca^{2+}, namely vaterite, aragonite, and calcite. The same cannot be done for say the K_{s0} of $Fe(OH)_{2(s)}$ and the K_{s0} of $FeCO_{3(s)}$ because the expressions for $\{Fe^{2+}\}$ in the solid POV are fundamentally different ($K_{s0}/\{OH^-\}^2$ *vs.* $K_{s0}/\{CO_3^{2-}\}$, respectively.)

11.4.5 Metal Hydroxide, Oxide, or Oxyhydroxide Equilibrated with Water and Some Value of $(C'_B - C'_A)$

11.4.5.1 Initially Pure Water: $(C'_B - C'_A) = 0$, $C_T = 0$

For the particular speciation that results when initially pure water is equilibrated with a given metal hydroxide, oxide, or oxyhydroxide, we need to consider the concentrations of all of the metal-containing species plus just two more, $[H^+]$ and $[OH^-]$. To compute values for this set of concentration unknowns, we have the following equations: (1) one $^*K_s$ for each of the metal-containing species; (2) K_w; and (3) the ENE. There is no MBE since we do not know *a priori* how much metal will dissolve.

With dissolution into initially pure water, there is no dissolved C_T. Also, there is no added strong acid like HCl, and no added strong base like NaOH; here we write that as

$$\left(C'_B - C'_A\right) = \left(\text{strong base like NaOH} - \text{strong acid like HCl}\right) = 0. \tag{11.47}$$

This quantity is called the "net simple strong base". We cannot say that $(C_B - C_A) = 0$ because the dissolution of a metal hydroxide, oxide, or oxyhydroxide introduces strong base into the solution by release of the anion of the solid OH^- (or O^{2-}, which reacts with H_2O to make to $2OH^-$), with the total metal that dissolves being the tracer, with appropriate coefficient, for the dissolved base of type ($C_B - C_A$). For dissolution of (am)$Fe(OH)_{3(s)}$ into initially pure water, the ENE is then

$$3[Fe^{3+}] + 2[FeOH^{2+}] + [Fe(OH)_2^+] + [H^+] = [Fe(OH)_4^-] + [OH^-]. \quad (11.48)$$

Neglecting activity corrections, we have

$$3^*K_{s0}[H^+]^3 + 2^*K_{s1}[H^+]^2 + {}^*K_{s2}[H^+] + [H^+] = \frac{^*K_{s4}}{[H^+]} + \frac{K_w}{[H^+]}. \quad (11.49)$$

The same ENE applies for (am)$Al(OH)_{3(s)}$. For $PbO_{(s)}$, the corresponding ENE is

$$2^*K_{s0}[H^+]^2 + {}^*K_{s1}[H^+] + [H^+] = \frac{^*K_{s3}}{[H^+]} + \frac{K_w}{[H^+]}. \quad (11.50)$$

Using the K values for 25 °C/1 atm in Table 11.4, for (am)$Fe(OH)_{3(s)}$ pH=7.00; this solid is very insoluble and has essentially no effect on the pH of initially pure water. For (am)$Al(OH)_{3(s)}$ added to initially pure water at 25 °C/1 atm, pH=6.85, so overall the solid is slightly more acidic than basic: the tendency to produce H^+ by production of $Al(OH)_4^-$ from the solid is slightly greater than the tendency to produce OH^- by production of Al^{3+}, $AlOH^{2+}$, and $Al(OH)_2^+$. For α-$PbO_{(s)}$ added to initially pure water at 25 °C/1 atm, pH=9.50, so overall, the solid is distinctly basic: the tendency to produce OH^- by production of Pb^{2+} and $PbOH^+$ from the solid is greater than the tendency to produce H^+ by production of $Pb(OH)_3^-$. (These results may be compared with the case, from Chapter 9, of $NaHCO_3$ dissolved in initially pure water: the solution is basic because HCO_3^- is a stronger base than it is an acid.) A dashed vertical line is drawn in each of the Figures 11.3 through 11.5 at the pH characterized by $(C'_B - C'_A) = 0$. For each solid, the speciation that results for dissolution into initially pure water is obtained by the intersections of that vertical line and with the various log concentration lines.

11.4.5.2 Non-Zero Net Simple Strong Base (*i.e.*, of the Types NaOH and HCl): $(C'_B - C'_A) \neq 0$ with $C_T \neq 0$

If there might be some added strong acid like HCl and/or some strong base like NaOH, and/or some dissolved C_T in an (am)$Fe(OH)_{3(s)}$/water system, then the ENE becomes

$$\begin{aligned}&[Na^+] + 3[Fe^{3+}] + 2[FeOH^{2+}] + [Fe(OH)_2^+] + [H^+] \\ &= [Fe(OH)_4^-] + [OH^-] + [Cl^-] + \alpha_1 C_T + 2\alpha_2 C_T.\end{aligned} \quad (11.51)$$

C_T here can be either fixed as in a closed system or allowed to vary with pH according to $C_T = K_H p_{CO_2}/\alpha_0$. With C'_A and C'_B as described above denoting a generic strong acid like HCl and a generic strong base like NaOH, respectively, Eq.(11.51) becomes

$$\begin{aligned}&(C'_B - C'_A) + 3[Fe^{3+}] + 2[FeOH^{2+}] + [Fe(OH)_2^+] + [H^+] \\ &= [Fe(OH)_4^-] + [OH^-] + \alpha_1 C_T + 2\alpha_2 C_T.\end{aligned} \quad (11.52)$$

Neglecting activity corrections,

$$\begin{aligned}&(C'_B - C'_A) + 3^*K_{s0}[H^+]^3 + 2^*K_{s1}[H^+]^2 + {}^*K_{s2}[H^+] + [H^+] \\ &= \frac{^*K_{s4}}{[H^+]} + \frac{K_w}{[H^+]} + \alpha_1 C_T + 2\alpha_2 C_T.\end{aligned} \quad (11.53)$$

Eq.(11.53) also applies for (am)$Al(OH)_{3(s)}$. There are three variables in Eq.(11.53), $(C_B' - C_A')$, C_T, and pH. Knowing two allows calculation of the third. For $PbO_{(s)}$, we use

$$(C_B' - C_A') + 2\,{}^*K_{s0}[H^+]^2 + {}^*K_{s1}[H^+] + [H^+]$$
$$= \frac{{}^*K_{s3}}{[H^+]} + \frac{K_w}{[H^+]} + \alpha_1 C_T + 2\alpha_2 C_T. \tag{11.54}$$

Example 11.6 Solubility of (am)$Fe(OH)_{3(s)}$ in Water with CO_2

Consider that some (am)$Fe(OH)_{3(s)}$ is equilibrated at 25 °C/1 atm in an open system with initially pure water and $p_{CO_2} = 10^{-3.4}$ atm. Compute the pH and the total dissolved iron, Fe_T. Neglect activity corrections.

Solution

Solving Eq.(11.52) with $C_T = K_H p_{CO_2}/\alpha_0$ gives pH = 5.61, the same pH as when (am)$Fe(OH)_{3(s)}$ is not present (Chapter 9); this solid is so insoluble that it has essentially no effect on the pH, the pH being determined nearly exclusively by the dissolution of $CO_{2(g)}$. $Fe_T = 7.96 \times 10^{-9}$ *M*.

Example 11.7 Dissolving a Stain of (am)$Fe(OH)_{3(s)}$ with HCl

Consider that 10^{-4} FW (= 0.0107 g) of (am)$Fe(OH)_{3(s)}$ is equilibrated with initially pure water at 25 °C/1 atm, as with an Fe(III) solid on a porcelain sink. If the sink is filled with 1 L of initially pure water, and neglecting contributions of CO_2 to the chemistry, what minimum value of $(C_B' - C_A')$ will be required to dissolve all the stain with addition of HCl? Neglect activity corrections. (This example has the solid POV.) Concentrated HCl can be purchased by consumers as "muriatic acid", but it is a dangerous chemical and must be handled exceedingly carefully (see Box 11.2). Neglect activity corrections.

Solution

With 1 L of water, then $Fe(III)_{T,sys} = 10^{-4}$ *M*. The low pH root of Eq.(11.44) is then 2.59, so $[H^+] = 2.57 \times 10^{-3}$ *M*. Solving Eq.(11.53) gives $(C_B' - C_A') = -2.79 \times 10^{-3}$ eq/L. The negative value indicates the need for a positive addition of HCl. We can check if this number makes sense. At pH = 2.59, from Figure 10.3 the dominant dissolved Fe(III) species is $FeOH^{2+}$. So, most of the dissolution occurs according to (am)$Fe(OH)_{3(s)} + 2H^+ = FeOH^{2+} + 2H_2O$. So ~2 H^+ were used per (am)$Fe(OH)_{3(s)}$ dissolved. With another 2.57×10^{-3} eq/L for the H^+ in solution, then the total added strong acid is estimated as $(2 \times 10^{-4} + 2.57 \times 10^{-3})$ eq/L = 2.77×10^{-3} eq/L $\approx -(C_B' - C_A') = 2.79 \times 10^{-3}$ eq/L.

Example 11.8 Calcium, Lime, & Rust Remover

Acidic consumer cleaning products less hazardous than HCl have been developed for removal (by acid dissolution) of mineral deposits. One example is CLR® (*Calcium, Lime & Rust Remover*, Jelmar, LLC, Skokie, IL). ("Calcium" and "lime" here both refer to precipitated calcium carbonate.) The 2016 material safety data sheet (MSDS) for CLR® only gives composition ranges (by weight) for the two main ingredients, lactic acid (5–18%) and lauramine oxide (1.5–7.5%). Lactic acid is a monoprotic organic acid with $pK_a = 3.86$ at 25 °C/1 atm; it is strong enough dissolve the target deposits, but not so strong that it is dangerous to handle, which is the case for HCl (see Box 11.2). Lauramine oxide is a surfactant, to provide a soapy effect. Modify the ENE that is Eq.(11.55) as needed, and calculate the minimum total lactic acid concentration that would be required to dissolve $Fe(III)_{T,sys} = 10^{-4}$ *M* worth of (am)$Fe(OH)_{3(s)}$. (This example has the solid POV.) Neglect activity corrections.

Solution

As in Example 11.6, neglecting activity corrections, the low pH root to dissolve $Fe(III)_{T,sys} = 10^{-4}$ *M* is pH = 2.59, so $[H^+] = 2.57 \times 10^{-3}$ *M*. Starting with Eq.(11.47), we remove $[Na^+]$ and $[Cl^-]$, and add $\alpha_1 L_T$ on the RHS for the concentration of lactate ion: $3[Fe^{3+}]+2[FeOH^{2+}]+[Fe(OH)_2^+]+[H^+] = [Fe(OH)_4^-]+[OH^-]+\alpha_1 L_T$. With $pK_a = 3.86$ and pH = 2.59, solving gives $L_T = 5.48 \times 10^{-2}$ *M*. This value is much higher than $Fe(III)_{T,sys}$, so the pH is mostly being set by the level of lactic acid, and not by the dissolution of the (am)$Fe(OH)_{3(s)}$. If the CLR® product is 10% by weight lactic acid, *i.e.*, ~100 g/L, then at MW = 90.08 g/mol this is ~1.1 *F*. To provide the needed L_T for the 1 liter of water in the sink, 1.1 $F \times x$ liters of CLR® = 5.48×10^{-2} $M \times 1$ liters, so $x \approx 0.05$ L (or ~50 mL).

BOX 11.2 Cautionary Statement to General Consumers about Handling "Muriatic Acid" (HCl)

Rust stains on porcelain, concrete, etc. provide an everyday context for understanding the solubility chemistry of Fe(III) hydroxide/oxide/oxyhydroxide solids. As defined here, $(C'_B - C'_A)$ is the concentration of net strong base of the NaOH/HCl type. $(C'_A - C'_B) >> 0$ for stain removal can be attained by addition of some concentrated HCl to a solution. Concentrated HCl can be purchased by consumers as "muriatic acid" (HCl, usually 31.5% by weight HCl, with solution density 1.16 g/mL, which all works out to be 11.0 *F* HCl). *Muria* is the Latin noun for *salt brine*, so muriatic acid is the acid related to NaCl, namely HCl. The average consumer needs to be mindful that "muriatic acid" is a hazardous chemical because: 1) it is a very strong acid; and 2) in concentrated form, it gives off HCl fumes. The following sober caution was found as part of an article entitled on removing rust stains (Robillard, n.d.):

> Do-it-yourselfer's should avoid muriatic acid where possible. Only use muriatic acid after exhausting other cleaning methods like TSP, or less caustic concrete stain removers. Muriatic acid is not the first choice for masonry cleaning but the last resort. Do not use this dangerous chemical unless you are sure you have no other choice. Muriatic acid is a highly reactive liquid acid, and one of the most dangerous chemicals you can buy for home use. It is an industrial-strength solution of hydrogen chloride …, also known as hydrochloric acid.

REFERENCES

Aqion (2017) *Solubility Product Constants* K_{sp} *at* 25 °C. Available at www.aqion.de/site/16 (Accessed 22 November 2017).

Ball JW, Nordstrom DK (1991) *User's Manual For WATEQ4F, with Revised Thermodynamic Data Base and Test Cases for Calculating Speciation of Major, Trace, and Redox Elements in Natural Waters.* U.S. Geological Survey Open-File Report 91-183.

Cheng YJ, Brittin HC (1991) Iron in food – effect of continued use of iron cookware. *Journal of Food Science*, **56**, 584–585.

Edzwald JK (2018) Personal communication, November 16, 2018.

Jang JH, Dempsey BA, Burgos WD (2007) Solubility of hematite revisited: Effects of hydration. *Environmental Science and Technology*, **41**, 7303–7308.

Pernitsky DJ, Edzwald JK (2003) Solubility of polyaluminum coagulants. *Journal of Water Supply: Research and Technology – Aqua*, **52**, 395–406.

Plummer NL, Parkhurst DL, Fleming GW, Dunkle SA (1988) *A Computer Program Incorporating Pitzer's Equations for Calculation of Geochemical Reactions in Brines*. U.S. Geological Survey Water-Resources Investigations Report 88-4153.

Robillard, R (n.d.) How to remove a rust stain off concrete or bluestone. A Concord Carpenter. Available at: https://www.aconcordcarpenter.com/how-to-remove-a-rust-stain-off-concrete-or-bluestone.html (Accessed 6 June 2019).

Sacks LA (1996) *Geochemical and Isotopic Composition of Ground Water with Emphasis on Sources of Sulfate in the Upper Floridan Aquifer in parts of Marion, Sumter, and Citrus Counties, Florida*. U.S. Geological Survey, Water-Resources Investigations Report 95-4251, Tallahassee, FL.

Smith RM, Martell AE (1976) *Critical Stability Constants, Volume 4, Inorganic Complexes*, New York: Plenum Press.

Stanforth R, Qiu J (2001) Effect of phosphate treatment on the solubility of lead in contaminated soil. *Environmental Geology*, **41**, 1–10.

Stumm W, Morgan JJ (1996) *Aquatic Chemistry: Chemical Equilibria and Rates in Natural Waters*, 3rd edn, Wiley Publishing Co., New York.

Wesolowski DJ, Palmer DA (1994) Aluminum speciation and equilibria in aqueous solution: V. Gibbsite solubility at 50 °C and pH 3-9 in 0.1 molal NaCl solutions (a general model for aluminum speciation; analytical methods). *Geochimica et Cosmochimica Acta*, **58**, 2947–2969.

12 Solubility Behavior of Calcium Carbonate and Other Divalent Metal Carbonates in Closed and Open Systems

12.1 INTRODUCTION

The ion-ion electrostatic attractions in mineral lattices are overarchingly important forces for mineral formation. **Trivalent** metals ions ($z = 3+$) have a large positive charge, and so are highly attracted to OH^- and O^{2-}. They are very stable (and insoluble) as hydroxide, oxide, or oxyhydroxide solids. They do not tend to form carbonate solids because if CO_3^{2-} is even moderately high in a system, OH^- will also be high so that a hydroxide can form, then ultimately dehydrate to form an oxide with the much more densely charged O^{2-}, for strong +3/−2 interactions. **Divalent** metal ions ($z = +2$) infrequently form solid metal hydroxides or oxides *e.g.*, $Fe(OH)_{2(s)}$ or $FeO_{(s)}$ are too soluble to precipitate over the corresponding metal carbonate $FeCO_{3(s)}$ unless the pH is high and C_T is limited, as in some closed systems. (The +2/−2 interactions in $CaCO_{3(s)}$ and $FeCO_{3(s)}$ are generally more favorable than the +2/−2 interactions in $CaO_{(s)}$ and $FeCO_{(s)}$, respectively.) **Univalent** metal ions ($z = +1$, as with Na^+) can also form solid carbonates under some conditions (as in alkaline playa lakes with high C_T), but these carbonates are fairly soluble, offering only +1/−2 interactions. Solids like $NaOH_{(s)}$ with +1/−1 interactions virtually never form in nature. To summarize, ambient minerals of trivalent metal ions are predominantly oxides, hydroxides, and oxyhydroxides; ambient minerals of divalent metal ions are predominantly carbonates; and, ambient minerals of univalent metal ions, when they do form in alkaline systems, are predominantly carbonates.

Examples of important divalent metal carbonates are:

1. $CaCO_{3(s)}$ in the hexagonal crystal form, known as calcite, and found in nature as deposits of marble, chalk, and limestone.
2. $CaCO_{3(s)}$ in the orthorhombic crystal form, known as aragonite, and found in nature in some shells and in geologically young deposits of $CaCO_{3(s)}$. Old deposits of $CaCO_{3(s)}$ are usually not aragonite, because aragonite is less stable than calcite and is converted to calcite over time.
3. $MgCa(CO_3)_{2(s)}$, known as dolomite (as in the Dolomite mountain range in the east Alps in northeast Italy).
4. $FeCO_{3(s)}$, known as siderite, and found in nature in hydrothermal veins, often as a result of the action of Fe^{2+}-containing solutions on calcium carbonate minerals.

If $CaCO_{3(s)}$ is equilibrated with initially pure water, then the ENE is

$$2[Ca^{2+}] + [CaOH^+] + [H^+] = [HCO_3^-] + 2[CO_3^{2-}] + [OH^-]. \qquad (12.1)$$

$CaCO_{3(s)}$ does not dissolve as a simple salt; some of the Ca^{2+} that goes into solution becomes $CaOH^+$, and some of the CO_3^{2-} that goes into solution becomes HCO_3^- and $H_2CO_3^*$. Dissolved metal complexes with more than one OH^- are important for many divalent metals (*e.g.*, Mn(II) and Fe(II)); for calcium, all we have to consider is $CaOH^+$. If some strong acid like HCl and/or some strong base like NaOH is also added, the ENE becomes

$$[Na^+] + 2[Ca^{2+}] + [CaOH^+] + [H^+] = [HCO_3^-] + 2[CO_3^{2-}] + [OH^-] + [Cl^-] \qquad (12.2)$$

or

$$(C'_B - C'_A) + 2[Ca^{2+}] + [CaOH^+] + [H^+] = [HCO_3^-] + 2[CO_3^{2-}] + [OH^-]. \qquad (12.3)$$

The concept of $(C'_B - C'_A)$ (the net simple strong base) was introduced in Chapter 11. Briefly, C'_B is strong base from chemicals like NaOH that do not involve the metal ion whose solubility is under consideration (so in this case, not $Ca(OH)_2$); C'_A is strong acid as from chemicals such as HCl. $(C'_B - C'_A)$ is one part of Alk. The other part is the contribution from dissolution of the solid (see Eq.(12.11)). For equilibrium with the solid,

$$CaCO_{3(s)} = Ca^{2+} + CO_3^{2-} \qquad K_{s0} = \frac{\{Ca^{2+}\}\{CO_3^{2-}\}}{\{CaCO_{3(s)}\}}. \qquad (12.4)$$

Taking $\{CaCO_{3(s)}\} = 1$ and neglecting activity corrections for the aqueous phase species (or substituting ${}^cK$ values to consider them),

$$[Ca^{2+}] = \frac{K_{s0}}{[CO_3^{2-}]} = \frac{K_{s0}}{\alpha_2^C C_T} \qquad (12.5)$$

$$(C'_B - C'_A) + \frac{2K_{s0}}{\alpha_2^C C_T} + [CaOH^+] + [H^+] = \alpha_1^C C_T + 2\alpha_2^C C_T + \frac{K_w}{[H^+]}. \qquad (12.6)$$

The addition of one OH^- to Ca^{2+} occurs according to

$$Ca^{2+} + OH^- = CaOH^+ \qquad K_{H1} = \frac{[CaOH^+]}{[Ca^{2+}][OH^-]}. \qquad (12.7)$$

We add the equilibrium

$$H_2O = H^+ = OH^- \qquad K_w = \left[H^+\right]\left[OH^-\right] \qquad (12.8)$$

to obtain

$$Ca^{2+} + H_2O = CaOH^+ + H^+ \qquad {}^*K_{H1} = K_{H1}K_w = \frac{[CaOH^+][H^+]}{[Ca^{2+}]}. \qquad (12.9)$$

By Eq.(12.6) with Eqs.(12.5) and (12.9), then neglecting activity corrections here and henceforth in this chapter,

$$(C'_B - C'_A) + \frac{2K_{s0}}{\alpha_2^C C_T} + \frac{K_{s0}}{\alpha_2^C C_T}\frac{{}^*K_{H1}}{[H^+]} + [H^+] = \alpha_1^C C_T + 2\alpha_2^C C_T + \frac{K_w}{[H^+]}. \qquad (12.10)$$

This chapter considers three case types for Eq.(12.10), each with both zero and nonzero $(C'_B - C'_A)$:

Case I. Closed system, $C_T = Ca_T$ (*i.e.*, $CaCO_{3(s)}$+water);
Case II. Closed system, $C_T = Ca_T + y$ (*i.e.*, $CaCO_{3(s)}$ + water + y dissolved CO_2);
Case III. Open system, $C_T = K_H p_{CO_2}/\alpha_0^C$ (*i.e.*, $CaCO_{3(s)}$+water + fixed p_{CO_2}).

Relevant example scenarios for the cases are given in Table 12.1.

TABLE 12.1
Example Scenarios in Closed and Open Systems for Equilibrium with $CaCO_{3(S)}$; the Noted Constraint Is Used to Solve Eq. (12.10)

	$(C'_B - C'_A)$	Example Scenarios
Case I	= 0	Water with little dissolved CO_2 percolates into a $CaCO_{3(s)}$ rock formation.
closed $C_T = Ca_T$	≠ 0	Water with little dissolved CO_2 but with some geologically produced strong acid percolates into a $CaCO_{3(s)}$ rock formation (*e.g.*, Carlsbad Caverns).
Case II closed	= 0	Pristine rainwater percolates through soil and absorbs significant CO_2 from carbon-respiring organisms, then percolates into a $CaCO_{3(s)}$ rock formation.
$C_T = Ca_T + y$	≠ 0	Pristine rainwater percolates through soil and absorbs significant CO_2, as well as some geologically produced strong acid, then percolates into a $CaCO_{3(s)}$ rock formation.
Case III open	= 0	Pristine rainwater falls on exposed $CaCO_{3(s)}$ bedrock, or on limestone structures or statues; or, limestone cave water equilibrates with cave air with specific p_{CO_2}.
p_{CO_2} = constant	≠ 0	Rainwater containing strong acid (as with "acid rain") falls on exposed $CaCO_{3(s)}$ bedrock, or on limestone structures or statues.

12.2 ALKALINITY (ALL CASES)

In all of the solutions of interest in this chapter, there are two possible sources of Alk: dissolution of carbonate rock, and addition of non-zero $(C'_B - C'_A)$. For the first, every FW of metal carbonate that dissolves introduces two equivalents of strong base. Thus, for the total Alk,

$$\text{Alk} = 2\text{Ca}_T + (C'_B - C'_A) = 2[\text{Ca}^{2+}] + 2[\text{CaOH}^+] + (C'_B - C'_A). \quad (12.11)$$

Eq.(12.3) is valid as the ENE in all of the systems in this chapter. Moving $[H^+]$ to the RHS gives

$$2[\text{Ca}^{2+}] + [\text{CaOH}^+] + (C'_B - C'_A) = [\text{HCO}_3^-] + 2[\text{CO}_3^{2-}] + [\text{OH}^-] - [\text{H}^+]. \quad (12.12)$$

The LHS of Eq.(12.12) differs from Alk as given by Eq.(12.11) only by 1×$[CaOH^+]$, so adding $[CaOH^+]$ to both sides of Eq.(12.12) gives

$$2[\text{Ca}^{2+}] + 2[\text{CaOH}^+] + (C'_B - C'_A) = [\text{HCO}_3^-] + 2[\text{CO}_3^{2-}] + [\text{OH}^-] + [\text{CaOH}^+] + -[\text{H}^+]. \quad (12.13)$$

The LHS of Eq.(12.13) is the same as the RHS of Eq.(12.11), so

$$\text{Alk} = [\text{HCO}_3^-] + 2[\text{CO}_3^{2-}] + [\text{OH}^-] + [\text{CaOH}^+] - [\text{H}^+]. \quad (12.14)$$

Compared to the expression for Alk given in Chapter 9, $CaOH^+$ is one more species that can carry Alk. If $Ca(OH)_2^o$ and $Ca(OH)_3^-$ were important in this system, then $+2[Ca(OH)_2^o]$ and $+3[Ca(OH)_3^-]$ would also be components of Alk.

12.3 CASE I: CLOSED SYSTEM, $C_T = Ca_T$ (CALCIUM CARBONATE + WATER + VARIABLE $(C'_B - C'_A)$)

12.3.1 SOLUBILITY AS A FUNCTION OF pH

If $CaCO_{3(s)}$ dissolves into water that initially contains no Ca^{2+} or CO_3^{2-}, and the dissolution occurs as a simple salt, then at saturation the MBE would be

$$[\text{Ca}^{2+}] = [\text{CO}_3^{2-}] = x \quad \text{(if the dissolution were simple)}. \quad (12.15)$$

Neglecting activity corrections (or substituting ${}^{c}K_{s0}$ to consider them),

$$K_{s0} = [Ca^{2+}][CO_3^{2-}] = x^2 \qquad \text{(if the dissolution were simple)} \tag{12.16}$$

$$[Ca^{2+}] = [CO_3^{2-}] = \sqrt{K_{s0}} \qquad \text{(if the dissolution were simple).} \tag{12.17}$$

But the dissolution is not simple. As noted above, some of the Ca^{2+} released from the solid becomes complexed with OH^- to form $CaOH^+$, and some of the CO_3^{2-} released from the solid becomes HCO_3^- and $H_2CO_3^*$. We will, however, still be able to write the MBE that

$$C_T = Ca_T. \tag{12.18}$$

The dissolution is not simple because Ca^{2+} and CO_3^{2-} are being pulled into other species, so more solid dissolves and $C_T = Ca_T > \sqrt{K_{s0}}$. Exactly how much greater? The answer is obtained as Eq.(12.23) below.

The equilibrium in Eq.(12.9) reveals the acidic nature of a metal ion that can add OH^- as a complexing ligand (Chapter 10). In fact, we can see that Ca^{2+} in a solution is indeed an acid that can lose a proton if we understand that every metal cation actually is a species that includes a group of closely associated water molecules of solvation (about six on average). The underlying representation of the equilibrium in Eq.(12.9) is then (Chapter 10)

$$Ca(H_2O)_6^{2+} = Ca(H_2O)_5OH^+ + H^+. \tag{12.19}$$

Conveniently, this means we can treat Ca^{2+} as an acid species, so that of the Ca_T in the solution,

$$[Ca^{2+}] = \alpha_0^{Ca} Ca_T. \tag{12.20}$$

Ca^{2+} does not add more than one OH^- at natural water pH values, so only one acidity constant is needed:

$$\alpha_0^{Ca} = \frac{1}{1 + \dfrac{{}^*K_{H1}}{[H^+]}}. \tag{12.21}$$

For equilibrium between $CaCO_{3(s)}$ and water, continuing to neglect activity corrections (or substituting ${}^{c}K$ values to consider them), by Eqs.(12.5) and (12.20)

$$\alpha_0^{Ca} Ca_T = \frac{K_{s0}}{\alpha_2^{C} C_T}. \tag{12.22}$$

From Eq.(12.18),

$$C_T = Ca_T = \sqrt{\frac{K_{s0}}{\alpha_2^{C}\alpha_0^{Ca}}} = \sqrt{K_{s0}}\left(\frac{1}{\sqrt{\alpha_2^{C}}}\,\frac{1}{\sqrt{\alpha_0^{Ca}}}\right). \tag{12.23}$$

The factor $1/\sqrt{\alpha_2^{C}}$ gives the increase in solubility due to CO_3^{2-} not acting as a spectator ion. The factor $1/\sqrt{\alpha_0^{Ca}}$ 1 gives the increase in solubility due to Ca^{2+} not acting as a spectator ion.

Eq.(12.23) gives C_T (= Ca_T) as a simple function of $[H^+]$. After calculating C_T (= Ca_T) for a given pH, then $[H_2CO_3^*]$, $[HCO_3^-]$, $[CO_3^{2-}]$, $[Ca^{2+}]$, and $[CaOH^+]$ can be obtained using α_0^{C}, α_1^{C}, α_2^{C}, α_0^{Ca} and α_1^{Ca}.

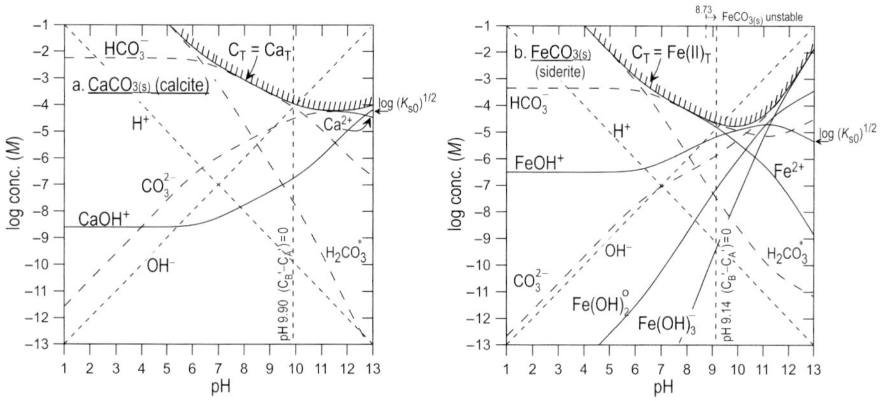

FIGURE 12.1 Log concentration (M) *vs.* pH diagrams with $C_T = M_T$ (closed system) for two divalent metal carbonate solids (calcite and siderite) in equilibrium with water at 25 °C/1 atm. Activity corrections neglected (not valid at very low or very high pH). It is assumed that a strong base like NaOH and/or strong acid like HCl is used to change the pH from that at $(C'_B - C'_A) = 0$. In this system, as discussed later (Section 14.4), above pH = 8.73, siderite is not stable relative to conversion to $Fe(OH)_{2(s)}$ because $\log p_{CO_2}$ falls below −5.42 atm, the value for equilibrium between $FeCO_{3(s)}$ and $Fe(OH)_{2(s)}$. The dashed vertical line gives the equilibrium composition at $(C'_B - C'_A) = 0$, *i.e.*, the metal carbonate is equilibrated with initially pure water. For both solids, at $(C'_B - C'_A) = 0$, for the dissolving CO_3^{2-}, the strong base approximation is roughly valid so that $\log C_T \approx \log[OH^-] \approx \log[HCO_3^-]$ approximates the conditions at $(C'_B - C'_A) = 0$. When the pH becomes low, the solubility increases because of the factor $1/\sqrt{\alpha_2^C}$; when the pH becomes high, it increases because of the factor $1/\sqrt{\alpha_0^{Ca}}$. The location of $\log(K_{s0})^{1/2}$ is noted, indicating the solubility that would result if simple salt dissolution occurred; this is approximately satisfied at pH values for which both $\alpha_2^C \approx \alpha_0^{Ca} \approx 1$.

Figure 12.1 contains log[ ] *vs.* pH diagrams for $CaCO_{3(s)}$ (calcite) and $FeCO_{3(s)}$ (siderite) at 25 °C/1 atm. The constants used to prepare the diagrams are provided in Table 12.2. Values for K_{s0} for calcite and aragonite as a function of temperature are given in Table 12.3.

For calcite and siderite in Figure 12.1, *only one* of the pH values is the one for the system in which the dissolution has taken place into initially-pure water, *i.e.*, when $(C'_B - C'_A) = 0$. From there, with no effect on the condition $C_T = Ca_T$, the addition of some strong acid like HCl (so that $(C'_B - C'_A) < 0$) gives access to lower pH values, and addition of some strong base like NaOH (so that $(C'_B - C'_A) > 0$) gives access to higher pH values.

The plots in Figure 12.1 extend over the full pH range. As discussed further in Chapter 14, not all metal carbonates in closed systems where CO_2 is limited are stable over this entire range. At high pH values, the in-situ p_{CO_2} value may become so low that a metal hydroxide solid becomes more stable than the carbonate. Thus, at a certain "transition pH", generically the equilibrium can be $MCO_{3(s)} + H_2O = M(OH)_{2(s)} + CO_{2(g)}$: if solids that form are pure, there is equilibrium between the two solids at only one p_{CO_2} value; below that value, if some $MCO_{3(s)}$ was present, it would be converted to $M(OH)_{2(s)}$ and vice versa.

12.3.2 The Particular pH When $(C'_B - C'_A) = 0$

To find the pH when $(C'_B - C'_A) = 0$, we solve Eq.(12.10) for that condition. With Eq.(12.23) giving C_T as function of [H⁺], the only unknown in Eq.(12.10) is [H⁺]. We can solve as usual by seeking the pH value that gives LHS − RHS = 0. For a metal carbonate for which the $M(OH)_3^-$ can be important, then $[M(OH)_3^-]$ $(= \alpha_3^M M_T)$ is included on the RHS. Table 12.2 gives the pH values computed

TABLE 12.2
Equilibrium Constant Values for Various Divalent Metal Ions, with the Closed-System pH (and Calculated Log p_{CO_2}) Value for a Solution in Equilibrium of Each Metal Carbonate Solid at 25 °C/1 atm When ($C'_B - C'_A$) = 0 and Added CO_2 Is Zero

Metal	$\log {}^*K_{H1}$	$\log {}^*K_{H2}$	$\log {}^*K_{H3}$	Metal Carbonate	$\log K_{s0}$	pH for $(C'_B - C'_A) = 0$	$\log p_{CO_2}$
Mg^{2+}	−11.42	a	a	$MgCO_{3(s)}$(magnesite)	−7.46	10.19	−6.13
Ca^{2+}	−12.7	a	a	$CaCO_{3(s)}$ (aragonite)	−8.34	9.94	−6.17
Ca^{2+}	−12.7	a	a	$CaCO_{3(s)}$ (calcite)	−8.48	9.90	−6.17
Sr^{2+}	−13.18	a	a	$SrCO_{3(s)}$	−9.03	9.73	−6.18
Mn^{2+}	−10.6	−10.6	−13.0	$MnCO_{3(s)}$	−9.30	9.62	−6.11
Fe^{2+}	−9.50	−11.1	−11.4	$FeCO_{3(s)}$ (siderite)	−10.68	9.14	−6.00

[a] Very small, and not well known; the associated species can be neglected.
Note: pH and log p_{CO_2} values calculated neglecting activity corrections.

TABLE 12.3
K_{s0} Values for Calcite and Aragonite at 1 atm as a Function of Temperature for Infinite Dilution Solution Conditions

	log K_{s0}	
t (°C)	aragonite	calcite
0	−8.24	−8.38
5	−8.26	−8.39
15	−8.30	−8.43
25	−8.36	−8.48
30	−8.39	−8.51
35	−8.43	−8.54

Note: Based on data of Plummer and Busenberg (1992); with T in degrees K, $\log K_{s0}^{\text{aragonite}} = -171.9773 - 0.077993\ T + 2903.293/T + 71.595\ \log T$; and $\log K_{s0}^{\text{calcite}} = -171.9065 - 0.077993\ T + 2839.319/T + 71.595\ \log T$.

at 25 °C/1 atm for several metal carbonates. The resulting values of $p_{CO_2} = \alpha_0^C C_T/K_H$ are very low relative to atmospheric levels.

Example 12.1 Solubility of Calcite in Initially Pure Water, Case I, $Ca_T = C_T$, $(C'_B - C'_A) = 0$

a. For calcite rock at 15 °C/1 atm and with $(C'_B - C'_A) = 0$, calculate pH, $C_T = Ca_T$, $[OH^-]$, $[HCO_3^-]$, and p_{CO_2}. **b.** Is the dissolving CO_3^{2-} largely acting to create a solution that follows the "strong base approximation" as discussed in Chapter 5? **c.** If this closed system solution were opened to the atmosphere with $p_{CO_2} = 10^{-3.4}$ atm, would some CO_2 enter the solution from the air, or would some CO_2 leave from the solution and go into the air? 15 °C is chosen for the problem because 25 °C is a little warm for most subsurface situations.

Solution

At 15 °C, from Table 9.2, we have log K_1 = −6.42, log K_2 = −10.43, log K_H = −1.32, and log K_w = −14.35. At 15 °C, from Table 12.3, for calcite, log K_{s0} = −8.43. From Table 12.2, log$^*K_{H1}$ = −12.7; this is for 25 °C/1 atm, but the resulting error will be negligible because $CaOH^+$ is not very important at the pH that will result. **a.** We solve for the value of pH that gives LHS − RHS = 0 for Eq.(12.10) while using Eq.(12.23) for C_T. Results: pH = 10.17; $Ca_T = C_T = 1.03\times10^{-4}$ *M*; $[OH^-] = 0.66\times10^{-4}$ *M*; $[HCO_3^-] = 0.66\times10^{-4}$ *M*; $p_{CO_2} = 10^{-6.61}$ atm. **b.** The "strong base approximation" is approaching acceptability here, as $[OH^-]$ is similar to C_T, with about 65% of the dissolving CO_3^{2-} forming HCO_3^- and OH^-. The equilibrium pH for $(C_B' - C_A') = 0$ is consequently given roughly by the general intersection of three lines, log C_T, log$[OH^-]$, and log$[HCO_3^-]$. (In Figure 12.1, which is for 25 °C/1 atm, this three-way nexus occurs near pH = 9.90.) **c.** Since $p_{CO_2} < 10^{-3.4}$ atm, CO_2 would enter the solution from the atmosphere if this solution was opened to ambient air.

12.3.3 pH When $(C_B' - C_A') \neq 0$

Cases when $CaCO_{3(s)}$ + water and $(C_B' - C_A') \neq 0$ certainly occur in the natural environment. An important case is that of Carlsbad Caverns in New Mexico for which negative values of $(C_B' - C_A')$ brought about much dissolution of a large limestone formation.

Example 12.2 Solubility of Calcite in Initially Pure Water Plus Some Strong Acid, Case I, $Ca_T = C_T$, $(C_B' - C_A') \neq 0$

Carlsbad Caverns in New Mexico was formed by the action of volcanic waters on a limestone formation. The volcanic waters contained H_2S which became H_2SO_4 upon oxidation. The H_2SO_4 reacted with the limestone and led to gypsum deposits in the cavern ($CaCO_{3(s)} + H_2SO_4 + H_2O = CaSO_4{\cdot}2H_2O + CO_2$). H_2SO_4 is a strong acid of the type that would need to be considered for $(C_B' - C_A')$; in this example, $2[SO_4^{2-}]$ is an accurate tracer for C_A' because $[HSO_4^-]$ is negligible at the pH that will result. Consider the case when water with 1.0×10^{-3} *F* H_2SO_4 equilibrates with calcite at 15 °C/1 atm; for this water, $(C_B' - C_A') = -2\times10^{-3}$ eq/L. **a.** Calculate pH, $Ca_T = C_T$, $[Ca^{2+}]$, $[CO_3^{2-}]$, $[HCO_3^-]$, and p_{CO_2}. **b.** Why is the dissolving CO_3^{2-} ending up in the solution mostly as HCO_3^-? **c.** Would the solution formed be supersaturated with gypsum at 15 °C/1 atm? $K_{s0} = 10^{-4.61}$ (Klimchouk, 1996). **d.** If, prior to any precipitation, this closed system solution were opened to the atmosphere with $p_{CO_2} = 10^{-3.4}$ atm, would some CO_2 enter the solution from the air, or would some CO_2 leave from the solution and go into the air?

Solution

We use the K_{s0} for gypsum as given, plus the same *K* values for 15 °C/1 atm as in Example 12.1. **a.** We solve for the value of pH that gives LHS − RHS = 0 for Eq.(12.10) with $(C_B' - C_A') = -2\times10^{-3}$ eq/L, while using Eq.(12.23) for C_T. Results: pH = 7.50, $Ca_T = C_T = 1.86\times10^{-3}$ *M* (almost 20 times more than in Example 12.1); $[Ca^{2+}] = 1.86\times10^{-3}$ *M* ($\approx Ca_T$); $[CO_3^{2-}] = 10^{-8.43}/[Ca^{2+}] = 2.00\times10^{-6}$ *M*, $[HCO_3^-] = [CO_3^{2-}][H^+]/K_2 = 1.71\times10^{-3}$ *M* (most of C_T); and $p_{CO_2} = [HCO_3^-][H^+]/(K_1K_H) = 10^{-2.52}$ atm. **b.** C_T is mostly HCO_3^-, because the pH is approximately midway between pK_1 and pK_2. **c.** $[Ca^{2+}][SO_4^{2-}] = (1.86\times10^{-3}\ M)(1.0\times10^{-3}\ M) = 1.86\times10^{-6} < K_{s0} = 10^{-4.61}$, the water formed will not be supersaturated with gypsum: no gypsum will precipitate for this level of added H_2SO_4. **d.** Since p_{CO_2} is greater than $10^{-3.4}$ atm, some CO_2 would leave the solution if it were opened to ambient air. This would cause the pH and thus $\alpha_2^C C_T$ to rise, and the solution would become supersaturated with $CaCO_{3(s)}$ leading to precipitation of "speleothems" (e.g., stalactites and stalagmites) of $CaCO_{3(s)}$, (see also Example 12.4).

12.4 CASE II: CLOSED SYSTEM, $C_T = Ca_T + y$ (CALCIUM CARBONATE + WATER + INITIAL y DISSOLVED CO_2 + VARIABLE $(C_B' - C_A')$)

12.4.1 The Particular pH When $(C_B' - C_A') = 0$

An extremely common case type for natural waters interacting with carbonate rock in a closed system involves rainwater that initially absorbs CO_2 to some C_T concentration y (as from the atmosphere or from soil gas), but nothing else of consequence. The water then percolates into a carbonate rock layer, and therein becomes a closed system because it has been cut off from the atmosphere. Because the solution starts off with some C_T before dissolution of rock begins, after equilibrium with the rock, we cannot write $C_T = Ca_T$. Rather, we have

$$C_T = Ca_T + y. \tag{12.24}$$

The relation for saturation is still

$$\alpha_0^{Ca}Ca_T = \frac{K_{s0}}{\alpha_2^C C_T} \tag{12.22}$$

so

$$\alpha_0^{Ca}Ca_T = \frac{K_{s0}}{\alpha_2^C(Ca_T + y)}. \tag{12.25}$$

This yields a second order polynomial in Ca_T:

$$(Ca_T)^2 + yCa_T - \frac{K_{s0}}{\alpha_2^C\alpha_0^{Ca}} = 0 \tag{12.26}$$

Eq.(12.26) has the form $ax^2+bx+c=0$, so we can use the "quadratic equation" to solve for Ca_T:

$$x = \frac{-b \pm \sqrt{b^2 - 4ac}}{2a} \tag{12.27}$$

$$Ca_T = \frac{-y \pm \sqrt{y^2 + \dfrac{4K_{s0}}{\alpha_2^C\alpha_0^{Ca}}}}{2}. \tag{12.28}$$

Since $-y$ is always ≤ 0, and because Ca_T must be positive, there is no ambiguity as to which of the two roots we want. We must take

$$Ca_T = \frac{-y + \sqrt{y^2 + \dfrac{4K_{s0}}{\alpha_2^C\alpha_0^{Ca}}}}{2}. \tag{12.29}$$

When $y = 0$, Eq.(12.29) reduces to Eq.(12.23), as required. Eq.(12.29) gives Ca_T and $C_T = Ca_T + y$ as functions of pH. Substituting Eq.(12.24) in Eq.(12.10) gives

$$\begin{aligned}(C_B' - C_A') + \frac{2K_{s0}}{\alpha_2^C(Ca_T + y)} + \frac{K_{s0}}{\alpha_2^C(Ca_T + y)}\frac{{}^*K_{H1}}{[H^+]} + [H^+] \\ = \alpha_1^C(Ca_T + y) + 2\alpha_2^C(Ca_T + y) + \frac{K_w}{[H^+]}.\end{aligned} \tag{12.30}$$

which can be solved to obtain pH after specifying y and $(C_B' - C_A')$, and using Eq.(12.29) for Ca_T.

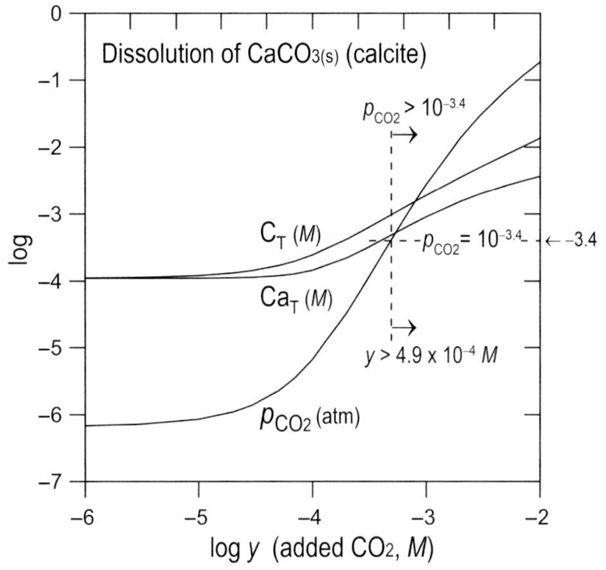

FIGURE 12.2 Log C_T, log Ca_T, and log p_{CO_2} as functions of log y (added CO_2, M) for calcite at 25 °C/1 atm. Activity corrections neglected. The value of p_{CO_2} becomes greater than $10^{-3.4}$ atm when $y > 4.9 \times 10^{-4}$ M.

12.4.2 The Effect of Added CO_2 (y) on Dissolution of Calcite When $(C'_B - C'_A) = 0$

Figure 12.2 illustrates the effect of y on the dissolution of calcite for a range of y values at 25 °C/1 atm. For $y = 10^{-6}$ M, the added dissolved CO_2 is very low and the solution chemistry is nearly the same as for dissolution of calcite into initially pure water, and log p_{CO_2} = −6.16, almost the same value as in Table 12.2. When $y = 4.9 \times 10^{-4}$ M, then log p_{CO_2} = −3.40. So, if water enters a calcite formation with $y > 4.9 \times 10^{-4}$ M, equilibrates with the rock at 25 °C/1 atm, then is opened to a gas phase with log p_{CO_2} = −3.40, some CO_2 will outgas (and some $CaCO_{3(s)}$ will precipitate). The value of y that gives log p_{CO_2} = −3.40 is temperature dependent.

Example 12.3 Solubility of Calcite in Initially Pure Water Plus Some CO_2, Case II, $C_T = Ca_T + y$, $(C'_B - C'_A) = 0$

Consider the case when water with an initial C_T from dissolved CO_2 of $y = 1.5 \times 10^{-3}$ M equilibrates in the subsurface with calcite rock at 15 °C/1 atm. For this water, $(C'_B - C'_A) = 0$. **a.** Calculate pH, Ca_T, $[Ca^{2+}]$, $[CO_3^{2-}]$, $[HCO_3^-]$, and p_{CO_2}. **b.** Why does the CO_3^{2-} that dissolves from the solid end up in the solution mostly as HCO_3^-? **c.** If this closed system solution were opened to the atmosphere with $p_{CO_2} = 10^{-3.4}$ atm, would some CO_2 enter the solution from the air, or would some CO_2 leave from the solution and go into the air?

Solution

We use the same K values for 15 °C/1 atm as in Example 12.1. **a.** We solve for the value of pH that gives LHS − RHS = 0 for Eq.(12.30) with $y = 1.5 \times 10^{-3}$ M and using Eq.(12.29) for Ca_T. Results: pH = 7.48, $[Ca^{2+}] = 1.28 \times 10^{-3}$ M (≈ Ca_T, about 12 times more than in Example 12.1); $C_T = Ca_T + y = 2.78 \times 10^{-3}$ M; $[CO_3^{2-}] = 10^{-8.43}/[Ca^{2+}] = 2.90 \times 10^{-6}$ M, $[HCO_3^-] = [CO_3^{2-}][H^+]/K_2 = 2.56 \times 10^{-3}$ M (most of C_T); and $p_{CO_2} = [HCO_3^-][H^+]/(K_1 K_H) = 10^{-2.34}$ atm. **b.** C_T is mostly HCO_3^- because the pH is about midway between pK_1 and pK_2. **c.** Since p_{CO_2} is greater than $10^{-3.4}$ atm, some CO_2 would leave the solution if it were opened to ambient air. This would cause the pH to rise, and make the solution supersaturated with $CaCO_{3(s)}$, leading to precipitation of $CaCO_{3(s)}$ (see also Example 12.4).

12.5 CASE III: OPEN SYSTEM, CONSTANT p_{CO_2}: CALCIUM CARBONATE + WATER + VARIABLE ($C'_B - C'_A$)

12.5.1 Solubility as a Function of pH

Natural waters at or near ground surface can equilibrate with both the rock and atmospheric levels of p_{CO_2} in an open system fashion. Even though limestone and dolomite rock are rather insoluble, over hundreds and thousands of years rainwater will slowly dissolve the rock, as assisted by the atmospheric CO_2 (Figures 12.3 and 12.4). Similarly, after water percolates through carbonate rock then emerges as drips from the ceilings of caves, it will tend to equilibrate with cave air at the p_{CO_2} value that is present.

In a system in which CO_2 can exchange between water and a gas phase at some p_{CO_2} value, there is no particular MBE constraining C_T and Ca_T as there was in both Cases I and II (Sections 12.3 and 12.4). There is, however, the constraint given by the K_H equilibrium:

FIGURE 12.3 Solution "epikarst" grooves ("karren") on large, steeply sloping limestone hill on Vancouver Island, British Columbia. (Photo: Paul Griffiths.)

FIGURE 12.4 Observer in blue jeans standing on gaping "epikarst" dissolution holes on limestone on Vancouver Island, British Columbia. (Photo: Paul Griffiths.)

$$CO_{2(g)} + H_2O = H_2CO_3^* \qquad K_H = \frac{[H_2CO_3^*]}{p_{CO_2}}. \tag{12.31}$$

For any situation involving solubility equilibrium of a metal carbonate, when neglecting activity corrections (or substituting ${}^cK_{s0}$ to consider them), Eq.(12.5) applies:

$$[Ca^{2+}] = \frac{K_{s0}}{[CO_3^{2-}]}. \tag{12.5}$$

Substituting for $[CO_3^{2-}]$ gives

$$[Ca^{2+}] = \frac{K_{s0}}{K_1K_2K_Hp_{CO_2}/[H^+]^2}. \tag{12.32}$$

for which $d\log[Ca^{2+}]/dpH = -2$. Therefore,

$$[CaOH^+] = [Ca^{2+}]\frac{{}^*K_{H1}}{[H^+]} = \frac{K_{s0}\,{}^*K_{H1}}{K_1K_2K_Hp_{CO_2}/[H^+]}. \tag{12.33}$$

for which $d\log[CaOH^+]/dpH = -1$. For Ca_T, from Eq.(12.32),

$$Ca_T = \frac{K_{s0}}{\alpha_0^{Ca}K_1K_2K_Hp_{CO_2}/[H^+]^2}. \tag{12.34}$$

Eq.(12.34) is a simple function of the equilibrium constants, p_{CO_2}, and $[H^+]$. Figure 12.5 illustrates the manner in which the solubility changes with pH for calcite and $FeCO_{3(s)}$ when $p_{CO_2} = 10^{-3.4}$ atm at 25 °C/1 atm. Figure 12.6 illustrates the case of calcite for three different values of p_{CO_2}, namely $10^{-4.4}$, $10^{-3.4}$, and $10^{-2.4}$ atm.

12.5.2 The Particular pH When $(C'_B - C'_A) = 0$

Neglecting activity corrections (or substituting ${}^cK$ values to consider them), Eq.(12.10) becomes

$$\begin{aligned}(C'_B - C'_A) + \frac{2K_{s0}}{K_1K_2K_Hp_{CO_2}/[H^+]^2} + \frac{K_{s0}\,{}^*K_{H1}}{K_1K_2K_Hp_{CO_2}/[H^+]} + [H^+] \\ = \frac{K_1K_Hp_{CO_2}}{[H^+]} + \frac{2K_1K_2K_Hp_{CO_2}}{[H^+]^2} + \frac{K_w}{[H^+]}.\end{aligned} \tag{12.35}$$

In each panel of Figures 12.5 and 12.6, one of the pH values is the one attained when $(C'_B - C'_A) = 0$, which, as usual, we can solve for pH once p_{CO_2} is specified. Once the pH is determined, the values of the concentrations of all the other individual species can be calculated. Table 12.4 gives the equilibrium pH values for 25 °C/1 atm when $p_{CO_2} = 10^{-4.4}$, $10^{-3.4}$, and $10^{-2.4}$ atm for calcite and aragonite, and for several other metal carbonates. (For some metal carbonates, the term for $[M(OH)_3^-]$ needs to be included in the solution.)

Table 12.5 contains the Alk values for calcite as the dissolving solid when $(C'_B - C'_A) = 0$ and $p_{CO_2} = 10^{-4.4}$, $10^{-3.4}$, and $10^{-2.4}$ atm, calculated according to Eq.(12.14), or equivalently according to $Alk = 2Ca_T$. In the absence of a basic solid like a metal carbonate, increasing CO_2 does not increase Alk. But increasing CO_2 in the presence of a carbonate solid causes Alk to increase, because more of the solid dissolves even as the pH drops. In other words, CO_2 intrinsically cannot bring in any alkalinity when it dissolves, but it does promote dissolution of any carbonate solid, which will bring

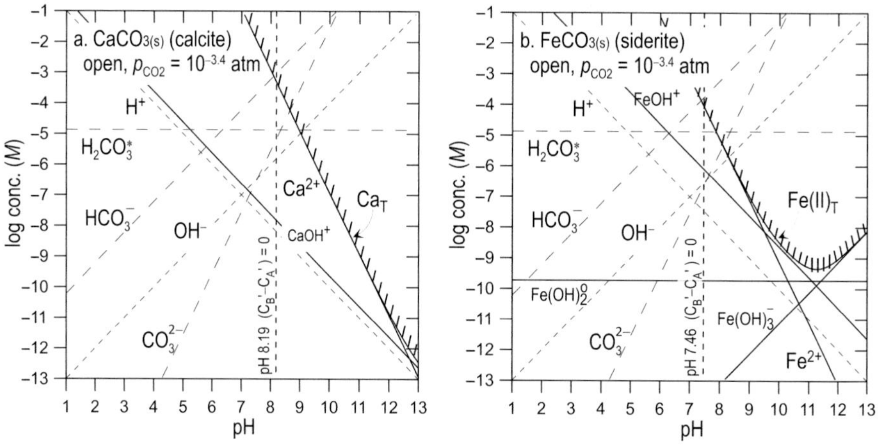

FIGURE 12.5 Log concentration (M) *vs.* pH diagrams with $p_{CO_2} = 10^{-3.4}$ atm (open system) for $CaCO_{3(s)}$ (calcite) and $FeCO_{3(s)}$ in equilibrium with water at 25 °C/1 atm. (The dihydroxo and trihydroxo complexes need to be considered for $FeCO_{3(s)}$.) Activity corrections neglected (not valid at very low or very high pH). Strong base (*e.g.*, NaOH) or strong acid (*e.g.*, HCl) is used to change the pH. In each case, a dashed vertical line gives the equilibrium composition when the metal carbonate is equilibrated with the gas phase and initially pure water, *i.e.*, when $(C'_B - C'_A) = 0$. Note the proximity of the $(C'_B - C'_A) = 0$ line to the pH at which $2[M^{2+}] \simeq [HCO_3^-]$ (approximate ENE at $(C'_B - C'_A) = 0$).

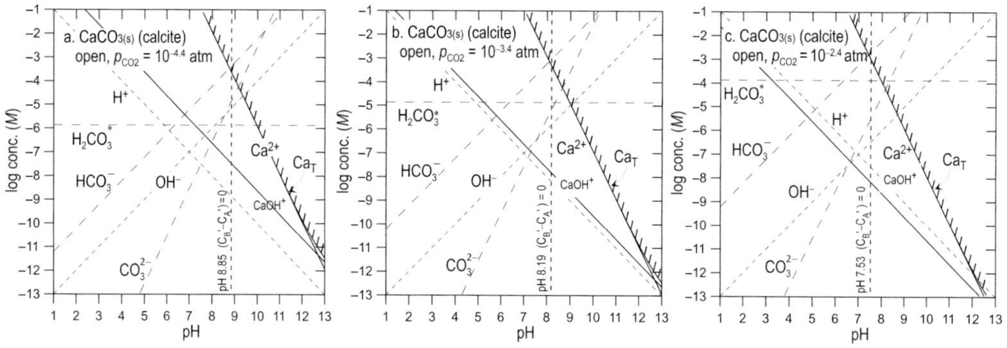

FIGURE 12.6 Log concentration (M) *vs.* pH diagrams for $CaCO_{3(s)}$ (calcite) for $p_{CO_2} = 10^{-4.4}$, $10^{-3.4}$, and $10^{-2.4}$ atm (open system) in equilibrium with water at 25 °C/1 atm. Activity corrections neglected (not valid at very low or very high pH). Strong base (*e.g.*, NaOH) or strong acid (*e.g.*, HCl) is used to change the pH. In each case, a dashed vertical line gives the equilibrium composition obtained when metal carbonate is equilibrated with the gas phase and initially pure water, *i.e.*, when $(C'_B - C'_A) = 0$. Note the proximity of each dashed vertical line to the pH at which $2[M^{2+}] \simeq [HCO_3^-]$ (the approximate ENE for the $(C'_B - C'_A) = 0$ case).

in alkalinity. Conversely, if a solution is saturated with a metal carbonate, lowering p_{CO_2} will cause the metal carbonate to precipitate and Alk for the solution will decrease (as when speleothems form in caves). Indeed, unfortunately, CO_2-exhaling visitors to karst caves can raise p_{CO_2} values in cave air significantly above undisturbed conditions, which can cause erosive redissolution of speleothems. Baker and Genty (1998) comment as follows:

TABLE 12.4
Equilibrium Constant Values for Various Divalent Metal Ions, with the Open-System pH Value for a Solution in Equilibrium of Each Metal Carbonate Solid at 25 °C/1 atm When $(C'_B - C'_A) = 0$

	log K Values				pH for $(C'_B - C'_A) = 0$ at p_{CO_2} (atm)[a]		
Metal Carbonate	log K_{s0}	log $*K_{H1}$	log $*K_{H2}$	log $*K_{H3}$	$10^{-4.4}$	$10^{-3.4}$	$10^{-2.4}$
$MgCO_{3(s)}$ (magnesite)	−7.46	−11.42	–	–	9.18	8.53	7.87
$CaCO_{3(s)}$ (aragonite)	−8.34	−12.7	–	–	8.90	8.24	7.58
$CaCO_{3(s)}$ (calcite)	−8.48	−12.7	–	–	8.85	8.19	7.53
$SrCO_{3(s)}$	−9.03	−13.18	–	–	8.67	8.01	7.35
$MnCO_{3(s)}$	−9.30	−10.6	−10.6	−13.0	8.58	7.92	7.26
$FeCO_{3(s)}$ (siderite)	−10.68	−9.5	−11.1	−11.4	8.13	7.46	6.80

Note: pH values calculated neglecting activity corrections.

[a] These p_{CO_2} values are sufficiently high that all of the metal carbonates are thermodynamically stable relative to their corresponding hydroxides.

TABLE 12.5
Open System Ca_T, Alk and pH Values for Solutions with $(C'_B - C'_A) = 0$ and in Equilibrium with $CaCO_{3(s)}$ (calcite) at 25 °C/1 atm for Three p_{CO_2} Values

	p_{CO_2} (atm)		
Parameter	$10^{-4.4}$	$10^{-3.4}$	$10^{-2.4}$
Ca_T (*M*)	0.23×10^{-3}	0.48×10^{-3}	1.02×10^{-3}
Alk (eq/L) = $2Ca_T$	0.46×10^{-3}	0.96×10^{-3}	2.05×10^{-3}
pH	8.85	8.19	7.53

It has been suggested that such elevated CO_2 may cause the destruction of speleothems within the caves, as increased CO_2 leads to a higher equilibrium concentration of calcium within the drip waters feeding the speleothems, and hence causes dissolution of existing features, although it has to be noted that there is a significant natural variability of cave air CO_2 and it is against this that the anthropogenic effects of visitors has to be judged.

Example 12.4 Equilibration of a Calcite + Water System at Fixed p_{CO_2}, Case III, $(C'_B - C'_A) = 0$

Assume that water from Example 12.3 is dripping off a cave ceiling and so equilibrates at 15 °C/1 atm with the cave gas at $p_{CO_2} = 10^{-3.2}$ atm. CO_2 will outgas from the drops during equilibration. Since $CaCO_{3(s)}$ is less soluble as p_{CO_2} decreases, some $CaCO_{3(s)}$ will precipitate. Assume that the form which precipitates is calcite. **a.** Calculate the value of Ca_T after equilibration, and ΔCa_T. **b.** If

1 FW of $CaCO_{3(s)}$ weighs ~100 g, calculate the number of liters needed to drip and equilibrate to cause the formation of a 1 metric ton stalactite.

Solution

We use the same K values for 15 °C/1 atm as in Example 12.1. **a.** For the dripping water, $(C_B' - C_A') = 0$. Solving Eq.(12.35) for pH when $p_{CO_2} = 10^{-3.2}$ atm gives pH = 8.06 and $Ca_T = 6.64\times10^{-4}$ M. In Example 12.3, $Ca_T = 1.28\times10^{-3}$ M. $\Delta Ca_T = -6.16\times10^{-4}$ M. Therefore, 6.16×10^{-4} FW of $CaCO_{3(s)}$ precipitates per liter of water that drips. At 100 g/FW, then 6.16×10^{-5} kg of $CaCO_{3(s)}$ precipitates per liter. The required volume of water to form the stalactite is calculated as 1000 kg/(6.16×10^{-5} kg/L) = 16.2 million liters.

Example 12.5 Weathering of Limestone Rock by Rain, Case III, $(C_B' - C_A') = 0$

There are many places around the world where carbonate bedrock (e.g., limestone, marble, dolomite) may be found as "outcrops" at the surface of the Earth, where it is subject to being dissolved by the action of precipitation as aided by atmospheric CO_2. Outcroppings on Vancouver Island in British Columbia, Canada, the Burren region in western Ireland, and Dent de Crolles, France are three examples. There are also many ancient man-made structures of limestone such as The Great Temple of Amman in Jordan (Paradise, 1998) that are similarly subjected to weathering. For an annual rainfall of ~1 m/y and a temperature of 15 °C, and assuming rain with $(C_B' - C_A') = 0$, what would be the velocity in cm/y at which flat calcite rock would be weathered away by dissolution? Assume equilibrium between the rock, rainwater, and atmospheric CO_2 at the pre-industrial value of p_{CO_2}(=$10^{-3.55}$ atm at sea level, 280 ppm). The FW of calcite is 100 g/FW and the density of calcite is 2.71 g/cm^3, which give 37.0 cm^3/FW.

Solution

For the sake of the calculation, assume 100 cm^2 area of rock. 1 m/y of rain places 10 L/y of water on that area. We use the same K values for 15 °C/1 atm as in Example 12.1. Solving Eq.(12.35) when $p_{CO_2} = 10^{-3.55}$ atm gives pH = 8.29, and $Ca_T = 5.09\times10^{-4}$ M. Thus, in the 10 L, 5.09×10^{-3} FW/y of $CaCO_{3(s)}$ will have dissolved (assuming equilibrium). 5.09×10^{-3} FW/y$\times$37.0 cm^3/FW = 0.19 cm^3/y. Denoting the weathering velocity as v (cm/y), then $v\times100$ cm^2 = 0.19 cm^3/y, or v = 0.0019 cm/y (or 19 cm in 10,000 y). This rough value is quite comparable to rates from field observations: 1) for exposed limestone rock in the Burren region in Ireland, Nagle and Spencer (2001) note an observed weathering velocity of 0.0008 cm/y (or 8 cm in 10,000 years); and 2) for limestone tombstones, Coleman (1981) cites data of Cann (1974) showing a dissolution velocity of 0.001 cm/y (or 10 cm in 10,000 years).

12.5.3 pH as Affected by $(C_B' - C_A') \neq 0$, as in Acid Rain

Whether $(C_B' - C_A')$ is zero or not, Eq.(12.35) is still easily solved. Note that Eq.(12.35) can be rearranged to give an explicit expression for $(C_B' - C_A')$, so for a range of conditions it is easier to consider multiple final pH values and calculate the corresponding $(C_B' - C_A')$ values, than to choose a range of $(C_B' - C_A')$ values and calculate each pH by an iteratively-based solution. Example 12.6 provides an example of the dissolution enhancement provided by a circumstance of "acid rain". Figure 12.7 illustrates the consequences of "acid rain", as enhanced by "dry deposition" of acidic aerosol particles.

FIGURE 12.7 Statue at Lichfield Cathedral, UK, that has been severely damaged by a combination of "acid rain" as well as, importantly, "dry deposition" of acidic, micron-sized and smaller atmospheric aerosol particles/droplets. (Shutterstock Stock Image ID: 2257879a.)

Example 12.6 Weathering of Limestone by "Acid Rain", Case III, $(C_B' - C_A') \neq 0$

Rain with pH values of 3 and even lower have not been uncommon in the last few decades in areas affected by sulfuric acid from the combustion of fossil fuels (especially coal) and nitric acid from vehicles. Consider rain at 15 °C with a pH of 3.5. **a.** Calculate the Alk value of that rain (which will be negative because the rain contains net mineral acidity). **b.** Equate the Alk value obtained in part a with $(C_B' - C_A')$ to determine the dissolution velocity of limestone (assuming equilibrium as in Example 12.5) for rainfall of ~1 m/y and a temperature of 15 °C. The FW of calcite is 100 g/FW and the density of calcite is 2.71 g/cm³, which give 37.0 cm³/FW. To simplify comparison with Example 12.5, assume equilibrium between the rock, rainwater, and atmospheric CO_2 at the pre-industrial sea-level value of $p_{CO_2} = 10^{-3.55}$.

Solution

a. We use the same K values for 15 °C/1 atm as in Example 12.1. If pH = 3.5, then Alk will be nearly exactly $-10^{-3.0}$ eq/L. Prior to the dissolution of any calcite, $(C_B' - C_A')$ = Alk = $-10^{-3.0}$ eq/L. For the sake of the calculation, assume 100 cm² area of rock. 1 m/y of rain places 10 L/y of water on that area. For $(C_B' - C_A') = -10^{-3.0}$ eq/L, solving Eq.(12.35) when $p_{CO_2} = 10^{-3.55}$ atm gives pH = 8.17 and $Ca_T = 8.85 \times 10^{-4}$ *M*. Thus, in the 10 L, 8.85×10^{-3} FW/y of $CaCO_{3(s)}$ will have dissolved (assuming equilibrium). 8.85×10^{-3} FW/y × 37.0 cm³/FW = 0.33 cm³/y. Denoting the weathering velocity as v (cm/y), then $v \times 100$ cm² = 0.33 cm³/y, or v = 0.0033 cm/y. (The current $p_{CO_2} = 10^{-3.40}$ gives pH = 8.08 and v = 0.0035 cm/y.)

REFERENCES

Baker A, Genty D (1998) Environmental pressures on conserving cave speleothems: effects of changing surface land use and increased cave tourism. *Journal of Environmental Management*, **53**, 165–175.

Cann JH (1974) A field investigation into rock weathering and soil forming processes. *Journal of Geological Education*, **22**, 226–230.

Coleman SM (1981) Rock-weathering rates as functions of time. *Quaternary Research*, **15**, 250–264.

Klimchouck A (1996) The dissolution and conversion of gypsum and anyhydrite. *International Journal of Speleology*, **25**, 21–36.

Nagle G, Spencer K (2001) *AS and A Level Geography Through Diagrams*, 3rd edn, Oxford University Press, Oxford, UK, p.16.

Paradise TR (1998) Limestone weathering and rate variability, Great Temple of Amman, Jordan. *Physical Geography*, **19**, 133–146.

Plummer LN, Busenberg E (1992) The solubilities of calcite, aragonite and vaterite in CO_2-H_2O solutions between 0 and 90 °C, and an evaluation of the aqueous model for the system $CaCO_3$-CO_2-H_2O. *Geochimica et Cosmochimica Acta*, **46**, 1011–1040.

13 Metal Phosphates

13.1 GENERAL

From Coulomb's Law, the higher the charge on some ion, the greater the force of attraction to an oppositely charged ion (a.k.a "counterion"). As a result, as the magnitude of the charge on the ion increases, the greater the likely stability of a solid formed with a given counterion. For Na^+, with Cl^- in $NaCl_{(s)}$ (halite), the electrostatic forces holding the lattice together are not very strong; this +1/−1 mineral is rather soluble (Chapter 10). (The low solubility of $AgCl_{(s)}$ (chlorargyrite), another +1/−1 mineral, is provided by *covalent* interactions between Ag^+ and Cl^-.) For Na^+ with SO_4^{2-} in $NaSO_4^{2-}\cdot(H_2O)_{10(s)}$ (mirabilite), the electrostatic forces holding the lattice together are much stronger; this +1/−2 mineral is less soluble than $NaCl_{(s)}$ (Chapter 10). For a higher charge cation, with Ca^{2+} combining with SO_4^{2-} in $CaSO_4{\cdot}2H_2O_{(s)}$ (gypsum, a +2/−2 mineral, Chapter 10), and Ca^{2+} with CO_3^{2-} in $CaCO_{3(s)}$ (calcite, another +2/−2 mineral, Chapter 11), the solubilities are lower than for mirabilite. Further, for Fe^{3+} in oxide and oxyhydroxide solids, the solubilities are quite low (with the consequence that dissolved Fe(III) concentrations are very low in natural waters at typical pH values, Chapter 10).

By the above discussion: (1) with a few exceptions (like $AgCl_{(s)}$), +1/−1 minerals are generally of high solubility; and (2) all solids that are +2 and +3 cation combinations with −2 and −3 anions will unquestionably be of low, if not exceedingly low, solubility. Among −3 anions, we have phosphate, PO_4^{3-}, a critically important aquatic/environmental nutrient and component of teeth, bones, and antlers. Solids of phosphate with Fe^{3+}, Al^{3+}, Fe^{2+}, Pb^{2+}, Zn^{2+}, Cd^{2+}, Zn^{2+}, Mg^{2+}, etc. are all of low to exceedingly low solubility. As result, mineral phosphates are important in many different environmental settings. Table 13.1 provides a summary.

SO_4^{2-} and CO_3^{2-}(in combination with its identity alters HCO_3^- and $H_2CO_3^*$) can be major species in natural waters and wastewaters. The same is not true for PO_4^{3-}, a.k.a "orthophosphate" with its alters HPO_4^{2-}, $H_2PO_4^-$, and H_3PO_4. Except as in some highly polluted surface waters and wastewaters, aquatic total orthophosphate values are typically <<1 mg/L as P, *i.e.*, <<3 mg/L as PO_4^{3-} (Domagalski and Johnson, 2012). Therefore, an ENE-based mathematical analysis of solid dissolution wherein all the dissolving ions from the solid may play important roles in the ENE (*e.g.*, Eq.(10.52) for (am)$FeOH_{3(s)}$, and Eq.(12.3) for calcium carbonate) is often not the best framework for presenting cases involving the dissolution of phosphate solids. It is, nevertheless, relevant in some cases, and provides pedagogical connection to Chapters 10 and 11. We use it in Sections 13.2 and 13.3 in consideration of the dissolution of natural hydroxyapatite and fluoroapatite in aquifers in such rock. For consideration of the precipitation/dissolution of struvite in waste treatment systems (Section 13.4), the ENE is not pertinent as a constraint.

13.2 HYDROXYAPATITE – $Ca_5(PO_4)_3(OH)_{(S)}$

For hydroxyapatite,

$$Ca_5(PO_4)_3OH_{(s)} = 5Ca^{2+} + 3PO_4^{3-} + OH^- \qquad K_{s0}^{HA} = \{Ca^{2+}\}^5\{PO_4^{3-}\}^3\{OH^-\}. \qquad (13.1)$$

For fluoroapatite,

TABLE 13.1
Situations Pertaining of Metal Phosphate Solids/Minerals in Environmental Settings and Human/Animal Physiological Systems

System	
A. Rocks/aquifers	The sedimentary mineral "phosphorite" contains various forms of apatite $Ca_5(PO_4)_3X_{(s)}$: hydroxyapatite, X = OH^-; fluoroapatite, X = F^-; chloroapatite, X = Cl^- (*e.g.*, see Dorozhkin, 2014).
B. Surface water, sediments, and soils	Control of low-level phosphate concentrations in natural water systems was thought for many years to be exclusively a simple matter of mineral solubility. The only question was, which phosphate solid is controlling in given system? *e.g.*, some iron (III) phosphate (Fox, 1989) or some calcium phosphate (Stumm and Morgan, 1996)? This can be the correct view in some circumstances. However, because of its high affinity for the mineral surfaces, phosphate will often simply *adsorb* on (and be solubility-controlled by) solid surfaces such as with (am)$Fe(OH)_{3(s)}$ and α-$FeOOH_{(s)}$ (goethite) (Barrow, 1983). If the phosphate begins to diffuse/migrate into the solid and replace O^{2-} or OH^-, then one may then be able to speak of formation of a phosphate solid, or at least a solid solution of a phosphate solid. Overall, however, controlling chemical processes in surface waters, sediments, and soils are significantly more complicated than can be described by simple application of a K_{s0} (Barrow, 1983; Barrow, 2015; Jarvis, 2015; Staunton et al., 2015). This limits applicability of the K_{s0} approach used in this chapter to the types of clear cases of mineral solubility considered here.
C. Contaminated soils and groundwaters	"Heavy metal" ions like Pb^{2+} and Cd^{2+} in contaminated moist soils can be precipitated and thus immobilized by addition of phosphate (*e.g.*, Stanforth and Qiu, 2001), using addition of sodium phosphate solutions and $Ca(H_2PO_4)_{(s)}$ fertilizer ("triple super phosphate", TSP); Cao et al. (2002), using solutions of H_3PO_4 combined with either $Ca(H_2PO_4)_2$ or with phosphate rock ("mainly of $Ca_{10}(PO_4)_6F_2$ [fluoroapatite] with substantial CO_3^{2-} substitution in the structure"); Thawornchaisit and Polprasert (2009) using addition of TSP, "diammonium phosphate" (DAP) ($(NH_4)_2HPO4_{(s)}$, or phosphate rock (hydroxy/fluoroapatite)). Wright et al. (2011) investigated the efficacy of forming calcium phosphate solids for co-precipitation/scavenging of dissolved metal contaminants in Ca^{2+}-containing groundwater by addition of NaH_2PO_4 solutions.
D. Municipal wastewater treatment plants	In municipal waste water treatment plants, $MgNH_4PO_4{\cdot}2H_2O_{(s)}$ (struvite, see Section 13.2) can precipitate in pumps, pipes, and valves, leading to costly repairs. However, by addition of some process chemistry, struvite precipitation can be controlled and the material recovered for beneficial use as an agricultural nutrient (*e.g.*, see Booker et al., 1999; Grooms et al., 2015).
E. *In vivo* waste processing (*e.g.*, kidneys) organs in humans and other animals	Struvite and calcium phosphates can form in human and animal waste treatment "plants" (*e.g.*, the kidneys), leading to kidney stones (*e.g.*, see Burns and Finlayson, 1982). Struvite stones are associated with urinary tract infections because bacteria can produce urease which hydrolyzes urea to ammonia, potentiating formation of struvite stones (Bichlera et al., 2002).
F. Teeth, bones, and antlers	$Ca_5(PO_4)_3OH_{(s)}$ (hydroxyapatite) and $Ca_5(PO_4)_3F_{(s)}$ (fluoroapatite) provide the mechanical strength needed in teeth, bones, and antlers (*e.g.*, ten Cate et al., 2009; Bonjour, 2011; Dorozhkin, 2014); substitution of F^- for OH^- converts hydroxyapatite to fluoroapatite, which greatly reduces the solubility of tooth enamel under acidic conditions and is highly effective in reducing caries (see Section 13.3).

$$Ca_5(PO_4)_3F_{(s)} = 5Ca^{2+} + 3PO_4^{3-} + F^- \qquad K_{s0}^{FA} = \{Ca^{2+}\}^5\{PO_4^{3-}\}^3\{F^-\}. \tag{13.2}$$

Here, and in the subsequent cases in this chapter, the expression used for K_{s0} assumes that the solid is pure, namely mole fraction = 1.

In hydroxyapatite, the repeating unit in the crystal lattice is actually $Ca_{10}(PO_4)_6(OH)_2$, and analogously for fluoroapatite. Some treatments of the solubilities of these two solids therefore choose to write the dissolution reactions with $Ca_{10}(PO_4)_6(OH)_{2(s)}$ and $Ca_{10}(PO_4)_6F_{2(s)}$, with the corresponding K_{s0} expressions and values being the squares of those given above. The computed chemistry is not affected by this choice: consider that representing the dissociation of water as $2H_2O = 2H^+ + 2OH^-$ with $K = (1.01 \times 10^{-14})^2$ at 25 °C/1 atm will give pH = 7 for pure water just as with $H_2O = H^+ + OH^-$ with $K = 1.01 \times 10^{-14}$.

Estimates of the $\log K_{s0}$ values found in the literature for particular apatites vary by several orders of magnitude, being dependent not only on the degree of purity/crystallinity, but also undoubtedly the adequacy of the experimental methods used. For example, for fluoroapatite at 37 °C, values of $\log K_{s0}$ that have been reported are −55.82 (Zhu et al., 2009); −60.15 (McCann, 1968); and −60.50 (Moreno et al., 1974). The value extrapolated to 37 °C from Ball and Nordstrom (1991/2001) is −55.37. The problem here is that whenever a solid is rather insoluble, equilibrium solid/solution equilibrium tends to be attained only slowly. *And*, since many different calcium phosphates have very low solubilities (Stumm and Morgan, 1996), subtle and difficult-to-detect mineral interconversions at the mineral/water interface can occur over the course of an experimental effort, as with "incongruent" dissolution, *i.e.*, dissolution of one solid with subsequent formation of another solid (see Shellis, 1996, for a discussion of these problems in the study of the solubilities of hydroxyapatite-related phases in teeth). Here we use the values in Table 13.2, but the reader should always remember that while calculated results can be obtained with unlimited precision (as with the pH values for the condition $(C'_B - C'_A) = 0$) in Figures 13.1 and 13.2, all such results are subject to significant caveats related to the existence of equilibrium and the correctness of the values of the K values utilized.

While there may be uncertainty regarding true, pure mineral K_{s0} values, all available evidence indicates that the values at 25 °C for crystalline hydroxyapatite and fluoroapatite are similar, and that the same is true at 37 °C. Therefore, the much lower solubility of fluoropatite *vs.* apatite at low pH (see Figure 13.3 below) is due mostly to the lower acidity of aqueous F^- as compared to aqueous OH^-. In particular, F^- reacts with available H^+ much less readily than does OH^-. Thus, fluoroapatite has less "willingness" to yield F^- to an acidic solution than hydroxyapatite has to yield OH^-. This is why low levels of fluoride in drinking water and relatively higher levels in dental care products are stunningly beneficial in preventing dental caries (Ripa, 1993). This author acquired many cavities as a child and teen; my children acquired none. Upon fluoridation, F^- exchanges for OH^- in tooth mineral according to F^- + hydroxyapatite = OH^- + fluoroapatite. In fact, because the hydroxyapatite structure easily accommodates F^-, it has been argued that the whole range of "solid solutions" of hydroxyapatite with fluoroapatite is possible, according to the formula $Ca_5(PO_4)_3OH_{1-x}F_{x(s)}$ (Moreno et al., 1974) where x can vary between 0 and 1.

Ca^{2+} can form complexes with phosphate species and with OH^-, and this will affect the solubility of calcium phosphate solids. Effects of metal complexes are also considered in prior chapters (for OH^- in Chapters 10 and 11, and for SO_4^{2-} in Chapter 11, Box 11.1). For phosphate complexes, we have

$$Ca^{2+} + H_2PO_4^- = CaH_2PO_4^+ \qquad K_{CaH_2PO_4^+} = K^+ = \frac{\{CaH_2PO_4^+\}}{\{Ca^{2+}\}\{H_2PO_4^-\}} = 10^{1.41} \tag{13.3}$$

$$Ca^{2+} + HPO_4^{2-} = CaHPO_4^o \qquad K_{CaHPO_4^o} = K^o = \frac{\{CaHPO_4^o\}}{\{Ca^{2+}\}\{HPO_4^{2-}\}} = 10^{2.74} \tag{13.4}$$

$$Ca^{2+} + PO_4^{3-} = CaPO_4^- \qquad K_{CaPO_4^-} = K^- = \frac{\{CaPO_4^-\}}{\{Ca^{2+}\}\{PO_4^{3-}\}} = 10^{6.46}. \tag{13.5}$$

TABLE 13.2
Selected Equilibrium Constants at 25 °C/1 atm and 37 °C/1 atm Relevant for Some Phosphate Minerals

		log K_{s0}	
Solubility		**25 °C**	**37 °C**
hydroxyapatite	$Ca_5(PO_4)_3OH_{(s)} = 5Ca^{2+} + 3PO_4^{3-} + OH^- \quad (K_{s0}^{HA})$	−54.92[a]	−55.27[a]
fluorapatite	$Ca_5(PO_4)_3F_{(s)} = 5Ca^{2+} + 3PO_4^{3-} + F^- \quad (K_{s0}^{FA})$	−55.10[a]	−55.37[a]
struvite	$MgNH_4PO_4 \cdot (H_2O)_{6(s)} = Mg^{2+} + NH_4^+ + PO_4^{3-} + 6H_2O \quad (K_{s0}^{Struv})$	−13.17[b]	−13.06[b]
Ca struvite[c]	$CaNH_4PO_4 \cdot (H_2O)_{n(s)} = Ca^{2+} + NH_4^+ + PO_4^{3-} + nH_2O \quad (K_{s0}^{CaStruv})$	unknown	unknown
Complexation		**log K**	
$Ca^{2+} + H_2PO_4^- = CaH_2PO_4^+ \quad (K_{Ca}^+)$		1.41[e]	1.50[e]
$Ca^{2+} + HPO_4^{2-} = CaHPO_4^o \quad (K_{Ca}^o)$		2.74[e]	2.83[e]
$Ca^{2+} + PO_4^{3-} = CaPO_4^- \quad (K_{Ca}^-)$		6.46[e]	6.54[e]
$Mg^{2+} + H_2PO_4^- = MgH_2PO_4^+ \quad (K_{Mg}^+)$		1.51[a]	1.61[a]
$Mg^{2+} + HPO_4^{2-} = MgHPO_4^o \quad (K_{Mg}^o)$		2.87[a]	2.96[a]
$Mg^{2+} + PO_4^{3-} = MgPO_4^- \quad (K_{Mg}^-)$		6.59[a]	6.68[a]
$Ca^{2+} + H_2O = CaOH^+ + H^+ \quad ({}^*K_{H1,Ca})$		12.7	12.25
$Mg^{2+} + H_2O = MgOH^+ + H^+ \quad ({}^*K_{H1,Mg})$		11.42	10.97
		pK_a	
Acid/Base		**25 °C**	**37 °C**
$H_3PO_4 = H^+ + H_2PO_4^- \quad (K_1^P)$		2.16[f]	2.32[f]
$H_2PO_4^- = H^+ + HPO_4^{2-} \quad (K_2^P)$		7.20[f]	7.11[f]
$HPO_4^{2-} = H^+ + PO_4^{3-} \quad (K_3^P)$		12.32[f]	12.18[f]
$NH_4^+ = H^+ + NH_3 \quad (K_a^N)$		9.24[g]	8.89[g]
$HF = H^+ + F^- \quad (K_a^F)$		3.18[f]	3.29[f]
$Ca^{2+} + H_2O = CaOH^+ + H^+ \quad ({}^*K_{H1,Ca})$		12.7	12.25
$Mg^{2+} + H_2O = MgOH^+ + H^+ \quad ({}^*K_{H1,Mg})$		11.42	10.97

[a] Ball and Nordstrom (1991/2001).
[b] Hanhoun et al. (2011); value at 37 °C is extrapolated.
[c] "Ca struvite" = Ca^{2+} analog of struvite. Very little is known about phosphate solids with Ca^{2+}+NH_4^+, including the number of molecules of water n that may be present in such solids.
[d] Fordham and Schwertmann (1977).
[e] Chughtai et al. (1968).
[f] Perrin (1982).
[g] Emerson et al. (1975); value at 37 °C is extrapolated.

Note the use of the abbreviated symbols K^+, K°, and K^-; the values are for 25 °C/1 atm. For complexation with OH^-, with the value given for 25 °C/1 atm,

$$Ca^{2+} + H_2O = CaOH^+ + H^+ \qquad {}^*K_{H1,Ca} = \frac{\{CaOH^+\}\{H^+\}}{\{Ca^{2+}\}} = 10^{-12.7} \tag{13.6}$$

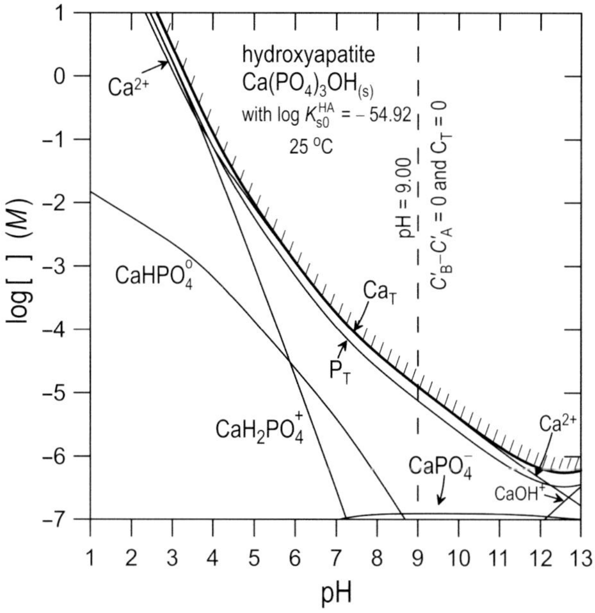

FIGURE 13.1 Solubility of hydroxyapatite *vs.* pH at 25 °C for the condition $Ca_T = \frac{5}{3}P_T$ (as with any case involving the solid + initially pure water + $(C'_B - C'_A) + C_T$).

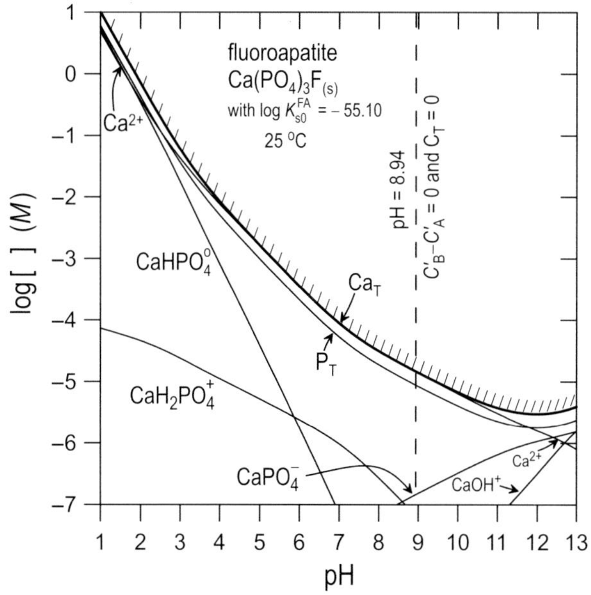

FIGURE 13.2 Solubility of fluoroapatite *vs.* pH at 25 °C for the conditions $Ca_T = \frac{5}{3}P_T$ and $F_T = \frac{1}{3}P_T$ (as with any case involving the solid + initially pure water + $(C'_B - C'_A) + C_T$).

In solution then, both calcium and phosphate can exist as multiple species. At saturation equilibrium for hydroxyapatite, neglecting activity corrections,

$$K_{s0}^{HA} = [Ca^{2+}][PO_4^{3-}][OH^-] = (\alpha^{Ca^{2+}} Ca_T)(\alpha_3^{P} P_T)[OH^-]. \tag{13.7}$$

The quantity $\alpha^{Ca^{2+}}$ is distinct from α_0^{Ca} as used in Chapter 12 (see below).

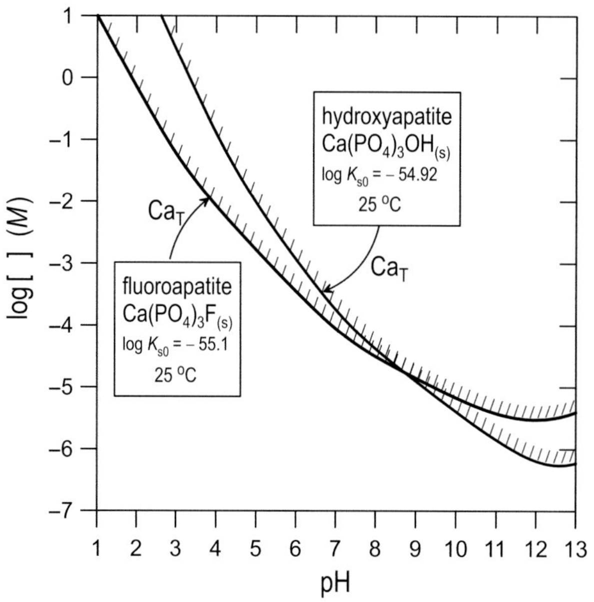

FIGURE 13.3 Total Ca_T solubilities of hydroxyapatite and fluoroapatite *vs.* pH at 25 °C. Conditions: for both solids, solid + initially pure water + $(C'_B - C'_A) + C_T$) so that $Ca_T = \frac{5}{3}P_T$, and also for fluoroapatite that $F_T = \frac{1}{3}P_T$.

For PO_4^{3-}, assuming there is little Mg^{2+} in the system to form complexes with phosphate,

$$\alpha^{PO_4^{3-}} \equiv \frac{[PO_4^{3-}]}{[H_3PO_4]+[H_2PO_4^-]+[HPO_4^{2-}]+[PO_4^{3-}]+[CaH_2PO_4^+]+[CaHPO_4^o]+[CaPO_4^-]} \quad (13.8)$$

$$= \frac{1}{\frac{[H_3PO_4]}{[PO_4^{3-}]}+\frac{[H_2PO_4^-]}{[PO_4^{3-}]}+\frac{[HPO_4^{2-}]}{[PO_4^{3-}]}+1+\frac{[CaH_2PO_4^+]}{[PO_4^{3-}]}+\frac{[CaHPO_4^o]}{[PO_4^{3-}]}+\frac{[CaPO_4^-]}{[PO_4^{3-}]}}. \quad (13.9)$$

Neglecting activity corrections,

$$\alpha^{PO_4^{3-}} = \frac{1}{\left(\frac{[H_3PO_4]}{[PO_4^{3-}]}+\frac{[H_2PO_4^-]}{[PO_4^{3-}]}+\frac{[HPO_4^{2-}]}{[PO_4^{3-}]}+1+\frac{K_{Ca}^{+}[Ca^{2+}][H_2PO_4^-]}{[PO_4^{3-}]} + \frac{K_{Ca}^{o}[Ca^{2+}][HPO_4^{2-}]}{[PO_4^{3-}]}+\frac{K_{Ca}^{-}[Ca^{2+}][PO_4^{3-}]}{[PO_4^{3-}]}\right)} \quad (13.10)$$

$$= \frac{1}{\frac{[H^+]^3}{K_1^P K_2^P K_3^P}+\frac{[H^+]^2}{K_2^P K_3^P}+\frac{[H^+]}{K_3^P}+1+[Ca^{2+}]\left(K_{Ca}^{+}\frac{[H^+]^2}{K_2^P K_3^P}+K_{Ca}^{o}\frac{[H^+]}{K_3^P}+K_{Ca}^{-}\right)}. \quad (13.11)$$

At 25 °C/1 atm, $pK_1^P = 2.16$, $pK_2^P = 7.20$, and $pK_3^P = 12.32$.

In Chapter 11, for Ca^{2+} the only complexing agent was H_2O ($Ca^{2+} + H_2O = CaOH^+ + H^+$) and so we used the purely acid/base analog that α_0 for the fraction of Ca_T in solution that is Ca^{2+}. Here, complexes with phosphate species are also present, so we introduce $\alpha^{Ca^{2+}}$ with

$$\alpha^{Ca^{2+}} \equiv \frac{[Ca^{2+}]}{[Ca^{2+}]+[CaOH^{+}]+[CaH_2PO_4^{+}]+[CaHPO_4^{o}]+[CaPO_4^{-}]} \quad (13.12)$$

$$= \frac{1}{1+\frac{[CaOH^{+}]}{[Ca^{2+}]}+\frac{[CaH_2PO_4^{+}]}{[Ca^{2+}]}+\frac{[CaHPO_4^{o}]}{[Ca^{2+}]}+\frac{[CaPO_4^{-}]}{[Ca^{2+}]}}. \quad (13.13)$$

Neglecting activity corrections,

$$\alpha^{Ca^{2+}} = \frac{1}{1+\frac{{}^{*}K_{H1,Ca}}{[H^{+}]}+K_{Ca}^{+}[H_2PO_4^{-}]+K_{Ca}^{o}[HPO_4^{2-}]+K_{Ca}^{-}[PO_4^{-}]} \quad (13.14)$$

$$= \frac{1}{1+\frac{{}^{*}K_{H1,Ca}}{[H^{+}]}+[PO_4^{-}]\left(K_{Ca}^{+}\frac{[H^{+}]^2}{K_2^PK_3^P}+K_{Ca}^{o}\frac{[H^{+}]}{K_3^P}+K_{Ca}^{-}\right)}. \quad (13.15)$$

Consider now the system hydroxyapatite + initially pure water + $(C_B' - C_A')$ + C_T; the amount of hydroxyapatite that dissolves and the pH depend on $(C_B' - C_A')$ and C_T. The MBE requirement is

$$Ca_T = \tfrac{5}{3}P_T. \quad (13.16)$$

Both Ca_T and P_T are measures of the total solubility. By Eq.(13.2), neglecting activity corrections,

$$K_{s0}^{HA} = [Ca^{2+}]^5[PO_4^{3-}]^3[OH^{-}] = (\alpha^{Ca^{2+}}Ca_T)^5(\alpha^{PO_4^{3-}}P_T)^3[OH^{-}]. \quad (13.17)$$

By Eq.(13.16),

$$P_T = \sqrt[8]{\frac{K_{s0}^{HA}}{(\frac{5}{3}\alpha^{Ca^{2+}})^5(\alpha^{PO_4^{3-}})^3(K_w/[H^{+}])}}. \quad (13.18)$$

Eq.(13.18) is a close conceptual relative of Eq.(12.23) for dissolution of $CaCO_{3(s)}$, which uses the acid/base α value that is α_0^{Ca} for Ca^{2+}. Here we use the symbol $\alpha^{Ca^{2+}}$ because the fraction $[Ca^{2+}]/Ca_T$ depends not only on pH, but also on $[PO_4^{3-}]$; and, $\alpha^{PO_4^{3-}}$ depends on pH and $[Ca^{2+}]$. So, to calculate the dissolved P_T $(= \frac{3}{5}Ca_T)$ at a given pH, one needs $[PO_4^{3-}]$ and $[Ca^{2+}]$ which are not known until the problem is solved, or at least iterated towards the solution. A numerical solution can be obtained in an overarching geochemical equilibrium algorithm such as MINEQL+. However, this text seeks to frame as many problems as possible for solution using spreadsheets, inasmuch as: (1) general algorithms have their own significant use-proficiency overhead; (2) most problems of interest are not so complicated that they require an algorithm such as MINEQL+; and (3) it is highly useful to develop the skills required to set up one's own *ad hoc* iterative schemes, hence the detail provided here.

The solubility of hydroxyapatite can be determined for each pH of interest in a spreadsheet according to the following steps.

1) For each pH row, a cell is established to initiate a guess value $\log[PO_4^{3-}]_o$. (Note: we guess and iterate on $\log[PO_4^{3-}]_o$ rather than $[PO_4^{3-}]_o$ so that the iterations can move more easily through many orders of magnitude in $[PO_4^{3-}]$.)
2) With $\log[PO_4^{3-}]_o$, K_{s0}^{HA} is used to compute a corresponding value of $[Ca^{2+}]$.

3) $[PO_4^{3-}]_o$ is used with Eq.(13.15) to estimate $\alpha^{Ca^{2+}}$, and $[Ca^{2+}]$ is used with Eq.(13.11) to estimate $\alpha^{PO_4^{3-}}$.
4) The estimates of $\alpha^{Ca^{2+}}$ and $\alpha^{PO_4^{3-}}$ are used to estimate P_T by Eq.(13.18).
5) A new estimate $\log[PO_4^{3-}]_1 = \log(\alpha^{PO_4^{3-}} P_T)$ is computed, but will not be identical to the preceding estimate of $\log[PO_4^{3-}]_o$ until the iteration has converged. The difference comparison that is $|\log[PO_4^{3-}]_1 - \log[PO_4^{3-}]_o|$ is computed, and Solver in Excel is invoked to vary $\log[PO_4^{3-}]_o$ until $|\log[PO_4^{3-}]_1 - \log[PO_4^{3-}]_o| = 0$, with ≤ 0.01 being adequate as the convergence criterion.

Iteration for a range of pH values can be carried out simultaneously (*en masse*) in Excel by placing the sum $\sum(|\log[PO_4^{3-}]_1 - \log[PO_4^{3-}]_o|)$ for the pH range in a particular cell, then invoking Solver with the request that the *sum* be brought close to 0 by varying the *range of cells* holding $\log[PO_4^{3-}]_o$ values. (Caveat: completely unreasonable initial guess values will not help Solver succeed.) For a pH range of 1 to 13 in steps of 0.1, there are 121 elements in the sum. If the convergence criterion is 1, then each $|\log[PO_4^{3-}]_1 - \log[PO_4^{3-}]_o|$ can be brought to <0.01. Unnecessarily taxing Solver by asking that the sum be brought to say <0.001 will only require more time and iterations and may cause the Solver scheme to quit before convergence is attained. Second, note that if absolute values were not used in the sum, negative differences would cancel out positive differences, leading to a the possibility that the sum $\sum(\log[PO_4^{3-}]_1 - \log[PO_4^{3-}]_o)$ could be made close to zero even when some of the components are not individually adequately small. Figure 13.1 is plot of the solubility of hydroxyapatite *vs.* pH. Which pH is extant in a system depends on $(C_B' - C_A')$ and C_T. Note the rapid increase in solubility for increasingly acidic solutions. A detailed consideration of the saturation state of hydroxyapatite in groundwaters in the Floridian aquifer system in Florida, Georgia, South Carolina, and Alabama is provided by Sprinkle (1989).

As usual, $(C_B' - C_A')$ is the component of Alk due to [strong base of the NaOH type] minus [strong acid of the HCl type], and C_T is the dissolved carbonate carbon, *e.g.*, as from CO_2 dissolved in rain. The solubility of hydroxyapatite for particular case values of $(C_B' - C_A')$ and C_T can be computed using the ENE:

$$\begin{aligned} &2[Ca^{2+}] + [CaOH^+] + [CaH_2PO_4^+] + [H^+] + (C_B' - C_A') \\ &\quad = [H_2PO_4^-] + 2[HPO_4^{2-}] + 3[PO_4^{3-}] \\ &\quad + [CaPO_4^-] + \alpha_1^C C_T + 2\alpha_2^C C_T + [OH^-] \end{aligned} \tag{13.19}$$

or

$$\begin{aligned} &\left([H_2PO_4^-] + 2[HPO_4^{2-}] + 3[PO_4^{3-}] + [CaPO_4^-] + [OH^-]\right) \\ &- \left(2[Ca^{2+}] + [CaOH^+] + [CaH_2PO_4^+] + [H^+]\right) \\ &= (C_B' - C_A') - \left(\alpha_1^C C_T + 2\alpha_2^C C_T\right) \end{aligned} \tag{13.20}$$

or

$$\left(\alpha^{PO_4^{3-}}P_T\left(\frac{[H^+]^2}{K_2^P K_3^P}+2\frac{[H^+]}{K_3^P}+3+\left(\alpha^{Ca^{2+}}\tfrac{5}{3}P_T\right)K_{Ca}^-\right)+\frac{K_w}{[H^+]}\right)$$

$$-\left(\alpha^{Ca^{2+}}\tfrac{5}{3}P_T\left[2+\frac{{}^*K_{H1}}{[H^+]}+\alpha^{PO_4^{3-}}P_T\frac{[H^+]^2}{K_2^P K_3^P}K_{Ca}^+\right]+[H^+]\right) \tag{13.21}$$

$$=(C_B'-C_A')-\left(\alpha_1^C C_T+2\alpha_2^C C_T\right).$$

For each pH in the spreadsheet, both the LHS and RHS of Eq.(13.20/13.21) can be computed. For specific values of $(C_B' - C_A')$ and C_T, the task is simply to search for the pH in the spreadsheet (with interpolation) that gives LHS = RHS. This second search could be carried out numerically, but for our purposes that level of additional complexity is perhaps not worth the effort.

Example 13.1 Solubility of Hydroxyapatite in Initially Pure Water

Initially pure water is equilibrated with hydroxyapatite at 25 °C/1 atm. At equilibrium, what is the pH, P_T, and Ca_T? Neglect activity corrections.

Solution

$(C_B' - C_A') = 0$ and $C_T = 0$. We want the pH at which the LHS of Eq.(13.21) is zero. In spreadsheet output for the problem, this occurs at pH = 9.00, as indicated in Figure 13.1; $P_T = 7.52\times10^{-6}$ *M*; $Ca_T = \tfrac{5}{3}P_T = 1.25\times10^{-5}$ *M*.

Example 13.2 Solubility of Hydroxyapatite in Infiltrated Pristine Rain Water

Pristine rain water ($(C_B' - C_A') = 0$) with $C_T = 10^{-5}$ *M* infiltrates into and equilibrates in an aquifer containing hydroxyapatite at 25 °C/1 atm. At equilibrium, what is the pH, P_T, and Ca_T? Neglect activity corrections.

Solution

For this case, $(C_B' - C_A') = 0$, and $C_T = 10^{-5}$ *M*. We thus want the pH at which the LHS of Eq.(13.21) equals $\left(\alpha_1^C+2\alpha_2^C\right)10^{-5}$. In spreadsheet output for the problem, this occurs between pH = 8.6 and 8.7. Using interpolation, pH = 8.67, $P_T = 1.11\times10^{-5}$ *M*, and $Ca_T = \tfrac{5}{3}P_T = 1.85\times10^{-5}$ *M*.

13.3 FLUOROAPATITE – $Ca_5(PO_4)_3F_{(S)}$

For fluoroapatite,

$$K_{s0}^{FA}=[Ca^{2+}]^5[PO_4^{3-}]^3[F^-]=(\alpha^{Ca^{2+}}Ca_T)^5(\alpha^{PO_4^{3-}}P_T)^3(\alpha_1^F F_T). \tag{13.22}$$

Since F^- does not interact with PO_4^{3-}, nor does it appreciably complex Ca^{2+}, Eq.(13.11) for $\alpha^{PO_4^{3-}}$ and Eq.(13.15) for $\alpha^{Ca^{2+}}$ remain applicable here. As with hydroxyapatite,

$$Ca_T = \tfrac{5}{3}P_T. \tag{13.16}$$

For fluoroapatite, we also have

$$\mathrm{F_T} = \tfrac{1}{3}\mathrm{P_T} \tag{13.23}$$

with

$$[\mathrm{F^-}] = \alpha_1^{\mathrm{F}}\mathrm{F_T}. \tag{13.24}$$

Neglecting activity corrections,

$$\alpha_1^{\mathrm{F}} \equiv \frac{[\mathrm{F^-}]}{[\mathrm{F^-}]+[\mathrm{HF}]} = \frac{K_a^{\mathrm{F}}}{[\mathrm{H^+}]+K_a^{\mathrm{F}}}. \tag{13.25}$$

By Eqs.(13.16), (13.22), (13.23), and (13.24),

$$\mathrm{P_T} = \sqrt[9]{\frac{K_{s0}^{\mathrm{FA}}}{(\tfrac{5}{3}\alpha^{\mathrm{Ca^{2+}}})^5(\alpha^{\mathrm{PO_4^{3-}}})^3(\tfrac{1}{3}\alpha_1^{\mathrm{F}})}}. \tag{13.26}$$

Eq.(13.26) is similar to Eq.(13.18) but involves a $\frac{1}{9}$th root, and contains $(\frac{1}{3}\alpha_1^{\mathrm{F}})$ for $[\mathrm{F^-}]$ rather than $K_w/[\mathrm{H^+}]$ for $[\mathrm{OH^-}]$.

As in Section 13.2, we avoid resorting to a geochemical equilibrium algorithm such as MINEQL+, and design our own iteration scheme which has the following steps.

1) For each pH row, a cell is established to initiate a guess value for $\log[\mathrm{PO_4^{3-}}]_o$. However, here we cannot immediately calculate a corresponding estimate for $[\mathrm{Ca^{2+}}]$ for computing $\alpha^{\mathrm{Ca^{2+}}}$ then $\alpha^{\mathrm{PO_4^{3-}}}$, then iterating on $\log[\mathrm{PO_4^{3-}}]_o$ because here there are three solubility-related variables, $[\mathrm{PO_4^{3-}}]$, $[\mathrm{Ca^{2+}}]$, and $[\mathrm{F^-}]$. (As in Section 13.2, the assumption is that there is negligible $\mathrm{Mg^{2+}}$ in the system to form complexes with phosphate.) After making the guess for $\log[\mathrm{PO_4^{3-}}]_o$ we need to make another guess, for $\log[\mathrm{F^-}]_o$.
2) A cell is established to initiate a guess value for $\log[\mathrm{F^-}]_o$.
3) $[\mathrm{PO_4^{3-}}]_o$, $[\mathrm{F^-}]_o$, and K_{s0}^{FA} are used to compute a corresponding estimate for $[\mathrm{Ca^{2+}}]$.
4) $[\mathrm{PO_4^{3-}}]_o$ is used with Eq.(13.15) to estimate $\alpha^{\mathrm{Ca^{2+}}}$, and $[\mathrm{Ca^{2+}}]$ is used with Eq.(13.11) to estimate $\alpha^{\mathrm{PO_4^{3-}}}$.
5) The estimates of $\alpha^{\mathrm{Ca^{2+}}}$ and $\alpha^{\mathrm{PO_4^{3-}}}$ plus α_1^{F} for the pH are used to estimate $\mathrm{P_T}$ by Eq.(13.24);
6) $\log[\mathrm{PO_4^{3-}}]_1 = \log(\alpha^{\mathrm{PO_4^{3-}}}\mathrm{P_T})$ is computed.
7) $|\log[\mathrm{PO_4^{3-}}]_1 - \log[\mathrm{PO_4^{3-}}]_o|$ is computed. Solver in Excel could be invoked at this point to vary $\log[\mathrm{PO_4^{3-}}]_o$ until $|\log[\mathrm{PO_4^{3-}}]_1 - \log[\mathrm{PO_4^{3-}}]_o| = 0$ with adequate convergence, but the first converged value for $\log[\mathrm{PO_4^{3-}}]_o$ will not be the correct overall value because the first guess for $\log[\mathrm{F^-}]_o$ undoubtedly was not adequately accurate. So, invoking Solver has to wait.
8) With $\mathrm{P_T}$ from step 5, $\mathrm{F_T}$ is computed as $\frac{1}{3}\mathrm{P_T}$ according to Eq.(13.23), then $\log[\mathrm{F^-}]_1$ is computed according to Eq.(13.24) as $\log(\alpha_1^{\mathrm{F}}\mathrm{F_T})$.
9) $|\log[\mathrm{F^-}]_1 - \log[\mathrm{F^-}]_o|$ is computed.
10) Solver in Excel is now invoked to vary both $\log[\mathrm{PO_4^{3-}}]_o$ and $\log[\mathrm{F^-}]_o$ until the sum that is $|\log[\mathrm{PO_4^{3-}}]_1 - \log[\mathrm{PO_4^{3-}}]_o| + |\log[\mathrm{F^-}]_1 - \log[\mathrm{F^-}]_o| = 0$ with adequate convergence.

The linked iteration can be done en masse for a range of pH values, as introduced above, though there is currently a limit in Solver for how many variable cells can be tolerated; for Excel in Office 2016, for pH in 121 steps of 0.1 from 1 to 13 the result is $2\times121 = 242$ variable cells which is too many.

Figure 13.2 is plot of the solubility of fluoroapatite *vs.* pH. What pH exists in a particular system depends on $(C'_B - C'_A)$ and C_T. Figure 13.3 provides a comparison of hydroxyapatite and fluoroapatite. For both solids, there is an increase in solubility for increasingly acidic solutions, but the increase for fluoroapatite is markedly less than for hydroxyapatite. This difference is the reason for the enormous advantages fluoridated of drinking water and tooth paste in preventing dental caries; as discussed by Ripa (1993), dental caries is due to formation of organic acids in the dental environment by biotic degradation of organic compounds in food residues. For the ambient environment, a detailed consideration of the saturation state of fluoroapatite in groundwaters in the Floridian aquifer system in Florida, Georgia, South Carolina, and Alabama is provided by Sprinkle (1989).

The fluoroapatite analogs of the ENE-based Eqs.(13.19) and (13.21) include $[F^-]$ $(=\alpha_1^F \frac{1}{3}P_T)$ as the only change:

$$\begin{aligned}2[Ca^{2+}] + [CaOH^+] + [CaH_2PO_4^+] + [H^+] + (C'_B - C'_A)\\ = [H_2PO_4^-] + 2[HPO_4^{2-}] + 3[PO_4^{3-}] + [CaPO_4^-] + [F^-]\\ + \alpha_1^C C_T + 2\alpha_2^C C_T + [OH^-]\end{aligned} \quad (13.25)$$

$$\begin{aligned}&\left(\alpha^{PO_4^{3-}} P_T\left(\frac{[H^+]^2}{K_2K_3} + 2\frac{[H^+]}{K_3} + 3 + \left(\alpha^{Ca^{2+}}\tfrac{5}{3}P_T\right)K_{Ca}^-\right) + \alpha_1^F\tfrac{1}{3}P_T + \frac{K_w}{[H^+]}\right)\\ &- \left(\alpha^{Ca^{2+}}\tfrac{5}{3}P_T\left[2 + \frac{{}^*K_{H1,Ca}}{[H^+]} + \alpha^{PO_4^{3-}}P_T\frac{[H^+]^2}{K_2K_3}K_{Ca}^+\right] + [H^+]\right)\\ &= (C'_B - C'_A) - \left(\alpha_1^C C_T + 2\alpha_2^C C_T\right).\end{aligned} \quad (13.26)$$

As with Eq.(13.20/13.21), for each pH in the spreadsheet solution, both the LHS and RHS can be computed. For specific values of $(C'_B - C'_A)$ and C_T, the task is to find the pH in the spreadsheet (with interpolation) that gives LHS = RHS.

Example 13.3 Solubility of Fluoropatite in Initially Pure Water

Initially pure water is equilibrated with fluoroapatite at 25 °C/1 atm. At equilibrium, what is the pH, P_T, and Ca_T? Neglect activity corrections.

Solution

For this case, $(C'_B - C'_A) = 0$ and $C_T = 0$. We thus want the pH at which the LHS of Eq.(13.26) is zero. This occurs at pH = 8.94, as indicated in Figure 13.2; $P_T = 8.84\times10^{-6}$ *M*; $Ca_T = \frac{5}{3}P_T = 1.47\times10^{-5}$ *M*. The pH of this solution is only very slightly lower than for Example 13.1 (hydroxyapatite) in which pH = 9.00 and $P_T = 7.56\times10^{-6}$ *M*. Fluoroapatite (containing F^-) is less basic than hydroxyapatite (containing OH^-), so the pH is indeed lower, but its $\log K_{s0}$ (as used here) is slightly larger, making the overall effect slight (consider $P_T = 7.56\times10^{-6}$ *M* for hydroxyapatite *vs.* $P_T = 8.92\times10^{-6}$ *M* for fluoroapatite).

Example 13.4 Solubility of Fluoroapatite in Infiltrated Pristine Rain Water

Pristine rainwater ($C_B' - C_A'$) = 0) with $C_T = 10^{-5}$ *M* infiltrates into and equilibrates in an aquifer containing fluoroapatite at 25 °C/1 atm. At equilibrium, what is the pH, P_T, and Ca_T? Neglect activity corrections.

Solution

For this case, $(C_B' - C_A') = 0$, and $C_T = 10^{-5}$ *M*. We thus want the pH at which the LHS of Eq.(13.26) equals $\left(\alpha_1^C + 2\alpha_2^C\right)10^{-5}$. This occurs at pH = 8.48. $P_T = 1.26\times10^{-5}$ *M*, and $Ca_T = \frac{5}{3}P_T = 2.10\times10^{-5}$ *M*.

Example 13.5 Saturation State of Hydroxyapatite and Fluoroapatite in Oral Fluids

Ten Cate et al. (2009, their Table 13.3) discuss estimated solution conditions for fluid within dental plaque in caries-positive individuals after rinsing with a 5% sucrose solution to induce "fermenting" with low pH conditions: pH, 5.29, Ca_T, 0.0082 *M*, P_T, 0.0135 *M*; and F^-, 0.005 *M*. Tatevossian (1987) reported data for plaque fluid suggesting about 50% of Ca_T is ionized and about 75% of P_T is inorganic (as considered in this chapter). So, for this problem take pH = 5.29, Ca_T = 0.004 *M*, P_T = 0.007 *M*; and F_T = 0.005 *M*. Evaluate the saturation index = $\log\Omega$ = log(ion activity product/K_{s0}) for both hydroxyapaptite and fluroapatite for 37 °C under the following assumptions: 1) the mineralogically relevant $\log K_{s0}$ values in Table 13.2 are applicable; the only forms comprising Ca_T and P_T are those considered above: *e.g.*, there are no organic complexes of Ca, and no organic forms of P; and 3) Mg_T levels are sufficiently low that phosphate complexes with Mg^{2+} can be neglected. Neglect activity corrections.

Solution

We know Ca_T and P_T, but we do not know $[Ca^{2+}]$ and $[PO_4^{3-}]$ because *a priori*, the extent of complex formation is unknown: an iterative solution is required. A spreadsheet for pH = 5.29 can be set up to carry out the following steps:

1) A cell is established to initiate a guess value for $\log[PO_4^{3-}]_o$.
2) $\log[PO_4^{3-}]_o$ with Eq.(13.15) is used to calculate $\alpha^{Ca^{2+}}$.
3) $\alpha^{Ca^{2+}}$ is used with Ca_T to calculate $[Ca^{2+}]$.
4) $[Ca^{2+}]$ with Eq.(13.11) is used to calculate $a^{PO_4^{3-}}$.
5) $\alpha^{PO_4^{3-}}$ with P_T is used to calculate $\log[PO_4^{3-}]_1$.
6) $\log[PO_4^{3-}]_o$ is iterated to convergence such that $|\log[PO_4^{3-}]_1 - \log[PO_4^{3-}]_o| < 0.01$.

With pH = 5.29, the results are: $\log[PO_4^{3-}]=10^{-11.15}$; $[PO_4^{3-}] = 7.03\times10^{-12}$; $[Ca^{2+}] = 3.21\times10^{-3}$; $\alpha_1^F = 0.992$, and $[F^-] \approx 5\times10^{-3}$ *M*.

For hydroxyapatite: ion activity product (IAP) = $[Ca^{2+}]^5[PO_4^{3-}]^3[OH^-] = 10^{-54.63}$. $IAP/K_{s0} = 10^{-54.63}/10^{-55.27} = 4.39$. $\log IAP/K_{s0} = \log\Omega = 0.64 \approx 0$. The assumptions and data lead to the conclusion that the conditions are somewhat supersaturated with crystalline hydroxyapatite. However, dental caries are known to form with non-fluoridated teeth under the ostensible experimental conditions of this problem, so while the analytical data may be problematic, the applicability of the mineralogical K_{s0} value used is also suspect (*too small*). This problem illustrates the difficulty in understanding solubility conditions when low-solubility materials are involved, especially if laid down biogenically. Indeed, the solubilities of hydroxyapapite-related materials in teeth are subject to considerable uncertainty (Aoba, 2004), are known to vary within the tooth

TABLE 13.3
Concentration and Phosphate Complexation Assumptions for Cases in Figure 13.4

Case	Phosphate Complexes Considered	$Mg(II)_T$	$Ca(II)_T$	$Ammonia_T$ (as NH_3)	$Phosphate_T$ (as PO_4^{3-})	$Mg(II)_T$	$Ca(II)_T$	$Ammonia_T$ (as NH_3)	$Phosphate_T$ (as PO_4^{3-})
	Y/N	mg/L	mg/L	mg/L	mg/L	*M*	*M*	*M*	*M*
A	Y	30	30	220	160	0.0012	0.00075	0.0127	0.0017
B	Y	80	30	220	160	0.0033	0.00075	0.0127	0.0017
C	Y	400	30	220	160	0.0165	0.00075	0.0127	0.0017
D	Y	80	30	220	800	0.0033	0.00075	0.0127	0.0084
E	Y	80	30	1100	160	0.0033	0.00075	0.0636	0.0017
E′	N	80	30	1100	160	0.0033	0.00075	0.0636	0.0017

Note: $Ca(II)_T$ value is representative of background Ca. Case A is a "Natural" water treatment case. For Case B, $Mg(II)_T$, ammonia, and phosphate values are from Grooms et al. (2015) as representative for the ostara pearl™ process. $Mg(II)_T$, ammonia, or phosphate are increased by 5× from Case B in cases C, D, E, and E′.

structure because of variations in composition (Aoba, 2004), and because of imperfections can be significantly *more soluble* than highly crystalline mineral apatite (Shellis, 1996).

For fluoroapatite: ion activity product (IAP) = $[Ca^{2+}]^5[PO_4^{3-}]^3[F^-]$ = $10^{-48.22}$. IAP/K_{s0} = $10^{-48.22}/10^{-55.37}$ = 1.4×10^7 >>> 1. Log IAP/K_{s0} = logΩ = 7.15 >> 0. The solution is calculated to be highly supersaturated with respect to fluoroapatite. The enormous benefits of fluoridation of drinking water and toothpaste relative to caries prevention are amply demonstrated.

13.4 STRUVITE – $MgNH_4PO_4{\cdot}6H_2O_{(S)}$

Struvite and Wastewater. Domestic wastewater contains compounds that break down and release phosphate and ammonia, both of which are nutrients in natural waters. And, ammonia is toxic to fish, so the general goal is to greatly reduce both constituents during treatment/prior to discharge. On a molar basis, there is usually much more ammonia than phosphorus in wastewater: the former originates predominantly from the large amounts of protein taken in by humans (and excreted as urea and fecal protein), while the latter originates from the lesser amounts of phosphate in excreted phospholipids and DNA. (In the U.S. and E.U., phosphate "builders" in detergents are mostly no longer in use.)

Biological phosphorus removal (BPR) from wastewater can be efficiently accomplished using the aerobic "activated sludge" process. However, during anaerobic digestion of the sludge (to reduce solids levels and obtain biogas), phosphate is released to the digest "liquor" and must be removed before the liquor can be discharged with water from other flows in the system. If the municipal source water contains significant magnesium (Mg^{2+}), struvite can precipitate spontaneously in/on the pipes, valves, and pump impellers handling the liquid. For struvite precipitation/dissolution, we have

$$MgNH_4PO_4{\cdot}6H_2O_{(s)} = Mg^{2+} + NH_4^+ + PO_4^{3-} + 6H_2O \qquad K_{s0} = \{Mg^{2+}\}\{NH_4^+\}\{PO_4^{3-}\}. \qquad (13.27)$$

Having six molecules of water in the crystal lattice per unit of $MgNH_4PO_4$, struvite is a "hexahydrate". Uncontrolled struvite precipitation in a wastewater treatment plant is an enormous and costly nuisance. Many astonishing photographs of pipes occluded with struvite can be found online; such occlusion will require complete teardown of systems (or attack with struvite dissolution products) leading to hugely problematic downtimes (sewage does not stop coming).

Ammonia and phosphate are high-value agricultural fertilizers, and struvite is sufficiently soluble to dissolve after application to soils. Thus, struvite is a valuable commodity if it can be harvested in a controlled manner, even if it does not tend to form on its own in a given system. One method is the Ostara Pearl™ process (Grooms et al., 2015): Mg^{2+} is added to anaerobic digest liquor, and pellets of struvite are the result. Mg^{2+} is added because significant ammonia and phosphate are always present in wastewater; the same is not true for Mg^{2+}, and varying the level of added Mg^{2+} allows for maximizing the struvite recovery. This recovery of struvite from wastewater is representative of an absolutely necessary and extraordinarily important movement towards increased commodity recovery (including stored energy) from human waste streams, see for example Batstone et al. (2015).

For both phosphate and ammonia, levels in digest liquor can be as high as 500 to 1000 mg/L, computed as mg PO_4^{3-}/L for phosphate and as mg NH_3/L for ammonia (Booker et al., 1999). On a molar basis, this is 0.005 to 0.01 *M* for P_T, and 0.03 to 0.06 *M* for N_T. Struvite precipitation can be used with high efficiency in removing phosphate. Since the N:P molar stoichiometry is 1:1 in struvite, given the generally $N_T >> P_T$ condition in municipal wastewater, then prior to discharge, residual ammonia will usually have to be removed by an additional method, such as stripping or nitrification/dentrification.

At saturation equilibrium for struvite, neglecting activity corrections, with the K_{s0} given pertaining to 25 °C/1 atm,

$$K_{s0} = [Mg^{2+}][NH_4^+][PO_4^{3-}] = (\alpha^{Mg^{2+}} Mg_T)(\alpha_0^N N_T)(\alpha^{PO_4^{3-}} P_T) = 10^{-13.26}. \quad (13.28)$$

The saturation index (Chapter 10) is

$$\log \Omega = \log \frac{(\alpha^{Mg^{2+}} Mg_T)(\alpha_0^N N_T)(\alpha^{PO_4^{3-}} P_T)}{K_{s0}}. \quad (13.29)$$

As usual, the conditions of supersaturation, equilibrium, and undersaturation are given by $\log \Omega > 0$, 0, and <0, respectively. Stumm and Morgan (1996) discuss rearranging Eq.(13.28) to obtain

$$Mg_T N_T P_T = \frac{K_{s0}}{\alpha^{Mg^{2+}} \alpha_0^N \alpha^{PO_4^{3-}}} = K_{s0}^{cond}. \quad (13.29)$$

K_{s0}^{cond} is a "conditional solubility product", used for a direct examination of the product $Mg_T N_T P_T$ when considering the saturation state Ω. Doyle and Parsons (2002) and others follow this approach. It has some advantage in helping build intuition regarding what types of Mg_T, N_T, and P_T values can lead to struvite precipitation, especially when Mg_T and/or P_T levels are low enough that complexes between Mg^{2+} and phosphate species can be neglected so that the three α values can be computed based simply on pH (as assumed by Stumm and Morgan (1996)). However, when such complexes cannot be neglected (the usual case in wastewater treatment), because $\alpha^{Mg^{2+}}$ and $\alpha^{PO_4^{3-}}$ depend implicitly on P_T and Mg_T, then given pH, P_T, and Mg_T values K_{s0}^{cond} cannot be evaluated prior to solving for the overall speciation. So, Doyle and Parsons (2002) used a numerical chemical speciation model (MINTEQA2) to calculate K_{s0}^{cond}, then the saturation state as $\Omega = K_{s0}^{cond}/(Mg_T N_T P_T)$. In other studies of struvite precipitation, Münch and Barr (2001) used MINTEQ, and Muster et al. (2013) used PHREEQC. In this chapter, as with the preceding calculations, we will use a simple, problem-focused, iterative spreadsheet approach to handle the numerical calculations.

Struvite formation is both well understood and common in the wastewater treatment domain, but the same is definitely not true for the Ca^{2+} analog $CaNH_4PO_4 \cdot nH_2O_{(s)}$ with $n = 6$, or with some other number. When precipitation of calcium phosphates in wastewater treatment is discussed, it is nearly always as hydroxyapatite (*e.g.*, see Rittmann et al., 2011), not as some $CaNH_4PO_4$ form. This seems a bit surprising given that: (1) molar levels for Ca^{2+} are often similar/greater than for Mg^{2+} in fresh waters (Hem, 1985); (2) phosphate species combine as well with Ca^{2+} as with Mg^{2+} (see values of K^+, K°, and K^- in Table 13.2); and, (3) for other mineral types, both Ca^{2+} and Mg^{2+} forms are well known (for carbonates, for Ca^{2+} we have calcite and aragonite for Ca^{2+}, and for Mg^{2+} there is magnesite). It seems then that it is the stability of the metal-ammonium-phosphate crystal lattice itself that is favored for Mg^{2+}. In eight-fold coordination, the Ca^{2+} ionic radius is 0.112 nm (nanometers), while for Mg^{2+} it is 0.089 nm. Konzett (2018) argues on this topic that: (1) it is very possible that within the struvite structure the coordination polyhedron that hosts Mg^{2+} in struvite is too small to easily accommodate the larger Ca^{2+}; and so (2) insertion of Ca^{2+} into the struvite polyhedron will destabilize the entire structure. Konzett (2018) notes further that in silicate structures, by comparison, Mg^{2+} and Ca^{2+} often occupy different coordination polyhedra, *i.e.*, force different numbers of next oxygen neighbors because of their different ionic radii. Overall, in the municipal wastewater treatment context, the net result seems to be that unless ammonium levels happen to be unusually high, calcium ammonium phosphate solids are too soluble to form, given that hydroxyapatite offers an alternative means to relieve the solution concentration of Ca^{2+} in the face of a significant PO_4^{3-} concentration. But, when ammonia levels *are* high, precipitation of $CaNH_4PO_4 \cdot 4H_2O_{(s)}$ has been described by Li et al. (2007) as achievable for removing ammonia from mining wastewaters involving ammonia-based lixiviant solutions, and by Quan et al. (2010) for removing ammonia from manure wastewaters from confined high-animal-density feeding operations. (Some refer to this type of enterprise as a form of agricultural "animal husbandry", but that is a truly cruel oxymoron if there ever was one: animals are raised under horrific, torturous, caged conditions, slaughtered one after the other in an abattoir, then mostly eaten by *Homo sapiens*.)

More α Values. In the above sections, $\alpha^{Ca^{2+}}$ is given by Eq.(13.12). Analogously for $\alpha^{Mg^{2+}}$ we have

$$\alpha^{Mg^{2+}} \equiv \frac{[Mg^{2+}]}{[Mg^{2+}]+[MgOH^{+}]+[MgH_2PO_4^{+}]+[MgHPO_4^{o}]+[MgPO_4^{-}]} \tag{13.30}$$

$$= \frac{1}{1+\dfrac{{}^{*}K_{H1,Mg}}{[H^{+}]}+K_{Mg}^{+}[H_2PO_4^{-}]+K_{Mg}^{o}[HPO_4^{2-}]+K_{Mg}^{-}[PO_4^{-}]}. \tag{13.31}$$

For $\alpha^{PO_4^{3-}}$, in Sections 13.2 and 13.3 for dissolution of hydroxyapatite and fluoroapatite there is only one 2+ metal species (Ca^{2+}) that forms complexes with phosphate species. In the general wastewater treatment context, however, both Mg^{2+} and Ca^{2+} are likely to be present at non-negligible levels, so for this section we have

$$\alpha^{PO_4^{3-}} = \frac{1}{\left[\dfrac{[H^+]^3}{K_1^PK_2^PK_3^P}+\dfrac{[H^+]^2}{K_2^PK_3^P}+\dfrac{[H^+]}{K_3^P}+1+[Ca^{2+}]\left(K_{Ca}^{+}\dfrac{[H^+]^2}{K_2^PK_3^P}+K_{Ca}^{o}\dfrac{[H^+]}{K_3^P}+K_{Ca}^{-}\right)+[Mg^{2+}]\left(K_{Mg}^{+}\dfrac{[H^+]^2}{K_2^PK_3^P}+K_{Mg}^{o}\dfrac{[H^+]}{K_3^P}+K_{Mg}^{-}\right)\right]}. \tag{13.32}$$

For α_0^N, as usual,

$$[NH_4^+] = \alpha_0^N N_T \tag{13.33}$$

$$\alpha_0^N \equiv \frac{[NH_4^+]}{[NH_4^+]+[NH_3]}. \tag{13.34}$$

Neglecting activity corrections,

$$\alpha_0^N = \frac{[H^+]}{K_a^N+[H^+]}. \tag{13.35}$$

Calculating Struvite Saturation. We now address how to answer whether struvite can precipitate from a water with particular composition values for Mg_T, Ca_T, N_T, and P_T, with pH being adjustable, as in a wastewater treatment operation. This is a solution POV problem. We are interested in how the saturation index for the solution will change as the pH may change. This problem is closely analogous with the case considered in Example 13.5, except that here we are interested in a range of pH values. We assume for the calculations that activity corrections can be neglected; they could be considered by substituting ${}^{c}K$ values for all K values.

For a given pH, at least we can calculate $[NH_4^+]$ directly. But while we know pH, Mg_T, Ca_T, and P_T, we cannot directly calculate $[Mg^{2+}]$, $[Ca^{2+}]$, and $[PO_4^{3-}]$. *A priori*, the extents of metal complex

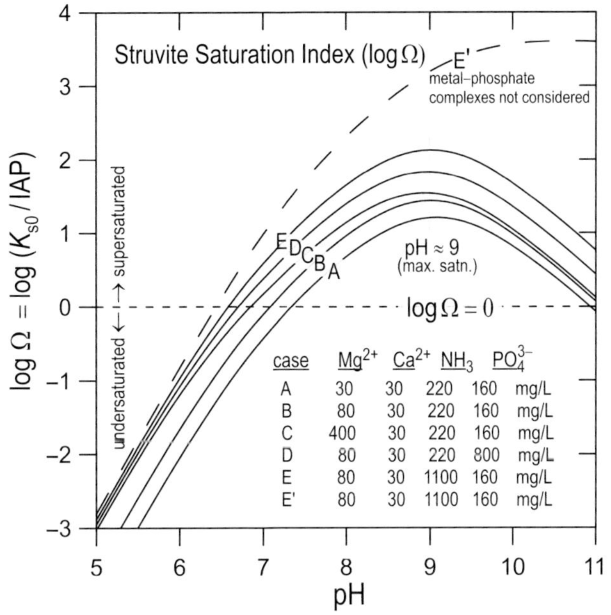

FIGURE 13.4 Struvite saturation index (log Ω) *vs.* pH for six cases: A to E consider metal-phosphate complexes; E′ does not.

formation are unknown; again, we require an iterative solution. We use a spreadsheet to carry out the following steps:

1) For each pH row, a cell is established to initiate a guess value for $\log[PO_4^{3-}]_o$.
2) With $\log[PO_4^{3-}]_o$, $\alpha^{Ca^{2+}}$ is calculated using Eq.(13.15), and $\alpha^{Mg^{2+}}$ is calculated using Eq.(13.31).
3) a) Using $\alpha^{Ca^{2+}}$ and Ca_T, $[Ca^{2+}]$ is calculated; b) Using $\alpha^{Mg^{2+}}$ and Mg_T, $[Mg^{2+}]$ is calculated; c) simply using pH, α_0^N is calculated using Eq.(13.35) then $[NH_4^+]$ is calculated using Eq.(13.33).
4) Using $[Ca^{2+}]$ and $[Mg^{2+}]$, $\alpha^{PO_4^{3-}}$ is calculated using Eq.(13.32).
5) Using $\alpha^{PO_4^{3-}}$, $[PO_4^{3-}]_1$ is calculated as $\alpha^{PO_4^{3-}} P_T$.
6) Iteration by varying $\log[PO_4^{3-}]_o$ to convergence is carried out for each pH such that $|\log[PO_4^{3-}]_1 - \log[PO_4^{3-}]_o| < 0.01$.

The iteration can be carried out en masse for a range of pH values, as described above for other problems. Curves for the saturation index = $\log\Omega$ as calculated using Eq.(13.29) by this method are plotted in Figure 13.4. The conditions for the six cases A–E and E′ are summarized in Table 13.3. For A–E, the $\log\Omega$ curves were calculated including the metal-phosphate complexes, but for E′ by not including those complexes. The downturns for curves A-E past pH ≈ 9 are due to: (1) decreasing α_0^N; and (2) decreasing $\alpha^{Mg^{2+}}$ (because of increasing complexation of Mg^{2+} with phosphate; $MgOH^+$ does not become important until pH approaches $p^*K_{H1,Mg}$). A comparison of the curves for E and E′ makes it clear that the complexes cannot be ignored for these types of $Mg_T/Ca_T/P_T$ conditions.

Case A is a case that might encountered in wastewater treatment, *i.e.*, absent any added Mg(II). Case B is from Grooms et al. (2015) as representative of the Ostara Pearl™ process in which Mg(II) is added. For a pilot study of that process, Vancouver (B.C.) Metro Wastewater Reclamation District (2011) comments as follows: "The [struvite] saturation point of a solution is strongly influenced by pH, hence if the feed stream does not have sufficient alkalinity, sodium hydroxide is added. The process is generally operated at a pH of between 7 and 8." Going from case B to cases C, D, and E, the Mg_T, ammonia, or phosphate level is raised by a factor of five, so that the curves for cases C, D, and E all lie above the case B curve; supersaturation is more easily achieved relative to case B.

Numerous authors, including Doyle and Parsons (2002) and references therein, have discussed that pH ≈ 9 gives maximum struvite supersaturation for wastewater treatment conditions; this may also be concluded based on Figure 13.4. For each curve in Figure 13.4, based on Eq.(13.29), the values of $\log(Mg_T N_T P_T)$ provide the vertical offsets from the corresponding curves (not plotted) for $\log(\alpha^{Mg^{2+}} \alpha_0^{N} \alpha^{PO_4^{3-}})$.

Example 13.6 Struvite Saturation

For the N_T, P_T, and Ca_T conditions for case B, and taking pH = 7.60, determine the value of Mg_T required to just achieve $\log\Omega = 0$ for struvite, *i.e.*, Mg_T such that any greater value would cause $\log\Omega > 0$ and allow precipitation. This is a solution POV problem. Neglect activity corrections.

Conditions: pH = 7.60, 25 °C/1 atm.

Constituent:	mg/L	molarity (*M*)
N_T: total ammonia (as NH_3)	220	0.01294
P_T: total orthophosphate (as PO_4)	160	0.001685
Ca_T: total Ca	30	0.000749

Solution

a. The same spreadsheet approach as discussed for the calculations in Figure 13.4 can be used here. In this case, iteration with Solver to convergence that $|\log[PO_4^{3-}]_1 - \log[PO_4^{3-}]_o| < 0.01$ by having Solver change two variable cells, the one for $\log[PO_4^{3-}]_o$ *and* one for $\log Mg_T$, all while subject to the constraint that the cell for $\log\Omega$ equals 0. The result is Mg_T = 14.1 mg/L.

REFERENCES

Aoba T (2004) Solubility properties of human tooth mineral and pathogenesis of dental caries. *Oral Diseases*, **10**, 249–257.

Ball JW, Nordstrom DK (1991/2001) *User's Manual for WATEQ4F, with Revised Thermodynamic Data Base and Test Cases for Calculating Speciation of Major, Trace, and Redox Elements in Natural Waters.* U.S. Geological Survey, Open-File Report 91-183, Menlo Park, CA, 1991 Revised and reprinted, April 2001.

Barrow, NJ (1983) A mechanistic model for describing the sorption and desorption of phosphate by soil. *Journal of Soil Science*, **34**, 733–750.

Barrow, NJ (2015) Reflections by N. J. Barrow. *European Journal of Soil Science*, **66**, 2–3.

Batstone DJ, Hülsen T, Mehta CM, Keller J (2015) Platforms for energy and nutrient recovery from domestic wastewater: A review. *Chemosphere*, **140**, 2–11.

Bichlera K-H, Eipper E, Naber K, Braunc V, Zimmermann R, Lahmea S (2002) Urinary infection stones. *International Journal of Antimicrobial Agents*, **19**, 488–498.

Bonjour J-P (2011) Calcium and phosphate: A duet of ions playing for bone health. *Journal of the American College of Nutrition*, **30**, Supplement 5, 438S–448S.

Booker NA, Priestley AJ, Fraser IH (1999) Struvite formation in wastewater treatment plants; opportunities for nutrient recovery. *Environmental Technology*, **20**, 777–782.

Burns JR, Finlayson B (1982) Solubility product of magnesium ammonium phosphate hexahydrate at various temperatures. *Journal of Urology*, **128**, 426–428.

Cao X, Ma LQ, Chen M., Singh SP, Harris WG (2002) Impacts of phosphate amendments on lead biogeochemistry at a contaminated site. *Environmental Science and Technology*, **36**, 5296–5304.
Chughtai, A, Marshall R, Nancollas GH (1968) Complexes in calcium phosphate solutions. *Journal of Physical Chemistry*, **72**, 208–211.
Domagalski JL and Johnson H (2012) Phosphorus and groundwater: Establishing links between agricultural use and transport to streams: U.S. Geological Survey Fact Sheet 2012–3004.
Dorozhkin SV (2014) Calcium orthophosphates: Occurrence, properties and major applications. *Bioceramics Development and Applications*, **4**, 081.
Doyle JD, Parsons SA (2002) Struvite formation, control and recovery. *Water Research*, **36**, 3925–3940.
Emerson K, Russo RC, Lund RE, Thurston RV (1975) Aqueous ammonia equilibrium calculations: Effect of pH and temperature. *Journal of the Fisheries Research Board of Canada*, **32**, 2379–2383.
Fox LE (1989) A model for inorganic control of phosphate concentrations in river waters. *Geochimica et Cosmochimica Acta*, **53**, 417–428.
Fordham AW, Schwertmann U (1977) Composition and reactions of liquid manure (Gülle), with particular reference to phosphate: II. Solid phase components. *Journal of Environmental Quality*, **6**, 136–140.
Grooms A, Reusser S, Dose, A, Britton A, Prasad, R (2015) Operating experience with Ostara struvite harvesting process. *Proceedings of the Water Environment Federation*, WEFTEC 2015, Chicago, IL, 26–30 September 2015, Session 204, 2162–2177.
Hanhoun M, Montastruc L, Azzaro-Pantel C, Biscans B, Frèche M, Pibouleau L (2011) Temperature impact assessment on struvite solubility product: A thermodynamic modeling approach. *Chemical Engineering Journal*, **167**, 50–58.
Hem JD (1985) *Study and Interpretation of the Chemical Characteristics of Natural Waters*, 3rd edn. U.S. Geological Survey Water-Supply Paper 2254. United States Government Printing Office. Available at: https://pubs.usgs.gov/wsp/wsp2254/pdf/wsp2254a.pdf (accessed 27 May 2019).
Jarvis S (2015) Landmark Papers: No. 4. *European Journal of Soil Science*, **66**, 1–1.
Konzett J (2018) Personal communication, September 21, 2018. Jürgen Konzett, Institut für Mineralogie und Petrographie, Universitat Innsbruck, Innrain 52, A-6020 Innsbruck, Austria.
Li Y, Yi L, Ma P, Zhou L (2007) Industrial wastewater treatment by the combination of chemical precipitation and immobilized microorganism technologies. *Environmental Engineering Science*, **24**, 736–744.
McCann HG (1968) The solubility of fluoroapatite and its relationship to that of calcium fluoride. *Archives of Oral Biology*, **13**, 987–1001.
Moreno EC, Kresak M, Zahradnik RT (1974) Fluoridated hydroxyapatite solubility and caries formation. *Nature*, **247**, 64–65.
Münch EV, Barr K (2001) Controlled struvite crystallisation for removing phosphorus from anaerobic digester sidestreams. *Water Research*, **35**, 151–159.
Muster TH, Douglas GB, Sherman N, Seeber A, Wright N, Güzükara Y (2013) Towards effective phosphorus recycling from wastewater: Quantity and quality. *Chemosphere*, **91**, 676–684.
Perrin DD (1982) *Ionisation Constants of Inorganic Acids and Bases in Aqueous Solution*, 2nd edn. IUPAC Chemical Data Series, No. 29, Pergamon Press, Oxford, UK.
Quan X, Ye C, Xiong Y, Xiang J, Wang F (2010) Simultaneous removal of ammonia, P and COD from anaerobically digested piggery wastewater using an integrated process of chemical precipitation and air stripping. *Journal of Hazardous Materials*, **178**, 326–332.
Ripa, LW (1993) A half-century of community water fluoridation in the United States: Review and commentary. *Journal of Public Health Dentistry*, **53**, 17–44.
Rittmann BE, Mayer B, Westerhoff P, Edwards M (2011) Capturing the lost phosphorus. *Chemosphere*, **84**, 846–853.
Shellis RP (1996) A scanning electron-microscopic study of solubility variations in human enamel and dentine. *Archives of Oral Biology*, **41**, 473–484.
Sprinkle, CL (1989) *Geochemistry of the Floridan Aquifer System in Florida and in Parts of Georgia, South Carolina, and Alabama*. U.S. Geological Survey Professional Paper 1403-I, United States Government Printing Office, Washington, D.C.
Stanforth R, Qiu J (2001) Effect of phosphate treatment on the solubility of lead in contaminated soil. *Environmental Geology*, **41**, 1–10.
Staunton S, Ludwig B, Torrent J (2015) Commentary on the impact of Barrow (1983). *European Journal of Soil Science*, **66**, 4–8.
Stumm W, Morgan JJ (1996) *Aquatic Chemistry: Chemical Equilibria and Rates in Natural Waters*, 3rd edn. Wiley Publishing Co., New York.

ten Cate JM, Larsen MJ, Pearce EIF, Fejerskov O (2009) Chemical interactions between the tooth and oral fluids, Chapter 12. In: *Dental Caries: The Disease and its Clinical Management*, 2nd edn, Fejerskov O and Kidd E with Nyvad B and Baelum V (Eds.), pp. 210–231. John Wiley & Sons, Chichester, UK.

Tatevossian A (1987) Calcium and phosphate in human dental plaque and their concentrations after overnight fasting and after ingestion of a boiled sweet. *Archives of Oral Biology*, **32**, 201–205.

Thawornchaisit U, Polprasert C (2009) Evaluation of phosphate fertilizers for the stabilization of cadmium in highly contaminated soils. *Journal of Hazardous Materials*, **165**, 1109–1113.

Vancouver Metro Wastewater Reclamation District (2011) Nutrient recovery by the Metro Wastewater Reclamation District, Pearl Nutrient Recovery Process Pilot Study Report, 13 April 2011. 690–1199 West Pender Street, Vancouver, BC, V6E 2R1.

Wright KE, Hartmann T, Fujita Y (2011) Inducing mineral precipitation in groundwater by addition of phosphate. *Geochemical Transactions*, **12**, 8.

Zhu Y, Zhang X, Chen Y, Xie Q, Lan J, Qian M, He N (2009) A comparative study on the dissolution and solubility of hydroxylapatite and fluorapatite at 25°C and 45°C. *Chemical Geology*, **268**, 89–96.

14 Which Solid Is Solubility Limiting? *Examples with Fe(II) for $FeCO_{3(s)}$ vs. $Fe(OH)_{2(s)}$ Using Log p_{CO_2} vs. pH Predominance Diagrams*

14.1 INTRODUCTION

Chapters 11–13 discuss how metal ions can form a variety of types of solids, simple salts, hydroxides, oxides, oxyhydroxides, carbonates, and phosphates. Other types of solids are possible. How, then, does one know which solid(s) to include when computing the equilibrium chemistry of a given system? With ferrous iron, for example, if it is known that there is enough Fe(II) in a certain system that either $Fe(OH)_{2(s)}$ or $FeCO_{3(s)}$ will be present, how does one know which solid to use in the calculations? A comprehensive geochemistry computer model loaded with a large geochemical equilibrium constant database (*e.g.*, Ball and Nordstrom, 1991/2001) can run through all the possibilities. However, we still want to understand this at a fundamental level. Besides, usually the systems of interest are not so complicated that such models are needed to solve problems, once we understand which solid(s) need to be considered.

The bottom line here is that for any specific conditions, the solid that will *limit* the solubility is the one that, if present, would specify the lowest value for the saturation activity of the metal ion of interest. Any other solid that, if present, would specify a higher saturation value cannot become saturated and form. To actually *control* the solubility of a metal ion, the solid has to be present, and that requires that the total system metal level per L of liquid system water ($M_{T,sys}$, see also Chapter 11) is high enough for some of the solid to have precipitated.

To determine solubility limitation for Fe^{2+} with $Fe(OH)_{2(s)}$ *vs.* $FeCO_{3(s)}$, we would determine which is lowest: (a) $\{Fe^{2+}\} = K_{s0}^{hyd}/\{OH^-\}^2$; or (b) $\{Fe^{2+}\} = K_{s0}^{car}/\{CO_3^{2-}\}$. Because these two solids have different chemistries (*i.e.*, they are not two different forms of $Fe(OH)_{2(s)}$, or similarly, not two different forms of $FeCO_{3(s)}$), it is not a matter of which solid simply has the lowest K_{s0}, but rather which solid gives the lowest $\{Fe^{2+}\}$ per the calculations indicated above in a *vs.* b.

Further, say that some Fe^{2+} or present in a solution at some pH so that we are interested in the equilibrium

$$Fe^{2+} + OH^- = FeOH^+ \qquad K_{H1} = \frac{\{FeOH^+\}}{\{Fe^{2+}\}\{OH^-\}}. \tag{14.1}$$

Whichever solid gives the lowest value for $\{Fe^{2+}\}$ in the matchup of a *vs.* b will also specify the lowest value for $\{FeOH^+\} = K_{H1}\{Fe^{2+}\}\{OH^-\}$, and by extension for all other dissolved Fe(II) species, and so also for the total dissolved Fe(II).

14.2 EQUILIBRIUM COEXISTENCE OF TWO SOLIDS

We combine the appropriate equilibria to obtain an equilibrium constant for the interconversion of $FeCO_{3(s)}$ and $Fe(OH)_{2(s)}$. Using the values for the various constants at 25 °C/1 atm, we have

$$FeCO_{3(s)} = Fe^{2+} + CO_3^{2-} \qquad K_{s0}^{car} = 10^{-10.68} \tag{14.2}$$

$$Fe^{2+} + 2OH^- = Fe(OH)_{2(s)} \qquad (K_{s0}^{hyd})^{-1} = 1/(10^{-15.1}) \tag{14.3}$$

$$2H_2O = 2H^+ + 2OH^- \qquad K_w^2 = 10^{-27.99} \tag{14.4}$$

$$2H^+ + CO_3^{2-} = H_2CO_3^* \qquad 1/(K_1K_2) = 1/10^{-16.68} \tag{14.5}$$

$$+H_2CO_3^* = H_2O + CO_{2(g)} \qquad 1/K_H = 1/10^{-1.47} \tag{14.6}$$

$$\boxed{FeCO_{3(s)} + H_2O + Fe(OH)_{2(s)} + CO_{2(g)} \qquad K = p_{CO_2} = 10^{-5.42}. \quad \text{gives one horizontal boundary line}} \tag{14.7}$$

We have assumed that when the two solids are present, they are both pure (*i.e.*, mole fraction $x = 1$, as with H_2O), so that their activities are unity, and we need not write them (or $\{H_2O\}$) in the expression for the final K. By the value of that K, the above two solids can co-exist at 25 °C/1 atm only when $p_{CO_2} = 10^{-5.42}$ atm. If $p_{CO_2} > 10^{-5.42}$ atm, any $Fe(OH)_{2(s)}$ present would tend to pick up CO_2 and be completely converted to $FeCO_{3(s)}$. If $p_{CO_2} < 10^{-5.42}$ atm, then any $FeCO_{3(s)}$ present will tend to release CO_2 and become $Fe(OH)_{2(s)}$. Further, on the basis of the preceding discussion, at 25 °C/1 atm, then

- For $p_{CO_2} > 10^{-5.42}$ atm, $FeCO_{3(s)}$ specifies a lower $\{Fe^{2+}\}$ than does $Fe(OH)_{2(s)}$ for all pH;
- For $p_{CO_2} = 10^{-5.42}$ atm, since both $Fe(OH)_{2(s)}$ and $FeCO_{3(s)}$ are stable, for every pH, they specify the same $\{Fe^{2+}\}$;
- For $p_{CO_2} < 10^{-5.42}$ atm, $Fe(OH)_{2(s)}$ specifies a lower $\{Fe^{2+}\}$ than does $FeCO_{3(s)}$ for all pH.

Figure 12.1.b shows the solubility of $FeCO_{3(s)}$ (siderite) *vs.* pH in a closed system such that the only source of C_T for the solution is dissolution of the solid. The caption there states that at pH > 8.73, $FeCO_{3(s)}$ becomes unstable in that system relative to conversion to $Fe(OH)_{2(s)}$. This is because at pH > 8.73, for the computed C_T, then $p_{CO_2} = [H_2CO_3^*]/K_H = \alpha_0 C_T/K_H < 10^{-5.42}$ atm. For pH > 8.73, Figure 14.1 for $Fe(OH)_{2(s)}$ then becomes the relevant equilibrium solubility diagram. Figure 14.1 is the ferrous hydroxide analog of Figure 11.3, for (amorphous ferric hydroxide). So, at pH = 8.73, $[Fe^{2+}]$ as given in Figure 12.1.b equals $[Fe^{2+}]$ as given in Figure 14.1, and similarly for the other Fe(II) species. At pH > 8.73, $[Fe^{2+}]$ as given in Figure 14.1 is less than $[Fe^{2+}]$ as given in Figure 12.1.b, and similarly for the other Fe(II) species. Figure 14.1 includes lines for three $Fe(II)_{T,sys}$ values of interest in this chapter.

14.3 Log p_{CO_2} *VS.* pH PREDOMINANCE DIAGRAMS WITH REGIONS FOR $FeCO_{3(s)}$, $Fe(OH)_{2(s)}$, AND DISSOLVED Fe(II) SPECIES

14.3.1 $Fe(II)_{T,sys} = 10^{-5}$ *M*

14.3.1.1 $FeCO_{3(s)}/Fe(OH)_{2(s)}$ Boundary Line

The *y* and *x*-axes of the "predominance diagrams" we are going to draw here are $\log p_{CO_2}$ and pH. By Eq.(14.7), for $FeCO_{3(s)}$ and $Fe(OH)_{2(s)}$ to be in equilibrium at 25 °C/1 atm, $\log p_{CO_2} = -5.42$. Since H^+ does not appear in the equilibrium, the coexistence of these two solids requires only that $\log p_{CO_2} = -5.42$: there is no *explicit* dependence on pH. So, without worrying for the moment exactly where it starts or stops, one can start by drawing a horizontal (*i.e.*, pH-independent) line at

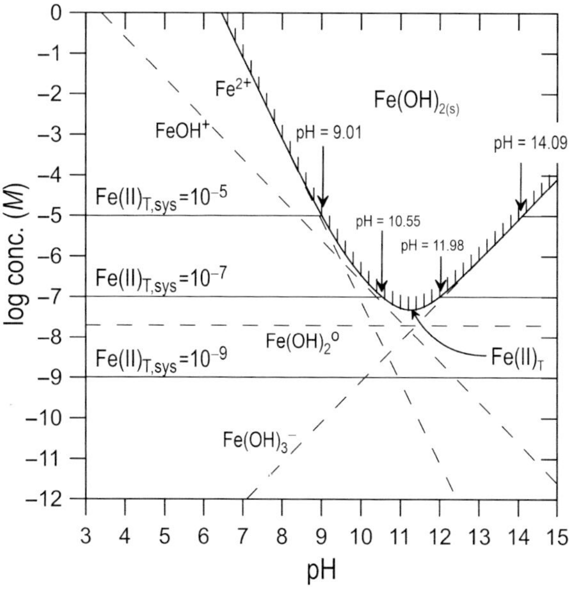

FIGURE 14.1 Solubility of $Fe(OH)_{2(s)}$ at 25 °C/1 atm for ${}^{*}K_{s0}^{hyd} = 10^{12.89}$, ${}^{*}K_{s1}^{hyd} = 10^{-3.40}$, $K_{s2}^{hyd} = 10^{-7.7}$, and ${}^{*}K_{s3}^{hyd} = 10^{-19.10}$. If $Fe(II)_{T,sys} = 10^{-5}$ *M*, then $Fe(OH)_{2(s)}$ can form when $9.01 < pH < 14.09$. If $Fe(II)_{T,sys} = 10^{-7}$ *M*, then $Fe(OH)_{2(s)}$ can form when $10.55 < pH < 11.98$. If $Fe(II)_{T,sys} = 10^{-9}$ *M*, then $Fe(OH)_{2(s)}$ cannot form at any pH.

$\log p_{CO_2} = -5.42$ (Figure 14.2). Above the line, $FeCO_{3(s)}$ is solubility limiting and is possible, and $Fe(OH)_{2(s)}$ is not possible and would be converted to $FeCO_{3(s)}$. *Below* the line, $Fe(OH)_{2(s)}$ is solubility limiting and is possible, and $FeCO_{3(s)}$ is not possible, and would be converted to $Fe(OH)_{2(s)}$. Along the line itself, both solids are possible at equilibrium; none, or one, or both may be present, depending on the particulars of the situation. For example, if $Fe(II)_{T,sys} = 0$, then neither solid can be present, regardless of which may set the limit for solubility.

14.3.1.2 $Fe^{2+}/Fe(OH)_{2(s)}$ and $Fe(OH)_{2(s)}/Fe(OH)_3^-$ Boundary Lines

$Fe^{2+}/Fe(OH)_{2(s)}$ – Solution/Solid BL. Consider that we are at some initial point A on the left side of the Figure 14.2 diagram, with $\log p_{CO_2} < -5.42$. $Fe(II)_{T,sys}$ for the diagram is 10^{-5} *M*, but no $Fe(OH)_{2(s)}$ is present at point A. As one moves from point A to higher pH, $Fe(II)_{T,sys} = 10^{-5}$ *M* is high enough that eventually the solution will become saturated with $Fe(OH)_{2(s)}$.

By analogy with the consideration in Figure 11.3 of the solubility of $(am)Fe(OH)_{3(s)}$, here we write

$$[Fe^{2+}] + [FeOH^{+}] + [Fe(OH)_2^{o}] + [Fe(OH)_3^{-}] = Fe(II)_T. \quad (14.8)$$

Neglecting activity corrections, using ${}^{*}K_s$ constants as introduced in Chapter 11, then for the pH that is just saturated without any $Fe(OH)_{2(s)}$ yet present,

$$\boxed{{}^{*}K_{s0}^{hyd}[H^{+}]^2 + {}^{*}K_{s1}^{hyd}[H^{+}] + K_{s2}^{hyd} + \frac{{}^{*}K_{s3}^{hyd}}{[H^{+}]} = Fe(II)_{T,sys}. \quad \text{can give two vertical BLs (lower pH, and higher pH)}} \quad (14.9)$$

Eq.(14.9) can give two BLs. At 25 °C/1 atm, $K_{s0}^{hyd} = 10^{-15.1}$, $K_{s1}^{hyd} = 10^{-10.6}$, $K_{s2}^{hyd} = 10^{-7.7}$, and $K_{s3}^{hyd} = 10^{-5.1}$, so with $Fe(II)_{T,sys} = 10^{-5}$ *M*, solving by LHS–RHS = 0 with a low pH initial guess gives pH = 9.01. This pH is basic, but it is still the lower pH root of the two roots for $Fe(II)_{T,sys} = 10^{-5}$ *M*. The second root is pH = 14.09 (see Figure 14.1). The ionic strength *I* in a solution with $pH \geq 13$ would be very high, so having neglected activity corrections to obtain the result for the second root

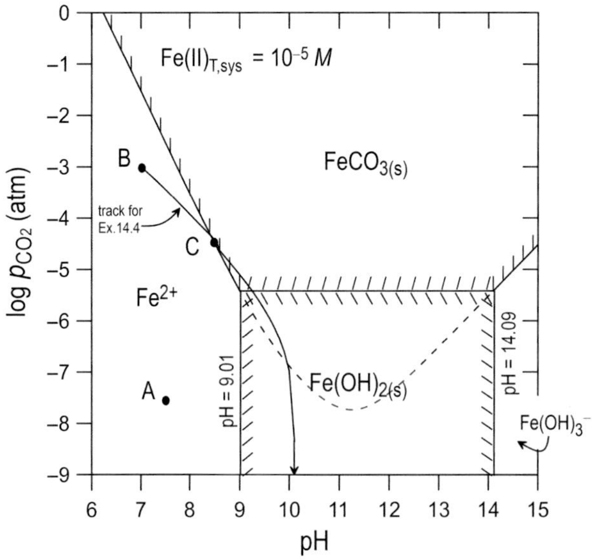

FIGURE 14.2 Log p_{CO_2}-pH predominance diagram for ferrous iron when $Fe(II)_{T,sys} = 10^{-5}$ *M*. Below log $p_{CO_2} = -5.42$, $FeCO_{3(s)}$ is not stable, so a dashed line is used for the solution/$FeCO_{3(s)}$ boundary line in that region. Points A, B, and C are discussed in the text.

and the representation of the very high pH portion of Figure 14.1 is not exactly correct. One could include activity corrections here, but it is definitely not worth the work: here the goal is to outline the overall chemistry, and pH values ≥13 are not found naturally anyway.

For the first root of Eq.(14.9), in Figure 14.2 we draw a vertical line at pH = 9.01 that starts at log $p_{CO_2} = -5.42$ and proceeds downward. The starting point identifies the left endpoint of the first line we discussed, log $p_{CO_2} = -5.42$. (At pH < 9.01, we cannot precipitate $Fe(OH)_{2(s)}$, so we cannot be interested in that part of an $FeCO_{3(s)}$/$Fe(OH)_{2(s)}$ equilibrium line.)

When log $p_{CO_2} < -5.42$ and pH ≤ 9.01, there is no $Fe(OH)_{2(s)}$ present. When log $p_{CO_2} < -5.42$ and 9.01 < pH < 14.09, some $Fe(OH)_{2(s)}$ is present. The amount of $Fe(OH)_{2(s)}$ present per liter of system water is given by $Fe(II)_{T,sys} - Fe(II)_T$, with $Fe(II)_T$ calculated by Eq.(14.8). The region below log $p_{CO_2} = -5.42$ with 9.01 < pH < 14.09 is labeled $Fe(OH)_{2(s)}$ because, *along with the solution phase*, the solid is present everywhere in that region, and possibly on the log $p_{CO_2} = -5.42$ line. Very close to pH = 9.01, there is less $Fe(OH)_{2(s)}$ per liter of system water than there is of dissolved Fe(II). All of the region is nevertheless labeled with $Fe(OH)_{2(s)}$; for solids, the region labeling rule is "if a solid is present in some portion of a predominance diagram, even though it will not predominate over other forms near some of the BLs, the *entire* region is labeled with the formula for that solid". It would therefore be more accurate to call these diagrams "predominance and/or presence diagrams" rather than "predominance diagram", but many years of precedence have established use of the latter.

Which dissolved Fe(II) species is dominant at pH = 9.01? Since we are on the line of saturation with $Fe(OH)_{2(s)}$, we can determine which species gives the largest term in Eq.(14.9) at pH = 9.01; the answer is Fe^{2+}, at 7.53×10^{-6} *M*, for ~75% of $Fe(II)_{T,sys}$. For pH values to the left of pH = 9.01, by Eq.(14.1), Fe^{2+} becomes more dominant over $FeOH^+$, so for all pH ≤ 9.01 we label the region with Fe^{2+}. To determine the particular distribution amongst the dissolved Fe(II) species for some pH < 9.01, one would proceed in a manner analogous with that used in Example 11.5 for dissolved Fe(III); anywhere inside the $Fe(OH)_{2(s)}$ region, or on its lines, we can use the equilibria with the solid.

$Fe(OH)_{2(s)}$/$Fe(OH)_3^-$. To the right of point A, as the pH moves higher from 9.01, the amount of $Fe(OH)_{2(s)}$ increases, then eventually starts to decrease towards zero. At the higher pH root of

Eq.(14.9), the solution is saturated, but solid is no longer present. Neglecting activity corrections, the higher pH root is 14.09 (see Figure 14.1). So, for the second BL from Eq.(14.9), in Figure 14.2 we draw a vertical line at pH = 14.09 downward from log $p_{CO_2} = -5.42$. Which dissolved species is dominant at pH = 14.09? Since we are on the line of saturation with $Fe(OH)_{2(s)}$, the dominant species is the one with the largest concentration term in Eq.(14.9); at pH = 14.09, that is $Fe(OH)_3^-$, at 9.98×10^{-6} *M*, or ~100% of $Fe(II)_{T,sys}$. For pH values to the right of 14.09, $Fe(OH)_3^-$ becomes even more dominant, and so for all pH ≥ 14.09, we label the region with $Fe(OH)_3^-$. Figure 14.2 goes out to pH = 15 because we want to see all this behavior, not because the pH = 14 to 15 range is highly relevant for natural systems.

14.3.1.3 $Fe^{2+}/FeCO_{3(s)}$ and $FeCO_{3(s)}/Fe(OH)_3^-$ Boundary Lines

$Fe^{2+}/FeCO_{3(s)}$. Consider now that we are at some initial point B on the left side of the Figure 14.2 diagram at which log $p_{CO_2} > -5.42$. As one moves from point B to higher pH, $Fe(II)_{T,sys} = 10^{-5}$ *M* is high enough that eventually the solution will become saturated with $FeCO_{3(s)}$. As with $Fe(OH)_{2(s)}$, for a given value of $Fe(II)_{T,sys}$, the "low-pH" collection of points at which saturation occurs will make a line. Along that line, Eq.(14.8) again applies, but now each of the dissolved Fe(II) species is set by $FeCO_{3(s)}$. Instead of Eq.(14.9), we now have

$$\frac{K_{s0}^{car}}{[CO_3^{2-}]}+\frac{K_{s0}^{car}}{[CO_3^{2-}]}\times\frac{{}^*K_{H1}}{[H^+]}+\frac{K_{s0}^{car}}{[CO_3^{2-}]}\times\frac{{}^*K_{H1}{}^*K_{H2}}{[H^+]^2}+\frac{K_{s0}^{car}}{[CO_3^{2-}]}\times\frac{{}^*K_{H1}{}^*K_{H2}{}^*K_{H3}}{[H^+]^3}=Fe(II)_{T,sys} \quad (14.10)$$

where activity corrections are again neglected. From Chapter 9, we have $[CO_3^{2-}] = K_1K_2K_Hp_{CO_2}/[H^+]^2$, so

$$\frac{K_{s0}^{car}}{K_1K_2K_Hp_{CO_2}}[H^+]^2\left(1+\frac{{}^*K_{H1}}{[H^+]}+\frac{{}^*K_{H1}{}^*K_{H2}}{[H^+]^2}+\frac{{}^*K_{H1}{}^*K_{H2}{}^*K_{H3}}{[H^+]^3}\right)=Fe(II)_{T,sys}. \quad (14.11)$$

can give two curved BLs
(lower pH, and higher pH)

At 25 °C/1 atm, $K_1K_2 = 10^{-16.68}$ and $K_H = 10^{-1.47}$; from Table 12.1, $K_{s0}^{car} = 10^{-10.68}$, ${}^*K_{H1} = 10^{-9.50}$, ${}^*K_{H2} = 10^{-11.10}$, ${}^*K_{H3} = 10^{-11.40}$. So,

$$\frac{10^{7.47}}{p_{CO_2}}[H^+]^2\left(1+\frac{10^{-9.50}}{[H^+]}+\frac{10^{-20.59}}{[H^+]^2}+\frac{10^{-31.99}}{[H^+]^3}\right)=Fe(II)_{T,sys}. \quad (14.12)$$

For log $p_{CO_2} = -5.42$ and $Fe(II)_{T,sys} = 10^{-5}$ *M*, solving by LHS–RHS = 0 with a low pH initial guess gives the root pH = 9.01. This is the same pH value that was obtained for the low pH root of Eq.(14.9) because at this log p_{CO_2} and pH, the two solids can be in equilibrium, so they specify the same value for each of the dissolved Fe(II) species: if one of them gives $Fe(II)_{T,sys}$ as the total dissolved Fe(II) at some pH, the other does as well. So, the point log $p_{CO_2} = -5.42$, pH = 9.01 is at the end of three different lines in the diagram: the horizontal $FeCO_{3(s)}/Fe(OH)_{2(s)}$ (solid/solid) boundary line (BL), the vertical $Fe^{2+}/Fe(OH)_{2(s)}$ (solution/solid) BL, and now the $Fe^{2+}/FeCO_{3(s)}$ (solution/solid) BL. Similarly, for log $p_{CO_2} = -5.42$ and $Fe(II)_{T,sys} = 10^{-5}$ *M*, a high pH initial guess with Eq.(14.12) gives the same root (pH = 14.09, within roundoff) as for the low pH root of Eq.(14.9). So, the point log $p_{CO_2} = -5.42$, pH = 14.10 is at the end of three different lines in the diagram: the horizontal $FeCO_{3(s)}/Fe(OH)_{2(s)}$ (solid/solid) BL, the vertical $Fe(OH)_{2(s)}/Fe(OH)_3^-$ (solid/solution) BL, and now the $FeCO_{3(s)}/Fe(OH)_3^-$ (solid/solution) BL.

To obtain more points for the $Fe^{2+}/FeCO_{3(s)}$ BL(we just have one so far), and more points for the $FeCO_{3(s)}/Fe(OH)_3^-$ BL, we could pick many more values of $\log p_{CO_2} = -5.42$ and for each solve numerically for the lower pH and higher pH roots. However, it is much easier to input pH values and compute the corresponding $\log p_{CO_2}$ values, then just keep the points for which $\log p_{CO_2} \geq -5.42$. These will be in two sets, a lower pH set (the lower pH roots) and a higher pH set (the higher pH roots). The lines are plotted in Figure 14.2. The $Fe^{2+}/FeCO_{3(s)}$ line slants to lower pH from pH = 9.01, so we know that the whole solution field on the left side is labeled with Fe^{2+}, not just the part to the left of the $Fe(OH)_{2(s)}$ region. Analogously, the whole solution field on the right side is labeled with $Fe(OH)_3^-$.

14.3.1.4 "Old-School" Approximations for Drawing Boundary Lines as Straight Lines

Before computers facilitated simplified geochemical calculations, locating solid/solution BLs in diagrams such as Figure 14.2 usually proceeded using simplifying assumptions as to which solution species dominated the speciation in the various regions. Many of the diagrams resulting from that era are still consulted today, so a brief word on how they were drawn is warranted.

The Two Solution/$Fe(OH)_{2(s)}$ Boundary Lines (BLs). For Figure 14.2, with Fe^{2+} still dominant by the time the pH is high enough to precipitate $Fe(OH)_{2(s)}$, the $Fe^{2+}/Fe(OH)_{2(s)}$ BL in earlier days would have been located using a simplified version of Eq.(14.9), namely $[Fe^{2+}] = K_{s0}^{hyd}/[OH^-]^2 \approx Fe(II)_{T,sys} = 10^{-5}$ *M*. This gives pH ≈ 8.94 rather than 9.01. That was close enough. On the other side of the $Fe(OH)_{2(s)}$ region, the simplified version of Eq.(14.9) would have been $[Fe(OH)_3^-] = K_{s3}^{hyd}[OH^-] \approx Fe(II)_{T,sys} = 10^{-5}$ *M*. That gives pH ≈ 14.10 rather than 14.09.

The Two Solution/$FeCO_{3(s)}$ Boundary Lines (BLs). For Figure 14.2, with Fe^{2+} still dominant in the upper part of the figure by the time the pH is high enough to precipitate $FeCO_{3(s)}$, the $Fe^{2+}/FeCO_{3(s)}$ BL would have been located using a simplified version of Eq.(14.11), namely $[Fe^{2+}] = K_{s0}^{car}[H^+]^2/(K_1K_2K_Hp_{CO_2}) \approx Fe(II)_{T,sys} = v10^{-5}$ *M*, which gives a straight log-log line with slope −2 that falls essentially on top of the corresponding BL in Figure 14.2 (which is slightly curved). On the other side of the $FeCO_{3(s)}$ region, the simplified version of Eq.(14.9) would have been $[Fe(OH)_3^-] = K_{s0}^{car}K_{H1}K_{H2}K_{H3}K_w^3/(K_1K_2K_Hp_{CO_2}[H^+]) \approx Fe(II)_{T,sys} = 10^{-5}$ *M*, which gives a straight log-log line with slope +1 that falls essentially on top of the corresponding BL in Figure 14.2.

Example 14.1 Calculating Solution Speciation When a Solid Is Not Present Using Metal Ion α Values

At point B in Figure 14.2, $\log p_{CO_2} = -3.0$ and pH = 7. If $Fe(II)_{T,sys} = 10^{-5}$ *M*, at 25 °C/1 atm, what is the concentration of each dissolved Fe(II) species? (Section 10.3.1 discusses the use of acid/base α values for metal ions.)

Solution

For dissolved Fe(II),

$$\alpha_0^{Fe(II)} = (1+{}^*K_{H1}/[H^+]+{}^*K_{H1}{}^*K_{H2}/[H^+]^2+{}^*K_{H1}{}^*K_{H2}{}^*K_{H3}/[H^+]^3)^{-1}.$$

At pH = 7.00, $\alpha_0^{Fe(II)} = 0.997$. $[Fe^{2+}] = 0.997 \times 10^{-5}$ *M*.

$\alpha_1^{Fe(II)} = \alpha_0^{Fe(II)}\,{}^*K_{H1}/[H^+] = 3.15 \times 10^{-3}$. At pH = 7, $[FeOH^+] = 3.15 \times 10^{-8}$ *M*.

$\alpha_2^{Fe(II)} = \alpha_1^{Fe(II)}\,{}^*K_{H2}/[H^+] = 2.50 \times 10^{-7}$. At pH = 7, $[Fe(OH)_2^o] = 2.15 \times 10^{-12}$ *M*.

$\alpha_3^{Fe(II)} = \alpha_2^{Fe(II)}\,{}^*K_{H3}/[H^+] = 9.97 \times 10^{-12}$. At pH = 7, $[Fe(OH)_3^-] = 9.97 \times 10^{-17}$ *M*.

Example 14.2 Calculating Solution Speciation When a Solid is Present Using Equilibrium with the Solid.1

Consider a system for which $Fe(II)_{T,sys} = 10^{-5}$ *M*, $\log p_{CO_2} = -3.5$, and pH = 9.5. According to Figure 14.2, $FeCO_{3(s)}$ is present. Compute the solution concentration of all the dissolved Fe(II) species, and also determine $[FeCO_{3(s)}]$, which is the amount of precipitated $FeCO_{3(s)}$ per liter of system water. Neglect activity corrections.

Solution

By equilibrium with $FeCO_{3(s)}$, then $[Fe^{2+}] = K_{s0}^{car}/[CO_3^{2-}]$. Substituting for $[CO_3^{2-}]$ gives

$[Fe^{2+}] = K_{s0}^{car}/(K_1K_2K_Hp_{CO_2}/[H^+]^2) = 9.33\times10^{-9}$ *M*.

$[FeOH^+] = [Fe^{2+}]^*K_{H1}/[H^+] = 9.33\times10^{-9}$ *M*.

$[Fe(OH)_2^0] = [Fe^{2+}]^*K_{H1}{}^*K_{H2}/[H^+]^2 = [FeOH^+]^*K_{H2}/[H^+] = 2.34\times10^{-10}$ *M*.

$[Fe(OH)_3^-] = [Fe^{2+}]^*K_{H1}{}^*K_{H2}{}^*K_{H3}/[H^+]^3 = [Fe(OH)_2^0]^*K_{H3}/[H^+] = 2.95\times10^{-12}$ *M*.

$[FeCO_{3(s)}] = Fe(II)_{T,sys} - [Fe^{2+}] - [FeOH^+] - [Fe(OH)_2^0] - [Fe(OH)_3^-] = 9.98\times10^{-6}$ *M*, so 99.8% of the system Fe(II) is precipitated.

Example 14.3 Calculating Solution Speciation When a Solid is Present Using Equilibrium with the Solid.2

Consider a system for which $Fe(II)_{T,sys} = 10^{-5}$ *M*, $\log p_{CO_2} = -7.0$, and pH = 9.1. According to Figure 14.2, $Fe(OH)_{2(s)}$ is present. Compute the solution concentration of all the dissolved Fe(II) species, and also determine $[Fe(OH)_{2(s)}]$, which is the amount of precipitated $Fe(OH)_{2(s)}$ per liter of system water. Neglect activity corrections.

Solution

By equilibrium with $Fe(OH)_{2(s)}$, then $[Fe^{2+}] = K_{s0}^{hyd}/[OH^-]^2$. Substituting for $[OH^-]^2$ gives

$[Fe^{2+}] = K_{s0}^{hyd}/(K_w^2/[H^+]^2) = 4.91\times10^{-6}$ *M*.

$[FeOH^+] = [Fe^{2+}]^*K_{H1}/[H^+] = 1.96\times10^{-6}$ *M*.

$[Fe(OH)_2^0] = [Fe^{2+}]^*K_{H1}{}^*K_{H2}/[H^+]^2 = [FeOH^+]^*K_{H2}/[H^+] = 1.98\times10^{-8}$ *M*.

$[Fe(OH)_3^-] = [Fe^{2+}]^*K_{H1}{}^*K_{H2}{}^*K_{H3}/[H^+]^3 = [Fe(OH)_2^0]^*K_{H3}/[H^+] = 9.80\times10^{-11}$ *M*.

$[Fe(OH)_{2(s)}] = Fe(II)_{T,sys} - [Fe^{2+}] - [FeOH^+] - [Fe(OH)_2^0] - [Fe(OH)_3^-] = 3.11\times10^{-6}$ *M*, so 31.1% of the system Fe(II) is precipitated.

Example 14.4 Moving in a $\log p_{CO_2}$ *vs.* pH Diagram with Conservation of System Alkalinity

Consider the chemistry at point B in Figure 14.2 where $\log p_{CO_2} = -3.0$, pH = 7, and $Fe(II)_{T,sys} = 10^{-5}$ *M* at 25 °C/1 atm. Neglect activity corrections for all parts of the problem. **a.** Compute Alk and C_T at point B; since $Fe(II)_{T,sys}$ is rather low relative to Alk, one can neglect terms for the hydroxo complexes of Fe(II) in Alk. **b.** Given the value of Alk, what is the highest the pH could reach if all the CO_2 was stripped from the solution and no solids precipitated? **c.** Assume that the system goes through a series of equilibrium states as the CO_2 outgasses so that $\log p_{CO_2}$ becomes lower, and lower pH becomes higher and higher. $FeCO_{3(s)}$ is the first solid to precipitate. This is a conservation of Alk situation for the solution up until that solid starts to precipitate (precipitating either $FeCO_{3(s)}$ or $Fe(OH)_{2(s)}$ removes Alk from the solution). For pH ≥ 7, compute C_T, $\log p_{CO_2}$, and the ion products for the two solids as a function of pH and verify that $FeCO_{3(s)}$ is the solid to become saturated first. Note the conditions (pH, C_T, and $\log p_{CO_2}$) when that happens. **d.** Continue the part c calculations to higher pH by neglecting the effect of solid precipitation on the solution Alk. What do the results indicate about whether an Fe(II) solid will be present once all the CO_2 was stripped from the solution? **e.** Re-solve the part d problem without making the simplifying assumption that Alk in the solution stays constant. This can be done by considering that there is a conservation of *system alkalinity,* with Alk_{sys} computed as a sum of the dissolved Alk + precipitated Alk. Alk_{sys} has units of eq per liter of system water. So, first develop an equation for Alk_{sys} which considers the presence of $FeCO_{3(s)}$. Then compute the pH- $\log p_{CO_2}$ track as the system moves through the $FeCO_{3(s)}$ region in Figure 14.2. What is the pH when the system arrives at the boundary with $Fe(OH)_{2(s)}$ at $\log p_{CO_2} = -5.42$? **f.** Continue to compute the pH and $\log p_{CO_2}$ track through the $Fe(OH)_{2(s)}$ region now using an expression for Alk_{sys} that considers the presence of $Fe(OH)_{2(s)}$.

Solution

a. Knowing pH and $\log p_{CO_2}$, Alk $= K_1K_Hp_{CO_2}/[H^+] + 2K_1K_2K_Hp_{CO_2}/[H^+]^2 + K_w/[H^+] - [H^+] = 1.51\times10^{-4}$ eq/L. Then, $C_T = K_Hp_{CO_2}/\alpha_0^C = 1.85\times10^{-4}$ *M*.
b. When no CO_2 is in the solution, then Alk = $[OH^-] - [H^+]$. With $[H^+]$ negligible relative to $[OH^-]$ at this Alk value, $1.515\times10^{-4} = 1.01\times10^{-14}/[H^+]$, so $[H^+] = 6.69\times10^{-11}$, pH = 10.176.
c. As CO_2 outgasses, the pH starts to go up, the race is on to see if either product, $[Fe^{2+}][CO_3^{2-}]$ or $[Fe^{2+}][OH^-]^2$, can reach its K_{s0} value for the corresponding solid, $FeCO_{3(s)}$ or $Fe(OH)_{2(s)}$ respectively. Up until one of the solids starts to precipitate, Alk in the solution stays constant at 1.51×10^{-4} eq/L, and all the calculations can be made as explicit functions proceeding from pH in a spreadsheet. $C_T = (\text{Alk} - [OH^-] + [H^+])/(\alpha_1^C + 2\alpha_2^C)$. For $FeCO_{3(s)}$, $K_{s0} = 10^{-10.68} = 2.09\times10^{-11}$. For $Fe(OH)_{2(s)}$, $K_{s0} = 10^{-15.10} = 7.94\times10^{-16}$. As indicated in the abbreviated table below, $FeCO_{3(s)}$ becomes saturated at pH = 8.54, with $Fe(OH)_{2(s)}$ not saturated at that point $(1.10\times10^{-16} < 7.94\times10^{-16})$.
d. $Fe(OH)_{2(s)}$ becomes saturated at pH = 9.01, but $\log p_{CO_2}$ at that point is still above −5.42, so for pH values somewhat greater than 9.01 we remain in the $FeCO_{3(s)}$ domain. The maximum pH is 10.176 when $p_{CO_2} = 0$. But, at pH somewhat above 10, $FeCO_{3(s)}$ becomes unsaturated, and $Fe(OH)_{2(s)}$ is saturated, the system will have crossed into and terminate in the $Fe(OH)_{2(s)}$ domain.
e. Once inside the $FeCO_{3(s)}$ domain, some of the solid has precipitated, and now Alk has been lowered from the initial value so we can no longer use it directly to compute the solution chemistry as a function of pH. The amount of precipitated $FeCO_{3(s)}$ per liter of water is given by $[FeCO_{3(s)}] = Fe(II)_{T,sys} - [Fe^{2+}] - [FeOH^+] - [Fe(OH)_2^o] - [Fe(OH)_3^-]$. Since the solid is present, we can use K_{s0}^{car} to obtain $[Fe^{2+}]$, and then use the $^*K_H$ values to obtain the hydroxo species from $[Fe^{2+}]$:

$$[Fe^{2+}] = K_{s0}^{car}/(K_1K_2K_Hp_{CO_2}/[H^+]^2);$$

$$[FeOH^+] = [Fe^{2+}]^*K_{H1}/[H^+];$$

$$[Fe(OH)_2^o] = [Fe^{2+}]^*K_{H1}{}^*K_{H2}/[H^+]^2;$$

$$[Fe(OH)_3^-] = [Fe^{2+}]^*K_{H1}{}^*K_{H2}{}^*K_{H3}/[H^+]^3;\ \text{so:}$$

$$[FeCO_{3(s)}] = Fe(II)_{T,sys} - K_{s0}^{car}/(K_1K_2K_Hp_{CO_2}/[H^+]^2)\times(1 + {}^*K_{H1}/[H^+] + {}^*K_{H1}{}^*K_{H2}/[H^+]^2 + {}^*K_{H1}{}^*K_{H2}{}^*K_{H3}/[H^+]^3)$$

which is a function of pH, but also a function of p_{CO_2}. We need a second equation to constrain p_{CO_2}. This is obtained from

$Alk_{sys} = \alpha_1^CC_T + 2\alpha_2^CC_T + [OH^-] - [H^+] + 2[FeCO_{3(s)}] = 1.515\times10^{-4}$ eq/L where C_T as usual is the dissolved-only total CO_2. Since $\alpha_0^CC_T = K_Hp_{CO_2}$, then $C_T = K_Hp_{CO_2}/\alpha_0^C$ so

Eq.(★): $Alk_{sys} = \alpha_1^CK_Hp_{CO_2}/\alpha_0^C + 2\alpha_2^CK_Hp_{CO_2}/\alpha_0^C + [OH^-] - [H^+] + 2[FeCO_{3(s)}] = 1.515\times10^{-4}$ eq/L.

Example 14.4 Some results for parts c and d. Ion products calculated for $FeCO_{3(s)}$ and $Fe(OH)_{2(s)}$ assuming no solids have precipitated)

A	B	C	D	E	F	G	H	I	J	K
									$FeCO_{3(s)}$ (K_{s0} = 2.09E−11) ion product	$Fe(OH)_{2(s)}$ (K_{s0} = 7.94E−16) ion product
pH	$[H^+]$	α_0^C	α_1^C	α_2^C	C_T	$\log p_{CO_2} = \log \alpha_0^C C_T/K_H$	$\alpha_0^{Fe(II)}$	$[Fe^{2+}] = \alpha_0^{Fe(II)} Fe(II)_{T,sys}$	$[Fe^{2+}][CO_3^{2-}] = \alpha_0^{Fe(II)} Fe(II)_{T,sys} \alpha_2^C C_T$	$[Fe^{2+}][OH^-]^2 = \alpha_0^{Fe(II)} Fe(II)_{T,sys}[OH^-]^2$
7.0	1.00E−07	0.183	0.817	3.82E−04	**1.85E−04**	−3.00	0.997	9.97E−06	7.06E−13	1.02E−19
7.5	3.16E−08	0.066	0.933	1.38E−03	1.62E−04	−3.44	0.990	9.90E−06	2.21E−12	1.01E−18
8.0	1.00E−08	0.022	0.974	4.55E−03	1.53E−03	−3.92	0.969	9.69E−06	6.76E−12	9.89E−18
c. 8.54	2.88E−09	0.0063	0.978	1.59E−02	1.47E−04	−4.46	0.901	9.01E−06	**2.09E−11 = K_{s0}**	1.10E−16
8.6	2.51E−09	5.49E−03	0.976	1.81E−02	1.46E−04	−4.52	0.888	8.88E−06	2.35E−11 > K_{s0}	1.44E−16
8.8	1.58E−09	3.43E−03	0.968	2.86E−02	1.42E−04	−4.73	0.833	8.33E−06	3.37E−11 > K_{s0}	3.38E−16
d. 9.01	9.77E−10	2.08E−03	0.952	4.56E−02	1.35E−04	−4.94 (> −5.42)	0.754	7.54E−06	4.65E−11 > K_{s0}	8.05E−16 > K_{s0}
9.5	3.16E−10	6.16E−04	0.871	0.129	1.06E−04	−5.47	0.494	4.94E−06	6.74E−11 > K_{s0}	5.05E−15 > K_{s0}
10.05	8.91E−11	1.31E−03	0.656	0.344	2.84E−05	−6.15	0.206	2.06E−06	2.09E−11 = K_{s0}	2.64E−14 > K_{s0}
10.176	6.67E−11	8.77E−05	0.588	0.412	2.17E−08	−6.32	0.159	1.59E−06	1.42E−14 < K_{s0}	**3.64E−14 > K_{s0}**

~fin~

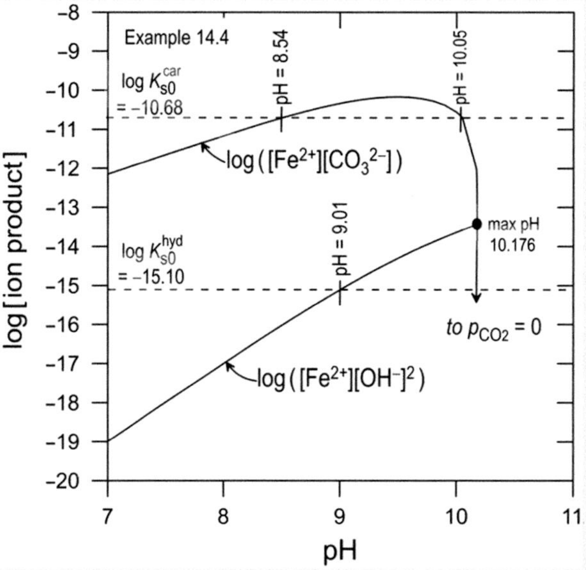

EXAMPLE 14.4 Parts a–d. Ion product values for $FeCO_{3(s)}$ and $Fe(OH)_{2(s)}$ *vs*. pH. Starting at point B in Figure 14.2 with $Fe(II)_{T,sys} = 10^{-5}$ *M*, CO_2 is gradually removed and pH rises. $FeCO_{3(s)}$ is the first solid to become saturated, at pH = 8.54. Continuing the calculations assuming Alk in solution remains at 1.515×10^{-4} eq/L, $Fe(OH)_{2(s)}$ becomes saturated at pH = 9.01, but log p_{CO_2} at that point is still above −5.42, so for pH values somewhat greater than 9.01 we remain in the $FeCO_{3(s)}$ domain. The maximum pH is 10.176 when $p_{CO_2} = 0$.

So, for each given pH value past 8.54 (*i.e.*, for each row) until $FeCO_{3(s)}$ disappears, we seek the log p_{CO_2} value that gives LHS−RHS = 0 (by Excel/Solver) for the Alk_{sys} equation, with $[FeCO_{3(s)}]$ computed as indicated above. Once pH = 9.25, the system is about to drop below log p_{CO_2} = −5.42, and thus into the $Fe(OH)_{2(s)}$ domain.

f. Once inside the $Fe(OH)_{2(s)}$ domain, some of that solid is precipitated, and as in part e, Alk is lower than the initial value. The amount of precipitated $Fe(OH)_{2(s)}$ per liter of water is given by $[Fe(OH)_{2(s)}] = Fe(II)_{T,sys} - [Fe^{2+}] - [FeOH^{+}] - [Fe(OH)_2^o] - [Fe(OH)_3^-]$. Since the solid is present, we can use K_{s0}^{hyd} to obtain $[Fe^{2+}]$, and then use the $^*K_H$ values to obtain the hydroxo species from $[Fe^{2+}]$:

$[Fe^{2+}]=K_{s0}^{hyd}/[OH^-]^2$;

$[FeOH^+] = [Fe^{2+}]{}^*K_{H1}/[H^+]$;

$[Fe(OH)_2^o] = [Fe^{2+}]{}^*K_{H1}{}^*K_{H2}/[H^+]^2$;

$[Fe(OH)_3^-] = [Fe^{2+}]{}^*K_{H1}{}^*K_{H2}{}^*K_{H3}/[H^+]^3$; so:

$$[Fe(OH)_{2(s)}] = Fe(II)_{T,sys} - K_{s0}^{hyd}/[OH^-]^2 \times (1 + {}^*K_{H1}/[H^+] + {}^*K_{H1}{}^*K_{H2}/[H^+]^2 + {}^*K_{H1}{}^*K_{H2}{}^*K_{H3}/[H^+]^3)$$

which is only a function of pH. We also have

Example 14.4 Some results for part e, $FeCO_{3(s)}$ is present

A	B	C	D	E	F	G	H	I	J	K
pH	$[H^+]$	α_0^C	α_1^C	α_2^C	log p_{CO_2}	p_{CO_2} (atm)	$[FeCO_{3(s)}]$	1E6*(LHS-RHS) with Eq.(★) for Alk_{sys} (for use with Solver using Column F value)	$FeCO_{3(s)}$ ion product $[Fe^{2+}][CO_3^{2-}]$ = $\alpha_0^{Fe(II)}\,Fe(II)_{T,sys}\,\alpha_2^C\,C_T$	$Fe(OH)_{2(s)}$ ion product $[Fe^{2+}][OH^-]^2$ = $\alpha_0^{Fe(II)}\,Fe(II)_{T,sys}[OH^-]^2$
8.6	2.51E–09	5.49E–03	0.976	1.82E–02	–4.63	2.33E–05	9.91E–07	~0	2.09E–11 = K_{s0}	1.29E–16 < K_{s0}
8.8	1.58E–09	3.43E–03	0.968	2.86E–02	–4.86	1.37E–05	3.49E–06	~0	2.09E–11 = K_{s0}	2.20E–16 < K_{s0}
9.0	1.00E–09	2.13E–03	0.953	4.46E–02	–5.10	7.93E–06	5.09E–06	~0	2.09E–11 = K_{s0}	3.80E–16 < K_{s0}
9.2	6.31E–10	1.31E–03	0.930	6.89E–02	–5.35	4.48E–06	6.05E–06	~0	2.09E–11 = K_{s0}	6.72E–16 < K_{s0}
9.25	5.62E–10	1.16E–03	0.922	7.67E–02	–5.41	3.86E–06	6.20E–06	~0	2.09E–11 = K_{s0}	7.80E–16 < K_{s0}

$Alk_{sys} = \alpha_1^C C_T + 2\alpha_2^C C_T + [OH^-] - [H^+] + 2[Fe(OH)_{2(s)}] = 1.515\times10^{-4}$ eq/L where C_T as usual is the dissolved-only total CO_2. Therefore, as was the case in part c, we can calculate C_T explicitly from pH, here using the dissolved Alk value as equal to $Alk_{sys} - 2[Fe(OH)_{2(s)}]$, so C_T is calculable explicitly as

$C_T = (Alk_{sys} - 2[Fe(OH)_{2(s)}] - [OH^-] + [H^+])/(\alpha_1^C + 2\alpha_2^C)$. Values up to the maximum pH are given in the table. The maximum pH (10.11) available is less than that computed in part b, because some of the total Alk has been removed from solution. The overall pH-log p_{CO_2} track, including the effects of solid precipitation on the dissolved Alk, is plotted in Figure 14.2 (starting at point B).

14.3.2 Dependence of the Diagram on $Fe(II)_{T,sys}$

For any element, lowering its amount per liter of system water will reduce the size(s) of any domain(s) in which the element can precipitate as a solid. If the level is lowered far enough, no solids will be able to form. Applied here, this means that if we reduce $Fe(II)_{T,sys}$ from 10^{-5} *M* to some lower number, the size of both the $Fe(OH)_{2(s)}$ and $FeCO_{3(s)}$ domains will decrease. For $Fe(OH)_{2(s)}$ this means that the two roots for Eq.(14.9) are moving closer (cf. the chemistry of (am)$Fe(OH)_{3(s)}$ in Figure 11.7), so the length of the $FeCO_{3(s)}/Fe(OH)_{2(s)}$ BL is becoming shorter. For $FeCO_{3(s)}$, the Eq.(14.11) curve is moving upwards. Once $Fe(II)_{T,sys}$ is reduced to 4.85×10^{-8} *M*, the two roots for Eq.(14.9) become identical (pH = 11.25) and: 1) the $FeCO_{3(s)}/Fe(OH)_{2(s)}$ BL is limited to that one point; 2) the minimum in the Eq.(14.11) log p_{CO_2} *vs.* pH curve for $FeCO_{3(s)}$ just touches that point; and 3) $Fe(OH)_{2(s)}$ can become saturated at that single pH, but will not form (cf. the circumstance of $Fe(III)_{T,sys} = 2.19\times10^{-10}$ *M* in Figure 11.7).

14.3.3 Log p_{CO_2} *vs.* pH Predominance Diagram for $Fe(II)_{T,sys} = 10^{-7}$ *M*

When $Fe(II)_{T,sys}$ is reduced from 10^{-5} *M* to 10^{-7} *M*, the two pH roots of Eq.(14.9) for $Fe(OH)_{2(s)}$ move towards each other to pH = 10.55 and pH = 11.98 (25 °C/1 atm). The $FeCO_{3(s)}/Fe(OH)_{2(s)}$ BL at log $p_{CO_2} = -5.42$ now extends only between those values. $Fe(OH)_3^-$ is dominant at pH = 11.98, so the entire solution region pH > 11.98 is labeled $Fe(OH)_3^-$. At pH 10.55, $FeOH^+$ is the dominant solution species in Eq.(14.8), so that the solution immediately to the left of pH = 10.55 is labeled $FeOH^+$. Since K_{H1} of Eq.(14.1) is $10^{4.5}$ at 25 °C/1 atm, neglecting activity corrections, we have a new vertical BL at pH = 9.50 on which $[Fe^{2+}] = [FeOH^+]$; for pH < 9.50, the solution region is labeled Fe^{2+}. Last, when $Fe(II)_{T,sys}$ is reduced from 10^{-5} *M* to 10^{-7} *M*, the Eq.(14.12) line for $FeCO_{3(s)}$ moves up by two log units. The result is Figure 14.3.

14.3.4 Log p_{CO_2} *vs.* pH Predominance Diagram for $Fe(II)_{T,sys} = 10^{-9}$ *M*

As seen in Figure 14.4, a number of things happen when $Fe(II)_{T,sys}$ is reduced from 10^{-7} *M* to 10^{-9} *M*. First, Eq.(14.9) for $Fe(OH)_{2(s)}$ no longer has any real positive roots at 25 °C/1 atm, so there no longer is a stability field for $Fe(OH)_{2(s)}$. Second, with no stability field for $Fe(OH)_{2(s)}$, there can no longer be a BL between $FeCO_{3(s)}$ and $Fe(OH)_{2(s)}$. ($Fe(OH)_{2(s)}$ is no longer possible in this system.) Third, the absence of the stability field for $Fe(OH)_{2(s)}$ fully unmasks what is going on in the solution, and we now have a solution dominance field for $Fe(OH)_2^o$; with log K_{H1}, log K_{H2}, and log K_{H3} values of 4.5, 2.9, and 2.6 at 25 °C/1 atm, the vertical BLs for transition in dominance between the various

Example 14.4 Some results for part f, $Fe(OH)_{2(s)}$ is present

A	B	C	D	E	F	G	H	I	J
pH	$[H^+]$	α_0^C	α_1^C	α_2^C	$[Fe(OH)_{2(s)}]$	dissolved Alk = 1.515E-04 −2$[FeCO_{3(s)}]$	C_T	$\log p_{CO_2}$	$Fe(OH)_{2(s)}$ ion product $[Fe^{2+}][OH^-]^2$
9.3	5.01E−10	1.03E−03	0.914	0.085	6.79E−06	1.38E−04	1.09E−04	−5.48	7.94E−16 = K_{s0}
9.5	3.16E−10	6.16E−04	0.871	0.129	8.42E−06	1.35E−04	9.10E−05	−5.78	7.94E−16 = K_{s0}
9.7	2.00E−10	3.62E−04	0.810	0.190	9.18E−06	1.33E−04	6.94E−05	−6.13	7.94E−16 = K_{s0}
9.9	1.26E−10	2.05E−04	0.729	0.271	9.55E−06	1.32E−04	4.11E−05	−6.60	7.94E−16 = K_{s0}
10.11	7.76E−11	1.08E−04	0.624	0.376	9.74E−06	1.32E−04	1.38E−06	−8.35	7.94E−16 = K_{s0}

~fin~

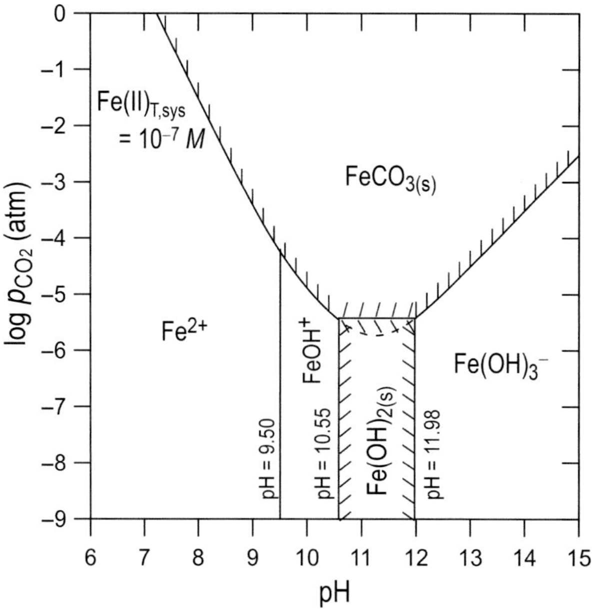

FIGURE 14.3 Log p_{CO_2}-pH predominance diagram for ferrous iron when $Fe(II)_{T,sys} = 10^{-7}$ *M*. Below log $p_{CO_2} = -5.42$, $FeCO_{3(s)}$ is not stable, so a dashed line is used for the solution/$FeCO_{3(s)}$ boundary line in that region.

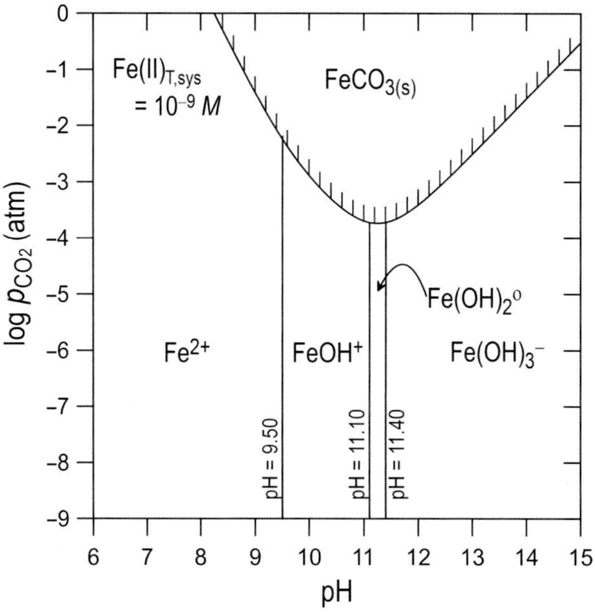

FIGURE 14.4 Log p_{CO_2}-pH predominance diagram for ferrous iron when $Fe(II)_{T,sys} = 10^{-9}$ *M*. The solution/$FeCO_{3(s)}$ boundary line lies above log $p_{CO_2} = -5.42$ for the entire pH range, so there is no stability field for $Fe(OH)_{2(s)}$, which unmasks the Fe(II) solution chemistry in the region around pH = 11.

dissolved Fe(II) species are pH = 9.50 for $[Fe^{2+}] = [FeOH^{+}]$, pH = 11.10 for $[FeOH^{+}] = [Fe(OH)_2^{o}]$, and pH = 11.40 $[Fe(OH)_2^{o}] = [Fe(OH)_3^{-}]$. Fourth, as compared to Figure 14.3, the Eq.(14.12) line for $FeCO_{3(s)}$ has moved up two log units.

REFERENCE

Ball JW, Nordstrom DK (1991/2001) *User's Manual for WATEQ4F, with Revised Thermodynamic Data Base and Test Cases for Calculating Speciation of Major, Trace, and Redox Elements in Natural Waters.* U.S. Geological Survey, Open-File Report 91-183, Menlo Park, CA, 1991. Revised and reprinted, April 2001.

15 The Kelvin Effect

The Effect of Particle Size on Dissolution and Evaporation Equilibria

15.1 INTRODUCTION

We commonly view the water solubility equilibrium constant for a crystalline solid such as calcite or sucrose as being a single particular value at some specific T and P; from Chapter 12, for calcite, $K_{s0} = 10^{-8.48}$ at 25 °C/1 atm. That singularity, however, requires that the crystals are macroscopic, not microscopic. This is because as crystal size becomes very small, thermodynamic solubility begins to increase. So, when one obtains a reference value for the water solubility of some solid material, it can generally be assumed that the value found pertains to the condition of macroscopic-sized crystals having been equilibrated with water, as for calcite that $K_{s0} = 10^{-8.48}$ at 25 °C/1 atm. The same applies to the solubility of liquids in water, and to the volatility into air of liquids and solids.

Many readers of this text have encountered particle size-dependent solubility in a chemistry laboratory course. Such a course often includes some type of experiment which leads to crystallization of a solid from a liquid medium. During the initial precipitation event, there may be rapid formation of numerous small particles. Although small crystals are relatively more soluble than larger crystals of the same solid, very small particles can nevertheless initially form, since the liquid medium in which the precipitation occurs is initially grossly supersaturated. Only later, as the liquid/solid system is allowed to sit for a week or two, and the gross supersaturation is relieved, can the numerous small crystals morph into fewer larger crystals (Figure 15.1). This is called "Ostwald ripening" after the German chemist Wilhelm Ostwald, the 1909 recipient of the Nobel Prize for work on chemical equilibrium and chemical reaction velocities. Ostwald ripening occurs because, by the Kelvin effect (after Lord Kelvin, 1824–1907, Scots–Irish physicist and engineer), a small crystal is more soluble in a liquid than a somewhat larger crystal that is nearby (the crystals will never all be the same size). This solubility difference creates a Fickian diffusion gradient in the dissolved concentration (Figure 15.2). The concentration halo around the smaller crystal causes the dissolved concentration near the larger crystal to be higher than the saturation value for that size, so mass begins to precipitate on the larger crystal causing the larger crystal to grow. This lowers the concentration around the smaller crystal, so it starts to dissolve, becoming even smaller. The result is a feedback loop that continuously and increasingly favors the loss of smaller crystals and growth of larger crystals.

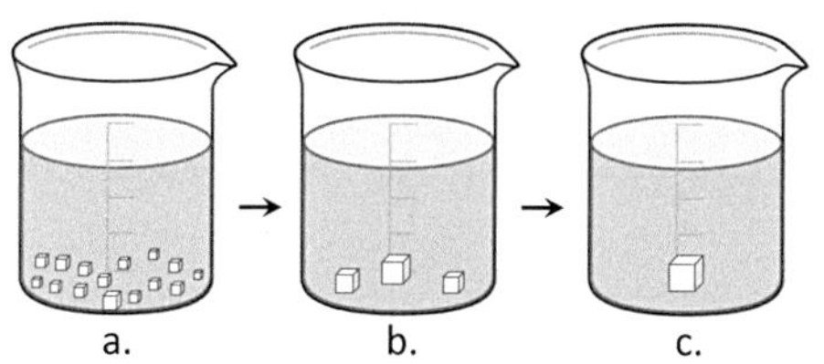

FIGURE 15.1 Process of crystal growth during "Ostwald ripening" in a beaker of liquid from: a) many very small, possibly amorphous particles; to b) fewer larger crystals; to, finally, c) one relatively large crystal.

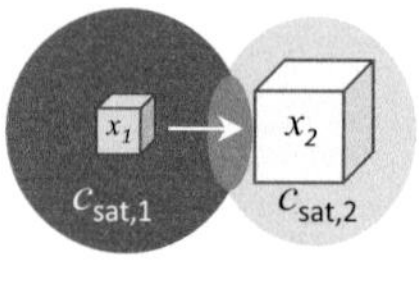

FIGURE 15.2 The two neighboring small crystals are of characteristic dimensions x_1 and x_2. $x_1 < x_2$. The concentration in the saturation halo around the x_1 crystal is $c_{sat,1}$. The concentration in the saturation halo around the x_2 crystal is $c_{sat,2}$. $c_{sat,1} > c_{sat,2}$. The concentration gradient (arrow) between the two crystals drives diffusion from around the small crystal towards the larger crystal. This leads to dissolution of the smaller crystal and precipitation on the larger crystal, *i.e.*, "Ostwald ripening".

In case some readers have not had the above experience in a chemistry laboratory course, let's talk a bit about the familiar "long-term storage of ice cream in the freezer" experiment. The phenomenon exactly as pictured in Figure 15.1 occurs when ice cream sits in a freezer. Initially, the ice crystals in the frozen cream system are very small, and the ice cream is nice and creamy. But the ice crystals are not all the same size. So, a small ice crystal is more soluble in the cream system than a somewhat larger ice crystal that is nearby. This creates a gradient in the water concentration in the cream system so that the water concentration is higher around the smaller crystals and lower around the larger crystals. In time, the ice cream becomes a mess of large ice crystals. Or, what about the "long-term storage of bread in the freezer" experiment. Say a loaf of bread in a plastic bag is placed in the freezer and left there for several months. Initially, the water in the bread is distributed throughout the bread; when the bread initially freezes, many very small crystals of ice are formed. By operation of the above effect, this time on the vapor pressure of water above the ice particles, the very small ice crystals will be lost and morph into a smaller number of large ice crystals. The concentration gradient that is moving the water is operating in the gas phase of the bread pores: small crystals of water have a higher equilibrium vapor pressure than do larger crystals. After several months, there is an astonishing amount of visible ice inside the plastic bag.

In natural water systems, conditions can change abruptly and a mineral can rapidly precipitate under conditions of high supersaturation. An example is the formation of $CaCO_{3(s)}$ particles during a "whiting event" in a lake. The small particles can settle into the sediments and there undergo Ostwald ripening (Morse and Casey, 1988). If left undisturbed long enough, a mass of small mineral crystals can morph into a very few large crystals (Figure 15.1c), which is exactly why crystal specimens that have formed over geologic time scales can be so large and beautiful. The effects of Ostwald ripening in geochemistry continue to be studied (*e.g.*, Poonoosamy et al., 2016; Steefel and Van Cappellen, 1990), and in the biological sciences as with hydroxyapatite in animal bone (Wong et al., 1995). Steefel and Van Cappellen (1990) include Ostwald ripening in a model of chemical weathering of granite.

This chapter considers the thermodynamics underlying the Kelvin effect: we need to know how small particle sizes need to be for Kelvin effect to influence solubility calculations. That said, most natural water chemists do not frequently make corrections for the Kelvin effect because: 1) very small particle sizes are more often the exception than the rule (because they are unstable); and 2) it is difficult to make such calculations reliably because there is never any single size, we typically do not know the sizes or shape uniformity very well, and the value of the interfacial tension σ between the two phases that causes the effect is material dependent, and usually not well known.

15.2 THE INTERFACIAL TENSION σ

15.2.1 The Origin of Interfacial Tension

For mineral solubility in water, the interface is between the mineral solid and the water. For ice crystals in bread, the interface is between the ice and the pore air in the bread. Some authors use γ for interfacial tension, and some call it the "surface tension". This text uses γ for activity coefficient on the molal scale, so a different symbol is needed here; we chose σ, and always call it the "interfacial tension".

The value of σ depends on the identities of the two adjacent phases, be they solid/liquid, solid/gas, liquid/liquid, or liquid/gas. So, the value of σ for $Fe(OH)_{3(s)}$/water is different from that for $Al(OH)_{3(s)}$/water. And, when liquid water is one of the phases, changing the nature of the water will have an effect, so σ for $CaCO_{3(s)}$/fresh water is different from that for $CaCO_{3(s)}$/seawater. Common units for σ are ergs/cm^2.

Dimensionally, σ has units of work (or energy) per unit area: it is a measure of how many ergs are required to increase the interfacial area by one cm^2. Energy per unit area is equivalent to force per unit length, so σ can also be expressed with units of dyne/cm (1 erg = 1 dyne-cm): σ can thus be thought of as the force (as with dynes) being exerted along the edge of a surface, per unit of edge length (as with cm), acting to reduce the surface area. The parcel of material in Figure 15.3a is composed of 64 cube sub-units. The parcel is surrounded by some host phase, so the interfacial area with the host is $(4\times4)\times6 = 96$ faces. The parcel is held together because each internal sub-unit face is attracted to the adjacent cube face, as for example by van der Waals forces, hydrogen bonding, etc. Let w_{inside} represent the work required to pull apart two sub-units interacting on one mutual adjacent internal face. There are 144 internal face interactions. Let w_{outside} represent the work required to completely separate one outside (*i.e.*, interfacial) sub-unit face from the adjacent host phase.

Consider now that the 64 cubic sub-units in Figure 15.3a are rearranged to form the configuration in Figure 15.3b. There is then an interfacial area with the host phase of 112 faces, and 136 internal adjacent face interactions. Going from 144 to 136 internal adjacent face interactions requires that positive work be done, specifically $8w_{\text{inside}}$. Going from 96 to 112 outside faces involves negative work (energy released), namely $-16w_{\text{outside}}$, because $+16w_{\text{outside}}$ would be the amount of work needed to *break* that number of attractive outside interfacial interactions. The net Δ work done is then

$$\Delta\,\text{work} = 8w_{\text{inside}} - 16w_{\text{outside}}. \tag{15.1}$$

This Δ work gave us a net increase in surface area:

$$\Delta\,\text{interfacial area} = 112\text{ faces} - 96\text{ faces} = 16\text{ faces}. \tag{15.2}$$

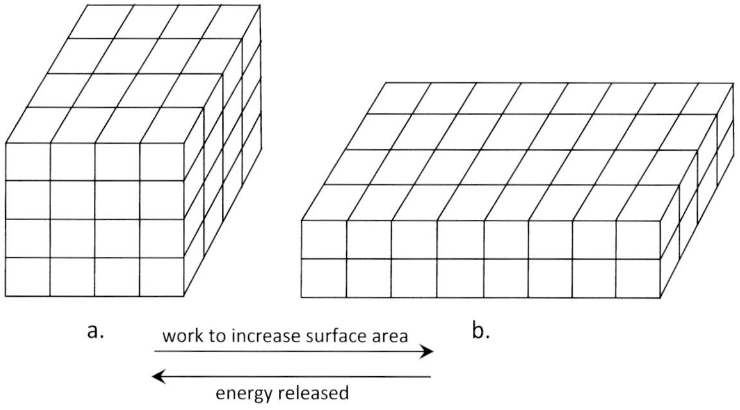

FIGURE 15.3 a. A cubic parcel of material composed of 64 cubic sub-units with 144 interior adjacent-face interactions, an interfacial area of 96 faces, and so 96 outside face interactions with the host phase; b. The 64 sub-units now arranged so that there are 136 interior adjacent-face interactions, an interfacial area of 112 faces, and so 112 outside face interactions with the host phase.

Each face has some value in cm². So, for the interfacial tension, a "finite difference" estimate of the work required per unit increase in area is

$$\sigma = \frac{\Delta\,\text{work}}{\Delta\,\text{area}} = \frac{8w_{\text{inside}} - 16w_{\text{outside}}}{16\text{ faces (cm}^2)}. \tag{15.3}$$

More precisely, σ is defined as the *differential* amount of work required to increase the surface area, that is,

$$\sigma \equiv \frac{dW}{dA}. \tag{15.4}$$

Since work per unit area is equivalent to force exerted by the interface per unit length along some edge, on an edge length x the interface exerts a force given by

$$\text{interfacial force on edge of length } x = \sigma x. \tag{15.5}$$

The tension force acting in an interface is analogous with the tension in the skin of a drum, in the skin of a balloon, etc. We know neither a drum skin nor a balloon is in favor of increasing its interfacial area. In fact, each is pulling to decrease its area. The interfacial tension thus causes a given volume of material to *pull in on itself*, to *reduce the surface area*, if possible. A blob of liquid water in zero gravity thus tends to adopt a spherical shape because that is the shape that has the least surface area for the given volume. This relates to how a water strider insect can "walk on water" (Figure 15.4): its legs are hydrophobic and cannot easily penetrate the water surface. Gravity pulling it downward leads to a slight concave depression at each leg, increasing the air/water interfacial surface area at the end of each leg, causing balancing upward forces on the insect.

15.2.2 The Effects of Surfactants on Interfacial Tension

A chemical that when added to a system changes the interfacial tension of some interface is a "surfactant", a portmanteau of surface + actant. Soaps and detergents are surfactants; they reduce the interfacial tension that water has with air, and with hydrophobic (a.k.a. "oily") phases.

For a clean water/air interface, at 25 °C/1 atm the interfacial tension is σ = 72.0 dynes/cm. For a soapy water/air interface, σ can be much lower, ~30 dynes/cm. When σ is lowered, less work is required to increase the surface area and create a bubble. Similarly, once a bubble is formed, when a surfactant is present, the energy driving force for collapse of the bubble has been lowered: soap bubbles can drift through air for some period before they burst due to desiccation, and/or thinning of the upper portion of the bubble due to gravity. Remarkably, epithelial cells in the alveoli of our lungs

FIGURE 15.4 Water strider on water. (Credit: Shutterstock. Photo ID number 128581670.)

secrete a mixture of lipid and protein compounds that act as a "pulmonary surfactant" to lessen the work required of us when we breathe and inflate the *~500 million* tiny alveolar "bubbles" in adult lungs (Ochs et al., 2004).

We understand now why, in the absence of a surfactant, it is difficult to disperse an oily phase into small water-suspended droplets: one large mass is the preferred state. But, when a surfactant is added, the work required to disperse the oil is greatly reduced so that completely cleaning up a greasy pot in the kitchen sink (or removing oily materials from clothes) is greatly facilitated: the oil/grease is suspendable for removal as innumerable fine, emulsified droplets. Surfactants, then, are extensively used in dealing with oil spills. Consider the Deepwater Horizon underwater crude oil well blowout event that began on April 20, 2010. The U.S. Coast Guard (2011) estimated a total spill volume of 210 million U.S. gallons, and that during the release a total of at least 1.8 million U.S. gallons of the "dispersants" Corexit 9500 and Corexit 9527 were applied on the water surface and injected subsea at the wellhead into the rapidly flowing oil (Chakraborty et al., 2012). While this application greatly reduced the formation of slicks on the sea surface, and promoted biological degradation, it also led to much oil being dispersed subsurface into the water column (and reducing dissolved oxygen), and depositing on the sea floor. Studies carried out later that year showed that deposition of a brown, oil-laden floc had greatly damaged deepwater coral communities as far as 11 km from the wellhead (White et al., 2012). So, sure, surfactants can be used to disperse oil spills, but the benefits can be ambiguous: there will be shift to another environmental compartment to deal with the release. When we use dishwashing liquid to clean our greasy pots and pans in the kitchen the dispersed grease still needs to be degraded someplace … in a water treatment plant, a septic tank/field system, or wherever the washwater ends up.

15.3 THE INTERFACIAL TENSION AND THE PRESSURE INCREASE ACROSS A CURVED INTERFACE

For a mass of material pulled into a spherical shape by action of the interfacial tension, the pressure inside the sphere is greater than the pressure outside. The magnitude of the pressure differential can be determined using a force balance on the sphere as divided into two hemispheres (Figure 15.5). The force generated by the interfacial tension for the upper hemisphere is exactly counterbalanced by the force generated for the lower hemisphere. Along the mutual circular edge length $2\pi r$, the upper hemisphere is pulling up on the lower hemisphere with force $2\pi r\sigma$, and the lower hemisphere is pulling down on the upper hemisphere with force $2\pi r\sigma$. The two balanced hemispheres compress the material contained within them.

Pressure has units of force per unit area. The force creating the pressure increase inside the sphere is $2\pi r\sigma$. We do not add the forces from both hemispheres to obtain $4\pi r\sigma$ because counterbalancing

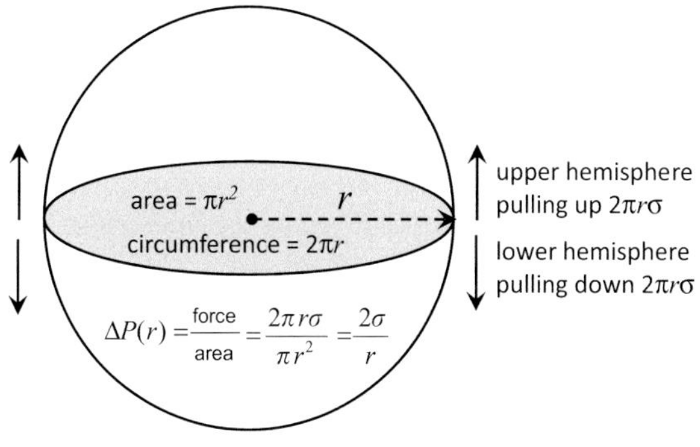

FIGURE 15.5 Interfacial tension balance across two hemispheres of some material surrounded by another phase.

forces do not increase the pressure; they just make for a static system. As an analogy, if we are swimming underwater in a pool, the pressure we are under is being generated by the water column above us. The fact that the water below us is also pressing up with equal force does not increase the local pressure, it just counterbalances it, and prevents the upper water from being pushed deeper.

For the sphere in Figure 15.5, the area over which the force $2\pi r\sigma$ is acting is the circular area on the bottom of the hemisphere, that is, πr^2. Thus,

$$\Delta P(r) = \frac{\text{force}}{\text{area}} = \frac{2\pi r\sigma}{\pi r^2} = \frac{2\sigma}{r}. \tag{15.6}$$

As $r \to \infty$ so that the surface becomes flat, then $\Delta P(r) \to 0$. $\Delta P(r)$ increases rapidly as $r \to 0$, in fact then $\Delta P(r) \to \infty$. For a water droplet in air, σ is the water/air interfacial tension; for a solid in water, σ is the solid/water interfacial tension. With σ expressed in units of ergs/cm^2, to obtain units for pressure from Eq.(15.6), we need a conversion factor between ergs and units for PV work as (pressure × volume). For PV work in L-atm, the conversion factor is 9.869×10^{-10} L-atm/erg.

Example 15.1 ΔP for Small Droplets of Water

At 20 °C, σ measured for a clean water/air interface is 72.8 dyne/cm = 72.8 ergs/cm^2. Calculate ΔP for a water droplets radii of 10, 100, 500, and 1000 Å.

Solution

1 Å = 10^{-8} cm, so values of 10, 100, 500, and 1000 Å correspond to $r = 10^{-7}$, 10^{-6}, 5×10^{-6}, and 10^{-5} cm, respectively. By Eq.(15.6) and using the conversion factor 9.869×10^{-10} L-atm/erg,

$$\Delta P(r) = \frac{2(72.8\ \text{erg/cm}^2)(9.8695 \times 10^{-10}\ \text{L} - \text{atm/erg})(1000\ \text{cm}^3/\text{L})}{r\ (\text{cm})}.$$

For radii of 10, 100, 500, and 1000 Å, then ΔP = 1440, 144, 28.7, and 14.4 atm respectively. Hydrostatic pressure in a column of fresh water increases at the rate of 1 atm for every 10.33 m of depth. So, to achieve 1440 atm added pressure by going underwater, one would have to be about 14 km deep. For a 10 Å droplet, though, it should be added that it is somewhat sketchy to have assumed that a value for σ measured using macroscopic bulk liquid water (72.8 dyne/cm) applies meaningfully to such a small cluster of molecules; the general trend is here is still valid, even if the exact values computed at very small sizes like 10 Å are questionable.

15.4 EFFECT OF $\Delta P(r)$ ON CHEMICAL POTENTIAL AND EQUILIBRIUM

15.4.1 General

The chemical potential of a chemical substance μ is defined as the free energy G per mol of that substance. The dependence of μ pressure P and T is given by the fundamental thermodynamic equation

$$d\mu = \bar{V}dP - \bar{S}dT \tag{15.7}$$

where $\bar{V}$ is the volume per mol (*i.e.*, "molar volume"), and $\bar{S}$ is the entropy per mol (*i.e.*, "molar entropy"), both as evaluated at T and P. At constant temperature ($dT = 0$),

$$d\mu = \bar{V}dP. \tag{15.8}$$

The quantity $\overline{V}$ is always positive, so increasing the pressure on any material at constant T will always increase μ. As an analogy, when you try to compress any physical materials, the material acts like a spring and resists. The compression work is stored in the increasing internal repulsion, and μ goes up. So, when you increase the molar free energy, the tendency to exit the phase by moving across the interface is increased: the vapor pressure for water inside a small droplet is greater than for water with a flat surface; the solubility of a tiny droplet of TCE in water is larger than that of a large droplet; the solubility of a small crystal of solid is greater than that for a large crystal.

We integrate Eq.(15.8):

$$\int d\mu = \int \overline{V} dP. \tag{15.9}$$

For this equation, all of the kinds of phases in which we are interested are either liquids or solids, *i.e.*, condensed phases. For condensed phases, $\overline{V}$ can generally be considered to be nearly independent of P for pressures less than a few thousand atm. Under that condition, Eq.(15.9) becomes

$$\int d\mu = \overline{V} \int dP. \tag{15.10}$$

Integration and then substitution of Eq.(15.6) give

$$\Delta\mu = \overline{V}\Delta P = \overline{V}\frac{2\sigma}{r}. \tag{15.11}$$

Expressed in words, $\Delta\mu$ is the difference computed as [μ for the material under a curved interface of radius r] minus [μ for the material under a flat interface]. Note that since $2\sigma/r$ has units of pressure, and $\overline{V}$ has units of volume per mol, so $\overline{V}2\sigma/r$ has units of work/mol (*i.e.*, energy/mol), as expected.

Example 15.2 $\Delta\mu$ for Small Droplets of Water

σ for a clean water/air interface is 72.8 dyne/cm = 72.8 ergs/cm^2. 1 L = 1000 cm^3 of liquid water weighs very close to 1000 g at 20 °C, so 1 cm^3 weighs 1 g. The MW of water is 18.01 g/mol. Obtain $\overline{V}$ for liquid water at 20 °C, and compute $\Delta\mu$ in kJ/mol for water droplet sizes of 10, 100, 500, and 1000 Å. 1 kJ = 10^{10} erg.

Solution

For 1 mol of water, at 1 g/cm^3, we require 18.01 cm^3: $\overline{V}$ = 18.01 cm^3/mol.

$$\Delta\mu(r) = (18.01\,\text{cm}^3/\text{mol})\frac{2(72.8\text{ ergs/cm}^2)}{r(\text{cm})}\left(\frac{1\,\text{kJ}}{10^{10}\,\text{ergs}}\right)$$

10, 100, 500, and 1000 Å correspond to r = 10^{-7}, 10^{-6}, 5×10^{-6}, and 10^{-5} cm, respectively. These values give $\Delta\mu$ = 2.6, 0.26, 0.052, and 0.026 kJ/mol, respectively.

15.4.2 Equilibrium across a Flat Interface

As noted above, particle size affects multiple types of interfacial equilibria. It is perhaps easiest to present the applicable math for the case of volatility of water droplets into air, and that case is very important anyway as regards how droplets of rain form in the atmosphere. Figure 15.6a illustrates the vaporization of liquid water from a flat interface into air. The vaporization reaction is

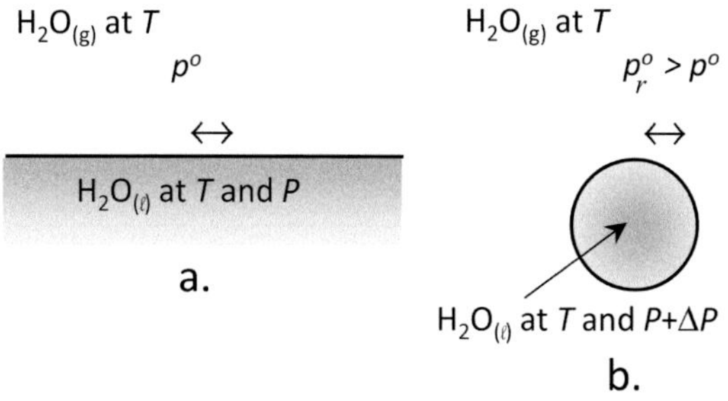

FIGURE 15.6 a. Vaporization of liquid water from a flat interface into air. b Vaporization of liquid water from a spherical interface of radius r into air.

$$\left[H_2O_{(\ell)}\right]_{\text{flat}} = H_2O_{(g)}. \tag{15.12}$$

Under the condition of chemical equilibrium (saturation), the values of the chemical potential for water in the two phases are equal:

$$\left[\mu \text{ of } H_2O_{(\ell)}\right]_{\text{flat}} = \mu \text{ of } H_2O_{(g)}. \tag{15.13}$$

The value of T in the two phases is the same. And, since the interface between the two phases is flat, across the interface $\Delta P = 0$, so P in both phases is also the same. We now expand the μ for each phase as the sum of a standard state chemical potential μ° plus an $RT\ln$(activity) term. The concentration scale most convenient for the liquid water is the mol fraction scale x_1. For the gas phase water, we use pressure p:

$$\mu_\ell^\text{o}(T,P) + RT\ln x_\ell = \mu_\text{g}^\text{o}(T) + RT\ln p. \tag{15.14}$$

Ideality is assumed in the liquid water phase, so that the mole-fraction-scale liquid-phase activity coefficient ζ_1 is unity: $a_\ell = x_\ell\zeta_\ell = x_\ell$. We do not need to specify a P value for $\mu_\text{g}^\text{o}(T)$ because μ_g depends on total P only insofar as non-idealities in the gas phase need to be considered, and these are negligible at ambient P values. Since p is the gas phase pressure of water, if $P > p$, then in the gas phase, there are other gases besides water present to provide the added pressure.

For the case of vaporization of essentially pure water, then $x_\ell \approx 1$, so $p = p^\circ$, the saturation vapor pressure of pure liquid water at T and P. Thus,

$$\mu_\ell^\text{o}(T,P) = \mu_\text{g}^\text{o}(T) + RT\ln p^\text{o} \tag{15.15}$$

or

$$\mu_\ell^\text{o}(T,P) - \mu_\text{g}^\text{o}(T) = RT\ln p^\text{o}. \tag{15.16}$$

Since we have taken for the liquid water that $a_\ell = 1$, we recognize p° as being the *equilibrium constant K* for the vaporization of pure liquid water across a flat water/air interface. And, the LHS of Eq.(15.17) is $-\Delta G^\circ$ (kJ/mol) for the vaporization. Thus, Eq.(15.16) has the form

$$\Delta G^\text{o}(T,P) = -RT\ln K(T,P). \tag{15.17}$$

So we see that the equilibrium constant for $H_2O_{(\ell)} = H_2O_{(g)}$ as we would have calculated it prior to encountering this chapter (as based on ΔG°) pertains to equilibrium across a flat interface.

15.4.3 Equilibrium across a Spherical Interface of Radius *r*

For $H_2O_{(\ell)} = H_2O_{(g)}$ equilibrium across a spherical interface of radius r (see Figure 15.6b), with p_r denoting the gas phase water saturation pressure for x_l and a spherical interface of radius r, Eqs.(15.12–15.14) become

$$\left[H_2O_{(\ell)}\right]_r = H_2O_{(g)} \tag{15.18}$$

$$\left[\mu \text{ of } H_2O_{(\ell)}\right]_r = \mu \text{ of } H_2O_{(g)} \tag{15.19}$$

$$\mu_\ell^o(T, P + \Delta P) + RT \ln x_\ell = \mu_g^o(T) + RT \ln p_r. \tag{15.20}$$

Now, for any essentially pure liquid ($x_\ell \approx 1$), then $p_r = p_r^o$ and

$$\mu_\ell^o(T, P + \Delta P) = \mu_g^o(T) + RT \ln p_r^o. \tag{15.21}$$

Subtracting Eq.(15.15) from Eq.(15.21), the $\mu_g^o(T)$ terms cancel, leaving

$$\mu_\ell^o(T, P + \Delta P) - \mu_\ell^o(T, P) = RT \ln p_r^o - RT \ln p^o. \tag{15.22}$$

The LHS is $\Delta\mu$ as given by Eq.(15.11). Thus,

$$\frac{2\bar{V}_\ell \sigma_\ell}{r} = RT \ln \frac{p_r^o}{p^o} \tag{15.23}$$

or

$$\boxed{\ln \frac{p_r^o}{p^o} = \frac{2\bar{V}_\ell \sigma_\ell}{rRT}} \quad \text{Kelvin Equation for liquid vaporization} \tag{15.24}$$

where σ_ℓ is the liquid/air interfacial tension, and $\bar{V}_\ell$ is the molar volume of the liquid. p_r^o always increases as r decreases.

Just as p° is the equilibrium constant for vaporization of pure water from a flat interface, p_r^o is the equilibrium constant for vaporization of pure water from an interface of radius r. We then deduce the general result (including for dissolution of a mineral into water) that

$$\boxed{\ln \frac{K_r}{K} = \frac{2\bar{V}\sigma}{rRT}} \quad \text{Kelvin Equation-general form.} \tag{15.25}$$

K_r always increases as r decreases. Expressed with units that are convenient here, $R = 8.31441$ J/K-mol. (1 J = 10^7 ergs.)

If σ has units of dyne/cm (ergs/cm^2), $\bar{V}$ has units of cm^3/mol, and r has units of cm, then Eq.(15.25) yields

$$\log K_r = \log K + \frac{1}{2.303} \times \frac{2\,(\bar{V}\text{ cm}^3/\text{mol})(\sigma \text{ ergs/cm}^2)}{(r\text{ cm})(8.31441\text{ J/K-mol})(T\text{ K})(10^7\text{ ergs/J})} \tag{15.26}$$

$$\boxed{\log K_r = \log K + 1.04 \times 10^{-8} \frac{\bar{V}\sigma}{rT}.} \tag{15.27}$$

Example 15.3 p_r^o for Small Droplets of Water

At 20 °C, the vapor pressure of pure water p^o = 17.5 Torr. σ for a clean water/air interface is 72.8 dyne/cm = 72.8 ergs/cm^2. $\bar{V}$ = 18.01 cm^3/mol. Using Eq.(15.27), compute the vapor pressure of water p_r^o in Torr for droplet sizes of 10, 100, 500, and 1000 Å.

Solution

We rewrite Eq.(15.27) as

$$\log p_r^o = \log p^o + 1.04 \times 10^{-8} \times (18.01) \times (72.8)/(r \times 293.15).$$

Droplet radii of 10, 100, 500, and 1000 Å correspond to $r = 10^{-7}$, 10^{-6}, 5×10^{-6}, and 10^{-5} cm, respectively. With p^o = 17.5 Torr (flat surface), these values give p_r^o = 51.1, 19.5, 17.9, and 17.7 Torr, respectively. Relative to a flat surface, the corresponding RH values are 290, 111, 102, and 101%. The p_r^o value for a droplet with a radius of 1000 Å is close to p^o. The results indicate why rain formation in the atmosphere requires the presence of cloud condensation nuclei (CNC): 1) if no CNC are present, nascent droplets have to start growing from r = 0; 2) at very small r, the required vapor pressure of water is unrealistically far above the ambient water p value: even RH = 110% is rarely attained in the atmosphere. So, small droplets (at say 10 Å) that do form will just re-evaporate. But, if dust or other types of particles at ~1000 Å are present in the atmosphere, then water can condense onto those particles. In effect, condensing water gets a "wormhole" to droplet sizes that are big enough such that RH values only slightly above 100% can drive continued growth. This is the principle behind "cloud seeding", including any possible attempts at climate change mitigation by seeding the "boundary-layer" with particles made from sea salt as discussed by Rasch et al. (2009).

For mineral particles, an estimate of the typical r might be obtained by electron microscopy. Alternatively, an average measure of the particle size could be obtained through an estimate of the molar surface area $\overline{\mathrm{SA}}$ (m^2/mol), as by application of the BET gas adsorption isotherm method. For such an application, assuming roughly spherical particles and defining n_p as the number of particles making up one mol of the material, and with r continuing to carry units of cm, then

$$\bar{V} = n_p \tfrac{4}{3}\pi r^3 \tag{15.28}$$

$$\overline{\mathrm{SA}}\,(\mathrm{m}^2/\mathrm{mol}) = (10^{-4}\ \mathrm{m}^2/\mathrm{cm}^2) n_\mathrm{p} 4\pi r^2. \tag{15.29}$$

Dividing Eq.(15.28) by (15.29) gives

$$r = \frac{3\bar{V}}{(10^4\ \mathrm{cm}^2/\mathrm{m}^2)\,\overline{\mathrm{SA}}}. \tag{15.30}$$

15.4.4 Cubic Particles of Dimension 2*r*

While small roughly spherical mineral particles do occasionally form in the environment (*e.g.*, see Gschwend and Reynolds, 1987), for non-spherical particles of a mineral solid there is some ambiguity as to how Eqs.(15.25) and (15.30) should be applied. Consider a small cubic crystal of mineral whose dimension $x = 2r$. We can proceed in a manner similar to that used in Figure 15.5, and slice the cube in half as in Figure 15.7. The different types of faces on a crystal will usually have different

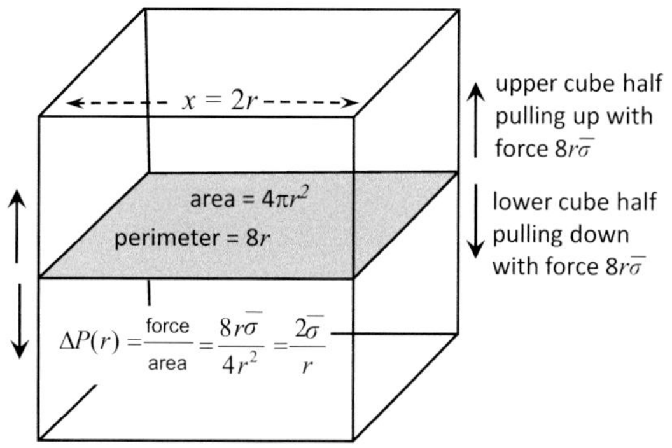

FIGURE 15.7 Interfacial tension balance across two halves of a cube of dimension $x = 2r$.

characteristic arrangements of the constituent species, so in general, the different faces will have different interfacial tension values with water. We denote the average interfacial tension for the different faces as $\bar{\sigma}$. The perimeter over which that surface tension will be acting is $4x = 8r$ According to the same type of reasoning used for Figure 15.5, the interfacial tension-generated force acting on the contents of the cubic crystal is $8r\bar{\sigma}$. The surface area over which that force will be acting is $x^2 = 4r^2$. So,

$$\Delta P = \frac{8r\bar{\sigma}}{4r^2} = \frac{2\bar{\sigma}}{r} \tag{15.31}$$

as before. And, for a collection of monodisperse cubes, we have

$$\bar{V} = n_p x^3 \tag{15.32}$$

$$\overline{\mathrm{SA}}\,(\mathrm{m^2/mol}) = (10^{-4}\,\mathrm{m^2/cm^2})n_p 6x^2. \tag{15.33}$$

Therefore, by dividing Eq.(15.32) by Eq.(15.33),

$$x = 2r = \frac{6\bar{V}}{(10^4\,\mathrm{cm^2/m^2})\,\overline{\mathrm{SA}}} \tag{15.34}$$

$$r = \frac{3\bar{V}}{(10^4\,\mathrm{cm^2/m^2})\,\overline{\mathrm{SA}}} \tag{15.35}$$

again as before. So, while uncertainty remains as to how $\bar{\sigma}$ might be evaluated for some particular mineral material, we certainly understand in general how and why mineral materials become more soluble are particle size decreases.

Example 15.4 Effect of Particle Size on Solubility

Schindler et al. (1965) obtained data for $Cu(OH)_{2(s)}$ at 25 °C/1 atm indicating $\bar{\sigma} = 410 \pm 130$ ergs/cm². For that value, what value of r will cause $K_{s0,r}$ to be twice the large-crystal K_{s0}? For computing $\bar{V}$, $Cu(OH)_{2(s)}$ has a FW of 97.56 g/FW, and a density of 3.93 g/cm³.

Solution

$\bar{V}$ = (97.56 g/FW)×(1 cm³/3.93 g) = 24.82 cm³/FW. By Eq.(15.27),

$$\log K_{s0,r} = \log K_{s0} + 1.04 \times 10^{-8}(\overline{V\sigma}/rT).$$

$$\log(K_{s0,r}/K_{s0}) = \log 2 = 0.301 = 1.04 \times 10^{-8}(\overline{V\sigma}/rT).$$

$$0.301 = 1.04 \times 10^{-8} \times 24.82 \times 410/(r \times 298.15).\ r = 1.18 \times 10^{-6}\ \text{cm}\ (= 118\ \text{Å}).$$

REFERENCES

Chakraborty R, Borglin SE, Dubinsky EA, Andersen GL, Hazen TC (2012) Microbial response to the MC-252 Oil and Corexit 9500 in the Gulf of Mexico. *Frontiers in Microbiology*, **3**, 357.

Gschwend PM, Reynolds MD (1987) Monodisperse ferrous phosphate colloids in an anoxic groundwater plume. *Journal of Contaminant Hydrology*, **1**, 309–327.

Morse JW, Casey WH (1988). Ostwald processes and mineral paragenesis in sediments. *American Journal of Science*.

Ochs M, Nyengaard JR, Jung A, Knudsen L, Voigt M, Wahlers T, Richter J, Gundersen HJ (2004) The number of alveoli in the human lung. *American Journal of Respiratory Critical Care Medicine*, **169**, 120–124.

Poonoosamy J, Curti E, Kosakowski G, Grolimund D, Van Loon LR, Mader U (2016) Barite precipitation following celestite dissolution in a porous medium: A SEM/BSE and mu-XRD/XRF study. *Geochimica et Cosmochimica Acta*, **182**, 131–144.

Rasch PJ, Latham L, Chen CC (2009) Geoengineering by cloud seeding: influence on sea ice and climate system. *Environmental Research Letters*, **4**, 045112 (8 pp).

Schindler P, Althaus H, Hofer F, Minder W (1965) Löslichkeitsprodukte von Metalloxiden und -hydroxiden. 10. Mitteilung. Löslichkeitsprodukte von Zinkoxid, Kupferhydroxid und Kupferoxid in Abhängigkeit von Teilchengrösse und molarer Oberfläche. Ein Beitrag zur Thermodynamik von Grenzflächen fest-flüssig. *Helvetica Chimica Acta*, **48**, 1204–1215.

Steefel CI, Van Cappellen P (1990) A new kinetic approach to modeling water-rock interaction: The role of nucleation, precursors, and Ostwald ripening. *Geochimica et Cosmochimica Acta*, **54**, 2657–2677.

U.S. Coast Guard (2011) *On Scene Coordinator Report: Deepwater Horizon Oil Spill*, September 2011, Washington, DC, U.S. Library of Congress Control Number 2012427375.

White HK, Hsing PY, Choc W, et al. (2012) Impact of the Deepwater Horizon oil spill on a deep-water coral community in the Gulf of Mexico. *Proceedings of the National Academy of Sciences*, **109**, 20303–20308.

Wong FSL, Elliott JC, Anderson P, Davis GR (1995) Mineral concentration in rat femoral diaphyses measured by X-ray microtomography. *Calcified Tissue International*, **56**, 62–70.

16 Solid/Solid and Liquid/Liquid Solution Mixtures

16.1 INTRODUCTION

Up to now in this text, whenever solids have been participants in equilibrium reactions, the only possibility considered has been that for any solid present, the mol fraction is 1.0. Since the standard state for all solids has been chosen to be the pure solid, this has meant that we have so far only considered cases when the solid activities are unity. However, given that many elements are present in natural systems, natural mineral solids are rarely highly pure. We therefore need to understand why non-pure solids form and how they behave. A side benefit will be that the general thermodynamics of solid solutions applies equally well to solutions of liquids in one another, *e.g.*, trichloroethylene (TCE) dissolved in a mostly water phase, and vice versa, TCE dissolved mix of other chlorinated solvents, etc.

When a condensed mass containing A is not pure A, then there are obviously other materials in that mass. This might happen in either of two ways:

1. The regions where A is present are pure, but there are cavities within or coatings on the surface that contain other compounds or elements.
2. Compounds or elements other than those comprising A are dispersed within A at a molecular or atomic level, *i.e.*, actually *dissolved* within it as a true solution. This is the case of interest in this chapter.

In the second case, alloys of metals like copper and zinc (*e.g.*, brass) and tin and lead (*e.g.*, lead-based solder) are examples of solid pairs that are miscible over all proportions at ambient T and P, so there is no solubility limit. For ionic solid pairs, a few are miscible over all proportions as with hydroxyapatite and fluoroapatite (Moreno et al., 1974). Being generally similar compositionally, divalent metal carbonate solids are prone to solid solution formation, as with substitution of large amounts of Mg^{2+} in $CaCO_{3(s)}$; the thermodynamic data compilation by Apps and Wilkin (2015) for metal carbonate solutions is particularly comprehensive. For less similar solids, say $NaCl_{(s)}$ and $CaCO_{3(s)}$, little (but not zero) interdissolution will occur.

16.2 THERMODYNAMIC EQUATIONS GOVERNING THE FORMATION OF SOLID AND LIQUID SOLUTIONS

16.2.1 General Equations

For any chemical process, we have

$$\Delta G = \Delta H - T\Delta S. \tag{16.1}$$

For the **mixing** (mxg) of two gases, two liquids, or two solids, we have

$$\Delta G_{\text{mxg}} = \Delta H_{\text{mxg}} - T\Delta S_{\text{mxg}}. \tag{16.2}$$

Mixing will occur when ΔG_{mxg} is less than zero. The generic reason why solid and liquid solutions can form is the same as the reason why gases mix.

Let us say that we are mixing some A with some B. *If the A–B interactions are of the same strength as the average of the A–A and B–B interactions, then mixing of A and B will neither take up or release any heat.* That is, $\Delta H_{mxg} = 0$. Analogously, let us say that we are mixing some people of region A with some people of region B, where A and B are nearby, and culturally very similar. A people like B people as well as they like themselves, and vice versa. In this case, we will say $\Delta H_{mxg} \approx 0$.

The entropy S is a measure of the disorder in a system. Mixing things always increases the system S: when mixing red marbles with blue marbles, $\Delta S_{mxg} > 0$. So, whenever $\Delta H_{mxg} = 0$, or nearly so, two pure liquids or two pure solids will always mix because by Eq(16.2) then $\Delta G_{mxg} < 0$.

Although $\Delta H_{mxg} \approx 0$ is possible, we also have $\Delta H_{mxg} > 0$ ("endothermic", heat taken up during the mixing), and $\Delta H_{mxg} < 0$ ("exothermic", heat released during the mixing). Continuing the people analogy started above, a case where $\Delta H_{mxg} > 0$ would be mixing of people from two traditionally competitive regions, say New York City and Boston, or, England and France, or India and Pakistan, etc. Strong differences of opinion about things like who had the best baseball team, who made the best cheese, or who had the best cricket team, might mean that the two groups would prefer not to socialize/mix. For even some mixing to proceed (required for equilibrium), some source energy for the endothermic mixing will need to be provided. As noted, $\Delta H_{mxg} > 0$ inhibits mixing. If $\Delta H_{mxg} >>> 0$, the two groups will be only slightly miscible, that is, only slightly soluble in one another. Consider an annual New York–Boston Baseball History Convention held in a neutral city. The New York delegates could perhaps absorb 1% Boston delegates in their midst, and vice versa. "Phase separation" will be present at the convention: a mostly New Yorker phase hanging out with one another talking about Babe Ruth and Willie Mays, and a mostly Bostonian phase talking about Ted Williams and Pedro Martinez (not that this author knows anything about baseball, just trying here to give an analogy). Say the entire Bostonian group leaves town, and takes the few poor New Yorkers in their midst with them back to Boston. Thereafter, the expectation of finding New York delegates walking in the neutral city will be similar to what would happen if no Bostonians were in their midst: calcite that is 99% pure will have a K_{s0} for Ca^{2+} and CO_3^{2-} at 25 °C/1 atm that is very close to the value for pure calcite, $10^{-8.48}$. The remaining Boston delegates are about 1% of the mostly New Yorker mix. But the expectation of finding remaining Bostonian delegates taking a walk in the neutral city will be much higher than 1% of the pure Bostonian value because they are so uncomfortable staying in the convention center with the New York group: a metal carbonate $MeCO_{3(s)}$ present at 1% in calcite as a hostile phase will act with an ion product that is much greater than the K_{s0} for $MeCO_{3(s)}$.

With $\Delta H_{mxg} < 0$, heat is released during mixing. In an anthropomorphic analogy, for $\Delta H_{mxg} < 0$, in 2018 we might mix some civilians from South Korea with their relatives in North Korea. Having not been together for decades, strong favorable interactions will favor the mixing. After mixing, they are going to want to be together and not do anything but chat. Sulfuric acid (H_2SO_4) and H_2O are two liquids that give $\Delta H_{mxg} \ll 0$. As a college student in the early 1970s, the author of this text learned first-hand just how exothermic that process is when he mixed four liters of each of these liquids in a tall cylindrical Pyrex pipette cleaning vessel. The mixture rapidly became exceedingly hot, and the bottom of the vessel cracked and broke out in a nice clean circle. The freshly combined H_2SO_4/water mixture poured onto the lab bench, the author, then the floor. It is the very strong interactions between the protons in the H_2SO_4 and H_2O that make this mixing process very exothermic. In a 1:1 molar mixture of H_2SO_4 and H_2O, the gas phase pressure of H_2O above the mixture will be much less than half the vapor pressure of pure water at the experimental temperature.

To summarize, for every case involving two initially pure materials, the portion of ΔG_{mxg} that is due to the entropy change on mixing is always negative: entropy effects always favor mixing. Depending on the nature of the chemical interactions between the two mixing components, the mixing can either be inhibited by enthalpic effects ($\Delta H_{mxg} > 0$) or promoted by enthalpic effects ($\Delta H_{mxg} < 0$).

Gases always mix because the intermolecular distances are large enough that the ΔS_{mxg} term always overwhelms any unfavorable interactions. Liquids and solids, being more condensed, are much more subject to molecule-molecule interactions. There are innumerable pairs of liquids and innumerable pairs of solids that are not completely miscible with one another, that is to say, not miscible over all proportions. Nevertheless, even in the event of immiscibility for some given proportion of materials, *some* interdissolution (*i.e.*, dissolution of two liquids in each other, or dissolution of two solids in one another) must *always* occur. No matter how insoluble a given organic liquid might be in water, the solubility can *never* be zero; the degree of dissolution (*i.e.*, mixing) that does result at equilibrium will represent the minimization of the system G over the sum of the opposing entropy (favorable) and enthalpic (repulsive) interactions. Note: when talking about an experimental determination of solubility, the presumption is that there are ample amounts of the two constituents to yield two separate phases at equilibrium.

Consider now the detailed thermodynamics of the mixing of two chemicals, A and B. The two chemicals might be two solids, or two liquids. Let n_A mols of pure A be mixed with n_B mols of pure B. Before the A and B are mixed, the free energy of the system is

$$G = n_A\mu_A^o + n_B\mu_B^o. \tag{16.3}$$

The mixing process is

$$n_A A + n_B B = A_{n_A}B_{n_B}. \tag{16.4}$$

When we are dealing with phases that have the possibility of being pure, it is convenient to use the mol fraction scale for activity. Here ζ_i will represent the activity coefficient of species i on the mol fraction scale. In general

$$\mu_i = \mu_i^o + RT\,\ln\zeta_i x_i \tag{16.5}$$

x_i is mol fraction of i. For pure i, $x_i = 1$, $\zeta_i = 1$, and $\mu_i = \mu_i^o$.

After mixing, we have

$$x_A = \frac{n_A}{n_A + n_B} \qquad x_B = \frac{n_B}{n_A + n_B} \qquad x_A + x_B = 1. \tag{16.6}$$

The overall G of the system, that is, the G of the mixture will be given by:

$$G_{mxt} = \left(n_A\mu_A^o + n_A RT\ln x_A + n_A RT\ln\zeta_A\right) + \left(n_B\mu_B^o + n_B RT\ln x_B + n_B RT\ln\zeta_B\right). \tag{16.7}$$

The first parenthetical term represents the G of the n_A mols of A in the solution; the second represents the G of the n_B mols of B in the solution. For the mixing process described by Eq.(16.4), the change in free energy of mixing is given by the difference between the G values as per Eq.(16.7) and Eq.(16.3):

$$\begin{aligned}\Delta G_{mxg} &= n_A\mu_A^o + n_A RT\ln x_A + n_A RT\ln\zeta_A \\ &\quad + n_B\mu_B^o + n_B RT\ln x_B + n_B RT\ln\zeta_B - \left(n_A\mu_A^o + n_B\mu_B^o\right).\end{aligned} \tag{16.8}$$

$$= n_A RT\ln x_A + n_A RT\ln\zeta_A + n_B RT\ln x_B + n_B RT\ln\zeta_B. \tag{16.9}$$

If ΔG_{mxg} is divided by the total number of mols ($n_A + n_B$), then the ΔG per mol of mixture, (the *molar* ΔG of mixing $\Delta\bar{G}_{mxg}$), is

$$\Delta\bar{G}_{mxg} = x_A RT\ln x_A + x_A RT\ln\zeta_A + x_B RT\ln x_B + x_B RT\ln\zeta_B. \tag{16.10}$$

The sum of the terms involving just the mol fractions represents the entropic contribution to $\Delta\overline{G}_{\text{mxg}}$. We know this because for A, in the absence of interactions that are different than those which A feels in pure A, then $\zeta_A = 1.0$, and only entropic effects exist; a similar statement can be made for B. The sum of the two terms in Eq.(16.10) involving activity coefficients represents the enthalpic contribution to $\Delta\overline{G}_{\text{mxg}}$.

$$-T\Delta\overline{S}_{\text{mxg}} = x_A RT \ln x_A + x_B RT \ln x_B. \quad (16.11)$$

$$\Delta\overline{H}_{\text{mxg}} = x_A RT \ln \zeta_A + x_B RT \ln \zeta_B. \quad (16.12)$$

16.2.2 Ideal Solutions

In the special case in which all $\zeta = 1$ over all possible compositions, the solution is ideal (idl): chemical A feels equally at home in the mixture as it does in pure A, and similarly for B: $\zeta_A = \zeta_B = 1$ for all mixture proportions. In the ideal case, Eq.(16.10) then reduces to the entropy terms alone:

$$\Delta\overline{G}_{\text{mxg}} = \Delta\overline{G}_{\text{mxg,idl}} = x_A RT \ln x_A + x_B RT \ln x_B. \quad (16.13)$$

In general then,

$$\Delta\overline{G}_{\text{mxg}} = \Delta\overline{G}_{\text{mxg,idl}} + x_A RT \ln \zeta_A + x_B RT \ln \zeta_B. \quad (16.14)$$

With $R = 8.314$ J/mol-K, $\Delta\overline{G}_{\text{mxg,idl}}$ is easily calculated for $0 < x_A < 1$. The results are presented in Table 16.1 and plotted in Figure 16.1. $\Delta\overline{G}_{\text{mxg,idl}} \leq 0$ for all x_A. There is perfect symmetry in $\Delta\overline{G}_{\text{mxg,idl}}$: the same value is obtained for $x_A = 0.1$ and 0.9, the same for $x_A = 0.2$ and 0.8, etc.

The quantity $\ln x_A$ cannot be evaluated when $x_A = 0.0$ ($x_B = 1.0$). And, $\ln x_B$ cannot be evaluated when $x_B = 0.0$ ($x_A = 1$). This does not cause a problem when evaluating $\Delta\overline{G}_{\text{mxg}}$ since when $x_A = 0$ or $x_B = 0$, there is no mixture. When no mixing has taken place, $\Delta\overline{G}_{\text{mxg}}$ must be zero. (Even from a

TABLE 16.1
$\Delta\overline{G}_{\text{mxg,idl}}$ (kJ/mol) as a Function of Composition at $T = 298$ K

x_A	x_B	$\Delta\overline{G}_{\text{mxg,idl}}$ (kJ/mol)
0.00	1.00	0
0.01	0.99	–0.14
0.05	0.95	–0.49
0.10	0.90	–0.81
0.20	0.80	–1.24
0.30	0.70	–1.51
0.40	0.60	–1.67
0.50	0.50	–1.72
0.60	0.40	–1.67
0.70	0.30	–1.51
0.80	0.20	–1.24
0.90	0.10	–0.81
0.95	0.05	–0.49
0.99	0.01	–0.14
1.00	0.00	0

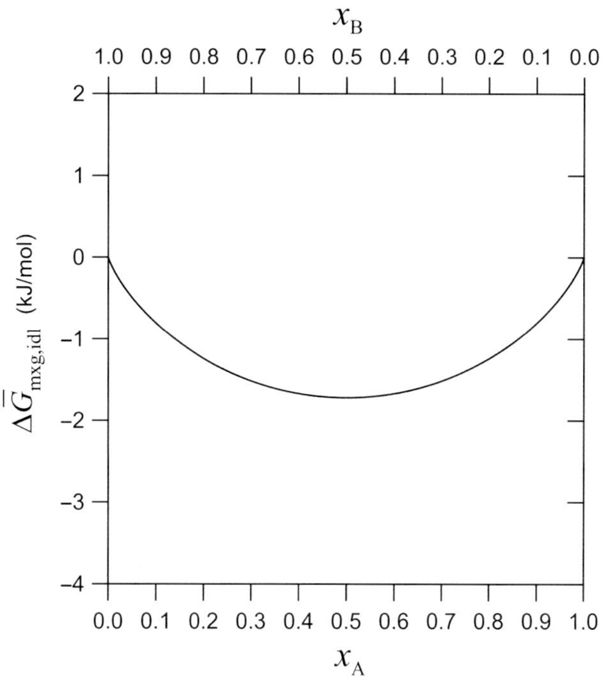

FIGURE 16.1 $\Delta\overline{G}_{mxg,idl}$ (kJ/mol) as a function of x_A for an ideal solid solution at 298.15 K.

strict mathematical point of view, there is no problem since the products $x_A \ln x_A$ and $x_B \ln x_B$ approach zero as x_A and x_B approach zero, respectively. Taking the former as an example, x_A becomes small faster than $\ln x_A$ approaches negative ∞. Consider that for x_A = 0.1, 0.01, and 0.001; $x_A \ln x_A$ = −0.23, −0.046, −0.0069, respectively.)

16.2.3 Non-Ideal Solutions

The values of $\Delta\overline{G}_{mxg,idl}$ in Table 16.1 are small. The maximum (in absolute terms) is only 1.7 kJ/mol. In contrast, among dissolved entities, hydrogen bonds can be as strong as ~170 kJ/mol, and the favorable ionic attractions in minerals can be much larger. So, if A can form hydrogen bonds with itself and other chemicals, and B cannot, then in a 1:1 molar mixture of A and B ($x_A = x_B = 0.5$), neither would feel comfortable at all: $\zeta_A >> 1$ and $\zeta_B >> 1$. (A molecules like to hold hands with one another, B cannot.) A measly −1.7 kJ/mol will not be favorable enough to hold the mixture as one phase: phase separation will occur. The bottom line here is that unless A and B are pretty similar, they are not going to dissolve in one another in all proportions. This is the origin of the well-known axiom "Like dissolves like". Fluoroapatite can dissolve well in hydroxyapatite. Do not expect the same for $SiO_{2(s)}$ and hydroxyapatite.

Over the last many decades, functions have been proposed by physical chemists for representing how ζ_A and ζ_B might behave in mixtures with different proportions of A and B. In such representations, as chemical boundary conditions, we must have that $\zeta_A = 1.0$ when $x_A = 1.0$, and ζ_B = 1.0 when $x_B = 1.0$. Component A must be represented as behaving ideally in pure A, as that is its ideality reference state, and similarly for B. The empirical "two-suffix Margules" parameterization (Prausnitz 1969) gives this kind of behavior:

$$\ln \zeta_A = (A/RT)x_B^2 \tag{16.15}$$

$$\ln \zeta_B = (A/RT)x_A^2 \tag{16.16}$$

A is a constant that has units of J/mol. The quantity A/RT is *dimensionless*. The value of A/RT will depend on which two chemicals are being mixed. Eqs.(16.15)/(16.16) have been used with success in describing the thermodynamics of binary liquid mixtures, and with mineral solid solutions as with, among others, alkali feldspars (Thompson and Waldbaum, 1969), pyroxenes (Lindsley et al., 1981), and Fe-Ti-oxides (Andersen and Lindsley, 1981).

When $A/RT = 0$, the solution is ideal: $\zeta_A = \zeta_B = 1$ for all x_A. When $A/RT > 0$, then $\Delta H_{mxg} > 0$, and mixing is inhibited. When $A/RT < 0$, then $\Delta H_{mxg} < 0$, and mixing is favored. Substituting Eqs. (16.15)/(16.16) into Eq.(16.14) gives

$$\Delta\bar{G}_{mxg} = \Delta\bar{G}_{mxg,idl} + x_A A x_B^2 + x_B A x_A^2. \quad (16.17)$$

$\Delta\bar{G}_{mxg,exs}$ is the "excess" (exs) free energy in the system due to non-ideality. For Eq.(16.17),

$$\Delta\bar{G}_{mxg,exs} = x_A A x_B^2 + x_B A x_A^2. \quad (16.18)$$

$$\Delta\bar{G}_{mxg} = \Delta\bar{G}_{mxg,idl} + \Delta\bar{G}_{mxg,exs}. \quad (16.19)$$

For an ideal mixture, $\Delta G_{mxg,exs} = 0$.

Eq.(16.18) gives

$$\Delta\bar{G}_{mxg,exs} = A x_A x_B (x_A + x_B) \quad (16.20)$$

$$= A x_A x_B. \quad (16.21)$$

With the two-suffix Margules representation, $\left|\Delta\bar{G}_{mxg,exs}\right|$ maximizes when $x_A = x_B = 0.5$.

Combining Eqs.(16.19) and (16.21),

$$\Delta\bar{G}_{mxg} = x_A RT \ln x_A + x_B RT \ln x_B + A x_A x_B. \quad (16.22)$$

Using $x_B = 1 - x_A$, Eq.(16.22) is easily evaluated as a function of x_A for any A/RT.

Eqs.(16.15)/(16.16) for the ζ values as functions of x_A are mirror images of one another. Since $\Delta\bar{G}_{mxg,idl}$ is symmetrical, then when the two-suffix Margules equations describe ζ_A and ζ_B, the $\Delta\bar{G}_{mxg}$ curve as given by Eq.(16.22) will be symmetrical (as in the ideal case). A real $\Delta\bar{G}_{mxg}$ curve will never be exactly symmetrical.

Figure 16.2 presents curves for A/RT values of –5, –3, –1, 0 (ideal solution), 1, 3, and 5. As A/RT becomes more and more positive, the solution process becomes less and less favored due to both the enthalpic term. As A/RT increases above 0, eventually a maximum in the $\Delta\bar{G}_{mxg}$ curve appears. Whenever the $\Delta\bar{G}_{mxg}$ curve has a maximum, then A and B will not be miscible over all proportions, and *separation into two phases* will tend to occur if similar molar amounts of A and B were forced together (as by some kind of hypothetical intense mixing) to give $x_A \approx x_B \approx 0.5$.

A fundamental aspect of mixing thermodynamics is that no matter how repulsively nonideal a given system is, very near $x_A = 0$ and very near $x_B = 0$, $\Delta\bar{G}_{mxg}$ will always be negative due to dominance by the entropy portion of $\Delta\bar{G}_{mxg}$. So, some interdissolution will *always* take place regardless of how repulsively non-ideal a given pair of chemicals are: **there is no such thing as a zero solubility**. Considered in the two-suffix Margules context, no matter how large A/RT becomes, the slope $d\Delta\bar{G}_{mxg}/dx_A$ will always be negative near $x_A = 0$:

$$\frac{d\Delta\bar{G}_{mxg}}{dx_A} = RT \ln x_A - RT \ln x_B + A - 2Ax_A. \quad (16.23)$$

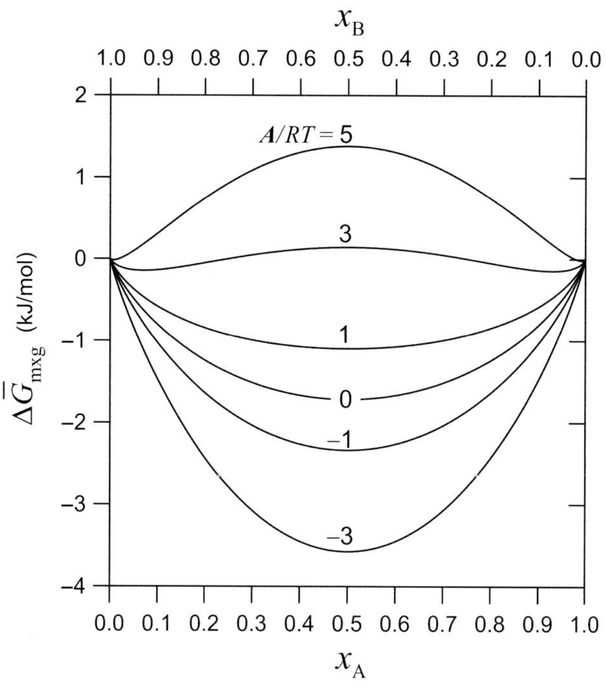

FIGURE 16.2 $\Delta\bar{G}_{mxg}$ (kJ/mol) at 298.15 K as a function of x_A for several values of the two-suffix Margules parameter A/RT. For an ideal solid solution, $A/RT = 0$.

As $x_A \to 0$, the derivative tends to $-\infty$. The same can be shown for $d\Delta\bar{G}_{mxg}/dx_B$ near $x_B = 0$. Thus, any pure chemical is an infinitely strong chemical magnet for other things to dissolve in it. Figure 16.3 plots $\Delta\bar{G}_{mxg}$ near $x_A = 0$ for the two-suffix Margules case with $A/RT = 4$, 5, and 6. In every case, $\Delta\bar{G}_{mxg}$ is negative for x_A values very close to zero. As A/RT becomes larger, the local minimum moves closer to $x_A = 0$.

16.2.4 Chemical A When Present in a Solution *vs.* When Pure

When a chemical A (liquid or solid) is pure, it will exert the full thermodynamic activity that corresponds to its mol fraction of 1.0. If it contains some dissolved B *and the system is at equilibrium*, then the thermodynamic activity of A must be less than or equal to that for pure A. It is clear that this is true when attractive non-ideality is active with $\Delta\bar{H}_{mxg} < 0$. In that case, then not only will the mol fraction x_A be less than 1.0, but so too will ζ_A, and therefore $x_A\zeta_A < 1$. When repulsive non-ideality is active with $\Delta\bar{H}_{mxg} > 0$, then while the mol fraction $x_A < 1$, we know that $\zeta_A > 1$. If the system is at equilibrium, we must still have $x_A\zeta_A \leq 1$, otherwise pure A would tend to form. For example, take chemical A to be some divalent metal carbonate $MeCO_{3(s)}$. For equilibrium dissolution into water at 25 °C/1 atm, say

$$\frac{\{Me^{2+}\}\{CO_3^{2-}\}}{\{MeCO_{3(s)}\}} = 10^{-9.00}. \tag{16.24}$$

$$\{Me^{2+}\}\{CO_3^{2-}\} = 10^{-9.00}\{MeCO_{3(s)}\}. \tag{16.25}$$

When the $MeCO_{3(s)}$ is pure,

$$\{Me^{2+}\}\{CO_3^{2-}\} = 10^{-9.00} \times 1 = 10^{-9.00}. \tag{16.26}$$

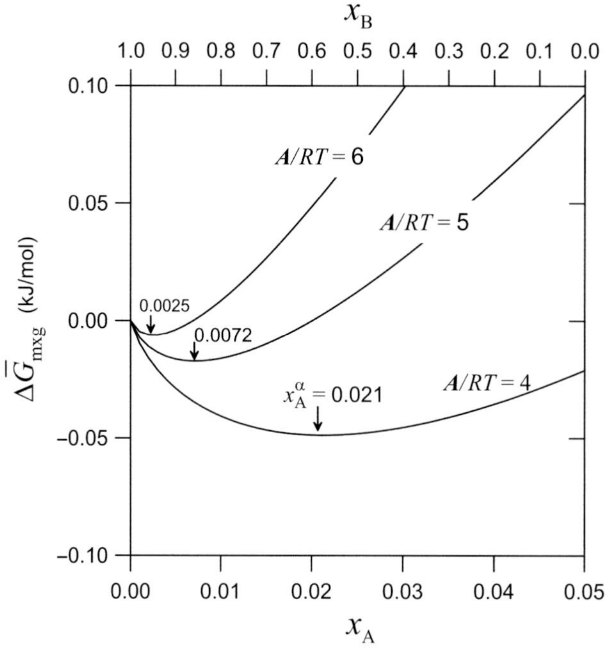

FIGURE 16.3 $\Delta\overline{G}_{mxg}$ (kJ/mol) near $x_A = 0$ for the two-suffix Margules case with A/RT = 4, 5, and 6. $\Delta\overline{G}_{mxg}$ (kJ/mol) is negative for x_A values close to zero.

When there is a solid solution of $MeCO_{3(s)}$ in $CaCO_{3(s)}$, then if the $MeCO_{3(s)}$ mol fraction were 0.01 and *if* the activity coefficient were 1000 ($\Delta\overline{H}_{mxg} > 0$), then $\{MeCO_{3(s)}\} = 10$, and dissolution into water would occur according to

$$\{Me^{2+}\}\{CO_3^{2-}\} = 10^{-9.00} \times 10 = 10^{-8.00}. \tag{16.27}$$

This is greater than the K_{s0} of the pure material, so the pure material would have to precipitate from the solution: the system could not have been at equilibrium. Overall, while the activity coefficients of minerals in a mineral solid solution are usually greater than 1, for equilibrium to be possible, it must always be true for the solution ion activity product (IAP) that

$$\text{IAP} \leq K_{s0}. \tag{16.28}$$

This discussion can be extended to the case when chemical A is an organic liquid (*e.g.*, chlorobenzene) dissolved in a liquid solution. Now, rather than Eq.(16.28), when the organic solution is exposed to water, and some A dissolves giving concentration c_A, then if the system is at equilibrium we must have

$$c_A \leq c_A^o \tag{16.29}$$

where c_A^o solubility concentration for pure A in water.

16.3 PHASE SEPARATION IN NON-IDEAL SOLUTIONS

16.3.1 The $\Delta\overline{G}_{mxg}$ Curve and Phase Separation

As noted above, when a binary system becomes more repulsively non-ideal (*i.e.*, for the two-suffix Margules case, that means A/RT becomes larger), at some point the system will not be miscible over all proportions. That is, although one can always succeed in dissolving *some* A in B, and vice versa, it will not be possible, for example, to obtain a stable mixture composed of equal amounts of A and B ($x_A = x_B = 0.5$); such a system will give two phases. One of the two phases will be predominantly A in composition, and the other phase will be predominantly B. The amount of A in the predominately B phase will be the "saturation" amount for A in B. Specifically, if we equilibrate 1 mol of trichloroethylene (TCE) with 1 mol of water, one of the phases will be mostly water with a little TCE dissolved in it, and the other phase will be mostly TCE with a little water dissolved in it.

As mentioned above, incomplete miscibility occurs for A and B pairs that are characterized by a $\Delta\bar{G}_{mxg}$ curve that has a maximum, as in Figure 16.4 for a general *non-symmetrical* case. For a mixture initially near $x_{A,a}$, phase separation will occur. The compositions of the resultant two equilibrium phases will be those that minimize the overall G of the system. These compositions are x_A^α and x_A^β. Phase separation moves a system to an overall minimum G value. When phase separation occurs, the same two values of x_A^α and x_A^β are always obtained, regardless of the initial values of n_A and n_B. Taking 1 mol of TCE and combining it with 2 mols of water gives the same phase compositions as 1 mol of TCE and combined with 3 mols of water. More details regarding this matter are provided in Appendix 16.A.

16.3.2 Tendency for Correlation between x_A^α and x_B^β

In a system in which phase separation can occur, the more repulsively non-ideal that a given A/B pair becomes, the more the value of x_A^α tends to move towards 0, and the more the value of x_A^β tends to move towards 1. So, when phase separation can occur, as the system becomes more repulsively non-ideal, then both the solubility of A in B and the solubility of B in A go down.

As discussed in Section 16.2.3, very near $x_A = 0$ and 1, the $\Delta\bar{G}_{mxg}$ curve is always negative, no matter how repulsively non-ideal the system. However, as a system becomes more and more

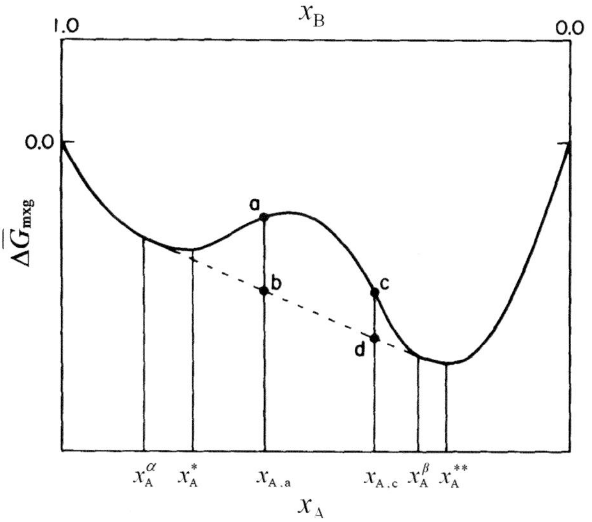

FIGURE 16.4 $\Delta\bar{G}_{mxg}$ as a function of x_A for a general, asymmetric, repulsively non-ideal mixing case. x_A^α and x_A^β are the compositions for the solubility limit amounts of B as dissolved in A, and A in B. Thus, x_A^α and x_A^β are the compositions that result after phase separation occurs for any forced mixture having an initial composition that is between these two values, and therefore unstable. x_A^* and x_A^{**} are the compositions at the two minima. For an asymmetric, repulsively non-ideal case such as this, $x_A^\alpha \neq x_A^*$ and $x_A^\beta \neq x_A^{**}$. For a symmetric case (*e.g.*, two-suffix Margules), $x_A^\alpha = x_A^*$ and $x_A^\beta = x_A^{**}$. The dashed line gives the net $\Delta\bar{G}_{mxg}$ case at equilibrium for any overall initial (*i.e.*, pre-phase separation) combination of n_A and n_B inside that range. The interval between x_A^α and x_A^β is often referred to as the "miscibility gap".

TABLE 16.2
Solubilities of Various Organic Liquid Compounds in Water and Vice Versa at 20 to 25 °C

		concentration of organic in the primarily water phase		concentration of water in the primarily organic phase	
Compound	**MW**	**wt %**	**$\log x_A^\alpha$**	**wt %**	**$\log x_B^\beta$**
heptane	100.2	0.00034	−6.21	0.0067	−3.43
hexane	86.2	0.00095	−5.70	0.009	−3.37
1,2,4-trichlorobenzene	181.4	0.0049	−5.31	0.02	−2.70
pentane	72.1	0.0038	−5.02	0.0101	−3.39
cyclohexane	84.2	0.0055	−4.93	0.00797	−3.43
cyclopentane	70.1	0.0156	−4.40	0.0189	−3.13
chlorobenzene	112.5	0.0498	−4.10	0.0446	−2.56
carbon tetrachloride	153.8	0.0793	−4.03	0.00857	−3.14
toluene	92.1	0.0526	−3.99	0.0552	−2.55
trichlorethylene (TCE)	131.3	0.128	−3.76	0.32	−1.64
1-chlorobutane	92.6	0.11	−3.67	0.08	−2.39
benzene	78.1	0.179	−3.38	0.0691	−2.52
methyl isoamyl ketone	114.1	0.54	−3.07	1.445	−1.07
chloroform	119.3	0.795	−2.92	0.0887	−2.23
n-butylacetate	118.1	0.84	−2.89	1.189	−1.14
1,2-dichoroethane	96.9	0.86	−2.79	0.182	−2.01
dichloromethane	84.9	1.3	−2.56	0.24	−1.95
methylisobutylketone (MIBK)	100.1	1.9	−2.46	2.056	−0.98
methylpropylketone	86.1	4.3	−2.03	4.118	−0.77
methyl-*t*-butylether (MTBE)	88.1	5.1	−1.96	1.5	−1.16
diethyl ether	90.1	6.04	−1.90	1.3	−1.21
1-butanol	74.1	6.32	−1.79	20	−0.29
ethyl acetate	88.1	8.7	−1.72	3.3	−0.84
isobutyl alcohol	74.1	8.5	−1.66	16	−0.36
2-butanol	74.1	12.5	−1.47	44	−0.12
methyl ethyl ketone (MEK)	72.1	22.3	−1.17	10	−0.51

repulsively non-ideal (higher and higher A/RT values), the $\Delta\overline{G}_{\text{mxg}}$ curve is pulled upwards, the height of the maximum increases, and the positions of the solubility mol fractions x_A^α and x_A^β are pulled closer to 0 and 1, respectively. Experimental values x_A^α and x_B^β values will tend to be correlated. As chemical A becomes more unlike and inhospitable to B, then chemical B will become more unlike and inhospitable to A. While there is not much data for solid solutions that can be examined in search of this correlation, there are data for liquid systems. Table 16.2 presents some data for a variety of organic compounds and water; Figure 16.5 plots the data in a $\log x_B^\beta$ *vs.* $\log x_A^\alpha$ manner. For these data, A is the organic compound, B is water. The α phase is the primarily water (aqueous) phase, and the β phase is the primarily organic phase. x_B^β represents the concentration of water in the primarily organic phase, and x_A^α represents the concentration of organic in the primarily water phase. There is a clear trend relating the two sets of mol fractions. The solubility of water in methylethylketone (MEK) is comparatively high, as is the solubility of MEK in water. The solubility of water in heptane is very low, as is the solubility of heptane in water.

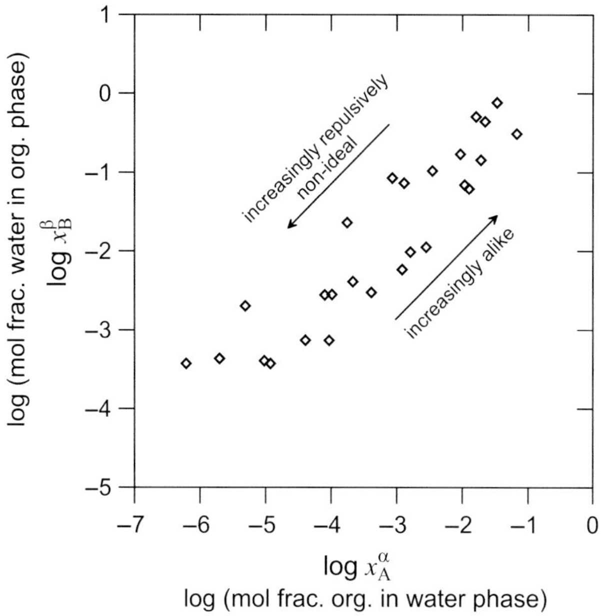

FIGURE 16.5 Log x_B^β *vs.* log x_A^α where A is an organic compound, B is water. These are the mutual solubility limit values.

APPENDIX 16.A

16.A.1 Minimization of the Free Energy G by Phase Separation

The two minima in Figure 16.4 occur at x_A^* and x_A^{**}. In the α phase, we have $x_A^\alpha + x_B^\alpha = 1$, and in the β phase we have $x_A^\beta + x_B^\beta = 1$. In this non-symmetrical case, the points x_A^* and x_A^{**} are not the same as the minima x_A^α and x_A^β. For given, constant total amounts of A and B, that is, for given values of

$$n_A^\alpha + n_A^\beta = n_A \quad \text{(constant)} \tag{16.30}$$

and

$$n_B^\alpha + n_B^\beta = n_B \quad \text{(constant),} \tag{16.31}$$

the *two* variables with respect to which G must be minimized are n_A^α and n_B^α. These two quantities represent the number of mols of A and B in the α phase. (By Eqs.(16.30) and (16.31), knowledge of n_A^α and n_B^α gives n_A^β and n_B^β.) After the phase separation of the initially-mixed $x_{A,a}$ system, the G of the two-phase system will be given by

$$G = (n_A^\alpha + n_B^\alpha)\bar{G}_{mxt}^\alpha + (n_A^\beta + n_B^\beta)\bar{G}_{mxt}^\beta \tag{16.32}$$

where $\bar{G}_{mxt}^\alpha$ and $\bar{G}_{mxt}^\beta$ represent the molar free energies for mixtures of the compositions of the α and β phases, respectively. By Eq.(16.7),

$$\bar{G}_{mxt}^\alpha = x_A^\alpha(\mu_A^o + RT\ln x_A^\alpha + RT\ln\zeta_A^\alpha) + x_B^\alpha(\mu_B^o + RT\ln x_B^\alpha + RT\ln\zeta_B^\alpha) \tag{16.33}$$

and

$$\bar{G}^{\beta}_{\mathrm{mxt}} = x^{\beta}_{\mathrm{A}}(\mu^{\mathrm{o}}_{\mathrm{A}} + RT\ln x^{\beta}_{\mathrm{A}} + RT\ln\zeta^{\beta}_{\mathrm{A}}) + x^{\beta}_{\mathrm{B}}(\mu^{\mathrm{o}}_{\mathrm{B}} + RT\ln^{\beta}_{\mathrm{B}} + RT\ln\zeta^{\beta}_{\mathrm{B}}). \tag{16.34}$$

Subtracting the total initial free energy of the unmixed components from the expression given by Eq.(16.32) gives the overall ΔG_{mxg}. That is,

$$\Delta G_{\mathrm{mxg}} = (n^{\alpha}_{\mathrm{A}} + n^{\alpha}_{\mathrm{B}})\bar{G}^{\alpha}_{\mathrm{mxt}} + (n^{\beta}_{\mathrm{A}} + n^{\beta}_{\mathrm{B}})\bar{G}^{\beta}_{\mathrm{mxt}} - n_{\mathrm{A}}\mu^{\mathrm{o}}_{\mathrm{A}} - n_{\mathrm{B}}\mu^{\mathrm{o}}_{\mathrm{B}}. \tag{16.35}$$

Since G differs from ΔG_{mxg} only by the quantity $(n_{\mathrm{A}}\mu^{\mathrm{o}}_{\mathrm{A}} + n_{\mathrm{B}}\mu^{\mathrm{o}}_{\mathrm{B}})$, a constant for any given case, the conditions that minimize ΔG_{mxg} for the formation of the two phases from the initially pure A and B will also minimize G. We can therefore investigate when ΔG_{mxg} is minimized.

When phase separation does occur, ΔG_{mxg} can be viewed as the result of the formation of two separate mixtures from the original n_{A} mols of A and the n_{B} mols of B. Thus, in addition to Eq.(16.35),

$$\Delta G_{\mathrm{mxg}} = (n^{\alpha}_{\mathrm{A}} + n^{\alpha}_{\mathrm{B}})\Delta\bar{G}^{\alpha}_{\mathrm{mxg}} + (n^{\beta}_{\mathrm{A}} + n^{\beta}_{\mathrm{B}})\Delta\bar{G}^{\beta}_{\mathrm{mxg}}. \tag{16.36}$$

By Eqs.(16.30) and (16.31),

$$\partial n^{\alpha}_{\mathrm{A}} = -\partial n^{\beta}_{\mathrm{A}} \tag{16.37}$$

$$\partial n^{\alpha}_{\mathrm{B}} = -\partial n^{\beta}_{\mathrm{B}}. \tag{16.38}$$

To find the minimum, we differentiate ΔG_{mxg} (as given by Eq.(16.36)) with respect to n^{α}_{A} and set the result equal to zero:

$$\begin{aligned}\partial\Delta G_{\mathrm{mxg}}/\partial n^{\alpha}_{\mathrm{A}} &= (n^{\alpha}_{\mathrm{A}} + n^{\alpha}_{\mathrm{B}})\partial\Delta\bar{G}^{\alpha}_{\mathrm{mxg}}/\partial n^{\alpha}_{\mathrm{A}} + \Delta\bar{G}^{\alpha}_{\mathrm{mxg}}\\ &\quad - (n^{\beta}_{\mathrm{A}} + n^{\beta}_{\mathrm{B}})\partial\Delta\bar{G}^{\beta}_{\mathrm{mxg}}/\partial n^{\beta}_{\mathrm{A}} - \Delta\bar{G}^{\beta}_{\mathrm{mxg}} = 0.\end{aligned} \tag{16.39}$$

By the chain rule,

$$\partial\Delta\bar{G}^{\alpha}_{\mathrm{mxg}}/\partial n^{\alpha}_{\mathrm{A}} = (\partial\Delta\bar{G}^{\alpha}_{\mathrm{mxg}}/\partial x^{\alpha}_{\mathrm{A}})(\partial x^{\alpha}_{\mathrm{A}}/\partial n^{\alpha}_{\mathrm{A}}) \tag{16.40}$$

$$\partial\Delta\bar{G}^{\beta}_{\mathrm{mxg}}/\partial n^{\beta}_{\mathrm{A}} = (\partial\Delta\bar{G}^{\beta}_{\mathrm{mxg}}/\partial x^{\beta}_{\mathrm{A}})(\partial x^{\beta}_{\mathrm{A}}/\partial n^{\beta}_{\mathrm{A}}) \tag{16.41}$$

Also,

$$(\partial x^{\alpha}_{\mathrm{A}}/\partial n^{\alpha}_{\mathrm{A}}) = n^{\alpha}_{\mathrm{B}}/(n^{\alpha}_{\mathrm{A}} + n^{\alpha}_{\mathrm{B}})^2 \tag{16.42}$$

$$(\partial x^{\beta}_{\mathrm{A}}/\partial n^{\beta}_{\mathrm{A}}) = n^{\beta}_{\mathrm{B}}/(n^{\beta}_{\mathrm{A}} + n^{\beta}_{\mathrm{B}})^2. \tag{16.43}$$

Substituting Eqs.(16.40)–(16.43) into Eq.(16.39) gives

$$\partial\Delta G_{\mathrm{mxg}}/\partial n^{\alpha}_{\mathrm{A}} = x^{\alpha}_{\mathrm{B}}\partial\Delta\bar{G}^{\alpha}_{\mathrm{mxg}}/\partial x^{\alpha}_{\mathrm{A}} + \Delta\bar{G}^{\alpha}_{\mathrm{mxg}} - x^{\beta}_{\mathrm{B}}\partial\Delta\bar{G}^{\beta}_{\mathrm{mxg}}/\partial x^{\beta}_{\mathrm{A}} - \Delta\bar{G}^{\beta}_{\mathrm{mxg}} = 0. \tag{16.44}$$

Differentiation of ΔG_{mxg} with respect to n^{α}_{B} and proceeding analogously gives

$$\partial\Delta G_{\mathrm{mxg}}/\partial n^{\alpha}_{\mathrm{B}} = x^{\alpha}_{\mathrm{A}}\partial\Delta\bar{G}^{\alpha}_{\mathrm{mxg}}/\partial x^{\alpha}_{\mathrm{B}} + \Delta\bar{G}^{\alpha}_{\mathrm{mxg}} - x^{\beta}_{\mathrm{A}}\partial\Delta\bar{G}^{\beta}_{\mathrm{mxg}}/\partial x^{\beta}_{\mathrm{B}} - \Delta\bar{G}^{\beta}_{\mathrm{mxg}} = 0. \tag{16.45}$$

Subtracting Eq.(16.45) from Eq.(16.44), then noting that

$$\partial x_B^\alpha = -\partial x_A^\alpha \quad \text{and} \quad \partial x_B^\beta = -\partial x_A^\beta \tag{16.46}$$

as well as $x_A^\alpha + x_B^\alpha = 1$ and $x_A^\beta + x_B^\beta = 1$, we obtain the result that the G of the two-phase system is minimized when

$$\partial\Delta\bar{G}_{mxg}^\alpha/\partial x_A^\alpha = \partial\Delta\bar{G}_{mxg}^\beta/\partial x_A^\beta. \tag{16.47}$$

Once phase separation is completed, and there is a single value for x_A^α and a single value for x_A^β, this analysis shows that the slope of the $\Delta\bar{G}_{mxg}$ *vs.* x_A curve at x_A^α is equal to the slope of the curve at x_A^β.

Application of Eq.(16.46) again in Eq.(16.45) along with Eq.(16.47) leads to

$$\partial\Delta\bar{G}_{mxg}^\alpha/\partial x_A^\alpha = \partial\Delta\bar{G}_{mxg}^\beta/\partial x_A^\beta = (\Delta\bar{G}_{mxg}^\beta - \Delta\bar{G}_{mxg}^\alpha)/(x_A^\beta - x_A^\alpha). \tag{16.48}$$

So, not only are the two slopes equal, but also given by the line connecting $(x_A^\alpha, \Delta\bar{G}_{mxg}^\alpha)$ and $(x_A^\beta, \Delta\bar{G}_{mxg}^\beta)$. The line is given in Figure 16.4 as a dashed line.

16.A.2 The Linear Mixing Curve for $\Delta\bar{G}_{mxg}$ *vs.* x_A in the Region Where Phase Separation Occurs

The dashed line in Figure 16.4 represents the "linear mixing curve" giving the *actual* $\Delta\bar{G}_{mxg}$ of any system whose *overall* composition falls between x_A^α and x_B^β. That is, we throw away the hump in Figure 16.4 and follow the dashed line instead. "Overall" composition means the composition that would result if there was only one phase.

$$\Delta G_{mxg} = (n_A + n_B)\Delta\bar{G}_{mxg}. \tag{16.49}$$

Consider an unstable, one phase mixture with composition $x_{A,a}$ for point a, as located in Figure 16.4. As the mixture separates into the α and β phases having compositions x_A^α and x_A^β, then ΔG_{mxg} can reach a lower value since the ΔG_{mxg} values for both x_A^α and x_A^β, are lower than at point a. The overall $\Delta\bar{G}_{mxg}$ of the system will be given by point b. Alternatively, if the initial composition of the one phase mixture is a point c and corresponding to $x_{A,c}$, then the mixture will again separate into two phases. The compositions of the two phases will again be x_A^α and the other x_A^β. The overall $\Delta\bar{G}_{mxg}$ will be given by point d. In this case, the majority of the material ends up in the β phase since $x_{A,c}$ is closer to x_A^β than it is to x_A^α. Thus, even though some of the material has to rise to a $\Delta\bar{G}_{mxg}$ that is higher than $\Delta\bar{G}_{mix,c}$, the fact that most of the material can lower its $\Delta\bar{G}_{mxg}$ is what drives the phase separation.

That it would even be possible to separate *any* one phase system with known overall composition X_A into two phases that have the specific compositions x_A^α and x_A^β may be understood as follows. There are *four* unknowns in the system, n_A^α, n_B^α, n_A^β, and n_B^β. Equations (16.30) and (16.31) are two equations that govern any system of this type. The thermodynamics of the system provide the other two equations, *i.e.*, that for every case, x_A^α and x_A^β are the two equilibrium compositions of the two phases.

16.A.3 When Phase Separation Occurs, the Same Two Values of x_A^α and x_A^β Are Obtained Regardless of the Values of n_A and n_B

That the *same* two phases would form whenever a phase separation occurs has been shown above by demonstrating that a specific pair of compositions will minimize the overall G of the system.

In addition, the following argument may be advanced. At equilibrium between the two phases, the chemical potentials of the two components must be equal in both phases:

$$\mu_A^\alpha = \mu_A^o + RT\ln\zeta_A^\alpha x_A^\alpha = \mu_A^\beta = \mu_A^o + RT\ln\zeta_A^\beta x_A^\beta \qquad (16.50)$$

$$\mu_B^\alpha = \mu_B^o + RT\ln\zeta_B^\alpha x_B^\alpha = \mu_B^\beta = \mu_B^o + RT\ln\zeta_B^\beta x_B^\beta. \qquad (16.51)$$

Since the same standard state μ values are used for both the α and β phases, Eqs.(16.50) and (16.51) reduce to an equating of the two activities of A and B in the two phases, that is,

$$\zeta_A^\alpha x_A^\alpha = \zeta_A^\beta x_A^\beta \qquad (16.52)$$

$$\zeta_B^\alpha x_B^\alpha = \zeta_B^\beta x_B^\beta. \qquad (16.53)$$

Eq.(16.52) and (16.53) may be used to solve any phase separation problem for a given pair of components. They contain the four unknown mol fractions. (The four activity coefficients are implicit functions of the mol fractions and do not therefore constitute additional unknowns.) The two additional equations needed are provided by the statements that the mol fractions sum to one for each phase. Thus, when phase separation occurs, the four equations used to solve the problem will always be exactly the same. Therefore, the two phases which form will always have the same composition no matter what the absolute amounts n_A and n_B are used to create the system. The *absolute* amounts of the two phases, will, of course, depend on n_A and n_B. Thus if we equilibrate 2 mols of TCE with 1 mol of water, we will get the same saturation concentrations of TCE in water and of water in TCE that we would get if we equilibrated 1 mol of TCE with 1 mol of water.

For a non-symmetrical $\Delta\bar{G}_{mxg}$ curve, the compositions x_A^α and x_A^β are not equal to the compositions which occur at the two minima at $\overset{*}{x}_A$ and $\overset{**}{x}_A$. This may be shown as follows. Consider the expression for the ΔG_{mxg} for fixed values of n_A and n_B:

$$\Delta G_{mxg} = (n_A^\alpha + n_B^\alpha)\Delta\bar{G}_{mxg}^\alpha + (n_A^\beta + n_B^\beta)\Delta\bar{G}_{mxg}^\beta. \qquad (16.36)$$

Thus, at equilibrium,

$$\begin{aligned}\partial\Delta G_{mxg}/\partial n_A^\alpha &= (n_A^\alpha + n_B^\alpha)\partial\Delta\bar{G}_{mxg}^\alpha/\partial n_A^\alpha + \Delta\bar{G}_{mxg}^\alpha \\ &\quad - (n_A^\beta + n_B^\beta)\partial\Delta\bar{G}_{mxg}^\beta/\partial n_A^\beta - \Delta\bar{G}_{mxg}^\beta = 0.\end{aligned} \qquad (16.39)$$

Substitution of Eqs.(16.40) and (16.41) yields

$$\begin{aligned}\partial\Delta G_{mxg}/\partial n_A^\alpha &= (n_A^\alpha + n_B^\alpha)(\partial\Delta\bar{G}_{mxg}^\alpha/\partial x_A^\alpha)(\partial x_A^\alpha/\partial n_A^\alpha) + \Delta\bar{G}_{mxg}^\alpha \\ &\quad - (n_A^\beta + n_B^\beta)(\partial\Delta\bar{G}_{mxg}^\beta/\partial x_A^\beta)(\partial x_A^\beta/\partial n_A^\beta) - \Delta\bar{G}_{mxg}^\beta = 0.\end{aligned} \qquad (16.54)$$

$(\partial\Delta\bar{G}_{mxg}^\alpha/\partial x_A^\alpha)$ and $(\partial\Delta\bar{G}_{mxg}^\beta/\partial x_A^\beta)$ are identically zero at $\overset{*}{x}_A$ and $\overset{**}{x}_A$, respectively. Since $(\partial x_A^\alpha/\partial n_A^\alpha)$ and $(\partial x_A^\beta/\partial n_A^\beta)$ are finite at the same points, respectively, then if the minima at $\overset{*}{x}_A$ and $\overset{**}{x}_A$ *did* represent the composition at equilibrium, then

$$\partial\Delta\bar{G}_{mxg}/\partial n_A^\alpha = \Delta\bar{G}_{mxg}^\alpha - \Delta\bar{G}_{mxg}^\beta = 0. \qquad (16.55)$$

The quantities $\Delta\bar{G}_{mxg}^\alpha$ and $\Delta\bar{G}_{mxg}^\beta$ will not *in general* be equal, so $\partial\Delta G/\partial n_A^\alpha$ will not be zero, and the x_A values for the minima in the $\Delta\bar{G}_{mxg}$ curve cannot represent the equilibrium compositions of the two-phase system in any non-symmetrical case. A similar conclusion can be reached based upon examination of $\partial\bar{G}_{mxg}/\partial n_B^\alpha$.

In the special case of *symmetrical* non-ideality (*e.g.*, that described by the two suffix Margules equations), the minima in the $\Delta\bar{G}_{\text{mxg}}$ curve *will* be the equilibrium compositions. At those points: 1) the slopes are equal; and 2) we satisfy the requirement of zero slopes that $\Delta\bar{G}^{\alpha}_{\text{mxg}} = \Delta\bar{G}^{\beta}_{\text{mxg}}$.

REFERENCES

Andersen DJ, Lindsley DH (1981) A valid Margules formulation for an asymmetric ternary solution: Revision of the olivine-ilmenite thermometer, with applications. *Geochimica et Cosmochimica Acta*, **45**, 847–853.

Apps JA, Wilkin RT (2015) *Thermodynamic Properties of Aqueous Carbonate Species and Solid Carbonate Phases of Selected Elements pertinent to Drinking Water Standards of the U.S. Environmental Protection Agency.* Lawrence Berkeley National Laboratory, LBNL-190825, 3 November 2015.

Lindsley D H, Grover JE, Davidson PM (1981) The thermodynamics of $Mg_2Si_2O_6$ – $CaMgSi_2O_6$: A review and improved model. In: *Thermodynamics of Minerals and Melts*, Newton RC, Navrotsky A, Wood BJ (Eds.), Springer, New York, pp.149–197.

Moreno EC, Kresak M, Zahradnik RT (1974) Fluoridated hydroxyapatite solubility and caries formation. *Nature*, **247**, 64–65.

Prausnitz JM (1969) *Molecular Thermodynamics of Fluid Phase Equilibria*, Prentice Hall, Englewood Cliffs, NJ, p.193.

Thompson JB Jr, Waldbaum DR (1969) Mixing properties of sanidine crystalline solutions. III. Calculations based on two-phase data. *American Mineralogist*, **54**, 811–838.

Part V

Redox Chemistry

17 Redox Reactions, E_H, and pe

17.1 INTRODUCTION

A "redox" reaction is a reaction involving electrons, and the word redox is a shortened portmanteau of "reduction/oxidation." A generalized redox "half-reaction" is

$$\overset{\text{reduction} \longrightarrow}{\underset{\longleftarrow \text{ oxidation}}{OX_1 + ne^- = RED_1}} \tag{17.1}$$

where OX_1 represents some oxidized species, RED_1 represents the reduced species created when OX_1 accepts n electrons. Reduction of a species involves the acceptance of one or more electrons by that species, and the oxidation of a species involves the loss of one or more electrons.

The word "reduction" relates to the fact that when chemical species accepts one or more electrons, which are negatively charged, the fundamental charge on some atomic unit in the species is reduced. The word "oxidation" relates to the fact that in the early days of chemistry, all known oxidation reactions involved *oxygen* as a reactant as in: 1) converting elemental iron to the oxide that is rust; 2) converting sulfur to the oxide SO_2, then to sulfuric acid; and 3) converting hydrogen gas to hydrogen oxide, namely water.

Eq.(17.1) is a half-reaction since it can only tell half of the story for a full redox reaction. This is because there never are many totally free electrons hanging about in chemical systems to supply electrons for Eq.(17.1). Rather, if the reaction in Eq.(17.1) is going to proceed to any significant degree, the electrons have to be provided by some other half-reaction proceeding according to

$$\overset{\text{oxidation} \longrightarrow}{\underset{\longleftarrow \text{ reduction}}{RED_2 = OX_2 + ne^-.}} \tag{17.2}$$

Summing Eq.(17.2) with Eq.(17.1) gives

$$RED_2 + OX_1 = OX_2 + RED_1. \tag{17.3}$$

For Eq.(17.3) in the forward direction, the electrons from RED_2 are reducing OX_1 to RED_1. Expressed another way, OX_1 is taking electrons from RED_2 and thereby oxidizing RED_2 to OX_2. RED_2 is a reductant because it is reducing OX_1 to RED_1. OX_1 is an oxidant because it is oxidizing RED_2 to OX_2. One specific example of a redox half-reaction of interest in natural water systems is

$$Fe^{3+} + e^- = Fe^{2+}. \tag{17.4}$$

17.2 ASSIGNING OXIDATION STATES

Working with redox chemistry requires being able to determine what the oxidation state is for each element of interest as it is found in a given atomic, molecular, and/or ionic species. When the species has bonds between atoms, the determination involves assigning control of the electrons in each bond to one or the other of the atoms: because oxidation means loss of electrons, the fewer the assigned electrons, the higher the assigned oxidation state. For covalent bonds, there is always some "sharing" of control. To assign oxidation states to atoms, and to account for the exchange of electrons in redox reactions, the electrons in each covalent bond must be assigned between the atoms in the bond. For a covalent bond between two identical atoms, the electrons are assigned equally between the two atoms. For a covalent bond between two dissimilar atoms, the electrons are wholly assigned to the atom with the greater electron affinity.

Atomic H has one electron. It is neither oxidized nor reduced relative to the atomic state, so the assigned oxidation state of atomic H is by definition 0, represented as H(0). Two atoms of H can combine to form a molecule of H_2 with a pair of shared electrons in the covalent bond holding the atoms together. In a molecule of H_2, since the two atoms are identical, the sharing of the electron pair is necessarily equal. For determination of the oxidation state, each H is therefore assigned one electron. So, as with atomic H, the oxidation state of H in H_2 is H(0). This is true for whatever phase the H_2 is in, whether in a gas phase, dissolved in water, or elsewhere.

Atomic O has six outer-shell electrons. As with atomic H, the assigned oxidation state of atomic O is defined as being 0, represented as O(0). Two atoms of O can combine to form a molecule of O_2, with two pairs of shared electrons in the two covalent bonds holding the atoms together. The two atoms are identical so (as in H_2), so the sharing of electrons is necessarily equal, and the oxidation state of O in O_2 is O(0). The same line of thinking leads to the result that for every element in the elemental state, the oxidation state is zero: N in N_2 is N(0); S in elemental sulfur is S(0) (whether represented as $S_{(s)}$, or sometimes as $S_{8(s)}$); C in graphite is C(0); Fe in iron metal is Fe(0), Hg in liquid mercury is Hg(0), etc.

When a bond is formed between two different atoms, then for purposes of assigning oxidation states, we need to know which element has the greater affinity for the electrons in the bond. The chemical term for electron affinity is "electronegativity". Figure 17.1 provides electronegativity values for a large portion of the periodic table. Electronegativity is a relative scale that goes from 0 to about 4, and does not have particular units. Elements that are very good oxidizing agents ("want electrons and want to become reduced") have high electronegativity values, as with O, F, and Cl. In fact, elemental F and Cl are such good oxidizing agents that, in nature, fluorine and chlorine are virtually always found in their reduced forms F^- and Cl^-, which are F(–I) and Cl(–I) respectively; similarly, oxygen is mostly found as O(–II). Elements that are good reducing agents (easily "give up" their electrons and become oxidized) have very low electronegativity values, as with H, Li, and Na. In fact, elemental Li and Na are such good reducing agents that lithium and sodium in nature are always found as Li^+ and Na^+, which are Li(I) and Na(I), respectively, and similarly, hydrogen is mostly found as H(I). As indicated, Roman numerals are usually used for oxidation states, unless a decimal is required.

Consider now the "Lewis diagram" in Figure 17.2a, which shows the outer shell bonding electrons in H_2O. Two hydrogen atoms bring one electron each (x) for bonding to the neutral oxygen atom with its six outer-shell electrons (•). The electronegativity value for O is 3.44, which is greater than that for H at 2.20. The oxygen is therefore assigned both of the electrons in both bonds. The result O with two more assigned electrons than in atomic O, so the O in H_2O is in the O(–II) oxidation state. For H, with one less electron than in atomic H, each H in H_2O is in the H(I) oxidation state.

Rule 1 – H in H_2 is H(0), but when bonded to *other* elements in nature, with its single bond, H is always H(I). H has an electronegativity value of 2.20, which is less than that of O (3.44), carbon (2.55), nitrogen (3.04), and chlorine (3.16).

1 **H** 2.20																	2 **He** --
3 **Li** 0.98	4 **Be** 1.57											5 **B** 2.04	6 **C** 2.55	7 **N** 3.04	8 **O** 3.44	9 **F** 3.98	10 **Ne** --
11 **Na** 0.93	12 **Mg** 1.31											13 **Al** 1.61	14 **Si** 1.90	15 **P** 2.19	16 **S** 2.58	17 **Cl** 3.16	18 **Ar** --
19 **K** 0.82	20 **Ca** 1.00	21 **Sc** 1.36	22 **Ti** 1.54	23 **V** 1.63	24 **Cr** 1.66	25 **Mn** 1.55	26 **Fe** 1.83	27 **Co** 1.88	28 **Ni** 1.91	29 **Cu** 1.90	30 **Zn** 1.65	31 **Ga** 1.81	32 **Ge** 2.01	33 **As** 2.18	34 **Se** 2.55	35 **Br** 2.96	36 **Kr** --
37 **Rb** 0.82	38 **Sr** 0.95	39 **Y** 1.22	40 **Zr** 1.33	41 **Nb** 1.6	42 **Mo** 2.16	43 **Tc** 1.9	44 **Ru** 2.2	45 **Rh** 2.28	46 **Pd** 2.20	47 **Ag** 1.93	48 **Cd** 1.69	49 **In** 1.78	50 **Sn** 1.96	51 **Sb** 2.05	52 **Te** 2.1	53 **I** 2.66	54 **Xe** 2.6
55 **Cs** 0.79	56 **Ba** 0.89	57 **La** 1.1	72 **Hf** 1.3	73 **Ta** 1.5	74 **W** 2.36	75 **Re** 1.9	76 **Os** 2.2	77 **Ir** 2.20	78 **Pt** 2.28	79 **Au** 2.54	80 **Hg** 2.00	81 **Tl** 1.62	82 **Pb** 2.33	83 **Bi** 2.02	84 **Po** 2.0	85 **At** 2.2	86 **Rn** --

FIGURE 17.1 Electronegativity values of the elements for a portion of the periodic table.

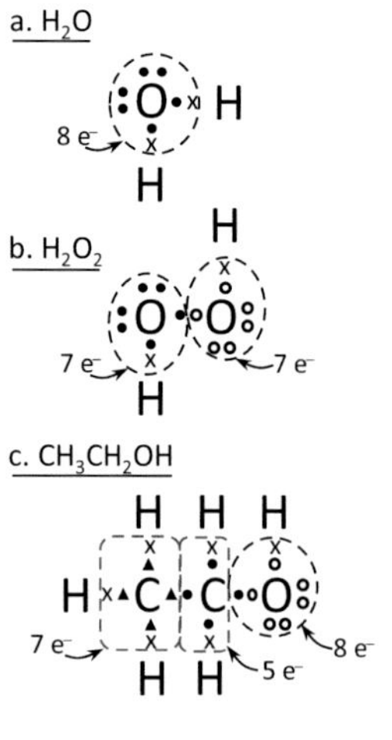

FIGURE 17.2 Lewis diagrams showing outer shell bonding electrons; electrons are assigned to atoms based on largest electronegativity value. a. water, H_2O; b. hydrogen peroxide, H_2O_2; c. ethanol.

Rule 2. O in O_2 is O(0), and in H_2O_2 it is O(–I), but in nature when both bonds are to *other* elements, O is always O(–II). O has an electronegativity value of 3.44, which is greater than that of H (2.20), and all other elements except fluorine (3.98). We essentially never need to talk about O–F bonds in water chemistry.

Rule 3.

$$\sum_i \nu_i o_i = \text{species charge.} \quad (17.5)$$

Any neutral molecule or ion is some combination of atoms with assignable outer-shell electrons and thus assignable oxidation states. Since the oxidation state indicates how many electrons the atom has relative to the neutral atom, the atom-by-atom sum of the oxidations states equals the charge on the species (zero for a neutral molecule). In Eq.(17.5), ν_i is the number of atoms of type i in the species, and o_i is the oxidation state of the type i atom. Examples: a) for H_2O, then 2(I) + 1(–II) = 0; b) for Na^+, then 1(I) = +1; c) for Cl^-, then 1(–I) = –1; d) for OH^-, 1(–II) + 1(I) = –1.

Example 17.1 Oxidation States of Oxygen

The Lewis diagram for hydrogen peroxide is given in Figure 17.2b. What is the assigned oxidation state of O in H_2O_2? Prepare a table ranking the oxidation states of O in O_2, H_2O_2, H_2O, and in OH^-.

Solution

In H_2O_2, each oxygen is assigned a total of 7 electrons (the electrons in the bond between the two oxygens are divided equally). This gives one more electron to each O than for the neutral atom, giving O(–I) for each O.

oxygen species	oxidation state
O_2	O(0)
H_2O_2	O(–I)
H_2O, OH^-	O(–II)

Example 17.2 Oxidation States of Inorganic Carbon

Use Rules 1, 2, and 3 to determine the assigned oxidation states of carbon in the following forms: **a.** CO_2; **b.** $H_2CO_3^*$; **c.** HCO_3^-; **d.** CO_3^{2-}; and **e.** CH_4 (methane). Order the chemicals from highest carbon oxidation state to lowest, and include elemental carbon as graphite ($C_{(s)}$) in the list at C(0).

carbon species	oxidation state	
CO_2, $H_2CO_3^*$, HCO_3^-, CO_3^{2-}	C(IV)	
$C_{(s)}$ (graphite)	C(0)	Δ = 8
CH_4	C(−IV)	

Solution

a. For CO_2, if o is the oxidation state of C, then $o + 2\times(-II) = 0$, $o = 4$, so we have C(IV). **b.** For $H_2CO_3^*$, then $2\times(I) + o + 3\times(-II) = 0$, $o = 4$, giving C(IV). **c.** For HCO_3^-, then $1\times(I) + o + 3\times(-II) = -1$, $o = 4$, giving C(IV). **d.** For CO_3^{2-}, then $o + 3\times(-II) = -2$, $o = 4$, giving C(IV). **e.** For CH_4, then $o + 4\times(I) = -4$, giving C(−IV).

Example 17.3 Oxidation States of Nitrogen

Use Rules 1, 2, and 3 to determine the assigned oxidation states of nitrogen in the following forms: **a.** HNO_3; **b.** NO_3^-; **c.** HNO_2; **d.** NO_2^-; **e.** NH_3; and **f.** NH_4^+ Order the chemicals from highest nitrogen oxidation state to lowest, and include N as N_2 in the list at N(0).

nitrogen species	oxidation state	
HNO_3,NO_3^-	N(V)	
HNO_2, NO_2^-	N(III)	Δ = 8
N_2	N(0)	
NH_3	N(−III)	

Solution

a. For HNO_3, if o is the oxidation state of N, then $1\times(I) + o + 3\times(-II) = 0$, $o = 5$, giving N(V). **b.** For NO_3^-, then $o + 3\times(-II) = -1$, $o = 5$, giving N(V). **c.** For HNO_2, then $1\times(I) + o + 2\times(-II) = 0$, $o = 3$, giving N(III). **d.** For NO_2^-, then $o + 2\times(-II) = -1$, $o = 3$, giving N(III). **e.** For NH_3, then $o + 3\times(I) = 0$, $o = -3$, giving N(−III). **f.** For NH_4^+, then $o + 4\times(I) = 1$, $o = -3$, so we have N(−III). The results are summarized in the accompanying table.

Example 17.4 Oxidation States of Sulfur

Use Rules 1, 2, and 3 to determine the assigned oxidation states of sulfur in the following forms: **a.** H_2SO_4; **b.** HSO_4^-; **c.** SO_4^{2-}; **d.** H_2SO_3; **e.** HSO_3^-; **f.** SO_3^{2-}; **g.** SO_2; **h.** S_2^{2-} (*e.g.*, as it appears in pyrite, $FeS_{2(s)}$); **i.** H_2S; **j.** HS^-; and **k.** S^{2-}. Order the chemicals from highest sulfur oxidation state to lowest, and include elemental sulfur $S_{(s)}$ in the list at S(0).

sulfur species	oxidation state	
$H_2SO_4, HSO_4^-, SO_4^{2-}$	S(VI)	$\Delta = 8$
H_2SO_3, HSO_3^-, SO_3^{2-}, SO_2	S(IV)	
$S_{(s)}$	S(0)	
S_2^{2-}	S(−I)	
H_2S, HS^-, S^{2-}	S(−II)	

Solution

a. For H_2SO_4, if o is the oxidation state of S, then 2×(I) + o + 4×(−II) = 0, o = 6, so we have S(VI). **b.** For HSO_4^-, then 1×(I) + o + 4×(−II) = −1, o = 6, giving S(VI). **c.** For SO_4^{2-}, then o + 4×(−II) = −2, o = 6, giving S(VI). **d.** For H_2SO_3, then 2×(I) + o + 3×(−II) = 0, o = 4, giving S(IV). **e.** For HSO_3^-, then 1×(I) + o + 3×(−II) = −1, o = 4, giving S(IV). **f.** For SO_3^{2-}, then o + 3×(−II) = −2, o = 4, giving S(IV). **g.** For SO_2, then o + 2×(−II) = 0, o = 4, giving S(IV). **h.** For S_2^{2-} 2o = −2, 0, o = −2, giving S(−I). **i.** For H_2S, then 2×(I) + o = 0, o = −2, giving S(−II). **j.** For HS^-, then 1×(I) + o = −1, o = −2, giving S(−II). **k.** For S^{2-}, then o = −2, giving S(−II).

All the assigned oxidation states considered above pertain to a single form of each element in the various species considered. For organic molecules, however, it is often the case that the different carbon atoms in a given molecule are bonded to different sets of other atoms, so that the different carbons technically would have different assigned oxidation states. Consider the ethanol molecule (Figure 17.2c). One C has three bonds to H atoms and one bond to the other another C. On the basis of the different atomic electronegativies, that C is C(−III). The other C has two bonds to H atoms, one bond to an O atom, and one bond to the other C. That C is C(−I). The average oxidation state for the two carbons is then [(−III) + (−I)]/2 = (−II). The same average oxidation state result can be obtained using Rules 1 to 3 as usual: for C_2H_6O (ethanol), if $\bar{o}$ is the average oxidation state of C in the molecule, then 2$\bar{o}$ + 6(I) + 1(−II) = 0, $\bar{o}$ = −2, so as the average, we have C(−II).

Example 17.5 Oxidation States of Inorganic and Organic Carbon

Use Rules 1, 2, and 3 to determine the assigned oxidation state of carbon in the following forms (compute the average value for compounds with more than one C): **a.** formic acid (HCOOH); **b.** formaldehyde (CH_2O); **c.** glucose ($C_6H_{12}O_6$); **d.** cellulose and starch (both are polymers of units with the formula $C_6H_{10}O_5$); **e.** benzene (C_6H_6); **f.** glyceryl tripalmitate ($C_3H_5O_6$-$(C_{16}H_{31}O_3)_3$, a triglyceride lipid oil); **g.** ethanol (CH_3CH_2OH); and **h.** octane (C_8H_{18}). Order the chemicals from the highest carbon oxidation state to the lowest, and include the forms of carbon considered in Example 17.1.

carbon species	oxidation state	
the four CO_2 species	C(IV)	$\Delta = 8$
HCOOH	C(II)	
$C_{(s)}$ (graphite)	C(0)	
CH_2O, $C_6H_{12}O_6$, cellulose, starch	C(0)	
glyceryl tripalmitate	C(−1.69)	
ethanol C_2H_6O	C(−II)	
C_8H_{18}	C(−2.25)	
CH_4	C(−IV)	

Solution

a. For formic acid (HCOOH), if o is the oxidation state of C, then $2(\mathrm{I}) + o + 2(-\mathrm{II}) = 0$, $o = 2$, giving C(II).
b. Formaldehyde (CH_2O), $o + 2(\mathrm{I}) + 1(-\mathrm{II}) = 0$, $o = 0$, giving C(0).
c. Glucose ($C_6H_{12}O_6$), for the average oxidation state, $6\bar{o} + 12(\mathrm{I}) + 6(-\mathrm{II}) = 0$, $\bar{o} = 0$, giving C(0).
d. Cellulose and starch ($C_6H_{10}O_5$), $6\bar{o} + 10(\mathrm{I}) + 5(-\mathrm{II}) = 0$, $\bar{o} = 0$, giving C(0).
e. Benzene (C_6H_6), $6o + 6(\mathrm{I}) = 0$, $o = -1$, giving C(−I).
f. Glyceryl tripalmitate ($C_3H_5O_6$-$(C_{16}H_{31}O_3)_3$), $51\bar{o} + 98(\mathrm{I}) + 6(-\mathrm{II}) = 0$, $\bar{o} = -1.68$, giving C(−1.68).
g. Ethanol, $2\bar{o} + 6(\mathrm{I}) + 1(-\mathrm{II}) = 0$, $\bar{o} = -2$, giving C(−II).
h. Octane (C_8H_{18}), $8\bar{o} + 18(\mathrm{I}) = 0$, $\bar{o} = -2.25$, giving C(−2.25).

For the compounds in the table for Example 17.4, the lower the carbon oxidation state, the more energy can be released per carbon upon combustion with oxygen to CO_2: oxidatively, CH_4 has the furthest to travel. But for C(IV), already being CO_2, the stored energy is by definition zero. For a lipid oil, the amount is similar to petroleum oil, which is why plants and animals use lipids in fat tissue and seeds to store chemical energy that is more portable than carbohydrates. Also, from Example 17.4, CH_2O, $C_6H_{12}O_6$, cellulose, and starch all contain C(0), as in graphite. In the formulas of each of these four chemicals, there is some number of C atoms, plus some integral number of H_2O units: for formaldehyde, 1 C and 1 H_2O. For glucose, 6 C atoms and 6 H_2O. All four chemicals are then literally hydrates of carbon, *i.e.*, "carbohydrates". So, "burning toast" and turning it totally black is not really "burning toast" at all, if burning means "oxidizing". "Burning toast" in fact is mostly "dehydrating toast": water is driven off the bread by the heat, turning it into carbon. (Of course, if the bread is set on fire in the toaster, that is another matter.) For bread approximated as starch, the dehydration reaction is $C_6H_{10}O_5 = 6C + 5H_2O$. This is the same chemistry involved when forming "biochar" (a.k.a., literally charcoal) from organic plant materials. Carbohydrate dehydration can also be carried out chemically, by the action of concentrated H_2SO_4. The acid vigorously draws out water from a carbohydrate so as to obtain water to hydrate the ions in the acid. The result is a carbon residue, as can be seen in online videos.

Example 17.6 Oxidation States of Metals

Use Rules 1, 2, and 3 to determine the assigned oxidation state of the metal in each of the following: **a.** Fe^{3+}; **b.** $FeOH^{2+}$; **c.** $Fe(OH)_{3(s)}$ (ferrihydrite); **d.** $FeOOH_{(s)}$ (goethite); **e.** $Fe_3O_{4(s)}$ (magnetite); **f.** Fe^{2+}; **g.** $FeOH^{+}$; **h.** MnO_4^- (permanganate ion, not found in nature, but has been used in some groundwater remediation efforts to oxidatively remove contaminants); **i.** $MnO_{2(s)}$; **j.** Mn^{2+}; **k.** $PbO_{2(s)}$; **l.** $PbO_{(s)}$; **m.** Pb^{2+}; and **n.** $Pb_{(s)}$.

metal species	ox. state	metal species	ox. state
Fe^{3+}	Fe(III)	MnO_4^-	Mn(VII)
$FeOH^{2+}$	Fe(III)	$MnO_{2(s)}$	Mn(IV)
$Fe(OH)_{3(s)}$	Fe(III)	Mn^{2+}	Mn(II)
$FeOOH_{(s)}$	Fe(III)	$PbO_{2(s)}$	Pb(IV)
$Fe_3O_{4(s)}$	Fe(2.67)	$PbO_{(s)}$	Pb(II)
Fe^{2+}	Fe(II)	Pb^{2+}	Pb(II)
$FeOH^{+}$	Fe(II)	$Pb_{(s)}$	Pb(0)

Solution

a. Fe^{3+}, $o = 3$, so we have Fe(III).
b. $FeOH^{2+}$, $o + 1(-II) + 1(I) = 2$, $o = 3$, giving Fe(III).
c. $Fe(OH)_{3(s)}$, $o + 3(-II) + 3(I) = 0$, $o = 3$, giving Fe(III).
d. $FeOOH_{(s)}$, $o + 2(-II) + 1(I) = 0$, $o = 3$, giving Fe(III).
e. $Fe_3O_{4\ (s)}$, $3\bar{o} + 4(-II) = 0$, $\bar{o} = 2.67$, giving Fe(2.67); as magnetite contains both Fe(II) and Fe(III).
f. Fe^{2+}, $o = 2$, giving Fe(II).
g. $FeOH^{+}$, $o + 1(-II) + 1(I) = 1$, $o = 2$, giving Fe(II).
h. MnO_4^-, $o + 4(-II) = -1$, $o = 7$, giving Mn(VII).
i. $MnO_{2(s)}$, $o + 2(-II) = 0$, $o = 4$, giving Mn(IV).
j. Mn^{2+}, $o = 2$, giving Mn(II).
k. $PbO_{2(s)}$, $o + 2(-II) = 0$, $o = 4$, giving Pb(IV).
l. $PbO_{(s)}$, $o + 1(-II) = 0$, $o = 2$, giving Pb(II).
m. Pb^{2+}, $o = 2$, giving Pb(II).
n. $Pb_{(s)}$, $o = 0$, giving Pb(0).

Example 17.7 Oxidation States of Chlorine

The chlorine oxidants HOCl, OCl^-, and Cl_2 are very important in disinfection of both drinking waters and waste waters. They are potent, and moreover are broadly lethal (bacteria, viruses, etc.) because they can oxidize cellular organic compounds in any kind of organism. Use Rules 1, 2, and 3 to determine the assigned oxidation state of Cl in each of the following: **a.** HOCl; **b.** OCl^-; **c.** Cl_2; **d.** HCl; and **e.** Cl^-.

Cl species	oxidation state
HOCl, OCl^-	Cl(I)
Cl_2	Cl(0)
HCl, Cl^-	Cl(−I)

Solution

a. HOCl, 1(I) + 1(–II) + *o* = 0, *o* = 1, giving Cl(I).
b. OCl^-, 1(–II) + *o* = –1, *o* = 1, giving Cl(I).
c. Cl_2, 2*o* = 0, giving Cl(0).
d. HCl, 1(I) + *o* = 0, *o* = –1, giving Cl(–I).
e. Cl^-, *o* = –1, giving Cl(–I).

17.3 E_H AND pe: EQUIVALENT WAYS OF HANDLING REDOX CALCULATIONS (BUT USING pe HAS ADVANTAGES)

17.3.1 E_H Equations

Consider the overall electrochemical cell in Figure 17.3. Let us assume for the moment that when mixed together, the chemicals in the two half cells would react in a forward direction according to

$$H_{2(g)} + Cu^{2+} = 2H^+ + Cu_{(s)} \tag{17.6}$$

which relates to Eq.(17.3) (*i.e.*, $RED_2 = H_{2(g)}$, $OX_1 = Cu^{2+}$, $OX_2 = 2H^+$, and $RED_1 = Cu_{(s)}$). Electrons are being passed from $H_{2(g)}$ (via H_2 in solution) to Cu^{2+}, becoming $2H^+$ and $Cu_{(s)}$, respectively. In an electrochemical cell, however, the two half cells are separated. This means that direct reaction cannot occur: if there is to be an exchange of electrons, electrons have to move through the wire that connects the two half cells. The meter in Figure 17.3 gives a reading of in units of volts which relates to the driving force for movement of electrons from the left half cell to the right half cell. This is a battery. For context, given their enormous importance in our lives, the common lead-acid battery and the alkaline battery and their electrochemistries are described in Boxes 17.1 and 17.2, respectively.

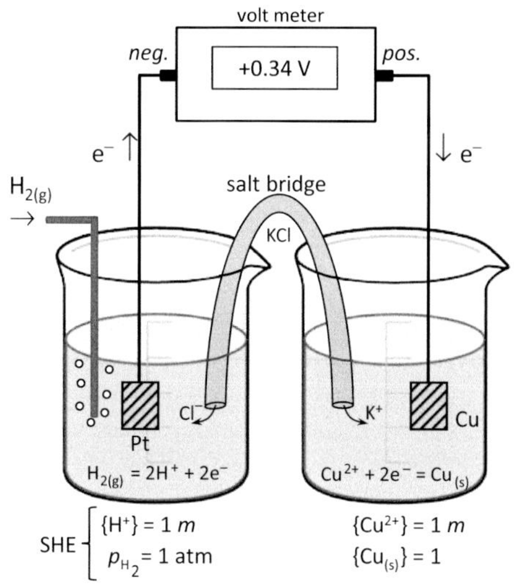

FIGURE 17.3 Electrochemical cell with a standard hydrogen electrode (SHE) as the left electrode, and a $Cu^{2+} + 2e^- = Cu_{(s)}$ half cell on the right. By convention, the negative terminal of the meter is on the left, and the positive terminal is on the right. $H_{2(g)}$ at 1 atm pressure is being bubbled through the solution to keep the dissolved H_2 activity corresponding to that for $H_{2(g)}$ at 1 atm. H_2 is being oxidized to $2H^+$ on the left so that electrons are entering the Pt electrode. On the right, Cu^{2+} is picking up those electrons, and being reduced to form more $Cu_{(s)}$ at the $Cu_{(s)}$ electrode. For the conditions indicated, the meter reads +0.34 V, which is E_H^o for the copper half cell.

The meter that is used to measure the voltage difference E between the electrodes of the two half cells is designed to give a positive reading ($E_{cell} > 0$) when electrons are flowing into the negative (black) terminal of the meter on the left, and out of the positive (red) terminal on the right. Electrons are flowing from the negative side towards the positive side, which makes sense since electrons are negative. The reaction at the left electrode is an oxidation, since it provides the electrons that are flowing from left to right. The reaction at the right electrode is a reduction since it is consuming electrons. If the battery is allowed to run long enough that the concentrations in the half cells change so that electrons no longer tend to flow, the battery is dead.

If electrons are moving from left to right in Figure 17.3, then the right side will start to develop a net negative charge as the Cu^{2+} ions are being removed from solution. The left side will start to develop a positive charge due to the H^+ ions that are being produced there. If the system is just the two half cells connected with a wire and meter between the two electrodes, the electron flow will soon stop because a build-up of negative charge on the right will repel entry of further electrons on the right. This problem is solved by adding a "salt bridge" which contains a concentrated solution of a salt like KCl, usually somehow physically stabilized (as in an agar gel, or a porous frit) so that salt solution does not all leak out into the half cells. If the salt is KCl, then K^+ ions can move from the bridge into the right cell to neutralize developing negative charge on the right, and an equal number

BOX 17.1 The Lead-Acid Battery

The lead-acid battery involves very simple, robust, and reliable chemistry which accounts for its myriad applications, not least of which is as the battery design-of-choice for almost all non-electric vehicles. The reactions shown are all for "discharge" conditions. For charge conditions, the directions are reversed. A single cell gives about 2V, so the common 12V battery requires 6 cells connected in series. $E°$ for the single cell is 2.041V, which means that in the ~5 M H_2SO_4 solution, $\{H^+\}^4\{SO_4^{2-}\}^2 \approx 1$.

Single-Cell Lead-Acid Battery

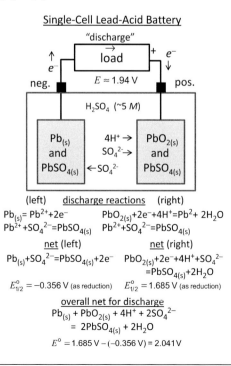

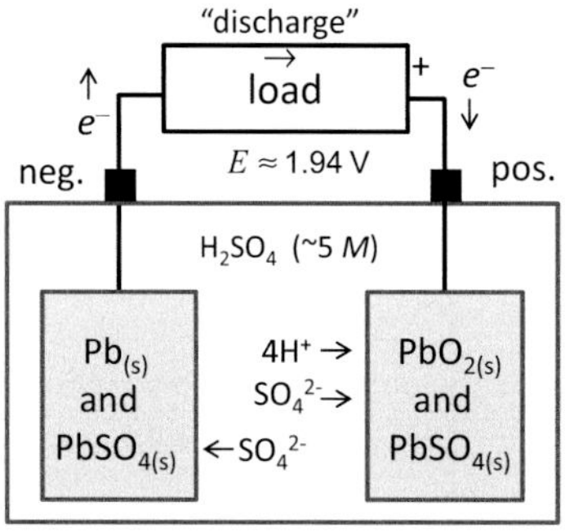

(left) discharge reactions (right)

(left)	(right)
$Pb_{(s)} = Pb^{2+} + 2e^-$	$PbO_{2(s)} + 2e^- + 4H^+ = Pb^{2+} + 2H_2O$
$Pb^{2+} + SO_4^{2-} = PbSO_{4(s)}$	$Pb^{2+} + SO_4^{2-} = PbSO_{4(s)}$
net (left)	net (right)
$Pb_{(s)} + SO_4^{2-} = PbSO_{4(s)} + 2e^-$	$PbO_{2(s)} + 2e^- + 4H^+ + SO_4^{2-} = PbSO_{4(s)} + 2H_2O$
$E^o_{1/2} = -0.356$ V (as reduction)	$E^o_{1/2} = 1.685$ V (as reduction)

overall net for discharge

$$Pb_{(s)} + PbO_{2(s)} + 4H^+ + 2SO_4^{2-} = 2PbSO_{4(s)} + 2H_2O$$

$$E^o = 1.685\text{ V} - (-0.356\text{ V}) = 2.041\text{ V}$$

BOX 17.2 The Alkaline Battery

Like the lead-acid battery, the alkaline battery is simple, robust, and reliable. The overall reaction involves only solids with no net consumption of OH^-. E for a single cell is ~1.5V, so multiple cells are not needed to provide 1.5V. If the caustic contents do leak, the KOH will soon be neutralized by atmospheric CO_2 to form bicarbonate/carbonate solids, which are not hazardous.

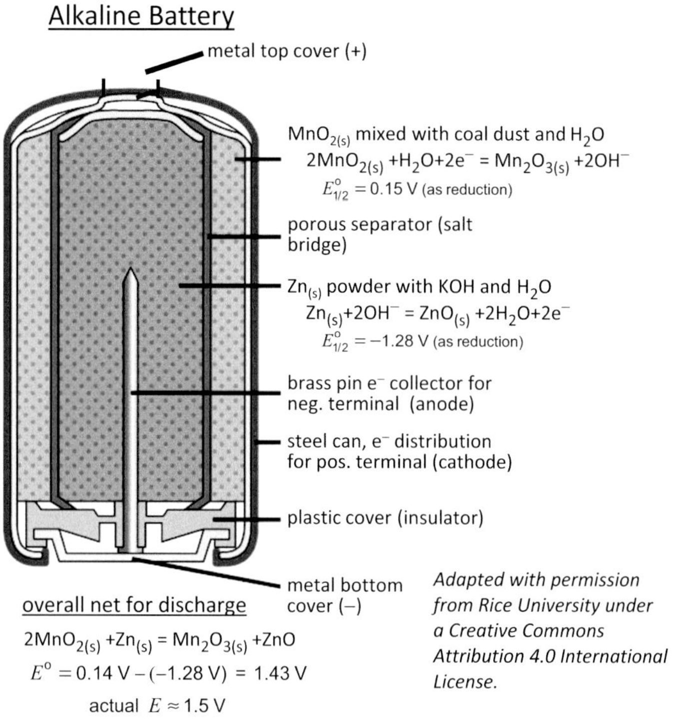

of Cl^- ions can move into the left cell to neutralize developing positive charge on the left. Overall, the two half cells remain neutrally charged, as does the salt bridge.

Figure 17.3 describes a system where the right electrode is made of elemental copper and the left electrode is made of platinum (Pt). The Pt does not participate directly in the reaction at the left electrode, but it does provide a conducting metal surface for the exchange of electrons. $H_{2(g)}$ is being bubbled in the solution near the left electrode to saturate the solution with $H_{2(g)}$ at 1 atm according to the K_H (M/atm) value for $H_{2(g)}$. It is then dissolved H_2 that is being oxidized at the left electrode to H^+. Cu^{2+} is being reduced at the right electrode forming more $Cu_{(s)}$ there. At 25 °C/atm, the meter will read +0.34 V. In practice, these kinds of measurements are made with a potentiometer in place of a simple meter. A potentiometer measures and shows the voltage as generated by the two half cells, and generates an equal and opposite voltage so that no current flows. That way there are no changes in the relevant conditions in the solution, in this case just $\{H^+\}$ and $\{Cu^{2+}\}$; if electrons *were* to flow from left to right, $\{H^+\}$ would increase and $\{Cu^{2+}\}$ would decrease.

The fundamental thermodynamic equivalence between the electrochemical reaction ΔG for an *overall cell* (two half cells) electrochemical reaction and the electrical work done by the reaction is

$$\Delta G = -nFE_{cell} \tag{17.7}$$

where n (mol/mol) is the number of mols of electrons exchanged per full mol unit of the reaction, F is the "Faraday" constant = 96,444.6 C/mol (C = coulombs of electrical charge), and E is the measured difference in electrical potential between the two electrodes. F is the amount of electrical charge held on one mol (= N_A = Avogadro's Number = 6.023×10^{23}) of elementary charges e_o. The charge on a proton is $+e_o$ and the charge on an electron is $-e_o$. The value of e_o is (96,444.6 C/mol)/N_A = 1.60×10^{-19} C.

ΔG is usually expressed in kJ/mol, so C/mol are not the most convenient units for F. But

$$1\ \text{C-V} = 10^{-3}\ \text{kJ} \tag{17.8}$$

so

$$F = \frac{96{,}485.33\ \text{C}}{\text{mol}} \times \frac{10^{-3}\ \text{kJ}}{\text{C-V}} = \frac{96{,}485.33\ \text{kJ}}{\text{V-mol}}. \tag{17.9}$$

1 C-V is a unit of energy as is one electron-volt ("eV"). One eV is the energy released when one positive elementary charge e_o falls through an electrical potential drop of 1 volt. 1 C-V is the energy released when one coulomb worth of positive charge falls through an electrical potential drop of 1 volt; One can make an exact analogy between electrical potential energy and gravitational potential energy.

From Chapter 2,

$$\Delta G = -RT\ \text{In}\frac{K}{Q}. \tag{17.10}$$

When all of the reactant and product species are at unit activity so that $Q = 1$, or if Q otherwise equals 1, then

$$\Delta G = \Delta G^{\text{o}} = -RT\ln K \tag{17.11}$$

By analogy, under the same conditions, for an electrochemical cell,

$$E_{\text{cell}} = E^{\text{o}}_{\text{cell}} \qquad (Q = 1). \tag{17.12}$$

In terms of the OX/RED notation, for Eq.(17.3) and the cell in Figure 17.3,

$$Q = \frac{\{\text{OX}_2\}\{\text{RED}_1\}}{\{\text{RED}_2\}\{\text{OX}_1\}} = \frac{\{\text{H}^+\}^2\{\text{Cu}_{(s)}\}}{p_{\text{H}_2}\{\text{Cu}^{2+}\}}. \tag{17.13}$$

If $E^{\text{o}}_{\text{cell}}$ is the potential when $Q = 1$, then

$$\Delta G^{\text{o}} = -nFE^{\text{o}}_{\text{cell}}. \tag{17.14}$$

Combining Eqs.(17.7), (17.10–17-11), and (17.14) gives

$$-nFE_{\text{cell}} = -nFE^{\text{o}}_{\text{cell}} + RT\ln Q \tag{17.15}$$

or

$$E_{\text{cell}} = E^{\text{o}}_{\text{cell}} + \frac{2.303\ RT}{nF}\log Q^{-1} \qquad \underline{\text{Nernst Equation.}} \tag{17.16}$$

R can be expressed using many different units, the choice depends on what is most convenient. Here, we use R = 8.314… C-V/mol-K. So, at T = 298.15 K (25 °C),

$$\begin{aligned}\frac{(2.303...)RT}{nF} &= \frac{(2.303...)(8.314...\text{C-V/mol-K})(298.15K)}{(n\ \text{mol/mol})(96{,}485.33...\text{C/mol})}\\ &= \boxed{\frac{0.05916}{n}\text{V.}}\end{aligned} \tag{17.17}$$

The units for n of mol/mol are kind of tricky and require a bit of thought to understand. In Eq.(17.7), we need: 1) the mol on top to multiply with FE_{cell} (C-V/mol) to get C–V (or kJ) of energy, and we need the mol on the bottom to get back to C–V/mol (or kJ/mol) for final units of energy/mol for ΔG, which actually is energy (usually kJ) *per* full "mol-based" unit of reaction as written; the stoichiometric coefficients for the reaction (which are pure numbers without units) *in that case* lead to the number of mols involved. And, in Eq.(17.17), with n having units of mol/mol, given the units of R and F, we end up with units of just V for E_{cell}. Moral: units are always important; if they do not work out, something is amiss.

For the reaction in Eq.(17.6),

$$E_{cell} = E^{o}_{cell} + \frac{2.303\,RT}{2F}\log\frac{p_{H_2}\{Cu^{2+}\}}{\{H^+\}^2\{Cu_{(s)}\}}. \tag{17.18}$$

When $E_{cell} > 0$, electrons are flowing left to right; when $E_{cell} < 0$, electrons are flowing right to left. E_{cell} increases with increasing p_{H_2} because the thermodynamic source of electrons on the left increases. E_{cell} increases with increasing $\{Cu^{2+}\}$ because the thermodynamic sink for electrons on the right increases. E decreases with increasing $\{H^+\}^2$ because the thermodynamic sink for electrons on the left increases. E decreases with increasing $\{Cu_{(s)}\}$ because the thermodynamic source for electrons on the right increases. ($\{Cu_{(s)}\}$ could be less than one in an alloy of Cu, but for pure $Cu_{(s)}$, then $\{Cu_{(s)}\} = 1$.)

At the atomic level in the Figure 17.3 system, there is an electron tug of war going on between the two electrodes. At the right electrode, there is some desire for Cu^{2+} to pull $2e^-$ out of the electrode and become metallic $Cu_{(s)}$. At the left electrode, there is a counter desire for $2H^+$ to pull $2e^-$ out of the electrode and become dissolved H_2, which then equilibrates with the bubbling $H_{2(g)}$. Which side wins the tug of war depends on: 1) the inherent tendencies of the two reaction to occur, which is built into E^{o}_{cell}; and 2) the values of the activity terms in Q, as elaborated in the previous paragraph regarding how the activities affect E_{cell}. Since the direction that the electrons actually flow is determined by which side wants the electrons more, this must be some kind of difference between the reduction tendencies of the two sides. So, we can divide each term on the RHS of Eq.(17.18) into two subterms, one for the right, one for the left:

$$E_{cell} = \overbrace{E^{o}_{Cu^{2+}+2e^-\rightarrow Cu_{(s)}} + \frac{2.303\,RT}{2F}\log\frac{\{Cu^{2+}\}}{\{Cu_{(s)}\}}}^{E_R} - \overbrace{E^{o}_{2H^+ +2e^-\rightarrow H_{2(g)}} + \frac{2.303\,RT}{2F}\log\frac{\{H^+\}^2}{p_{H_2}}}^{E_L}. \tag{17.19}$$

The single log term in Eq.(17.18) leads directly to the two log terms in Eq.(17.19); each is straightforward to evaluate. For the division of the single E^{o}_{cell} term in Eq.(17.18) into the two $E°$ terms in Eq.(17.19), it makes sense that when both of the log terms are zero so that $E_{cell} = E^{o}_{cell}$, that E^{o}_{cell} is made up of the difference between the two half cells tending to proceed as reductions under standard conditions, with the difference taken as R minus L because $E_{cell} > 0$ when the right side is winning the tug of water and proceeding as reduction. So, it makes sense that we can divide E^{o}_{cell} for the Figure 17.3 cell into two parts according to

$$E^{o}_{cell} = \overbrace{E^{o}_{Cu^{2+}+2e^-\rightarrow Cu_{(s)}}}^{E^{o}_{R}} - \overbrace{E^{o}_{2H^+ +2e^-\rightarrow H_{2(g)}}}^{E^{o}_{L}}. \tag{17.20}$$

But what is the value of each of the terms in Eq.(17.20)? One cannot measure the tendency of the half reaction $Cu^{2+} + 2e^- \rightarrow Cu_{(s)}$ to occur *all by itself* because there is no source of electrons in that context. The same can be said for the half reaction $2H^+ + 2e^- \rightarrow H_{2(g)}$. In order to evaluate either of the two half-cell E° values, one has to measure it as an E_{cell} with some standardized availability of electrons provided to the wire on the left, and the half cell of interest on the right. OK, but exactly what "standardized availability of electrons"? We could phone up God and ask that question, or maybe look for the book *All Secrets of the Universe*, but neither option is practical, at least at the moment. We are going to have "wing" this one as best we can and pick *something*. The interesting thing is that it does not really matter what is picked, because whatever is picked, after the two measurements, the difference for Eq.(17.20) will always be the same (because each different pick affects both measurements by the same amount). Cool! Now we can move this along. OK, what to pick? More exactly we should ask: "What *was* picked by chemists when this field was developed?" The answer is the hydrogen electrode on the left in Figure 17.3, which thereby is named the "standard hydrogen electrode" (SHE). Why this choice? Because early chemists had a very good understanding of how to prepare solutions of strong acids so as to obtain known H^+ activities in solution, and a good understanding of how to prepare $H_{2(g)}$, being that H was one of the very first elements discovered and studied (1766 in Britain, by Henry Cavendish).

Looking more closely at Eq.(17.20), $E^o_{2H^+ + 2e^- \rightarrow H_{2(g)}}$ pertains exactly to the SHE. But evaluating $E^o_{2H^+ + 2e^- \rightarrow H_{2(g)}}$ relative to itself is going to give exactly zero, so $E^o_{2H^+ + 2e^- \rightarrow H_{2(g)}} \equiv 0$. And, since it has been decided to measure half cell E° values relative to the SHE, then we add in a subscript H, so for the Figure 17.3 cell,

$$E^o_{cell} = \overbrace{E^o_{H,\,Cu^{2+} + 2e^- \rightarrow Cu_{(s)}}}^{E^o_R} - \overbrace{0}^{E^o_L\ \mathrm{SHE}} = E^o_{H,Cu^{2+} + 2e^- \rightarrow Cu_{(s)}} \tag{17.21}$$

Half cell E° values measured relative to the SHE are designated E^o_H values; they are named "standard reduction potentials" (see Table 17.1).

An E_H value is a type of E_{cell}. It is a measure of the redox state that a solution would have if it is *at equilibrium*. In particular, E_H is the E_{cell} when there is some kind of electrode capable of sensing the availability of electrons in the solution of interest as the right half cell (on the positive side of the meter/potentiometer), and the SHE is the left half cell (on the negative side). The solution of interest in the right cell could have been prepared with some known OX/RED chemistry, or it might be a natural water sample in which we do not know the OX/RED pair operating at the electrode. Before proceeding, two important comments are needed:

Comment 1. We can (and mostly do) talk about the E_H of a system, even if there is no electrode used: E_H is a property of the solution that might be measureable with an electrode, but exists regardless, just like pH.

Comment 2. The caveat "*at equilibrium*" is a big one, because most redox reactions do not come to equilibrium quickly, and this most certainly includes natural waters. This is distinctly unlike the situation with acid/base reactions, which equilibrate exceedingly rapidly. However, we should be grateful that redox equilibria are generally slow unless specifically catalyzed; otherwise, life as we know it would be impossible. For example, if all such reactions were fast, organic compounds in cellular solutions containing oxygen would be directly and immediately stoichiometrically oxidized by the dissolved O_2 present, and that would be the end. Even so, the very small relative degree of such reactions of this type causes much damage and aging in cells. The matter of the kinetics of redox reactions is addressed somewhat further in Section 17.3.2.5.

Okay, back to the subject directly at hand. With $E_L = 0$ for this system, for any OX + ne^- = RED half-reaction in the solution of interest, then Eq.(17.19) gives

$$E_H = E^o_{H,\,OX+ne^- \rightarrow RED} + \frac{2.303\,RT}{nF}\log\frac{\{OX\}}{\{RED\}}. \tag{17.22}$$

TABLE 17.1
Data for Selected Redox Reactions at 25 °C/1 atm in Order of Increasing pe° and E_H^o

Reduction Half-Reaction	E_H^o	log *K*	pe°	pe°(W)
$OX + ne^- = RED$	$= 0.05916\ pe^\circ$	$= \frac{\{RED\}}{\{OX\}\{e^-\}^n} = \frac{E_H^o}{2.303RT/(nF)}$	$= \frac{1}{n}\log K = \frac{E_H^o}{2.303RT/F}$	$= pe^\circ - \frac{n_H}{n_e}7$
$Na^+ + e^- = Na_{(s)}$	−2.71	−46.0	−46.0	−46.0
$2H_2O + 2\ e^- = H_{2(g)} + 2OH^-$	−0.83	−27.99	−14.00	−7.00
$2H_2O + 2\ e^- = H_2 + 2OH^-$	−0.92	−31.10	−15.55	−8.55
$Zn^{2+} + 2e^- = Zn_{(s)}$	−0.76	−26.0	−13.0	−13.0
$FeCO_{3(s)} + 2e^- = Fe_{(s)} + CO_3^{2-}$	−0.76	−25.58	−12.79	−12.79
$Fe^{2+} + 2e^- = Fe_{(s)}$	−0.44	−14.9	−7.45	−7.45
$CO_{2(g)} + H^+ + 2e^- = HCOO^-$	−0.29	−9.66	−4.83	−8.33
$H_2CO_3^* + H^+ + 2e^- = HCOO^- + H_2O$	−0.24	−8.19	−4.10	−7.60
$CO_{2(g)} + 4H^+ + 4e^- = CH_2O + H_2O$	−0.071	−4.8	−1.2	−8.2
$S_{(s)} + H^+ + 2e^- = HS^-$	−0.065	−2.2	−1.1	−4.6
$H_2CO_3^* + 4H^+ + 4e^- = CH_2O + 2H_2O$	−0.049	−3.33	−0.83	−7.83
$CO_{2(g)} + 4H^+ + 4e^- = \frac{1}{6}C_6H_{12}O_6(glucose) + H_2O$	−0.012	−0.8	−0.2	−7.2
$H_2CO_3^* + 4H^+ + 4e^- = \frac{1}{6}C_6H_{12}O_6(glucose) + 2H_2O$	0.010	0.67	0.17	−6.83
$2H^+ + 2e^- = H_{2(g)}$	0.00	0.0	0.0	−7.00
$2H^+ + 2e^- = H_2$	−0.092	−3.11	−1.55	−8.55
$HSO_4^- + 3H^+ + 2e^- = H_2SO_3^* + H_2O$	0.012	0.40	0.20	−10.3
$S_{(s)} + H^+ + e^- = \frac{1}{2}H_2S_2$	0.027	0.45	0.45	−6.55
$N_{2(g)} + 6H^+ + 6e^- = 2NH_3$	0.093	9.5	1.58	−5.42
$N_2 + 6H^+ + 6e^- = 2NH_3$	0.125	12.69	2.11	−4.89
$S_{(s)} + 2H^+ + 2e^- = H_2S$	0.14	4.8	2.4	−4.6
$Cu^{2+} + e^- = Cu^+$	0.16	2.7	2.7	2.7
$H_2CO_3^* + 8H^+ + 8e^- = CH_4 + 3H_2O$	0.16	21.62	2.70	−4.30
$HCOO^- + 3H^+ + 2e^- = CH_2O + H_2O$	0.17	5.64	2.82	−7.68
$CO_{2(g)} + 8H^+ + 8e^- = CH_{4(g)} + 2H_2O$	0.17	23.0	2.87	−4.13
$AgCl_{(s)} + e^- = Ag_{(s)} + Cl^-$	0.22	3.7	3.7	3.7
$CH_2O + 2H^+ + 2e^- = CH_3OH$	0.24	8.0	4.0	−3.0
$SO_4^{2} + 9H^+ + 8e^- = HS^- + 4H_2O$	0.25	34.0	4.25	−3.63
$\frac{1}{2}H_2S_2 + H^+ + e^- = H_2S$	0.26	4.35	4.35	−2.65
$Hg_2Cl_{2(s)} + 2e^- = 2Hg_{(l)} + 2Cl^-$	0.268	9.06	4.53	4.53
$N_{2(g)} + 8H^+ + 6e^- = 2NH_4^+$	0.28	28.1	4.68	−4.65
$SO_4^{2} + 10H^+ + 8e^- = H_2S + 4H_2O$	0.30	41.0	5.13	−3.62
$N_2 + 8H^+ + 6e^- = 2NH_4^+$	0.31	31.3	5.21	−4.12
$\frac{1}{6}C_6H_{12}O_6(glucose) + 4H^+ + 4e^- = CH_4 + H_2O$	0.31	20.95	5.24	−1.76
$Cu^{2+} + 2e^- = Cu_{(s)}$	0.34	11.4	5.7	5.7
$HSO_4^- + 7H^+ + 6e^- = S_{(s)} + 4H_2O$	0.34	34.2	5.7	−2.47
$SO_4^{2-} + 8H^+ + 6e^- = S_{(s)} + 4H_2O$	0.36	36.2	6.03	−3.3
$CH_2O + 4H^+ + 4e^- = CH_4 + H_2O$	0.37	24.95	6.24	−0.76
$CH_2O + 4H^+ + 4e^- = CH_{4(g)} + H_2O$	0.41	27.8	6.94	−0.06

(Continued)

TABLE 17.1 (CONTINUED)
Data for Selected Redox Reactions at 25 °C/1 atm in Order of Increasing pe° and E_H^o

Reduction Half-Reaction	E_H^o	log *K*	pe°	pe°(W)
$OX + ne^- = RED$	$= 0.05916\ pe^\circ$	$= \frac{\{RED\}}{\{OX\}\{e^-\}^n} = \frac{E_H^o}{2.303\,RT\,/\,(nF)}$	$= \frac{1}{n}\log K = \frac{E_H^o}{2.303RT\,/\,F}$	$= pe^\circ - \frac{n_H}{n_e}7$
$H_2SO_3^* + 4H^+ + 4e^- = S_{(s)} + 3H_2O$	0.50	33.8	8.45	1.45
$CH_3OH + 2H^+ + 2e^- = CH_4 + H_2O$	0.50	16.95	8.48	1.48
$Cu^+ + e^- = Cu_{(s)}$	0.52	8.8	8.8	8.8
$H_3AsO_4 + 2H^+ + 2e^- = H_3AsO_3 + H_2O$	0.575	9.72	19.44	5.44
$CH_3OH + 2H^+ + 2e^- = CH_{4(g)} + H_2O$	0.58	19.8	9.88	2.88
$Fe^{3+} + e^- = Fe^{2+}$	0.77	13.0	13.0	13.0
$\alpha\text{-}FeOOH_{(s)} + HCO_3^- + 2H^+ + e^- = FeCO_{3(s)} + 2H_2O$	0.78	13.15	13.15	–0.85
$\alpha\text{-}FeOOH_{(S)} + 3H^+ + e^- = Fe^{2+} + 2H_2O$	0.80	13.5	13.5	–7.5
$Ag^+ + e^- = Ag_{(s)}$	0.80	13.5	13.5	13.5
$NO_2^- + 7H^+ + 6e^- = NH_3 + 2H_2O$	0.80	81.5	13.58	5.41
$NO_3^- + 2H^+ + 2e^- = NO_2^- + H_2O$	0.84	28.3	14.15	7.15
$NO_3^- + 10H^+ + 8e^- = NH_4^+ + 3H_2O$	0.88	119.2	14.9	6.15
$NO_2^- + 8H^+ + 6e^- = NH_4^+ + 2H_2O$	0.90	90.8	15.14	5.82
$MnO_{2(S)} + HCO_3^- + 3H^+ + 2e^- = MnCO_{3(s)} + 2H_2O$	0.94	25.8	15.9	5.4
$(am)Fe(OH)_{3(s)} + 3H^+ + e^- = Fe^{2+} + 3H_2O$	0.96	16.2	16.2	–4.8
$O_{2(g)} + 4H^+ + 4e^- = 2H_2O$	1.23	83.1	20.78	13.78
$NO_3^- + 6H^+ + 5e^- = \frac{1}{2}N_2 + 3H_2O$	1.23	103.7	20.74	12.34
$NO_3^- + 6H^+ + 5e^- = \frac{1}{2}N_{2(g)} + 3H_2O$	1.25	105.3	21.05	12.65
$O_2 + 4H^+ + 4e^- = 2H_2O$	1.27	86.0	21.50	14.50
$MnO_{2(s)} + 4H^+ + 2e^- = Mn^{2+} + 2H_2O$	1.29	43.6	21.8	7.8
$Fe^{3+} + CO_3^{2-} + e^- = FeCO_{3(S)}$	1.40	23.68	23.68	23.68
$Cl_2 + 2e^- = 2Cl^-$	1.40	47.2	23.6	23.6
$NO_2^- + 4H^+ + 3e^- = \frac{1}{2}N_2 + 2H_2O$	1.49	75.4	25.13	15.8
$(am)Fe(OH)_{3(s)} + CO_3^{2-} + 3H^+ + e^- = FeCO_{3(s)} + 3H_2O$	1.59	26.88	26.88	5.88
$HOCl + H+ + e– = ½Cl_2 + H_2O$	1.59	26.9	26.9	19.9
$ClO^- + 2H^+ + 2e^- = Cl^- + H_2O$	1.71	57.8	28.9	21.8
$H_2O_2 + 2H^+ + 2e^- = 2H_2O$	1.76	59.6	29.80	22.80

Note: Values are from Bard et al. (1985) and Stumm and Morgan (1996). Under standard conditions {OX}/{RED} = 1), the reducing strength of the RED species is highest at the top of the table, and the oxidizing strength of the OX species is the highest at the bottom of the table.

Eq.(17.22) is a highly important variation of the Nernst Equation, *i.e.*, as applied to an electrochemical half cell rather than an overall cell with two half reactions.

The higher the value of E_H, the higher the {OX}/{RED} ratio, and the solution becomes more oxidizing. This means it is more inclined to consume electrons as they may be made available from reduced species: as the ratio {OX}/{RED} increases, the {OX} activity term, which promotes consumption of electrons, is increasing, and/or the {RED} activity term, which promotes release of electrons, is decreasing. Conversely, the lower the value of E_H, the lower the {OX}/{RED} ratio, and the solution becomes more reducing, *i.e.*, more inclined to release electrons to oxidants as they may be made available. Compare this with: 1) the higher the pH, the more basic the solution, and the more inclined the solution is to consume protons as acids may be made available; 2) the lower the pH, the more acidic the solution, and the more inclined the solution is to release protons to bases as bases may be made available. E_H is thus a redox analog of pH, though pe (as discussed below) is a more direct analog.

Example 17.8 Writing Expressions for {OX} and {RED}

For a half reaction, the {OX} term is the product of the activities of all the species except electrons on the OX side. The {RED} term is the product of the activities of all the species on the RED side. $\{H_2O\}$ is taken as equal to 1. Three examples for illustration are:

$$2H^+ + 2e^- = H_{2(g)}, \qquad \{OX\} = \{H^+\}^2 \text{ and } \{RED\} = p_{H_2};$$

$$Cu^{2+} + 2e^- = Cu_{(s)}, \qquad \{OX\} = \{Cu^{2+}\} \text{ and } \{RED\} = \{Cu_{(s)}\} = 1 \text{ for pure } Cu_{(s)};$$

$$NO_3^- + 6H^- + 5e^- = \tfrac{1}{2}N_{2(g)} + 3H_2O, \qquad \{OX\} = \{NO_3^-\}\{H^-\}^6, \qquad \{RED\} = p_{N_2}^{\frac{1}{2}}.$$

What are {OX} and {RED} for each of the additional half reactions? Take $\{H_2O\} = 1$	Solution
a. $O_{2(g)} + 4H^+ + 4e^- = 2H_2O$	$\{OX\} = p_{O_2}\{H^+\}^4$, $\{RED\} = 1$.
b. $NO_3^- + 6H^+ + 5e^- = \frac{1}{2}N_2 + 3H_2O$ (N_2 is dissolved)	$\{OX\} = \{NO_3^-\}\{H^+\}^6$, $\{RED\} = \{N_2\}^{1/2}$.
c. $Fe^{3+} + e^- = Fe^{3+}$	$\{OX\} = \{Fe^{3+}\}$, $\{RED\} = \{Fe^{2+}\}$.
d. $SO_4^{2-} + 10H^+ + 8e^- = H_2S + 3H_2O$	$\{OX\} = \{SO_4^{2-}\}\{H^+\}^{10}$, $\{RED\} = \{H_2S\}$.
e. $H_2CO_3^* + 4H^+ + 4e^- = \frac{1}{6}C_6H_{12}O_6\text{(glucose)} + 2H_2O$	$\{OX\} = \{H_2CO_3^*\}\{H^+\}^4$, $\{RED\} = \{C_6H_{12}O_6\}^{1/6}$.

Example 17.9 Computing E_H Values

Compute E_H at 25 °C/1 atm for the half reactions in parts a–e of Example 17.8, given the following conditions and the E_H^o values. Neglect activity corrections. Per Eq.(17.17), $2.303RT/F = 0.05916$ at 25 °C.

Conditions	E_H^o (V)
a. $p_{O_2} = 0.21$ atm, pH = 5.0.	1.23
b. $[NO_3^-] = 10^{-5}$ *M*, $[N_2] = 0.0005$ *M*, pH = 5.0.	1.23
c. $[Fe^{3+}] = 10^{-5}$ *M*, $[Fe^{2+}] = 10^{-3}$ *M*.	0.77
d. $[SO_4^{2-}] = 10^{-5}$ *M*, $[H_2S] = 10^{-5}$ *M*, pH = 5.0.	0.30
e. $[H_2CO_3^*] = 10^{-4.87}$ *M*, [glucose] $= 10^{-3}$ *M*, pH = 5.0.	0.010

Solution.

$$\text{a. } E_H = 1.23 + \frac{0.05916}{4}\log\frac{0.21(10^{-5.0})^4}{1} = 0.924\text{ V}$$

$$\text{b. } E_H = 1.23 + \frac{0.05916}{5}\log\frac{10^{-5}(10^{-5.0})^6}{(0.0005)^{0.5}} = 0.835\text{ V}$$

$$\text{c. } E_H = 0.77 + \frac{0.05916}{1}\log\frac{10^{-5}}{10^{-3}} = 0.652\text{ V}$$

$$\text{d. } E_H = 0.30 + \frac{0.05916}{8}\log\frac{10^{-5}(10^{-5.0})^{10}}{10^{-5}} = -0.070\text{ V}$$

$$\text{e. } E_H = 0.010 + \frac{0.05916}{4}\log\frac{10^{-4.87}(10^{-5.0})^4}{10^{-3}} = -0.313\text{ V}$$

Example 17.10 Computing Predicted Equilibrium Chemistry Based on E_H

Example 17.4 considers the forms of S in different oxidation states, including S(VI) and S(−II). For S(VI), pK_2 for H_2SO_4 is at least 2, so at pH = 5.0 $\alpha_2^{S(VI)} \approx 1$, and $S(VI)_T \approx [SO_4^{2-}]$. For S(−II), pK_1 for H_2S is about 7, so at pH = 5.0 $\alpha_0^{S(-II)} \approx 1$, and $S(-II)_T \approx [H_2S]$. Therefore, in Example 17.9.d, when E_H = −0.070 V, then $S(VI)_T \approx S(-II)_T$ and $S_T \approx 2 \times 10^{-5}$ *M*. For pH = 5.0 and $S_T \approx 2 \times 10^{-5}$ *M*, use the same pH-based approximations to estimate equilibrium values of $[SO_4^{2-}]$ and $[H_2S]$ for E_H values of: **a.** −0.040 V; and **b.** −0.100 V. Neglect activity corrections.

Solution

$$\textbf{a. } E_H = 0.30 + \frac{0.05916}{8}\log\frac{[SO_4^{2-}](10^{-5.0})^{10}}{[H_2S]} = -0.040\text{ V}$$

$$\log\frac{[SO_4^{2-}](10^{-5.0})^{10}}{[H_2S]} = \frac{(-0.040 - 0.30)8}{0.05916} = -45.97\text{, so } \frac{[SO_4^{2-}]}{[H_2S]} = \frac{10^{-45.97}}{(10^{-5.0})^{10}} = 1.05 \times 10^4$$

Essentially all of the S_T is present as SO_4^{2-}, so $[SO_4^{2-}] = 2 \times 10^{-5}$ *M*, and $[H_2S] = (2 \times 10^{-5}\ M)/1.09 \times 10^4 = 1.83 \times 10^{-9}$ *M*.

$$\textbf{b. } E_H = 0.30 + \frac{0.05916}{8}\log\frac{[SO_4^{2-}](10^{-5.0})^{10}}{[H_2S]} = -0.100\text{ V}$$

$$\log\frac{[SO_4^{2-}](10^{-5.0})^{10}}{[H_2S]} = \frac{(-0.100 - 0.30)8}{0.05916} = -54.09\text{ , so } \frac{[SO_4^{2-}]}{[H_2S]} = \frac{10^{-54.09}}{(10^{-5.0})^{10}} = 8.12 \times 10^{-5}$$

Essentially all of the S_T is present as H_2S, so $[H_2S] = 2 \times 10^{-5}$ *M*, and $[SO_4^{2-}] = 8.12 \times 10^{-5} \times (2 \times 10^{-5}\ M) = 1.62 \times 10^{-9}$ *M*.

17.3.2 pe Equations

17.3.2.1 The Concept of pe

With pH = −log {H⁺}, we have a direct means for expressing and evaluating activities and thereby concentrations of species that participate in acid/base reactions. For example, for a monoprotic acid HA system of total concentration A_T, then $[HA] = \alpha_0^A A_T$, and $[A^-] = \alpha_1^A A_T$, where α_0^A and

α_1^A are simple functions of pH and the pK_a for HA. For a redox half reaction, {OX} and {RED} are determined by E_H, but the route between an E_H value and {OX} and {RED} involves several steps: in Example 17.10, a subtraction to obtain ($E_H - E_H^o$), division by RT/F (= 0.05916 at 298.15 K), multiplication by n, and an exponentiation. The pe approach reduces the number of operations by introducing the concept of $\{e^-\}$, the activity of the electron in solution, with pe = $-\log\{e^-\}$. This approach allows equilibrium constants to be used in parameterizing redox reactions: for the reaction in Eq.(17.4), we write

$$Fe^{3+} + e^- = Fe^{2+} \qquad K = \frac{\{Fe^{2+}\}}{\{Fe^{3+}\}\{e^-\}} = \frac{\{Fe^{2+}\}}{\{Fe^{3+}\}10^{-pe}} = 10^{13.0} \tag{17.23}$$

and the particular value for K pertains to 25 °C/1 atm.

As with standard reduction potentials for half cells, redox K values are written as reductions: an oxidized species combines with electrons to produce a reduced species. This approach has the *stability* K perspective, as with a metal complexation reaction wherein a metal ion combines with a ligand to produce a metal-ligand complex (*e.g.*, $Fe^{3+} + OH^- = FeOH^{2+}$, $K_{H1} = \{FeOH^{2+}\}/(\{Fe^{3+}\}\{OH^-\})$). The higher the K for formation of a metal-ligand complex, the more stable is the metal-ligand complex. Analogously, the higher the K for a redox half-reaction, the more the stable is the RED form (and the stronger the OX form is as an oxidant). By comparison, an acid K_a value take the *instability* perspective: the stronger the acid, the higher its K_a, and the more it tends to *dissociate*.

Most chemistry texts do not discuss the pe (with redox K) approach, using instead only the E_H (with Nernst equation) approach. But, some geochemistry references, including this book, mainly use the pe approach because it provides a somewhat faster way to work redox problems, makes it simple to combine redox half reactions with non-redox reactions, and is entirely equivalent (as it must be) to the E_H approach. Perhaps the biggest advantage of the pe approach is that it simplifies the combining of redox half reactions with non-redox reactions, as in Example 17.11.

Example 17.11 Combining Redox and Non-Redox Equilibria

Combine the equilibrium reaction and K value for Eq.(17.23) with

$$Fe^{2+} + H_2O = FeOH^+ + H^+ \qquad {}^*K_{H1} = \frac{\{FeOH^+\}}{\{Fe^{2+}\}\{OH^-\}} = 10^{-9.50} \qquad (25°C/1\,atm)$$

to obtain the redox K value for $Fe^{3+} + H_2O + e^- = FeOH^+ + H^+$ for 25 °C/1 atm.

Solution

$$Fe^{3+} + e^- = Fe^{2+} \qquad K = \frac{\{Fe^{2+}\}}{\{Fe^{3+}\}\{e^-\}} = 10^{13.0}$$

$$+\,Fe^{2+} + H_2O = FeOH^+ + H^+ \qquad {}^*K_{H1} = \frac{\{FeOH^+\}\{H^+\}}{\{Fe^{2+}\}\{H_2O\}} = 10^{-9.50}$$

$$Fe^{3+} + H_2O + e^- = FeOH^+ + H^+ \qquad K = \frac{\{FeOH^+\}\{H^+\}}{\{Fe^{3+}\}\{H_2O\}\{e^-\}} = 10^{4.5}$$

17.3.2.2 pe and the Redox K as Developed from the Nernst Equation

For any half reaction OX + ne^- = RED, the redox K is defined as

$$K \equiv \frac{\{\text{RED}\}}{\{\text{OX}\}\{\text{e}^-\}^n} \tag{17.24}$$

$$= \frac{\{\text{RED}\}}{\{\text{OX}\}(10^{-\text{pe}})^n}. \tag{17.25}$$

Starting with Eq.(17.22), shortening the subscript on E_{H}^{o}, we have

$$nFE_{\text{H}} = nFE_{\text{H}}^{\text{o}} + 2.303\,RT \log \frac{\{\text{OX}\}}{\{\text{RED}\}} \tag{17.26}$$

$$\frac{nFE_{\text{H}}}{2.303\,RT} = \frac{nFE_{\text{H}}^{\text{o}}}{2.303RT} + \log \frac{\{\text{OX}\}}{\{\text{RED}\}} \tag{17.27}$$

$$10^{\frac{nFE_{\text{H}}}{2.303\,RT}} = 10^{\frac{nFE_{\text{H}}^{\text{o}}}{2.303\,RT}} \frac{\{\text{OX}\}}{\{\text{RED}\}} \tag{17.28}$$

$$\frac{\{\text{RED}\}}{\{\text{OX}\}} 10^{\frac{nFE_{\text{H}}}{2.303\,RT}} = 10^{\frac{nFE_{\text{H}}^{\text{o}}}{2.303\,RT}} \tag{17.29}$$

$$\frac{\{\text{RED}\}}{\{\text{OX}\}\left(10^{\frac{-FE_{\text{H}}}{2.303\,RT}}\right)^n} = 10^{\frac{nFE_{\text{H}}^{\text{o}}}{2.303\,RT}} \tag{17.30}$$

Comparing terms between Eq.(17.24) and Eq.(17.29), then it is seen that the E_{H}^{o} with Nernst equation approach is interconvertible with the pe with redox K approach, with

$$\boxed{\text{pe} = \frac{FE_{\text{H}}}{2.303\,RT}} \tag{17.31}$$

$$\boxed{K = 10^{\frac{nFE_{\text{H}}^{\text{o}}}{2.303\,RT}}} \tag{17.32}$$

Exactly as with E_{H}, pe is a measure of the redox character of the solution, *i.e.*, how oxidizing/reducing the solution is. pe is directly proportional to E_{H}, and vice versa, so they have a common zero: when $E_{\text{H}} = 0$, then pe $= 0$. From Eq.(17.31), at 25 °C, the proportionality relationship is

$$\text{pe} = \frac{E_{\text{H}}}{0.05916} \qquad T = 298.15\ \text{K} \tag{17.33}$$

And, as noted above, K increases as E_{H}^{o} increases, with Eq.(17.32) providing the specific relationship. Further, for standard conditions, when {OX}/{RED} $= 1$, then pe $=$ pe° and from Eq.(17.31)

$$\boxed{\text{pe}^{\text{o}} = \frac{FE_{\text{H}}^{\text{o}}}{2.303\,RT}} \tag{17.34}$$

$$\text{pe}^{\text{o}} = \frac{E_{\text{H}}^{\text{o}}}{0.05916} \qquad T = 298.15\ \text{K} \tag{17.35}$$

By Eqs.(17.32) and (17.34),

$$\boxed{pe^o = \frac{1}{n}\log K} \quad (17.36)$$

By Eqs.(17.24) and (17.36), the pe analog of Eq.(17.22) is obtained:

$$\boxed{pe = pe^o + \frac{1}{n}\log\frac{\{OX\}}{\{RED\}}} \quad (17.37)$$

Example 17.12 Computing pe Values: Method 1, Using *K*

For comparison with the E_H calculations in Example 17.9, compute pe at 25 °C/1 atm for the half reactions in parts a–d of Example 17.8, given the following conditions and the log *K* values. Neglect activity corrections.

Conditions	**log *K***
a. $p_{O_2} = 0.21$ atm, pH = 5.0.	83.1
b. $[NO_3^-] = 10^{-5}$ *M*, $[N_2] = 0.0005$ *M*, pH = 5.0.	103.7
c. $[Fe^{3+}] = 10^{-5}$ *M*, $[Fe^{2+}] = 10^{-3}$ *M*.	13.0
d. $[SO_4^{2-}] = 10^{-5}$ *M*, $[H_2S] = 10^{-5}$ *M*, pH = 5.0.	41.0
e. $[H_2CO_3^*] = 10^{-4.87}$ *M*, [glucose] = 10^{-3} *M*, pH = 5.0.	0.67

Solution

a. $O_{2(g)} + 4H^+ + 4e^- = 2H_2O$

$$K = 10^{83.1} = \frac{1}{p_{O_2}[H^+]^4\{e^-\}^4} = \frac{1}{0.21(10^{-5.0})^4\{e^-\}^4}, \text{ so pe} = 15.61\ (E_H = 0.923\text{ V}).$$

b. $NO_3^- + 6H^+ + 5e^- = \frac{1}{2}N_2 + 3H_2O$ (N_2 is dissolved)

$$K = 10^{103.7} = \frac{\{N_2\}^{0.5}}{\{NO_3^-\}[H^+]^6\{e^-\}^5} = \frac{(0.0005)^{0.5}}{10^{-5.0}(10^{-5.0})^6\{e^-\}^5}, \text{ so pe} = 14.07\ (E_H = 0.832\text{ V}).$$

c. $Fe^{3+} + e^- = Fe^{3+}$

$$K = 10^{13.0} = \frac{\{Fe^{2+}\}}{\{Fe^{3+}\}\{e^-\}} = \frac{10^{-3.0}}{10^{-5.0}\{e^-\}}, \text{ so pe} = 11.0\ (E_H = 0.651\text{ V}).$$

d. $SO_4^{2-} + 10H^+ + 8e^- = H_2S + 3H_2O$

$$K = 10^{41.0} = \frac{\{H_2S\}}{\{SO_4^{2-}\}[H^+]^{10}\{e^-\}^8} = \frac{10^{-5.0}}{10^{-5.0}(10^{-5.0})^{10}\{e^-\}^8}, \text{ so pe} = -1.13\ (E_H = -0.067\text{ V}).$$

e. $H_2CO_3^* + 4H^+ + 4e^- = \frac{1}{6}C_6H_{12}O_6(\text{glucose}) + 2H_2O$

$$K = 10^{0.67} = \frac{\{\text{glucose}\}}{\{H_2CO_3^*\}[H^+]^4\{e^-\}^4} = \frac{10^{-3.0}}{10^{-4.87}(10^{-5.0})^4\{e^-\}^4}, \text{ so pe} = -5.30\ (E_H = -0.314\text{ V}).$$

The corresponding E_H values have been calculated by Eq.(17.33); the values agree with those in Example 17.9 to within small differences caused by round-off between the E_H^o and log *K* values.

Example 17.13 Computing pe Values: Method 2, Using pe° (Easier)

Compute pe at 25 °C/1 atm for the half reactions in parts a–e of Example 17.8, given the following conditions and the pe° values. Neglect activity corrections.

Conditions	pe°
a. p_{O_2} = 0.21 atm, pH = 5.0.	20.78
b. $[NO_3^-]$ = 10^{-5} *M*, $[N_2]$ = 0.0005 *M*, pH = 5.0.	20.74
c. $[Fe^{3+}]$ = 10^{-5} *M*, $[Fe^{2+}]$ = 10^{-3} *M*.	13.0
d. $[SO_4^{2-}]$ = 10^{-5} *M*, $[H_2S]$ = 10^{-5} *M*, pH = 5.0.	5.13
e. $[H_2CO_3^*]$ = $10^{-4.87}$ *M*, [glucose] = 10^{-3} *M*, pH = 5.0.	0.17

Solution

pe = pe° + (1/*n*)log [{OX}/{RED}].

a. $O_{2(g)} + 4H^+ + 4e^- = 2H_2O$

$n = 4$, pe = 20.78 + $\frac{1}{4}\log(p_{O_2}[H^+]^4)$ = 20.78 + $\frac{1}{4}\log p_{O_2}$ − pH = 15.61.

b. $NO_3^- + 6H^+ + 5e^- = \frac{1}{2}N_2 + 3H_2O$ (N_2 is dissolved)

$n = 5$, pe = 20.74 + $\frac{1}{5}\log([NO_3^-][H^+]^6/[N_2]^{0.5})$ = 20.74 + 0.20 log $[10^{-5}(10^{-5})^6/(0.0005)^{0.5}]$ = 14.07.

c. $Fe^{3+} + e^- = Fe^{3+}$

$n = 1$, pe = 13.0 + $\frac{1}{1}\log([Fe^{3+}]/[Fe^{2+}])$ = 13.0 + log$(10^{-5}/10^{-3})$ = 11.0.

d. $SO_4^{2-} + 10H^+ + 8e^- = H_2S + 3H_2O$

$n = 8$, pe = 5.13 + $\frac{1}{8}\log([SO_4^{2-}][H^+]^{10}/[H_2S])$ = 5.13 + 0.125 log $[10^{-5}(10^{-5})^{10}/(10^{-5})]$ = −1.12.

e. $H_2CO_3^* + 4H^+ + 4e^- = \frac{1}{6}C_6H_{12}O_6$ (glucose) + $2H_2O$

$n = 4$, pe = 0.17 + $\frac{1}{4}\log([H_2CO_3^*][H^+]^4/[\text{glucose}])$ = 0.17 + 0.25 log $[10^{-4.87}(10^{-5})^4/(10^{-3})]$ = −5.30.

The corresponding pe values between Example 17.12 and here agree to within small differences caused by round-off between the log *K* and pe° values.

17.3.2.3 pe Is as Meaningful as E_H

A few geochemists and natural water chemists dislike pe, noting that electrons do not exist in aqueous solution at the types of *concentrations* that would be given by 10^{-pe} values that can be calculated for real circumstances for an {OX}, {RED} pair and the corresponding tabulated redox *K* value. For example, for the $2H^+/H_{2(g)}$ half cell, $E_H^o = 0$, so by Eq.(17.32) $K = 1$:

$$2H^+ + 2e^- = H_{2(g)} \qquad K = \frac{p_{H_2}}{\{H^+\}^2\{e^-\}^2} = 1 \tag{17.38}$$

In the SHE, $\{H^+\} = 1$ and $p_{H_2} = 1$, so $\{e^-\} = 1$. Obviously, the concentration of electrons is not 1 *M*, but something many orders of magnitude smaller. However, per the definition of activity,

$$\{e^-\} \equiv \gamma_e[e^-] \tag{17.39}$$

where γ_e is the activity coefficient of the hydrated electron. The value of $[e^-]$ in water is very small even when it is at its largest. (In contrast, in solutions of alkali metals in ammonia, the solvated electron can reach appreciable concentrations.) As a result, it can be concluded that γ_e in water, whatever it is (and it *is* something) is enormous. So, since $\{e^-\}_{SHE} = 1$, whatever the tiny, non-zero value for $[e^-]_{SHE}$ is, then in water in an SHE, $\gamma_e = 1/[e^-]_{SHE}$. Fortunately, we do not ever have to worry about what γ_e and $[e^-]$ are individually, only what their product is, per Eq.(17.39). To summarize, $10^{-pe} \neq [e^-]$; rather, $10^{-pe} = \{e^-\}$.

17.3.2.4 At Equilibrium, All Redox Half Reactions Specify the Same pe (and E_H)

We know that at equilibrium, an aqueous solution has only one pH: this is *the* pH of the solution. If more than one conjugate acid/conjugate base system is present (acetic acid, HF, $H_2CO_3^*$, whatever), because acid/base reactions equilibrate rapidly, all of the pairs are in equilibrium with the same $\{H^+\}$. For example, for a solution that contains some mix of monoprotic acids HA, HB, HC, plus maybe some strong acid, and maybe some strong base, then at equilibrium we have

$$\text{pH} = \left(\text{p}K_{\text{HA}} + \log\frac{\{\text{A}^-\}}{\{\text{HA}\}}\right) = \left(\text{p}K_{\text{HB}} + \log\frac{\{\text{B}^-\}}{\{\text{HB}\}}\right) = \left(\text{p}K_{\text{HC}} + \log\frac{\{\text{C}^-\}}{\{\text{HC}\}}\right). \quad (17.40)$$

We have applied this concept extensively in previous chapters when solving problems based on the ENE, in which we assumed that species concentrations could be represented as a product of an α value and a total acid concentration, as with $\alpha_1^A A_T$ for $[A^-]$, and all the different acid/base α values were functions of the same $\{H^+\}$ value. The situation with pe is exactly analogous in that there can be only one pe (and so one E_H) in any given solution at any given time, and if the solution is in full equilibrium, all the redox pairs are in equilibrium with that single pe (E_H) value. Under those conditions, for redox half cell reactions 1, 2, and 3 involving n_1, n_2, and n_3, electrons, respectively,

$$\text{pe} = \left(\text{pe}_1^\circ + \frac{1}{n_1}\log\frac{\{\text{OX}_1\}}{\{\text{RED}_1\}}\right) = \left(\text{pe}_2^\circ + \frac{1}{n_2}\log\frac{\{\text{OX}_2\}}{\{\text{RED}_2\}}\right) = \left(\text{pe}_3^\circ + \frac{1}{n_3}\log\frac{\{\text{OX}_3\}}{\{\text{RED}_3\}}\right). \quad (17.41)$$

At full equilibrium then, a knowledge of the pe given by one redox half-reaction gives us the pe for all the other redox half-reactions, and we can calculate the {OX}/{RED} ratio for all other half-reactions. However, as mentioned above in the context of cellular fluids, many electron exchange processes are not fast. For natural waters, some dissolved organic carbon (DOC) is always present, even though organic carbon is wholly unstable relative to oxidation to CO_2 at atmospheric levels of O_2. In general, then, if one were to obtain a real-world water sample and were to measure $\{OX_1\}$ and $\{RED_1\}$, $\{OX_2\}$ and $\{RED_2\}$, $\{OX_3\}$ and $\{RED_3\}$, etc. for the various redox pairs present, then calculate the pe indicated by each pair, the result will be different pe values.

17.3.2.5 Reference Electrodes, the E_H (pe) Combination Electrode, and the pH Combination Electrode

Standard Hydrogen Electrode (SHE). The standard hydrogen electrode (SHE) has many advantages as a reference electrode for electrochemical measurements: it possesses adequately fast kinetics at the Pt electrode, and the chemistry is simple, well-defined, and well-understood. The SHE has one big disadvantage: it uses a continuous flow of $H_{2(g)}$, which is far too dangerous for routine measurements in laboratories, not to mention the need for a tank of compressed $H_{2(g)}$. As a result, alternative reference electrodes have been developed. The two most important ones are the SCE, and the SSCE.

Saturated Calomel Electrode (SCE). All measurement electrodes discussed in the remainder of this section are used with a potentiometer so that the potential difference across the two sides of the potentiometer is measured, but virtually no actual current is allowed to flow. If current were to flow, it would alter and thus damage the electrode.

In strict chemical terms, the SCE is an excellent reference electrode, being both robust and very stable. As such, the SCE remains in use in research laboratories. However, because it contains Hg, it is now in complete disfavor as a routine-use reference electrode. Completely supplanting the SCE in the latter setting is the silver/silver chloride electrode (SSCE). We nevertheless discuss the SCE because of its great historical importance in electrochemical measurements.

The SCE utilizes calomel, $Hg_2Cl_{2(s)}$, which is the chloride salt of the Hg(I) "polycation" Hg_2^{2+}. There is a very long and not-so-pleasant history of calomel being used in large doses in human and veterinary medicine for problems ranging from constipation to syphilis (Schmid, 2009). The redox half-reaction for the SCE is

$$Hg_2Cl_{2(s)} + 2e^- = 2\,Hg_{(\ell)} + 2\,Cl^- \qquad E_H^o = 0.268\text{ V}. \tag{17.42}$$

Therefore,

$$E_H = 0.268 + \frac{2.303\,RT}{2\,F}\log\frac{\{Hg_2Cl_{2(s)}\}}{\{Hg_{(l)}\}^2\{Cl^-\}^2}. \tag{17.43}$$

A schematic diagram of an SCE is given in Figure 17.4a. A Pt wire is immersed in a paste of $Hg_2Cl_{2(s)}$, liquid mercury ($Hg_{(\ell)}$), and often some KCl. That paste is wet and is in contact with an internal solution that is saturated with KCl. The porous frit on the side of the SCE (or similar such arrangement) isolates the chemicals in the SCE from those in the test solution in which the SCE might be immersed, and also provides the needed electrical connection (salt bridge) between the SCE and the test solution. As such, it could be placed directly in the solution of a SHE, both then connected to a meter or potentiostat. With the $Hg_2Cl_{2(s)}$ and $Hg_{(\ell)}$ both being individually pure (mol fraction x =1) in the SCE, then

$$E_{H,SCE} = 0.268 + \frac{2.303\,RT}{2\,F}\log\frac{1}{\{Cl^-\}^2} = 0.241\text{ V}. \tag{17.44}$$

By measurement against the SHE, at 25 °C/1 atm the SCE gives a reading of 0.241 V. By Eq.(17.44) this gives $\{Cl^-\} = 2.86\ m$.

Silver/Silver Chloride Electrode (SSCE). The chemistry of the SSCE is very similar to that of the SCE, but is based on

$$AgCl_{(s)} + e^- = Ag_{(s)} + Cl^- \qquad E_H^o = 0.222\text{ V} \tag{17.45}$$

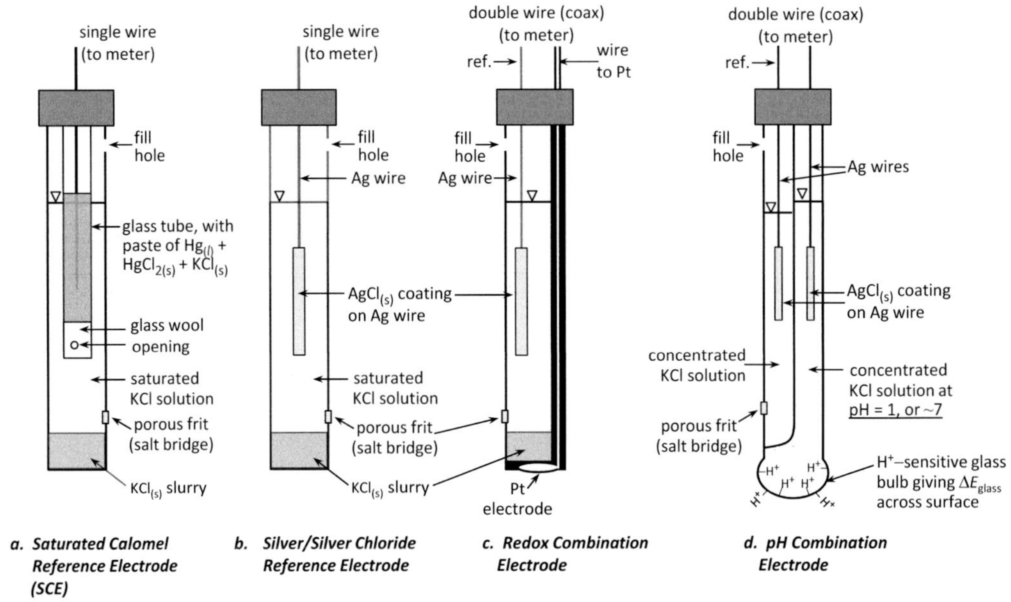

FIGURE 17.4 a. Saturated calomel electrode (SCE). b. Silver/silver chloride electrode (SSCE). c. Redox combination electrode. d. pH combination electrode.

$$E_H = 0.222 + \frac{2.303\,RT}{F}\log\frac{\{AgCl_{(s)}\}}{\{Ag_{(s)}\}\{Cl^-\}}. \tag{17.46}$$

A schematic diagram of an SSCE is given in Figure 17.4b. A portion of an $Ag_{(s)}$ wire is coated with $AgCl_{(s)}$, with the coated portion in contact with an internal solution that is saturated with KCl. The porous frit on the side of the SSCE (or similar such arrangement) serves the exact same purposes as the porous frit on the SCE. With the $AgCl_{(s)}$ and $Ag_{(s)}$ both being individually pure (mol fraction $x = 1$) in the SSCE, then

$$E_{H,SSCE} = 0.222 + \frac{2.303\,RT}{F}\log\frac{1}{\{Cl^-\}} = 0.197\text{ V}. \tag{17.47}$$

By measurement against the SHE, at 25 °C/1 atm the SSCE gives a reading of 0.197 V. By Eq.(17.47) this means $\{Cl^-\}$ in a saturated solution of KCl is 2.65 *m*, which is adequately close to 2.86 given experimental uncertainties in the E_H^o values for the two electrodes.

Redox Combination Electrode. Since E_H (pe) is such an important concept in natural waters, it is natural that many people would seek to make such measurements. Indeed, this is routinely attempted using a Pt electrode placed in a water sample and connected to the positive side of the meter, and some reference electrode in the sample connected to the negative side. Ideally, the meter then reads an overall E_{cell} value given by

$$E_{cell} = E_{H,sample} - E_{H,ref} \tag{17.48}$$

so

$$E_{H,sample} = E_{cell} + E_{H,ref} \tag{17.49}$$

where $E_{H,ref}$ is the E_H value of the reference electrode; 0.241 V for the SCE, and 0.197 V for the SSCE. A redox combination facilitates such a measurement by building both electrodes into one combination electrode (Figure 17.4c) with two wires, one from each electrode. The Pt electrode is on the outside of the combination electrode, visible usually as a shiny disk on the bottom. The reference electrode is housed inside the combination electrode, with a frit for the electrical connection (salt bridge) with the test solution. To obtain E_H, since the reference electrode is not the SHE, $E_{H,ref} \neq 0$, one must carry out the calculation in Eq.(17.49) either by hand, or as built into the instrument calibration.

Making and interpreting E_H measurements with environmental samples is a flawed and difficult business. One of the potential problems relates to the fact that in the case of an anoxic water sample, preventing atmospheric oxygen from contaminating the sample is exceedingly difficult. More problematic are the slow kinetics of most natural redox half reactions. The U.S.G.S. *National Field Manual for the Collection of Water-Quality Data* (*Section 6.5, Reduction-Oxidation Potential (Electrode Method)* by Nordstrom and Wilde, 2005) states:

> In contrast to other field measurements, the determination of the reduction-oxidation potential of water (referred to as redox) should not be considered a routine determination. Measurement of redox potential, described here as Eh [E_H] measurement, is not recommended in general because of the difficulties inherent in its theoretical concept and its practical measurement (see "Interferences and Limitations," Section 6.5.3.A).
>
> - Eh measurement[s] may show qualitative trends but generally cannot be interpreted as [giving] equilibrium values.
> - Determinations of redox using the platinum (or other noble metal) electrode method (Eh) are valid only when redox species are (a) electroactive, and (b) present in the solution at concentrations of about 10^{-5} molal and higher. Redox species in natural waters generally do not reach equilibrium with metal electrodes.

So, if the E_H (pe) of a natural system is very difficult/impossible to measure and use, then why are E_H (pe) calculations still worthwhile? Because we can most certainly use such calculations in the reverse context, *i.e.*, to predict *a priori* what, if any, redox chemistry *will tend to occur*. Examples:

1. Is organic carbon stable in water in the presence of significant dissolved oxygen? No, it will tend to be oxidized to CO_2, and the oxygen will be reduced to water: "biological oxygen demand" (BOD) in water will act reduce the dissolved O_2 level as fast as the organisms present can metabolize it.
2. When dissolved oxygen has become depleted, is organic carbon in water stable in the presence of significant nitrate? No, it will tend to be oxidized to CO_2, and the nitrate will be reduced to N_2: nitrate can be removed in sewage treatment by microbiologically-mediated dentrification.
3. When dissolved oxygen and nitrate have become depleted, is organic carbon stable all by itself? No, it will tend to undergo fermentation (a.k.a. disproportionation) by methanogens to a mix of compounds containing CO_2 and methane: dammed reservoirs feeding hydroelectric power plants can become totally anoxic in their bottom waters, leading to formation of amounts of methane (formed from the incoming organic matter in the river water) that can greatly reduce the CO_2 emission-reduction advantages of hydroelectric-derived *vs.* fossil-fuel derived power (Maeck et al., 2013).

pH Combination Electrode. The common pH electrode is actually a combination electrode (two electrodes in one unit), with design features as shown in Figure 17.4d. The pH-sensing functionality is not redox-based, but is a consequence of a pH-dependent adsorption of H^+ ions onto the inner and outer surfaces of the glass bulb at the tip of the electrode. The pH combination electrode utilizes two redox reference electrodes, one on the right (R) that is exposed to the solution that also covers the inside glass surface, and one on the left (L) in the solution that is electrically connected to the test solution via the porous frit (salt bridge). On the inside glass surface, the extent of H^+ adsorption remains constant because the internal pH is constant. On the outside glass surface, the extent of H^+ adsorption varies according to the pH in the test solution. There is then a differential degree of charging across the glass surface, and this leads to a measurable potential difference ΔE_{glass} from one side of the glass surface to the other. ΔE_{glass} is analogous with the temperature-dependent ΔE that develops at the interface between two dissimilar metals in a thermocouple, wherein electrons preferentially leave one side, and differentially charge the other. ΔE_{glass} contributes to the overall E_{cell} according to

$$E_{cell} = E_{H,ref\text{-}R} - E_{H,ref\text{-}L} + \Delta E_{glass} \tag{17.50}$$

with

$$\Delta E_{glass} = E_G + \frac{2.303\,RT}{(+1)F}\log\{H^+\} \tag{17.51}$$

where E_G is a constant determined by the properties of the glass and the pH on the inside of the electrode. The $2.303RT/F$ multiplier arises for the same reason as in the Nernst Equation; the +1 multiplying F arises because +1 is the charge z on the ion (H^+) which is causing ΔE_{glass} due to the role for zFE as the electrostatic energy of adsorption on the outside glass surface. Collecting all the constants into a single constant A,

$$E_{cell} = A + \frac{2.303\,RT}{F}\log\{H^+\} \tag{17.52}$$

$$= A - \frac{2.303\,RT}{F}\text{pH}. \tag{17.53}$$

A pH meter then measures a signal that is linearly related to pH, and thus can be calibrated with buffer solutions of known pH. At 25 °C, it is said that the electrode is behaving in a perfect "Nernstian" ("100% slope") manner when the slope $dE_{cell}/d\text{pH} = -0.05916$ V/pH unit. pH electrodes usually do not follow perfect Nernstian behavior. Calibrating a pH meter is then a matter of having the meter determine the E_{cell} values for two different buffer solutions (*e.g.*, pH = 4.0 and 7.0), and then the slope, for measurement of any other pH.

The more than 100 year history of the efforts to understand the details of the complex chemistry underlying Eq.(17.52) is reviewed in some detail by Graham et al. (2013). The story began when Cremer (1906) noted a dependence in the electrical response of glass membranes in solution to changing H^+ concentrations. After Arnold Beckman registered the first patent in 1934 for such an instrument, the glass electrode/potentiometer-based pH meter quickly became a universally used laboratory instrument. Remarkably, though, even after 100 years, numerous details are still not agreed upon as regards how electrical conductivity though the glass is accomplished; there is consensus nevertheless about the basic idea that it is adsorption and desorption of protons at the hydrated glass surface that is at the root of how glass pH electrodes work.

17.4 REDOX LADDER

17.4.1 Redox Ladder under Standard Conditions (Use pe° or E_H^o)

Say that we have first monoprotic acid $HA_\dagger$ for which $pK_a = 5$, and second monoprotic acid $HA_\ddagger$ for which $pK_a = 7$. In a solution in which $\{HA_\dagger\}/\{A_\dagger^-\} = 1$, then pH = 5. In a separate solution in which $\{HA_\ddagger\}/\{A_\ddagger^-\} = 1$, then pH = 7. If we could somehow instantaneously prepare a single solution in which $\{HA_\dagger\}/\{A_\dagger^-\} = 1$ and $\{HA_\ddagger\}/\{A_\ddagger^-\} = 1$, then protons on $HA_\dagger$ molecules would start to protonate the $A_\ddagger^-$; the final pH would be somewhere between 5 and 7. Therefore, the pK_a values of acids in increasing order provide an acid/base ladder according to which the conjugate base forms show increasing tendency to accept protons when all $\{HA\}/\{A^-\} = 1$ (the condition that is analogous to "standard conditions" for redox half reactions). It is exactly the same when comparing pe° (E_H^o) values. If we had two solutions under standard conditions in the two different half cells of an overall cell, the electrons would flow in the direction of the solution with the higher pe°.

Consider the half cells $Cu^{2+} + 2e^- = Cu_{(s)}$ and $2H^+ + 2e^- = H_{2(g)}$. The pertinent values from Table 17.1 are given in Table 17.2. So, by the above arguments, the $Cu^{2+} + 2e^- = Cu_{(s)}$ half cell wants electrons more under standard conditions than does the hydrogen half cell. This is why, in the Figure 17.2 cell (which is at standard conditions), that electrons flow from $H_{2(g)}$ to Cu^{2+}, and the meter reads 0.34 V. All the values in Table 17.1 can then be considered to yield a "redox ladder, with the ladder rungs ordered from lowest to highest pe° (E_H^o), according to which the OX forms show increasing tendency to accept electrons (under standard conditions).

17.4.2 Redox Ladder under Non-Standard Conditions

For natural waters, a simple comparison of two pe° (E_H^o) values as in Table 17.2 can be misleading. One reason is that many important reactions involve $\{H^+\}$ (usually in {OX}), and under standard conditions $\{H^+\} = 1$ *m* (pH = 0), which is hardly representative of conditions in natural water systems. To deal with such pH effects, the pe°(W) concept was developed as giving the pe value when all terms in {OX} and {RED} are unity, except H^+, for which $\{H^+\}$ is set to 10^{-7} *m* (pH = 7). The W is intended to invoke the idea of regular "water." For pe°(W), the number of protons in the reaction is of interest, so we have two *n* values to consider. For some general redox half-reaction, we add subscripts as follows:

$$aA + n_H H^+ + n_e e^- = cC + dD \qquad K = \frac{\{C\}^c\{D\}^d}{\{A\}^a\{e^-\}^{n_e}\{H^+\}^{n_H}} \tag{17.54}$$

TABLE 17.2
Data for the Two Half Reactions in the Cell in Figure 17.3 Giving a Portion the Redox Ladder for Standard Conditions

half reaction	pe°	E_H^o
$2H^+ + 2e^- = H_{2(g)}$	0.0	0.0
$Cu^{2+} + 2e^- = Cu_{(s)}$	5.7	0.34

$$\text{pe} = \text{pe}^o + \frac{1}{n_e}\log\frac{\{\text{OX}\}}{\{\text{RED}\}} \tag{17.55}$$

$$= \text{pe}^o + \frac{1}{n_e}\log\frac{\{\text{A}\}^a\{\text{H}^+\}^{n_H}}{\{\text{C}\}^c\{\text{D}\}^d} \tag{17.56}$$

$$= \text{pe}^o + \frac{1}{n_e}\log\frac{\{\text{A}\}^a}{\{\text{C}\}^c\{\text{D}\}^d} + \frac{n_H}{n_e}\log\{\text{H}^+\} \tag{17.57}$$

$$= \text{pe}^o + \frac{1}{n_e}\log\frac{\{\text{A}\}^a}{\{\text{C}\}^c\{\text{D}\}^d} - \frac{n_H}{n_e}\text{pH} \tag{17.58}$$

When we have standard conditions for everything but $\{H^+\}$ (which is at 10^{-7} *m*), or at least the quotient $\{A\}^a/(\{B\}^b\{C\}^c) = 1$, the remaining log term in Eq.(17.58) is zero, and pe = pe°(W) so

$$\text{pe}^o(\text{W}) = \text{pe}^o - \frac{n_H}{n_e}7.00. \tag{17.59}$$

pe°(W) can be of limited use because for many half cells of interest, conditions in natural waters do not typically lead to a value near 1 for the argument of the remaining log term, so it is usually best to just calculate the pe (or E_H) value for each half cell of interest, then make any needed comparison. In other words, do not rely on the positions of rungs under standard, or even pe°(W) conditions, but see exactly where the rungs are, then decide who is going to get electrons from whom.

Example 17.14 Comparing pe Values for Different Half Cells – Which Will Win Electrons?

When comparing two half-reactions, the one with the higher pe will take electrons and proceed as a reduction of its oxidized form; the half-reaction with the lower pe will yield electrons and proceed as an oxidation of its reduced form. Consider the conditions for the half cells in parts a–e of Example 17.13. The calculated pe values are summarized below. State what would tend to happen if some natural water had the concentration characteristics each of the given pairs.

Conditions	pe
a. p_{O_2} = 0.21 atm, pH = 5.0.	15.61
b. $[NO_3^-] = 10^{-5}$ *M*, $[N_2]$ = 0.0005 *M*, pH = 5.0.	14.07
c. $[Fe^{3+}] = 10^{-5}$ *M*, $[Fe^{2+}] = 10^{-3}$ *M*.	11.0
d. $[SO_4^{2-}] = 10^{-5}$ *M*, $[H_2S] = 10^{-5}$ *M*, pH = 5.0.	−1.12
e. $[H_2CO_3^*] = 10^{-4.87}$ *M*, [glucose] = 10^{-3} *M*, pH = 5.0.	−5.30

Pair	Solution
a and **b**	15.61 > 14.07, so under these conditions, O_2 can theoretically oxidize N_2 to NO_3^-, and thereby be reduced to water. However, oxidation of N_2 **does not occur naturally** except under extreme conditions that favor the kinetics, as in lightning. With its N≡N triple bond, N_2 is very stable and a "tough nut to crack".
a and **c**	15.61 > 11.0, so under these conditions, O_2 will oxidize Fe^{2+} to Fe^{3+}, and thereby be reduced to water. This can happen both abiotically as well as biotically.
a and **d**	15.61 > −1.12, so under these conditions O_2 will oxidize H_2S to SO_4^{2-}, and thereby be reduced to water. This can happen both abiotically as well as biotically.
d and **e**	−1.12 > −5.30, so under these conditions, SO_4^{2-} will oxidize glucose (and other organic compounds) to CO_2, and thereby be reduced to H_2S. This can happen biotically as carried out by sulfate-reducing bacteria which use sulfate as a "terminal electron acceptor" (oxidant) to respire organic compounds. This is how H_2S forms in sewers, sewage, and in tidal flats, etc.
e and **c**	11 > −5.30, so under these conditions, Fe^{3+} will oxidize glucose (and other organic compounds) to CO_2, and thereby be reduced to Fe^{2+}. This can happen biotically as carried out by iron-reducing bacteria which use Fe(III) as a "terminal electron acceptor" (oxidant) to respire organic compounds. This is how ferrous iron (Fe(II)) forms in anoxic groundwater systems. This will happen before the pe drops down to sulfate-reducing conditions, as in the case above. So, sulfate-reducing conditions will be reached if there is more than enough organic matter to exhaust all of the Fe(III). Then, if some sulfate is present, the system will go on to produce H_2S (which will usually lead to precipitation of $FeS_{(s)}$).

17.5 REDOX α VALUES

17.5.1 General

Acid/base α values, as pH-dependent functions, are basic for both: A) understanding acid/base chemistry; and B) solving every pH-related problem that involves acid and bases. Analogously with A, redox α values for pe (E_H)-dependent functions are basic for understanding redox chemistry, *i.e.*, understanding pe (E_H)-related chemistry as it involves oxidants and reductants. However, for B, the analogy does not pertain because redox α values are not very useful at all when solving redox problems. The reason is that in natural water chemistry, usually, the conditions in a solution are such that: 1) one redox α is essentially unity; 2) the others are very close to zero; and 3) we are far, far away from being in the pe neighborhood where the others will start to become important. For example, in anoxic groundwaters, virtually 100% of any iron present is Fe(II), and virtually 100% of any sulfur present will be S(−II). If the waters become even slightly oxic, virtually 100% of any iron present is Fe(III), and virtually 100% of any sulfur present will be S(VI). (For sulfur, an exception would be geothermally active areas such as Yellowstone National Park in the U.S., where $S_{(s)}$ can be found in abundance at the surface where it continually is being formed from H_2S.) Here we discuss the formulation of expressions for redox α values, and how they vary with pe.

17.5.2 $\alpha_{O_2}^{\text{remaining}}$ – A Fraction-Remaining α for O_2 (For Comparison with Actual Redox α Values)

How much oxygen is present in natural waters is obviously of great importance for all aerobic life. We could use $\alpha_{O(0)}$ as the fraction of total solution O that is in the O(0) state, as given by

$$\alpha_{O(0)} \equiv \frac{2[O_2]}{2[O_2] + \left([H_2O] + [OH^-]\right)} \tag{17.60}$$

where all concentrations are in M, $2[O_2]$ is the concentration of O(0), and ($[H_2O]+[OH^-]$) is the total concentration of O(–II). However, there is a tremendous amount of oxygen in all the water, so it is more useful to talk about how the fraction-remaining α for O will change as pe (E_H) drops from its initial value. If $[O_2]_{initial}$ dissolved concentration of O_2, then

$$\alpha_{O2}^{remaining} \equiv \frac{[O_2]}{[O_2]_{initial}} \tag{17.61}$$

From Table 17.1, for dissolved O_2,

$$K = \frac{1}{\{O_2\}\{H^+\}^4\{e^-\}^4} = 10^{86.0} \tag{17.62}$$

We neglect activity corrections (except $\{e^-\}$ Then,

$$[O_2] = \frac{1}{K[H^+]^4\{e^-\}^4} = \frac{1}{K10^{-4pH}10^{-4pe}} \tag{17.63}$$

and $\alpha_{O2}^{remaining}$ can be calculated accordingly with Eq.(17.61).

17.5.3 $\alpha_{N(V)}$, THE REDOX α FOR N(V) WHEN ALL N IS DISSOLVED AND CONSIDERING ONLY N(V) (AS NO_3^-) AND N(0) (AS N_2)

For dissolved nitrogen, neglecting N(III),

$$\alpha_{N(V)} \equiv \frac{[NO_3^-]}{N_T} = \frac{[NO_3^-]}{[NO_3^-]+2[N_2]} \tag{17.64}$$

From Table 17.1,

$$K = \frac{\{N_2\}^{1/2}}{\{NO_3^-\}\{H^+\}^6\{e^-\}^5} = 10^{103.7} \tag{17.65}$$

Neglecting activity corrections,

$$[N_2]^{1/2} = K[NO_3^-][H^+]^6\{e^-\}^5 \tag{17.66}$$

$$[N_2] = K^2[NO_3^-]^2[H^+]^{12}\{e^-\}^{10} \tag{17.67}$$

$$N_T = [NO_3^-] + \tfrac{1}{2}[N_2] \tag{17.68}$$

$$= [NO_3^-] + \tfrac{1}{2}K^2[NO_3^-]^2 10^{-12pH}10^{-10pe} \tag{17.69}$$

Rearranging gives a second order polynomial in $[NO_3^-]$:

$$\left(\tfrac{1}{2} \times K^2 10^{-12pH}10^{-10pe}\right)[NO_3^-]^2 + [NO_3^-] - N_T = 0 \tag{17.70}$$

which is of the form $ax^2 + bx + c = 0$, and so can be solved using the "quadratic equation":

$$[NO_3^-] = \frac{-1+\sqrt{1+2\times K^2 10^{-12pH}10^{-10pe}N_T}}{K^2 10^{-12pH}10^{-10pe}} \tag{17.71}$$

The positive sign is selected for the square root term so that $[NO_3^-]$ is positive. The value of $\alpha_{N(V)}$ can then be obtained as a function of pe and pH by Eqs.(17.64) and (17.71).

Let $2 \times K^2 10^{-12pH} 10^{-10pe} N_T = q$. When the solution conditions make q small relative to 1, as it will be at high pe where $[NO_3^-]$ will greatly predominate over ½$[N_2]$, then the mathematical approximation $\sqrt{1+q} \approx 1 + q/2$ applies (*e.g.*, $\sqrt{1+0.01} = 1.0049876\ldots \approx 1.005$), so

$$[NO_3^-] \approx \frac{-1 + (1 + K^2 10^{-12pH} 10^{-10pe} N_T)}{K^2 10^{-12pH} 10^{-10pe}} = N_T \tag{17.72}$$

and

$$\alpha_{N(V)} \approx 1. \tag{17.73}$$

When q becomes *really* small (at high pe), although mathematically Eq.(17.72) continues to be valid, a spreadsheet can deliver wonky results when accurate evaluation of $\sqrt{1+q}$ (and therefore $-1+\sqrt{1+q}$) cannot be accomplished: the level of binary precision needed for the calculation exceeds what is built into the spreadsheet code.

17.5.4 $\alpha_{Fe(III)}$, THE REDOX α FOR Fe(III) WHEN ALL Fe IS DISSOLVED

If all system Fe is dissolved,

$$\alpha_{Fe(III)} \equiv \frac{Fe(III)_T}{Fe_T} = \frac{Fe(III)_T}{Fe(II)_T + Fe(III)_T} \tag{17.74}$$

$$[Fe^{2+}] = \alpha_0^{Fe(II)} Fe(II)_T \tag{17.75}$$

$$[Fe^{3+}] = \alpha_0^{Fe(III)} Fe(III)_T. \tag{17.76}$$

Note that $\alpha_0^{Fe(II)}$ and $\alpha_0^{Fe(III)}$ are acid/base α values, not redox α values; they can be calculated using the relevant $^*K_H$ values (see Section 10.3.1). To simplify their expression, we introduce the nomenclature

$$^*\beta_{Hn} \equiv {}^*K_{H1}\ldots{}^*K_{Hn}. \tag{17.77}$$

For example, $^*\beta_{H1}^{Fe(II)} = {}^*K_{H1}^{Fe(II)}$, $^*\beta_{H2}^{Fe(II)} = {}^*K_{H1}^{Fe(II)}\,{}^*K_{H2}^{Fe(II)}$, etc. If the Fe(II) species considered are Fe^{2+}, $FeOH^+$, $Fe(OH)_2^o$, and $Fe(OH)_3^-$, then neglecting activity corrections,

$$\alpha_0^{Fe(II)} = \frac{1}{1 + \frac{^*\beta_{H1}^{Fe(II)}}{[H^+]} + \frac{^*\beta_{H2}^{Fe(II)}}{[H^+]^2} + \frac{^*\beta_{H3}^{Fe(II)}}{[H^+]^3}}. \tag{17.78}$$

Similarly, if the Fe(III) species considered are Fe^{3+}, $FeOH^{2+}$, $Fe(OH)_2^+$, $Fe(OH)_3^o$, and $Fe(OH)_4^-$, then

$$\alpha_0^{Fe(III)} = \frac{1}{1 + \frac{^*\beta_{H1}^{Fe(III)}}{[H^+]} + \frac{^*\beta_{H2}^{Fe(III)}}{[H^+]^2} + \frac{^*\beta_{H3}^{Fe(III)}}{[H^+]^3} + \frac{^*\beta_{H4}^{Fe(III)}}{[H^+]^4}}. \tag{17.79}$$

From Table 17.1,

$$K = \frac{\{Fe^{2+}\}}{\{Fe^{3+}\}\{e^-\}} = 10^{13.0}. \tag{17.80}$$

Neglecting activity corrections for the iron species, by Eqs.(17.75) and (17.76),

$$K = \frac{\alpha_0^{\text{Fe(II)}}\text{Fe(II)}_\text{T}}{\alpha_0^{\text{Fe(III)}}\text{Fe(III)}_\text{T}\{\text{e}^-\}} \tag{17.81}$$

$$\frac{\text{Fe(II)}_\text{T}}{\text{Fe(III)}_\text{T}} = \frac{K\alpha_0^{\text{Fe(III)}}\{\text{e}^-\}}{\alpha_0^{\text{Fe(II)}}} = \frac{K\alpha_0^{\text{Fe(III)}}10^{-\text{pe}}}{\alpha_0^{\text{Fe(II)}}}. \tag{17.82}$$

From Eq.(17.74),

$$\alpha_{\text{Fe(III)}} = \frac{1}{\dfrac{\text{Fe(II)}_\text{T}}{\text{Fe(III)}_\text{T}} + 1} \tag{17. 83}$$

$$= \frac{1}{\dfrac{K\alpha_0^{\text{Fe(III)}}10^{-\text{pe}}}{\alpha_0^{\text{Fe(II)}}} + 1} \tag{17.84}$$

which can be evaluated as a function of pe and pH.

17.5.5 $\alpha_{S(VI)}$, the Redox α for S(VI) When All S Is Dissolved and Considering Only S(VI) (as HSO_4^- and SO_4^{2-}) and S(–II) (as H_2S and HS^-)

If all system S is dissolved, and considering only S(VI) and S(–II) species,

$$\alpha_{\text{S(VI)}} \equiv \frac{\text{S(VI)}_\text{T}}{\text{S}_\text{T}} = \frac{\text{S(VI)}_\text{T}}{\text{S(-II)}_\text{T}+\text{S(VI)}_\text{T}} \tag{17.85}$$

$$[\text{H}_2\text{S}] = \alpha_0^{\text{S(-II)}}\text{S(-II)}_\text{T} \tag{17.86}$$

$$[\text{SO}_4^{2-}] = \alpha_2^{\text{S(VI)}}\text{S(VI)}_\text{T} \tag{17.87}$$

where $\alpha_0^{\text{S(-II)}}$ and $\alpha_2^{\text{S(VI)}}$ are acid/base α values. From Table 17.1,

$$K = \frac{\{\text{H}_2\text{S}\}}{\{\text{SO}_4^{2-}\}\{\text{H}^+\}^{10}\{\text{e}^-\}^8} = 10^{41.0}. \tag{17.88}$$

Neglecting activity corrections, by Eqs.(17.86) and (17.87),

$$K = \frac{\alpha_0^{\text{S(-II)}}\text{S(-II)}_\text{T}}{\alpha_2^{\text{S(VI)}}\text{S(VI)}_\text{T}[\text{H}^+]^{10}\{\text{e}^-\}^8} \tag{17.89}$$

$$\frac{\text{S(-II)}_\text{T}}{\text{S(VI)}_\text{T}} = \frac{K\alpha_2^{\text{S(VI)}}[\text{H}^+]^{10}\{\text{e}^-\}^8}{\alpha_0^{\text{S(-II)}}} = \frac{K\alpha_2^{\text{S(VI)}}10^{-10\text{pH}}10^{-8\text{pe}}}{\alpha_0^{\text{S(-II)}}}. \tag{17.90}$$

From Eq.(17.85),

$$\alpha_{\text{S(VI)}} = \frac{1}{\dfrac{\text{S(-II)}_\text{T}}{\text{S(VI)}_\text{T}} + 1} \tag{17.91}$$

$$= \frac{1}{\dfrac{\alpha_2^{\text{S(VI)}}K10^{-10\text{pH}}10^{-8\text{pe}}}{\alpha_0^{\text{S(-II)}}} + 1} \tag{17.92}$$

which can be evaluated as a function of pe and pH.

17.5.6 $\alpha_{C(IV)}$, The Redox α for C(IV) When All C Is Dissolved and Considering Only C(IV) (the CO_2 Species) and C(−IV) (as CH_4)

If all system C is dissolved, and considering only C(IV) and C(−IV) species,

$$\alpha_{C(IV)} \equiv \frac{C(IV)_T}{C_T} = \frac{C(IV)_T}{C(-IV)_T + C(IV)_T}. \tag{17.93}$$

C(−IV) (methane) has no acid base chemistry, so

$$[CH_4] = C(-IV)_T. \tag{17.94}$$

For C(IV),

$$[H_2CO_3^*] = \alpha_0^{C(IV)} C(IV)_T \tag{17.95}$$

where $\alpha_0^{C(IV)}$ is an aid base α value in the CO_2 system. From Table 17.1,

$$K = \frac{\{CH_4\}}{\{H_2CO_3^*\}\{H^+\}^8\{e^-\}^8} = 10^{21.62}. \tag{17.96}$$

Neglecting activity corrections, by Eqs.(17.94) and (17.95),

$$K = \frac{C(-IV)_T}{\alpha_0^{C(IV)} C(IV)_T [H^+]^8 \{e^-\}^8} \tag{17.97}$$

$$\frac{C(-IV)_T}{C(IV)_T} = K\alpha_0^{C(IV)}\{H^+\}^8\{e^-\}^8 = K\alpha_0^{C(IV)} 10^{-8pH} 10^{-8pe}. \tag{17.98}$$

From Eq.(17.93),

$$\alpha_{C(IV)} = \frac{1}{\dfrac{C(-IV)_T}{C(IV)_T} + 1} \tag{17.99}$$

$$= \frac{1}{K\alpha_0^{C(IV)} 10^{-8pH} 10^{-8pe} + 1} \tag{17.100}$$

which can be evaluated as a function of pe and pH.

Example 17.16 Plots of Redox α Values for Oxidized Forms as pe Drops

α values for five extremely important redox systems, O, N, Fe, S, and C are provided by Eq.(17.61) with Eq.(17.63), Eq.(17.64) with Eq.(17.71), Eq.(17.84), Eq.(17.92), and Eq.(17.100), respectively. Consider three systems, all at 25 °C/1 atm, but with pH values of 6, 7, and 8. We are interested in what happens as pe drops, as caused perhaps by the addition of biological carbon ("organic matter"), and how that is affected by pH. In each system, the water is initially saturated with O_2 for p_{O_2} = 0.21 atm (the atmospheric value). For O_2, $K_H = 10^{-2.88}$ (*M*/atm). Initially, $[NO_3^-] \approx N_T$ = 5.0×10^{-4} *M*. For Fe, S, and C, we do not need to specify Fe_T, S_T, and C_T other than to say that the levels are low enough we do not precipitate any solids with those elements: here we are interested in seeing at what pe values the α values for the oxidized forms start to decline significantly from 1, and not how precipitation of solids affects things. (*E.g.*, formation of one or more solids will complicate the situation once the initial Fe(III) is mostly reduced to Fe(II), then tending to precipitate as

$FeS_{(s)}$ once S(VI) starts to be reduced significantly to S(−II).) **a.** Compute $[O_2]_{initial}$ and the initial pe for each case. **b.** Compute and plot the five α values as evaluated on the *x*-axis, with decreasing pe on the *y*-axis.

Solution

a. Initially, for all three systems, $[O_2]_{initial} = K_H p_{O_2} = 10^{-2.88} 0.21 = 2.77 \times 10^{-4}\ M$. As used in Example 17.13, at 25 °C/1 atm, $pe = 20.78 + \frac{1}{4}\log p_{O_2} - pH = 20.78 + \frac{1}{4}\log(0.21) - pH$. At pH = 6.0, pe = 14.61. At pH = 7.0, pe = 13.61. At pH = 6.0, pe = 12.61.

b. In the figure for this example, there is one panel each for pH = 6, 7, and 8. More e^- in a reaction gives sharper disappearance as pe drops. Each curve delineates the fraction remaining: for the O_2 curve, this is because it is the definition of the α plotted, and for the others it is because what is shown is a transition from essentially 100% of the more oxidized form. The relative order of reduction (first O_2, then NO_3^-, etc.) is the same at all three pH values. As pH rises, all five curves shift downwards (higher $\{e^-\}$, lower pe) because the driving force for uptake of e^- is lowered. For O_2, NO_3^-, SO_4^{2-}, and $H_2CO_3^*$, this is because H^+ is a reactant in the half reaction, so lower $\{H^+\}$ leads to higher $\{e^-\}$ (lower pe), all other things remaining equal. For Fe(III)T this is due to the nature of the implicit dependence on H^+: higher pH moves more dissolved Fe(III) into hydroxo complexes than is the case of dissolved Fe(II) (the $^*K_{H1}$ values are larger for Fe(III)). This makes it harder to reduce Fe(III).

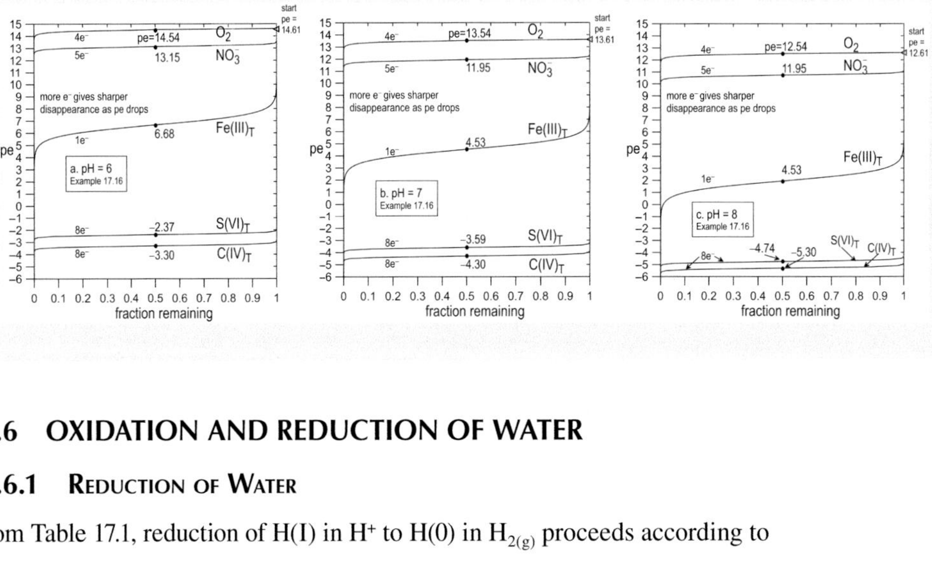

17.6 OXIDATION AND REDUCTION OF WATER

17.6.1 Reduction of Water

From Table 17.1, reduction of H(I) in H^+ to H(0) in $H_{2(g)}$ proceeds according to

$$2H^+ + 2e^- = H_{2(g)} \qquad K = 1, \qquad pe^o = 0. \tag{17.101}$$

We also have

$$2H_2O = 2H^+ + 2OH^- \qquad K = K_w^2 = (1.01 \times 10^{-14})^2. \tag{17.102}$$

For the reduction of H(I) in H_2O to H(0) in $H_{2(g)}$, adding the reactions in Eqs.(17.101) and (17.102) gives

$$2H_2O + 2e^- = H_{2(g)} + 2OH^- \qquad K = (1.01 \times 10^{-14})^2 \qquad pe^o = -14.00. \tag{17.103}$$

The *K* values given are for 25 °C/1 atm.

Because H^+ is in direct equilibrium with H_2O, either Eq.(17.101) or Eq.(17.103) may be used to describe how H(I) in aqueous solutions will accept electrons. Eq.(17.101) gives

$$\log p_{H_2} = 0.0 - 2\text{pH} - 2\text{pe} \quad (17.104)$$

$$\text{pe} = -\tfrac{1}{2}\log p_{H_2} - \text{pH} \quad (17.105)$$

$$\left(\frac{\partial \log p_{H_2}}{\partial \text{pe}}\right)_{\text{pH}} = -2. \quad (17.106)$$

As indicated, the partial derivative takes pH to be constant.

17.6.2 Oxidation of Water

From Table 17.1,

$$O_{2(g)} + 4H^+ + 4e^- = 2H_2O \qquad \log K = 83.1 \qquad \text{pe}^\circ = 20.78. \quad (17.107)$$

Oxidation of O(–II) in H_2O to O(0) in $O_{2(g)}$ occurs when the reaction proceeds backwards. We know

$$4H_2O = 4H^+ + 4OH^- \qquad \log K = K_w^4 = (1.01 \times 10^{-14})^4. \quad (17.108)$$

Subtracting Eq.(17.108) from Eq.(17.107) gives

$$O_{2(g)} + 2H_2O + 4e^- = 4OH^- \qquad \log K = 27.12 \qquad \text{pe}^\circ = 6.78. \quad (17.109)$$

which describes the oxidation of O(–II) in hydroxide to O(0) in O_2. The K values given are for 25 °C/1 atm. Because OH^- is in direct equilibrium with H_2O, either Eq.(17.107) or Eq.(17.109) may be used to describe how O(–II) in aqueous solutions will release electrons. Eq.(17.107) gives

$$\log p_{O_2} = -83.1 + 4\text{pH} + 4\text{pe} \quad (17.110)$$

$$\text{pe} = 20.78 + \tfrac{1}{4}\log p_{O_2} - \text{pH} \quad (17.111)$$

$$\left(\frac{\partial \log p_{O_2}}{\partial \text{pe}}\right)_{\text{pH}} = 4. \quad (17.112)$$

As indicated, the partial derivative takes pH to be constant. Based on Eqs.(17.104) and (17.110), lines for $\log p_{H_2}$ and $\log p_{O_2}$ *vs.* pe are drawn in Figure 17.5 for pH = 4, 7, and 10 and 25 °C/1 atm.

17.6.3 Redox Stability Limits for Water: The Redox Analog of Thermal Boiling

If a system of interest is assumed to be at 1 atm, neither p_{H_2} nor p_{O_2} can exceed 1 atm. If either the *in-situ* p_{H_2} or *in-situ* p_{O_2} did exceed 1 atm, the water would start to boil with one or the other gas. Not boil thermally, which would mean the vapor pressure of water $p_{H_2O} > 1$ atm as in Figure 17.6a, but boil for redox reasons because either: 1) so many electrons are available in solution that H_2O is reducible to H_2 to such a degree that the *in-situ* $p_{H_2} > 1$ atm; or 2) so few electrons were available in solution that the H_2O is oxidizable to O_2 and can release electrons to such a degree that enough O_2 is left in the solution and the *in-situ* $p_{O_2} > 1$ atm. So, at a system pressure of 1 atm, the thermodynamic stability range for H_2O is limited to conditions between ≤ 1 atm p_{O_2} on the oxidizing side (Figure 17.6b) and $p_{H_2} \leq 1$ atm on the reducing side (Figure 17.6c). This is called the stability field for water, and is demarcated on pe-pH (E_H-pH) diagrams, as will be shown in the next few chapters.

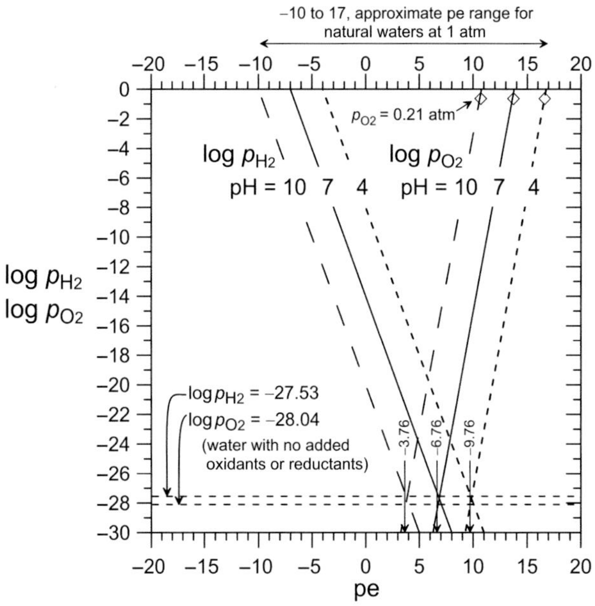

FIGURE 17.5 Lines for log p_{H_2} for pH = 4, 7, and 10 and similarly for and log p_{O_2}. For pH = 10, p_{H_2}= 1 atm when pe = −10. For pH = 4, p_{O_2}= 1 atm when pe ≈ 17. With pH = 4 and 10 as the range most natural waters, the pe range as limited by stability of water at 1 atm is then pe = −10 to 17. Water with no added oxidants or reductants will be characterized by $[O_2] = \frac{1}{2}[H_2]$. At 25 °C/1 atm this leads to log p_{H_2} = −27.53 and log p_{O_2} = −28.04. For pure water at pH = 7.00, then pe = −6.76; this is the redox analog of pH = 7.00 in pure water because then $[H^+] = [OH^-] = \sqrt{K_w}$. For water adjusted to pH = 4 with non-redox active strong acid and no added oxidants or reductants, pe = −3.76; for water adjusted to pH = 10 with non-redox active strong base, pe = −9.76.

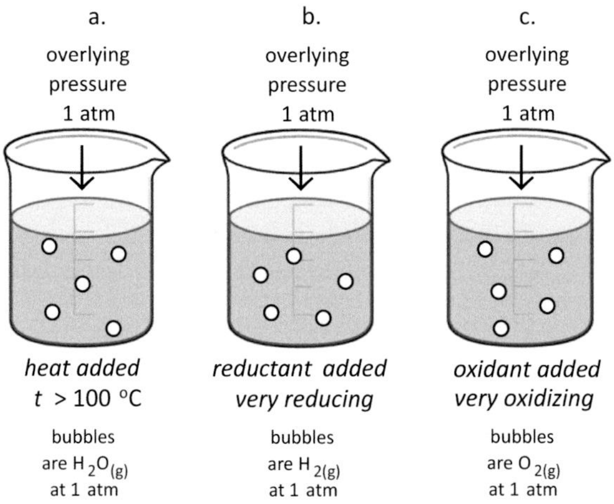

FIGURE 17.6 All systems at 1 atm total pressure. **a.** Thermal boiling when the temperature is slightly above 100 °C because p_{H_2O}>1 atm. **b.** Boiling in a very reducing environment because p_{H_2}>1 atm. **c.** Boiling in a very oxidizing environment because p_{O_2} > 1 atm.

17.6.4 pe Range for Most Natural Waters at 1 atm

The vast bulk of natural waters have pH values that fall between about 4 and 10, hence the choice of pH = 4, 7, and 10 for the lines in Figure 17.5. On the oxidizing side, with p_{O_2} limited at 1 atm, then p_{O_2} = 1 atm and these pH values give pe = 16.78, 13.78, and 10.78, respectively. So the high-side pe limit for natural waters at 1 atm is about 17. On the reducing side, with p_{H_2} limited at 1 atm, then p_{H_2} = 1 atm and these pH values give pe = –4, –7, and –10, respectively. So the low-side pe limit for natural waters at 1 atm is about –10. As shown in Figure 17.5, with atmospheric levels of p_{O_2} = 0.21 atm, and thus not-so-far from 1 atm, water in equilibrium with atmospheric levels of p_{O_2} are quite oxidizing. So, as noted above, the only thing that saves life as we know it from going up in a big puff of CO_2 is the slow kinetics of many redox reactions.

17.6.5 The pe (E_H) of Pure Water with No Added Oxidants of Reductants – The Redox Analog of pH = ½ log K_w for Pure Water (Optional)

Given the importance that pH = 7 has as a reference value in acid/base chemistry of aqueous solutions, it is natural to consider the pe (E_H) of pure water. Knowing the answer does not carry much practical utility, but it is perhaps interesting, and not that much work.

First, from Eq.(17.101),

$$2H^+ + 2e^- = H_{2(g)} \qquad \log K = 0. \tag{17.113}$$

For half of Eq.(17.107), and reversing the direction,

$$H_2O = \tfrac{1}{2}O_{2(g)} + 2H^+ + 2e^- \qquad \log K = -41.55. \tag{17.114}$$

Adding Eqs.(17.113) and (17.114) gives

$$H_2O = H_{2(g)} + \tfrac{1}{2}O_{2(g)} \qquad \log K = -41.55. \tag{17.115}$$

So, at 25 °C/1 atm, for every aqueous solution

$$p_{H_2}p_{O_2}^{\frac{1}{2}} = 10^{-41.55}. \tag{17.116}$$

In pure water there are no added oxidants or reductants. This means the only way to produce H_2 and O_2 is by dissociation of water. The stoichiometry of Eq.(17.115) then gives

$$[O_2] = \tfrac{1}{2}[H_2]. \tag{17.117}$$

By application of the two Henry's Law constants, $[O_2] = K_H^{O_2}p_{O_2}$ and $[H_2] = K_H^{H_2}p_{H_2}$. Substitution for p_{O_2} in Eq.(17.116) gives

$$p_{H_2}\left(\frac{1}{2}\frac{K_H^{H_2}}{K_H^{O_2}}p_{H_2}\right)^{\frac{1}{2}} = 10^{-41.55} \tag{17.118}$$

From Table 9.3, at 25 °C, $K_H^{H_2} = 10^{-3.11}$ and $K_H^{O_2} = 10^{-2.90}$, giving

$$p_{H_2} = 2.95 \times 10^{-28} \tag{17.119}$$

and so

$$p_{O_2} = 9.11 \times 10^{-29}. \tag{17.120}$$

For pH = 7.0, substituting Eq.(17.119) into Eq.(17.105) gives pe = 6.76. Assuming that HCl and NaOH would not be redox active, it would be possible to adjust the pH without affecting the validity of Eqs.(17.117–17.120). So, pe = 9.76 at pH = 4.0, and pe = 3.76 at pH 10.0. Each of these circumstances is marked in Figure 17.5 by a short vertical line drawn at the pe value where the log p_{H_2} line is 0.301 log units (= log ½) above the corresponding log p_{O_2} line.

REFERENCES

Bard AJ, Parsons R, Jordan J (1985) *Standard Potentials in Aqueous Solution*. Marcel Dekker, New York.

Cremer M (1906) Über die Ursache der elektromotorischen Eigenschaften der Gewebe, zugleich ein Beitrag zur Lehre von den polyphasischen Elektrolytketten. *Zeitschrift für Biologie*, **47**, 562–607.

Graham DJ, Jaselskis B, Moore CE (2013) Development of the glass electrode and the pH response. *Journal of Chemical Education*, **90**, 345–351.

Maeck A, DelSontro T, McGinnis DF, Fischer H, Flury S, Schmidt M, Fietzek P, Lorke A (2013) Sediment trapping by dams creates methane emission hot spots. *Environmental Science and Technology*, **47**, 8130–8137.

Rice University (n.d.) *Chemisty*. Available at: https://opentextbc.ca/chemistry/chapter/17-5-batteries-and-fuel-cells/ (Accessed 15 October 2019).

Schmid J (2009) Beautiful black poison. Available at: www.westonaprice.org/health-topics/environmental-toxins/beautiful-black-poison/ (Accessed 29 December 2017).

Stumm W, Morgan JJ (1996) *Aquatic Chemistry: Chemical Equilibria and Rates in Natural Waters*, 3rd Edition. Wiley Publishing Co., New York.

18 Introduction to pe–pH Diagrams

The Cases of Aqueous Chlorine, Hydrogen, and Oxygen

18.1 INTRODUCTION

pe (or equivalently E_H) and pH are the two most useful variables for understanding the chemistry of a natural water system, so these two are sometimes called "master variables". An important way to look at the chemistry of an element is, therefore, to examine how that chemistry varies, at equilibrium, as one moves around in a pe–pH field. This leads to the preparation of figures with pe on the *y*-axis, and pH on the *x*-axis. Such a figure is referred to as a "pe–pH predominance diagram" (or "E_H-pH predominance diagram") because for dissolved forms of an element, the convention used to label the different regions in the figure involves labeling each region according to which form is dominant (>50%) in that region. However, if it happens that a particular solid form of the element is predicted to be present in some region of the diagram, then throughout that region the identity of the solid is the label, even though the amount of solid will not comprise >50% of the element per liter of system water *everywhere* in the region. As such, a more accurately descriptive name would be "pe–pH predominance/and or solid-presence diagram", but that is a little long and not used in the field, so we will stick with the shorter name and just keep this caveat in mind.

Prior discussion in this text of the concept of predominance diagrams occurred with Figures 14.2-14.4, wherein the variables were $\log p_{CO_2}$ and pH. The species having predominance regions included Fe^{2+}, $FeOH^+$, $Fe(OH)_2^o$, $Fe(OH)_3^-$, $FeCO_{3(s)}$, and $Fe(OH)_{2(s)}$. At any given point in a region labeled Fe^{2+}, the species Fe^{2+} is present at a concentration higher than that of any other dissolved Fe(II)-containing species, and, no solid is present. In a region labeled with one of the solids, that solid is present, and is controlling the aqueous chemistry of the element; at a given point in the region, the solid may not represent the form holding most of the element as summed over the total system, per L of water (some of the element is in solution throughout every solid-labeled region). When adjacent to a solution-labeled region, *immediately* inside one of the solid regions there will be more of the element in solution (per liter of water) than in the solid (per liter of water). But the further one moves inside a solid region, the more of the element is present in the solid, and the less is in the solution. Because predominance diagrams are constructed using a log/log format, the change occurs quickly, *i.e.*, often within a fraction of a log unit; deep inside a solid region, the vast majority of the element is in the form of that solid.

With pe being an important master variable in natural waters, we now see that Figures 14.1–14.4 were prepared assuming that all Fe in the system is in the Fe(II) oxidation state: the pe is low enough that Fe(III) is not important, but not so low that Fe(0) can be present. (Fe(I) is not stable under any conditions.) If we add another axis for pe to Figure 14.2, then conceptually, we obtain Figure 18.1. A $\log p_{CO_2}$ *vs.* pH diagram thus represents a $\log p_{CO_2}$ *vs.* pH slice in Figure 18.1 at some constant pe; so, Figure 14.2 is a slice at fairly low pe where Fe(II) dominates; a $\log p_{CO_2}$ *vs.* pH slice where Fe(III) dominates, would be a slice at higher pe.

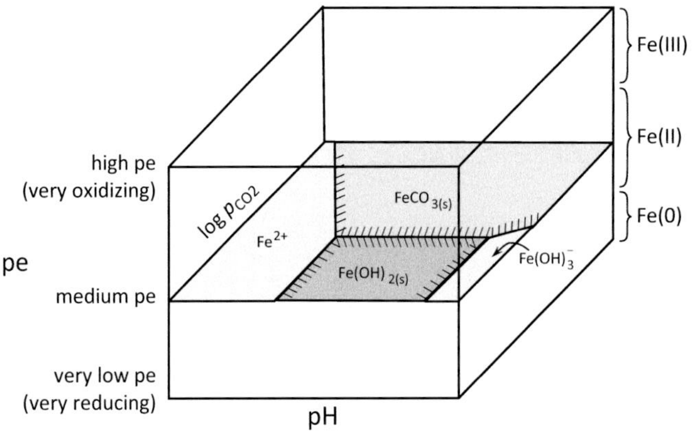

FIGURE 18.1 Three-dimensional pe–pH-log p_{CO_2} diagram for $Fe_{T,sys} = 10^{-4}$ *M* (mols Fe/L of system water). At high pe, the Fe(III) oxidation state will dominate (very low levels of Fe(II) will be present). At intermediate pe, the Fe(II) oxidation state will dominate (very low levels of Fe(III) will be present). At very low pe, solid Fe(0) will be present (with low levels of Fe(II) and exceedingly low levels of Fe(III)). Figure 14.2 for $Fe_T = 10^{-4}$ *M* is shown as a horizontal, log p_{CO_2} *vs.* pH slice.

pe *vs.* pH predominance diagrams can be used to study the manner in which the chemistry of an element changes in response to changes in the redox and pH conditions in the solution. This chapter considers pe–pH diagrams for chlorine, hydrogen and oxygen. For all three elements, the value of log p_{CO_2} will be considered to be sufficiently low that CO_2 has no effect on the chemistry. For chlorine, we follow the approach of Stumm and Morgan (1996).

18.2 pe–pH DIAGRAM FOR AQUEOUS CHLORINE

18.2.1 Preliminary pe–pH Diagram for Aqueous Chlorine

Chlorine has several oxidations states that are well-known, including Cl(VII), as in perchlorate (ClO_4^-) which has become a groundwater contaminant in some locations because of its use as a powerful oxidant in solid fuel rocket engines ($ClO_4^- = Cl^- + 4O_2$). We limit our discussion here to three oxidation states, I, 0, and –I. Cl(0) is elemental chlorine, which may exist in aqueous solution as Cl_2 and is the reference point from which the other oxidation state values are measured. Cl in the I oxidation state has lost one electron relative to Cl in Cl_2. Chlorine in the –I oxidation state has gained one electron relative Cl in Cl_2. Two Cl(I) species are HOCl and OCl^-. Two Cl(–I) species are HCl and Cl^-, though we know that in aqueous solutions, HCl is such a strong acid that it is never important as a molecular species.

For a first take on the problem, the pe–pH diagram for chlorine will look *something* like Figure 18.2. In that diagram, all of the acid/base boundary lines (*i.e.*, the HOCl/OCl^- and HCl/Cl^- boundary lines) are simple vertical lines: they do not depend on pe. All of the boundary lines separating different redox forms (*i.e.*, the HOCl/Cl_2, OCl^-/Cl_2, Cl_2/HCl, and Cl_2/Cl^- boundary lines) have been drawn as simple horizontal lines, though as we know, many redox half-reactions depend on pe as well as pH, so many redox boundary lines will not be horizontal with slope zero.

18.2.2 Actual pe–pH Diagram for Aqueous Chlorine for Some Specific Cl_T

We seek to determine equations for the various boundary lines for aqueous chlorine. We neglect activity corrections (except for the electron here and henceforth). Then, from Table 17.1, the following equilibrium data are available at 25 °C/1 atm:

$$HOCl + H^+ + e^- = \tfrac{1}{2}Cl_2 + H_2O \qquad K = \frac{[Cl_2]^{\frac{1}{2}}}{[HOCl][H^+]\{e^-\}} = 10^{26.9} \qquad (18.1)$$

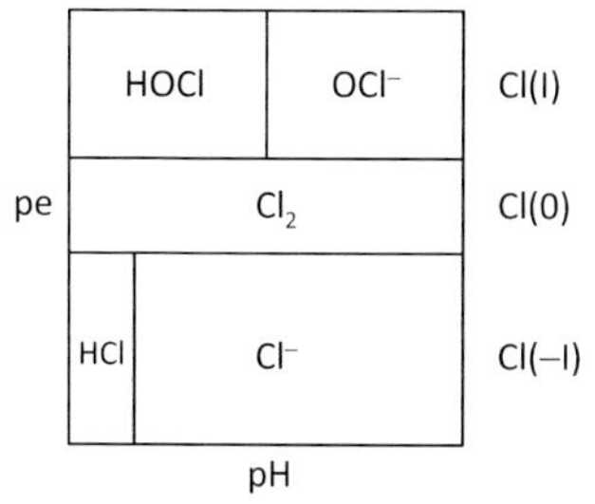

FIGURE 18.2 Preliminary pe–pH predominance diagram for aqueous chlorine at a Cl_T value large enough for a region for Cl_2 to be present. A line between regions for species in the same oxidation state (*e.g.*, HOCl and OCl^-) is a simple vertical line at a specific pH. For HOCl and OCl^-, this is pH=pK for HOCl. A line between predominance regions for two species not in the same oxidation state may be horizontal, or may have a non-zero slope. The Cl_2/Cl^- boundary line does not involve H^+ and so it is a horizontal line for as far as the contribution of HOCl to Cl_T can be neglected. The region for HCl is included to illustrate the analogous behavior of HOCl and OCl^- for Cl(I) and HCl and Cl^- for Cl(–I) (in practice, HCl has no predominance region in the range pH=0 to 14 because its pK<<0).

$$\tfrac{1}{2}Cl_2 + e^- = Cl^- \qquad K = \frac{[Cl^-]}{[Cl_2]^{\frac{1}{2}}\{e^-\}} = 10^{23.6}. \tag{18.2}$$

For the acid dissociation of HOCl,

$$HOCl = H^+ + OCl^- \qquad K = \frac{[H^+][OCl^-]}{[HOCl]} = 10^{-7.3}. \tag{18.3}$$

The equilibrium in Eq.(18.1) gives

line	equation	depends on	boundary	
(1)	$pe = 26.9 + \log[HOCl] - \tfrac{1}{2}\log[Cl_2] - pH$	pe and pH	$\frac{HOCl}{Cl_2}$	(18.4)

Along a pe–pH boundary line, in general there are two goals: 1) to obtain a simple function between pe and pH, *e.g.*, for Eq.(18.4) one way to do this would be to eliminate the portion $(\log[HOCl] - \tfrac{1}{2}\log[Cl_2])$ by setting $\log[HOCl] = \tfrac{1}{2}\log[Cl_2]$, which would mean $[HOCl] = [Cl_2]^{\frac{1}{2}}$); 2) to have the element in the species present at equal concentrations in the two forms, so that if we cross the boundary line, we shift from dominance by one form to dominance by the other form. For the second goal, for HOCl which contains one Cl, and Cl_2 which contains two Cl, this suggests $\tfrac{1}{2}[HOCl] = [Cl_2]$. However, this is in conflict with the first goal in that $[HOCl] = [Cl_2]^{\frac{1}{2}}$ does not give $\tfrac{1}{2}[HOCl] = [Cl_2]$. *E.g.* if $[HOCl] = 10^{-3}$ M, then $[Cl_2] = 10^{-6}$ M. To accomplish the latter, more important goal, assume a Cl_T concentration that we can divide amongst the different Cl species. Following Stumm and Morgan (1996), we will choose Cl_T=0.04 M. This is a very high value, and is not particularly relevant *per se* for fresh waters, but as long as we keep that in mind, there are advantages in doing so, which will be discussed in due course. We now have

$$Cl_T = 0.04\ M = [HOCl] + [OCl^-] + 2[Cl_2] + [Cl^-]. \tag{18.5}$$

Along the boundary line between HOCl and Cl_2: 1) OCl^- will not contribute significantly to Cl_T (because that boundary line will be found to end before we get close to the pK for HOCl); and 2)

Cl^- will not contribute significantly to Cl_T (at least for most of the range for which the boundary line will be drawn). Therefore,

$$Cl_T = 0.04\ M \approx [HOCl] + 2[Cl_2]. \tag{18.6}$$

Now, for the second goal mentioned above (equal amounts of Cl in the two forms along the line), we specify

$$[Cl_2] = \tfrac{1}{2}[HOCl] \tag{18.7}$$

so

$$[HOCl] = 0.02\ M \quad \text{and} \quad [Cl_2] = 0.01\ M. \tag{18.8}$$

Thus,

$$pe = 26.9 - 1.70 + 1.00 - pH \tag{18.9}$$

or

$$\boxed{\text{line 1: } pe = 26.2 - pH \qquad \frac{dpe}{dpH} = -1.} \tag{18.10}$$

Line 1 is drawn in Figure 18.3.

For the boundary line between the regions for Cl_2 and Cl^-, we use

line	equation	depends on	boundary	
(2)	$pe = 23.6 + \frac{1}{2}\log[Cl_2] - \log[Cl^-]$	pe only	$\frac{Cl_2}{Cl^-}$.	(18.11)

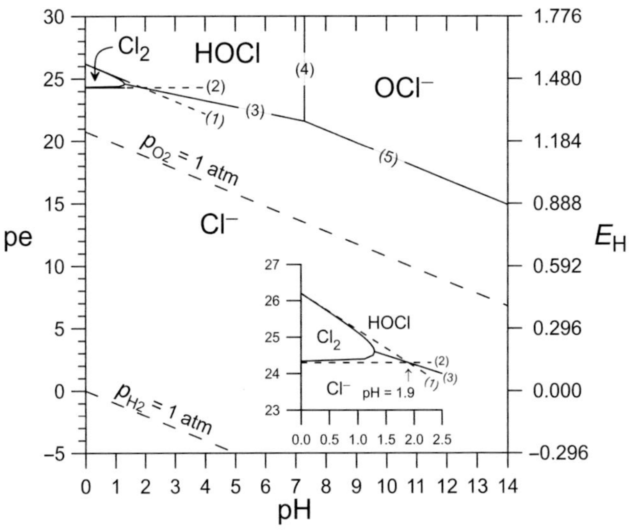

FIGURE 18.3 pe–pH predominance diagram for aqueous chlorine at 25 °C/1 atm and Cl_T=0.04 *M*. Each straight boundary line assumes equal solution mass concentrations of the element in the two adjacent forms. This assumption leads to a contradiction at the intersection of lines 1 and 2. In actuality, the Cl_2 region is not a sharp wedge, but a blunt wedge defined by the curve computed according to Box 18.1 (see also inset figure). Cl_2, HOCl, and Cl^- are all strong oxidants in solution: their predominance regions lie far above the water stability line for p_{O_2} = 1 atm; this accounts for their utility as disinfection agents for drinking water and waste water.

To be consistent with the conditions used to obtain Eq.(18.9), we again let $Cl_T = 0.04\ M$, but this time we presume that Cl_2 and Cl^- are the major species. Assuming equal amounts of chlorine in the two forms along the boundary line, we set

$$[Cl_2] = \tfrac{1}{2}[Cl^-] \tag{18.12}$$

$$[Cl_2] = 0.01\ M \qquad [Cl^-] = 0.02\ M \tag{18.13}$$

and so

$$\boxed{\text{line 2: pe} = 24.3 \qquad \frac{d\text{pe}}{d\text{pH}} = 0.} \tag{18.14}$$

Lines 1 and 2 cross at pH = 1.9. This poses two problems.

The first problem is that Cl_2 cannot predominate past this intersection. As presented schematically in Figure 18.4, this can be demonstrated considering what the relative locations of the dashed portions of the lines imply for when we are past the wedge tip. (The dashed lines do not exactly trace the points for the conditions $[Cl_2] = ½[HOCl]$ and $[Cl_2] = ½[Cl^-]$, because both lines assume $[Cl_2] = 0.01\ M$, and $[Cl_2]$ will not be that large anywhere past the wedge tip.) *Nota bene*: for pH values past the wedge tip, one *could* determine where the $[Cl_2] = ½[HOCl]$ line actually falls, and similarly for the $[Cl_2] = ½[Cl^-]$ line. Since the latter will be above the former, the conclusions reached in the discussion below will be valid.

In region *A*, all points are above line 2 so $[Cl_2] > ½[Cl^-]$, and all points are above line 1 so $½[HOCl] > [Cl_2]$; the net result is $½[HOCl] > [Cl_2] > ½[Cl^-]$, and HOCl dominates. In region *C*, all points are below line 1 so $[Cl_2] > ½[HOCl]$, and all points are below line 2 so $½[Cl^-] > [Cl_2]$; the net result is $½[Cl^-] > [Cl_2] > ½[HOCl]$, and Cl^- dominates. In region *B*, all points are above line 1 so $½[HOCl] > [Cl_2]$, but also all points are below line 2 so $½[Cl^-] > [Cl_2]$. So, Cl_2 cannot dominate in region B, and that region will have to be **divided** between HOCl and Cl^-. Up to some pH approaching ~1.9, we can draw both lines 1 and 2, but at pH = 1.9 and beyond, neither line is relevant because Cl_2 cannot dominate. The larger the value of Cl_T, the bigger the wedge is for Cl_2 dominance, and the higher the pH at the wedge point. This is why a large Cl_T value was chosen for our diagram: we want to see the wedge. As Cl_T decreases, the wedge recedes off the diagram to the left.

The second problem is that lines 1 and 2 cannot both be valid even at the intersection: for line 1 it was assumed that $[HOCl] = 0.02\ M$ and $[Cl_2] = 0.01\ M$, and for line 2 that $[Cl_2] = 0.01\ M$ and $[Cl^-] = 0.02\ M$. If both lines were valid at the intersection, this gives $Cl_T = 0.06\ M$ while it has been

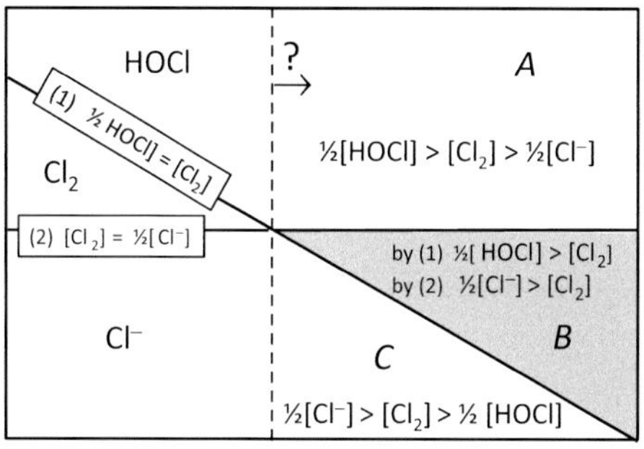

FIGURE 18.4 In the three regions to the left of the vertical dashed line, the relative dominance of the different Cl species is clear: HOCl dominates above line 1; Cl_2 dominates below line 1; but above line 2; and Cl^- dominates below line 2. To the right of wedge tip for Cl_2, in region *A*, for all points $½[HOCl] > [Cl_2] > -[Cl^-]$ so HOCl dominates. In region *C*, for all points $[Cl^-] > [Cl_2] > ½[HOCl]$ and Cl^- dominates. In region *B*, $½[HOCl] > [Cl_2]$ and $[Cl^-] > [Cl_2]$. Cl_2 cannot dominate past the wedge tip, so region *B* will have to be divided between HOCl and Cl^-.

assumed that $Cl_T = 0.04\ M$. The tip of the Cl_2 region must therefore end at some pH < 1.9. This difficulty is usually ignored when dealing with pe–pH diagrams, because the main purpose of such diagrams is to give a "pretty good" understanding of where the different regions lie: if the map is off by some fraction of a pH unit or some fraction of a pe unit here and there, well, the whole pe–pH domain is so big in comparison that this is not a big deal. This is kind of like you having a map of the United States for which up at the northwest corner of California, the state lines between California, Oregon, and Nevada are somehow off by ~100 miles. If by your readings and map consultation, you decide you are in the vicinity of that corner (in any of the states), you still have a pretty good idea of your location relative to the three states and the rest of the United States, even if do not know exactly what state you are in. Moreover, for redox maps, a "pretty good" idea where you are is all that matters, given that environmental systems are generally not a full equilibrium anyway. Nevertheless, Box 18.1 discusses how one can fix the Cl "map" in this regard, and that approach will be used for the analogous wedge that exists for S(0) in the sulfur pe–pH diagram. The map analogy, by the way, provides a perfect description of how these diagrams are understood and used. In fact, compilations of such diagrams for scores of elements for different conditions have been assembled into "atlases", as with the landmark book by Pourbaix (1974), and the lengthy work by Takeno (2005) which

BOX 18.1 The Exact Boundary Line for the Cl_2 Region as a Function of Cl_T

For this boundary line,

$$[Cl_2] = \tfrac{1}{4} Cl_T,$$

the other half of Cl_T is distributed over the other two species

$$[HOCl] + [Cl^-] = \tfrac{1}{2} Cl_T.$$

By Eq.(18.1),

$$\frac{[Cl_2]^{\frac{1}{2}}}{[HOCl]10^{-pH}10^{-pe}} = 10^{26.9}$$

$$[HOCl] = \frac{[Cl_2]^{\frac{1}{2}}}{10^{26.9}10^{-pH}10^{-pe}} = \frac{0.1}{10^{26.9}10^{-pH}10^{-pe}}.$$

By Eq.(18.2),

$$\frac{[Cl^-]}{[Cl_2]^{\frac{1}{2}}10^{-pe}} = 10^{23.6}$$

$$[Cl^-] = 10^{23.6}[Cl_2]^{\frac{1}{2}}10^{-pe} = 10^{23.6}(0.1)10^{-pe}.$$

Substituting into the second equation for [HOCl] and $[Cl^-]$ gives

$$\frac{(\frac{1}{4}Cl_T)^{\frac{1}{2}}}{10^{26.9}10^{-pe}(\frac{1}{2}Cl_T - 10^{23.6}(\frac{1}{4}Cl_T)^{\frac{1}{2}}10^{-pe})} = 10^{-pH}.$$

Substituting a range of pe values produces the corresponding pH values. When $Cl_T = 0.04\ M$, the function blows up for pe ≤ 24.3; per Eq.(18.14), for that Cl_T, that pe is the lower limit for the Cl_2 region.

compares diagrams prepared using different *K*-value datasets. The text by Brookins (2012) provides E_H-pH diagrams for 75 elements found at the surface of the earth, including transuranic and other radioactive species. A smaller but highly relevant group of useful diagrams with discussion for the most common elements can be found in the geochemistry text by Garrels and Christ (1965).

Once the Cl_2 region ends, we must find the boundary line between HOCl and Cl^-. We derive the equation for this line by adding together the chemical equilibrium equations underlying lines 1 and 2. In particular,

line	equation	depends on	boundary	
(1)	$pe = 26.9 + \log[HOCl] - \frac{1}{2}\log[Cl_2] - pH$	pe and pH	$\frac{HOCl}{Cl_2}$	(18.4)
+ (2)	$pe = 23.6 + \frac{1}{2}\log[Cl_2] - \log[Cl^-]$	pe	$\frac{Cl_2}{Cl^-}$	(18.10)
(3)	$2pe = 50.5 + \log[HOCl] - \log[Cl^-] - pH$	pe and pH	$\frac{HOCl}{Cl^-}$.	(18.13)

In contrast to the cases with lines 1 and 2, the two species appearing in the equation for line 3 both have one Cl per species. Now, when equal amounts of chlorine are present in the two species along the boundary line, $[HOCl]=[Cl^-]$ so for 25 °C/1 atm

$$\text{line 3: } pe = 25.25 - \tfrac{1}{2}pH \qquad \frac{dpe}{dpH} = -\tfrac{1}{2}. \tag{18.14}$$

Line 3 gives the locus of all points for which $[HOCl]=[Cl^-]$ for any Cl_T, even in regions wherein HOCl and Cl^- are present at low concentrations relative to other species. As Cl_T varies, the wedge point for the Cl_2 region slides along line 3.

Example 18.1 Calculating the Concentration of a Species Outside of Its Dominance Region, Part A

Calculate $[Cl_2]$ when $Cl_T=0.04$ *M* for the pe and pH at pH=5 and on the HOCl/Cl^- line (line 3, Eq.18.14).

Solution

For line 3, $pe = 25.25 - \frac{1}{2}(5.0) = 22.75$. As the two equi-important species along the line, $[HOCl]=[Cl^-]=½\ Cl_T$. If we have a system with $Cl_T=0.04$ *M*, then $[HOCl]=[Cl^-] \approx 0.02$ *M*. Eq.(18.4) gives $pe = 26.9 + \log[HOCl] - \frac{1}{2}\log[Cl_2] - pH$, so $22.75 = 26.9 + \log(0.02) - \frac{1}{2}\log[Cl_2] - 5.0$. $\log[Cl_2] = 10^{-5.10} = 7.98 \times 10^{-6}$ *M*.

For the boundary line between the regions for HOCl and OCl^-, we use the equilibrium given in Eq.(18.3). Neglecting activity corrections, at 25 °C/1 atm,

line	equation	depends on	boundary	
(4)	$pH = 7.3 - \log[HOCl] + \log[OCl^-]$	pH	HOCl | OCl^-.	(18.15)

The pK for HOCl is 7.3. At pH = 7.3, at all pe, [HOCl] = [OCl^-] and predominance between HOCl and OCl^- switches moving through pH = 7.3.

$$\text{line 4: } pH = 7.3 \qquad \text{independent of pe.} \tag{18.16}$$

Also, line 4 (like line 3): 1) is independent of Cl_T because both species have the same number of chlorine atoms (one); and 2) gives the locus of all points for which the species for the boundary line are present at equal concentrations, even those lying in regions of the diagram wherein the species for the line are at concentrations that are low relative to other species.

Last, we now obtain the boundary line between OCl^- and Cl^- by adding up the equations underlying lines 1, 2, and 4:

line	equation	depends on	boundary	
(1)	$pe = 26.9 + \log[HOCl] - \frac{1}{2}\log[Cl^-] - pH$	pe and pH	$\frac{HOCl}{Cl_2}$	(18.4)
(2)	$pe = 23.6 + \frac{1}{2}\log[Cl_2] - \log[Cl^-]$	pe	$\frac{Cl_2}{Cl^-}$	(18.10)
+ (4)	$pH = 7.3 - \log[HOCl] + \log[OCl^-]$	pH	HOCl | OCl^-	(18.15)
(5)	$2pe = 57.8 + \log[OCl^-] - \log[Cl^-] - 2pH$	pe and pH	$\frac{OCl^-}{Cl^-}$.	(18.17)

Along the boundary line, [OCl^-] = [Cl^-], so

$$\text{line (5): } pe = 28.9 - pH \qquad \frac{dpe}{dpH} = -1. \tag{18.18}$$

As with lines 3 and 4, line 5 is independent of Cl_T. Line 5 gives the locus of all points for which [OCl^-] = [Cl^-], even those lying in regions of the diagram wherein OCl^- and Cl^- are present at low concentrations relative to other species. All the lines are drawn in Figure 18.3 for the portions of the pe–pH domain over which they are needed as predominance boundary lines.

In addition to having the various lines for Cl species in Figure 18.3, we need to draw the two lines that describe the stability limits of water when the system pressure is 1 atm. For the oxidation of water to $O_{2(g)}$ at high pe, at 25 °C/1 atm, we have

$$pe = 20.78 + \tfrac{1}{4}\log p_{O_2} - pH. \tag{17.111}$$

When p_{O_2} = 1 atm, which is the upper limit when the system pressure is 1 atm,

$$pe = 20.78 - pH \qquad (p_{O_2} = 1 \text{ atm}). \tag{18.19}$$

Above this line, $O_{2(s)}$ bubbles will start forming in the water.

Example 18.2 Calculating the Concentration of a Species Outside of Its Dominance Region, Part B

Calculate [OCl^-] when: **a.** Cl_T = 0.04 M; and **b.** Cl_T = 0.001 M for the pe and pH along the p_{O_2} = 1 line (Eq.18.19) when pH = 8.0.

Solution

Here, pe = 20.78 – (8.0) = 12.78. In this portion of the diagram. We are far inside the [Cl^-] region, so [Cl^-] ≈ Cl_T. **a.** If we have a system with Cl_T = 0.04 *M*, then [Cl^-] ≈ 0.04 *M*. Eq.(18.17) gives 2pe = 57.8 + log[OCl^-] – log[Cl^-] – 2pH, so 2(12.78) = 57.8 + log[OCl^-] – log(0.04) – 2(8). log [OCl^-] = $10^{-17.64}$ = 2.30 × 10^{-18} *M*. **b.** If we have a system with Cl_T = 0.001 *M*, then [Cl^-] ≈ 0.001 *M*. Eq.(18.17) gives 2pe = 57.8 + log[OCl^-] – log[Cl^-] – 2pH, so 2(12.78) = 57.8 + log[OCl^-] – log(0.001) – 2(8). log[OCl^-] = $10^{-19.24}$ = 5.75 × 10^{-20} *M*.

For the reduction of water to $H_{2(g)}$ at 25 °C/1 atm, we have

$$pe = -\tfrac{1}{2}\log p_{H_2} - pH. \qquad (17.105)$$

When p_{H_2} = 1 atm, which is the upper limit when the system pressure is 1 atm,

$$pe = -pH \qquad (p_{H_2} = 1\ \text{atm}). \qquad (18.20)$$

Below that line, $H_{2(g)}$ bubbles can start forming if the system pressure is 1 atm. With the lines for Eqs.(18.19) and (18.20) included in Figure 18.3, it can be concluded that Cl^- is the dominant equilibrium Cl species for natural waters at a system pressure of 1 atm, and also for pressures quite a bit higher than that given how high lines 1, 2, 3, and 5 lie in the diagram.

18.2.3 Aqueous Cl_2: Disproportionation and Disinfection

Because Cl_2 cannot predominate for Cl at moderate pH unless Cl_T is unrealistically high, if it is placed in water at moderate pH, it will tend strongly to decompose into those forms that can predominate. Like Fe(I), Cl(0) is not stable. If we were to try and make some Fe(I), it would just react according to 2Fe(I) = Fe(0) + Fe(II). Putting subscripts on the two Fe to see what is happening, Fe_a(I) + Fe_b(I) = Fe_a(0) + Fe_b(II). Neither Fe wants to stay as Fe(I), so Fe_a(I) takes an electron from Fe_b(I), and both escape the Fe(I) oxidation state. This type of reaction is termed a disproportionation: two Fe(I) dissociate proportionally according to the number of electrons exchanged; one becomes Fe(0) and the other becomes Fe(II). This is what happens with most Cl_2 added to water (unless we are in the Cl_2 dominance wedge because pH is very low and Cl_T is very, very high). The two Cl initially share a bond pair of electrons. As illustrated in Figure 18.5, one of the Cl(0) takes both electrons in the bond and goes down to Cl(–I), sending the other Cl(0) up to Cl(I). The net reaction for the process is obtained as

$$\tfrac{1}{2}Cl_2 + e^- = Cl^- \qquad K = 10^{23.6} \qquad (18.2)$$

$$-\left(HOCl + H^+ + e^- = \tfrac{1}{2}Cl_2 + H_2O \qquad K = 10^{26.9}\right) \qquad (18.1)$$

$$Cl_2 + H_2O = HOCl + H^+ + Cl^- \qquad K = \frac{[HOCl][H^+][Cl^-]}{[Cl_2]} = 10^{-3.3} \qquad (18.21)$$

where the expression for the equilibrium constant neglects activity corrections. (Reaction Eq.(18.21) in reverse is an important step in the activation of trapped Cl in the atmosphere (as $ClONO_2$, chlorine nitrate) during ozone hole formation in polar latitudes; the Cl_2 formed is photolyzed by the increasing Spring sunlight to Cl atoms, and these then catalyze ozone depletion.)

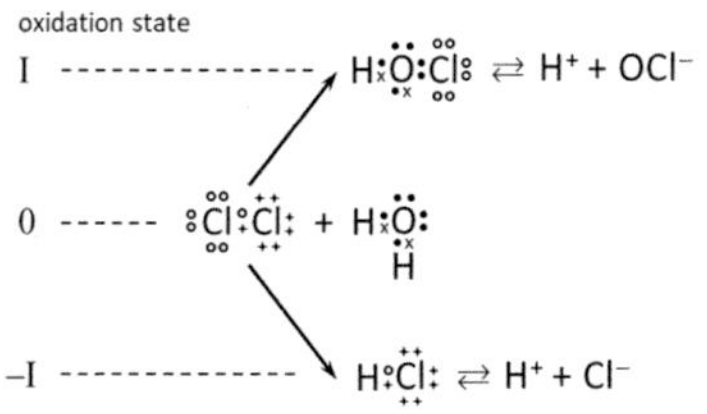

FIGURE 18.5 Disproportionation of Cl_2 (which is in the Cl(0) oxidation state) to one Cl(I) and one Cl(–I). An equilibrium is established between the Cl(0) state and the Cl(I) and Cl(–I) states. The two oxidation states formed in the disproportionation are symmetrically distributed around the originating oxidation state, so equal amounts of chlorine move to the I and –I states.

While the K value for the disproportionation reaction given in Eq.(18.21) may look small, the value is deceiving. If the pH is say ~7, and if $[Cl^-] < 0.003\ M$ (as for most fresh waters), then

$$\frac{[HOCl]}{[Cl_2]} > 10^{6.2}. \tag{18.22}$$

So, when small amounts of Cl_2 are added to drinking water, swimming pool water, wastewater, etc. to accomplish "chlorine disinfection", disproportionation to Cl(I) and Cl(–) is strongly favored. Alternatively, Cl(I) can be added directly (and frequently is with swimming pools) in the form of powdered $NaOCl_{(s)}$ (sodium hypochlorite), or as dissolved NaOCl. The latter is also extensively used as "bleach" for laundry because it removes stains. It is clear from Figure 18.3 why disinfection and bleaching by Cl(I) occurs: HOCl and OCl^- are strongly oxidizing species, with predominance regions far above the $p_{O_2} = 1$ atm line (even $p_{O_2} = 1$ is pretty oxidizing). HOCl and OCl^- have a strong desire to pick up electrons and become Cl^-. They will thus oxidize any readily available organic molecule. If pathogens are present in water, portions of their cells will be oxidized (and thus seriously damaged) by the HOCl and OCl^-, without any organism specificity: Cl_2 and $HOCl/OCl^-$ are broad-spectrum disinfectants. And, if there is some Cl_2 and $HOCl/OCl^-$ that is left over, drinking water in a distribution pipe will be protected if some contaminated water is accidentally pulled into the pipe should there be a temporary loss of line pressure. When used as bleach, C(I) oxidizes the offending stain molecules so that they no longer absorb light. Unfortunately, it will also degrade and weaken the fabric, as someone who has created some artistic bleach spots on jeans may attest: first some nice white spots, then eventually some nice holes, which can also be in style.

18.3 pe–pH DIAGRAM FOR AQUEOUS OXYGEN

18.3.1 Preliminary pe–pH Diagram for Aqueous Oxygen

For oxygen, we consider three oxidation states, O(0) as in O_2, O(–I) as in H_2O_2, and O(–II) as in H_2O (mostly, with minor additional amounts as OH^-). A preliminary version of what the pe–pH diagram for O might look like is then given by Figure 18.6a. It will turn out (see Section 18.3.2) that H_2O_2 does not have a predominance region. Because it is universally familiar as a disinfectant, it is worth discussing why this is, and what implications it holds for H_2O_2 as an oxidant.

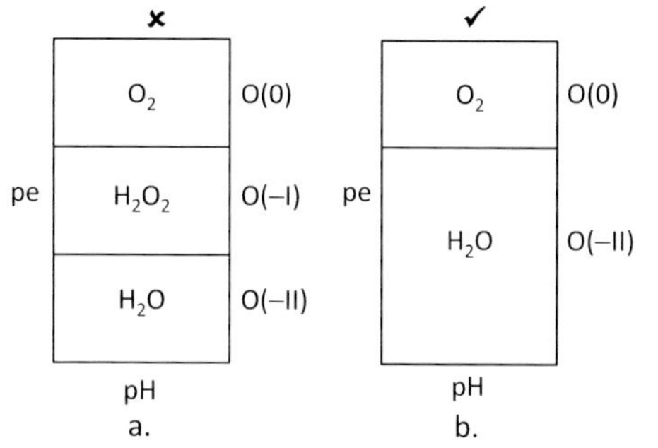

FIGURE 18.6 **a.** Preliminary pe–pH predominance diagram for aqueous oxygen if H_2O_2 has a region, which is not (✗) the case. **b.** Preliminary pe–pH predominance diagram for aqueous oxygen with just dissolved O_2 and H_2O. Redox boundary lines in the final pe–pH diagram may not be horizontal as they are above.

18.3.2 Why H_2O_2 Does Not Have a Predominance Region for Aqueous Oxygen (Optional)

From Table 17.1, at 25 °C/1 atm,

$$H_2O_2 + 2H^+ + 2e^- = 2H_2O \qquad K = \frac{1}{[H_2O_2][H^+]^2\{e^-\}^2} \quad K = 10^{59.6}. \tag{18.23}$$

where the expression for K takes $x_{H_2O} \approx 1$.

For an equation for the H_2O_2/H_2O boundary line, this gives

$$pe = \tfrac{1}{2}59.8 + \tfrac{1}{2}\log[H_2O_2] - pH \qquad H_2O_2 \big/ H_2O \; @ \; x_{H_2O} \approx 1^{\cdot} \tag{18.24}$$

At $x_{H_2O} \approx 1$, waters is about 55.5 M. If for a boundary line the amount of O in H_2O_2 and in water were taken to be equal, then $[H_2O_2]$ would have to be so high that one would no longer be talking about "water". So for an H_2O_2/H_2O boundary line we will leave x_{H_2O}at ~1 and just take $[H_2O_2] = 1$ M (that is still *a lot*). That gives

$$pe = 29.9 - pH \qquad H_2O_2 \; @ \; 1\,M \big/ H_2O \; @ \; x_{H_2O} \approx 1^{\cdot} \tag{18.25}$$

As shown in Figure 18.7, this plots far above the water-stability-upper-limit line for a total system pressure of 1 atm, which is

$$pe = 20.78 - pH \qquad (p_{O_2} = 1\text{ atm}). \tag{18.19}$$

Eq.(18.25) also plots far above the line along which $[O_2] = 1$ M (see Eq.(18.30 below). So, if pe was increased starting from moderate levels, we would first pass the line for $p_{O_2} = 1$ atm, then pass the line for $[O_2] = 1$ M (if the system pressure was high enough to keep it from bubbling out), then finally reach the line for $[H_2O_2] = 1$ M, which is not even the 1:1 line for O(–I):O(–II). So, H_2O_2 does not have a predominance region for water, not even close.

If we do put some H_2O_2 in water, it will tend strongly to disproportionate, analogous with the behavior of Cl_2. With H_2O_2, the disproportionation takes O(–I) in H_2O_2 and moves equal amounts:

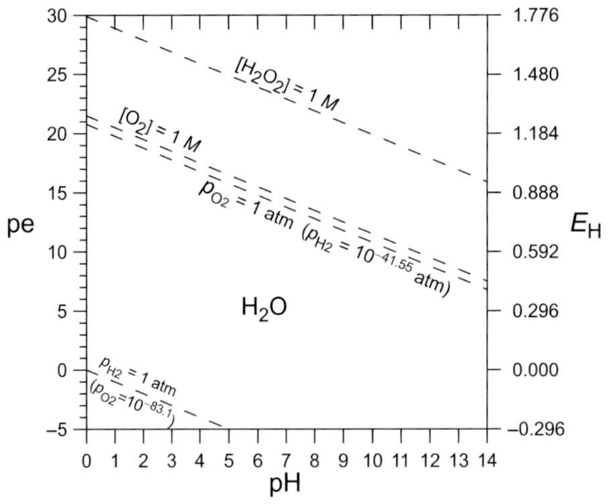

FIGURE 18.7 pe–pH predominance diagram for aqueous oxygen at 25 °C/1 atm. H_2O is the predominant form of oxygen within the stability limit lines for a system pressure of 1 atm.

1) up to O(0) in O_2, and 2) down to O(−II) in H_2O. From Table 17.1, neglecting activity corrections for O_2 and H_2O_2, and noting that $[O_2]$ is the dissolved O_2 concentration, the net reaction can be obtained according to

$$-\left(O_2 + 4H^+ + 4e^- = 2H_2O \qquad K = \frac{1}{[O_2][H^+]^4\{e^-\}^4} = 10^{86.0}\right) \tag{18.26}$$

$$+2\left(H_2O_2 + 2H^+ + 2e^- = 2H_2O \qquad K = \frac{1}{[H_2O_2][H^+]^2\{e^-\}^2}\ K = 10^{59.6}\right) \tag{18.27}$$

$$2H_2O_2 = 2H_2O + O_2 \qquad K = \frac{[O_2]}{[H_2O_2]^2} = 10^{33.2}. \tag{18.28}$$

H_2O_2 solutions that are ~3% by weight in water have long been used for cleaning wounds. The vigorous bubbling that takes place when such a solution is placed in a wound is due to release of $O_{2(g)}$ by disproportionation (Eq.(18.28), as catalyzed by the cellular enzyme catalase found in the wound). This has medicinal worth because the bubbling can help clean a wound. In addition, at levels like 3% the oxidative character of the H_2O_2 tends to kill some types of bacteria, especially Gram-positive bacteria (*e.g.*, *Streptococcus* and *Staphylococcus*). However, the oxidizing power of H_2O_2 is nonspecific, and so it has also been argued that use of ~3% H_2O_2 can retard healing because it can cause collateral damage to body cells involved in wound repair (Wilgus et al., 2005; Loo et al., 2012). In any case, H_2O_2 at 3% does not kill all bacteria, and so it is not a broad-spectrum antibiotic. At higher concentrations, 6 to 35%, H_2O_2 can kill a broad range of microbes: essentially a wide range of chemicals and cellular structures are chemically burned, and so such H_2O_2 solutions can be used as a biocide for sterilizing equipment and surfaces in the medical treatment environment (Linley et al., 2012). Also, hair bleaching formulations as 12% H_2O_2 work by oxidatively destroying hair pigment molecules. Certainly one of the most interesting things about the biochemistry of H_2O_2 is that in animals, protective phagocytes use a respiratory burst to create sufficient H_2O_2 to disable/kill foreign cells, H_2O_2 being a by-product of aerobic metabolism (Juven and Pierson, 1996; Wittman et al., 2012). When stabilized with some phosphoric acid and left undisturbed in an opaque bottle, an H_2O_2 solution can have a shelf-life of years. Opaque bottles are used because light photolytically induces the disproportionation.

18.3.3 Actual pe–pH Diagram for Aqueous Oxygen

Continuing as above, from Table 17.1, at 25 °C/1 atm,

$$O_2 + 4H^+ + 4e^- = 2H_2O \qquad K = \frac{1}{[O_2][H^+]^4\{e^-\}^4} = 10^{86.0} \tag{18.26}$$

where the O_2 is dissolved. For an equation related to the O_2/H_2O boundary line, the equilibrium in Eq.(18.26) gives

$$pe = 21.5 + \tfrac{1}{4}\log[O_2] - pH \qquad {}^{O_2}\!/_{H_2O\ @\ x_{H_2O}\ \approx\ 1}. \tag{18.29}$$

As pe increases, $[O_2]$ increases very rapidly. For the same reason used for setting $[H_2O_2] = 1\ M$ to obtain Eq.(18.25), we take $[O_2] = 1\ M$, and leave $x_{H_2O} \approx 1$. This gives

$$pe = 21.5 - pH \qquad {}^{O_2\ @\ 1\ M}\!/_{H_2O\ @\ x_{H_2O}\ \approx\ 1}. \tag{18.30}$$

This line is plotted in Figure 18.7, and lies above the line for Eq.(18.19), so dissolved O_2 cannot reach 1 M inside the water-stability limit posed by a system pressure of 1 atm. Indeed, consider that $[O_2] = K_H^{O_2} p_{O_2}$. At 25 °C, $K_H^{O_2}$ =-$10^{-2.88}$ M/atm, so if $[O_2]$ were 1 M, then $p_{O_2} = 1\ M/K_H^{O_2} = 10^{2.88}$ atm, clarifying our understanding that the line for Eq.(18.30) lies above that the 1 atm pressure limit line (Eq.(18.19)). We finish the diagram (Figure 18.7) by also including, as we always do, the water-stability lower-limit line for a system pressure of 1 atm,

$$pe = -pH \qquad (p_{H_2} = 1\ \text{atm}). \tag{18.20}$$

18.4 pe–pH DIAGRAM FOR AQUEOUS HYDROGEN

For hydrogen, we consider two oxidation states, H(I) as in H_2O (mostly, with minor additional amounts as H^+ and $-OH^-$), and H(0) as in dissolved H_2. Accordingly, Figure 18.8 gives a preliminary-pe–pH diagram for hydrogen.

From Table 17.1, at 25 °C/1 atm, neglecting activity corrections,

$$2H^+ + 2e^- = H_2 \qquad K = \frac{[H_2]}{[H^+]^2\{e^-\}^2} = 10^{-3.11} \tag{18.31}$$

where $[H_2]$ is the dissolved $[H_2]$ concentration. For an equation related to the H_2O/H_2 boundary line, the equilibrium in Eq.(18.31) gives

$$pe = -3.11 - \log[H_2] - pH \qquad {}^{H_2O\ @\ x_{H_2O}\ \approx\ 1}\!/_{H_2}. \tag{18.32}$$

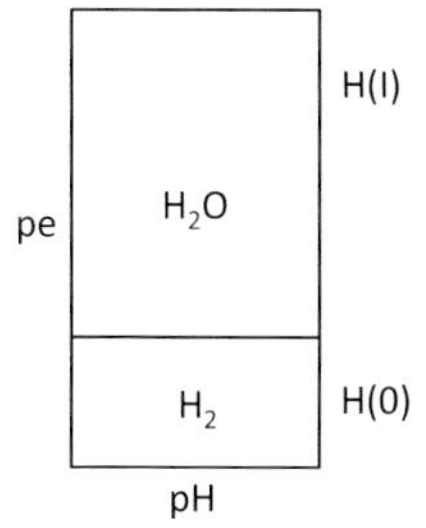

FIGURE 18.8 Preliminary pe–pH predominance diagram for aqueous hydrogen. The redox boundary line shown may not in fact be horizontal. In the final diagram, the H_2O/H_2 boundary line will be found to lie below the lower water stability limit line for a total pressure of 1 atm, and $p_{H_2} = 1$ atm.

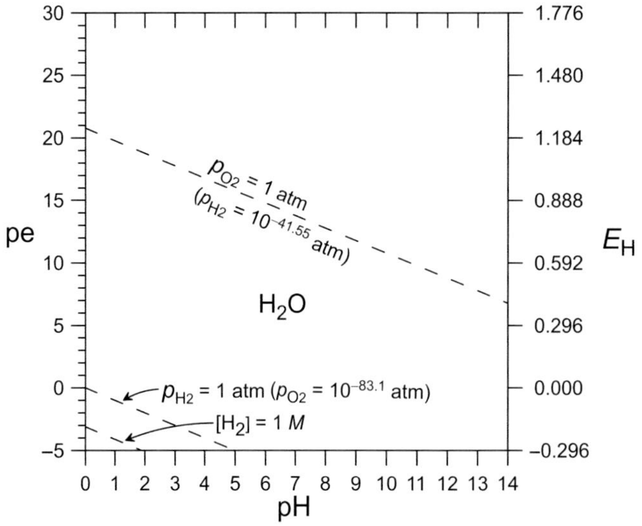

FIGURE 18.9 pe–pH predominance diagram for aqueous hydrogen at 25 °C/1 atm. H_2O is the predominant form of hydrogen within the stability limit lines for a system pressure of 1 atm.

As pe decreases, $[H_2]$ increases very rapidly. For the same reason used for setting $[H_2O_2] = 1\ M$ to obtain Eq.(18.25), we take $[H_2] = 1\ M$, and leave $x_{H_2O} \approx 1$. This gives

$$\text{pe} = -3.11 - \text{pH} \qquad H_2O\ @\ x_{H_2O} \approx 1 \Big/ H_2\ @\ 1\ M. \tag{18.33}$$

This line is plotted in Figure 18.9, and lies below the line for Eq.(18.19), so dissolved H_2 cannot reach 1 M inside the water-stability limit posed by a system pressure of 1 atm. Indeed, consider that $[H_2] = K_H^{H_2} p_{H_2}$. Since $K_H^{H_2}$ - $10^{-3.11}$ M/atm at 25 °C, if $[H_2]$ were 1 M, then $p_{H_2} = 1\ M/K_H^{H_2} = 10^{3.11}$ atm, clarifying our understanding that the line for Eq.(18.33) lies below that for Eq.(18.20). We finish the diagram (Figure 18.9) by including, as we always do, the water-stability upper-limit line for a system pressure of 1 atm,

$$\text{pe} = 20.78 - \text{pH} \qquad (p_{O_2} = 1\ \text{atm}). \tag{18.19}$$

18.5 THE MUTUAL EXCLUSION OF O_2 AND H_2

The relationship between $[O_2]$ and $[H_2]$ at 25 °C/1 atm is obtained by combining

$$2\left(2H^+ + 2e^- = H_{2(g)} \qquad \log K = -0\right) \tag{18.34}$$

$$-\left(O_{2(g)} + 4H^+ + 4e^- = 2H_2O \qquad \log K = 83.1\right) \tag{18.26}$$

$$2H_2O = 2H_{2(g)} + O_{2(g)} \qquad \log K = -83.1. \tag{18.35}$$

With the activity of liquid water as usual taken as $x_{H_2O} \approx 1$, Eq.(18.35) yields

$$p_{H_2}^2 p_{O_2} = 10^{-83.1} \tag{18.36}$$

$$2\log p_{H_2} + \log p_{O_2} = -83.1. \tag{18.37}$$

When $p_{O_2} = 1$ atm, then at equilibrium at 25 °C/1 atm $p_{H_2} = 10^{-41.55}$ atm. When $p_{O_2} = 0.21$ atm (the ambient atmospheric value), at equilibrium $p_{H_2} = 10^{-41.21}$ atm; it is not surprising then that there is little H_2 in the earth's atmosphere. When $p_{H_2} = 1$ atm, at equilibrium at 25 °C/1 atm, $p_{O_2} = 10^{-83.1}$ atm. This is why liquid oxygen and liquid hydrogen are used to propel rockets: the driving force to combine is very high.

Example 18.3 Mutual Exclusion of H_2 and O_2

At pH = 7.0, the difference in pe between the upper- and lower-water-stability limit lines is (20.775 – 7.0) – (–7.0) = 20.775. (Three figures beyond the decimal are used in 20.775, which originates as 83.1/4.) Divide the distance into 5 fractional intervals. Make a table of $\log p_{H_2}$ and $\log p_{O_2}$ at the points corresponding to the ends of each of those intervals.

Solution

We use $\log p_{O_2} = -83.1 + 4\text{pH} + 4\text{pe}$, and
$2\log p_{H_2} + \log p_{O_2} = -83.1$ (Eq.18.37).

	fraction	pe	$\log p_{O_2}$	$\log p_{H_2}$
a.	0	13.775	0.0	–41.55
b.	0.2	9.620	–16.62	–33.24
c.	0.4	5.465	–33.24	–24.93
d.	0.6	1.310	–49.86	–16.62
e.	0.8	–2.845	–66.48	–8.31
f.	1.0	–7.000	–83.10	0.0

REFERENCES

Brookins DG (2012) *Eh-pH Diagrams for Geochemistry*, Springer Science & Business Media, New York.

Garrels R, Christ CL (1965) *Solutions, Minerals, and Equilibria*, Harper and Row, New York, 450 pages.

Juven BJ, Pierson MD, (1996) Antibacterial effects of hydrogen peroxide and methods for its detection and quantitation. *Journal of Food Protection*, **59**, 1233–1241.

Linley E, Denyer SP, McDonnell G, Simons C, Maillard J-Y (2012) Use of hydrogen peroxide as a biocide: new consideration of its mechanisms of biocidal action. *Journal of Antimicrobial Chemotherapy*, **67**, 1589–1596.

Loo, AEK, Wong YT, Ho R, Wasser M, Du T, Ng WT, Halliwell B, Sastre J (2012) Effects of hydrogen peroxide on wound healing in mice in relation to oxidative damage. *PLoS ONE*, **7**, e49215.

Pourbaix M (1974) *Atlas of Electrochemical Equilibria in Aqueous Solutions*, 2nd English edn, National Association of Corrosion Engineers, Houston, TX.

Stumm W, Morgan JJ (1996) *Aquatic Chemistry: Chemical Equilibria and Rates in Natural Waters*, 3rd edn, Wiley Publishing Co., New York.

Takeno N (2005) *Atlas of Eh-pH Diagrams Intercomparison of Thermodynamic Databases*, Geological Survey of Japan, Open File Report No.419, National Institute of Advanced Industrial Science and Technology Research Center for Deep Geological Environments.

Wilgus TA, Bergdall VK, Dipietro LA, Oberyszyn TM (2005) Hydrogen peroxide disrupts scarless fetal wound repair. *Wound Repair and Regeneration*, **13**, 513–529.

Wittmann C, Chockley P, Singh SK, Pase L, Lieschke GJ, Grabher C (2012) Hydrogen peroxide in inflammation: Messenger, guide, and assassin (Review Article). *Advances in Hematology*, **2012**, Article ID 541471.

19 pe–pH Diagrams for Lead (Pb) with Negligible Dissolved CO_2

19.1 INTRODUCTION

For the pe–pH diagram for chlorine considered in Chapter 18, we did not consider the possibility of solid forms of chlorine, only dissolved forms. Solid salts of hypochlorite and chloride (*e.g.*, NaOCl and $NaCl_{(s)}$ respectively) are too soluble to precipitate, except as can occur for NaCl under the strongly evaporative conditions of salt flats in closed basin systems. In contrast, for metal elements that have +II, +III, and +IV oxidation states, one or more solids may well form under natural conditions, and increasing the oxidation state greatly increases the likelihood that a solid will form.

Lead (Pb) is an environmentally relevant, toxic, redox-active metal element that can form solids. A question of considerable importance pertains then to the historical use of lead pipes for connecting water mains to points of service: if a $Pb_{(s)}$ pipe corrodes, how soluble will the lead corrosion products be under the conditions in the water?

Here we discuss preparation of the pe–pH predominance diagrams for Pb at low C_T for $Pb_{T,sys}$ values of 10^{-2}, 10^{-4}, and 10^{-7} *M*, where for $Pb_{T,sys}$ values *M* means mols of Pb (solid + dissolved) per liter of system water. As in Chapter 10, the subscript "T" *without* the added subscript "sys" refers to a dissolved-only value. We here consider the low C_T situation first, so as to not tackle too much all at once; Chapter 20 will address the added complexity that results when dissolved CO_2 is present so that carbonate solids of Pb(II) may form, and when phosphate may also be present so that phosphate solids of Pb(II) may form.

General treatments of the redox chemistry of Pb are provided by Garrels and Christ (1965), Hem (1976), and Stumm and Morgan (1996). Treatments addressing the Pb-lead-pipe context are provided by Schock et al. (1996) and Lytle and Schock (2005). A lengthy review of the environmental chemistry of numerous metals is provided by Langmuir et al. (2004).

19.2 pe–pH DIAGRAM FOR $Pb_{T,sys} = 10^{-2}$ *M*

19.2.1 General – Pb Has Three Important Oxidation States IV, II, and 0

The relevant chemical equilibria are given below. The *K* values pertain to 25 °C/1 atm. Whenever it appears in a *K* expression, we will take $\{H_2O\} = 1$.

$$PbO_{2(s)} + 4H^+ + 2e^- = 2H_2O + Pb^{2+} \qquad \log K = 49.2 \quad \text{(redox: Pb(IV) to Pb(II))} \tag{19.1}$$

$$PbO_{2(s)} + H_2O = PbO_3^{2-} + 2H^+ \qquad \log K = -31.03 \quad \text{(solubility of Pb(IV))} \tag{19.2}$$

$$Pb^{2+} + 2e^- = Pb_{(s)} \qquad \log K = -4.26 \quad \text{(redox: Pb(II) to Pb(0))} \tag{19.3}$$

$$\alpha\text{-}PbO_{(s)} + 2H^+ = Pb^{2+} + H_2O \qquad \log {}^*K_{s0} = 12.60 \quad \text{(solubility of Pb(II))} \tag{19.4}$$

$$Pb^{2+} + H_2O = PbOH^+ + H^+ \qquad \log {}^*K_{H1} = -7.22 \quad \text{(complexation of dissolved Pb(II))} \tag{19.5}$$

$$PbOH^+ + H_2O = Pb(OH)_2^o + H^+ \quad \log{}^*K_{H2} = -9.69 \quad \text{(complexation of dissolved Pb(II))} \tag{19.6}$$

$$Pb(OH)_2^o + H_2O = Pb(OH)_3^- + H^+ \quad \log{}^*K_{H3} = -11.17 \quad \text{(complexation of dissolved Pb(II))} \tag{19.7}$$

$$Pb(OH)_3^- + H_2O = Pb(OH)_4^{2-} + H^+ \quad \log{}^*K_{H4} = -11.64 \quad \text{(complexation of dissolved Pb(II))} \tag{19.8}$$

As in Chapters 17 and 18, we will neglect activity corrections for dissolved species (except for electron, for which activity corrections are built into the definition of pe); they could be included by substituting ${}^cK$ values.

Pb(IV), Pb(II), and Pb(0) are the only oxidation states to be considered. If any amount of Pb(III) were synthesized and placed in an aqueous system, virtually all of it would disproportionate to become Pb(IV) and Pb(II). An analogous statement can be made about Pb(I).

For Pb(0) we have $Pb_{(s)}$, which is essentially insoluble: in solution, the species Pb^o, $PbOH^-$, etc. are of negligible importance.

For Pb(II) (which is the "first stop" in the corrosion of Pb(0)) the species Pb^{2+}, $PbOH^+$, $Pb(OH)_2^o$, $Pb(OH)_3^-$, and $Pb(OH)_4^{2-}$ are of interest. Their concentrations can be limited by the solubility Pb(II) oxides and hydroxides. The Pb(II) solid emphasized in this chapter is α-$PbO_{(s)}$ (litharge, a.k.a. "red $PbO_{(s)}$"). pe–pH diagrams could also be drawn for Pb considering β-$PbO_{(s)}$ (massicot, a.k.a. "yellow $PbO_{(s)}$"), but at 25 °C/1 atm α-$PbO_{(s)}$ is slightly less soluble than β-$PbO_{(s)}$ and so is the thermodynamically more stable form.

For Pb(IV), two forms of $PbO_{2(s)}$ are known, plattnerite (β-$PbO_{2(s)}$) and scrutinyite (α-$PbO_{2(s)}$). In lead pipes, both have been reported as forming: 1) directly on the $Pb_{(s)}$ (Lytle and Schock, 2005); and 2) on Pb(II) mineral solids that directly overlie the $Pb_{(s)}$ (Wang et al., 2010). It is believed that chlorinated drinking water containing "free chlorine" ($HOCl/OCl^-$) can favor $PbO_{2(s)}$ formation in some circumstances because of the highly oxidizing character of free chlorine (Lytle and Schock, 2005; Wang et al., 2010). Both α- and β-$PbO_{2(s)}$ are very insoluble as Pb(IV), except at very high pH where PbO_3^{2-} can become important (see Eq.(19.2)). Other dissolved Pb(IV) species (Pb^{4+}, $PbOH^{3+}$, etc. are generally thought to be unimportant. The species PbO_3^{2-} can be thought of as forming from $Pb(OH)_6^{2-}$ by loss of $3H_2O$. (Actually, pH considerations alone would not allow PbO_3^{2-} and $Pb(OH)_6^{2-}$ to be distinguished from one another, but we do not have to worry about such fine points here.)

For intermediate pH, and under oxidizing conditions where Pb(IV) is dominant, it has been argued that the low solubility of α- and β-$PbO_{2(s)}$ scales inside lead plumbing is helpful in minimizing the dissolution of lead into tap water (Lytle and Schock, 2005; Wang et al., 2010; Triantafyllidou et al., 2015). Here, "low solubility" of $PbO_{2(s)}$ refers to: 1) its negligible dissolution as Pb(IV) except at high pH; and 2) when the pe is high, it low reductive solubility as Pb(II). However, if highly oxidizing conditions are not maintained, and organic compounds are present and thus capable of serving as reductants, both α- and β-$PbO_{2(s)}$ may undergo reductive dissolution (Eq.(19.1)) thereby forming the more soluble Pb(II)). The relative thermochemical stabilities (ΔG_f^o values) of α-$PbO_{2(s)}$ *vs.* β-$PbO_{2(s)}$ are not well understood, so this text will not particularize the form when referring to $PbO_{2(s)}$, and will rely on generic thermochemical data for "$PbO_{2(s)}$".

In a pe–pH diagram with solution/solid boundary lines, the total amount of element in the system is the T,sys value. We begin by considering $Pb_{T,sys} = 10^{-2}$ M. This seems like a high value, but not when a piece of lead pipe is being exposed to adjacent water. $Pb_{T,sys} = 10^{-2}$ M is high enough that $PbO_{2(s)}$, α-$PbO_{(s)}$, and $Pb_{(s)}$ all have regions inside the water-stability domain (*i.e.*, between the lines for $p_{O_2} = 1$ atm and $p_{H_2} = 1$ atm).

19.2.2 The Two Vertical Solution $|\alpha\text{-}PbO_{(s)}|$ Solution Boundary Lines and the Vertical Solution/Solution Boundary Line for Pb(II)

We focus first on intermediate pe values where α-$PbO_{(s)}$ will be found when $Pb_{T,sys} = 10^{-2}$ *M*. At intermediate pe, essentially all the dissolved Pb is Pb(II) because Pb(0) is considered insoluble, and because dissolution of Pb(IV) from $PbO_{2(s)}$ to form $Pb_2O_3^-$ by Eq.(19.3) can be neglected:

$$Pb_T = Pb(II)_T = [Pb^{2+}] + [PbOH^+] + [Pb(OH)_2^o] + [Pb(OH)_3^-] + [Pb(OH)_4^{2-}]. \quad (19.9)$$

As discussed in Section 10.3 and Box 19.1, $^*K_H$ values act like acidity *K* values for metal ions. Therefore,

$$\alpha_0^{Pb(II)} = \frac{[Pb^{2+}]}{Pb(II)_T} = \frac{1}{1+\frac{^*K_{H1}}{[H^+]}+\frac{^*K_{H1}{}^*K_{H2}}{[H^+]^2}+\frac{^*K_{H1}{}^*K_{H2}{}^*K_{H3}}{[H^+]^3}+\frac{^*K_{H1}{}^*K_{H2}{}^*K_{H3}{}^*K_{H4}}{[H^+]^4}}. \quad (19.10)$$

From Eqs.(19.5)–(19.8), at 25 °C/1 atm, $^*K_{H1} = 10^{-7.22}$, $^*K_{H2} = 10^{-9.69}$, $^*K_{H3} = 10^{-11.17}$, and $^*K_{H4} = 10^{-11.64}$. Eq.(19.10) is written more economically as

$$\alpha_0^{Pb(II)} = \frac{[Pb^{2+}]}{Pb(II)_T} = \frac{1}{1+\frac{^*\beta_{H1}}{[H^+]}+\frac{^*\beta_{H2}}{[H^+]^2}+\frac{^*\beta_{H3}}{[H^+]^3}+\frac{^*\beta_{H4}}{[H^+]^4}} \quad (19.11)$$

where at 25 °C/1 atm, $^*\beta_1 \equiv {}^*K_{H1} = 10^{-7.22}$, $^*\beta_2 \equiv {}^*K_{H1}{}^*K_{H2} = 10^{-16.91}$, $^*\beta_3 \equiv {}^*K_{H1}{}^*K_{H2}{}^*K_{H3} = 10^{-28.08}$, and $^*\beta_4 \equiv {}^*K_{H1}{}^*K_{H2}{}^*K_{H3}{}^*K_{H4} = 10^{-39.72}$. If $[Pb^{2+}]$ is set by saturation equilibrium with α-$PbO_{(s)}$, then by Eq.(19.4),

$$Pb(II)_T = {}^*K_{s0}[H^+]^2 \frac{1}{\alpha_0^{Pb(II)}}. \quad (19.12)$$

As noted above, at intermediate pe and/or pH values, in solution essentially all dissolved Pb is Pb(II). On the boundary line, $Pb(II)_T = Pb_T = Pb_{T,sys} = 10^{-2}$ *M*. There is no movement of electrons when Pb(II) in α-$PbO_{(s)}$ dissolves, or precipitates from Pb(II) in solution: a boundary line between Pb(II) in solution and Pb(II) in α-$PbO_{(s)}$ does not depend on pe. In the pe–pH diagram at intermediate pe (where PbO_3^{2-} can be neglected), the solid has boundary lines with the solution at two pH values:

$$Pb_{T,sys} = {}^*K_{s0}[H^+]^2 \frac{1}{\alpha_0^{Pb(II)}} \quad (19.13)$$

or equivalently,

$$Pb_{T,sys} = {}^*K_{s0}[H^+]^2 + {}^*K_{s1}[H^+] + K_{s2} + \frac{^*K_{s3}}{[H^+]} + \frac{^*K_{s4}}{[H^+]^2} \qquad \text{solution} \overset{pH=7.55}{|} \alpha\text{-}PbO_{(s)} \overset{pH=12.53}{|} \text{solution}. \quad (19.14)$$

From Eq.(19.4), at 25 °C/1 atm, for α-$PbO_{(s)}$, $^*K_{s0} = 10^{12.60}$. From Eq.(19.5–19.8), then: $^*K_{s1} = {}^*K_{s0}{}^*\beta_{H1} = 10^{5.38}$; $K_{s2} = {}^*K_{s0}{}^*\beta_{H2} = 10^{-4.31}$; $^*K_{s3} = {}^*K_{s0}{}^*\beta_{H3} = 10^{-15.48}$, $^*K_{s4} = {}^*K_{s0}{}^*\beta_{H4} = 10^{-27.12}$. Determining the two pH values proceeds as in the conceptually identical cases in Chapter 11 for the low and high pH roots for dissolution/precipitation of a metal oxide (or hydroxide, see Figure 11.7). For $Pb_{T,sys} = 10^{-2}$ *M*, the two roots of Eq.(19.13/19.14) are pH = 7.55 and pH = 12.53.

We need to label the solution regions on either side of the α-$PbO_{(s)}$ region. Consider the acid/base α values introduced in Chapter 11, and given specifically for Pb(II) in Box 19.1 when neglecting

carbonate-related species. Figure 19.1 is a plot for the five $\alpha^{\mathrm{Pb(II)}}$ *vs.* pH. At pH = 7.55, $PbOH^+$ is the dominant solution species, so for the region in the pe–pH diagram to the immediate left of pH = 7.55, $PbOH^+$ is the labeled species (at intermediate pe). For the region pH > 12.53, $Pb(OH)_4^{2-}$ is the labeled species for the diagram (at intermediate pe).
For equilibrium between Pb^{2+} and $PbOH^+$, from Eq.(19.5),

$$\log[\mathrm{PbOH^+}] = -7.22 + \mathrm{pH} + \log[\mathrm{Pb^{2+}}]. \quad (19.15)$$

For the predominance boundary line between Pb^{2+} and $PbOH^+$, we take $[Pb^{2+}] = [PbOH^+]$ so that

$$\mathrm{Pb^{2+}} \;\overset{\mathrm{pH}=7.22}{\Big|}\; \mathrm{PbOH^+}. \quad (19.16)$$

Pb^{2+} has a dominance region for pH < 7.22. The pe–pH diagram for $\mathrm{Pb_{T,sys}} = 10^{-2}$ *M* and low C_T will thus appear roughly as shown in Figure 19.2 (There is no wedge limit on pH for dominance by Pb(II) because Pb(II) does not disproportionate to Pb(IV) + Pb(0), as would be analogous to the disproportionation of Cl(0) to Cl(I) + Cl (–I)). Line (3) is pH = 7.22, line (4) is "pH = 7.55", and line (5) is "pH = 12.53". These lines are plotted in Figure 19.3. Directly on line (4) and on line (5), there is no α-$PbO_{(s)}$ present, but the solution is saturated for $\{\alpha\text{-}PbO_{(s)}\} = 1$. From either line, movement by the slightest amount towards the interior the α-$PbO_{(s)}$ region will cause some α-$PbO_{(s)}$ to precipitate.

BOX 19.1 Acid/Base $\alpha^{\mathrm{Pb(II)}}$ Values

Functions for the $\alpha^{\mathrm{Pb(II)}}$ values (neglecting activity corrections for dissolved species). All *K* values at 25 °C/1 atm:

$^*K_{H1} = 10^{-7.22}$, $^*K_{H2} = 10^{-9.69}$, $^*K_{H3} = 10^{-11.17}$, $^*K_{H4} = 10^{-11.64}$.
$^*\beta_{H1} = 10^{-7.22}$, $^*\beta_{H2} = 10^{-16.91}$, $^*\beta_{H3} = 10^{-28.08}$, $^*\beta_{H4} = 10^{-39.72}$.
$^*\beta_{H1} = {}^*K_{H1} = 10^{-7.22}$, $^*\beta_{H1} = {}^*K_{H1}{}^*K_{H2} = 10^{-16.91}$, $^*\beta_{H3} = {}^*K_{H1}{}^*K_{H2}{}^*K_{H3} = 10^{-28.08}$,
$^*\beta_{H4} = {}^*K_{H1}{}^*K_{H2}{}^*K_{H3}{}^*K_{H4} = 10^{-39.72}$.

definition	expression
$[\mathrm{Pb^{2+}}] = \alpha_0^{\mathrm{Pb(II)}}\mathrm{Pb(II)_T}$	$\alpha_0^{\mathrm{Pb(II)}} = \dfrac{[\mathrm{Pb^{2+}}]}{[\mathrm{Pb^{2+}}]+[\mathrm{PbOH^+}]+[\mathrm{Pb(OH)_2^o}]+[\mathrm{Pb(OH)_3^-}]} = \left(1+\dfrac{^*\beta_{H1}}{[\mathrm{H^+}]}+\dfrac{^*\beta_{H2}}{[\mathrm{H^+}]^2}+\dfrac{^*\beta_{H3}}{[\mathrm{H^+}]^3}+\dfrac{^*\beta_{H4}}{[\mathrm{H^+}]^4}\right)^{-1}$
$[\mathrm{PbOH^+}] = \alpha_1^{\mathrm{Pb(II)}}\mathrm{Pb(II)_T}$	$\alpha_1^{\mathrm{Pb(II)}} = \alpha_0^{\mathrm{Pb(II)}}\dfrac{^*K_{H1}}{[\mathrm{H^+}]}$
$[\mathrm{Pb(OH)_2^o}] = \alpha_2^{\mathrm{Pb(II)}}\mathrm{Pb(II)_T}$	$\alpha_2^{\mathrm{Pb(II)}} = \alpha_1^{\mathrm{Pb(II)}}\dfrac{^*K_{H2}}{[\mathrm{H^+}]}$
$[\mathrm{Pb(OH)_3^-}] = \alpha_3^{\mathrm{Pb(II)}}\mathrm{Pb(II)_T}$	$\alpha_3^{\mathrm{Pb(II)}} = \alpha_2^{\mathrm{Pb(II)}}\dfrac{^*K_{H3}}{[\mathrm{H^+}]}$
$[\mathrm{Pb(OH)_4^{2-}}] = \alpha_4^{\mathrm{Pb(II)}}\mathrm{Pb(II)_T}$	$\alpha_4^{\mathrm{Pb(II)}} = \alpha_3^{\mathrm{Pb(II)}}\dfrac{^*K_{H4}}{[\mathrm{H^+}]}$

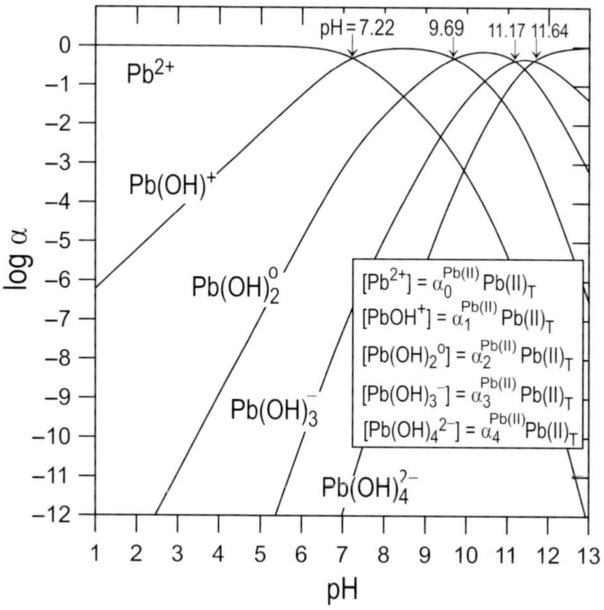

FIGURE 19.1 Log α values *vs.* pH for dissolved Pb(II) species at 25 °C/1 atm (see Box 19.1). These α apply for dissolved Pb(II) regardless of whether any solid is present. Pb^{2+} predominates for pH < 7.22; $PbOH^+$ predominates for 7.22 < pH < 9.69; $Pb(OH)_2^0$ predominates for 9.69 < pH < 11.17; $Pb(OH)_3^-$ predominates for 11.17 < pH < 11.64; and $Pb(OH)_4^{2-}$ predominates for pH > 11.64.

We have not yet identified the pe values at which lines (3), (4), and (5) each start and stop. For each line, the upper pe will be given by its intersection with the line above, and the lower pe will be given by its intersection with the line below. We need to know where the other lines lie.

Example 19.1 Saturation with Respect to a Solid on a Solution/Solid Boundary Line (A Solution POV Example)

Considering Figure 19.3 (25 °C/1 atm), compute $[Pb^{2+}]$ at point E where pH = 7.55 and for $Pb_{T,sys} = 10^{-2}$ *M*, and show that the solution is exactly saturated with α-$PbO_{(s)}$ at that pH according to Eq.(19.4). Neglect activity corrections for dissolved species.

Solution

Point E is on the boundary line at pH = 7.55. $\alpha_0^{Pb(II)} = 0.317$. $Pb(II)_T = Pb_{T,sys}$. $[Pb^{2+}] = \alpha_0^{Pb(II)} Pb_{T,sys} = 0.317 \times 10^{-2}\ M = 3.17 \times 10^{-3}\ M$. At saturation with α-$PbO_{(s)}$, $[Pb^{2+}]/[H^+]^2 = {}^*K_{s0} = 10^{12.60}$. At pH = 7.55, $[Pb^{2+}]/[H^+]^2 = 3.17 \times 10^{-3}/(10^{-7.55})^2 = 3.99 \times 10^{12} = 10^{12.60}$. The solution is exactly saturated with α-$PbO_{(s)}$. Raising the pH by any amount will cause some α-$PbO_{(s)}$ to form, causing Pb_T in solution to fall below 10^{-2} *M* (at equilibrium).

19.2.3 The Curved $PbO_{2(s)}$/Solution Boundary Line

In a manner analogous to the three non-redox solution phase/$FeCO_{3(s)}$ boundary lines in Figures 14.2 through 14.4, the three sections labeled as line (1) in Figure 19.2 are portions of the

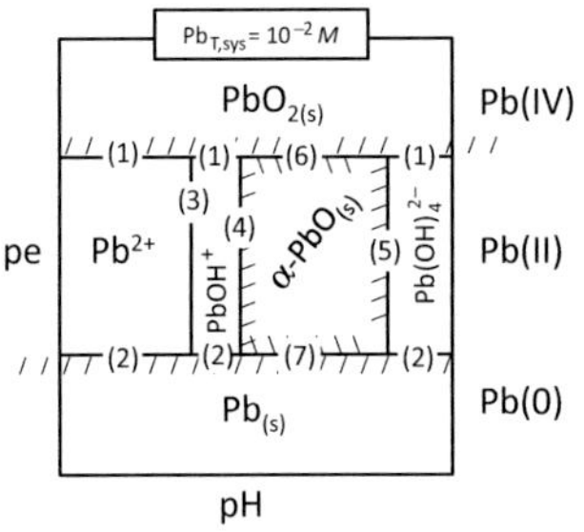

FIGURE 19.2 Preliminary pe–pH diagram for $Pb_{T,sys} = 10^{-2}$ *M* and 25 °C/1 atm assuming that $PbOH^+$ and $Pb(OH)_4^{2-}$ are the dominant, dissolved phase Pb(II) species immediately to the left and right of the region for α-$PbO_{(s)}$. Pb^{2+} is dominant at $pH < p^*K_{H1} = 7.22$; $Pb(OH)_4^{2-}$ is dominant at $pH > p^*K_{H4} = 11.64$.

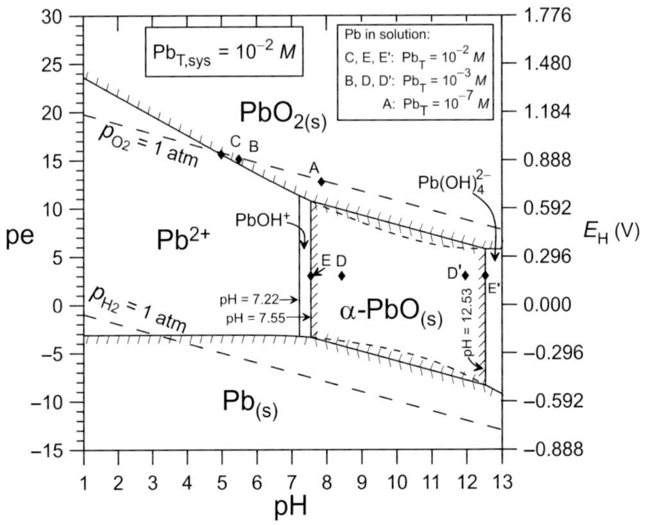

FIGURE 19.3 pe–pH predominance diagram for $Pb_{T,sys} = 10^{-2}$ *M* at 25 °C/1 atm and low C_T. Assumptions: 1) all solution-phase activity coefficients are unity; 2) when a solid is present, it has unit activity.

same, curved, $PbO_{2(s)}$/solution boundary line along which the dissolved $Pb_T = Pb_{T,sys} = 10^{-2}$ *M*. At high pe and pH, a term for $Pb(IV)_T$ may contribute to Pb_T:

$$Pb_T = Pb(II)_T + Pb(IV)_T. \tag{19.17}$$

Saturation equilibrium between $PbO_{2(s)}$ and Pb(II) is a reductive dissolution equilibrium where $\{e^-\}^2$ appears in the solubility product that corresponds to Eq.(19.1). At 25 °C/1 atm,

$$2pe = 49.2 + \log\{PbO_{2(s)}\} - \log[Pb^{2+}] - 4pH \tag{19.18}$$

Taking $\{PbO_{2(s)}\} = 1$, using Eq.(19.11) gives

$$2pe = 49.2 - \log Pb(II)_T - \log \alpha_0^{Pb(II)} - 4pH. \tag{19.19}$$

At high pe and high pH, PbO_3^{2-} (the one Pb(IV) species we are considering) cannot be neglected, so by Eq.(19.17) with $Pb(IV)_T = [PbO_3^{2-}]$,

$$Pb_T = Pb(II)_T + [PbO_3^{2-}]. \tag{19.20}$$

By the equilibrium in Eq.(19.2), at 25 °C/1 atm and taking $\{PbO_{2(s)}\} = 1$,

$$[PbO_3^{2-}] = \frac{10^{-31.03}}{[H^+]^2} \tag{19.21}$$

$$Pb(II)_T = Pb_T - \frac{10^{-31.03}}{[H^+]^2} \tag{19.22}$$

$$2pe = 49.2 - \log\left(Pb_T - \frac{10^{-31.03}}{[H^+]^2}\right) - \log \alpha_0^{Pb(II)} - 4pH. \tag{19.23}$$

Directly on line (1), $Pb_T = Pb_{T,sys}$, so $PbO_{2(s)}$ is not present, but the solution is saturated for $\{PbO_{2(s)}\} = 1$:

$$2pe = 49.2 - \log\left(Pb_{T,sys} - \frac{10^{-31.03}}{[H^+]^2}\right) - \log \alpha_0^{Pb(II)} - 4pH. \quad \frac{PbO_{2(s)}}{\text{solution}}. \tag{19.24}$$

Taking a range of pH values, one can solve for the corresponding pe values to obtain the desired, curved $PbO_{2(s)}$/solution boundary line. This is easier than taking a range of pe values and solving for the corresponding pH values. (When $Pb_{T,sys}$ is low enough (not 10^{-2} *M*), at pH < 14 the term $-10^{-31.03}/[H^+]^2$ can cause the line in the diagram to abruptly turn up as the $PbO_{2(s)}$ becomes soluble as $Pb_2O_3^-$.) Line (1) pertains to pH ≤ 7.55 and pH ≥ 12.53. These portions are plotted in Figure 19.3. The portion between the two pH values is plotted as a dashed line because it is inside the α-$PbO_{(s)}$ region where $PbO_{2(s)}$ does not exist at equilibrium. Between the 1/4/6 intersection and the 6/5/1 intersection, we need to plot a $PbO_{2(s)}/\alpha$-$PbO_{(s)}$ boundary line (line (6)).

Example 19.2 Saturation with Respect to a Solid inside a Solid Predominance Region (A Solid POV Example)

Considering Figure 19.3, compute $[Pb^{2+}]$ (*M*) and Pb_T (*M* and μg/L) expected at equilibrium for pH = 8.00, p_{O_2} = 0.21 atm, and 25 °C/1 atm. Neglect activity corrections for dissolved species.

Solution

Eq.(17.111) gives $pe = 20.78 - pH + \frac{1}{4}\log p_{O_2}$. $pe = 20.78 - 8.00 + \frac{1}{4}\log(0.21) = 12.61$. Eq.(19.18) gives $2pe = 49.2 + \log\{PbO_{2(s)}\} - \log[Pb^{2+}] - 4pH$. For $\{PbO_{2(s)}\} = 1$, then $[Pb^{2+}] = 9.55 \times 10^{-9}$ *M*. $Pb_T = [Pb^{2+}]/\alpha_0^{Pb(II)}$. At pH = 8.00, $\alpha_0^{Pb(II)} = 0.140$. $Pb_T = 6.83 \times 10^{-8}$ *M*, or 14.2 μg/L.

19.2.4 The Straight $PbO_{2(s)}/\alpha$-$PbO_{(s)}$ Boundary Line

Line (6) can be obtained at 25 °C/1 atm by combining

$$2pe = 49.2 + \log\{PbO_{2(s)}\} - \log[Pb^{2+}] - 4pH \quad (19.18) \quad \frac{PbO_{2(s)}}{Pb^{2+}}$$

$$-\left(\log\{Pb^{2+}\} = 12.60 - 2pH + \log\{\alpha\text{-}PbO_{(s)}\}\right) \quad (19.25) \quad \frac{Pb^{2+}}{\alpha\text{-}PbO_{(s)}}$$

$$2pe = 36.60 - 2pH + \log\{PbO_{2(s)}\} - \log\{\alpha\text{-}PbO_{(s)}\}. \quad (19.26) \quad \frac{PbO_{2(s)}}{\alpha\text{-}PbO_{(s)}}$$

Eq.(19.25) originates in Eq.(19.4). When $\{PbO_{2(s)}\}$ and $\{\alpha\text{-}PbO_{(s)}\}$ = 1, then Eq.(19.26) reduces to

$$pe = 18.30 - pH \qquad \frac{PbO_{2(s)}}{\alpha\text{-}PbO_{(s)}} \tag{19.27}$$

which is line (6) at 25 °C/1 atm and is plotted in Figure 19.3. The slope *d*pe/*d*pH = –1 because for every electron consumed in converting $PbO_{2(s)}$ to $\alpha\text{-}PbO_{(s)}$, one proton is consumed ($PbO_{2(s)} + 2e^- + 2H^+ = \alpha\text{-}PbO_{(s)} + H_2O$). Since no solution phase Pb species are involved, Eq.(19.27) is independent of $Pb_{T,sys}$; the plotted starting and stopping points do, however, depend on $Pb_{T,sys}$.

19.2.5 The Curved Solution/$Pb_{(s)}$ Boundary Line

Analogously with line (1) in Figure 19.2, line (2) is one curved, solution phase/$Pb_{(s)}$ boundary line along which the dissolved $Pb_T = Pb_{T,sys} = 10^{-2}$ *M*. The equation for that line can be developed by analogy with Section 19.2.3. For equilibrium between $Pb_{(s)}$ and the solution, at 25 °C/1 atm, by Eq.(19.3)

$$2pe = -4.26 + \log[Pb^{2+}] - \log\{Pb_{(s)}\}. \tag{19.28}$$

Along line (2), the solid is not present, but the solution is saturated (oxidative dissolution) for $\{Pb_{(s)}\} = 1$ and $Pb_T = Pb_{T,sys}$. Taking $Pb_T = Pb_{T,sys}$, combining Eq.(19.11) with Eq.(19.28) gives

$$2pe = -4.26 + \log Pb_{T,sys} + \log \alpha_0^{Pb(II)}. \qquad \frac{\text{solution}}{Pb_{(s)}} \tag{19.29}$$

For a range of pH values, one can solve for the corresponding pe values to obtain the solution/$Pb_{(s)}$ boundary line. For $Pb_{T,sys} = 10^{-2}$ *M*, the low and high pH ranges over which that boundary line will be applicable will be the same as discussed for Eq.(19.24). Eq.(19.29) with $Pb_{T,sys} = 10^{-2}$ *M* does not apply within the pH range that $\alpha\text{-}PbO_{(s)}$ is present.

19.2.6 The Straight α-$PbO_{(s)}$/$Pb_{(s)}$ Boundary Line

For the equilibrium pertaining to line (7), we combine:

$$2pe = -4.26 + \log[Pb^{2+}] - \log\{Pb_{(s)}\} \qquad (19.28) \qquad \frac{Pb^{2+}}{Pb_{(s)}}$$

$$+\log\{Pb^{2+}\} = 12.60 - 2pH + \log\{\alpha\text{-}PbO_{(s)}\} \qquad (19.25) \qquad \frac{\alpha\text{-}PbO_{(s)}}{Pb^{2+}}$$

$$2pe = 8.34 - 2pH + \log\{\alpha\text{-}PbO_{(s)}\} - \log\{Pb_{(s)}\}. \qquad (19.30) \qquad \frac{\alpha\text{-}PbO_{(s)}}{Pb_{(s)}}$$

When $\{\alpha\text{-}PbO_{(s)}\} = 1$ and $\{Pb_{(s)}\} = 1$, Eq.(19.30) reduces to

$$pe = 4.17 - pH. \qquad \frac{\alpha\text{-}PbO_{(s)}}{Pb_{(s)}} \tag{19.31}$$

which is line (7) at 25 °C/1 atm, and is plotted in Figure 19.3. As with Eq.(19.27), Eq.(19.31) does not depend on $Pb_{T,sys}$; its plotted starting and stopping points do depend on $Pb_{T,sys}$. The slope *d*pe/*d*pH = –1 because for every electron that is consumed in the conversion of $\alpha\text{-}PbO_{(s)}$ to $Pb_{(s)}$, one proton is also consumed: $\alpha\text{-}PbO_{(s)} + 2e^- + 2H^+ = Pb_{(s)} + H_2O$.

Example 19.3 Computing Solution Phase Concentrations Using the Solid POV

Considering Figure 19.3 for $Pb_{T,sys} = 10^{-2}$ *M*, the curved solution/$Pb_{(s)}$ boundary line goes through the point pe = –4.95, pH = 10.20, which is inside the region for α-$PbO_{(s)}$. **a.** What would be the dissolved Pb_T for equilibrium (25 °C/1 atm) at that point if it was set by $Pb_{(s)}$? **b.** What is the dissolved Pb_T if it was set by α-$PbO_{(s)}$? **c.** Explain why the values from parts **a** and **b** prove that $Pb_{(s)}$ is not stable relative to α-$PbO_{(s)}$ at that point. Neglect activity corrections for dissolved species.

Solution

a. Since the point is on the solution/$Pb_{(s)}$ boundary line for $Pb_{T,sys} = 10^{-2}$ *M*, the dissolved $Pb_T = 10^{-2}$ *M* if it was set by $Pb_{(s)}$.
b. By Eq.(19.4), as set by α-$PbO_{(s)}$, $\log[Pb^{2+}] = 12.60 - 2pH = 12.60 - 2(10.20) = -7.80$. $Pb_T = [Pb^{2+}]/\alpha_0^{Pb(II)}$. At pH = 10.20, $\alpha_0^{Pb(II)} = 2.28\times10^{-4}$. $Pb_T = 10^{-7.80}$ $M/2.28\times10^{-4} = 6.96\times10^{-5}$ *M*.
c. α-$PbO_{(s)}$ is prescribing a lower value for dissolved Pb_T than is $Pb_{(s)}$, so α-$PbO_{(s)}$ is more stable at this point than is $Pb_{(s)}$.

19.2.7 Implications of the Low C_T pe–pH Predominance Diagrams for Tap Water by Oxidative (Corrosive) Dissolution of Pb(0) in Lead Pipes, Lead-Containing Solder, and Lead-Containing Brass

Dissolved Pb(II) can enter drinking water by the oxidative (corrosive) dissolution of: 1) the lead "water service line" pipes historically used to connect water mains to points of use; 2) copper pipes joined with lead containing solder; and 2) brass fittings containing some lead. Corrosion of lead water service lines is what has caused the Pb-in-drinking-water crisis in Flint, Michigan (Pieper et al., 2017; Olson et al., 2017). The problem surfaced in early 2015 when 104 ppb of lead was found in a sample of tap water from a Flint home: the "ground-zero home". The abstract in the Pieper et al. (2017) paper merits in toto inclusion here:

> Flint, Michigan switched to the Flint River as a temporary drinking water source without implementing corrosion control in April 2014. Ten months later, water samples collected from a Flint residence revealed progressively rising water lead levels (104, 397, and 707 μg/L) coinciding with increasing water discoloration. An intensive follow-up monitoring event at this home investigated patterns of lead release by flow rate–all water samples contained lead above 15 μg/L and several exceeded hazardous waste levels (>5000 μg/L). Forensic evaluation of exhumed service line pipes compared to water contamination "fingerprint" analysis of trace elements, revealed that the immediate cause of the high water lead levels was the destabilization of lead-bearing corrosion rust layers that accumulated over decades on a galvanized iron pipe downstream of a lead pipe. After analysis of blood lead data revealed spiking lead in blood of Flint children in September 2015, a state of emergency was declared and public health interventions (distribution of filters and bottled water) likely averted an even worse exposure event due to rising water lead levels.
>
> **(Reprinted by permission of the authors and the American Chemical Society)**

In Portland, Oregon, most lead service lines had been removed by the late 1990s, but the corrosion of lead-based solder connecting copper pipes, and older brass fittings containing lead, has continued to cause problems of Pb in tap water (Renner, 2010). The Portland problem is caused in part because most drinking water for Portland is from the Bull Run watershed on the east side of the Mt. Hood, a dormant volcano. The rock underlying the watershed is mostly basalt, so the water tends to both low pH value (~7) and low Alk (~1.4×10^{-4} eq/L, 7 mg/L "$CaCO_3$") (City of Portland, 2011). With Alk low, C_T is low, so carbonate Pb(II) solids do not tend to form and thereby passivate Pb surfaces;

current protocols provide for NaOH addition to raise the pH to 8.0, before release to the distribution system (City of Portland, 2016).

Overall, the matter of Pb in drinking water greatly motivates the study of pe–pH diagrams for Pb. Relevant here is where the different regions lie relative to each other, as well as relative to the water stability lines for the oxidation and reduction of water. The latter are included in Figure 19.3 for 25 °C/1 atm, and are

$$pe = -\tfrac{1}{2}\log p_{H_2} - pH \tag{17.104}$$

and

$$pe = 20.78 - pH + \tfrac{1}{4}\log p_{O_2}. \tag{17.111}$$

For $p_{H_2} = 1$ atm, Eq.(17.104) becomes

$$pe = -pH \qquad (p_{H_2} = 1 \text{ atm}). \tag{19.32}$$

For $p_{O_2} = 1$ atm, Eq.(17.111) becomes

$$pe = 20.78 - pH \qquad (p_{O_2} = 1 \text{ atm}). \tag{19.33}$$

In Figure 19.3 for $Pb_{T,sys} = 10^{-2}$ *M*, within the stability field for water at 1 atm, a pe–pH domain for each of the three oxidation states can be found. The region for Pb(IV) as $PbO_{2(s)}$ is present in the upper portion of the field for pH values greater than about 5. Regions for dissolved and solid Pb(II) occupy the middle of field. Pb(0) can be present only under very reducing conditions, when pH is greater than about 3.5. Pb(0), as $Pb_{(s)}$ will certainly not be present at equilibrium in oxic waters: if some $Pb_{(s)}$ were added to an oxic system, it is strongly subject to oxidation. We know from daily life, however, that $Pb_{(s)}$ (as in fishing weights, scuba diving weights, lead-containing solder, and lead pipes) seems quite stable in the presence of air and water. But this is *metastability*: in the presence of oxygen, some oxidation of $Pb_{(s)}$ to Pb(II) and/or Pb(IV) solids occurs, and a layer of oxide, carbonate, and/or hydroxycarbonate solids can coat the $Pb_{(s)}$ surface, protecting ("passivating") the underlying $Pb_{(s)}$ from further oxidation. Scratching the surface exposes the shiny $Pb_{(s)}$ underneath, then the new surface becomes oxidized and thereby re-coated; an analogous process happens with $Fe_{(s)}$, $Al_{(s)}$, and many other elemental metal solids.

Consider points A, B, and C in Figure 19.3. The pe–pH coordinates are given in Table 19.1. For each, $p_{O_2} = 0.21$ atm, so these points are slightly below the line for Eq.(19.33). For the three points, assuming equilibrium with $\{PbO_{2(s)}\} = 1$ gives Pb_T (dissolved) values of 10^{-2} *M*, 10^{-3} *M*, and 10^{-7} *M*, respectively. For dissolved $Pb_T = 10^{-7}$ *M*, with the atomic weight of Pb at 207.2 g/mol, this is equivalent to ~21 µg/L Pb. If a $PbO_{2(s)}$ layer on $Pb_{(s)}$ is intact and effectively separates the $Pb_{(s)}$ from the solution, the very low reductive-dissolution solubility of $PbO_{2(s)}$ is predicted (Lytle and Schock, 2005) and observed (Triantafyllidou et al., 2015) to provide very good protection against contamination from Pb plumbing for oxic waters at neutral to basic pH values, as assisted by some kind of regular flow through the pipes and fittings to further lower the concentration: according to the "Lead and Copper Rule", the current U.S. EPA action level (AL) for Pb in drinking water is 15 µg/L (U.S. EPA, 1991). α-$PbO_{2(s)}$ (scrutinyite) was reported by Reuter et al. (2005) as being found as "scale" on the inside walls of a lead water service line, along with the two Pb(II) oxides litharge and massicot, plus hydrocerussite and cerussite which are carbonate-containing Pb(II) solids to be considered in Chapter 20. For lead scale samples obtained from 15 multiple lead service lines exhumed in Cincinnati over the period 2000 to 2014, Triantafyllidou et al. (2015) observed that in every case "the scale consisted of a nearly pure β-PbO_2 (plattnerite) layer at the scale/water contact, underlain by a thin layer of tetragonal PbO (α-PbO, litharge) adjoining the lead pipe."

TABLE 19.1
Seven Points in Pb pe–pH Diagrams with Three Levels of $Pb_{T,sys}$

point	pe	pH	for $Pb_{T,sys} = 10^{-2}$ *M* (Figure 19.3), then Pb_T (dissolved) =	for $Pb_{T,sys} = 10^{-3}$ *M* (Figure 19.4), then Pb_T (dissolved) =	for $Pb_{T,sys} = 10^{-7}$ *M* (Figure 19.5), then Pb_T (dissolved) =
A	12.758	7.853	10^{-7} *M*	10^{-7} *M*	
B	15.118	5.493	10^{-3} *M*		
C	15.621	4.990			
E	3.0	7.55	10^{-2} *M*		10^{-7} *M*
E'	3.0	12.53		10^{-3} *M*	
D	3.0	8.43	10^{-3} *M*		
D'	3.0	11.96			

It is not known whether the presence of significant dissolved O_2 is sufficiently oxidative to maintain a $PbO_{2(s)}$ surface layer within a lead service line, as would be predicted from Figure 19.3. A problem is that Pb(IV) is a good oxidant, so it is subject to reductive dissolution by dissolved organic compounds that are present in the water, even as dissolved O_2 might tend to re-form $PbO_{2(s)}$. However, "free chlorine" ($HOCl/OCl^-$, as in chlorinated drinking water) is strongly oxidizing, and can surely form protective layers of $PbO_{2(s)}$ (Lytle and Schock, 2005; Wang et al., 2010).

If all of a $PbO_{2(s)}$ passivating surface should dissolve, then a next possible line of defense would be a Pb(II) solid. But in the absence of significant dissolved C_T or phosphate, the only possibilities are oxides like α-$PbO_{(s)}$ and β-$PbO_{(s)}$, both of which are too soluble to provide adequate protection of drinking water: as noted, Figure 19.3 indicates that at pH = 7.55, α-$PbO_{(s)}$ (the less soluble of the two) has a saturation concentration of $Pb_T = 10^{-2}$ *M*, or ~2×10^6 μg/L, as at point E. At pH = 8.43, the saturation concentration of Pb_T is still ~200,000 μg/L (10^{-3} *M*). Even for the minimum solubility for α-$PbO_{(s)}$ at pH = 10.41, the Pb_T saturation concentration is ~14,000 μg/L ($10^{-4.17}$ *M*, see Figure 19.4).

Besides Pb in tap water, lead poisoning continues to be caused by consumption of "moonshine" produced using distillation ("still") equipment involving Pb. Pb arises in this setting when copper piping is connected with tin/lead solder, and/or an automobile radiator manufactured with tin/lead solder is used as a flow-through condenser for the gaseous water/alcohol/air mix created by boiling fermented "mash" (MMWR, 1992; Morgan et al., 2004). Distilling moonshine with flow-through condensation is non-reflux distillation in a "pot still", *i.e*, a "1-theoretical plate" en masse flow-through condensation of alcohol and water from the gaseous water/alcohol mix flowing through the condenser. (Obtaining high-proof moonshine in this manner requires sequential re-distillation.) Overall, drinking moonshine of dubious provenance is a risky undertaking, even without considering that poisonous levels of methanol may be present if the distiller did not discard the early distilling fraction (a.k.a. "head", "foreshot"). For one case study of lead poisoning from moonshine, Arnold and Morgan (2015) note:

> On **hospital day 16** [emphasis added], she was found to have a whole blood lead concentration of 148.2 μg/dL [1482 ppb] ... Testing of the moonshine acquired from the patient's home ... revealed a lead concentration of 15,000 μg/L-50 times the threshold set by the FDA's highest threshold for possible adverse health effects.

Last, ducks and geese are at great risk for lead poisoning when they ingest the lead ($Pb_{(s)}$) shot used in hunting with waterfowl with shotguns, and lead weights used in fishing; there is a strong argument for using replacement for lead in these applications (Goddard et al., 2008).

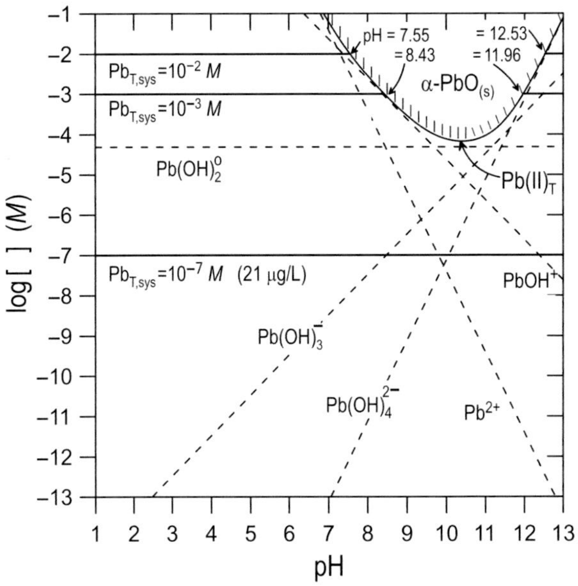

FIGURE 19.4 Log concentration (M) *vs.* pH lines for various individual Pb(II) species and for log $Pb(II)_T$ for equilibrium with α-$PbO_{(s)}$ at 25 °C/1 atm in the absence of significant C_T. At $Pb_{T,sys} = 10^{-2}$ M, α-$PbO_{(s)}$ is present only for $7.55 < pH < 12.53$. At $Pb_{T,sys} = 10^{-3}$ M, α-$PbO_{(s)}$ is present only for $8.43 < pH < 11.96$. At $Pb_{T,sys} = 10^{-7}$ M (21 ppb), α-PbO(s) cannot form at any pH. Assumptions: 1) all solution-phase activity coefficients are unity; 2) when a solid is present, it has unit activity.

Because a dissolved Pb_T concentration of 10^{-2} M corresponds to ~2×10^6 μg/L, much lower $Pb_{T,sys}$ values than 10^{-2} M are highly relevant both toxicologically and geologically. We need to consider pe–pH diagrams for lower values of $Pb_{T,sys}$.

19.3 pe–pH DIAGRAM FOR $Pb_{T,sys} = 10^{-3}$ M

The width of the region for α-$PbO_{(s)}$ in a pe–pH predominance diagram will decrease with decreasing $Pb_{T,sys}$, and can be reduced to zero: as $Pb_{T,sys}$ decreases, the value of the low pH root of Eq.(19.13/19.14) increases, and the value of the high pH root decreases. Eventually, at the minimum solubility, the system metal level is reduced to the point that the two roots are the same. The solution will be saturated at that one pH, but cannot form at that level, or at any level below that.

As indicated in the solubility *vs.* pH plot for α-$PbO_{(s)}$ in Figure 19.4, when $Pb_{T,sys} = 10^{-3}$ M, with increasing pH (from low values), α-$PbO_{(s)}$ can become saturated, but Eq.(19.13/19.14) now gives a low pH root of 8.61, and a high pH root of 11.85. From Figure 19.1, at pH = 8.61, $PbOH^+$ is the dominant dissolved species; at pH = 11.85, $Pb(OH)_4^{2-}$ is the dominant species. So far then, for Figure 19.5, we have that there is boundary line between regions for Pb^{2+} and $PbOH^+$ at pH = 7.22, a boundary line between regions for $PbOH^+$ and α-$PbO_{(s)}$ at pH = 8.61, and a boundary line between regions for α-$PbO_{(s)}$ and $Pb(OH)_4^{2-}$ at pH = 11.85. The $PbO_{2(s)}$/α-$Pb0_{(s)}$ and α-$Pb0_{(s)}$/$Pb_{(s)}$ boundary lines are in the same positions as in Figure 19.3, but shorter (because the α-$Pb0_{(s)}$ region is narrower.)

Next, we need the redox boundary line between $PbO_{2(s)}$ and the solution, and the boundary line between the solution and $Pb_{(s)}$. For the first, as with Figure 19.3, there is one overall line between $PbO_{2(s)}$ and the solution, as obtained from Eq.(19.24). The line is drawn in Figure 19.5 for $Pb_{T,sys} = 10^{-3}$ M. A dashed line is used for the portion that falls inside the region for α-$PbO_{(s)}$; for that range, Eq.(19.27) is drawn. For the second, there is one overall line between the solution and $Pb_{(s)}$, as obtained from Eq.(19.29). The line is drawn in Figure 19.5 for $Pb_{T,sys} = 10^{-3}$ M. A dashed line is used for the portion

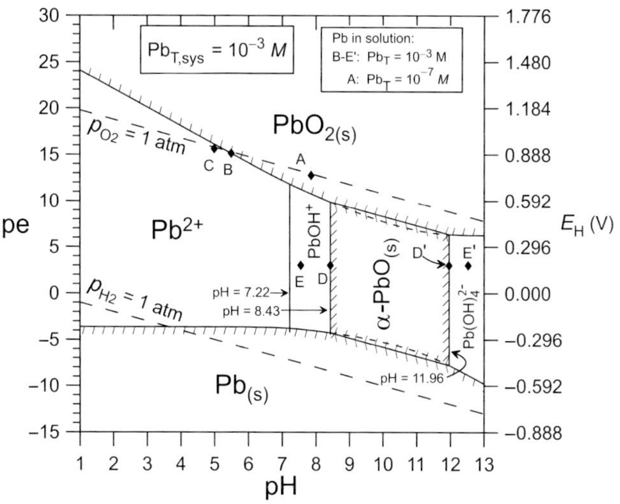

FIGURE 19.5 pe–pH predominance diagram for $Pb_{T,sys} = 10^{-3}$ M at 25 °C/1 atm and low C_T. Assumptions: 1) all solution-phase activity coefficients are unity; 2) when a solid is present, it has unit activity.

that falls inside the region for α-$PbO_{(s)}$; for that range, Eq.(19.31) is drawn. Compared to Figure 19.3, all the regions for the solids in Figure 19.5 are smaller: a lower value of $Pb_{T,sys}$ makes it harder to precipitate any solid. This is a general result for any solid region for any element in any pe–pH diagram.

19.4 pe–pH DIAGRAM FOR $Pb_{T,sys} = 10^{-7}$ M (21 ppb)

For $Pb_{T,sys} = 10^{-7}$ M, if all the Pb was dissolved, the level would be 21 μg/L = 21 ppb. For $Pb_{T,sys} = 10^{-7}$ M, Figure 19.4 indicates that α-$PbO_{(s)}$ cannot form at any pH, so α-$PbO_{(s)}$ no longer has a predominance region in the pe–pH diagram. At intermediate pe values, there are no regions for solids, only for the dissolved Pb species, namely Pb^{2+}, $PbOH^+$, $Pb(OH)_2^o$, $Pb(OH)_3^-$, and $Pb(OH)_3^{2-}$. (It is not that $Pb(OH)_2^o$ and $Pb(OH)_3^-$ are not present in the solution over the pH range for the α-$PbO_{(s)}$ region in Figure 19.5, but that a solid is present in the pH ranges where $Pb(OH)_2^o$ or $Pb(OH)_3^-$ can predominate the *solution* chemistry; by the rules of drawing these diagrams, everywhere a solid is present, then that solid is depicted. Calculation of the solution chemistry can then be done for all species using the solid POV.) Both the $PbO_{2(s)}$/solution boundary line and the solution/$Pb_{(s)}$ boundary line now extend uninterrupted from low to high pH, and we have Figure 19.6.

From Figure 19.1 at 25 °C/1 atm: 1) the $[Pb^{2+}] = [PbOH^+]$ boundary line is at pH = 7.22; 2) the $[Pb(OH)^+] = [Pb(OH)_2^o]$ boundary line is at pH = 9.69; 3) the $[Pb(OH)_2^o] = [Pb(OH)_3^-]$ boundary line is at pH = 11.17; and 4) the $[Pb(OH)_3^-] = [Pb(OH)_4^{2-}]$ boundary line is at pH = 11.64. Those considerations give four vertical boundary lines:

$$\begin{array}{ccccccccc} & \text{pH=7.22} & & \text{pH=9.69} & & \text{pH=11.17} & & \text{pH=11.64} & \\ Pb^{2+} & | & PbOH^+ & | & Pb(OH)_2^o & | & Pb(OH)_3^- & | & Pb(OH)_4^{2-}. \end{array} \qquad (19.34)$$

When comparing Figure 19.6 to Figure 19.5, not only has the region for α-$PbO_{(s)}$ been eliminated, but there is now also a sharp upturn of the $PbO_{2(s)}$/solution boundary line at pH = 12.0. This is because at high pH, at $Pb_{T,sys} = 10^{-7}$ M, the dissolved Pb(IV) species PbO_3^{2-} has become important in Eq.(19.20). This causes the term $-\log(Pb_{T,sys} - 10^{-31.03}/[H^+]^2)$ to sharply become large-positive as $10^{-31.03}/[H^+]^2$ becomes close to $Pb_{T,sys} = 10^{-7}$ (because $(Pb_{T,sys} - 10^{-31.03}/[H^+]^2) \rightarrow 0$. This occurs at $[H^+] = (10^{-31.03}/10^{-7})^{\frac{1}{2}} = 10^{-12.015}$. (At $Pb_{T,sys} = 10^{-3}$ M in Figure 19.5, a slight upturn is visible at

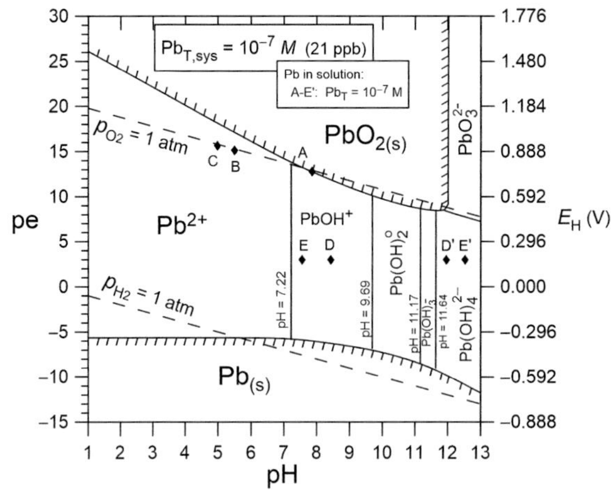

FIGURE 19.6 pe–pH predominance diagram for PbT,sys = 10^{-7} *M* (21 μg/L = 21 ppb) at 25 °C/1 atm and low CT. Assumptions: 1) all solution-phase activity coefficients are unity; 2) when a solid is present, it has unit activity.

pH ≈14.) This means for $Pb_{T,sys} = 10^{-7}\,M$ that there is a region for PbO_3^{2-} at high pH and high pe. So, we need one more, last boundary line, namely that between $Pb(IV)_T$ (= $[PbO_3^{2-}]$) and $Pb(II)_T$. (Recall here, from Section 19.1, that all Pb_T values without a subscript "sys" are *dissolved* values.) We combine

$$2pe = 49.2 + \log\{PbO_{2(s)}\} - \log[Pb^{2+}] - 4pH \qquad (19.18) \qquad \frac{PbO_{2(s)}}{Pb^{2+}}$$

$$-\left(\log[PbO_3^{2-}] = -31.03 + 2pH + \log\{PbO_{2(s)}\}\right) \qquad (19.35) \qquad \frac{PbO_3^{2-}}{PbO_{2(s)}}$$

$$2pe = 80.23 - 6pH + \log[PbO_3^{2-}] - \log[Pb^{2+}]. \qquad (19.36) \qquad \frac{PbO_3^{2-}}{Pb^{2+}}$$

Eq.(19.35) originates in Eq.(19.2). Using Eq.(19.11) and $Pb(IV)_T = [PbO_3^{2-}]$ with Eq.(19.36) gives

$$2pe = 80.23 - 6pH + \log Pb(IV)_T - \log Pb(II)_T - \log\alpha_0^{Pb(II)} \qquad (19.37)$$

Along the desired boundary line, $Pb(IV)_T = Pb(II)_T$, giving

$$2pe = 80.23 - 6pH - \log\alpha_0^{Pb(II)} \qquad \frac{PbO_3^{2-}}{Pb^{2+}}. \qquad (19.38)$$

This line, as plotted in Figure 19.6 from pH = 12.01 to 14.00, lies slightly below the $p_{O_{2(g)}} = 1$ atm water stability line.

Example 19.4 Saturation Index and Solid Formation

Consider Figure 19.6 and the system wherein $Pb_{T,sys} = 10^{-7}\,M$. For intermediate pe values where only Pb(II) is stable, use Eq.(19.4) and $\alpha_0^{Pb(II)}$ to compute and plot the Saturation Index (SI) *vs.* pH from pH = 8 to 12 for α-$PbO_{(s)}$. Eq.(11.6) provides the definition of SI; saturation occurs when SI = 0. Neglect activity corrections for dissolved species.

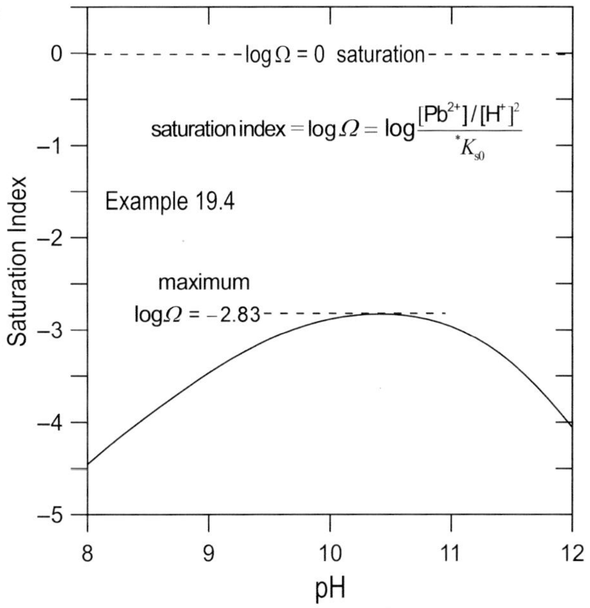

Solution

SI = $\log\Omega$ = log [(ion activity product)/K_s], with K_s being the solubility equilibrium constant corresponding to the ion activity product. When SI < 0, the solution is undersaturated; when SI = 0, a solution is exactly saturated with a solid; when SI > 0, a solution is supersaturated with a solid. When K_s is taken to be $^*K_{s0}$ for α-$PbO_{(s)}$, the ion activity product = $[Pb^{2+}]/[H^+]^2$, so $\Omega = \log[([Pb^{2+}]/[H^+]^2)/{}^*K_{s0}]$. $[Pb^{2+}] = \alpha_0^{Pb(II)}Pb(II)_T$. We take $Pb(II)_T = Pb_{T,sys}$ to see if this value of $Pb(II)_T$ can lead to saturation. The plot indicates it cannot because SI < 0 at all pH. No α-$PbO_{(s)}$ can form at $Pb_{T,sys} = 10^{-7}$ *M*.

19.5 USE OF "SIMPLIFYING" ASSUMPTIONS TO DRAW pe–pH (E_H-pH) DIAGRAMS

Before the 1980s advent of personal computers and their easy computational power, a curved boundary line such as that represented by Eq.(19.15) or (19.20) was not drawn as a single curve line, but rather approximated as a series of straight segments, each of which involved a locally applicable approximation. This approach was used for the non-blunt-wedge version of Figure 18.3. If we had used this approach for Figure 19.6, the $PbO_{2(s)}$/solution boundary would have been drawn: 1) taking $[Pb^{2+}] = 10^{-7}$ *M* from pH = 0 to pH = 7.7; 2) taking $[PbOH^+] = 10^{-7}$ *M* from pH = 7.7 to pH = 9.4; 3) taking $[Pb(OH)_2^o] = 10^{-7}$ *M* from pH = 9.4 to pH = 11.0; and last 4) taking $[Pb(OH)_3^-] = 10^{-7}$ *M* from pH = 11.0 to pH = 14.0. There are countless examples of this practice in books, reports, theses, and scientific articles. Even at the time of this writing, use of that approach in new work remains much more common than not. Its existence is pointed out here lest the reader be unnecessarily confused when pondering the work of others, even as it is hoped that the considerable conceptual and practical advantages of the curved approach have been communicated.

REFERENCES

Arnold J, Morgan B (2015) Management of lead encephalopathy with DMSA after exposure to lead-contaminated moonshine. *Journal of Medical Toxicology*, **11**, 464–467.

City of Portland (2011) Portland Water Bureau Treatment Variance Request, June 6, 2011, Section 2, Portland's Bull Run Watershed, Table 2-2, page 2–16. Available at: www.portlandoregon.gov/water/54913. (Accessed 10 February 2018).

City of Portland (2016) Portland Water Bureau Triannual Water Quality Analysis, November, 2016. Available at: www.portlandoregon.gov/water/article/484002. (Accessed 10 February 2018).

Lytle DA, Schock MR (2005) Formation of Pb(IV) oxides in chlorinated water. *Journal of the American Water Works Association*, **97**, 102–114.

Garrels RM, Christ CL (1965) *Solutions, Minerals, and Equilibria.* Harper & Row, New York.

Goddard CI, Leonard NJ, Stang DL, Wingate PJ, Rattner BA, Franson JC, Sheffield SR (2008) Management concerns about known and potential impacts of lead use in shooting and in fishing activities. *Fisheries*, **33**, 228–236.

Hem JD (1976) Inorganic chemistry of lead in water. In: *Lead in the Environment.* TG Levering (Ed.). U.S. Geological Survey Professional Paper 957, Washington, D.C., pp.5–11.

Langmuir D, Chrostowski P, Vigneault B, Chaney R (2004) *Issue Paper on the Environmental Chemistry of Metals. Report to U.S. Environmental Protection Agency*, Contract #68-C-02-060, Risk Assessment Forum, Washington, D.C.

Morgan BW, Parramore CS, Ethridge M (2004) Lead contaminated moonshine: A report of Bureau of Alcohol, Tobacco and Firearms analyzed samples. *Veterinary Human Toxicology*, **46**, 89–90.

MMWR (Morbidity and Mortality Weekly Report) (1992) Elevated blood lead levels associated with illicitly distilled alcohol – Alabama, 1990–1991, May 01, 1992/41(17), 294–295.

Olson TM, Wax M, Yonts J, Heidecorn K, Haig S-J, Yeoman D, Hayes Z, Raskin L, Ellis BR (2017) Forensic estimates of lead release from lead service lines during the water crisis in Flint, Michigan. *Environmental Science and Technology Letters*, **4**, 356–361.

Pieper KJ, Tang M, Edwards MA (2017) Flint water crisis caused by interrupted corrosion control: Investigating "ground zero" home. *Environmental Science and Technology*, **51**, 2007–2014.

Renner R (2010) Exposure on tap: Drinking water as an overlooked source of lead. *Environmental Health Perspectives*, **118**, A68–A74.

Reuter C, Conway J, Mast DB, Gerke TL, Maynard B (2005) The mineralogy of Pb scales in drinking water distribution systems as revealed by combined XRD and micro-raman spectroscopy. Presented at North-Central Section, GSA, Minneapolis, MN, May 19–20, 2005, Available at: https://cfpub.epa.gov/si/index.cfm (Accessed 19 February 2018).

Schock MR, Wagner I, Oliphant RJ (1996) Chapter 4, Corrosion and Solubility of Lead in Drinking Water. In: *Internal Corrosion Of Water Distribution Systems: Cooperative Research Report*, 2nd Edition, AWWA Research Foundation (Denver, CO) and DVGW-Technologiezentrum Wasser (Karlsruhe, Federal Republic of Germany).

Stumm W, Morgan JJ (1996) *Aquatic Chemistry: Chemical Equilibria and Rates in Natural Waters*, 3rd edn. Wiley Publishing Co., New York.

Triantafyllidou S, Schock MR, DeSantis MK, White C (2015) Low contribution of PbO_2-coated lead service lines to water lead contamination at the tap. *Environmental Science and Technology*, **49**, 3476–3754.

U.S. Environmental Protection Agency (1991) Maximum contaminant level goals and national primary drinking water regulations for lead and copper; final rule. *Federal Register*, **56**, 26460.

Wang Y, Xie J, Li W, Wang Z, Giammar DE (2010) Formation of lead(IV) oxides from lead(II) compounds. *Environmental Science and Technology*, **44**, 8950–8956.

20 pe–pH Diagrams for Lead (Pb) in the Presence of CO_2 with Fixed C_T, and Fixed C_T and Phosphate

20.1 INTRODUCTION

Chapter 19 considered pe–pH diagrams for aqueous Pb when both CO_2 and phosphate are low. Some natural water systems do exist in which little CO_2 is present, but most contain important amounts of dissolved CO_2. Thus, in a pe–pH diagram for a metal, if some CO_2 is present, then we need to consider that there may be predominance region for a metal carbonate solid, possibly more than one carbonate solid variation. This is especially true if the metal has a stable II oxidation state: while many metal ions with +2 charge are fairly soluble as hydroxides and oxides (Chapter 11), many form fairly insoluble carbonate solids (*e.g.*, $CaCO_{3(s)}$, $FeCO_{3(s)}$, $PbCO_{3(s)}$, etc., Chapter 12). Also, metal ions with +2 charge like Ca^{2+}, Mg^{2+}, and Pb^{2+} form insoluble phosphate solids, as with hydroxyapatite and struvite ($Ca_5(PO_4)_3OH_{(s)}$) and $MgNH_4PO_4{\cdot}6H_2O_{(s)}$, Chapter 13)), and hydroxypyromorphite ($Ca_5(PO_4)_3OH_{(s)}$), this chapter). So, if C_T and/or P_T are significant, *and* the T,sys value for the metal is high enough, then a Pb(II) carbonate and/or a Pb(II) phosphate will have a predominance region in the relevant pe–pH diagram. (As in Chapter 19, all Pb_T values without a subscript "sys" are *dissolved* values.) For metals in a III oxidation state or higher, generally carbonate and phosphate solids do not need to be considered, because hydroxide, oxide, and oxyhydroxide solids are, in general, more stable. For example: 1) Fe(III) will form $Fe(OH)_{3(s)}$, $FeOOH_{(s)}$, and $Fe_2O_{3(s)}$, but not the Fe(III) carbonate $Fe_2(CO_3)_{2(s)}$; that solid is unstable relative to conversion to $Fe_2O_{3(s)} + 2CO_2$; and 2) Pb(IV) will form PbO_2, but not Pb(IV) carbonate $Pb(CO_3)_{2(s)}$. A lengthy review of the environmental chemistry of numerous metals (including lead) is provided by Langmuir et al. (2004).

20.2 pe–pH DIAGRAM FOR $Pb_{T,sys} = 10^{-5}$ *M* AND $C_{T,free} = 10^{-3}$ *M*

20.2.1 General

Four points from Chapter 19 warrant review here:

1) Pb has three major oxidation states that must be considered: IV, II, and 0.
2) $PbO_{2(s)}$ is essentially insoluble as Pb(IV), except at very high pH, at which PbO_3^{2-} can form.
3) $PbO_{2(s)}$ is only very weakly soluble by reductive dissolution to Pb(II), so the high pe conditions induced by use of direct chlorination of drinking water disinfection favor formation of $PbO_{2(s)}$ and so can be beneficial in limiting dissolved Pb (Lytle and Schock, 2005).
4) $Pb_{(s)}$ is essentially insoluble as Pb(0) at all pH.
5) In the aqueous phase, multiple Pb(II) species need to be considered. (In this chapter, in addition to Pb^{2+}, $PbOH^+$, $Pb(OH)_2^o$, and $Pb(OH)_3^-$, complexes with carbonate- and phosphate-related species are relevant (see Eqs.(20.12–20.14) and Eqs.(20.40-20.41).)

As far as Pb(II) solids are concerned, we again consider α-$PbO_{(s)}$ as in Chapter 19. As carbonate solids, we add $PbCO_{3(s)}$ (cerussite) and $Pb_3(CO_3)_2(OH)_{2(s)}$ (hydrocerussite). Overall, with the K values pertaining to 25 °C/1 atm, we have

$$PbO_{2(s)} + 4H^+ + 2e^- = 2H_2O + Pb^{2+} \quad \log K = 49.2 \quad \text{redox: Pb(IV) to Pb(II)} \tag{20.1}$$

$$PbO_{2(s)} + H_2O = PbO_3^{2-} + 2H^+ \quad \log K = -31.03 \quad \text{solubility of Pb(IV)} \tag{20.2}$$

$$Pb^{2+} + 2e^- = Pb_{(s)} \quad \log K = -4.26 \quad \text{redox: Pb(II) to Pb(0)} \tag{20.3}$$

$$\alpha\text{-}PbO_{(s)} + 2H^+ = Pb^{2+} + H_2O \quad \log K = 12.6 \quad \text{solubility of Pb(II)} \quad \text{(Pb(II) solid 1)} \tag{20.4}$$

$$PbCO_{3(s)} = Pb^{2+} + CO_3^{2-} \quad \log K = -12.8 \quad \text{solubility of Pb(II)} \quad \text{(Pb(II) solid 2)} \tag{20.5}$$

$$Pb_3(CO_3)_2(OH)_{2(s)} + 2H^+ = 3Pb^{2+} + 2CO_3^{2-} + 2H_2O \quad \log K = -18.8 \quad \text{solubility of Pb(II)} \quad \text{(Pb(II) solid 3)} \tag{20.6}$$

$$PbO \cdot PbCO_{3(s)} + 2H^+ = 2Pb^{2+} + CO_3^{2-} + H_2O \quad \log K = -0.5 \quad \text{solubility of Pb(II)} \quad \text{(Pb(II) solid 4)} \tag{20.7}$$

$$Pb^{2+} + H_2O = PbOH^+ + H^+ \quad \log {}^*K_{H1} = -7.22 \quad \text{complexation of dissolved Pb(II)} \tag{20.8}$$

$$PbOH^+ + H_2O = Pb(OH)_2^o + H^+ \quad \log {}^*K_{H2} = -9.69 \quad \text{complexation of dissolved Pb(II)} \tag{20.9}$$

$$Pb(OH)_2^o + H_2O = Pb(OH)_3^- + H^+ \quad \log {}^*K_{H3} = -11.17 \quad \text{complexation of dissolved Pb(II)} \tag{20.10}$$

$$Pb(OH)_3^- + H_2O = Pb(OH)_4^{2-} + H^+ \quad \log {}^*K_{H4} = -11.64 \quad \text{complexation of dissolved Pb(II)} \tag{20.11}$$

$$Pb^{2+} + CO_3^{2-} = PbCO_3^o \quad \log K_{C1} = 7.10 \quad \text{complexation of dissolved Pb(II)} \tag{20.12}$$

$$PbCO_3^o + CO_3^{2-} = Pb(CO_3)_2^{2-} \quad \log K_{C2} = 3.23 \quad \text{complexation of dissolved Pb(II)} \tag{20.13}$$

$$Pb^{2+} + HCO_3^{2-} = PbHCO_3^+ \quad \log K_{HC1} = 2.26 \quad \text{complexation of dissolved Pb(II).} \tag{20.14}$$

Some of the above *K* values were informed by the summary of aqueous Pb chemistry by Schock et al. (1996). (The formation constants of $PbCO_3^o$ and $Pb(CO_3)_2^-$ were also studied by Easley and Byrne (2011).) Throughout this chapter, when equilibrium constants are deployed simply with concentrations rather than activities, it is being assumed that corresponding solution-phase activity corrections can be neglected; they could be included by using ${}^cK$ values.

20.2.2 Solution $\alpha^{Pb(II)}$ Values with Hydroxide and Carbonate-Related Complexes

For dissolved Pb(II), adding $PbCO_3^o$, $Pb(CO_3)_2^{2-}$, and $PbHCO_3^+$ to the five dissolved Pb(II) species considered in Chapter 19 gives

$$\alpha_0^{Pb(II)} \equiv \frac{[Pb^{2+}]}{Pb_T} = \frac{[Pb^{2+}]}{[Pb^{2+}]+[PbOH^+]+[Pb(OH)_2^o]+[Pb(OH)_3^-] + [Pb(OH)_4^{2-}]+[PbCO_3^o]+[Pb(CO_3)_2^{2-}]+[PbHCO_3^+]} \tag{20.15}$$

$$= \frac{1}{\left(1+\frac{[PbOH^+]}{[Pb^{2+}]}+\frac{[Pb(OH)_2^o]}{[Pb^{2+}]}+\frac{[Pb(OH)_3^-]}{[Pb^{2+}]}+\frac{[Pb(OH)_4^{2-}]}{[Pb^{2+}]} +\frac{[PbCO_3^o]}{[Pb^{2+}]}+\frac{[Pb(CO_3)_2^{2-}]}{[Pb^{2+}]}+\frac{[PbHCO_3^+]}{[Pb^{2+}]}\right)}. \tag{20.16}$$

The second to fifth terms in the denominator are obtained as in Box 19.1 For the sixth to eighth terms, from Eqs.(20.12–20.14),

$$\begin{aligned} \frac{[PbCO_3^o]}{[Pb^{2+}]} &= K_{C1}\alpha_2^C C_{T,free}, \\ \frac{[Pb(CO_3)_2^{2-}]}{[Pb^{2+}]} &= K_{C1}K_{C2}(\alpha_2^C C_{T,free})^2, \\ \frac{[PbHCO_3^+]}{[Pb^{2+}]} &= K_{HC1}\alpha_1^C C_{T,free} \end{aligned} \tag{20.17}$$

with $C_{T,free}$ as defined in Box 20.1. Thus, for Pb^{2+},

$$\alpha_0^{Pb(II)} = \frac{1}{\left[1+\left(\frac{{}^*\beta_{H1}}{[H^+]}+\frac{{}^*\beta_{H2}}{[H^+]^2}+\frac{{}^*\beta_{H3}}{[H^+]^3}+\frac{{}^*\beta_{H4}}{[H^+]^4}\right) +\left(K_{C1}\alpha_2^C C_{T,free}+K_{C1}K_{C2}(\alpha_2^C C_{T,free})^2+K_{HC1}\alpha_1^C C_{T,free}\right)\right]} \tag{20.18}$$

where the two sets of parentheses in the denominator collect the terms for the hydroxo complexes and the carbonate-related complexes, respectively. The α value expressions for the seven other Pb(II) species are given in Box 20.2. A log α *vs.* pH plot is provided in Figure 20.1 for $C_{T,free} = 10^{-3}$ *M*. These α values apply for the dissolved Pb(II) regardless of whether any solid(s) is(are) present.

BOX 20.1 Definitions of $C_{T,free}$ and $C_{T,all\ species}$

$$C_{T,free} \equiv [H_2CO_3^*] + [HCO_3^-] + [CO_3^{2-}]$$

$$C_{T,all\ species} \equiv [H_2CO_3^*] + [HCO_3^-] + [CO_3^{2-}] + [PbCO_3^o] + [Pb(CO_3)_2^{2-}] + [PbHCO_3^+]$$

Because the concentration of each of the Pb(II) complexes in $C_{T,all\ species}$ is proportional to $[Pb^{2+}]$ (see Eq.20.16), then as long as $[Pb^{2+}]$ does not become "too high", then

$$C_{T,free} \approx C_{T,all\ species}$$

and diagrams drawn can pertain to the solution constraint that

$$C_{T,free} = 10^{-3} M \approx C_{T,all\ species}.$$

BOX 20.2 α Values beyond α_0 for Hydroxo and Carbonate-Related Aqueous Pb(II) Species

Expressions for $\alpha_1^{Pb(II)}$, $\alpha_2^{Pb(II)}$, $\alpha_3^{Pb(II)}$, and $\alpha_4^{Pb(II)}$ as based on $\alpha_0^{Pb(II)}$ are as developed in Box 19.1. Recall that $^*\beta_{H1} = {}^*K_{H1}$, $^*\beta_{H1} = {}^*K_{H1}{}^*K_{H2}$, $^*\beta_{H3} = {}^*K_{H1}{}^*K_{H2}{}^*K_{H3}$, and $^*\beta_{H4} = {}^*K_{H1}{}^*K_{H2}{}^*K_{H3}{}^*K_{H4}$. Using $^*\beta$ notation, neglecting activity corrections, these are,

$$\alpha_1^{Pb(II)} = \alpha_0^{Pb(II)} \frac{^*\beta_{H1}}{[H^+]}, \qquad \alpha_2^{Pb(II)} = \alpha_0^{Pb(II)} \frac{^*\beta_{H2}}{[H^+]^2},$$

$$\alpha_3^{Pb(II)} = \alpha_0^{Pb(II)} \frac{^*\beta_{H3}}{[H^+]^3}, \qquad \alpha_4^{Pb(II)} = \alpha_0^{Pb(II)} \frac{^*\beta_{H4}}{[H^+]^4}.$$

For the complexes with CO_3^{2-} and HCO_3^-, using $\alpha_0^{Pb(II)} = [Pb^{2+}] / Pb_T$, Eqs.(20.12–20.14), and neglecting activity corrections in the second and third rows:

$\alpha_{C1}^{Pb(II)} \equiv \dfrac{[PbCO_3^o]}{Pb(II)_T}$	$\alpha_{C2}^{Pb(II)} \equiv \dfrac{[Pb(CO_3)_2^{2-}]}{Pb(II)_T}$	$\alpha_{HC1}^{Pb(II)} \equiv \dfrac{[PbHCO_3^+]}{Pb(II)_T}$
$= \alpha_0^{Pb(II)} \dfrac{[PbCO_3^o]}{[Pb^{2+}]}$	$= \alpha_0^{Pb(II)} \dfrac{[Pb(CO_3)_2^{2-}]}{[Pb^{2+}]}$	$= \alpha_0^{Pb(II)} \dfrac{[PbHCO_3^+]}{[Pb^{2+}]}$
$= \alpha_0^{Pb(II)} K_{C1} \alpha_2^C C_{T,free}$	$= \alpha_0^{Pb(II)} K_{C1} K_{C1} (\alpha_2^C C_{T,free})^2$	$= \alpha_0^{Pb(II)} K_{HC1} \alpha_1 C_{T,free}$

As discussed in Box 20.1, $C_{T,free}$ refers only to the sum $[H_2CO_3^*]+[HCO_3^-]+[CO_3^{2-}]$, and does not include any of the CO_3^{2-} or HCO_3^- that is present within the dissolved carbonate-related complexes of Pb(II).

20.2.3 Identification of the Pb(II) Solids That Limit Pb(II) Solubility

Chapter 19 discussed that Pb(IV) is very insoluble, except for the formation of PbO_3^{2-} at high pH. Therefore, even at rather low $Pb_{T,sys}$, the pe can generally be made high enough to form $PbO_{2(s)}$ at acidic to neutral pH, though this may require $p_{O_2} >> 1$ atm. Similarly, because $Pb_{(s)}$ is generally assumed to be totally insoluble as Pb(0), then even at rather low $Pb_{T,sys}$ the pe can be made low enough to form $Pb_{(s)}$ at any pH, though this may require $p_{H_2} >> 1$ atm.

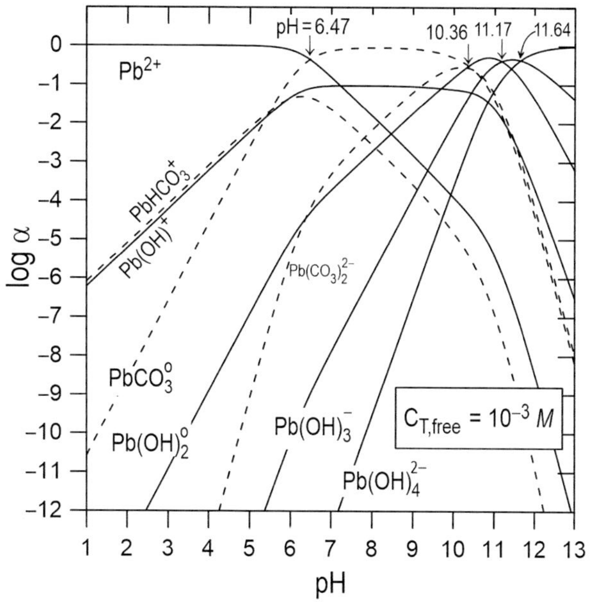

FIGURE 20.1 Log α *vs.* pH for dissolved Pb(II) species at 25 °C/1 atm with $C_{T,free} = 10^{-3}$ M (see Box 20.1). These α apply for dissolved Pb(II) regardless of whether any solid is present. Dashed lines are used to distinguish the carbonate-related Pb(II) complexes. Pb^{2+} predominates for pH < 6.47; $PbCO_3^o$ predominates for 6.47 < pH < 10.36; $Pb(OH)_2^o$ predominates for 10.36 < pH < 11.17; $Pb(OH)_3^-$ predominates for 11.17 < pH < 11.64; and $Pb(OH)_4^{2-}$ predominates for pH > 11.64. All solution-phase activity coefficients are assumed to be unity. Compare with Figure 19.1 for which carbonate-related complexes are not considered.

Pb(II) solids are generally more soluble than $PbO_{2(s)}$ and $Pb_{(s)}$. In Chapter 19 (no CO_2), it was discussed that it may not be possible to form $PbO_{(s)}$ at any pH when $Pb_{T,sys}$ is low (see Figure 19.6). For this chapter (similarly), for any given nonzero value of C_T, when $Pb_{T,sys}$ is sufficiently low it will not be possible at any pH to form $PbO_{(s)}$ or any Pb(II) carbonate solid. However, as $Pb_{T,sys}$ increases, predominance regions for one or more Pb(II) solids among the four in Eqs.(20.10–13) will eventually appear in the pe–pH diagram. But which one(s) will form at equilibrium? To answer that question, recall from Chapter 14 that:

1. The solid that prescribes the lowest concentration of any given dissolved metal species or total dissolved metal is the *solubility limiting* solid.
2. When a given solid is solubility-limiting, then at equilibrium, the solution phase does not have the ability to hold more of any given dissolved metal species (or total dissolved metal) than is prescribed by that solid.
3. When a given solid is solubility limiting as well as actually *solubility controlling*, then the solution is characterized by the concentration values for the individual dissolved metal species and the total dissolved metal as prescribed by equilibrium with that solid for the solution conditions of interest.
4. Two different solids can be present simultaneously if the concentrations that they prescribe are exactly the same for each individual dissolved metal species, as well as (therefore) for the total dissolved metal.

So, for a given $Pb_{T,sys}$, to determine which of the four Pb(II) solids have predominance regions and the pH ranges of those regions, one first needs to use a total solubility diagram to determine where the different solids are solubility limiting. *Then*, whether or not a given Pb(II) solid will be present so that it is solubility controlling (*i.e.*, have a pe–pH predominance region) can be determined from that same total solubility diagram based on the value of $Pb_{T,sys}$ for which the pe–pH diagram is being constructed. Specifically, as in Chapter 19, a given Pb(II)-solubility-limiting solid can only occur over the pH range for which that solid is actually solubility controlling, *i.e.*, prescribing that the dissolved $Pb(II)_T < Pb_{T,sys}$.

To investigate solubility limitation by the four different Pb(II) solids, we can compute $Pb(II)_T$ as it is prescribed by each of the solids *vs.* pH, while assuming that $C_{T,free}$ for the solution remains constant for all pH. Unlike $Pb(II)_T$, which can be reduced below $Pb_{T,sys}$, $C_{T,free}$ is considered to remain constant. We now develop the equations for $[Pb^{2+}]$ for each of the solids, then invoke $\alpha_0^{Pb(II)}$ to obtain the corresponding dissolved $Pb(II)_T$ functions.

For $PbO_{(s)}$,

$$[Pb^{2+}]_{PbO_{(s)}} = 10^{12.7}[H^+]^2 \tag{20.19}$$

$$\log[Pb^{2+}]_{PbO_{(s)}} = 12.7 - 2\text{ pH}. \tag{20.20}$$

For $PbO{\cdot}CO_{3(s)}$,

$$[Pb^{2+}]_{PbO\cdot CO_{3(s)}} = \left(\frac{10^{-0.5}[H^+]^2}{[CO_3^{2-}]}\right)^{1/2} \tag{20.21}$$

$$\log[Pb^{2+}]_{PbO\cdot CO_{3(s)}} = -0.25 - \text{pH} - \tfrac{1}{2}\log\alpha_2^C - \tfrac{1}{2}\log C_{T,free}. \tag{20.22}$$

For $PbCO_{3(s)}$,

$$[Pb^{2+}]_{PbCO_{3(s)}} = \frac{10^{-12.8}}{[CO_3^{2-}]} \tag{20.23}$$

$$\log[Pb^{2+}]_{PbCO_{3(s)}} = -12.8 - \log\alpha_2^C - \log C_{T,free}. \tag{20.24}$$

For $Pb_3(CO_3)_2(OH)_{2(s)}$,

$$[Pb^{2+}]_{Pb_3(CO_3)_2(OH)_{2(s)}} = \left(\frac{10^{-18.8}[H^+]^2}{[CO_3^{2-}]^2}\right)^{1/3} \tag{20.25}$$

$$\log[Pb^{2+}]_{Pb_3(CO_3)_2(OH)_{2(s)}} = -6.27 - \tfrac{2}{3}\text{pH} - \tfrac{2}{3}\log\alpha_2^C - \tfrac{2}{3}\log C_{T,free}. \tag{20.26}$$

From above,

$$Pb(II)_T = \frac{[Pb^{2+}]}{\alpha_0^{Pb(II)}} \tag{20.15}$$

where $\alpha_0^{Pb(II)}$ is a property of the solution only, given here by Eq.(20.18), and dependent only on pH and $C_{T,free}$, and not on whether any particular solid is present. If a particular Pb(II) solid is present and specifying $[Pb^{2+}]$ then Eq.(20.15) can be used with Eqs.(20.20), (20.22), (20.24), and (20.26) to obtain the corresponding dissolved $Pb(II)_T$ curves. As seen in Figure 20.2, for almost the entire pH range, $Pb_3(CO_3)_2(OH)_{2(s)}$ is the least soluble, and therefore limits Pb(II) solubility; whether it will be present depends on $Pb_{T,sys}$.

At low pH values, $Pb_3(CO_3)_2(OH)_{2(s)}$ is only slightly less soluble than $PbCO_{3(s)}$, but at slightly alkaline pH values it is more than an order of magnitude less soluble. The solubility limit curve

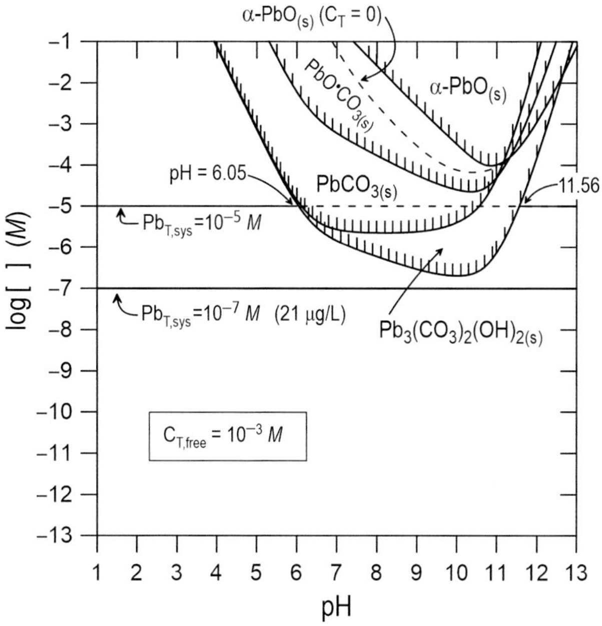

FIGURE 20.2 Log $Pb(II)_T$ for four solids *vs.* pH (solid lines) at 25 °C/1 atm with $C_{T,free} = 10^{-3}$ *M*. **a.** When $Pb_{T,sys} = 10^{-5}$ *M*, at equilibrium: 1) when pH ≤ 6.05 and pH ≥ 11.56, all $Pb(II)_{T,sys}$ is in solution; 2) when 6.05 < pH < 11.56, $Pb_3(CO_3)_2(OH)_{2(s)}$ is present and $Pb(II)_T < 10^{-5}$ *M*. **b.** When $Pb_{T,sys} = 10^{-7}$ *M*, no solid can form at any pH. The line for Log $Pb(II)_T$ for $PbO_{(s)}$ (dashed) for when $C_{T,free} = 0$ (from Figure 19.4) is included for comparison. Assumptions: 1) all solution-phase activity coefficients are unity; and 2) when a solid is present, it has unit activity.

for α–$PbO_{(s)}$ in Figure 20.2 for $C_{T,free} = 10^{-3}$ *M* lies above the line for $C_{T,free} = 0$ (dashed line, from Figure 19.4); this is because carbonate-related Pb(II) complexes reduce $\alpha_0^{Pb(II)}$ at all pH values, so that α–$PbO_{(s)}$ solubility is increased when $C_T > 0$. Because of their low solubilities, all along the curves plotted for $PbCO_{3(s)}$ and $Pb_3(CO_3)_2(OH)_{2(s)}$, $C_{T,free} = 10^{-3}$ *M* ≈ $C_{T,all\ species}$ (see Box 20.3).

The most important lesson from Figure 20.2 is how much lower the lines for $PbCO_{3(s)}$ and $Pb_3(CO_3)_2(OH)_{2(s)}$ are as compared to the line (dashed) for α–$PbO_{(s)}$ when $C_T = 0$. Thus, the presence of dissolved carbonate can be greatly advantageous in reducing Pb levels in drinking water delivered by Pb-containing plumbing, as follows: if Pb(II) is released oxidatively from $Pb_{(s)}$, then a Pb(II) carbonate can form as scale on the plumbing walls (and not, hopefully, as drinkable Pb-containing particles suspended in the water). However, despite the large solubility reduction for Pb(II) that is implied by $C_{T,free} = 10^{-3}$ *M* for $Pb_3(CO_3)_2(OH)_{2(s)}$, the solubility limit curve for that solid is still above the current U.S. EPA action level (AL) for drinking water of 15 μg/L (1991). This is why

BOX 20.3 $C_{T,free}$ *vs.* $C_{T,all\ species}$ for the Lines for $C_{T,free} = 10^{-3}$ *M* for the Four Solids in Figure 20.2

α–$PbO_{(s)}$ is so inherently soluble (see Box 20.1) that for its plotted Pb_T curve, $C_{T,free}$ does not comprise most of $C_{T,all\ species}$ until pH > ~10. If drawn, the curve for α–$PbO_{(s)}$ when $C_{T,all\ species} = 10^{-3}$ *M* would lie visibly lower for pH < ~10 than the line for α–$PbO_{(s)}$ for $C_{T,free} = 10^{-3}$ *M*. For the plotted Pb_T curve for $PbO{\cdot}CO_{3(s)}$, $C_{T,free}$ does not comprise most of $C_{T,all\ species}$ until pH > ~7.5.

orthophosphate (PO_4^{3-}) is sometimes added to finished drinking water: it allows formation of a very low solubility Pb(II) phosphate solid (see Section 20.4).

20.2.4 The Two Vertical Solution/Pb(II) Solid Boundary Lines and the Vertical Solution/Solution Boundary Lines for Pb(II)

From Figure 20.2, when $C_{T,free} = 10^{-3}$ *M*, then $Pb_{T,sys} = 10^{-5}$ *M* is high enough for $Pb_3(CO_3)_2(OH)_{2(s)}$ to form over some pH range. There are two pH values at which the log $Pb_{T,sys}$ line intersects the line for dissolved Pb_T as limited by $Pb_3(CO_3)_2(OH)_{2(s)}$. In particular, from Eqs.(20.15) and (20.26), for $Pb_{T,sys} = Pb_T$, then

$$\log \alpha_0^{Pb(II)} + \log Pb_{T,sys} = -6.27 - \tfrac{2}{3}\text{pH} - \tfrac{2}{3}\log \alpha_2^{C} - \tfrac{2}{3}\log C_{T,free}. \tag{20.27}$$

$$\text{solution} \mid Pb_3(CO_3)_2(OH)_{2(s)} \mid \text{solution}$$

When $C_{T,free} = 10^{-3}$ *M* and $Pb_{T,sys} = 10^{-5}$ *M*, and $\alpha_0^{Pb(II)}$ is given by Eq.(20.18), then the two pH roots of Eq.(20.27) are pH = 6.05 and 11.56. $Pb_3(CO_3)_2(OH)_{2(s)}$ will be present at intermediate pe when $6.05 < \text{pH} < 11.56$.

According to Figure 20.1, for all pH < 6.47, the dominant dissolved Pb(II) species is Pb^{2+}; for $11.17 < \text{pH} < 11.64$, the dominant species is $Pb(OH)_3^-$; for pH > 11.64, the dominant species is $Pb(OH)_4^{2-}$. Overall, through the Pb(II) region, the sequence of vertical lines is then

$$Pb^{2+} \overset{\text{pH}=6.05}{\Big|} Pb_3(CO_3)_2(OH)_{2(s)} \overset{\text{pH}=11.56}{\Big|} Pb(OH)_3^- \overset{\text{pH}=11.64}{\Big|} Pb(OH)_4^{2-}. \tag{20.28}$$

In generic terms then, we can expect that the pe–pH diagram desired will look something like Figure 20.3, though it will be complicated due to reduction of C(IV) at very low pe (see Section 20.2.5 and Eq.(20.30): C(IV), the redox state for $C_{T,free}$, is not stable at very low pe due to conversion to methane, C(–IV), as can occur by some methanogenic microorganisms (Oren, 1999).

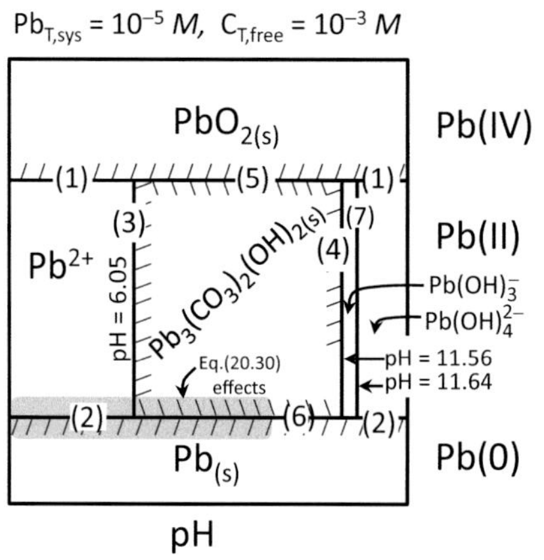

FIGURE 20.3 Generic pe–pH diagram for $Pb_{T,sys} = 10^{-5}$ *M* with $C_{T,free} = 10^{-3}$ *M*. At points below the line for Eq.(20.30) (not shown), reduction of C(IV) carbon to C(–IV) (*i.e.*, methane) at low pe values influences a region (shaded) where the lines are affected due to the essential absence of C(IV).

20.2.5 C(IV) Converted to C(−IV) (*i.e.*, CH_4) under Very Reducing Conditions

Like Pb, CO_2 species are subject to redox reactions. As discussed in Chapter 17, at full redox equilibrium, the two main carbon oxidation states of interest are: 1) C(IV) as in $H_2CO_3^*$, HCO_3^-, and CO_3^{2-}; and 2) C(−IV) in methane (CH_4). At 25 °C/1 atm, from Eqs.(17.94, (17.96) and (17.98),

$$\frac{\text{C(IV)}_T}{\text{C(−IV)}_T} = \frac{10^{-21.62}10^{8\text{pH}}10^{8\text{pe}}}{\alpha_0^{\text{C(IV)}}} \quad (20.29)$$

At any given pH, because of the factor 8 in 10^{8pe}, as pe drops through the $C(IV)_T$ to CH_4 transition, the loss of stability of $C(IV)_T$ relative to CH_4 occurs abruptly. The condition that $C(IV)_T/[CH_4] = 1$ therefore essentially defines the limits of the stability of $Pb_3(CO_3)_2(OH)_{2(s)}$. For $C(IV)_T/[CH_4] = 1$,

$$\text{pe} = 2.70 + \tfrac{1}{8}\alpha_0^{\text{C(IV)}} - \text{pH} \quad (20.30)$$

where $\alpha_0^{C(IV)}$ is the acid/base fraction of dissolved C(IV) present as $H_2CO_3^*$. Above the line given by Eq.(20.30), most of the dissolved C_T (both free and complexed) will be present as C(IV), and Pb lines drawn assuming that $[CO_3^{2-}] = \alpha_2^C C_{T,free}$ will be valid. Below that line, at full equilibrium, most of the dissolved C_T will be present as CH_4, and lines drawn assuming that $[CO_3^{2-}] = \alpha_2^C C_{T,free}$ will be wrong. The Eq.(20.30) line is included in the pe–pH diagrams for this chapter. Because of where that line falls, for $Pb_{T,sys} = 10^{-5}$ *M*, on the "low" pH side, two results are: 1) a wedge of extended solution predominance for Pb^{2+}; and 2) some vertical extension of the region for Pb(s) where line (6) in Figure 20.3 lies below Eq.(20.30) (see Figure 20.4).

20.2.6 The Curved $PbO_{2(s)}$/Solution Boundary Line

The left and right sides of line (1) in Figure 20.3 are parts of the same, curved $PbO_{2(s)}$/solution boundary line. From Eq.(19.24), for that line (which considers the contribution of PbO_3^{2-} to the Pb_T (dissolved)), when $Pb_T = Pb_{T,sys}$,

$$2\text{pe} = 49.2 - \log\left(\text{Pb}_{T,\text{sys}} - \frac{10^{-31.03}}{[\text{H}^+]^2}\right) - \log\alpha_0^{\text{Pb(II)}} - 4\text{pH} \quad \frac{\text{PbO}_{2(s)}}{\text{solution}} \quad (20.31)$$

where for this section $\alpha_0^{Pb(II)}$ is computed using Eq.(20.18) with $C_{T,free} = 10^{-3}$ *M*.

20.2.7 The Curved Solution/$Pb_{(s)}$ Boundary Line

The left and right sides of line (2) in Figure 20.3 are parts of the same, curved $Pb_{(s)}$/solution boundary line. From Eq.(19.29), for that line,

$$2\text{pe} = -4.26 + \log\text{Pb}_{T,\text{sys}} + \log\alpha_0^{\text{Pb(II)}}. \quad \frac{\text{solution}}{\text{Pb}_{(s)}} \quad (20.32)$$

In Chapter 19, $\alpha_0^{Pb(II)}$ was by Eq.(19.11) *i.e.*, without inclusion of carbonate-related complexes of Pb(II). In this chapter, for the line for Eq.(20.32): 1) at points below the line for Eq.(20.30), $\alpha_0^{Pb(II)}$ can be approximated by Eq.(19.11) because C(IV) is essentially gone); 2) at points above the line for Eq.(20.30), can be obtained for $C_{T,free} = 10^{-3}$ *M* using Eq.(20.18) which includes carbonate-related complexes of Pb(II). This two segment approximation approach is adequate; there is little payoff for going to the trouble to use Eq.(20.2) to exactly compute $C(IV)_{T,free}$ as a function of pe and pH to

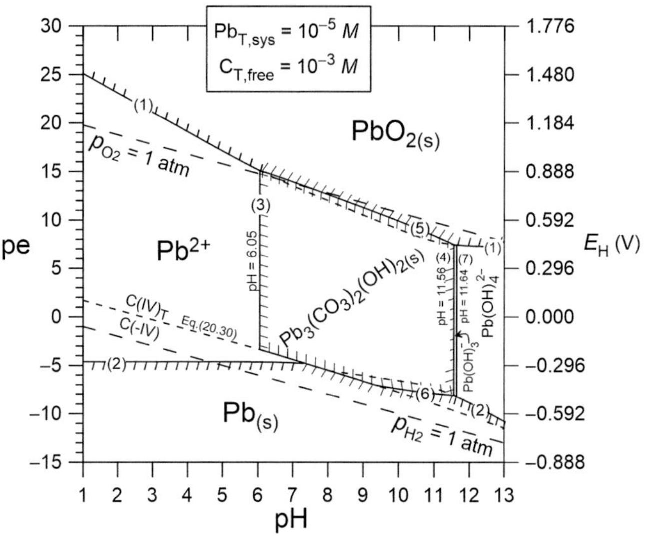

FIGURE 20.4 pe–pH predominance diagram for $Pb_{T,sys} = 10^{-5}$ *M* with $C_{T,free} = 10^{-3}$ *M* at 25 °C/1 atm. Assumptions: 1) all solution-phase activity coefficients are unity; and 2) when a solid is present, it has unit activity.

obtain the exact single line that covers both portions. Two summary comments can then be made about Eq.(20.32):

1) at points above the line for Eq.(20.30):
   1a) $\alpha_0^{Pb(II)}$ can be approximated in Eq.(20.32) including carbonate-related complexes of Pb(II) (Eq.20.18), with $C_{T,free} = 10^{-3}$ *M*;
   1b) $PbCO_3^o$ can be a dominant *solution* Pb(II) species;
2) at points below the line for Eq.(20.30), because C(IV) is essentially gone, then:
   2a) $\alpha_0^{Pb(II)}$ in Eq.(20.32) can be approximated by excluding carbonate-related complexes of Pb(II) (Eq.19.11);
   2b) $PbCO_3^o$ cannot be a dominant *solution* Pb(II) species, and the upper boundary of the added wedge for Pb^{2+} is (approximately), Eq.(20.30).

20.2.8 The Curved $PbO_{2(s)}/Pb_3(CO_3)_2(OH)_{2(s)}$ and $Pb_3(CO_3)_2(OH)_{2(s)}/Pb_{(s)}$ Solid/Solid Boundary Lines

For the solid/solid $PbO_{2(s)}/Pb_3(CO_3)_2(OH)_{2(s)}$ boundary line (line (5) in Figure 20.3),

$$PbO_{2(s)} + 4H^+ + 2e^- = 2H_2O + Pb^{2+} \qquad \log K = 49.2 \qquad (20.1)$$

$$-\tfrac{1}{3}\left[Pb_3(CO_3)_2(OH)_{2(s)} + 2H^+ = 3Pb^{2+} + 2CO_3^{2-} + 2H_2O \qquad \log K = -18.8\right] \qquad (20.6)$$

$$PbO_{2(s)} + \tfrac{10}{3}H^+ + \tfrac{2}{3}CO_3^{2-} + 2e^- = \tfrac{1}{3}Pb_3(CO_3)_2(OH)_{2(s)} + \tfrac{4}{3}H_2O \qquad \log K = 55.47. \qquad (20.33)$$

At 25 °C/1 atm, assuming unit activities for the solids, and unit activity coefficients in the solution phase, Eq.(20.33) yields

$$2\text{pe} = 55.47 + \tfrac{2}{3}\log\alpha_2^{\mathrm{C}} + \tfrac{2}{3}\log C_{\mathrm{T,free}} - \tfrac{10}{3}\,\text{pH}. \tag{20.34}$$

For the $Pb_3(CO_3)_2(OH)_{2(s)}/Pb_{(s)}$ boundary line (line (6) in Figure 20.3),

$$\tfrac{1}{3}\Big[Pb_3(CO_3)_2(OH)_{2(s)} + 2H^+ = 3Pb^{2+} + 2CO_3^{2-} + 2H_2O \qquad \log K = -18.8\Big] \tag{20.6}$$

$$+Pb^{2+} + 2e^- = Pb_{(s)} \qquad \log K = -4.26 \tag{20.3}$$

$$\tfrac{1}{3}Pb_3(CO_3)_2(OH)_{2(s)} + \tfrac{2}{3}H^+ + 2e^- = Pb_{(s)} + \tfrac{2}{3}CO_3^{2-} + \tfrac{2}{3}H_2O \qquad \log K = -10.53. \tag{20.35}$$

At 25 °C/1 atm, assuming unit activities for the solids, and unit activity coefficients in the solution phase, Eq.(20.35) yields

$$2\text{ pe} = -10.53 - \tfrac{2}{3}\log\alpha_2^{\mathrm{C}} + \log C_{\mathrm{T,free}} - \tfrac{2}{3}\,\text{pH}. \tag{20.36}$$

Eq.(20.36) applies for line (6) in Figure 20.3) only for the portion that lies above the line for Eq.(20.30). When below that line, then the line for Eq.(20.30) defines the lower boundary of the region for $Pb_3(CO_3)_2(OH)_{2(s)}$. Overall, for the final pe–pH diagram, Figure 20.4 contains the Pb boundary lines represented by Eqs.(20.28), (20.31), (20.32), (20.34), and (20.36).

Example 20.1 Effects of Changing pH and $C_{T,free}$ on the Solubility of $Pb_3(CO_3)_2(OH)_{2(s)}$

For a given C_T in water leaving a water distribution plant, Figure 20.2 shows how solubility limitation by $Pb_3(CO_3)_2OH_{2(s)}$ greatly reduces the solubility of Pb(II) at neutral to somewhat basic pH values. Raising $C_{T,free}$ will also affect the solubility as $Pb(II)_T$, but in two competing ways: lowering $[Pb^{2+}]$ by the action of the term $\log C_{T,free}$ with Eq.(20.26), but also lowering $\alpha_0^{Pb(II)}$ (which raises $Pb(II)_T = [Pb^{2+}]/\alpha_0^{Pb(II)}$). For both parts **a** and **b**, use Eq.(20.26) to obtain $[Pb^{2+}]$ for equilibrium with $Pb_3(CO_3)_2(OH)_{2(s)}$. Then, use Eq.(20.18) to obtain $\alpha_0^{Pb(II)}$ and then $Pb(II)_T$ in both molar and µg/L. The atomic weight of Pb is 207.2 g/mol. **a.** $C_{T,free} = 10^{-3}$ *M* at pH = 6.5 and 8.0. **b.** $C_{T,free} = 5 \times 10^{-3}$ *M* at pH = 6.5 and 8.0. Within the results for part **a**, and similarly for part **b**, what are the effects on $Pb(II)_T$ of raising the pH from 6.5 to 8.0? **c.** At pH = 6.5 (and 8.0), using the results for parts **a** and **b**, what are the effects on $Pb(II)_T$ of increasing $C_{T,free}$ from 1×10^{-3} *M* to 5×10^{-3} *M*.

Solution

	Equilibrium with $Pb_3(CO_3)_2OH_{2(s)}$		
$C_{T,free}$	pH = 6.5	pH = 8.0	Comment
a. 1×10^{-3} *M*	$[Pb^{2+}] = 10^{-5.90}$ *M*, $\alpha_0^{Pb(II)} = 0.419$, $Pb(II)_T = 10^{-5.52}$ *M* (630 µg/L)	$[Pb^{2+}] = 10^{-8.04}$ *M*, $\alpha_0^{Pb(II)} = 0.0154$, $Pb(II)_T = 10^{-6.23}$ *M* (123 µg/L)	raising pH from 6.5 to 8.0 at constant $C_{T,free}$ significantly reduces $Pb(II)_T$
b. 5×10^{-3} *M*	$[Pb^{2+}] = 10^{-6.36}$ *M*, $\alpha_0^{Pb(II)} = 0.139$, $Pb(II)_T = 10^{-5.50}$ *M* (649 µg/L)	$[Pb^{2+}] = 10^{-8.51}$ *M*, $\alpha_0^{Pb(II)} = 0.00327$, $Pb(II)_T = 10^{-6.02}$ *M* (196 µg/L)	
c. Comment	at pH = 6.5, raising $C_{T,free}$ increases $Pb(II)_T$ by 3% (reduction of $[Pb^{2+}]$ is almost completely offset by the reduction of $\alpha_0^{Pb(II)}$)	at pH = 8.0, raising $C_{T,free}$ increases $Pb(II)_T$ by ~60% (reduction of $[Pb^{2+}]$ is significantly smaller than the reduction of $\alpha_0^{Pb(II)}$)	

At constant $C_{T,free} = 1\times10^{-3}$ *M* (part a) and also for $C_{T,free} = 5\times10^{-3}$ *M* (part b), **raising the pH from 6.5 to 8.0 has a significant beneficial effect** in lowering $Pb(II)_T$. However, increasing $C_{T,free}$ from 1×10^{-3} *M* to 5×10^{-3} *M* at both pH values does not decrease $Pb(II)_T$, because in this range of $C_{T,free}$, the reduction of $[Pb^{2+}]$ is not as large as the reduction of $\alpha_0^{Pb(II)}$ (due to increased formation of carbonate-related Pb(II) complexes in solution). This indicates that $C_{T,free}$ is subject to some optimization when seeking to reduce $Pb(II)_T$ as limited by $Pb_3(CO_3)_2(OH)_{2(s)}$, though the dependence on $C_{T,free}$ is not strong.

Example 20.2 Effect of Changing C_T,free on Solubility of $Pb_3(CO_3)_2(OH)_2$(s) at pH = 8.0

For pH = 8.0 and $C_{T,free}$ ranging from 10^{-4} *M* to 10^{-2} *M*, compute $\alpha_0^{Pb(II)}$. Then, for equilibrium with $Pb_3(CO_3)_2(OH)_{2(s)}$, compute log $[Pb^{2+}]$ (*M*), log $Pb(II)_T$ (*M*), and $Pb(II)_T$ (μg/L).

Solution

Selected values are given in the accompanying table, as plotted in the accompanying figure. $Pb(II)_T$ varies by about a factor of 2 over two orders of magnitude in $C_{T,free}$.

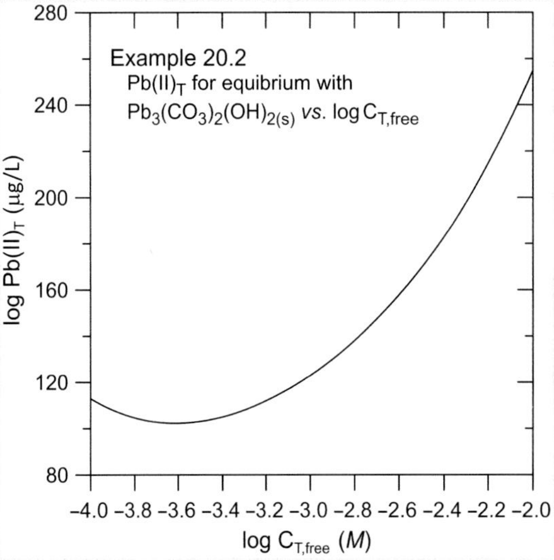

$C_{T,free}$ (*M*)	log $C_{T,free}$ (*M*)	$\alpha_0^{Pb(II)}$	log $[Pb^{2+}]$ (*M*)	log $Pb(II)_T$ (*M*)	$Pb(II)_T$ (μg/L)
0.0001	−4.00	7.75E-02	−7.37	−6.26	113
0.0002	−3.70	5.36E-02	−7.57	−6.30	103
0.0003	−3.52	4.09E-02	−7.69	−6.30	103
0.0004	−3.40	3.31E-02	−7.78	−6.30	105
0.0005	−3.30	2.78E-02	−7.84	−6.28	108
0.0007	−3.15	2.10E-02	−7.94	−6.26	114
0.0010	−3.00	1.54E-02	−8.04	−6.23	123
0.0020	−2.70	8.07E-03	−8.24	−6.15	147
0.0030	−2.52	5.44E-03	−8.36	−6.09	167
0.0050	−2.30	3.27E-03	−8.51	−6.02	198
0.0075	−2.12	2.16E-03	−8.62	−5.96	229
0.0100	−2.00	1.60E-03	−8.71	−5.91	255

20.3 pe–pH DIAGRAM FOR $Pb_{T,sys} = 10^{-7}$ *M* AND $C_{T,free} = 10^{-3}$ *M*

20.3.1 General

We follow the same approach as laid out in Section 20.2.1.

20.3.2 Solution $\alpha^{Pb(II)}$ Values with Hydroxide and Carbonate-Related Complexes

The same equations for the $\alpha^{Pb(II)}$ expressions with $C_T = 10^{-3}$ *M* apply as in Section 20.2.2.

20.3.3 Identification of the Pb(II) Solids That Limit Pb(II) Solubility

Based on Figure 20.2, when $C_{T,free} = 10^{-3}$ *M*, $Pb_{T,sys} = 10^{-7}$ *M* is not high enough for $Pb_3(CO_3)_2(OH)_{2(s)}$ or any of the other Pb(II) solids to form at any pH.

20.3.4 The Vertical Solution/Solution Boundary Lines for Pb(II)

As noted in Section 20.3.3, all through the intermediate pH range there is no predominance region for a Pb(II) solid: from low to high pH, dominance passes from one dissolved Pb(II) species to another. For $C_{T,free} = 10^{-3}$ *M*, the sequence is given in Figure 20.1. Thus for the vertical Pb(II) solution/solution boundary lines (lines (3), (4), (5), and (6) in Figure 20.5), then

$$Pb^{2+} \overset{pH=6.47}{\Big|} PbCO_3^o \overset{10.36}{\Big|} Pb(OH)_2^o \overset{11.17}{\Big|} Pb(OH)_3^- \overset{11.64}{\Big|} Pb(OH)_4^{2-}. \tag{20.37}$$

As reviewed below (Section 20.3.6), there is one more solution species that will have a predominance region when $Pb_{T,sys} = 10^{-7}$ *M*, namely PbO_3^{2-}, which is a Pb(IV) species. We expect this based on Figure 19.6 for $Pb_{T,sys} = 10^{-7}$ *M*. Overall, the pe–pH diagram will look something like Figure 20.5.

20.3.5 C(IV) Converted to C(–IV) (*i.e.*, CH_4) under Very Reducing Conditions

Due to the reductive loss of C(IV) from the system at low pe according to Eq.(20.30), the portion of the region for the solution species $PbCO_3^o$ below Eq.(20.30) is lost to area for non-carbonate-related

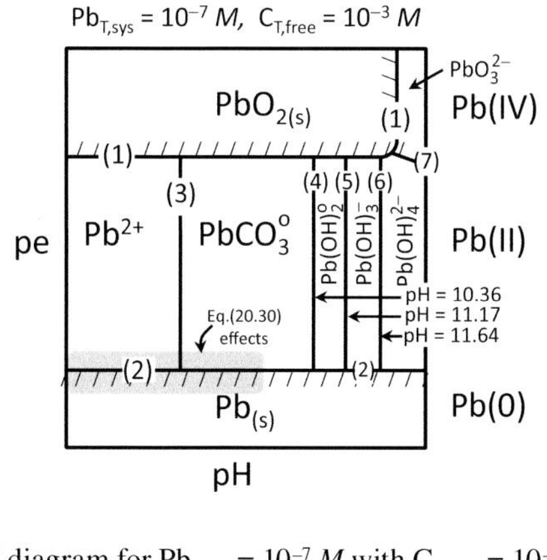

FIGURE 20.5 Generic pe–pH diagram for $Pb_{T,sys} = 10^{-7}$ *M* with $C_{T,free} = 10^{-3}$ *M*. At points below the line for Eq.(20.30) (not shown), reduction of C(IV) carbon to C(–IV) (*i.e.*, methane) influences a region (shaded) due to the essential absence of C(IV).

Pb(II) solution species, like Pb^{2+} and $PbOH^{+}$. Also, the portion of the curved solution $Pb_{(s)}$/solution boundary line that lies below Eq.(20.30) will be affected, as discussed in Section 20.3.7.

20.3.6 The Curved $PbO_{2(s)}$/Solution Boundary Line

As in Section 20.2.6, Eq.(20.31) gives the curved $PbO_{2(s)}$/solution boundary line (line (1) in Figure 20.5), and a region for PbO_3^{2-} is opened up at high pH and pe. It is not that PbO_3^{2-} reaches a higher concentration at high pe and high pH with $Pb_{T,sys} = 10^{-7}$ *M* than when $Pb_{T,sys} = 10^{-5}$ *M*; the opposite is largely the case. The region for PbO_3^{2-} appears when $Pb_{T,sys} = 10^{-7}$ *M* because then $PbO_{2(s)}$ is not present at high pe and pH > 12: the $PbO_{2(s)}$/solution boundary line ("coastline", if you will) for $PbO_{2(s)}$ has receded to the left, causing a solution region for PbO_3^{2-} to open up in the diagram at high pe and pH > 12. This means we will need a Pb(IV)/Pb(II) solution/solution boundary line (line (7) in Figure 20.5), as is now reviewed from Chapter 19.

20.3.7 The Curved $PbO_3^{2-}/Pb(II)_T$ Boundary Line

From Eq.(19.38), for the curved $PbO_3^{2-}/Pb(II)_T$ boundary line,

$$2pe = 80.23 - 6pH - \log \alpha_0^{Pb(II)} \frac{PbO_3^{2-}}{Pb^{2+}}. \qquad (20.38)$$

Eq.(20.38) does not depend on $Pb_{T,sys}$, but it does depend on $C_{T,free}$, by Eq.(20.18) for $\alpha_0^{Pb(II)}$.

20.3.8 The Curved Solution/$Pb_{(s)}$ Boundary Line

As in Section 20.2.7, Eq.(20.32) gives the curved $Pb_{(s)}$/solution boundary line (2) in Figure 20.5), and the two summary comments made in that section apply here as well. Overall, Figure 20.6 is obtained using Eqs.(20.31), (20.32), (20.37), and (20.38), with Eq.(20.30) superseding Eq.(20.32) as the approximate lower boundary of the region for $PbCO_3^o$ when Eq.(20.32) falls below Eq.(20.30).

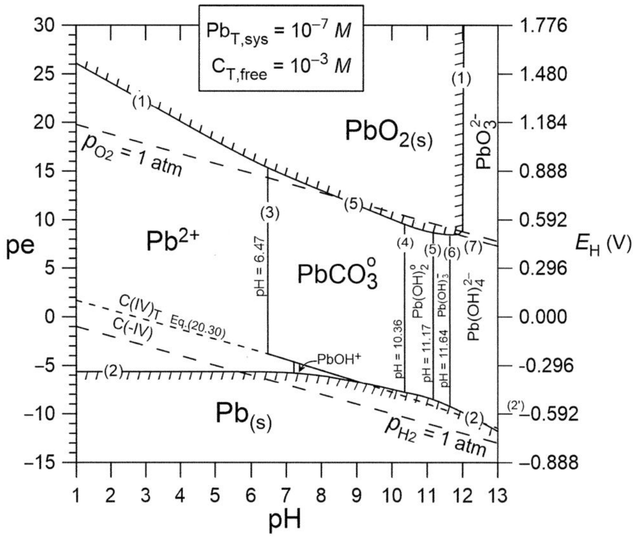

FIGURE 20.6 pe–pH predominance diagram for $Pb_{T,sys} = 10^{-7}$ *M* with $C_{T,free} = 10^{-3}$ *M* at 25 °C/1 atm. Assumptions: 1) all solution-phase activity coefficients are unity; and 2) when a solid is present, it has unit activity.

20.4 pe–pH DIAGRAM FOR $Pb_{T,sys} = 10^{-7}$ *M* WITH $C_{T,free} = 10^{-3}$ *M* AND $P_{T,free} = 10^{-5}$ *M* (~1 MG/L AS ORTHOPHOSPHATE)

20.4.1 General

Figure 20.2 illustrates that formation of and equilibrium with hydrocerussite ($Pb_3(CO_3)_2(OH)_{2(s)}$) is capable of considerable limitation of Pb(II) levels below what a Pb(II) oxide solid can accomplish, indeed to ~10^{-6} *M* Pb levels (~200 µg/L) at neutral to moderately basic pH values. Moreover, since some period of stagnancy will be required to reach equilibrium between water and hydrocerrusite scale within in a service line pipe (or with within indoor plumbing containing Pb), flushing clean water through the pipe and plumbing before consumption can be enormously helpful in further reducing Pb levels in drinking water, allowing conformance with the current U.S. EPA action level (AL) for drinking water of 15 µg/L (EPA, 1991). However, since rigorous water line flushing may not always occur with plumbing containing Pb, it is of interest to consider solids that are even less soluble than hydrocerrusite. Pb(II) solids with phosphate PO_3^{3-} come immediately to mind given that: 1) PO_3^{3-} bears an ionic charge of −3 (which will be more strongly attracted to a charge of +2 as on Pb^{2+} than an ionic charge of either −2 as on CO_3^{2-}; 2) hydroxyapatite ($Ca_5(PO_4)_3OH_{(s)}$) is known to have exceedingly low solubility, so its Pb analog hydroxypyromorphite ($Pb_5(PO_4)_3OH_{(s)}$) is likely to be is similarly insoluble. Also, phosphate is of generally low human toxicity, so adding a few mg/L of phosphate to finished drinking water is not problematic: Moser et al. (2015) report a median value of 324 mg/L (as orthophosphate) for 46 non-alcoholic beverages, while for 2001 survey data from 264 drinking water utilities carried out by the American Water Works Association, McNeill and Edwards (2002) state that 56% reported adding phosphate-based corrosion inhibitors at levels mostly in the range 0.2 to 3 mg/L (as orthophosphate), the median level being ~1 mg/L (~10^{-5} *M*). ~57% of those responding were using phosphate for the purpose of mitigating lead levels, and ~84% were using it to mitigate copper and/or lead levels. We now consider the nature of the pe–pH diagram for $Pb_{T,sys} = 10^{-7}$ *M*, $C_{T,free} = 10^{-3}$ *M*, and $P_{T,free} = 10^{-5}$ *M*.

Phosphate-related equilibria needed here are

$$Pb_5(PO_4)_3OH_{(s)} = 5Pb^{2+} + 3PO_4^{3-} + OH^- \qquad \log K_{s0} = -78.8 \qquad \text{solubility of Pb(II)} \quad \text{(Pb(II) solid 5)} \tag{20.39}$$

$$Pb^{2+} + HPO_4^{2-} = PbHPO_4^{o} \qquad \log K_{Pb}^{o} = 3.09 \qquad \text{complexation of dissolved Pb(II)} \tag{20.40}$$

$$Pb^{2+} + H_2PO_4^{-} = PbH_2PO_4^{+} \qquad \log K_{Pb}^{+} = 1.53 \qquad \text{complexation of dissolved Pb(II)} \tag{20.41}$$

$$H_3PO_4 = H^+ + H_2PO_4^- \qquad \log K_1^P = -2.16 \tag{20.42}$$

$$H_2PO_4^- = H^+ + HPO_4^{2-} \qquad \log K_2^P = -7.20 \tag{20.43}$$

$$HPO_4^{2-} = H^+ + PO_4^{3-} \qquad \log K_3^P = -12.32. \tag{20.44}$$

The values for the equilibrium constants pertain to 25 °C/1 atm. The values for the phosphate-related complexes (Eqs.(20.40) and (20.41) are from Schock et al., 1996), and not very large (~1000, and ~30, respectively). As a result, with $P_{T,free}$ low at 10^{-5} M, those complexes will not have a large effect on the α values dissolved for the Pb(II) species (*cf.* Figure 20.1 *vs.* Figure 20.7 below).

Two points can be made regarding Pb(II) phosphate solids. First, as with any extremely low solubility solid, there is some uncertainty in the $\log K_{s0}$ for $Pb_5(PO_4)_3OH_{(s)}$: Schock et al. (1996) cite literature data giving $\log K_{s0} = -76.8$, for 25 °C/1 atm; Zhu et al. (2016) report $\log K_{s0} = -80.8$. Here we take $\log K_{s0} = -78.8$, the geometric mean of the two K_{s0} values. Second, while $Pb_5(PO_4)_3Cl_{(s)}$ (chloropyromorphite) is known geologically, and the data compilation of Ball and Nordstrom (1991/2001) lists $\log K_{s0}$ −84.3 at 25 °C/1 atm, this value is suspect, and the solid will not be considered further here; Schock et al. (1996) note: "scant evidence has been uncovered that chloropyromorphite is an important solubility-controlling precipitate in potable waters".

20.4.2 Solution $\alpha^{Pb(II)}$ Values with Hydroxide, Carbonate-Related, and Phosphate-Related Complexes

Extension of Eq.(20.15) to include phosphate-related complexes gives

$$\alpha_0^{Pb(II)} = \frac{[Pb^{2+}]}{\begin{array}{c}[Pb^{2+}]+[PbOH^{+}]+[Pb(OH)_2^{o}]+[Pb(OH)_3^{-}]+[Pb(OH)_4^{2-}]\\ +[PbCO_3^{o}]+[Pb(CO_3)_2^{2-}]+[PbHCO_3^{+}]+[PbHPO_4^{o}]+[PbH_2PO_4^{+}]\end{array}} \quad (20.45)$$

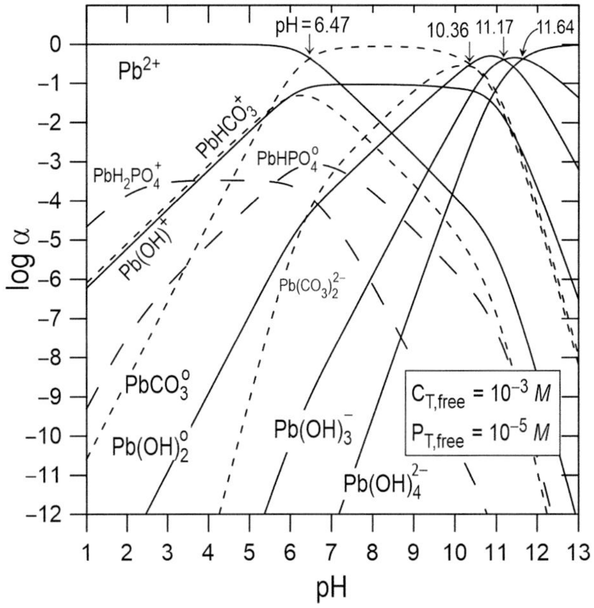

FIGURE 20.7 Log α *vs.* pH for dissolved Pb(II) species at 25 °C/1 atm, considering carbonate-related complexes with $C_{T,free} = 10^{-3}$ M and phosphate-related complexes with $P_{T,free} = 10^{-5}$ M. These α apply for dissolved Pb(II) regardless of whether any solid is present. Short-dashed lines are used to distinguish the carbonate-related Pb(II) complexes; long-dashed lines are used for the phosphate-related Pb(II) complexes. The phosphate-related complexes do not significantly affect the lines for the non-phosphate-related complexes so that as in Figure 20.1, Pb^{2+} predominates for pH < 6.47; $PbCO_3^o$ predominates for 6.47 < pH < 10.36; $Pb(OH)_2^o$ predominates for 10.36 < pH < 11.17; $Pb(OH)_3^-$ predominates for 11.17 < pH < 11.64; and $Pb(OH)_4^{2-}$ predominates for pH > 11.64. All solution-phase activity coefficients are assumed to be unity.

$$= \frac{1}{1+\frac{[PbOH^+]}{[Pb^{2+}]}+\frac{[Pb(OH)_2^o]}{[Pb^{2+}]}+\frac{[Pb(OH)_3^-]}{[Pb^{2+}]}+\frac{[Pb(OH)_4^{2-}]}{[Pb^{2+}]} +\frac{[PbCO_3^o]}{[Pb^{2+}]}+\frac{[Pb(CO_3)_2^{2-}]}{[Pb^{2+}]}+\frac{[PbHCO_3^+]}{[Pb^{2+}]}+\frac{[PbHPO_4^o]}{[Pb^{2+}]}+\frac{[PbH_2PO_4^+]}{[Pb^{2+}]}}. \tag{20.46}$$

The second to fifth terms in the denominator are obtained as in Box 19.1 The sixth to eighth terms are obtained as in Box 20.2. For the ninth and tenth terms, from Eqs.(20.40–20.41),

$$\frac{[PbHPO_4^o]}{[Pb^{2+}]} = K_{Pb}^o\alpha_2^P P_{T,free}, \qquad \frac{[PbH_2PO_4^+]}{[Pb^{2+}]} = K_{Pb}^+\alpha_1^P P_{T,free}. \tag{20.47}$$

$P_{T,free}$ is defined analogously with $C_{T,free}$ (Box 20.1). Thus, by extension of Eq.(20.18) to include phosphate-related complexes, for Pb^{2+},

$$\alpha_0^{Pb(II)} = \frac{1}{\left[1+\left(\frac{{}^*\beta_{H1}}{[H^+]}+\frac{{}^*\beta_{H2}}{[H^+]^2}+\frac{{}^*\beta_{H3}}{[H^+]^3}+\frac{{}^*\beta_{H4}}{[H^+]^4}\right) + \left(K_{C1}\alpha_2^C C_{T,free} + K_{C1}K_{C2}(\alpha_2^C C_{T,free})^2 + K_{HC1}\alpha_1^C C_{T,free}\right) + \left(K_{Pb}^o\alpha_2^P P_{T,free} + K_{Pb}^+\alpha_1^P P_{T,free}\right)\right]} \tag{20.48}$$

where the three sets of parentheses in the denominator collect the terms for the hydroxo complexes, the carbonate-related complexes, and the phosphate-related complexes, respectively. The α value expressions for the nine other Pb(II) species are given in Boxes 20.2 and 20.4. A $\log\alpha$ *vs.* pH plot is provided in Figure 20.7 for $C_{T,free} = 10^{-3}$ *M* and $P_{T,free} = 10^{-5}$ *M*. These α values apply for the dissolved Pb(II) regardless of whether any solid(s) is(are) present.

20.4.3 Identification of the Pb(II) Solids That Limit Pb(II) Solubility

From Section 20.2.3, for $C_{T,free} = 10^{-3}$ *M*, $Pb_3(CO_3)_2(OH)_{2(s)}$ is the only solid that could possibly compete with $Pb_5(PO_4)_3OH_{(s)}$. For $Pb_5(PO_4)_3OH_{(s)}$, by analogy with Eqs.(20.19–20.26), based on Eq.(20.39),

$$[Pb^{2+}]_{Pb_5(PO_4)_3OH_{(s)}} = \left(\frac{10^{-78.8}}{[PO_4^{3-}]^3[OH^-]}\right)^{1/5} = \left(\frac{10^{-78.8}[H^+]}{[PO_4^{3-}]^3 K_w}\right)^{1/5} = \left(\frac{10^{-64.8}[H^+]}{[PO_4^{3-}]^3}\right)^{1/5} \tag{20.49}$$

$$\log[Pb^{2+}]_{Pb_3(CO_3)_2(OH)_{2(s)}} = -12.96 - \tfrac{1}{5}pH - \tfrac{3}{5}\log\alpha_3^P - \tfrac{3}{5}\log P_{T,free}. \tag{20.50}$$

Using Eq.(20.15), Figure 20.8 provides a plot of $\log Pb(II)_T$ for the two solids using Eqs.(20.26) and (20.50). For both lines, $\alpha_0^{Pb(II)}$ is given by Eq.(20.48) with $C_{T,free} = 10^{-3}$ *M* and $P_{T,free} = 10^{-5}$ *M*. The lines indicate that $Pb_5(PO_4)_3OH_{(s)}$ is solubility limiting at pH < 9.90, and $Pb_3(CO_3)_2(OH)_{2(s)}$ is solubility limiting at pH > 9.90; the two solids could co-exist at pH = 9.90 when $Pb_{T,sys}$ is higher than the $Pb_T(II)$ for the intersection of the two lines ($10^{-6.69}$). Overall, for the important pH range of 6 to 9, there is a striking improvement in $Pb(II)_T$ solubility limitation accomplished by the presence of just $P_{T,free} = 10^{-5}$ *M* (~1 mg/L phosphate, as orthophosphate).

BOX 20.4 α Values beyond α_0 for Phosphate-Related Aqueous Pb(II) Species

For the phosphate-related complexes, using $\alpha_0^{Pb(II)} = [Pb^{2+}]/Pb_T$, and Eq.(20.40) and Eq.(20.41), and neglecting solution-phase activity corrections in the third row of each box of equations,

$$\alpha_{HP1}^{Pb(II)} \equiv \frac{[PbHPO_4^o]}{Pb(II)_T} = \alpha_0^{Pb(II)} \frac{[PbHPO_4^o]}{[Pb^{2+}]} = \alpha_0^{Pb(II)} K_{Pb}^{o} \alpha_2^{P} P_{T,free}$$

$$\alpha_{HC1}^{Pb(II)} \equiv \frac{[PbH_2PO_4^+]}{Pb(II)_T} = \alpha_0^{Pb(II)} \frac{[PbH_2PO_4^+]}{[Pb^{2+}]} = \alpha_0^{Pb(II)} K_{Pb}^{+} \alpha_1^{P} P_{T,free}$$

By analogy with $C_{T,free}$ (see Box 20.1), $P_{T,free}$ refers only to the sum $[H_3PO_4]+[H_2PO_4^-]+[HPO_4^{2-}]+[PO_4^{3-}]$, and does not include any of the HPO_4^{2-} or $H_2PO_4^-$ complexed to Pb(II).

20.4.4 The Two Vertical Solution/Pb(II) Solid Boundary Lines and the Vertical Solution/Solution Boundary Lines for Pb(II)

From Figure 20.8, when $C_{T,free} = 10^{-3}$ *M* and $P_{T,free} = 10^{-5}$ *M*, then $Pb_{T,sys} = 10^{-7}$ *M* is high enough for $Pb_5(PO_4)_3OH_{(s)}$ to form over some pH range, and is the only solid that can form. There are two pH values at which the log $Pb_{T,sys}$ line intersects the line for dissolved Pb_T as limited by $Pb_5(PO_4)_3OH_{(s)}$. From Eqs.(20.15) and (20.50), for $Pb_{T,sys} = Pb_T$, then

$$\log\alpha_0^{Pb(II)} + \log Pb_{T,sys} = -12.96 - \tfrac{1}{5}pH - \tfrac{3}{5}\log\alpha_3^{P} - \tfrac{3}{5}\log P_{T,free}. \quad (20.51)$$

$$\text{solution} \mid Pb_5(PO_4)_3OH_{(s)} \mid \text{solution}$$

When $C_{T,free} = 10^{-3}$ *M* and $P_{T,free} = 10^{-5}$ *M*, and $\alpha_0^{Pb(II)}$ given by Eq.(20.48), then the two pH roots of Eq.(20.51) are pH = 6.66 and 8.70. The solid $Pb_5(PO_4)_3OH_{(s)}$ will be present at intermediate pe when $6.66 < pH < 8.70$.

According to Figure 20.7, for all pH < 6.47, the dominant dissolved Pb(II) species is Pb^{2+}; for $6.47 < pH < 10.36$, the dominant species is $PbCO_3^o$; for $10.36 < pH < 11.17$, the dominant species is $Pb(OH)_3^o$; for $11.17 < pH < 11.64$, the dominant species is$Pb(OH)_3^-$; for pH > 11.64, the dominant species is $Pb(OH)_4^{2-}$. Overall, through the Pb(II) region, the sequence of vertical lines is

$$Pb^{2+} \overset{pH=6.47}{\mid} PbCO_3^o \overset{pH=6.66}{\mid} Pb_5(PO_4)_3OH_{(s)} \overset{pH=8.70}{\mid} PbCO_3^o \overset{pH=10.36}{\mid} Pb(OH)_2^o \overset{pH=11.17}{\mid} Pb(OH)_3^- \overset{pH=11.64}{\mid} Pb(OH)_4^{2-}. \quad (20.52)$$

Below the line for Eq.(20.30), the region for $PbCO_3^o$ on both sides of the region for $Pb_5(PO_4)_3OH_{(s)}$ will be truncated due to reduction of C(IV) and the region distributed to other Pb(II) solution species. Also, both boundary lines of the solution with $Pb_5(PO_4)_3OH_{(s)}$ will be curved, making the solid region somewhat larger, like roots at the bottom of a tree: when $C(IV)_{T,free} = 0$ and $P_{T,free} = 10^{-5}$ *M*, then $\alpha_0^{Pb(II)}$ is given by Eq.(20.18), and the low pH root of Eq.(20.51) is pH = 6.31 (rather than 6.66), and the high pH root is 10.28 (rather than 8.70).

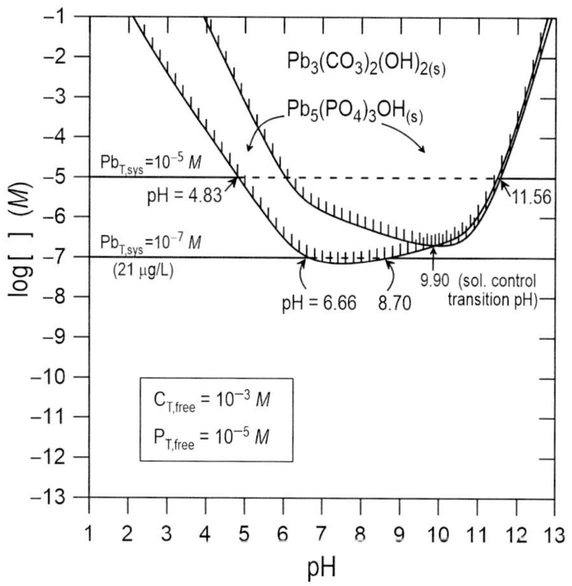

FIGURE 20.8 Log$Pb(II)_T$ for two solids *vs.* pH at 25 °C/1 atm with $C_{T,free} = 10^{-3}$ *M* and $P_{T,free} = 10^{-5}$ *M*. **a.** When $Pb_{T,sys} = 10^{-5}$ *M*, at equilibrium: 1) when pH ≤ 4.83 and pH ≥ 8.70), all $Pb(II)_{T,sys}$ is in solution; 2) when 4.83 < pH < 9.90, $Pb_5(PO_4)_3OH_{(s)}$ is present; when 9.90 < pH < 11.56, $Pb_3(CO_3)_2(OH)_{2(s)}$ is present; when pH = 9.90, both solids could be present, depending on how the system was attained; 3) $Pb(II)_T < 10^{-5}$ *M* when 4.83 < pH < 11.56. **b.** When $Pb_{T,sys} = 10^{-7}$ *M*, at equilibrium: 1) when pH ≤ 6.66 and pH ≥ 8.70, all $Pb(II)_{T,sys}$ is in solution; 2) when 6.66 < pH < 8.70, $Pb_5(PO_4)_3OH_{(s)}$ is present and $Pb(II)_T < 10^{-7}$ *M*. Assumptions: 1) all solution-phase activity coefficients are unity; and 2) when a solid is present, it has unit activity.

20.4.5 C(IV) Converted to C(–IV) (*i.e.*, CH_4) under Very Reducing Conditions

See Section 20.3.5; the same comments apply here.

20.4.6 The Curved $PbO_{2(s)}$/Solution Boundary Line

As in Sections 20.2.6 and 20.3.6, Eq.(20.31) gives the curved $PbO_{2(s)}$/solution boundary line. Here, $\alpha_0^{Pb(II)}$ is calculated by Eq.(20.48) using $C_{T,free} = 10^{-3}$ *M* and $P_{T,free} = 10^{-5}$ *M*. This line applies for pH < 6.66 and pH > 8.70.

20.4.7 The Curved $PbO_3^{2-}/Pb(II)_T$ Boundary Line

As in Section 20.3.7, Eq,(20.38) gives the curved $PbO_3^{2-}/Pb(II)_T$ boundary line with $\alpha_0^{Pb(II)}$ here calculated by Eq.(20.48) with $C_{T,free} = 10^{-3}$ *M* and $P_{T,free} = 10^{-5}$ *M*.

20.4.8 The Curved Solution/$Pb_{(s)}$ Boundary Line

As in Sections 20.2.7 and 20.3.8, when $Pb_{T,sys} = 10^{-7}$ *M*, then Eq.(20.32) again gives the curved $Pb_{(s)}$/solution boundary line (line (2) in Figure 20.5). In Chapter 19, $\alpha_0^{Pb(II)}$ was computed without inclusion of carbonate-related complexes of Pb(II). Here, for the line given by Eq.(20.32):

1) at points above the line for Eq.(20.30):
   1a) $\alpha_0^{Pb(II)}$ can be approximated computed including carbonate-related complexes of Pb(II), with $C_{T,free} = 10^{-3}$ *M*; and
   1b) $PbCO_3^o$ may be the dominant *solution* Pb(II) species;

2) at points below the line for Eq.(20.30), because C(IV) is essentially gone:
   2a) $\alpha_0^{\mathrm{Pb(II)}}$ can be approximated by excluding carbonate-related complexes of Pb(II); and
   2b) $PbCO_3^o$ cannot be the dominant *solution* Pb(II) species.

20.4.9 The Curved Solid/Solid $PbO_{2(s)}/Pb_5(PO_4)_3OH_{(s)}$ and $Pb_5(PO_4)_3OH_{(s)}/Pb_{(s)}$ Boundary Lines

For the solid/solid $PbO_{2(s)}/Pb_5(PO_4)_3OH_{(s)}$ boundary line,

$$PbO_{2(s)} + 4H^+ + 2e^- = 2H_2O + Pb^{2+} \qquad \log K = 49.2 \tag{20.1}$$

$$-\tfrac{1}{5}\left[Pb_5(PO_4)_3OH_{2(s)} = 5Pb^{2+} + 3PO_4^{3-} + OH^- \qquad \log K = -78.8\right] \tag{20.39}$$

$$+\left[\tfrac{1}{5}(H_2O = H^+ + OH^- \qquad \log K_w = -14.00\right] \tag{20.53}$$

$$PbO_{2(s)} + \tfrac{19}{5}H^+ + \tfrac{3}{5}PO_4^{3-} + 2e^- = \tfrac{1}{5}Pb_5(PO_4)_3OH_{(s)} + \tfrac{9}{5}H_2O \qquad \log K = 62.2. \tag{20.54}$$

At 25 °C/1 atm, assuming unit activities for the solids, and unit activity coefficients in the solution phase, Eq.(20.54) yields

$$2pe = 62.2 + \tfrac{3}{5}\log \alpha_3^P + \tfrac{3}{5}\log P_{T,free} - \tfrac{19}{5}pH. \tag{20.55}$$

For the $Pb_5(PO_4)_3OH_{(s)}/Pb_{(s)}$ boundary line,

$$\tfrac{1}{5}\left[Pb_5(PO_4)_3OH_{2(s)} = 5Pb^{2+} + 3PO_4^{3-} + OH^- \qquad \log K = -78.8\right] \tag{20.39}$$

$$+Pb^{2+} + 2e^- = Pb_{(s)} \qquad \log K = -4.26 \tag{20.3}$$

$$-\tfrac{1}{5}\left[H_2O = H^+ + OH^- \qquad \log K_w = -14.00\right] \tag{20.53}$$

$$\tfrac{1}{5}Pb_5(PO_4)_3OH_{(s)} + \tfrac{1}{5}H^+ + 2e^- = Pb_{(s)} + \tfrac{3}{5}PO_4^{3-} + \tfrac{1}{5}H_2O \qquad \log K = -17.22. \tag{20.56}$$

At 25 °C/1 atm, assuming unit activities for the solids, and unit activity coefficients in the solution phase, Eq.(20.56) yields

$$2pe = -17.22 - \tfrac{3}{5}\log \alpha_3^P - \tfrac{3}{5}\log P_{T,free} - \tfrac{1}{5}pH. \tag{20.57}$$

Two comments can be made about Eq (20.57):

1) at points above the line for Eq.(20.30):
   1a) $\alpha_0^{\mathrm{Pb(II)}}$ can be approximated in Eq.(20.48) including carbonate-and phosphate-related complexes of Pb(II), with $C_{T,free} = 10^{-3}$ *M* and $P_{T,free} = 10^{-5}$ *M*;
   1b) $PbCO_3^o$ can be the dominant *solution* Pb(II) species.

2) at points below the line for Eq.(20.30), because C(IV) is essentially gone, then:

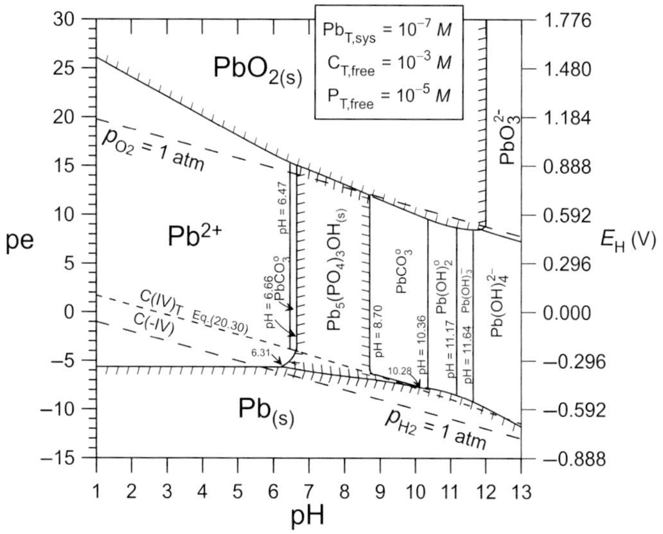

FIGURE 20.9 pe–pH predominance diagram for $Pb_{T,sys} = 10^{-7}$ *M* with $C_{T,free} = 10^{-5}$ *M* and $P_{T,free} = 10^{-5}$ *M* at 25 °C/1 atm. Assumptions: 1) all solution-phase activity coefficients are unity; and 2) when a solid is present, it has unit activity.

2a) $\alpha_0^{Pb(II)}$ in Eq.(20.57) can be approximated by excluding the Pb(II) carbonate-related complex terms in Eq.(20.48) but with $P_{T,free} = 10^{-5}$ *M*;

2b) $PbCO_3^o$ cannot be the dominant *solution* Pb(II) species, and the lower boundary of the region for $PbCO_3^o$ is, approximately, Eq.(20.30).

Overall, Figure 20.9 is obtained using Eqs.(20.52), (20.31), (20.32), (20.38), (20.55), and (20.57), with Eq.(20.30) superseding Eq.(20.32) as the approximate lower boundary of the region for $PbCO_3^o$ when Eq.(20.32) falls below Eq.(20.30).

Example 20.3 Effects of Phosphate on the Solubility of Pb(II)

Figure 20.9 shows how adding a low level of phosphate at pH values in the 6.5 to 8.5 range allows $Pb_5(PO_4)_3OH_{(s)}$ to greatly limit the solubility of $Pb(II)_T$. For pH = 7.5 and $P_{T,free} = 3\times10^{-5}$ *M* (~3 mg/L as orthophosphate), use Eq.(20.50) to obtain $[Pb^{2+}]$ for equilibrium with $Pb_5(PO_4)_3OH_{(s)}$. Then, if $C_{T,free} = 2\times10^{-3}$ *M*, use Eq.(20.48) to obtain $\alpha_0^{Pb(II)}$ and then $Pb(II)_T$ in both molar and μg/L. The atomic weight of Pb is 207.2 g/mol.

Solution

$[Pb^{2+}] = 10^{-8.75}$ *M*; $\alpha_0^{Pb(II)} = 10^{-1.58}$; $Pb(II)_T = 10^{-8.75}$ *M*$/10^{-1.58} = 10^{-7.17}$ *M* or 14 μg/L.

REFERENCES

Ball JW, Nordstrom DK (1991/2001) *User's Manual For WATEQ4F, With Revised Thermodynamic Data Base and Test Cases For Calculating Speciation Of Major, Trace, And Redox Elements In Natural Waters.* U.S. Geological Survey, Open-File Report 91-183, Menlo Park, CA, 1991. Revised and reprinted, April 2001.

Easley RA, Byrne RH (2011) The ionic strength dependence of lead (II) carbonate complexation in perchlorate media. *Geochimica et Cosmochimica Acta*, **75**, 5638–5647.

Langmuir D, Chrostowski P, Vigneault B, Chaney R (2004). Issue Paper on the Environmental Chemistry of Metals. Report to U.S. Environmental Protection Agency, Contract #68-C-02-060, Risk Assessment Forum, Washington, D.C.

Lytle DA, Schock MR (2005) Formation of Pb(IV) oxides in chlorinated water. *American Water Works Association Journal*, **97**, 102–114.

McNeill L, Edwards M (2002) Phosphate inhibitor use at US utilities. *American Water Works Association. Journal*, **94**, 57–63.

Moser M, White K, Henry B, Oh S, Miller ER, Anderson CA, Benjamin J, Charleston J, Appel LJ, Chang AR (2015) Phosphorus content of popular beverages. *American Journal of Kidney Diseases*, **65**, 967–971.

Oren A (1999) Bioenergetic aspects of halophilism. *Microbiology and Molecular Biology Reviews*, **63**, 334–348.

Schock MR, Wagner I, Oliphant RJ (1996) Chapter 4, Corrosion and solubility of lead in drinking water. In: *Internal Corrosion Of Water Distribution Systems: Cooperative Research Report*, 2nd edn. AWWA Research Foundation (Denver, CO) and DVGW-Technologiezentrum Wasser (Karlsruhe, Federal Republic of Germany).

U.S. EPA (1991) Title 40: Protection of Environment, Part 141—National Primary Drinking Water Regulations, Federal Register, 56 FR 26548, June 7, 1991.

Zhu Y, Huang B, Zhu Z, Liu H, Huang Y, Zhao X, Liang M (2016) Characterization, dissolution and solubility of the hydroxypyromorphite–hydroxyapatite solid solution [$(Pb_xCa_{1-x})_5(PO_4)_3OH$] at 25 °C and pH 2–9. *Geochemical Transactions*, **17**:2, 1–18.

21 pe and Natural Systems

21.1 REDOX-CONTROLLING ELEMENTS IN NATURAL WATERS

21.1.1 MAJOR *VS.* MINOR REDOX ELEMENTS

The list of elements that can play a controlling role in determining the redox conditions in natural waters is short, and is usually oxygen, nitrogen, iron, sulfur, carbon, and hydrogen. Adding in chlorine, which is heavily in used in the disinfection of drinking water and waste water, the corresponding mnemonic for this group is ClONFeSCH. The author had been using I for iron, but Cervarich (2019) aptly suggested using Fe, giving (roughly) "clownfish", which is perhaps more easily remembered. The element order gives the relative oxidizing strength of the elements when they are in their highest common oxidation states, *i.e.*, Cl(I) > O(0) > N(V) > Fe(III) > S(VI) > C(IV) > H(I). The order thus gives the redox-ladder succession according to which those oxidized forms will tend to be reduced as a given system becomes more reducing, as might occur by a titration with organic carbon.

Besides the ClONFeSCH group, other redox-active elements (*e.g.*, manganese, arsenic, lead, mercury, selenium, etc.) can also be present in natural systems, but usually they are very minor components so that their chemistries tend to "go along for the ride" as established by the dominant redox elements. For example, in a high-organic buried sediment, trace levels of Hg(II) will in general tend to be reduced to elemental mercury Hg(0), with hardly any depletion of the organic carbon that is present; and As(V) will be reduced to yield the more mobile As(III). This situation for overall redox chemistry is analogous with what happens with overall acid/base equilibrium chemistry in a solution as follows: in a given natural water situation, a small number of acidic and basic species will largely control the pH (*e.g.*, $H_2CO_3^*$ plus some positive or negative net strong base), and the α values for minor acid/base-active species will be set largely by the dominant chemistry.

Chapter 18 considered the pe–pH predominance diagrams of the elements Cl, O, and H. Chlorine was considered first because of its great importance in water treatment; oxygen and hydrogen were considered because of their roles in setting the limits of water stability for systems at 1 atm total pressure, as with the pe–pH line for p_{O_2} = 1 atm as the upper stability limit line, and the pe–pH line for p_{H_2} = 1 atm as the lower stability limit line. In this chapter, Sections 21.2–21.5 consider the pe–pH predominance diagrams of the remaining elements of the ClONFeSCH group, namely nitrogen, iron, sulfur, and carbon, following the order in ClONFeSCH. A collection of useful predominance diagrams for important elements is provided by Garrels and Christ (1965); a lengthy review of the chemistry of numerous metals is provided by Langmuir et al. (2004).

21.1.2 FULL REDOX EQUILIBRIUM RARELY OBTAINED IN NATURAL WATERS

Acid/base equilibrium in water usually reaches equilibrium in fractions of a second: proton exchange is very fast. So, as reviewed in Section 17.3.2.4, in natural waters one can always assume that all conjugate acid/base pairs HA/A⁻, HB/B⁻, HC/C⁻, etc. are describing the same pH according to

$$\mathrm{pH} = \left(\mathrm{p}K_{\mathrm{HA}} + \log\frac{\{\mathrm{A}^-\}}{\{\mathrm{HA}\}}\right) = \left(\mathrm{p}K_{\mathrm{HB}} + \log\frac{\{\mathrm{B}^-\}}{\{\mathrm{HB}\}}\right) = \left(\mathrm{p}K_{\mathrm{HC}} + \log\frac{\{\mathrm{C}^-\}}{\{\mathrm{HC}\}}\right). \quad (17.40)$$

Thus, the acid/base activity ratios in a given aqueous system, together with the various pK values, all indicate the same pH.

For redox reactions, the analog of Eq.(17.40) is Eq.(17.41)

$$\mathrm{pe} = \left(\mathrm{pe}_1^\circ + \frac{1}{n_1}\log\frac{\{\mathrm{OX}_1\}}{\{\mathrm{RED}_1\}}\right) = \left(\mathrm{pe}_2^\circ + \frac{1}{n_2}\log\frac{\{\mathrm{OX}_2\}}{\{\mathrm{RED}_2\}}\right) = \left(\mathrm{pe}_3^\circ + \frac{1}{n_3}\log\frac{\{\mathrm{OX}_3\}}{\{\mathrm{RED}_3\}}\right). \quad (17.41)$$

where each the {OX} term represents the product of the activities of all of the reactants (except e^-) raised to their respective stoichiometric powers, and {RED} represents the product of all of the activities of all of the products raised to their respective stoichiometric powers. Unlike H^+ exchange, abiotically the great majority of e^- exchange (redox) reactions are very slow. This is particularly true for reactions involving the exchange of multiple electrons. For example, while the removal of a single electron from Fe(II) to form Fe(III) can abiotically proceed relatively quickly, the oxidation of H_2S (S(–II)) to SO_4^{2-} to (S(VI)) involves the removal of eight electrons and abiotically is very slow. We can all be very grateful that at ambient temperatures, the abiotic multi-electron oxidation by oxygen of carbohydrates, lipids, proteins, DNA etc. to CO_2 is exceedingly slow. If it were very fast, we and all other life forms would be quickly oxidized (and maybe catch on fire): aerobic life forms are wildly out of equilibrium with their oxic environments. Moreover, happily, biologically mediated reactions can be kept under control so that things do not get out of hand. We can eat a piece of breakfast toast with jam in the morning, and then rather than just being respired to CO_2 in seconds or minutes, the organic material (fuel) can keep us going for hours; and, if we eat more than we need to keep going, some of the consumed organic calories can be stably held as fat to carry us through possible tough times ahead.

Multiple electron reactions are slow because they entail complex mechanisms, and because the actual physical concentration of the electron $[e^-]$ in aqueous systems is always very low ($\mathrm{pe} \equiv \log [e^-]\gamma_{e^-}$): if $[e^-]$ is low, it is hard to find e^- to make an exchange. So, in the ambient environment, in the absence of microorganisms, multiple electron redox reactions are very slow, and one cannot think about the application of equilibrium principles on time scales shorter than thousands to millions of years. By catalyzing redox reactions and moving nature towards equilibrium, organisms can obtain the energy that is stored in a nonequilibrium redox situation. Cells use this energy to continue living, build cell mass, and reproduce. In nature, virtually any redox reaction that can be out of equilibrium will have several strains of microorganism that can move such a system towards equilibrium: if there is an ecological niche that can be occupied, nature will fill it. Many of these bacteria have names that reflect what they do, as with *Thiobacillus ferrooxidans*. This bacterium can oxidize H_2S and $S_{(s)}$ to SO_4^{2-} (the "thio" part), and can oxidize Fe(II) to Fe(III) (the "ferro" part).

If the constant input of solar energy to the Earth were to cease, systems at the surface of the Earth would gradually attain full redox equilibrium and be fully dead; life is a nonequilibrium condition. The daily input of solar energy continuously frustrates the approach to total redox equilibrium at the surface of the Earth; solar energy provides the energy stream for most ecosystems. One exception is found at deep oceanic vents where reduced compounds like H_2S escape into the overlying seawater; a food web can be built upon the primary production of microorganisms that use O_2 in the seawater to oxidize those sulfur compounds (*e.g.*, Van Dover et al., 2002).

Overall, as noted, the ubiquitous absence of total redox equilibrium in natural systems means that Eq.(17.41) is never exactly applicable. If one were to obtain chemical analytical data for some aqueous system that allowed calculating the {OX}/{RED} ratios for a series of redox half-reactions, it is certain all of the resulting calculated pe values would be different. The virtually unknowable actual nonequilibrium pe value that a given water will have will be determined by the two redox half-reactions that are exchanging electrons with the greatest facility, and even those two half-reactions will likely not be in equilibrium. As a result, the main purpose of equilibrium redox calculations for natural waters is not to give us a means to predict exactly what the actual pe is for a given system, or conversely to use a pe to calculate exactly what all the {OX}/{RED} activity ratios are, but rather to indicate the general redox conditions in a system (*e.g.*, very oxidizing, somewhat oxidizing, somewhat reducing, or very reducing). For example, in a very reducing system, while calculations based on the different redox half reactions in Eq.(17.41) will not give the same pe, they will at least all give a very low pe: if H_2S is found in some groundwater, it is certain that there is no measurable O_2, nitrate, or Fe(III).

21.1.3 pe° AND pe°(W)

pe° values as introduced in Chapter 17 provide a somewhat useful reference scale for comparing the relative oxidation and reduction strengths of different redox half-reactions. What limits the utility of pe° values for natural systems is the fact that each such value pertains to the condition that all individual activities in {OX} and also in {RED} are unity, or at least that the overall {OX}/{RED} = 1. Consider the simple half-reaction

$$Fe^{3+} + e^- = Fe^{2+} \qquad pe = \frac{1}{1}\left(\log K + \log\frac{\{OX\}}{\{RED\}}\right) = pe^o + \log\frac{\{Fe^{3+}\}}{\{Fe^{2+}\}} \tag{21.1}$$

for which pe° = log K = 13.0 at 25 °C/1 atm. (All specific K values given in this chapter pertain to 25 °C/1 atm.) Knowing that the pe = pe° = 13.0 means that the pe is poised at the value when the two species exchanging the electrons are at equal activities: $\{Fe^{3+}\} = \{Fe^{2+}\}$. As seen in Table 17.1, many redox half-reactions involve more than just the two chemical species that are exchanging electrons. Consider the reduction of nitrate to nitrite:

$$NO_3^- + 2H^+ + 2e^- = NO_2^- + H_2O$$

$$pe = \frac{1}{2}\left(\log K + \log\frac{\{OX\}}{\{RED\}}\right) = pe^o + \frac{1}{2}\log\frac{\{NO_3^-\}\{H^+\}^2}{\{NO_2^-\}}. \tag{21.2}$$

With log$K$ = 28.3, when $\{NO_3^-\} = \{NO_2^-\}$, then pe ≠ pe° = 14.15 unless $\{H^+\} = 1$. $\{H^+\} = 1$ is not a very relevant environmental condition. In other words, if we know $\{NO_3^-\} = \{NO_2^-\}$, this will not tell us the pe that the nitrate/nitrite couple would give for environmentally relevant conditions.

The pe°(W) scale has been developed to mitigate the general incomparability of pe° values for environmental conditions. The pe°(W) scale gives the pe values for redox half-reactions when all species within the {OX}/{RED} expression combine to give a value of 1 (as with $\{NO_3^-\} = \{NO_2^-\}$), and $\{H^+\}$ is set at 10^{-7}, the value for neutral water. For the half-reaction in Eq.(21.1), there is no H^+ in the half reaction, so pe° = pe°(W) =13.0. For other half-reactions, the general relation between pe° and pe°(W) at 25 °C/1 atm is (see Section 17.4.2)

$$pe^o(W) = pe^o - \frac{n_H}{n_e}7.0 \tag{21.3}$$

where n_H is the number of H^+ on the LHS of the half reaction, and n_e is the number of electrons. For the half reaction in Eq.(21.2), $n_H = 2$ and $n_e = 2$, so $pe^o(W) = 14.15 - \frac{2}{2}7.0 = 7.15$.

Other texts in water chemistry discuss the use of pe°(W) values, *e.g.*, Stumm and Morgan (1996). However, as a practical matter, this author does not find pe°(W) to be worth the trouble. In many cases, one is not seeking the particular pe°(W) condition, because the electron exchanging species are not at similar activities, nor is pH = 7. Moreover, there is the added considerable complication that many redox half-reactions involve solids, as with

$$SO_4^{2-} + 8H^+ + 6e^- = S_{(s)} + 4H_2O$$

$$pe = \frac{1}{6}\left(\log K + \log\frac{\{OX\}}{\{RED\}}\right) = pe^o + \frac{1}{6}\log\frac{\{SO_4^{2-}\}\{H^+\}^8}{\{S_{(s)}\}}. \tag{21.4}$$

With log $K = 36.2$, now pe°(W) = $\frac{1}{6}36.2 - \frac{8}{6}7.0 = -3.30$. To use this pe°(W)value, $\{S_{(s)}\}$ is taken to be 1, which means that for $\{SO_4^{2-}\}/\{S_{(s)}\} = 1$, which means $\{SO_4^{2-}\} = 1$: so this pe°(W) and many others that involve solids are not directly relevant to environmental considerations. The approach in this text is then not to depend on tables of pe° or pe°(W) to characterize redox ladder positions in natural systems.

21.2 pe–pH DIAGRAM FOR NITROGEN

21.2.1 Redox Equilibria Governing Nitrogen Species

Solution-phase activity coefficients are assumed to be unity. Nitrogen is in the VA column of the periodic table, so atomic nitrogen has 5 valence electrons. Removing 5 electrons takes nitrogen to the relatively stable state of an empty outer shell, N(V). Adding 3 electrons takes nitrogen to a full outer shell of 8, N(–III). Inorganic nitrogen can exist in four oxidation states, V, III, 0, and –III (Table 21.1), with other inorganic nitrogen oxidation states being unstable relative to disproportionation. The redox half-reactions in Table 17.1 that involve the Table 21.1 species are collected in Table 21.2. Thus, inorganic nitrogen can range from the N(V) species NO_3^- down to the N(–III) species NH_4^+ and NH_3. Eight electrons are added to NO_3^- to convert it to NH_4^+ or NH_3 and vice versa. This difference of 8 in oxidation states is the same as the difference of 8 between C(IV) and C(–IV).

At sufficiently high pe, the predominant oxidation state will be N(V), as with HNO_3 and NO_3^-. HNO_3, however, is such a strong acid that it cannot have a predominance region even at very low pH, so for all pH there will be one N(V) region labeled NO_3^-. HNO_2 is only a modestly strong acid, with $pK_a = 3.25$. Thus, if there is predominance layer for N(III), it will be labeled HNO_2 for pH<3.25, and NO_2^- for pH>3.25. N_2 has no acid/base properties, so if there is a predominance layer for N(0), for all pH it will be labeled N_2. At sufficiently low pe, the predominant oxidation state will be N(III), as with NH_4^+ and NH_3. For NH_4^+, $pK_a = 9.24$. The predominance field for N(–III) will be labeled NH_4^+ for pH<9.24, and NH_3 for pH>9.24. With the four oxidation states and the preceding considerations, absent additional information, the pe–pH diagram for aqueous N may then look something like the diagram in Figure 21.1. Whether there are regions for N(III) and N(0) will depend on where lines (1), (2), and (3) actually fall. Arsenic (As) is also a group VA element, being directly below nitrogen in the periodic table.

21.2.2 Identification of the pe–pH Predominance Regions for Aqueous Nitrogen

Whether N(III) and N(0) will have predominance regions in the final diagram for nitrogen can be determined by examining the equations for lines (1), (2), and (3). In Figure 21.1, the vertical sequence

TABLE 21.1
Important Inorganic Forms of Nitrogen in Natural Water Systems and Their Oxidation States pK_a Values at 25 °C/1 atm

oxidation state	species	pK_a values
V	NO_3^-	
III	HNO_2 and NO_2^-	$pK_a = 3.25$
0	N_2	
–III	NH_4^+ and NH_3	$pK_a = 9.24$

TABLE 21.2
Data for Selected Nitrogen-Related Redox Reactions at 25 °C/1 atm in Order of Decreasing pe° and E_H^0

reduction half reaction	log K	pe°	pe°(W)	E_H^0(=0.05916 pe°)
$NO_2^- + 4H^+ + 3e^- = \frac{1}{2}N_2 + 2H_2O$	75.4	25.13	15.80	1.49
$NO_3^- + 6H^+ + 5e^- = \frac{1}{2}N_2 + 3H_2O$	103.7	20.74	12.34	1.23
$NO_2^- + 8H^+ + 6e^- = NH_4^+ + 2H_2O$	90.8	15.14	5.82	0.90
$NO_3^- + 10H^+ + 8e^- = NH_4^+ + 3H_2O$	119.2	14.9	6.15	0.88
$NO_3^- + 2H^+ + 2e^- = NO_2^- + H_2O$	28.3	14.15	7.15	0.84
$NO_2^- + 7H^+ + 6e^- = NH_3 + 2H_2O$	81.5	13.58	5.41	0.80
$N_2 + 8H^+ + 6e^- = 2NH_4^+$	31.3	5.21	–4.12	0.31
$N_2 + 6H^+ + 6e^- = 2NH_3$	12.69	2.11	–4.89	0.125

Note: N_2 is dissolved nitrogen.

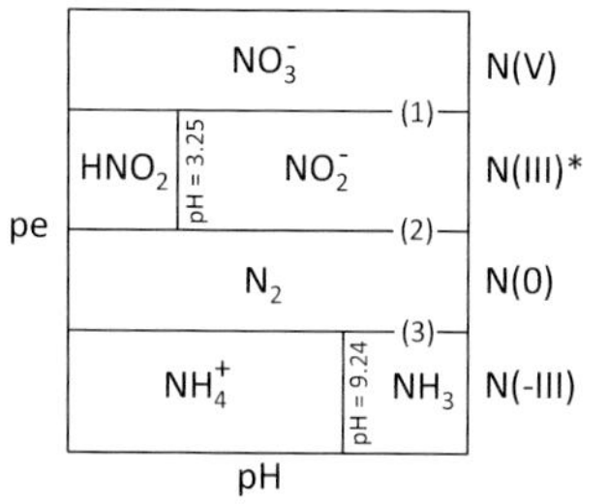

FIGURE 21.1 Preliminary pe–pH diagram for nitrogen. The asterisk on N(III) indicates that because of where lines (1), (2), and (3) actually fall, in the final diagram there will be no layer at equilibrium for N(III). At 25 °C/1 atm, for HNO_2, $pK_a = 3.25$; for NH_4^+, $pK_a = 9.24$.

is line (1) > line (2) > line (3). The equations are developed in Table 21.3; the two involving N_2 depend on the value of total dissolved nitrogen, which is set at 10^{-3} *M*. The lines are plotted in Figure 21.2, and the actual vertical sequence is line (2) > line (1) > line (3). This means that at equilibrium, N(III) cannot have a predominance region at any pH: above line (1), N(V) predominates over N(III); below line (2), N(0) predominates over N(III). These two conditions cover the whole diagram, so N(III) cannot predominate anywhere in the diagram. So, when N(III) *is* found in an aquatic system, *e.g.*, groundwater contaminated with nitrate fertilizers, N equilibrium cannot be present.

Consider now the region between lines (2) and (1) in Figure 21.2. N(V) predominates above line (2) because N(V) > N(III) everywhere above line (1). N(0) predominates below line (1) and down to line (3) because line (1) is below line (2). The region between lines (2) and (1) will be divided between predominance by N(V) for an upper portion, and N(0) for a lower portion: to demarcate the boundary between those two portions, we need the N(V)/N(0) line, which is Eq.(21.20) as developed in Table 21.3. This line is a hybrid of line (1) for N(V)/N(III) and line (2) for N(III)/N(0). This chemistry is a direct analog of the situation with chlorine in Chapter 18 for the region where the Cl(I)/Cl(0) line lies below the Cl(0)/Cl(–I) line: a hybridization is needed beyond the Cl_2 wedge to obtain the Cl(I)/Cl(–I) line.

The final diagram is provided in Figure 21.3. As catalyzed by common biochemical enzymes, the dashed arrows show the inter-valence state conversions that are kinetically difficult, and which

TABLE 21.3
Equations for pe–pH Boundary Lines for $N(V)/N(III)_T$, $N(III)_T/N(0)$, and $N(0)/N(-III)_T$, with Equilibrium Constant Values for 25 °C/1 Atm. As Needed, Total Dissolved N (= $N_{T,Sys}$) = 10^{-3} *M*

Couple	equations	
$\frac{N(V)}{N(III)_T}$ line (1)	$NO_3^- + 2H^+ + 2e^- = NO_2^- + H_2O \quad pe^o = 14.15$	(21.5)
	$pe = 14.15 + \frac{1}{2}\log\frac{\{NO_3^-\}\{H^+\}^2}{\{NO_2^-\}}$	(21.6)
	$N(III)_T = [HNO_2] + [NO_2^-]$; for HNO_2, $pK_a = 3.25$ for use in $\alpha_1^{N(III)}$	
For unit γ values, then:	$pe = 14.15 + \frac{1}{2}\log\frac{[NO_3^-]}{\alpha_1^{N(III)}N(III)_T} - pH.$	(21.7)
There are equal amounts of N in the two oxidation states when $[NO_3^-] = N(III)_T$. Then:		
	$pe = 14.15 + \frac{1}{2}\log\alpha_1^{N(III)} - pH \quad$ line (1)	(21.8)
$\frac{N(III)_T}{N(0)}$ line (2)	$NO_2^- + 4H^+ + 3e^- = \frac{1}{2}N_2 + 2H_2O \quad pe^o = 25.13$	(21.9)
	$pe = 25.13 + \frac{1}{3}\log\frac{\{NO_2^-\}\{H^+\}^4}{\{N_2\}^{\frac{1}{2}}}$	(21.10)
	$N(III)_T = [HNO_2] + [NO_2^-]$; for HNO_2, $pK_a = 3.25$ for use in $\alpha_1^{N(III)}$.	
For unit γ values:	$pe = 25.13 + \frac{1}{3}\log\frac{\alpha_1^{N(III)}N(III)_T}{[N_2]^{\frac{1}{2}}} - \frac{4}{3}pH.$	(21.11)
Neglecting N(V) and N(−III) in the region of the line, for $N_{T,sys} = 10^{-3}$ *M*, there are equal amounts of dissolved nitrogen in the two oxidation states when:		
	$N(III)_T = 2[N_2] = 0.5 \times 10^{-3} M.$	
Then:	$pe = 24.63 + \frac{1}{3}\log\alpha_1^{N(III)} - \frac{4}{3}pH \quad$ line (2)	(21.12)
$\frac{N(0)}{N(-III)_T}$ line (3)	$2N_2 + 6H^+ + 6e^- = 2NH_3 \quad pe^o = 2.11$	(21.13)
	$pe = 2.11 + \frac{1}{6}\log\frac{\{N_2\}\{H^+\}^6}{\{NH_3\}^2}$	(21.14)
	$N(-III)_T = [NH_4^+] + [NH_3]$; for NH_4^+, $pK_a = 9.24$ for use in $\alpha_1^{N(-III)}$.	
For unit γ values:	$pe = 2.11 + \frac{1}{6}\log\frac{[N_2]}{(\alpha_1^{N(-III)}N(-III)_T)^2} - pH.$	(21.15)
Neglecting N(V) and N(III) in the region of the line, for $N_{T,sys} = 10^{-3}$ *M*, there are equal amounts of dissolved nitrogen in the two oxidation states when:		
	$2[N_2] = N(-III)_T = 0.5 \times 10^{-3} M.$	
Then:	$pe = 2.61 - \frac{1}{3}\log\alpha_1^{N(-III)} - pH \quad$ line (3)	(21.16)

(Continued)

TABLE 21.3 (CONTINUED)
Equations for pe–pH Boundary Lines for $N(V)/N(III)_T$, $N(III)_T/N(0)$, and $N(0)/N(-III)_T$, with Equilibrium Constant Values for 25 °C/1 Atm. As Needed, Total Dissolved N (= $N_{T,Sys}$) = 10^{-3} *M*

Couple	equations	
$\frac{N(V)}{N(0)}$ line hybrid (1+2)	$NO_3^- + 6H^+ + 5e^- = \frac{1}{2}N_2 + 3H_2O \quad pe^\circ = 20.74$	(21.17)
	$pe = 20.74 + \frac{1}{5}\log\frac{\{NO_3^-\}\{H^+\}^6}{\{N_2\}^{\frac{1}{2}}}.$	(21.18)
For unit γ values:	$pe = 20.74 + \frac{1}{5}\log\frac{[NO_3^-]}{[N_2]^{\frac{1}{2}}} - \frac{6}{5}pH.$	(21.19)
Neglecting N(III) in the region of the line, for $N_{T,sys} = 10^{-3}$ *M*, there are equal amounts of dissolved nitrogen in the two oxidation states when:	$[NO_3^-] = 2[N_2] = 0.5 \times 10^{-3} M$	
	$pe = 20.44 - \frac{6}{5}pH$ line hybrid (1+2).	(21.20)

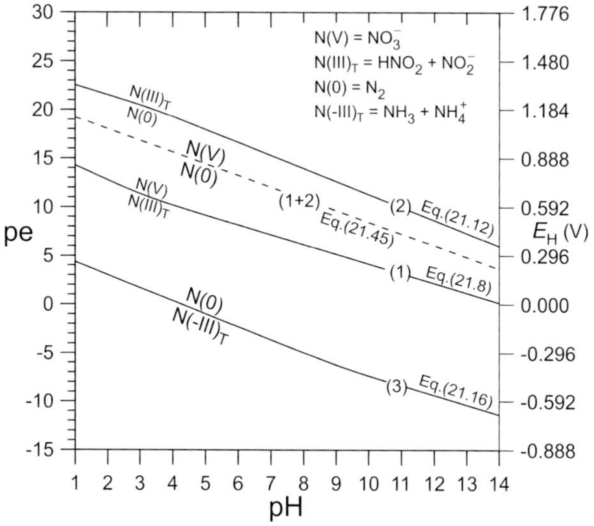

FIGURE 21.2 pe–pH predominance lines for nitrogen at 25 °C/1 atm for $N(V)/N(III)_T$, $N(III)_T/N(0)$, $N(0)/N(-III)_T$, and the hybrid of the first two for N(V)/N(0). For lines involving N(0), $N_{T,sys} = 10^{-3}$ *M*. Solution-phase activity coefficients are assumed to be unity.

are kinetically facile. In particular: 1) if redox conditions in a real system fall below the (1+2) hybrid line, NO_3^- that is present is easily reduced to N_2 by denitrifying bacteria; 2) if redox conditions in a real system move from below line (3) to above the (1+2) hybrid line, NH_4^+ and NH_3 that are present are easily oxidized to NO_3^- (via NO_2^-) by nitrifying bacteria; 3) under reducing conditions, reduction to NH_4^+/NH_3 is possible for NO_3^-, but very slow N_2; and 4) under oxidizing conditions, N_2 is kinetically very slow to be oxidized to NO_3^-. The kinetic inertness of N_2 is a result of the strong triple bond between the two nitrogen atoms; at 25 °C/1 atm, the bond strength is about 950 kJ/mol.

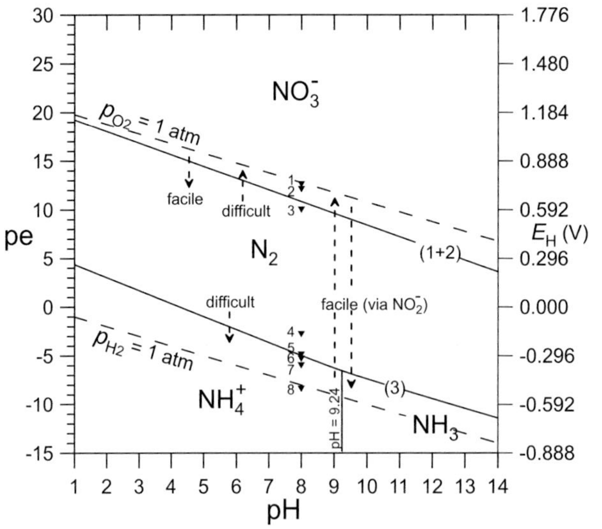

FIGURE 21.3 Final pe–pH predominance diagram for nitrogen at 25 °C/1 atm. For lines involving N(0) (N_2), $N_{T,sys} = 10^{-3}$ *M*. The size of the region for N_2 will increase if $N_{T,sys}$ is increased. As catalyzed by common biochemical enzymes, the dashed arrows show which inter-valence state conversions are kinetically difficult, and which are kinetically facile. Solution-phase activity coefficients are assumed to be unity. See Table 21.4 for description of points 1 to 8 at pH = 8.

By comparison, the bond strength of a typical carbon-carbon single bond is about 350 kJ/mol. Very few organisms can break the triple bond to convert N_2 to other nitrogen forms. Regarding point 4, note that the NO_3^-/N_2 boundary line lies well below the p_{O_2} = 1 atm for the O_2/H_2O redox couple. The line for p_{O_2} = 0.21 atm will be very close to the p_{O_2} = 1 atm line. Thus, the predominance region for NO_3^- extends well into the pe–pH region for typical oxygenated waters; the fact that N_2 typically remains abundant in the presence of oxygen demonstrates its redox inertness.

In anoxic municipal wastewaters, there is abundant ammonia (as NH_4^+ and NH_3). It is not the consequence of the reduction of N_2 (see above), but rather mostly due to the hydrolysis of abundant urea (from urine). Ammonia in wastewater can be removed during wastewater treatment by a two-step process: 1) nitrification by one set of bacteria to NO_3^-; and 2) denitrification of the NO_3^- by another set of bacteria to form N_2 (which will then simply off-gas to the atmosphere). Alternatively, in "ammonia stripping", ammonia is removed directly by shifting the pH to above the pK_a for NH_4^+ (so that NH_3 predominates), then "sparging" the wastewater with air bubbles to volatilize the NH_3.

Table 21.4 lists the pe values for 8 points at pH = 8. These are included in Figure 21.3. Point 1 is fully oxic with p_{O_2} = 0.21 atm. NO_3^- is predicted to be the predominant form of nitrogen. At point 2, 99% of the oxygen is gone, but NO_3^- is still dominant. For points 3–5, N_2 is predominant. For points 6–8, one would predict that nitrogen is present mostly as NH_4^+, though in real systems descending through the N_2 region does not lead to conversion to significant NH_4^+: N_2 is too inert.

Example 21.1 a. pe Threshold for Nitrate Reduction

A high level of NO_3^- for surface waters as from fertilizers would be 5 mg/L (as NO_3^-). At atmospheric levels of O_2, as noted for Figure 21.3, dissolved N_2 is unstable relative to oxidation to NO_3^-: further NO_3^- would form if the oxidation proceeded. Assume 25 °C/1 atm; neglect activity corrections.

Dissolution of $N_{2(g)}$: $N_{2(g)} = N_2$, $K_H = 10^{-3.20}$ M/atm.
Dissolution of $O_{2(g)}$: $O_{2(g)} = O_2$, $K_H = 10^{-2.90}$ M/atm.

a. If pH = 8.0, calculate the value to which the pe would need to drop in order for 5 mg/L NO_3^- to become the stable concentration assuming equilibrium with atmospheric levels of N_2 ($p_{N_2} = 0.78$ atm). It is below this pe that reduction of NO_3^- to N_2 becomes thermodynamically possible (absent the biochemical input of other energy). **b.** Calculate the corresponding p_{O_2} value, and the oxygen concentration in mg/L.

Solution

a. The FW for NO_3^- is 62.00 g/FW. $[NO_3^-]$ = 5 mg/L × (1 g/1000 mg) × (1 FW/62.00 g) = $10^{-7.09}$ FW/L (also M, in this case). For N_2, $[N_2]$ = 0.78 atm × $10^{-3.20}$ M/atm = $10^{-3.31}$ M. By Eq.(21.19),
$\text{pe} = 20.74 + \frac{1}{5}\log[NO_3^-]/[N_2]^{\frac{1}{2}} - \frac{6}{5}\text{pH}$. $\text{pe} = 20.74 + \frac{1}{5}\log\left(10^{-7.09}/(10^{-3.31})^{\frac{1}{2}}\right) - \frac{6}{5}(8.0)$.
pe = 10.05.

b. From Chapter 17, for the O_2/H_2O redox couple, $\text{pe} = 20.78 + \frac{1}{4}\log p_{O_2} - \text{pH}$. $\log p_{O_2} = 4(\text{pe} - 20.78 + \text{pH}) = -10.92$, p_{O_2} = **$10^{-10.92}$ atm.** $[O_2] = 10^{-10.92}$ atm × $10^{-2.90}$ M/atm = $10^{-13.82}$ M. $10^{-13.82}$ M × (32.00 g/1 MW) ×(1000 mg/g) = **$10^{-9.31}$ mg/L.** The O_2 is far, far gone by the time that reduction of NO_3^- to N_2 can begin.

21.3 pe–pH DIAGRAMS FOR IRON WITH $Fe_T = 10^{-5}$ M, $C_T \approx 0$

21.3.1 General – Fe Has Three Important Oxidation States: III, II, and 0

Solution-phase activity coefficients are assumed to be unity. The approach by which pe–pH diagrams for aqueous iron are prepared is similar to that discussed for aqueous lead in Chapters 19 and 20: 1) for the common oxidation states (here III, II, and 0, Table 21.5), no disproportion occurs, so that the vertical sequence is predictable (here as III, II, then 0); 2) hydroxide or oxide, solids need to be considered, and their formation will depend on T,sys for the metal; 3) the elemental, *i.e.*, zero-oxidation state form of the metal can be considered to be insoluble.

The redox chemistry of iron is simpler than that of nitrogen: for iron we can be confident about the oxidation states that have predominance layers, and we do not need to develop equations that ultimately do not appear in the final diagram (as with nitrogen, sulfur, and carbon). Thus for iron, the boundary line equations are developed in the body of the text rather than in a broader table as in Tables 21.3 (N), 21.10 and 21.11 (S), and 21.14 (C).

For iron solids, over the timeframe of interest, in the absence of significant carbonate carbon, it can be often assumed that (am)$Fe(OH)_{3(s)}$ and $Fe(OH)_{2(s)}$ are the only possible Fe(III) and Fe(II) solids. (The temporal metastability of (am)$Fe(OH)_{3(s)}$ relative to Fe(III) oxides and oxyhyroxides such as goethite and hematite is discussed in Chapter 11.) For dissolved iron species, we have Fe^{3+} and Fe^{2+} and their complexes with OH^-. Note regarding Fe^{3+}-related species that in contrast to aqueous Pb for which the solubility of the most oxidized state (IV) is very low at ~neutral to acidic pH, the solubility of Fe(III) at ~neutral to acidic pH certainly cannot be neglected. Also, when $Fe_{T,sys}$ is low (*e.g.*, $Fe_{T,sys} = 10^{-3.5}$ M or less), at equilibrium the dimer $Fe_2(OH)_2^{4+}$ will not be significant (see Chapter 10). The forms of Fe that will be considered here as possibly having pe–pH predominance regions are given in Table 21.5. Table 21.6 provides the equilibrium constants and equations for the acid/base chemistry of Fe(III) and Fe(II). Table 21.7 summarizes the relevant redox reactions for iron at 25 °C/1 atm.

Figure 21.4 is a preliminary pe–pH diagram for aqueous iron with $Fe_{T,sys} = 10^{-5}$ M and $C_T \approx 0$, as based on: 1) the above discussion (*e.g.*, Fe(I) can be ruled out because it would disproportionate

TABLE 21.4
Change in *In-Situ* Gas Pressure, Dominance, and Presence/Absence of Forms of Six ClONFeSCH Elements at pH = 8.0 as pe Is Lowered

point	pe	*in-situ* gas pressure p_{O_2} (atm)	$N_{T,sys} = 10^{-3}$ *M* predominant dissolved N form	$Fe_{T,sys} = 10^{-5}$ *M* $C_{T,free} = 0$ $S_{T,sys} = 0$ predom. Fe form	$Fe_{T,sys} = 10^{-5}$ *M* $C_{T,free} = 10^{-3}$ *M* $S_{T,sys} = 0$ solid Fe form	$Fe_{T,sys} = 10^{-5}$ *M* $C_{T,free} = 10^{-3}$ *M* $S_{T,sys} = 10^{-3}$ *M* solid Fe form	$S_{T,sys} = 10^{-3}$ *M* predominant dissolved S form	predominant dissolved C form	*in-situ* gas pressure p_{H_2} (atm)
1	12.61	0.21	NO_3^-	$Fe(OH)_{3(s)}$	$Fe(OH)_{3(s)}$	$Fe(OH)_{3(s)}$	SO_4^{2-}	HCO_3^-	6.3×10^{-42}
2	12.11	0.0021	NO_3^-	$Fe(OH)_{3(s)}$	$Fe(OH)_{3(s)}$	$Fe(OH)_{3(s)}$	SO_4^{2-}	HCO_3^-	6.3×10^{-41}
3	10.0	7.6×10^{-12}	N_2	$Fe(OH)_{3(s)}$	$Fe(OH)_{3(s)}$	$Fe(OH)_{3(s)}$	SO_4^{2-}	HCO_3^-	1.0×10^{-36}
4	–2.7	1.2×10^{-62}	N_2	$Fe(OH)_{3(s)}$ [a]	$FeCO_{3(s)}$	$FeCO_{3(s)}$	SO_4^{2-}	HCO_3^-	2.5×10^{-11}
5	–4.9	1.9×10^{-71}	N_2	Fe^{2+}	$FeCO_{3(s)}$	$FeS_{(s)}$	HS^-	HCO_3^-	6.3×10^{-7}
6	–5.3	4.8×10^{-73}	NH_4^+ [b]	Fe^{2+}	$FeCO_{3(s)}$	$FeS_{(s)}$	HS^-	HCO_3^-	4.0×10^{-6}
7	–6.0	7.6×10^{-76}	NH_4^+ [b]	Fe^{2+}	Fe^{2+} [c]	$FeS_{(s)}$	HS^-	CH_4 [d]	1.0×10^{-4}
8	–8.4	1.9×10^{-85}	NH_4^+ [b]	Fe^{2+}	Fe^{2+} [c]	$FeS_{(s)}$	HS^-	CH_4 [d]	6.3

Conditions: $Fe_{T,sys} = 10^{-5}$ *M*, $C_{T,free} = 10^{-3}$ *M*, and $S_{T,sys}$ 10^{-3} *M* (ClONFeSCH = chlorine, oxygen, nitrogen, iron, sulfur, carbon, and hydrogen).

[a] $Fe(OH)_{3(s)}$ is still present, but barely.

[b] Actually, N_2 would persist at this pe because NH_3 and NH_4^+ are not formed from N_2 at meaningful rates.

[c] Actually, $FeCO_{3(s)}$ would persist at this pe because C(IV) is not converted rapidly to CH_4.

[d] Actually, C(IV) would persist at this pe because CH_4 is not formed rapidly from C(IV).

TABLE 21.5
Important Forms of Iron in Natural Water Systems

oxidation state	solids, $C_T \approx 0$	solids, $C_T > 0$	dissolved species
III	$(am)Fe(OH)_{3(s)}$	$(am)Fe(OH)_{3(s)}$	Fe^{3+}, $FeOH^{2+}$, $Fe(OH)_2^+$, $Fe(OH)_3^o$, $Fe(OH)_4^-$
II	$Fe(OH)_{2(s)}$	$Fe(OH)_{2(s)}$, $FeCO_{3(s)}$	Fe^{2+}, $FeOH^+$, $Fe(OH)_2^o$, $Fe(OH)_3^-$
0	$Fe_{(s)}$	$Fe_{(s)}$	–

TABLE 21.6
Acid/Base Equilibrium Constant Values (25 °C/1 atm), α_0 Expressions for the Acid-Metal Ions Fe^{3+} and Fe^{2+}, and Solubility Constant Values (25 °C/1 atm)

Fe(II) Family

$Fe^{2+} + H_2O = FeOH^+ + H^+$ $\quad \log {}^*K_{H1}^{Fe(II)} = -9.5$

$FeOH^+ + H_2O = Fe(OH)_2^o + H^+$ $\quad \log {}^*K_{H2}^{Fe(II)} = -11.1$

$Fe(OH)_2^o + H_2O = Fe(OH)_3^- + H^+$ $\quad \log {}^*K_{H3}^{Fe(II)} = -11.4$

Fe(III) Family

$Fe^{3+} + H_2O = FeOH^{2+} + H^+$ $\quad \log {}^*K_{H1}^{Fe(III)} = -2.2$

$FeOH^{2+} + H_2O = Fe(OH)_2^+ + H^+$ $\quad \log {}^*K_{H2}^{Fe(III)} = -3.5$

$Fe(OH)_2^+ + H_2O = Fe(OH)_3^o + H^+$ $\quad \log {}^*K_{H3}^{Fe(III)} = -7.33$

$Fe(OH)_3^o + H_2O = Fe(OH)_4^- + H^+$ $\quad \log {}^*K_{H4}^{Fe(III)} = -8.57$

Summary	$p{}^*K_{Hn}^{Fe(\ )} = -\log {}^*K_{Hn}^{Fe(\ )}$				$p{}^*\beta_{Hn}^{Fe(\ )} = -\log {}^*\beta_{Hn}^{Fe(\ )}$			
	$n = 1$	2	3	4	$n = 1$	2	3	4
Fe(II)	9.50	11.1	11.4	–	9.50	20.5	31.9	–
Fe(III)	2.20	3.50	7.33	8.57	2.20	5.70	13.03	21.60

Expressions for α_0 (solution-phase activity coefficients assumed to be unity)

$[Fe^{2+}] = \alpha_0^{Fe^{2+}} Fe(II)_T$

$$\alpha_0^{Fe^{2+}} = \left(1 + {}^*\beta_{H1}/[H^+] + {}^*\beta_{H2}/[H^+]^2 + {}^*\beta_3/[H^+]^3\right)^{-1}$$

$[Fe^{3+}] = \alpha_0^{Fe^{3+}} Fe(III)_T$

$$\alpha_0^{Fe^{3+}} = \left(1 + {}^*\beta_{H1}/[H^+] + {}^*\beta_{H2}/[H^+]^2 + {}^*\beta_3/[H^+]^3 + {}^*\beta_4/[H^+]^4\right)^{-1}$$

Solubility Equilibria

$(am)Fe(OH)_{3(s)} + 3H^+ = Fe^{3+} + 3H_2O$ $\quad \log {}^*K_{s0}^{(am)Fe(OH)_{3(s)}} = 3.2$

$Fe(OH)_{2(s)} + 2H^+ = Fe^{2+} + 2H_2O$ $\quad \log {}^*K_{s0}^{Fe(OH)_{2(s)}} = 12.9$

$FeCO_{3(s)} = Fe^{2+} + CO_3^{2-}$ $\quad \log K_{s0}^{FeCO_{3(s)}} = -10.68$

to a mix of higher and lower oxidation states), and Fe(II) will not disproportionate to Fe(III) and Fe(0); 2) the assumption that $Fe_{T,sys}$ is high enough that both $(am)Fe(OH)_{3(s)}$ and $Fe(OH)_{2(s)}$ have predominance regions.

On both the left- and right-hand sides of the preliminary diagram in Figure 21.4, the solution phase has been divided into regions labeled simply as $Fe(III)_T$ and $Fe(II)_T$. Once the solution/solid boundary lines are located, it will be possible to subdivide those regions into sub-regions labeled as Fe^{3+}, $FeOH^{2+}$, etc., and Fe^{2+}, $FeOH^+$, etc., as necessary. Because high pH values generally favor

TABLE 21.7
Data for Selected Iron-Related Redox Reactions at 25 °C/1 atm in Order of Decreasing pe° and E_H^o

Redox Half-Reaction	log *K*	pe°	pe°(W)	E_H^o (=0.05916 pe°)
(am)$Fe(OH)_{3(s)} + CO_3^{2-} + 3H^+ + e^- = FeCO_{3(s)} + 3H_2O$	26.88	26.88	5.88	1.59
$Fe^{3+} + e^- + CO_3^{2-} = FeCO_{3(s)}$	23.68	23.68	23.68	1.40
(am)$Fe(OH)_{3(s)} + 3H^+ + e^- = Fe^{2+} + 3H_2O$	16.2	16.2	−4.8	0.96
$Fe^{3+} + e^- = Fe^{2+}$	13.0	13.0	13.0	0.77
$Fe^{2+} + 2e^- = Fe_{(s)}$	−14.9	−7.45	−7.45	−0.44
$FeCO_{3(s)} + 2e^- = Fe_{(s)} + CO_3^{2-}$	−25.58	−12.79	−12.79	−0.76

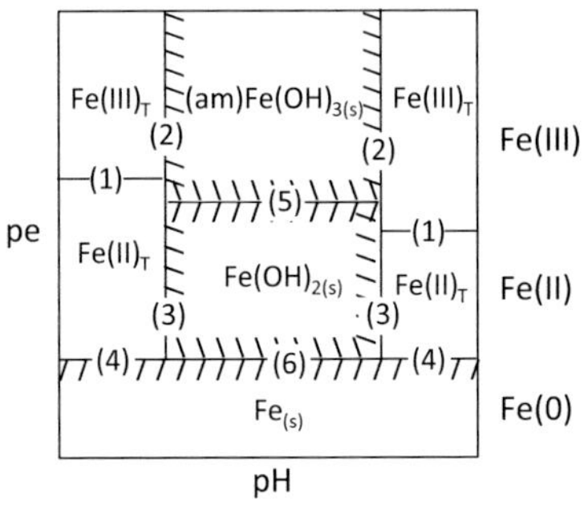

FIGURE 21.4 Preliminary pe–pH diagram for iron with $Fe_{T,sys} = 10^{-5}$ *M* and $C_{T,free} = 0$. For solid regions, boundaries are marked with hash marks.

the reduced forms of metals, reality has been anticipated in Figure 21.4 by placing the right-hand portion of line (1) below the left-hand portion.

21.3.2 $Fe(III)_T/Fe(II)_T$ Boundary Line

For the full pH range, there is one continuous $Fe(III)_T/Fe(II)_T$ boundary line, and Figure 21.4 anticipates that two portions of that line (1) will be plotted, one on each side. For solution-phase redox equilibrium between Fe(III) and Fe(II) species, from Table 21.6,

$$Fe^{3+} + e^- = Fe^{2+} \qquad pe^o = 13.0 \tag{21.21}$$

$$pe = 13.0 + \log\frac{\{Fe^{3+}\}}{\{Fe^{2+}\}}. \tag{21.22}$$

For unit γ values,

$$pe = 13.0 + \log\frac{[Fe^{3+}]}{[Fe^{2+}]}. \tag{21.23}$$

Invoking acid/base α values within each oxidation state for the (dissolved) $Fe(III)_T$ and $Fe(II)_T$,

$$[Fe^{3+}] = \alpha_0^{Fe(III)}Fe(III)_T, \qquad [Fe^{2+}] = \alpha_0^{Fe(II)}Fe(II)_1 \tag{21.24}$$

$$pe = 13.0 + \log \frac{\alpha_0^{Fe(III)} Fe(III)_T}{\alpha_0^{Fe(II)} Fe(II)_T}. \quad (21.25)$$

Table 21.6 reviews the expressions for the two α_0 values.

Along the boundary line,

$$Fe(III)_T = Fe(II)_T \quad (21.26)$$

so

$$pe = 13.0 + \log \frac{\alpha_0^{Fe(III)}}{\alpha_0^{Fe(II)}} \quad \text{line (1)}. \quad (21.27)$$

Line (1) is plotted in Figure 21.5 except for the portion that lies under the regions for $Fe(OH)_{3(s)}$ and $Fe(OH)_{2(s)}$; the visible portion at high pH immediately below pH = 14 is very short, serving as the boundary in the diagram between the regions for $Fe(III)_T$ and $Fe(II)_T$, as for $Fe(OH)_4^-$ and $Fe(OH)_3^-$.

21.3.3 Solution/(am)$Fe(OH)_{3(s)}$ Boundary Line

Figure 21.4 shows two pieces of the boundary line (2) between the solution and (am)$Fe(OH)_{3(s)}$. This boundary line is related to Eq.(11.44), for which the two pH roots (low and high) describe the pH values at which (am)$Fe(OH)_{3(s)}$ (neglecting $Fe(II)_T$) is saturated with $Fe(III)_T = Fe_{T,sys}$. (For $Fe_{T,sys} = 10^{-5}$ *M*, the two roots are pH = 3.09 and 13.40.) In this chapter, at high pe where Fe(III) dominates, the two asymptotically vertical portions of line (2) give those same pH values. As pe lowers, visible concave-up curvature develops because of the increasing importance of $Fe(II)_T$.

For line (2), considering both $Fe(III)_T$ and $Fe(II)_T$ as components of the dissolved Fe, we have

$$Fe(III)_T + Fe(II)_T = Fe_{T,sys}. \quad (21.28)$$

By Eq.(21.24),

$$\frac{[Fe^{3+}]}{\alpha_0^{Fe(III)}} + \frac{[Fe^{2+}]}{\alpha_0^{Fe(II)}} = Fe_{T,sys}. \quad (21.29)$$

By Eq.(21.23), for redox equilibrium, neglecting activity corrections,

$$[Fe^{2+}] = 10^{13.0} 10^{-pe} [Fe^{3+}]. \quad (21.30)$$

By Eq.(11.22), for solubility equilibrium with (am)$Fe(OH)_{3(s)}$,

$$[Fe^{3+}] = {}^*K_{s0}^{(am)Fe(OH)_{3(s)}} [H^+]^3 \quad (21.31)$$

where ${}^*K_{s0}$ (=$10^{3.2}$, 25 °C/1 atm) pertains. Combining Eqs.(21.29)–(21.31),

$$\frac{{}^*K_{s0}^{(am)Fe(OH)_{3(s)}} [H^+]^3}{\alpha_0^{Fe(III)}} + \frac{10^{13.0} 10^{-pe}\, {}^*K_{s0}^{(am)Fe(OH)_{3(s)}} [H^+]^3}{\alpha_0^{Fe(II)}} = Fe_{T,sys}. \quad (21.32)$$

Rearranging,

$$pe = -\log \frac{Fe_{T,sys} - \dfrac{{}^*K_{s0}^{(am)Fe(OH)_{3(s)}} [H^+]^3}{\alpha_0^{Fe(III)}}}{\dfrac{10^{13.0}\, {}^*K_{s0}^{(am)Fe(OH)_{3(s)}} [H^+]^3}{\alpha_0^{Fe(II)}}}. \quad \text{line (2)} \quad (21.33)$$

Line (2) can be plotted for any value of $Fe_{T,sys}$ larger than the minimum solubility of $Fe(III)_T$ from $(am)Fe(OH)_{3(s)}$ (as given by the curve with pH for $K_{s0}^{(am)Fe(OH)3(s)}[H^+]^3/\alpha_0^{Fe(III)}$). Line (2) is continuously concave up, and for $Fe_{T,sys} = 10^{-5}$ *M*, portions lie under the region for $Fe(OH)_{2(s)}$. It is plotted in Figure 21.5 for $Fe_{T,sys} = 10^{-5}$ *M*, dashed for the portion lying under the $Fe(OH)_{2(s)}$ region. Both the LHS and RHS arms of line (2) are asymptotically vertical because of the rapid diminishment of Fe(II).

21.3.4 Solution/$Fe(OH)_{2(s)}$ Boundary Line

Figure 21.4 shows two pieces of the boundary line (3) between the solution and $Fe(OH)_{2(s)}$. Eqs. (21.28)–(21.30) again apply. But now, instead of Eq.(21.31) (according to which $(am)Fe(OH)_{3(s)}$ controls Fe^{3+}), switching to $^*K_{s0}$ (= $10^{12.9}$, 25 °C/1 atm) for $Fe(OH)_{2(s)}$, neglecting activity corrections,

$$[Fe^{2+}] = {}^*K_{s0}^{Fe(OH)2(s)}[H^+]^2 \tag{21.34}$$

Combining Eqs.(21.29), (21.30) and (21.34) gives

$$\frac{{}^*K_{s0}^{Fe(OH)2(s)}[H^+]^2}{\alpha_0^{Fe(III)}10^{13.0}10^{-pe}} + \frac{{}^*K_{s0}^{Fe(OH)2(s)}[H^+]^2}{\alpha_0^{Fe(II)}} = Fe_{T,sys}. \tag{21.35}$$

Rearranging,

$$pe = \log\frac{Fe_{T,sys} - \dfrac{{}^*K_{s0}^{Fe(OH)2(s)}[H^+]^2}{\alpha_0^{Fe(II)}}}{\dfrac{10^{-13.0}\,{}^*K_{s0}^{Fe(OH)2(s)}[H^+]^2}{\alpha_0^{Fe(III)}}}. \quad \text{line (3)} \tag{21.36}$$

Line (3) can be plotted for any value of $Fe_{T,sys}$ larger than the minimum solubility of $Fe(II)_T$ from $Fe(OH)_{2(s)}$ (which is given by the curve with pH for ${}^*K_{s0}^{Fe(OH)2(s)}[H^+]^2/\alpha_0^{Fe(II)}$). Line (3) is continuously concave down. As plotted for $Fe_{T,sys} = 10^{-5}$ *M* in Figure 21.5, portions (dashed) lie under the

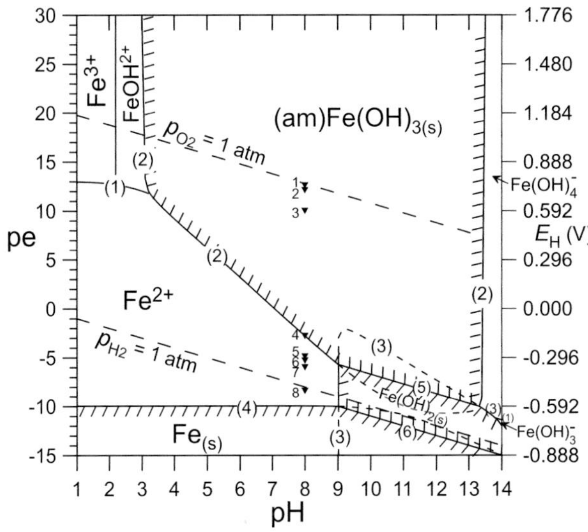

FIGURE 21.5 Final pe–pH predominance diagram for iron at 25 °C/1 atm for $Fe_{T,sys} = 10^{-5}$ *M* and $C_{T,free} = 0$. For each region for an Fe solid, the extent will increase as $Fe_{T,sys}$ is increased. Solution-phase activity coefficients are assumed to be unity. See Table 21.4 for description of points 1 to 8 at pH = 8.

regions for (am)$Fe(OH)_{3(s)}$ and $Fe_{(s)}$. Both the LHS and RHS arms of line (3) are asymptotically vertical because of the rapid diminishment of Fe(III) at very low pe. For the $Fe(OH)_{2(s)}$ analogy for Eq.(11.44), the two pH roots (low and high) describe the pH values at which $Fe(OH)_{3(s)}$ (neglecting $Fe(III)_T$) is saturated with $Fe(II)_T = Fe_{T,sys}$. For $Fe_{T,sys} = 10^{-5}$ *M*, the two roots are pH = 9.01 and 14.09.

21.3.5 Solution/$Fe_{(s)}$ Boundary Line

For the boundary between the solution and $Fe_{(s)}$, Eq.(21.28)–(21.30) again apply, but now instead of Eq.(21.34) (according to which $Fe(OH)_{2(s)}$ controls Fe^{2+}), from Table 21.6 for control by $Fe_{(s)}$,

$$[Fe^{2+}] = 10^{14.9}10^{2pe}. \tag{21.37}$$

Under the highly reducing conditions required for stability of $Fe_{(s)}$, $Fe(III)_T$ is negligible in Eq.(21.28) so with essentially no error,

$$Fe(II)_T = Fe_{T,sys}. \tag{21.38}$$

Then, by Eqs.(21.24) and (21.37),

$$\frac{10^{14.9}10^{2pe}}{\alpha_0^{Fe(II)}} = Fe_{T,sys} \tag{21.39}$$

$$pe = \tfrac{1}{2}\log\left(10^{-14.9}\alpha_0^{Fe(II)}Fe_{T,sys}\right). \qquad \text{line (4)} \tag{21.40}$$

Line (4) is continuously concave down. As plotted for $Fe_{T,sys} = 10^{-5}$ *M* in Figure 21.5, a portion (dashed) lies under the region for $Fe(OH)_{2(s)}$. As pH increases, line (4) trends downward because Fe^{2+} represents a smaller and smaller fraction of $Fe(II)_T$: a higher $\{e^-\}$ is required to precipitate $Fe_{(s)}$.

21.3.6 (am)$Fe(OH)_{3(s)}$/$Fe(OH)_{2(s)}$ Boundary Line

Taking *K* values from Tables 21.5 and 21.6, we combine

$$(am)Fe(OH)_{3(s)} + 3H^+ = Fe^{3+} + 3H_2O \qquad \log {}^*K_{s0}^{(am)Fe(OH)_{3(s)}} = 3.2 \tag{21.41}$$

$$+\ Fe^{3+} + e^- = Fe^{2+} \qquad \log K = 13.0 \tag{21.42}$$

$$-\ \left(Fe(OH)_{2(s)} + 2H^+ = Fe^{2+} + 2H_2O \qquad \log {}^*K_{s0}^{Fe(OH)_{2(s)}} = 12.9\right) \tag{21.43}$$

$$(am)Fe(OH)_{3(s)} + H^+ + e^- = Fe(OH)_{2(s)} + H_2O \qquad \log K = 3.3. \tag{21.44}$$

$$\frac{\{Fe(OH)_{2(s)}\}}{\{Fe(OH)_{3(s)}\}[H^+]\{e^-\}} = 10^{3.3}. \tag{21.45}$$

On the (am)$Fe(OH)_{3(s)}$/$Fe(OH)_{2(s)}$ boundary line, the solids are considered to be at unit activity. Thus, at 25 °C/1 atm,

$$pe = 3.3 - pH. \qquad \text{line (5)} \tag{21.46}$$

Line (5) is plotted in Figure 21.5, but only for the pH range over which (am)$Fe(OH)_{3(s)}$ and $Fe(OH)_{2(s)}$ can both be present, as given by the two intersections of line (2) with line (5).

21.3.7 $Fe(OH)_{2(s)}/Fe_{(s)}$ Boundary Line

Taking K values from Tables 21.5 and 21.6, we combine

$$Fe(OH)_{2(s)} + 2H^+ = Fe^{2+} + 2H_2O \qquad \log {}^*K_{s0}^{Fe(OH)_{2(s)}} = 12.9 \tag{21.47}$$

$$+ \qquad Fe^{2+} + 2e^- = Fe_{(s)} \qquad \log K = -14.9 \tag{21.48}$$

$$Fe(OH)_{2(s)} + 2H^+ + 2e^- = Fe_{(s)} + 2H_2O \qquad \log K = -2.0. \tag{21.49}$$

$$\frac{\{Fe_{(s)}\}}{\{Fe(OH)_{2(s)}\}[H^+]\{e^-\}^2} = 10^{-2.0}. \tag{21.50}$$

On the $Fe(OH)_{2(s)}/Fe_{(s)}$ boundary line, but solids are considered to be at unit activity. Thus, at 25 °C/1 atm,

$$pe = -1.0 - pH. \qquad \text{line (6)} \tag{21.51}$$

Line (6) is plotted in Figure 21.5, but only for the pH range over which $Fe(OH)_{2(s)}$ and $Fe_{(s)}$ can both be present, as given by the two intersections of line (3) with line (6).

21.3.8 Final pe–pH Diagram Comments

On the low pH side of Figure 21.5, the solution/(am)$Fe(OH)_{3(s)}$ boundary line (2) falls higher with respect to pH than p^*K_{H1} for Fe^{3+}, but lower than p^*K_{H2}. On the high pH side, line (2) falls higher than p^*K_{H4} for Fe^{3+}. This means the region for $Fe(III)_T$ will be divided into sub-regions that are labeled Fe^{3+} and $FeOH^{2+}$ on the left (separated by pH = p^*K_{H1} for Fe^{3+}), and $Fe(OH)_4^-$ on the right.

On the low pH side of Figure 21.5, the solution/(am)$Fe(OH)_{3(s)}$ boundary line (2) and the solution/$Fe(OH)_{2(s)}$ boundary line (3) fall lower with respect to pH p^*K_{H1} for Fe^{2+}. On the high pH side, line (3) falls higher than p^*K_{H3} for Fe^{2+}. This means the region for $Fe(II)_T$ will be divided into a subregion that is labeled Fe^{2+} on the left, and a tiny region for $Fe(OH)_3^-$ on the right. Last, wherever a portion of a solid/solution boundary line is largely straight, it is because one solution species predominates over all others. This fact is sometimes used to approximate a boundary line like (2) as a series of straight lines, taking Fe_T (dissolved) to be dominated sequentially by $FeOH^{2+}$ (for pH ≈3), Fe^{2+} (for pH ~3 to ~9), then $Fe(OH)_4^-$ (for pH ~13.4).

Overall, Fe(III) will be the dominant oxidation state of Fe in waters containing even small amounts of oxygen (note the position of the line for p_{O_2} = 1 atm): Fe(II) can only be formed in significant measure when the conditions are rather reducing. Last, when Fe(0) is present (as $Fe_{(s)}$), then unless $[Fe^{2+}]$ and/or the pH are very high, $Fe_{(s)}$ is capable of reducing water to H_2 with $p_{H_2} > 1$. Regarding points 1–8 in Table 21.4, while is NO_3^- gone at point 3 and below, when $Fe_{T,sys} = 10^{-5}$ M then (am)$Fe(OH)_{3(s)}$ persists down to point 4. Below point 4, all the Fe is solubilized as dissolved Fe(II) with the predominant species being Fe^{2+}. Along line (2) in this region, the pe for complete disappearance of (am)$Fe(OH)_{3(s)}$ decreases by only 1 for every order of magnitude increase in $Fe_{T,sys}$; if $Fe_{T,sys}$ is much higher at $Fe_{T,sys} = 10^{-3}$ M (as perhaps in a soil-aquifer/water system), all of the Fe will be solubilized as Fe(II) below about pe = −4.7 (rather than pe = −2.7 as at point 4).

In Bangladesh, organic-carbon-driven reduction of Fe(III) oxide and hydroxide solids to dissolved Fe(II) has been described by Neumann et al. (2010) as the cause of problematically high arsenic (As) levels in numerous small irrigation and public supply wells located in shallow aquifers. Mechanistically, natural As(V) is present as sorbed to Fe(III) oxide, hydroxide, and oxyhydroxide solids, and so has low mobility in groundwater systems when Fe(III) solids are stable. However, when reductive loss of the sorbing solids occurs, As is mobilized and drawn into the shallow wells by the pumping-induced groundwater flow. The reductive conversion of As(V) to As(III) also plays a role, As(III) being generally less strongly adsorbed to mineral surfaces than As(V).

Example 21.2 a. pe Threshold for Significant Reduction of (am)Fe(OH)$_{3(s)}$

Consider a system with $Fe_{T,sys} = 10^{-5}$ *M*, and a beginning pe (= 10.02) and pH (=8.0) as in Example 21.1. Take $C_{T,free}$ to be low enough that the possible formation of $FeCO_{3(s)}$ need not be considered. Assume that the pH stays constant at 8.00. **a.** With guidance from Figure 21.5 as to the dominant Fe(II) solution species at pH = 8.0, calculate the pe required to have brought 1% of the (am)Fe(OH)$_{3(s)}$ into solution as Fe(II)$_T$. **b.** If $N_{T,sys} \approx 5\times10^{-4}$ *M* as in Example 21.1, with guidance from Figure 21.3 as to the dominant N form at the calculated pe, calculate the predicted remaining concentration of NO_3^- at equilibrium.

Solution

a. At pH = 8.0 in the region for Fe^{2+} in Figure 21.5, there is no nearby region for $FeOH^+$: by far the dominant dissolved Fe(II) species is Fe^{2+}, so we seek the conditions that give $0.01\times Fe_{T,sys} = [Fe^{2+}] = 10^{-7}$ *M*. By Eq.(21.30), $[Fe^{2+}] = 10^{13.0}10^{-pe}[Fe^{3+}]$. By Eq.(21.31), for solubility equilibrium with (am)Fe(OH)$_{3(s)}$, $[Fe^{3+}] = {}^*K_{s0}^{(am)Fe(OH)3(s)}[H^+]^3$. ${}^*K_{s0}^{(am)Fe(OH)3(s)} = 10^{3.2}$. Thus, $10^{-7} = 10^{13.0}10^{-pe}10^{3.2}[H^+]^3$, and **pe = 0.80**.

b. At pe = 7.2, we are well into the region for N_2. Thus, for $N_{T,sys} = 5\times10^{-4}$ *M*, $[N_2] = 5\times10^{-4}$ *M*. By Eq.(21.19), $pe = 20.74 + \frac{1}{5}\log[NO_3^-]/[N_2]^{\frac{1}{2}} - \frac{6}{5}pH$. $0.80 = 20.74 + \frac{1}{5}\log[NO_3^-]/[5\times10^{-4}]^{\frac{1}{2}} - \frac{6}{5}(8.0)$. $[NO_3^-] = \mathbf{10^{-50.05}}$ *M*. For comparison, a single NO_3^- per liter would be $10^{-23.78}$ *M*. NO_3^- is far, far gone by the time that reduction of (am)Fe(OH)$_{3(s)}$ to Fe(II) is underway.

Example 21.3 Reduction of As(V) to As(III)

Compute the solution As(V)$_T$/As(III)$_T$ ratio at equilibrium for 25 °C/1 atm when all $Fe_{T,sys}$ (= 10^{-3} *M*) has been solubilized as Fe(II)$_T$ at pH = 8.0. Both arsenic acid (H_3AsO_4, As(V)) and arsenious acid (H_3AsO_3, As(III)) are triprotic acids. Assume that there is little dissolved C(IV), so that the possible formation of $FeCO_{3(s)}$ need not be considered. Neglect activity corrections. The following equilibrium data are available at 25 °C/1 atm:

As(V) to As(III): $H_3AsO_4 + 2H^+ + 2e^- = H_3AsO_3 + H_2O \qquad \log K = 19.44 \qquad pe^o = 9.72.$

As(III) acid/base: $H_3AsO_3 = H^+ + H_2AsO_3^- \qquad \log K_1^{As(III)} = -9.23 \qquad \log\beta_1^{As(III)} = -9.23.$

$H_2AsO_3^- = H^+ + HAsO_3^{2-} \qquad \log K_2^{As(III)} = -12.10 \qquad \log\beta_2^{As(III)} = -21.33.$

$HAsO_3^{2-} = H^+ + AsO_3^{3-} \qquad \log K_3^{As(III)} = -13.41 \qquad \log\beta_3^{As(III)} = -34.74.$

As(V) acid/base: $H_3AsO_4 = H^+ + H_2AsO_4^- \qquad \log K_1^{As(V)} = -2.24 \qquad \log\beta_1^{As(V)} = -2.24.$

$H_2AsO_4^- = H^+ + HAsO_4^{2-} \qquad \log K_2^{As(V)} = -6.76 \qquad \log\beta_2^{As(V)} = -9.00.$

$HAsO_4^{2-} = H^+ + AsO_4^{3-} \qquad \log K_3^{As(V)} = -11.60 \qquad \log\beta_3^{As(V)} = -20.60.$

As(V) is far more acidic than As(III).

Solution

As in Example 21.2, at pH = 8.0, by far the dominant dissolved Fe(II) species is Fe^{2+}, so we seek the conditions that give $Fe_{T,sys} = [Fe^{2+}] = 10^{-3}$ *M*. By Eq.(21.30), $[Fe^{2+}] = 10^{13.0}10^{-pe}[Fe^{3+}]$. By Eq.(21.31), for solubility equilibrium with (am)$Fe(OH)_{3(s)}$, $[Fe^{3+}] = {}^*K_{s0}^{(am)Fe(OH)3(s)}[H^+]^3$. ${}^*K_{s0}^{(am)Fe(OH)3(s)} = 10^{3.2}$. Thus, $10^{-3} = 10^{13.0}10^{-pe}10^{3.2}[H^+]^3$, and pe = −4.80. $pe = 9.72 + \frac{1}{2}\log\frac{\{H_3AsO_4\}\{H^+\}^2}{\{H_3AsO_3\}}$. Neglecting activity corrections, $4.80 = 9.72 + \frac{1}{2}\log\frac{\alpha_0^{As(V)}As(V)_T[H^+]^2}{\alpha_0^{As(III)}As(III)_T}$. At pH = 8.0, $\alpha_0^{As(III)} = 0.944$, and $\alpha_0^{As(V)} = 9.45\times10^{-8}$. $4.80 = 9.72 + \frac{1}{2}\log\frac{\alpha_0^{As(V)}As(V)_T[H^+]^2}{\alpha_0^{As(III)}As(III)_T}$. $As(V)_T/As(III)_T = 10^{-6.02}$. Reduction of As(V) to As(III) is essentially complete. For waters containing arsenic, the sorbing Fe(III) solid is gone, and the As is in the less strongly sorbing (As(III)) form.

21.4 pe–pH DIAGRAMS FOR IRON WITH $Fe_T = 10^{-5}$ *M*, $C_{T,free} = 10^{-3}$ *M*

21.4.1 General – Fe Oxidation States Are III, II, and 0; $FeCO_{3(s)}$ Needs to Be Considered

As in Section 21.4, Fe(III), Fe(II) and Fe(0) are considered. But now, with carbon present at a concentration of $C_{T,free}$ (= total dissolved, non-metal-complexed carbon) = 10^{-3} *M*, the possibility of the formation of $FeCO_{3(s)}$ needs to be considered. (Fe(III) carbonate solids are not stable; see Section 20.1.) Unlike in Chapter 20, where dissolved carbonate-related complexes of Pb(II) were considered, for Fe(II) such complexes are not believed to be important. Above the $C(IV)_T/C(-IV)$ line (*i.e.*, the $C(IV)_T/CH_4$ line), the majority of the dissolved carbon is C(IV) and $FeCO_{3(s)}$ may form. Below the $C(IV)_T/C(-IV)$ line, the majority of the dissolved carbon is CH_4, and $FeCO_{3(s)}$ will not form.

21.4.2 Solubility Limitation by $FeCO_{3(s)}$ *vs.* $Fe(OH)_{2(s)}$

As in Chapter 20 for Pb(II) in aqueous systems, to identify the pH regimes for solubility limitation by the (II) oxidation state forms (oxide or hydroxide solid *vs.* carbonate solid), the curves for the total solubility of each solid must be examined. For solubility control of Fe(II) by $Fe(OH)_{2(s)}$,

$$[Fe^{2+}]_{Fe(OH)2(s)} = {}^*K_{s0}[H^+]^2 = 10^{12.9}[H^+]^2. \tag{21.52}$$

Then, by Eq.(21.24),

$$\log Fe(II)_T = 12.9 - 2pH - \log\alpha_0^{Fe(II)} \qquad Fe(OH)_{2(s)}. \tag{21.53}$$

For solubility control of Fe(II) by $FeCO_{3(s)}$,

$$[Fe^{2+}]_{FeCO3(s)} = \frac{10^{-10.68}}{[CO_3^{2-}]}. \tag{21.54}$$

Then, by Eq.(21.24),

$$\log Fe(II)_T = -10.68 - \log\alpha_2^C - \log C_{T,free} - \log\alpha_0^{Fe(II)}. \qquad FeCO_{3(s)} \tag{21.55}$$

The lines for log $Fe(II)_T$ for the two solids are plotted in Figure 21.6 with $C_{T,free} = 10^{-3}$ *M*. $FeCO_{3(s)}$ is solubility limiting at pH ≤ 10.06. $Fe(OH)_{2(s)}$ is solubility limiting at pH ≥ 10.06. For $Fe_{T,sys} = 10^{-5}$ *M*: 1) $FeCO_{3(s)}$ is present at 9.01 < pH < 10.06; 2) $Fe(OH)_{2(s)}$ is present at 10.06 < pH < 14.10; and 3) one or both solids is/are present at pH = 10.06, depending on how the system was formed.

We can conclude that a preliminary version of the pe–pH diagram for this case is as given in Figure 21.7. For pH > 10.06, things are the same as when $C_{T,free} = 0$ in Figure 21.4. For pH < 10.06, there are four new boundary lines: these involve $FeCO_{3(s)}$. Line (3′) is an essentially vertical line at pH = 9.01. Line (7) is a vertical line at pH = 10.06. Lines (5′) and (6′) are developed as follows. If it

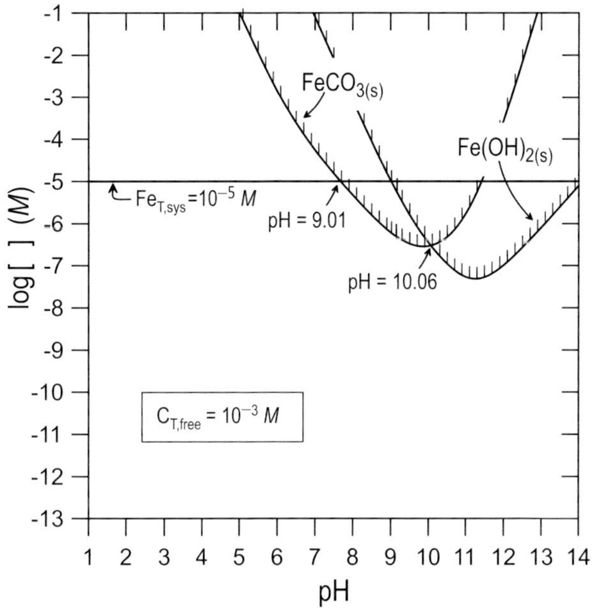

FIGURE 21.6 $Fe(II)_T$ *vs.* pH for solubility control by $FeCO_{3(s)}$ and $Fe(OH)_{2(s)}$ at 25 °C/1 atm for $C_{T,free} = 10^{-3}$ *M*. Solution-phase activity coefficients are assumed to be unity. $FeCO_{3(s)}$ is solubility limiting at pH ≤ 10.06. $Fe(OH)_{2(s)}$ is solubility limiting at pH ≥ 10.06. For $Fe_{T,sys} = 10^{-5}$ *M*: 1) $FeCO_{3(s)}$ is definitely present at 9.01 < pH < 10.06; 2) $Fe(OH)_{2(s)}$ is definitely present at 10.06 < pH < 14.10; and 3) one or both solids is/are present at pH = 10.06.

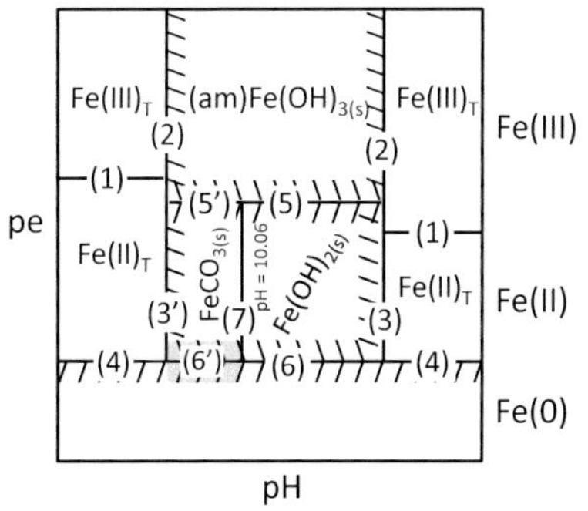

FIGURE 21.7 Preliminary pe–pH diagram for iron with $Fe_{T,sys} = 10^{-5}$ *M* and $C_{T,free} = 10^{-3}$ *M*. For solid regions, boundaries are marked with hash marks. The shaded region involving line (6′) indicates that line (6′) will not appear in the final diagram if it lies below the $C(IV)_T/CH_4$ line (it does): C(IV) decreases rapidly below the $C(IV)_T/CH_4$ line, so $FeCO_{3(s)}$ cannot be stable there. See Chapter 20 for the analogous circumstance with Pb(II) carbonate solids.

falls below the $C(VI)_T/CH_4$ line, line (6′) will not appear in the diagram, and the $Fe(OH)_{2(s)}$ region will show a leftward directed extension past the pH = 10.06 transition pH for $FeCO_{3(s)}$ | $Fe(OH)_{2(s)}$.

21.4.3 (am)$Fe(OH)_{3(s)}/FeCO_{3(s)}$ Boundary Line

We combine

$$\text{(am)Fe(OH)}_{3(s)} + 3H^+ = Fe^{3+} + 3H_2O \qquad \log {}^*K_{s0} = 3.2 \tag{21.41}$$

$$+Fe^{3+} + e^- = Fe^{2+} \qquad \log K = 13.0 \tag{21.42}$$

$$-\left(FeCO_{3(s)} = Fe^{2+} + CO_3^{2-} \qquad \log K = -10.68\right) \tag{21.56}$$

$$\text{(am)Fe(OH)}_{3(s)} + 3H^+ + CO_3^{2-} + e^- = FeCO_{3(s)} + 3H_2O \qquad \log K = 26.88. \tag{21.57}$$

$$pe = 26.88 + \log \alpha_2^C + \log C_{T,free} - 3pH. \qquad \text{line (5′)} \tag{21.58}$$

Eq.(21.58) is plotted in Figure 21.8, but only for the pH range over which both (am)$Fe(OH)_{3(s)}$ and $FeCO_{3(s)}$ are present.

21.4.4 $FeCO_{3(s)}/Fe_{(s)}$ Boundary Line

From Table 21.6,

$$FeCO_{3(s)} + 2e^- = Fe_{(s)} + CO_3^{2-} \qquad \log K = -25.58. \tag{21.59}$$

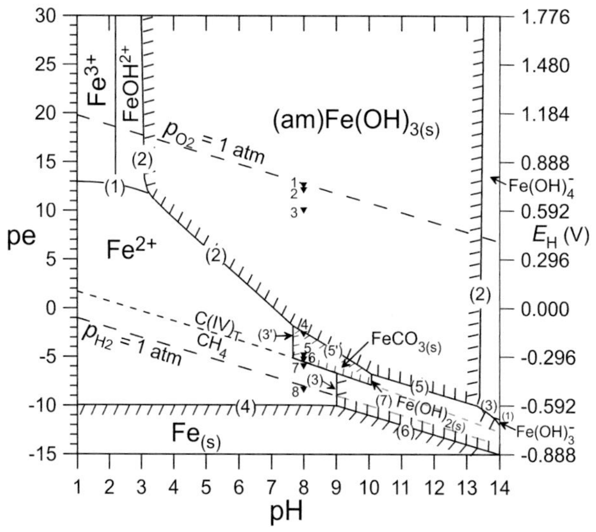

FIGURE 21.8 Final pe–pH predominance diagram for Fe at 25 °C/1 atm for $Fe_{T,sys} = 10^{-5}$ *M* and $C_{T,free} = 10^{-3}$ *M*. Solution-phase activity coefficients are assumed to be unity. The size of all regions for Fe solids will increase as $Fe_{T,sys}$ increased. The size of the region for $FeCO_{3(s)}$ will increase with increasing $C_{T,free}$, but will not in any case extend measurably below the $C(IV)_T/CH_4$ line if $C(IV)_T$ can be converted to CH_4 for the time scale of interest. See Table 21.4 for consideration of points 1 to 8 at pH = 8.

$$pe = -12.79 - \tfrac{1}{2}\log\alpha_2^{C} - \tfrac{1}{2}\log C_{T,free}. \quad \text{line } (6') \tag{21.60}$$

As developed in Section 20.2.5, and again in Table 21.14, the $C(IV)_T/CH_4$ line is

$$pe = 2.70 + \tfrac{1}{8}\alpha_0^{C(IV)} - pH. \tag{21.61}$$

The $C(IV)_T/CH_4$ line falls above line (6′), so line (6′) is not plotted in Figure 21.8.

21.4.5 Solution/$FeCO_{3(s)}$ Boundary Line

From Figure 21.7, we only draw line (3′) for the lower pH side boundary of the solution with $FeCO_{3(s)}$. In developing the concave-down equation for the solution/$Fe(OH)_{2(s)}$ boundary line (line (3), see Section 21.3.4), consideration of $Fe(III)_T$ as a component of the dissolved Fe only leads to some curvature on the high pH side of the diagram: on the lower pH side, line (3) is essentially vertical (pH = 9.01) because dissolved Fe(III) is unimportant there. Analogously, for line (3′), the pH of the boundary line is determined by the solubility of $FeCO_{3(s)}$:

$$Fe_{T,sys} = Fe(II)_T. \tag{21.62}$$

Because of equilibrium with $FeCO_{3(s)}$ along the line,

$$[Fe^{2+}] = \frac{K_{s0}}{\alpha_2^{C} C_T}. \tag{21.63}$$

where the K_{s0} is for $FeCO_{3(s)}$ (Table 21.5). By Eqs.(21.24), (21.62), and (21.63),

$$Fe_{T,sys} - \frac{K_{s0}}{\alpha_2^{C} C_T \alpha_0^{Fe}} = 0. \quad \text{line } (3') \tag{21.64}$$

The lower root of Eq.(21.64) is pH = 7.67, which serves as line (3′) in Figure 21.8.

21.4.6 Final pe–pH Diagram Comments

As compared to Figure 21.5 ($C_{T,free} = 0$), Figure 21.8 is very similar, though the sum total of the areas for Fe(II) solids in Figure 21.8 is larger than for Figure 21.5. This is because for $C_{T,free} = 10^{-3}$ *M*, $FeCO_{3(s)}$ is less soluble than $Fe(OH)_{2(s)}$ when pH < 10.06 (see Figure 21.6). Regarding points 1–8 in Table 21.4, when $C_{T,free} = 0$, (am)$Fe(OH)_{3(s)}$ persists down to point 4, and is completely dissolved as $Fe(II)_T$ (mostly Fe^{2+}) at points 5 and 6. In the presence of $C_{T,free} = 10^{-3}$ *M*, points 4–6 are inside the region for $FeCO_{3(s)}$, and the region for Fe^{2+} is entered for points 7 and 8 only if the $C(IV)_T$ is considered labile relative to reduction to CH_4.

Example 21.4 Alkalinity and Oxygen Demand Effects of Oxidation of Fe^{2+}

a. Some anoxic well water at pH = 7.0 is brought to ground surface. O_2 enters the solution and oxidizes the Fe(II) to Fe(III). Determine the major Fe(II) species by far at pH = 7.0, and develop the net reaction for its oxidation by O_2 of Fe^{2+} at pH = 7.0 to Fe(III). Fe(III) is very insoluble at pH = 7 (Chapter 11); assume for the reaction that all the Fe(III) reacts with H_2O to precipitate as (am)$Fe(OH)_{3(s)}$. **b.** In mg/L, how much oxygen will be consumed by the oxidation of $10^{-2.5}$ *M* of the $Fe(II)_T$ at pH = 7.0 to Fe(III)? **c.** What will be the effect on Alk for the net reaction?

Solution

a. From Table 21.6, $p^{*}K_{H1}^{Fe(II)} = 9.5$, which is 2.5 units from pH = 7: Fe^{2+} is the major dissolved Fe(II) species. For oxidation of Fe^{2+} by O_2, followed by the precipitation reaction, the overall net reaction is obtained according to

$$\tfrac{1}{4}\left(O_2 + 4H^+ + 4e^- = 2H_2O\right)$$

$$-\left(Fe^{3+} + e^- = Fe^{2+}\right)$$

$$+Fe^{3+} + 3H_2O = (am)Fe(OH)_{3(s)} + 3H^+$$

$$\tfrac{1}{4}O_2 + Fe^{2+} + \tfrac{5}{2}H_2O = (am)Fe(OH)_{3(s)} + 2H^+.$$

b. By the stoichiometry of the above reaction, to oxidize $10^{-2.5}$ *M* of Fe^{2+} to form $(am)Fe(OH)_{3(s)}$, the level of O_2 needed is $\tfrac{1}{4}10^{-2.5}$ $M = 7.91 \times 10^{-4}$ *M* of O_2. The MW of O_2 is 32.00 g/mol, so 32,000 mg/mol. 7.91×10^{-4} $M \times 32{,}000$ mg/mol = **25.3 mg/L**.

c. 2.0 equivalents of strong acid are produced per mol of Fe^{2+} oxidized, so the effect $\Delta Alk = -(2\ eq/mol) \times 10^{-2.5}$ $M =$ $\mathbf{-6.32 \times 10^{-3}}$ **eq/L**.

21.5 pe–pH DIAGRAM FOR SULFUR WITH $S_T = 10^{-3}$ *M*

21.5.1 Redox Equilibria Governing Sulfur Species

Sulfur is in the VIA column of the periodic table, so atomic sulfur has 6 valence electrons. Removing 6 electrons takes sulfur to the relatively stable state of an empty outer shell, S(VI). Adding 2 electrons takes sulfur to a full outer shell of 8, S(–II). Inorganic sulfur can exist in five main oxidation states, VI, IV, 0, –I, and –II (Table 21.8), with other oxidation states being very unstable relative to disproportionation. The redox half-reactions in Table 17.1 that involve the Table 21.8 species are collected in Table 21.9. Thus, inorganic sulfur ranges from the S(VI) species HSO_4^- and SO_4^{2-} down to the S(–II) species H_2S, HS^-, and S^{2-}. Eight electrons can be added to each S(VI) species, converting it to a S(–II) species, and *vice versa*. This difference of 8 in oxidation states is the same as the difference of 8 between C(IV) and C(–IV), and between N(V) and N(–III).

In the first edition of this text, the possibility of predominance by aqueous S(–I) was not included. In this edition, it is considered because of: 1) the sometimes presence of the aqueous species H_2S_2, HS_2^-, and S_2^{2-}, as well as higher polysulfide forms (see below); and 2) the equilibrium importance of pyrite ($FeS_{2(s)}$) under very reducing conditions in iron+sulfur systems. We will be able to rule out the possibility of predominance at equilibrium of dissolved S(–I), though doing so will require some thought (see below).

At very high pe, the highest oxidation state must predominate, namely S(VI), as with HSO_4^- and SO_4^{2-} (the S(VI) species H_2SO_4, is such a strong acid that it cannot have a predominance region at high pe even at very low pH). Overall, with $pK_2 = 1.99$ for H_2SO_4 (25 °C/1 atm), at high pe, the predominance layer will be divided into two regions, one for HSO_4^-, and the other for SO_4^{2-}.

For S(IV), analogously as with dissolved CO_2, dissolved SO_2 exists as the molecular species SO_2 and, to a much lesser extent, the hydrated species H_2SO_3. So, analogously as with dissolved CO_2, $[H_2SO_3^*] \equiv [SO_2] + [H_2SO_3]$. $H_2SO_3^*$ is diprotic; $pK_1 = 1.85$ and $pK_2 = 7.18$ (Table 21.8). Thus, if there were a predominance layer for S(IV), it would be divided into three regions labeled $H_2SO_3^*$, HSO_3^-, and SO_3^{2-}.

For S(0), $S_{(s)}$ is essentially insoluble, and has no acid/base properties. Thus, the S(0) layer (to the extent that it exists, which depends on $S_{T,sys}$), will have one region, for $S_{(s)}$. $S_{(s)}$ is composed

TABLE 21.8
Important Inorganic Forms of Sulfur in Natural Water Systems and Their Oxidation States

oxidation state	species	p*K* values
VI	HSO_4^-, SO_4^{2-}	$pK_1^{H_2SO_4} = <-3$, $pK_2^{H_2SO_4} = 1.99$
IV	$H_2SO_3^*, HSO_3^-, SO_3^{2-}$	$pK_1^{H_2SO_3} = 1.85$; $pK_2^{H_2SO_3} = 7.18$
0	$S_{(s)}$	
–I	H_2S_2, HS_2^-, S_2^{2-}	$pK_1^{H_2S_2} = 5.0$[a]; $pK_2^{H_2S_2} = 9.7$[a]
II	H_2S, HS^-, S^{2-}	$pK_1^{H_2S} = 7.0$[b]; $pK_2^{H_2S} = 12.92$[b]

Note: $H_2SO_3^*$ is the SO_2 analog of CO_2 so that $[H_2SO_3^*] \equiv [SO_2] + [H_2SO_3]$.
[a] Gun et al. (2000).
[b] Christensen et al. (1976).

predominantly of eight-membered, ringed molecules, that is, S_8. Smaller and larger rings can also be found in solid sulfur. Thus, solid elemental sulfur is a solid solution of S molecules of different ring sizes. Because the chemical potentials of sulfur in the various rings are all very similar, the solid can be denoted simply as $S_{(s)}$.

For S(–I), the species of interest here are H_2S_2, HS_2^-, and S_2^{2-}. This is the smallest "polysulfide" family, and can be thought of as a consequence of the dissolution of $S_{(s)}$ by action of H_2S (or by action of HS^- or S^{2-}) according to $H_2S + S_{(s)} = H_2S_2$ (or $HS^- + S_{(s)} = HS_2^{2-}$, or $S^{2-} + S_{(s)} = S_2^{2-}$). Polysulfides with more than 2 S per unit are known to form in natural waters (Gun et al., 2000). Each member of an H_2S_n family carries an average sulfur oxidation state of $-\frac{2}{n}$; for the H_2S family this gives $-\frac{2}{2} = -1$; for the H_2S_3 family, $-\frac{2}{3}$; for the H_2S_4 family, $-\frac{2}{4} = -\frac{1}{2}$; and so on towards zero. None of the families for H_2S_2, H_2S_3 H_2S_4, etc. can be dominant in a pe–pH diagram for S; the case of the H_2S_2 family at S(–I) is considered in the next section.

For S(–II), this the lowest oxidation state will be the predominant oxidation state at very low pe, as with H_2S, HS^- and S^{2-}. The S(–II) predominance layer will be divided into three regions according to $pK_1 = 7.00$ and $pK_2 = 12.92$ (Table 21.7). Overall, a very preliminary version of pe-diagram for S is as given in Figure 21.9.

21.5.2 Identification of the pe–pH Predominance Regions for Aqueous Sulfur

The important redox equilibria relating the Table 21.8 species are collected in Table 21.9. The equations for lines (1) to (4) in Figure 21.9 are developed in Table 21.10, with $S_{T,sys}$ taken to be 10^{-3} *M* for the lines with S(0) in $S_{(s)}$. The four lines are plotted in Figure 21.10.

The possibility of a predominance layer for $S(IV)_T$ is excluded as follows. In Figure 21.10, $S(VI)_T$ predominates over $S(IV)_T$ above line 1, and S(0) predominates over $S(IV)_T$ below line (2) (if $S(IV)_T$ were 10^{-3} *M*.) Since line (1) lies below line (2), this covers the whole diagram, and $S(IV)_T$ cannot have any predominance regions: *an $S(VI)_T/S(0)$ boundary line will be present in the diagram as a (1+2) line hybrid.* (As a detail, line (2) assumes equilibrium with $S_{(s)}$, so is not valid at pH values above the pH maximum of the $S_{(s)}$ wedge (see below), so exclusion of $S(IV)_T$ predominance below line (2) has actually not yet been shown for the whole pH range. However, the $S(IV)_T/S(-II)_T$ line, which can be shown to be $pe = 6.43 + \frac{1}{6}\log(\alpha_0^{S(IV)}/\alpha_0^{S(-II)}) - pH$, lies close to line (2) (see Figure 21.10), and so it remains that $S(IV)_T$ does not predominate below ~line (2).)

TABLE 21.9
Data for Selected Sulfur-Related Redox Reactions at 25 °C/1 atm in Order of Decreasing pe° and E^0_H

reduction half reaction	log *K*	pe°	pe°(W)	E^0_H(=0.05916 pe°)
$H_2SO_3^* + 4H^+ + 4e^- = S_{(s)} + 3H_2O$	33.8	8.45	1.45	0.50
$SO_4^{2-} + 8H^+ + 6e^- = S_{(s)} + 4H_2O$	36.2	6.03	−3.3	0.36
$HSO_4^- + 7H^+ + 6e^- = S_{(s)} + 4H_2O$	34.2	5.7	−2.47	0.34
$SO_4^{2-} + 10H^+ + 8e^- = H_2S + 4H_2O$	41.0	5.13	−3.62	0.30
$\frac{1}{2}H_2S_2 + H^+ + e^- = H_2S$	4.35[a]	4.35[a]	−2.65[a]	0.26[a]
$SO_4^{2-} + 9H^+ + 8e^- = HS^- + 4H_2O$	34.0	4.25	−3.63	0.25
$S_{(s)} + 2H^+ + 2e^- = H_2S$	4.8	2.4	−4.6	0.14
$SO_4^{2-} + 4H^+ + 2e^- = H_2SO_3^* + H_2O$	2.4	1.2	−12.8	0.07
$S_{(s)} + H^+ + e^- = \frac{1}{2}H_2S_2$	0.45[a]	0.45[a]	−6.55[a]	0.027[a]
$HSO_4^- + 3H^+ + 2e^- = H_2SO_3^* + H_2O$	0.4	0.2	−10.3	0.012
$S_{(s)} + H^+ + 2e^- = HS^-$	−2.2	−1.1	−4.6	−0.065

[a] Values obtained based on
1) Rickard and Luther (2007) that $S_2^{2-} + H_2O = 1.75HS^- + 0.25SO_4^{2-} + 0.25H^+ \quad \log K = 0.8893$;
2) p*K* values in Table 21.7 for H_2S_2 and H_2S; and, depending on the reaction,
3) either $SO_4^{2-} + 10H^+ + 8e^- = H_2S + 4H_2O \quad \log K = 41.0$, or $S_{(s)} + 2H^+ + 2e^- = H_2S$, $\log K = 4.8$.

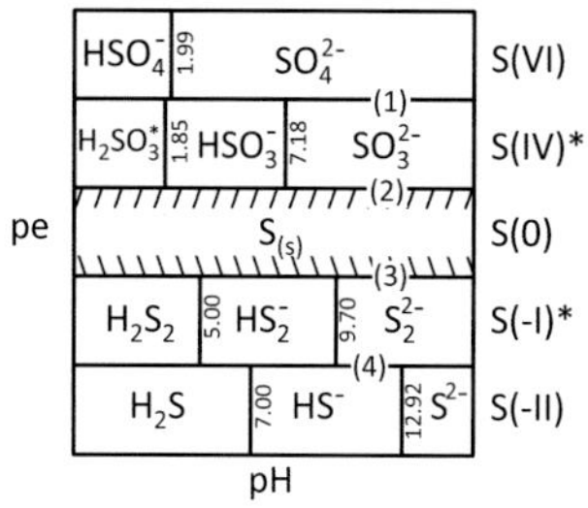

FIGURE 21.9 Preliminary pe–pH diagram for sulfur. The asterisk on S(IV) indicates that because of where lines (1) and (2) lie, in the final diagram there will not be a layer for S(IV). The asterisk on S(−I) indicates that because of where lines (3) and (4) lie, in the final diagram there will not be a layer for S(−I).

The possibility of a predominance layer for $S(-I)_T$ is excluded as follows. In Figure 21.10, S(0) predominates over $S(-I)_T$ above line (3), and $S(-II)_T$ predominates over $S(-I)_T$ below line (4). Since line (3) in Figure 21.10 lies below line (4) except at high pH, $S(-I)_T$ cannot have any predominance regions except perhaps at high pH: *an S(0)/S(−II)$_T$ boundary line will be present in the diagram as a (3 + 4) line hybrid.* (Because line (3) assumes equilibrium with $S_{(s)}$, it is not valid at pH values above the pH maximum of the $S_{(s)}$ wedge (see below), so exclusion of $S(-I)_T$ predominance above line (3) for most of the pH range has actually not yet been shown. However, the entirety of the $S(VI)_T/S(-I)_T$ line, which can be shown to be $\mathrm{pe} = 5.03 + \frac{1}{7}\log(\alpha_2^{S(VI)}/(\alpha_0^{S(-I)})^{\frac{1}{2}}) - \frac{9}{7}\mathrm{pH}$, lies below line (4) (see Figure 21.10), and so $S(-I)_T$ indeed cannot predominate at any pH.)

TABLE 21.10
Equations for pe–pH Boundary Lines for $S(VI)_T/S(IV)_T$, $S(IV)_T/S(0)$, $S(0)/S(-I)_T$, and $S(-I)_T/S(-II)_T$, with Equilibrium Constant Values for 25 °C/1 atm

Couple	Equations
$\frac{S(VI)_T}{S(IV)_T}$ line (1)	$SO_4^{2-} + 4H^+ + 2e^- = H_2SO_3^* + H_2O \quad pe^o = 1.20$ (21.65)

$$pe = 1.20 + \tfrac{1}{2}\log\frac{\{SO_4^{2-}\}\{H^+\}^4}{\{H_2SO_3^*\}} \quad (21.66)$$

$$S(VI)_T = \left[HSO_4^-\right] + \left[SO_4^{2-}\right]$$

$$S(IV)_T = \left[H_2SO_3^*\right] + \left[HSO_3^-\right] + \left[SO_3^{2-}\right]$$

For unit γ values, then:

$$pe = 1.20 + \tfrac{1}{2}\log\frac{\alpha_2^{S(VI)}S(VI)_T}{\alpha_0^{S(IV)}S(IV)_T} - 2pH \quad (21.67)$$

For H_2SO_4, $pK_1^{H_2SO_4} < -3$, so $[H_2SO_4] \approx 0$; $pK_2^{H_2SO_4} = 1.99$, for use in $\alpha_2^{S(VI)} = \dfrac{[SO_4^{2-}]}{[HSO_4^-] + [SO_4^{2-}]} = \dfrac{K_2^{H_2SO_4}}{[H^+] + K_2^{H_2SO_4}}$

For $H_2SO_3^*$, $pK_1^{H_2SO_3} = 1.85$ and $pK_2^{H_2SO_3} = 7.18$, for use in $\alpha_0^{S(IV)} = \dfrac{[H_2SO_3^*]}{[H_2SO_3^*] + [HSO_3^-] + [SO_3^{2-}]}$

There are equal amounts of sulfur in the two oxidation states when $S(VI)_T = S(IV)_T$. Then:

$$pe = 1.20 + \tfrac{1}{2}\log\frac{\alpha_2^{S(VI)}}{\alpha_0^{S(IV)}} - 2pH \quad \text{line (1)} \quad (21.68)$$

Couple	Equations
$\frac{S(IV)_T}{S(0)}$ line (2)	$H_2SO_3^* + 4H^+ + 4e^- = S_{(s)} + 3H_2O \quad pe^o = 8.45$ (21.69)

$$pe = 8.45 + \tfrac{1}{4}\log\frac{\{H_2SO_3^*\}\{H^+\}^4}{\{S_{(s)}\}} \quad (21.70)$$

$$S(IV)_T = \left[H_2SO_3^*\right] + \left[HSO_3^-\right] + \left[SO_3^{2-}\right]$$

For unit γ values and $\{S_{(s)}\} = 1$:

$$pe = 8.45 + \tfrac{1}{4}\log\left(\alpha_0^{S(IV)}S(IV)_T\right) - pH \quad (21.71)$$

For $H_2SO_3^*$, $pK_1^{H_2SO_3} = 1.85$ and $pK_2^{H_2SO_3} = 7.18$, for use in $\alpha_0^{S(IV)} = \dfrac{[H_2SO_3^*]}{[H_2SO_3^*] + [HSO_3^-] + [SO_3^{2-}]}$

With $S_{T,sys} = 10^{-3}$ *M*, along the presumed boundary line with $S_{(s)}$, neglecting the other oxidation states for dissolved S, the dissolved $S(IV)_T = 10^{-3}$ *M*. Then:

$$S(IV)_T = 10^{-3}M$$

Then:

$$pe = 7.70 + \tfrac{1}{4}\log\alpha_0^{S(IV)} - pH \quad \text{line (2)} \quad (21.72)$$

Couple	Equations
$\frac{S(0)}{S(-I)_T}$ line (3)	$S_{(s)} + H^+ + e^- = \tfrac{1}{2}H_2S_2 \quad pe^o = 0.45$ (21.73)

$$pe = 0.45 + \tfrac{1}{1}\log\frac{\{S_{(s)}\}\{H^+\}}{\{H_2S_2\}^{\frac{1}{2}}} \quad (21.74)$$

$$S(-I)_T = 2\left(\left[H_2S_2\right] + \left[HS_2^-\right] + \left[S_2^{2-}\right]\right) = 2\left[H_2S_2 \text{ related}\right]_T$$

(Continued)

TABLE 21.10 (CONTINUED)
Equations for pe–pH Boundary Lines for $S(VI)_T/S(IV)_T$, $S(IV)_T/S(0)$, $S(0)/S(-I)_T$, and $S(-I)_T/S(-II)_T$, with Equilibrium Constant Values for 25 °C/1 atm

Couple	Equations
For unit γ values and $\{S_{(s)}\} = 1$:	$pe = 0.45 - \frac{1}{2}\log\left(\alpha_0^{S(-I)}[H_2S_2 \text{ related}]_T\right) - pH$ (21.75)
For H_2S_2, $pK_1^{H_2S_2} = 5.0$; $pK_2^{H_2S_2} = 9.7$ for use in $\alpha_0^{S(-I)} = \frac{[H_2S_2]}{[H_2S_2]+[HS_2^-]+[S_2^{2-}]}$	
All H_2S_2-related species contain 2S per unit. With $S_{T,sys} = 10^{-3}$ *M*, along the presumed boundary line for $S_{(s)}$, the total dissolved S = 10^{-3} *M*. Thus, along the presumed boundary line, neglecting the other oxidation states for dissolved S, total dissolved S = $S(-I)_T = 2[H_2S_2\text{-related}]_T = 10^{-3}$ *M*. Then:	$[H_2S_2 \text{ related}]_T = 0.5\times10^{-3}\ M$
Then:	$pe = 2.10 - \frac{1}{2}\log\alpha_0^{S(-I)} - pH$ line (3) (21.76)
$\frac{S(-I)_T}{S(-II)_T}$ line (4)	$\frac{1}{2}H_2S_2 + H^+ + e^- = H_2S \quad pe° = 4.35$ (21.77)
	$pe = 4.35 + \frac{1}{1}\log\frac{\{H_2S_2\}^{\frac{1}{2}}\{H^+\}}{\{H_2S\}}$ (21.78)
	$S(-I)_T = 2\left([H_2S_2]+[HS_2^-]+[S_2^{2-}]\right) = 2[H_2S_2 \text{ related}]_T$
	$S(-II)_T = [H_2S]+[HS^-]+[S^{2-}]$
For unit γ values:	$pe = 4.35 + \log\frac{\left(\alpha_0^{S(-I)}[H_2S_2 \text{ related}]_T\right)^{\frac{1}{2}}}{\alpha_0^{S(-II)}S(-II)_T} - pH$ (21.79)
For H_2S_2, $pK_1^{H_2S_2} = 5.0$; $pK_2^{H_2S_2} = 9.7$ for use in $\alpha_0^{S(-I)} = \frac{[H_2S_2]}{[H_2S_2]+[HS_2^-]+[S_2^{2-}]}$	
For H_2S, $pK_1^{H_2S} = 7.0$; $pK_2^{H_2S} = 12.92$ for use in $\alpha_0^{S(-II)} = \frac{[H_2S]}{[H_2S]+[HS^-]+[S^{2-}]}$	
For total dissolved S = 10^{-3} *M*, there are equal amounts of sulfur in the two oxidation states when:	$2[H_2S_2 \text{ related}]_T = S(-I)_T = S(-II)_T = 0.5\times10^{-3}\ M$
Then:	$pe = 5.85 + \log\frac{\left(\alpha_0^{S(-I)}\right)^{\frac{1}{2}}}{\alpha_0^{S(-II)}} - pH$ line (4) (21.80)

Note: As needed, $S_{T,sys} = 10^{-3}$ *M*.

The $S(VI)_T/S(0)$ (*i.e.*, (1+2) hybrid) and $S(0)/S(-II)_T$ (*i.e.*, (3+4) hybrid) boundary line equations ((21.84) and (21.88)) are developed in Table 21.11 and plotted in Figure 21.11. Analogously with the Cl(I)/Cl(0) and Cl(0)/Cl(–I) lines in Figure 18.3, here these two lines cross (at pH = 4.88 for $S_{T,sys} = 10^{-3}$ *M*). $S_{(s)}$ will therefore disproportionate (and completely dissolve) beyond the pH maximum of the $S_{(s)}$ wedge, and a new, $S(VI)_T/S(-II)_T$ line will begin to be drawn, as the hybrid of (1+2) with (3+4). When organic matter or methane is present with SO_4^{2-}, the forward, reductive direction for the underlying reaction

$$SO_4^{2-} + 10H^+ + 8e^- = H_2S + 4H_2O \tag{21.89}$$

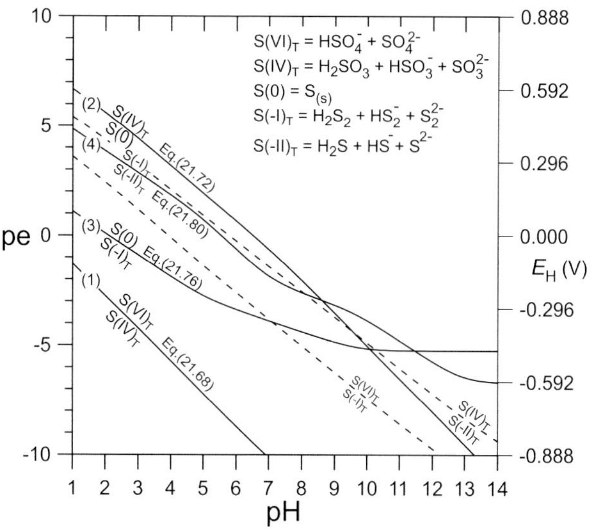

FIGURE 21.10 pe–pH predominance lines for sulfur at 25 °C/1 atm for $S(VI)_T/S(IV)_T$ (line 1), $S(IV)_T/S(0)$ (line 2), $S(0)/S(-I)_T$ (line 3), and $S(-I)/S(-II)_T$ (line 4). Also shown are lines (dashed) for $S(IV)_T/S(-II)_T$ and $S(VI)_T/S(-I)_T$. For lines involving S(0), $S_{T,sys} = 10^{-3}$ *M*. Solution-phase activity coefficients are assumed to be unity.

is mediated by a considerable number of different sulfate-reducing microorganism that includes sulfate-reducing bacteria, as well as sulfate-reducing archaea (archaea being a set of organisms distinct from, but resembling bacteria). The reverse, oxidative direction for the Eq.(21.89) is carried out under oxic conditions by sulfate oxidizing bacteria. The equation for the (1+2)+(3+4) hybrid based on Eq.(21.92) is developed in Table 21.11. Involving only dissolved species, this line is perfectly valid inside the $S_{(s)}$ wedge as the boundary along which $S(VI)_T = S(-II)_T$, but inside the wedge, it is not drawn. This follows the rule that where a solid is present in a pe–pH diagram, that region is labeled with the solid, and other lines are generally not drawn. As $S_{T,sys}$ is lowered or raised, the wedge for the solid will recede or grow, just as the wedge for Cl_2 recedes or grows with changing $Cl_{T,sys}$. For sulfur, the tip of the $S_{(s)}$ wedge tracks along the $S(VI)_T/S(-II)_T$ line. For chlorine, the tip of the Cl_2 wedge tracks along the Cl(I)/Cl(−I) line.

The intersection at pH = 4.88 is an approximation for the pH at the tip of the $S_{(s)}$ wedge. This is because along both lines (21.84) and (21.88), all of $S_{T,sys}$ is assumed to be dissolved and reside in a single dominant oxidation state. Along the $S(VI)_T/S(0)$ line, it was assumed that dissolved $S(VI)_T = 10^{-3}$ *M*, and along the $S(0)/S(-II)_T$ line, it was assumed that dissolved $S(-II)_T = 10^{-3}$ *M*. This would require that the total dissolved at the intersection would be 2×10^{-3} *M* even as the diagram is being drawn for $S_{T,sys} = 10^{-3}$ *M*. The single boundary line defining the $S_{(s)}$ wedge can be computed by calculating the set of pe and pH values for which $S_{(s)}$ specifies that the dissolved $S_T = 10^{-3}$ *M* as summed over $S(VI)_T$ and $S(-II)_T$ ($S(IV)_T$ and $S(-I)_T$ be negligible):

$$[HSO_4^-] + [SO_4^{2-}] + [H_2S] + [HS^-] + [S^{2-}] = S_{T,sys}. \qquad (21.93)$$

At 25 °C/1 atm, K_2 for H_2SO_4 is $10^{-2.0}$, K_1 for H_2S is $10^{-7.0}$, and K_1K_2 for H_2S is $10^{-19.92}$. Therefore, neglecting activity corrections, Eq.(21.93) becomes

$$[SO_4^-]\left(\frac{[H^+]}{10^{-1.99}}+1\right) + [H_2S]\left(1+\frac{10^{-7.00}}{[H^+]}+\frac{10^{-19.82}}{[H^+]^2}\right) = S_{T,sys}. \qquad (21.94)$$

TABLE 21.11
Equations for Sulfur Redox Half Reactions for S(VI)$_T$/S(0), S(0)/S(−II)$_T$, and S(VI)$_T$/S(−II)$_T$, with Equilibrium Constant Values for 25 °C/1 atm

Couple	Equations	
$\frac{S(VI)_T}{S(0)}$ line (1+2)	$SO_4^{2-} + 8H^+ + 6e^- = S_{(s)} + 4H_2O \quad pe^o = 6.03$	(21.81)
	$pe = 6.03 + \frac{1}{6}\log\frac{\{SO_4^{2-}\}\{H^+\}^8}{\{S_{(s)}\}}$	(21.82)
	$S(VI)_T = [HSO_4^-] + [SO_4^{2-}]$	
For unit γ values and $\{S_{(s)}\} = 1$:	$pe = 6.03 + \frac{1}{6}\log\left(\alpha_2^{S(VI)}S(VI)_T\right) - \frac{4}{3}pH$	(21.83)
For H_2SO_4, $pK_1 < -3$; $pK_2 = 1.99$ for use in $\alpha_2^{S(VI)} = \frac{[SO_4^{2-}]}{[HSO_4^-] + [SO_4^{2-}]}$		
With $S_{T,sys} = 10^{-3}$ *M*, along the boundary line with $S_{(s)}$, the dissolved $S(VI)_T = 10^{-3}$ *M*. Then:		
	$pe = 5.53 + \frac{1}{6}\log\alpha_2^{S(VI)} - \frac{4}{3}pH$ line (1+2)	(21.84)
$\frac{S(0)_T}{S(-II)_T}$ line (3 + 4)	$S_{(s)} + 2H^+ + 2e^- = H_2S \quad pe^o = 2.4$	(21.85)
	$pe = 2.4 + \frac{1}{2}\log\frac{\{S_{(s)}\}\{H^+\}^2}{\{H_2S\}}$	(21.86)
	$S(-II)_T = [H_2S] + [HS^-] + [S^{2-}]$	
For unit γ values and $\{S_{(s)}\} = 1$:	$pe = 2.4 - \frac{1}{2}\log\left(\alpha_0^{S(-II)}S(-II)_T\right) - pH$	(21.87)
With $S_{T,sys} = 10^{-3}$ *M*, along the boundary line with $S_{(s)}$, the dissolved $S(-II)_T = 10^{-3}$ *M*. Then: Then:		
	$pe = 3.9 - \frac{1}{2}\log\alpha_0^{S(-II)} - pH$ line (3 + 4)	(21.88)
$\frac{S(VI)_T}{S(-II)_T}$ line ((1 + 2) + (3 + 4))	$SO_4^{2-} + 10H^+ + 8e^- = H_2S + 4H_2O \quad pe^o = 5.13$	(21.89)
	$pe = 5.13 + \frac{1}{8}\log\frac{\{SO_4^{2-}\}\{H^+\}^{10}}{\{H_2S\}}$	(21.90)
	$S(VI)_T = [HSO_4^-] + [SO_4^{2-}]$	
	$S(-II)_T = [H_2S] + [HS^-] + [S^{2-}]$	
For unit γ values:	$pe = 5.13 + \frac{1}{8}\log\frac{\alpha_2^{S(VI)}S(VI)_T}{\alpha_0^{S(-II)}S(-II)_T} - \frac{10}{8}pH$	(21.91)
There are equal amount of sulfur in the two oxidation states when $S(VI)_T = S(-I)_T$. Then:		
	$pe = 5.13 + \frac{1}{8}\log\frac{\alpha_2^{S(VI)}}{\alpha_0^{S(-II)}} - \frac{10}{8}pH$ line ((1 + 2) + (3 + 4))	(21.92)

Note: As needed, $S_{T,sys} = 10^{-3}$ *M*.

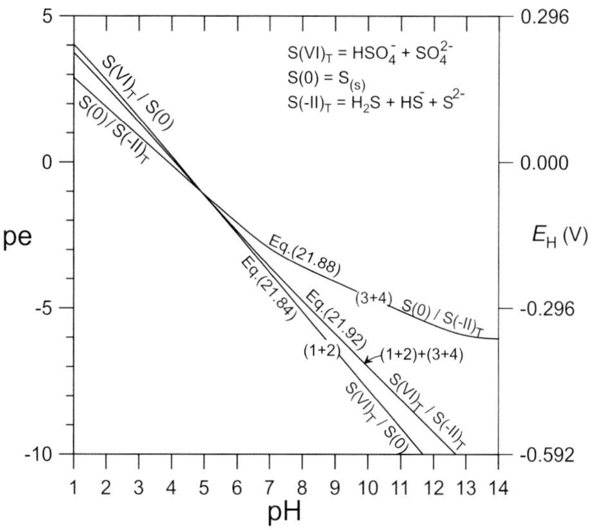

FIGURE 21.11 pe–pH predominance lines for sulfur at 25 °C/1 atm for $S(VI)_T/S(0)$, $S(0)/S(-II)_T$/, and the hybrid line $S(VI)/S(-II)_T$. For lines involving S(0), $S_{T,sys} = 10^{-3}$ *M*. Solution-phase activity coefficients are assumed to be unity.

From Table 21.8, we have for equilibrium with S(s),

$$SO_4^{2-} + 8H^+ + 6e^- = S_{(s)} + 4H_2O \qquad K = 10^{36.2} = \frac{1}{[SO_4^{2-}]10^{-8pH}10^{-6pe}} \tag{21.95}$$

$$S_{(s)} + 2H^+ + 2e^- = H_2S \qquad K = 10^{4.8} = \frac{[H_2S]}{10^{-2pH}10^{-2pe}} \tag{21.96}$$

where the *K* expressions neglect activity corrections for the S species. The curve for the $S_{(s)}$ wedge is then

$$\frac{1}{10^{36.2}10^{-8pH}10^{-6pe}}\left(\frac{[H^+]}{10^{-1.99}}+1\right) + 10^{4.8}10^{-2pH}10^{-2pe}\left(1+\frac{10^{-7.0}}{[H^+]}+\frac{10^{-19.82}}{[H^+]^2}\right) = S_{T,sys} \tag{21.97}$$

The $S_{(s)}$ wedge is angled downwards. Therefore, for the pH range over which the wedge extends, for each value of pH. Eq.(21.97) has two real solutions for pe: a higher pe root and a lower pe root. Near the tip, the two roots are very close to one another, so, obtaining both pe roots in an organized manner by a numerical algorithm can be challenging; when seeking one of the roots, a numerical algorithm can jump to the other.

21.5.3 Final pe–pH Diagram Comments

Figure 21.12 gives the final pe–pH diagram for $S_T = 10^{-3}$ *M* and 25 °C/1 atm. At equilibrium, the concentrations of S(IV) species are extremely low (though never zero) everywhere in the pe–pH domain. Equation (21.97) is plotted in Figure 21.12 as the blunt wedge defining the $S_{(s)}$ region. As discussed above, the blunting makes the maximum pH of the wedge pH = 4.40 rather than the pH of 4.88 that is obtained as the intersection of lines (21.84) and (21.88). Lines at pH = 1.99, pH = 7.0, and pH = 12.92 provide the HSO_4^-/SO_4^{2-}, H_2S/HS^-, and HS^-/S^{2-} boundary lines, respectively. The $S(VI)_T/S(-II)_T$ line given by Eq.(21.92) and is valid over its entire length, though not plotted inside the $S_{(s)}$ wedge.

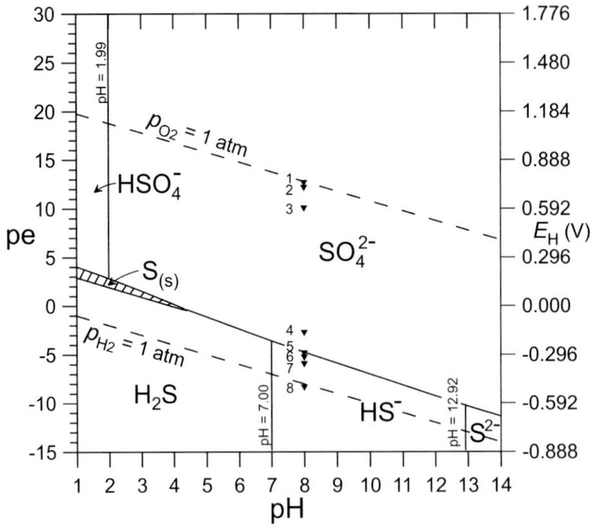

FIGURE 21.12 Final pe–pH predominance diagram for S at 25 °C/1 atm for $S_{T,sys} = 10^{-3}$ *M*. Solution-phase activity coefficients are assumed to be unity. The extent of the region for $S_{(s)}$ will increase as $S_{T,sys}$ is increased; at $S_{T,sys} = 10^{-3}$ *M*, the S(s) wedge extends only to pH 4.40. See Table 21.4 for description of points 1 to 8 at pH = 8.

Regarding points 1–8 in Table 21.4, for $Fe_{T,sys} = 10^{-5}$ *M*, when $C_{T,free} = 10^{-3}$ *M*, points 4–6 are inside the region for $FeCO_{3(s)}$, then the region for Fe^{2+} is entered for points 7 and 8 (Figure 21.8). In the presence of sulfur, the situation is unchanged for point 4 because at that point S is present predominantly as SO_4^{2-}. But at points 5–8, if $S_{T,sys} = 10^{-3}$ *M*, there will be appreciable S(–II) in the system. The solid of Fe^{2+} with S(–II), namely $FeS_{(s)}$, is extremely insoluble: at points 5–8, which are below the $S(VI)_T/S(-II)_T$ line, Fe chemistry will be dominated by $FeS_{(s)}$.

Example 21.5 H_2S in Sewers

Because of the abundant organic carbon that is present, considerable levels of H_2S can develop in sewers by reaction of the organic carbon in the sewage with sulfate, as mediated by sulfate-reducing microorganisms. This can happen even if the sewer gas contains significant oxygen, because accumulated solids at the *bottom* of the pipe can become anaerobic muck, cut off from an oxic environment above it, so that sulfate reduction can occur there. For some history, see Smith (2010).

Warning: H_2S in sewer gas can be absolutely deadly. Every year, numerous workers maintaining sewers around the world are killed by H_2S. In the U.S. alone, for the period 1993 to 1999, the U.S. Bureau of Labor Statistics Census for Fatal Occupational Injuries (CFOI) tabulated 52 worker deaths due to H_2S; ~50% of the fatalities occurred within the first year of employment (Hendrickson and Hamilton, 2004), apparently as a result of inadequate training and carelessness. In 21% of the cases, a co-worker died simultaneously, or in an effort to save a stricken colleague. It has been said that one breath at 1000 ppm (10^{-3} atm) can cause immediate collapse ("knock down"), and death (Hariharan et al., 2016).

a. Develop the net redox reaction for reduction of SO_4^{2-} to H_2S by the oxidation of a carbohydrate (with formula $C_6H_{12}O_6$) by sulfate-reducing microorganisms, forming C(IV) according to the reverse of $H_2CO_3^* + 4H^+ + 4e^- = \frac{1}{6}C_6H_{12}O_6$ (glucose) + $2H_2O$. **b.** H_2S can react corrosively (abiotically) with cast iron sewer pipe. The reaction involves the reduction of the H(I) in H_2S yielding H_2 and S^{2-} by action of the Fe(0), which is converted to Fe(II); precipitation of the very insoluble corrosion product $FeS_{(s)}$ follows immediately. Develop the net redox reaction. **c.** Discharge of industrial acid into a sewer can release massive amounts of H_2S to sewer gas by immediate solubilization by the acid of accumulated $FeS_{(s)}$. Assuming that the acid is HCl (though it could be any of many acids, including acetic acid), write the net reaction. **d.** The Henry's Gas Law constant for H_2S is 0.1 *M*/atm at 25 °C/1 atm. (The first "atm" in the preceding sentence refers to the gas pressure of H_2S; the second refers (as usual) to the total system pressure.) If sewer gas with 1000 ppm (10^{-3} atm) H_2S is at equilibrium with sewage flowing beneath it at 25 °C/1 atm, what is the concentration of H_2S in the sewage? **e.** An enormous problem caused by H_2S in sewers involves O_2 that is usually present at significant levels in sewer gas. Water condenses on the top of concrete sewer pipes. Gaseous H_2S dissolves in that water and is there oxidized by sulfate oxidizing microorganisms to H_2SO_4, which then proceeds to attack the concrete (which is an alkaline material). Develop the net redox reaction for oxidation of H_2S to SO_4^{2-} by the action of O_2, which in turn is reduced to H_2O.

Solution

a.

$$SO_4^{2-} + 10H^+ + 8e^- = H_2S + 4H_2O$$

$$-2\left(H_2CO_3^* + 4H^+ + 4e^- = \tfrac{1}{6}C_6H_{12}O_6\ \text{(glucose)} + 2H_2O\right)$$

$$SO_4^{2-} + \tfrac{1}{3}C_6H_{12}O_6 + 2H^+ = H_2S + 2H_2CO_3^*$$

b.

$$H_2S = 2H^+ + S^{2-}$$

$$2H^+ + 2e^- = H_2$$

$$-\left(Fe^{2+} + 2e^- = Fe_{(s)}\right)$$

$$+Fe^{2+} + S^{2-} = FeS_{(s)}$$

$$H_2S + Fe_{(s)} = FeS_{(s)} + H_2$$

c. $2H^+ + FeS_{(s)} = H_2S + Fe^{2+}$

d. $H_2S_{(g)} = H_2S$. $K_H = 0.1\ M/\text{atm} = [H_2S]/p_{H_2S}$. $p_{H_2S} = 10^{-3}$ atm. $[H_2S] = 10^{-4}\ M$.

e.

$$2\left(O_2 + 4H^+ + 4e^- = 2H_2O\right)$$

$$-\ SO_4^{2-} + 10H^+ + 8e^- = H_2S + 4H_2O$$

$$H_2S + 2O_2 = 2H^+ + SO_4^{2-}$$

Oxidation of H_2S by O_2 leads to H_2SO_4, which attacks the alkaline concrete.

Example 21.6 Formation of $FeS_{(s)}$ under Highly Reducing Conditions when Both Fe and S Are Present in the System

Consider a sediment system at 25 °C/1 atm for which $Fe_{T,sys} = 10^{-3}$ *M*, $C_{T,free} = 10^{-3}$ *M*, and $S_{T,sys} = 10^{-3}$ *M*. Point 5 in Table 21.4 and in the figures is pe = −4.9, pH = 8.0. The line between $(am)Fe(OH)_{3(s)}$ and $FeCO_{3(s)}$ in Figure 21.8 does not depend of $Fe_{T,sys}$, so since point 5 is below that line when $Fe_{T,sys} = 10^{-5}$ *M*, it is also below that line when $Fe_{T,sys} = 10^{-3}$ *M*. Neglect activity corrections. **a.** At pH = 8.0, compute the pe when the solution ratio $S(VI)_T/(S{-}II)_T = 1$. **b.** Compute $S(VI)_T/(S{-}II)_T$ at point 5. If all of the $S_{T,sys}$ is in solution, then calculate $S(-II)_T$ at point 5. **c.** For $[Fe^{2+}]$ at saturation with $FeCO_{3(s)}$, would the solution be supersaturated with $FeS_{(s)}$ if the $K_{s0}^{FeS} = 10^{-18}$?

Solution

a. From Eq.(21.91), pe $= 5.13 + \frac{1}{8}\log\frac{\alpha_2^{S(VI)}S(VI)_T}{\alpha_0^{S(-II)}S(-II)_T} - \frac{10}{8}$pH. At pH = 8.0, $\alpha_2^{S(VI)} = 1.00$ and $\alpha_0^{S(-II)} = 0.091$.

For $S(VI)_T/(S{-}II)_T = 1$, pe = −4.74.

b. pe = −4.9, from Eq.(21.91), $-4.9 = 5.13 + \frac{1}{8}\log\frac{\alpha_2^{S(VI)}S(VI)_T}{\alpha_0^{S(-II)}S(-II)_T} - \frac{10}{8}$pH . At pH = 8.0, $\alpha_2^{S(VI)} = 1.00$ and $\alpha_0^{S(-II)} = 0.091$. $S(VI)_T/(S{-}II)_T = 0.052$. The fraction of the solution S that is present as $S(-II)_T = S(-II)_T/(S(-II)_T + S(VI)_T) = (1 + S(VI)_T/(S{-}II)_T)^{-1} = 0.95$. $S(-II)_T = 0.95\times S_{T,sys} = 0.95\times10^{-3}$ *M*.

c. Calculate $[CO_3^{2-}]$: at pH = 8.0, $\alpha_0^{C(IV)} = 0.022$; $\alpha_2^{C(IV)} = \alpha_0^{C(IV)}K_1^{C(IV)}K_2^{C(IV)}/[H^+]^2 = 4.55\times10^{-3}$; $[CO_3^{2-}] = \alpha_2^{C(IV)}C_{T,free} = 4.55\times10^{-6}$ *M*. Calculate $[Fe^{2+}]$: for equilibrium with $FeCO_{3(s)}$, $[Fe^{2+}] = K_{s0}^{FeCO_{3(s)}}/[CO_3^{2-}] = 10^{-10.68}/4.55\times10^{-6} = 4.59\times10^{-6}$ *M*. Calculate $[S^{2-}]$: $\alpha_2^{S(-II)} = \alpha_0^{S(-II)}K_1^{H_2S}K_2^{H_2S}/[H^+]^2$; at pH = 8.0, $\alpha_2^{S(-II)} = 1.09\times10^{-5}$; $[S^{2-}] = \alpha_2^{S(-II)}S(-II)_T = 1.09\times10^{-5}\times0.95\times10^{-3}$ *M* $= 1.04\times10^{-8}$ *M*. Check saturation state with $FeS_{(s)}$: $[Fe^{2+}][S^{2-}] = 4.59\times10^{-6}$ *M* $\times 1.04\times10^{-8}$ *M* $= 4.75\times10^{-14}$ *M* >>> $K_{s0}^{FeS} = 10^{-18}$: if $S_{T,sys} = 10^{-3}$ *M*, $FeS_{(s)}$ will form at point 5 (and below, see Table 21.4).

21.6 pe–pH DIAGRAM FOR CARBON

21.6.1 Redox Equilibria Governing Carbon Species

The pe–pH diagrams for oxygen and hydrogen, were considered in Chapter 18. For carbon, there are innumerable possible carbon-containing molecules, and therefore many different average carbon oxidation states. The example carbon forms considered in this section are given in Table 21.1 with their oxidation states (averages when appropriate).

Carbon is in the IVA column of the periodic table, so atomic carbon has 4 valence electrons. Removing 4 electrons takes carbon to the relatively stable state of an empty outer shell, C(VI). Adding 4 electrons takes carbon to a full outer shell of 8 and C(–IV). As a result, carbon-containing species range from: 1) CO_2-type species, which are C(IV) species; to 2) the C(0) forms of graphite and diamond (graphite is more stable at 25 °C/1 atm; both are exceedingly slow to form at 25 °C/1 atm, and will be ignored in this chapter); and 3) carbohydrates such as glucose which is C(0) (on average); and finally to 4) CH_4, which is C(–IV). Other intermediate average oxidation states occur for C in organic compounds.

Some of the possible redox half-reactions involving the Table 21.12 compounds are given in Table 21.13. The question now is which of the different carbon oxidation states and species can have predominance regions in equilibrium pe–pH diagrams.

21.6.2 Identification of the pe–pH Predominance Regions for Aqueous Carbon

General Expectations. For the simple/limited collection of species in Table 21.12, a preliminary pe–pH diagram for carbon would be composed of five oxidation state layers. The layer for C(IV)

TABLE 21.12
Some Examples of Carbon in Natural Water Systems and Their Oxidation States (or Average Oxidation States)

oxidation state	Species	pK values
IV	$H_2CO_3^*$, HCO_3^-, CO_3^{2-}	$pK_1 = 6.35$; $pK_2 = 10.33$
II	HCOOH (formic acid), $HCOO^-$ (formate ion)	$pK_a = 3.75$
0	$C_6H_{12}O_6$ (glucose) and CH_2O (formaldehyde)	–
–II	CH_3OH (methanol)	–
–IV	CH_4 (methane)	–

Note: pK values at 25 °C/1 atm.

TABLE 21.13
Data for Selected Carbon-Related Redox Reactions at 25 °C/1 atm in Order of Decreasing pe° and E_H^o

reduction half reaction	log K	pe°	pe°(W)	E_H^o(=0.05916 pe°)
$CH_3OH + 2H^+ + 2e^- = CH_4 + H_2O$	16.95	8.48	1.48	0.50
$CH_2O + 4H^+ + 4e^- = CH_4 + H_2O$	24.95	6.24	–0.76	0.37
$\frac{1}{6}C_6H_{12}O_6(\text{glucose}) + 4H^+ + 4e^- = CH_4 + H_2O$	20.95	5.24	–1.76	0.31
$HCOO^- + 7H^+ + 6e^- = CH_4 + 2H_2O$	29.8	4.97	–3.20	0.29
$CH_2O + 2H^+ + 2e^- = CH_3OH$	8.0	4.0	–3.0	0.24
$HCOO^- + 3H^+ + 2e^- = CH_2O + H_2O$	5.64	2.82	–7.68	0.17
$H_2CO_3^* + 8H^+ + 8e^- = CH_4 + 3H_2O$	21.6	2.70	–4.30	0.16
$H_2CO_3^* + 6H^+ + 6e^- = CH_3OH + 2H_2O$	4.66	0.78	–6.22	0.046
$H_2CO_3^* + 4H^+ + 4e^- = \frac{1}{6}C_6H_{12}O_6\ (\text{glucose}) + 2H_2O$	0.67	0.17	–6.83	0.010
$H_2CO_3^* + 4H^+ + 4e^- = CH_2O + 2H_2O$	–3.33	–0.83	–7.83	–0.049
$H_2CO_3^* + H^+ + 2e^- = HCOO^- + H_2O$	–8.19	–4.10	–7.60	–0.24

Note: CH_4 is dissolved methane.

would be divided into three regions as for $H_2CO_3^*$, HCO_3^-, and CO_3^{2-}, as separated by pK_1 and pK_2 for $H_2CO_3^*$. The layer for C(II) (Formic$_T$) would be divided into two as for HCOOH and HCO_2^-, as separated by the pK for HCOOH.

Given the infinite number of average oxidation states for organic compounds comprised of C, H, and O, there would be an infinite number of oxidation state rows, but we will shortly show that no aqueous organic compound has a predominance region at equilibrium. Showing this is facilitated by first identifying the $C(IV)_T/C(-IV)_T$ line (a.k.a. the $C(IV)_T/CH_4$ line).

The $C(IV)_T/C(-IV)$ Line. From Table 21.13,

$$H_2CO_3^* + 8H^+ + 8e^- = CH_4 + 3H_2O \quad \log K = 21.60 \tag{21.98}$$

$$pe = 2.70 + \tfrac{1}{8}\log\frac{\{H_2CO_3^*\}}{\{CH_4\}} - pH. \tag{21.99}$$

CH_4 has no acid/base chemistry so $[CH_4]$ equals the total dissolved C(–IV) concentration. $H_2CO_3^*$ has acid/base chemistry. As usual, $[H_2CO_3^*] = \alpha_0^C C(IV)_T$ where $C(IV)_T$ represents only the C(IV) portion of the total dissolved carbon. Neglecting activity corrections in Eq.(21.99) (and here forward),

$$pe = 2.70 + \tfrac{1}{8}\log\frac{\alpha_0^C C(IV)_T}{[CH_4]} - pH. \quad (21.100)$$

When there are equal amounts of carbon in the two oxidation states, $C(IV)_T = [CH_4]$ so

$$pe = 2.70 + \tfrac{1}{8}\log\alpha_0^C - pH \qquad C(IV)_T/CH_4. \quad (21.101)$$

C(IV) *vs.* Other Oxidation State Forms: The $C(IV)_T$/C(II), $C(IV)_T$/C(0), $C(IV)_T$/C(–II) Lines. Table 21.14 develops pe–pH equations for C(IV) *vs.* other oxidation state forms: C(II) (as formic acid/formate ion), two forms of C(0) (glucose and formaldehyde), and C(–II) (methanol). These lines are given by Eqs.(21.105), (21.110), (21.114), and (21.118). *All four lines fall below the line for Eq.(21.101).* This means that none of these other oxidation state forms can have a predominance region in the pe–pH diagram for carbon. For example, consider the line for Eq.(21.105), $C(IV)_T/Formic_T$. Above that line, $C(IV)_T$ predominates over $Formic_T$. But, that line is below Eq.(21.101) in a region where CH_4 already predominates over $C(IV)_T$. This means that CH_4 already greatly predominates over $Formic_T$ along the Eq.(21.105) line. Since CH_4 is more reduced than $Formic_T$, going to pe values below the Eq.(21.105) makes the system more reducing, and so can only further increase the predominance of CH_4 over $Formic_T$. Overall, $Formic_T$ cannot predominate above line for Eq.(21.105) because $C(IV)_T > Formic_T$, and it cannot predominate below the line for Eq.(21.105) because $CH_4 > Formic_T$. The same rationale can be applied to rule out predominance regions for glucose, formaldehyde, and methanol. Indeed, all other $C(IV)_T$/organic lines (if developed) would fall below the line for Eq.(21.101). The net result is that if graphite and diamond forms of C(0) are ruled out as being kinetically excluded, the only forms having predominance regions in the pe–pH diagram for carbon are $C(IV)_T$ and CH_4. We can therefore expect that the pe–pH diagram for carbon will look something like Figure 21.13, with the C(IV) region divided into the three species making up $C(IV)_T$ according to the values of pK_1 and pK_2 for $H_2CO_3^*$.

21.6.3 Final pe–pH Diagram Comments

The exact diagram is given in Figure 21.14. Note, however, that while only C(IV) or C(–IV) (*i.e.*, CH_4) species can predominate in the pe–pH diagram for carbon (when graphite is excluded), with the needed K values in hand, one will still calculate non-zero (albeit miniscule) equilibrium concentrations for any particular organic compound. If graphite were considered, it would appear at a wedge as with Cl(0) and S(0) in the chlorine (Figure 18.3) and sulfur (Figure 21.12) systems. At equilibrium, C(VI) persists down to ~point 6, for which N(V), Fe(III), and S(VI) are far gone. At points 7 and 8, C(–IV) (CH_4) is the stable carbon form.

Example 21.7 The $C(IV)_T/CH_4$ Transition Zone

Consider Eqs.(21.7–21.8). **a.** For pH = 7.8, compute the pe on the $C(IV)_T/[CH_4] = 1$ boundary line (Eq.(21.101)) for 25 °C/1 atm. For that pe, also calculate pe ± 1. For the three pe values, calculate the ratio $C(IV)_T/[CH_4]$ for pH = 7.8. **b.** For total dissolved C = 10^{-3} *M*, and neglecting all compounds except those for $C(IV)_T$ and CH_4, calculate the redox α values (see Section 17.5) for $C(IV)_T$ and CH_4, then the concentrations $[C(IV)_T]$ and $[CH_4]$.

Solution

At pH = 7.8, $\alpha_0^{C} = 0.0342$. **a.** On the predominance boundary line, by Eq.(21.101), at pH = 7.8, pe = −5.28. The three desired pe values for the example are then −4.28, −5.28, and −6.28. By Eq.(21.100), the associated $C(IV)_T/[CH_4]$ ratios are obtained. **b.** For the desired redox α values, by Eq.(17.99), $\alpha_{C(IV)_T} = \left(CH_4/C(IV)_T + 1\right)^{-1}$. Thus, $\alpha_{CH4} = 1 - \alpha_{C(IV)_T}$. Moving to pe values above the line quickly moves to essentially total dominance by $C(IV)_T$; moving to pe values below the line quickly moves to essentially total dominance by CH_4.

	part a		part b			
pH	pe	$\frac{[C(IV)_T]}{[CH_4]}$	$\alpha_{C(IV)T}$	α_{CH_4}	$[C(IV)_T]$ (*M*)	$[CH_4]$ (*M*)
7.8	−4.28	1.00E+08	~1.0	1.00E-08	1.00E-03	1.00E-11
7.8	−5.28	1	0.5	0.5	5.00E-04	5.00E-04
7.8	−6.28	1.00E-08	1.00E-08	~1.0	1.00E-11	1.00E-03

21.6.4 Disproportionation of Aqueous Organic Carbon Compounds

C(IV) and C(−IV) being the only stable oxidation states for carbon at 25 °C/1 atm (neglecting graphite and diamond) necessarily means all organic compounds are unstable relative to disproportionation. This the reason why methanogenic organisms can transform glucose to CO_2 and CH_4 by disproportionation:

$$C_6H_{12}O_6 = 3CO_2 + 3CH_4. \tag{21.119}$$

Similarly, yeast cells (classified as members of the fungus kingdom) can ferment carbohydrates to CO_2 and alcohols. For glucose as the parent compound and ethanol as the alcohol, the reaction is

$$C_6H_{12}O_6 = 2CO_2 + 2C_2H_5OH\,(\text{ethanol}). \tag{21.120}$$

Alcohols are necessarily subject to subsequent disproportionation to CO_2 and CH_4; for ethanol the subsequent reaction is

$$2C_2H_5OH = CO_2 + 3CH_4. \tag{21.121}$$

In every redox disproportionation reaction, the parent compound *dissociates* into two different product oxidation states, in *proportions* that reflect the difference in oxidation states between each product and the parent compound. For the reaction in Eq.(21.119), glucose contains C(0) (on average), the CO_2 is C(+IV), and CH_4 is C(−IV). One product is $-4e^-$ above the parent, the other is $+4e^-$ below. These two differences being equal, the proportion of carbons going up equals the proportion going down. So, starting with 6 carbons in glucose, the products are $3CO_2$ and $3CH_4$. In this disproportionation of the parent molecule, one portion of the carbon is oxidizing the other portion.

For the reaction in Eq.(21.120), glucose contains C(0) (on average), the CO_2 is C(+IV), and C_2H_5OH is C(−II) (on average). One product is $-4e^-$ above the parent, the other is $+2e^-$ below. To take the electrons, there has to be more carbons going down than going up: there is a 2:1 ratio between the two differences, so the proportion of carbons going down is $\frac{2}{1} = 2$ times that going up. So, starting with 6 carbons in glucose, the products are $2CO_2$ and $2C_2H_5OH$.

TABLE 21.14
Equations for pe–pH Boundary Lines for $C(IV)_T$ *vs.* Other Carbon Oxidation State Forms in Aqueous Systems, with Equilibrium Constant Values for 25 °C/1 atm

couple	equations	
$\frac{C(IV)_T}{C(-IV)}\left(=\frac{C(IV)_T}{CH_4}\right)$	$H_2CO_3^* + 8H^+ + 8e^- = CH_4 + 3H_2O \quad pe^o = 2.70$	(21.98)
	$pe = 2.70 + \frac{1}{8}\log\frac{\{H_2CO_3^*\}\{H^+\}^8}{\{CH_4\}}$	(21.99)
For unit γ values, then:	$pe = 2.70 + \frac{1}{8}\log\frac{\alpha_0^C C(IV)_T}{[CH_4]} - pH$	(21.100)
There are equal amount of C in the two oxidation states when $C(IV)_T = [CH_4]$. Then:		
	$pe = 2.70 + \frac{1}{8}\log\alpha_0^C - pH$	(21.101)
$\frac{C(IV)_T}{C(II)}\left(=\frac{C(IV)_T}{Formic_T}\right)$	$H_2CO_3^* + H^+ + 2e^- = HCOO^- + H_2O \quad pe^o = -4.10$	(21.102)
	$pe = -4.10 + \frac{1}{2}\log\frac{\{H_2CO_3^*\}\{H^+\}}{\{HCOO^-\}}$	(21.103)
For unit γ values:	$pe = -4.10 + \frac{1}{2}\log\frac{\alpha_0^C C(IV)_T}{\alpha_1^F Formic_T} - \frac{1}{2}pH$	(21.104)
$Formic_T = [HCOOH] + [HCOO^-]$; for HCOOH, $pK_a = 3.75$ for use in α_1^F. There are equal amounts of C in the two oxidation states when $C(IV)_T = Formic_T$. Then:		
	$pe = -4.10 + \frac{1}{2}\log\alpha_0^C - \frac{1}{2}\log\alpha_1^F - \frac{1}{2}pH$ Lies below Eq.(21.101).	(21.105)
$\frac{C(IV)_T}{C(0)}$ Case 1, C(0) = glucose	$H_2CO_3^* + 4H^+ + 4e^- = \frac{1}{6}C_6H_{12}O_6(\text{glucose}) + 2H_2O \quad pe^o = 0.17$	(21.106)
	$pe = 0.17 + \frac{1}{4}\log\frac{\{H_2CO_3^*\}\{H^+\}^4}{\{C_6H_{12}O_6\}^{1/6}}$	(21.107)
For unit γ values:	$pe = 0.17 + \frac{1}{4}\log\frac{\alpha_0^C C(IV)_T}{[C_6H_{12}O_6]^{1/6}} - pH$	(21.108)
Unlike Eq.(21.7) and Eq.(21.11), obtaining a pe–pH line from Eq.(21.15) requires specification of the total amount dissolved C. For 10^{-3} *M*, there are equal amounts of C in the two oxidation states when:		
	$C(IV)_T = 6[C_6H_{12}O_6] = 5.0 \times 10^{-4}\,M$	(21.109)
Then:	$pe = -0.49 + \frac{1}{4}\log\alpha_0^C - pH.$ Lies below Eq.(21.101).	(21.110)
$\frac{C(IV)_T}{C(0)}$ Case 2, C(0) = formaldehyde	$H_2CO_3^* + 4H^+ + 4e^- = CH_2O + 2H_2O \quad pe^o = -0.83$	(21.111)
	$pe = -0.83 + \frac{1}{4}\log\frac{\{H_2CO_3^*\}\{H^+\}^4}{\{CH_2O\}}$	(21.112)

(Continued)

TABLE 21.14 (CONTINUED)
Equations for pe–pH Boundary Lines for $C(IV)_T$ *vs.* Other Carbon Oxidation State Forms in Aqueous Systems, with Equilibrium Constant Values for 25 °C/1 atm

couple	equations	
For unit γ values:	$pe = -0.83 + \frac{1}{4}\log\frac{\alpha_0^C C(IV)_T}{[CH_2O]} - pH$	(21.113)
There are equal amounts of C in the two oxidation states when $C(IV)_T = [CH_2O]$. Then:		
	$pe = -0.83 + \frac{1}{4}\log\alpha_0^C - pH$ Lies below Eq.(21.101).	(21.114)
$\frac{C(IV)_T}{C(-II)}\left(=\frac{C(IV)_T}{CH_3OH}\right)$	$H_2CO_3^* + 6H^+ + 6e^- = CH_3OH + 2H_2O \quad pe^o = 0.78$	(21.115)
	$pe = 0.78 + \frac{1}{6}\log\frac{\{H_2CO_3^*\}\{H^+\}^6}{\{CH_3OH\}}$	(21.116)
For unit γ values:	$pe = 0.78 + \frac{1}{6}\log\frac{\alpha_0^C C(IV)_T}{[CH_3OH]} - pH$	(21.117)
There are equal amounts of C in the two oxidation states when $C(IV)_T = [CH_3OH]$. Then:		
	$pe = 0.78 + \frac{1}{6}\log\alpha_0^C - pH$ Lies below Eq.(21.101).	(21.118)

Note: As needed, total dissolved C = 10^{-3} *M*.

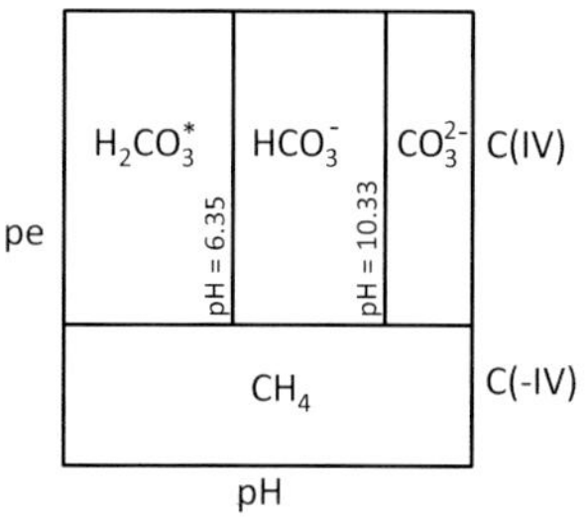

FIGURE 21.13 Generic pe–pH diagram for C after exclusion of organic forms of C (because they are unstable due to disproportionation). Graphite and diamond (solid C(0) forms) are assumed to be kinetically excluded in natural water systems. At 25 °C/1 atm, for $H_2CO_3^*$ $pK_1 = 6.35$ and $pK_2 = 10.33$.

For the reaction in Eq.(21.121), C_2H_5-OH contains C(–II) (on average), the CO_2 is C(+IV), and CH_4 is C(–IV). One product is $-6e^-$ above the parent, the other is $+2e^-$ below. There has to be more carbons going down than going up: there is a 3:1 ratio between the two differences, so the proportion of carbons going down is $\frac{3}{1} = 3$ times that going up. So, starting with 4 carbons in $2C_2H_5$-OH, the products are CO_2 and $3CH_4$. The same principles can be used to determine an overall disproportionation reaction for any organic compound, with water added as needed to get a balanced reaction; H and O not changing oxidation states means water has no effect on the proportions among the products.

The disproportionation of Cl_2 discussed in Chapter 18 (Figure 18.5) is exactly analogous to the story here for organic compounds. Beyond the Cl_2 wedge in the pe–pH diagram, Cl_2 is will mostly undergo disproportionation according to

$$Cl_2 + H_2O = HOCl + HCl. \qquad (21.122)$$

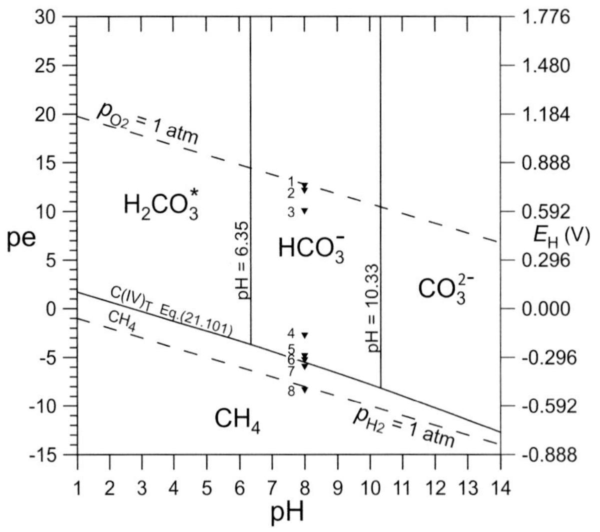

FIGURE 21.14 pe–pH predominance diagram for carbon at 25 °C/1 atm. Assumptions: 1) graphite and diamond as solid C(0) forms of carbon are not kinetically relevant in natural water systems; and 2) all solution-phase activity coefficients are unity. See Table 21.4 for description of points 1 to 8 at pH = 8. At equilibrium, C(VI) persists down to ~point 6, for which N(V), Fe(III), and S(VI) are far gone; C(–IV) (CH_4) is the stable carbon form at points 7 and 8.

The Cl_2 (oxidation state 0) dissociates to the +I and –I oxidation states in proportions that reflect the difference in oxidation states between the products HOCl and HCl and the reactant Cl_2 (Figure 18.5). The zero-valent wedge in the Cl diagram is analogous to what happens with carbon when the graphite form of $C_{(s)}$ is considered relevant, and for S(s). For C(s), if $C_{T,sys}$ is high enough, there is a $C_{(s)}$ wedge in the pe–pH diagram beyond which only CO_2 and CH_4 have predominance regions. For $S_{(s)}$, if $S_{T,sys}$ is high enough, there is an $S_{(s)}$ wedge in the pe-pH diagram beyond which only $S(VI)_T$ and $S(–II)_T$ have predominance regions.

Example 21.8 Determining Disproportionation Stoichiometry

a. Determine the overall disproportionation reaction for propanol (C_3H_7OH) going to CO_2 and CH_4.
b. Determine the overall disporprotionation reaction for benzene (C_6H_6) going to CO_2 and CH_4.

Solution

a. Based on the principles discussed in Section 17.2, C_3H_7OH contains C(–II). CO_2 is C(+IV), and CH_4 is C(–IV). One product is $-6e^-$ above the parent, the other is $+2e^-$ below. There has to be more carbons going down than going up; there is a 3:1 ratio between the two differences, so the proportion of carbons going down is $\frac{3}{1}$ = 3 times that going up. C_3H_7-OH contains 3 carbons. We need a multiple of 4 carbons so we can get 3:1 ratio in the products. 12 works. So we start with

$$4C_3H_7OH = 3CO_2 + 9CH_4 \quad \text{(not balanced)}$$

This reaction is not balanced with respect to H and O: 32 H and 4 O on the LHS, and 6 O and 36 H on the RHS. 2 H_2O is the difference, so the net reaction is

$$4C_3H_7OH + 2H_2O = 3CO_2 + 9CH_4 \quad \text{(balanced)}$$

b. C_6H_6 contains C(–I). The CO_2 is C(+IV), and CH_4 is C(–IV). One product is $-5e^-$ above the parent, the other is $+3e^-$ below. There has to be more carbons going down than going up; there is a 5:3 ratio between the two differences, so the proportion of carbons going down is $\frac{5}{3}$ times that going up. C_6H_6 contains 6 carbons. We need a multiple of 8 carbons so we can get 5:3 ratio in the products. 24 works. So we start with

$$4C_6H_6 = 9CO_2 + 15CH_4 \quad \text{(not balanced)}$$

There are 24 H and 0 O on the LHS, and 60 H and 18 O on the RHS. 18 H_2O is the difference, so the net reaction is

$$4C_6H_6 + 18H_2O = 9CO_2 + 15CH_4 \quad \text{(balanced)}$$

Example 21.9 Computing Concentrations of Minor Redox Species

For each row of the results table in Example 21.7, because the sum $C(IV)_T + [CH_4]$ dominates, use the pe, pH, and concentration values with the equations in Table 21.14 to compute the minor concentrations at equilibrium of: **a.** Formic_T; **b.** glucose; and **c.** methanol.

Solution

At pH = 7.8: 1) $\alpha_0^C = 0.0342$; 2) with pK = 3.75, $\alpha_1^F \approx 1.00$.

a. Formic_T. Eq.(21.11) gives $\text{pe} = -4.10 + \frac{1}{2}\log\frac{\alpha_0^C C(IV)_T}{\alpha_1^F \text{Formic}_T} - \frac{1}{2}\text{pH}$. $\text{Formic}_T = \frac{\alpha_0^C C(IV)_T}{\alpha_1^F 10^{2(\text{pe}+4.10+\frac{1}{2}\text{pH})}}$

b. glucose. Eq.(21.15) gives $\text{pe} = 0.17 + \frac{1}{4}\log\frac{\alpha_0^C C(IV)_T}{[C_6H_{12}O_6]^{1/6}} - \text{pH}$. $[C_6H_{12}O_6] = \left(\frac{\alpha_0^C C(IV)_T}{10^{4(\text{pe}-0.17+\text{pH})}}\right)^6$

c. methanol. Eq.(21.24) gives $\text{pe} = 0.78 + \frac{1}{6}\log\frac{\alpha_0^C C(IV)_T}{[CH_3OH]} - \text{pH}$. $[CH_3OH] = \frac{\alpha_0^C C(IV)_T}{10^{6(\text{pe}-0.78+\text{pH})}}$

inputs			**solution**		
pH	**pe**	**[C(IV)$_T$] (*M*)**	**a. Formic$_T$ (*M*)**	**b. glucose (*M*)**	**c. methanol (*M*)**
7.8	–4.28	1.00E-03	1.26E-12	7.60E-108	1.30E-21
7.8	–5.28	5.00E-04	6.30E-11	1.19E-85	6.49E-16
7.8	–6.28	1.00E-11	1.26E-16	7.60E-108	1.30E-17

REFERENCES

Cervarich AS (2019) Personal communication.

Christensen JJ, Hansen LD, Izatt RM (1976) *Handbook of Proton Ionization Heats and Related Thermodynamic Quantities*, J. Wiley and Sons, New York.

Garrels RM, Christ CL (1965) *Solutions, Minerals, and Equilibria*. Harper & Row, New York.

Gun J, Goifman A, Shkrob I, Kamyshny A, Ginzburg B, Hadas O, Dor I, Modestov AD, Lev O (2000) Formation of polysulfides in an oxygen rich freshwater lake and their role in the production of volatile sulfur compounds in aquatic systems. *Environmental Science and Technology*, **34**, 4741–4746.

Hendrickson RG, Hamilton R (2004) Co-worker fatalities from hydrogen sulfide. *American Journal of Industrial Medicine* **45**, 346–50.

Hariharan U, Bhasin N, Mittal V, Sood R (2016) A fatal case of septic tank gas poisoning: critical care challenges; case report. *Journal Anesthesia & Critical Care* (Open Access), **6**, 00228.

Neumann RB, Ashfaque KN, Badruzzaman ABM, Ali MA, Shoemaker JK, Harvey CF (2010) Anthropogenic influences on groundwater arsenic concentrations in Bangladesh. *Nature Geoscience.* **3**, 46–52.

Langmuir D, Chrostowski P, Vigneault B, Chaney R (2004). *Issue Paper on the Environmental Chemistry of Metals. Report to U.S. Environmental Protection Agency,* Contract #68-C-02-060, Risk Assessment Forum, Washington, D.C.

Pankow JF (1991) *Aquatic Chemistry Concepts* (First Edition). CRC Press, Boca Raton, FL.

Rickard D, Luther GW (2007) Chemistry of iron sulfides. *Chemical Reviews*, **107**, 514–562.

Smith RP (2010) A short history of hydrogen sulfide. *American Scientist*, **98**, 6–9.

Stumm W, Morgan JJ (1996) *Aquatic Chemistry: Chemical Equilibria and Rates in Natural Waters* 3rd edn, Wiley Publishing Co, New York

Van Dover CL, German CR, Speer KG, Parson LM, Vrijenhoek RC (2002) Evolution and biogeography of deep-sea vent and seep invertebrates. *Science*, **295**, 1253–1257.

22 Redox Succession (Titration) in a Stratified Lake during a Period of Summer Stagnation

22.1 INTRODUCTION

22.1.1 Lake Dynamics

Over the course of a year, a large lake (or reservoir) in a temperate latitude can provide the stage for dramatic and predictable changes in redox chemistry. This is because the bottom portion of such a lake can become quite isolated during the summer, allowing dissolved oxygen to become exhausted, followed by the sequential exhaustion of N(V) which is converted to N(0), Fe(III) which is converted to Fe(II), and S(VI) which is converted to S(–II), then disproportionation (fermentation) of organic carbon (OC) to C(IV) and C(–IV) (*i.e.*, to CO_2 species and CH_4), with some production of H_2.

Figure 22.1 shows how the density of pure water depends on temperature. In lakes of moderate depth, the interplay of temperature, density, and wind during the seasons of the year can produce a cycle of characteristic patterns of thermal stratification. Figure 22.2 illustrates a typical series of temperature and density gradients for such lakes. The next four paragraphs paraphrase and closely adopt the view of this cycle by Fair et al. (1968).

During winter, water immediately below the ice is essentially at 0 °C. The ice itself can be much colder than 0 °C at and near the air surface. An unusual but very fortunate characteristic of water is that its solid form (ice) is less dense than the liquid, and so ice floats. (For many compounds, their solid forms are more dense than their corresponding liquids.) The ice layer forms a thermal cover for the lake during the winter, limiting (though not eliminating) further freezing. If ice were more dense than liquid water near the freezing point, it would sink, cover the sediments and aquatic plants, and the lake might even fill up with ice: many forms of life would die.

During winter, the temperature of the water at the bottom of the lake is not far from 4 °C, which is the temperature at which water reaches its maximum density. Thus, the water column is in a state of stable physical equilibrium. It is inversely stratified in terms of temperature, but directly and stably stratified in terms of density. The ice cover shuts out the shear forces of the wind, and so there is little/no movement of water under the ice; advective "roll cells" that might provide vertical mixing and move surface water down and deep water up cannot develop. This is the period of *winter stagnation*.

In the spring, the ice begins to break up, and the water near the surface begins to warm. As the temperature of maximum density for water is approached, the surface waters also become denser and tend to sink. Movement of surface water due to winds begins, as well as diel fluctuations in the temperatures of the surface water. Once the water temperature is nearly uniform at all depths and close to the temperature of maximum density, vertical mixing becomes pronounced. This is the period of *spring overturn*. In a given lake, it varies in length during different years and can last several weeks.

With the onset of summer, the surface waters become progressively warmer. Soon, lighter water overlies denser water once again. As the temperature differences between the upper and lower waters increase, circulation becomes increasingly more confined to the upper waters. A second period of stable density equilibrium is established. The water column is now stratified in terms

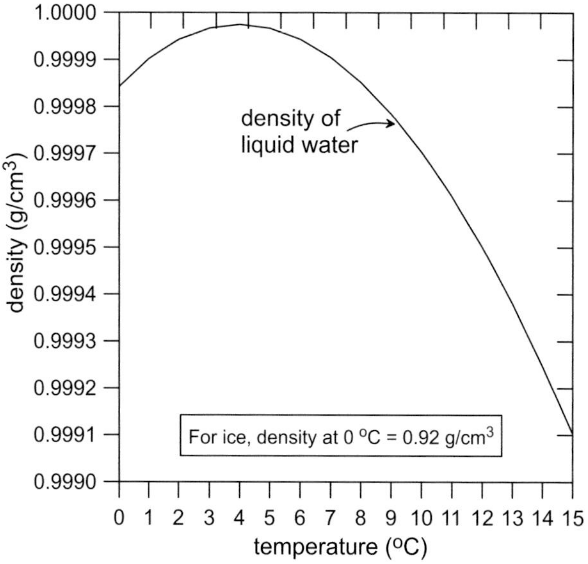

FIGURE 22.1 Density (g/cm^3) of pure water as a function of temperature (°C).

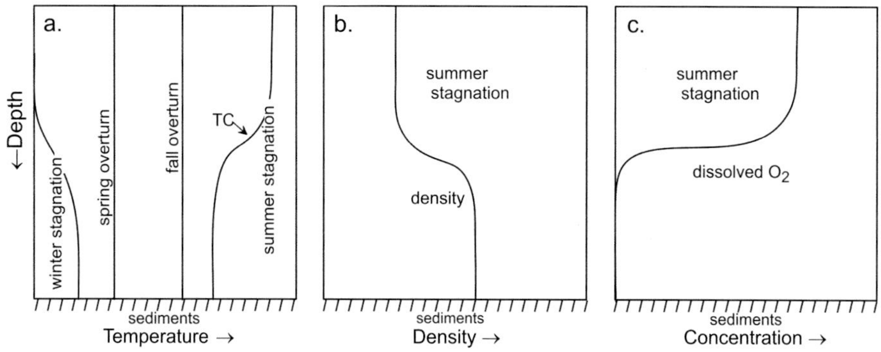

FIGURE 22.2 **a**. Temperature as a function of depth during the four seasons of the year in a lake or reservoir in the middle latitudes. The lake/reservoir mixes completely during both the spring and fall overturning. TC = thermocline. **b.** Density as a function of depth summer stagnation. **c.** Dissolved oxygen as a function of depth during summer stagnation. Adapted from Fair et al. (1968).

of both temperature and density. In comparatively deep lakes, often the water below ~8 m is now nearly stagnant, and the temperature at the bottom remains almost constant and at ~4 °C. This is the period of *summer stagnation*. For lakes in northern latitudes, it can extend from April to November and provides more than adequate time for dramatic changes in the chemistry of the bottom waters.

With the arrival of autumn, surface layers cool and sink, and the summer equilibrium is upset. The water is mixed to greater and greater depths, and a vertical temperature gradient eventually disappears completely. At this point, autumn winds easily induce vertical mixing, and the *great overturn* or *fall overturn* occurs. At this time, many lakes become fully vertically mixed. When winter arrives, the surface waters again freeze, and the condition of winter stagnation begins again. A lake that thus undergoes two periods of mixing (spring and autumn) is referred to as *dimictic*.

The model system for a hypothetical dimictic lake during the summer is represented in Figure 22.3. The "epilimnion" (upper portion) is warmer than the "hypolimnion" (lower portion). The temperature of the epilimnion increases steadily as the summer progresses, so the stratification in the lake becomes increasingly more stable. The bulk of the temperature change with depth occurs within the "mesolimnion", which is often fairly thin, and contains the "thermocline" (depth of maximum rate of temperature change). Persons who dive into a lake in the summer can encounter increasingly cold waters below a certain depth; this is an encounter with the mesolimnion/thermocline. (Because of the dependence of the refractive index of light on water density, the location of the thermocline can sometimes be seen using a dive mask.) The stability of the summer stratification is based in the fact that any downward movement of a parcel of water in the epilimnion towards the hypolimnion is resisted by buoyancy effects, and most of the parcel remains in the epilimnion. The greater the temperature difference between the epilimnion and the hypolimnion, the more stable is the stratification. As summer progresses, the thermocline gradually deepens as heat is slowly transported downward. Unless the lake is comparatively shallow, the fall overturn usually occurs before the thermocline can reach the bottom.

22.1.2 The Effects of Lake Dynamics on pe

Summer stratification can be a major problem in many lakes and reservoirs. Advective mixing across the mesolimnion is very limited during that period, and simple molecular diffusion is very slow: there is little ability for surface oxygen to move from the epilimnion into the hypolimnion. The oxygen supply in the hypolimnion is thus essentially limited to what is present at the beginning of the summer, prior to the onset of stratification. An oxygen level needed for aerobic life might not last until the fall overturn remixes the lake. Indeed, over the course of the summer, oxygen will be consumed due to the biological degradation of: 1) OC initially in the water column in the hypolimnion; 2) dead and/or live OC (*e.g.*, algal and diatom cells) in the epilimnion that sediments downwards through the mesolimnion; and 3) OC in the sediments. The production of OC in the epilimnion is more likely to occur in lakes that are nutrient rich (eutrophic) than in lakes that are nutrient poor (oligotrophic). During the course of the summer, the pe and the pH will change along a track that can be plotted on a pe–pH diagram.

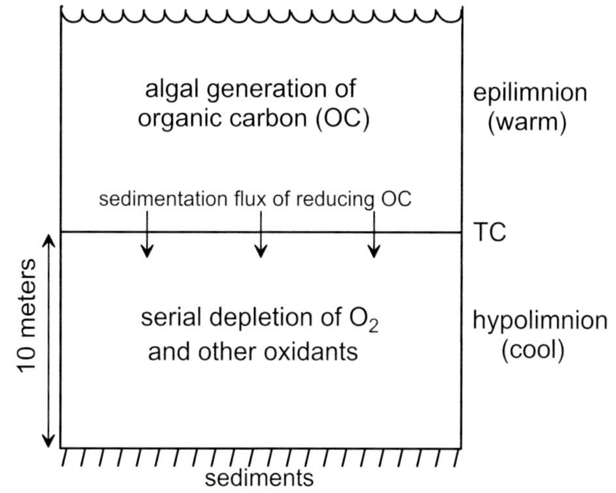

FIGURE 22.3 Schematic model of a lake during the summer. Organic carbon (OC) as a result of primary biological production in the epilimnion is sedimenting down through the thermocline into the 10 m thick hypolimnion. Oxygen and other oxidants in the hypolimnion are reduced sequentially as the summer progresses. Slow transport across the thermocline (TC) prevents replenishment of the dissolved oxygen.

22.2 MODEL CONSIDERATIONS FOR A HYPOTHETICAL LAKE

22.2.1 Initial Conditions

Figure 22.3 gives a schematic view of the model lake discussed here. The changing redox chemistry in the hypolimnion will be discussed in terms of an equilibrium model (except for reaction of N_2: see Section 21.3.2). Thus, as the chemistry of the system changes with time due to the reaction of extant OC, or reaction of newly arrived OC, a series of pe–pH equilibrium states for a series of discrete points in time will be computed. Analogously, in an acid/base titration curve, a series of equilibrium pH values is calculated and plotted *vs.* $(C_B - C_A)$, or *vs.* time if the time rate of addition of $(C_B - C_A)$ is known.

Table 22.1 gives the parameters assumed to remain constant for the lake. Table 22.2 summarizes the assumed initial concentrations of O_2 and other redox-active chemical species in the hypolimnion. Table 21.3 summarizes the relevant redox reactions. We will set up and solve all of the pertinent equilibrium-based equations for the problem. This will build chemical intuition that is not easily obtained by looking at output from a general equilibrium geochemistry computer application.

Many significant figures will be retained in the calculations. Although the accuracy with which we know equilibrium constants and initial conditions do not justify such a great number of significant figures, once we select our best values for the equilibrium constants and conditions, it is necessary to carry many significant figures because at certain points during the summer, premature rounding to fewer significant figures will have large effects on the calculated pe and pH values, and therefore on the concentrations of pe- and pH-dependent species.

An equilibrium model assumes that there are no kinetic limitations on the rates of the reactions of interest. While it is certainly never true that full redox equilibrium exists at any point in any real lake (or anywhere else in the environment), results obtained using an equilibrium model provide insight as to the sequence of events that can be expected, if not the exact timing. Since the main objective here is to illustrate some redox chemistry (and not to develop a detailed and perfect lake model), other assumptions used are as follows:

1. A one-box model for the hypolimnion is adequate even though complicated multi-layer models have long been used (*e.g.*, Imboden and Lerman, 1978).
2. The reducing role of OC in the sediments can be neglected.
3. The thickness of the hypolimnion is constant.
4. Equilibrium constants for 25 °C/1 atm apply even though hypolimnion temperatures are usually lower (corrections for pressure would be exceedingly small), and activity corrections can be neglected.
5. Prior to summer stratification, the water in the lake is well-mixed. The initial levels of the dissolved gases O_2, N_2, and $H_2CO_3^*$ (Table 22.2) are those for equilibrium at 25 °C/1 atm with an atmosphere for which $p_{N_2} = 0.78$ atm, $p_{O_2} = 0.21$ atm, and $p_{CO_2} = 10^{-3.5}$ atm, with $K_H = 10^{-3.20}$ *M*/atm, $K_H = 10^{-2.90}$ *M*/atm, and $K_H = 10^{-1.47}$ *M*/atm.

TABLE 22.1
Conditions in Hypolimnion Assumed to Remain Constant

Thickness (thermocline to sediments)	10 m
Organic carbon (OC) loading rate (as $C_6H_{12}O_6$)	$= 450$ mg C/m^2-day $= 3.746566 \times 10^{-2}$ mol/m^2-day $= 1.498626 \times 10^{-1}$ eq/m^2-day $= 3.746566 \times 10^{-6}$ mol/L-day $= 1.498626 \times 10^{-5}$ eq/L-day

TABLE 22.2
Initial Chemical Conditions in the Hypolimnion

Redox Species	mg/L	*M* or *F*	*n* (equivalents/mol)	eq/L
O_2	8.459662	2.643943×10^{-4}	4 if reduced to water	1.057497×10^{-3}
$Fe(OH)_{3(s)}$	0.50 (as Fe)	8.953032×10^{-6}	1 if reduced to Fe(II)	8.953032×10^{-6}
NO_3^-	0.20 (as N)	1.427888×10^{-5}	5 if reduced to $\frac{1}{2}N_2$	7.139440×10^{-5}
SO_4^{2-}	3.00 (as S)	9.357455×10^{-5}	8 if reduced to H_2S/HS^-	7.485964×10^{-4}
OC	1.0 (as C)	8.325701×10^{-5}	4 if oxidized to CO_2	3.330281×10^{-4}

Note: $Fe(OH)_{3(s)}$ is suspended in the water column; organic carbon (OC) is presumed to have the formula $C_6H_{12}O_6$.
pH = 8.00, $C(IV)_T = 4.915840\times10^{-4}$ *M*, $Alk = 4.840975\times10^{-4}$ eq/L, $N_T = 2[N_2] = 9.985723\times10^{-4}$ *M*.

6. The initial and subsequently sedimenting OC is in an average oxidation state of zero, and will be oxidized by O_2 and other oxidants according to

$$\tfrac{1}{6}C_6H_{12}O_6 + H_2O = CO_2 + 4H^+ + 4e^-. \qquad (22.1)$$

7. The initial presence of both OC and O_2 at the beginning of the stratification is not an equilibrium situation; an initial step will be to oxidize the initial OC.

22.2.2 Assumptions Governing Nitrogen Redox Chemistry

Given the annotations in Figure 21.3 about the high stability of N_2, no oxidation of N_2 to NO_3^- and no reduction of N_2 to NH_3/NH_4^+ will be allowed. Reduction of NO_3^- to N_2 will be allowed (as mediated by denitrifying microorganisms) as soon as it becomes favorable.

22.2.3 Equivalents per Liter (eq/L) as Units for Redox Reactions

The initial species concentrations and assumed OC flux rate are given in units involving mols, as in molarity (*M*) and mols/m^2-day. However, we know that when different redox-active species react, different numbers of mols of electrons are transferred: for one mol of Fe^{3+} to one mol of Fe^{2+}, one mol of electrons is gained; for one mol of O_2 to two mols of H_2O, four mols of electrons are gained. This is analogous to the situation with acids and bases, in which case more than one mol of protons may be lost or gained per mol of acid or base. As discussed in Chapter 7, to simplify acid/base calculations, the concept of equivalents of acid and base is introduced (see the review in Box 22.1).

BOX 22.1 Review of Equivalents as Used for Acid/Base Calculations

1.0 eq of strong acid will exactly neutralize 1.0 eq of strong base. For HCl, because it is monoprotic, 1.0 FW equals 1 eq, and a 1.0 *F* solution contains 1.0 eq/L solution. For H_2SO_4, because it is diprotic, 1.0 FW equals 2.0 eq, and a 1.0 *F* solution is a 2.0 eq/L solution. For the base NaOH, 1.0 FW equals 1 eq, and a 1.0 *F* solution is a 1.0 eq/L solution. For the base $Mg(OH)_2$, 1.0 FW equals 1.0 eq, and a 1.0 *F* solution is a 2.0 eq/L solution. In summary, multiplying the FW by the number of acidic or basic units per FW will convert units of FW to units of eq, and units of *F* to units of eq/L.

If the hypolimnion is assumed to remain well-mixed (this is the essence of a one-box model), dividing the flux of sedimenting OC by the thickness of the box yields the input rate of OC in units of M/time:

$$\frac{3.746566\times10^{-2}\ \text{mol/m}^2\text{-day}}{10\ \text{m}}\times\frac{1\ \text{m}^3}{1000\ \text{L}}=3.746566\times10^{-6}\ M/\text{day}. \tag{22.2}$$

Units of eq and eq/L are useful for redox-active species. Here, the number of electrons n being exchanged in the redox reaction of interest provides the conversion factor between FW (or mols) and eq, and between F (or M) and eq/L. As noted, for the oxidation of the OC considered here to CO_2, $n = 4$. Therefore, multiplying by the associated conversion factor of (4 eq C/mol C) leads to

$$\begin{aligned}\begin{matrix}\text{input rate of reduced}\\ \text{carbon to hypolimnion}\end{matrix}&=3.746566\times10^{-6}\,M/\text{day}\times\frac{4\ \text{eq C}}{\text{mol C}}\\&=1.498626\times10^{-5}\ \text{eq/L-day}.\end{aligned} \tag{22.3}$$

22.2.4 pe for Before any Reactions Occur (Period 0)

For the pe given by the initial oxygen concentration and the initial pH in the hypolimnion,

$$\text{pe} = 21.50 + \tfrac{1}{4}\log[O_2] - \text{pH}. \tag{22.4}$$

During the course of the summer, we will be keeping track of how much oxygen is present in terms of its concentration in units of eq/L. Thus, we recast Eq.(22.4) in a form where the O_2 concentration is expressed in eq/L. As an oxidant, O_2 takes 4 electrons per mol:

$$[O_2] = \tfrac{1}{4}[O_2,\text{eq/L}]. \tag{22.5}$$

Eq.(22.4) and Eq.(22.5) yield

$$\text{pe} = 21.35 + \tfrac{1}{4}\log[O_2,\text{eq/L}] - \text{pH}. \tag{22.6}$$

For $[O_2,\ \text{eq/L}] = 1.057497\times10^{-3}$ eq/L and pH = 8.00,

$$\text{pe} = 21.35 + \tfrac{1}{4}\log(1.057497\times10^{-3}\ \text{eq/L}) - 8.00 = 12.606. \tag{22.7}$$

The value pe = 12.606 does not represent initial equilibrium since it has been assumed that some initial OC is present in the water column along with the O_2. It will be necessary to calculate a second pe for $t = 0$ (beginning of Period 1) when oxidation of the initial OC by the initial O_2 is complete.

22.3 THE FOUR SEQUENTIAL REDOX EQUIVALENCE POINTS AND ADDITIONAL REDOX LANDMARKS

22.3.1 The Sequential Equivalence Point (EP) Landmarks for O_2, Nitrate, Iron(III), and Sulfate

Slow addition of OC to the hypolimnion will cause a near perfect step-wise reduction of the four initial oxidants (*i.e.*, electron acceptors) in the water. We only say near perfect because, just as in an alkalimetric titration in which a mixture of different acids is not titrated in a perfect stepwise manner, so too does a steady lowering of the pe continuously affect the levels of all electron acceptors. (Solids can, however, disappear abruptly.) The order of reduction will follow the ClONFeSCH acronym. The length of time to reach each successive equivalence point (EP) can be calculated.

For O_2 which is reduced to H_2O, there is the H_2O EP. For NO_3^- which is reduced to N_2, there is the N_2 EP. For (am)$Fe(OH)_{3(s)}$ which is reduced to Fe^{2+} (and $FeOH^+$ and other Fe(II) forms), there is the Fe(II) EP. For SO_4^{2-} which is reduced to H_2S (as well as HS^- and other S(−II) forms), there is the S(−II) EP.

For reduction of O_2,

$$\text{time to reach } H_2O \text{ EP} = \frac{(1.057497 \times 10^{-3} - 3.330281 \times 10^{-4}) \text{ eq/L}}{1.498626 \times 10^{-5} \text{ eq/L-day}} \tag{22.8}$$

$$= 48.342 \text{ days.}$$

The numerator in Eq.(22.8) is the difference between the initial O_2 level and the initial OC level. The net balanced reaction is

$$\tfrac{1}{6}C_6H_{12}O_6 + O_2 = CO_2 + H_2O. \tag{22.9}$$

The pH will fall during this period. With $[O_2]$ and pe dropping rapidly as $t \rightarrow 48.342$ days, $t \approx 48.34$ days is the time at which NO_3^- will become subject to reduction to N_2.

For reduction of NO_3^- to N_2, the number of equivalents of NO_3^- available at the beginning of the summer is 7.139440×10^{-5} eq/L. Thus,

$$\text{time to reach } N_2 \text{ EP} = 48.342 \text{ days} + \frac{7.139440 \times 10^{-5} \text{ eq/L}}{1.498626 \times 10^{-5} \text{ eq/L-day}} \tag{22.10}$$

$$= 53.106 \text{ days.}$$

In the stepwise view of the reaction sequence, between $t = 48.342$ and $t = 53.106$ days, the vast majority of the reduction occurring involves the conversion of NO_3^- to N_2. The net balanced redox reaction is

$$\tfrac{5}{6}C_6H_{12}O_6 + 4NO_3^- + 4H^+ = 5H_2CO_3^* + 2N_2 + 2H_2O. \tag{22.11}$$

Removing $4H^+$ from the solution will tend to raise the pH more than adding $5H_2CO_3^*$ will tend to lower the pH: the pH will rise between 48.342 and 53.106 days. With NO_3^- and therefore pe dropping rapidly as $t \rightarrow 53.11$ days, (am)$Fe(OH)_{3(s)}$ will become subject to significant reduction to Fe(II).

The amount of (am)$Fe(OH)_{3(s)}$ initially available for reduction is 8.953032×10^{-6} eq/L. Thus, the (am)$Fe(OH)_{3(s)}$ will not last beyond

$$\text{time to reach Fe(II) EP} = 53.106 \text{ days} + \frac{8.953032 \times 10^{-6} \text{ eq/L}}{1.498626 \times 10^{-5} \text{ eq/L-day}} \tag{22.12}$$

$$= 53.703 \text{ days.}$$

Between $t = 53.106$ and 53.703 days, the vast majority of the reduction occurring involves the conversion of (am)$Fe(OH)_{3(s)}$ to $Fe(II)_T$. The net balanced redox reaction is

$$\tfrac{1}{6}C_6H_{12}O_6 + 4(\text{am})Fe(OH)_{3(s)} + 8H^+ = H_2CO_3^* + 4Fe^{2+} + 10H_2O. \tag{22.13}$$

Removing 8H⁺ from solution will tend to raise the pH much more than adding one $H_2CO_3^*$ will tend to lower the pH: the pH will rise until the Fe(II) EP is reached.

If reduced to S(–II), the amount of SO_4^{2-} initially in the system is 7.485964×10^{-4} eq/L. Thus,

$$\text{time to reach S(–II) EP} = 53.703 \text{ days} + \frac{7.485964 \times 10^{-4} \text{ eq/L}}{1.498626 \times 10^{-5} \text{ eq/L-day}} \tag{22.14}$$

$$= 103.655 \text{ days.}$$

At pH < 7 where most of the dissolved $S(-II)_T$ is present as H_2S, and except for the short period when $FeS_{(s)}$ is precipitating, the net balanced redox reaction is

$$\tfrac{2}{6} C_6H_{12}O_6 + 2H^+ + SO_4^{2-} = 2H_2CO_3^* + H_2S. \tag{22.15}$$

Removing 2H⁺ from solution will tend to raise the pH more than adding two $H_2CO_3^*$ will tend to lower the pH: the pH will rise until the S(–II) EP is reached.

OC is a relatively weak reducing agent, and so will not be able to reduce significant H(I) in water to H_2. Thus, the S(–II) EP is the last redox EP, and there is not a rapid decline in pe when passing through the S(–II) EP. Once the pe has been reduced sufficiently that little SO_4^{2-} remains, with no further significant oxidants present, the added $C_6H_{12}O_6$ now begins to disproportionate to CO_2 and CH_4: C(IV) and C(–IV) are the only stable oxidation states for carbon (see Figure 21.14):

$$C_6H_{12}O_6 = [(CO_2)_3(CH_4)_3] = 3CO_2 + 3CH_4. \tag{22.16}$$

Some of the zero-oxidation-state carbon in $C_6H_{12}O_6$ is oxidized up to C(IV), and an equal amount is reduced down to C(–IV) as CO_2: the pH will fall. There will be no redox EP for the carbon because added $C_6H_{12}O_6$ will just continue to disproportionate.

22.3.2 Landmark Times Delineating the Different Periods

22.3.2.1 General

The specific governing equations that will be solved to obtain pe and pH as they change with time will depend on how redox-active solids disappear and appear in the system.

22.3.2.2 Period 0: Consumption of Initial OC

The 0th step in the model involves oxidation of the initial OC by initial O_2. This production of CO_2 causes the pH to drop from 8.0 to close to 7 (see below).

22.3.2.3 Period 1: Reduction of O_2 up to Activation of NO_3^-

During Period 1, O_2 is the main oxidant being lost. As it is being lost, it will be necessary to test for the point at which NO_3^- can begin to accept electrons (viz., when reduction of NO_3^- by OC can begin). That point will mark the end of Period 1.

22.3.2.4 Period 2: Reduction of NO_3^- up to Disappearance of (am)$Fe(OH)_{3(s)}$ At Fe(II) EP

By Eq.(22.11), the Fe(II) EP is reached at ~54 days. Since the stagnation that begins in the summer can last many months, the (am)$Fe(OH)_{3(s)}$ will disappear at ~54 days; Period 2 will end.

22.3.2.5 Period 3: Reduction of SO_4^{2-} up to Appearance of $FeS_{(s)}$

Will $FeCO_{3(s)}$ Form in This System? No. As dissolved Fe(II) builds up during reduction of the (am)$Fe(OH)_{3(s)}$, at some point $FeCO_{3(s)}$ might form. The K_{s0} for $FeCO_{3(s)}$ is $10^{-10.68}$. For $FeCO_{3(s)}$ to become supersaturated, when activity corrections in the solution can be neglected,

$$[Fe^{2+}][CO_3^{2-}] = \alpha_0^{Fe(II)}Fe(II)_T\,\alpha_2^{C(IV)}C(IV)_T > 10^{-10.68}. \quad (22.17)$$

With $Fe(II)_T$ as the total dissolved Fe(II), and $\alpha_{0,Fe(II)}$ as the fraction of Fe(II) present as Fe^{2+}, as usual,

$$\alpha_0^{Fe(II)} \equiv \frac{[Fe^{2+}]}{[Fe^{2+}]+[FeOH^+]+[Fe(OH)_2^o]+[Fe(OH)]_3^-}. \quad (22.18)$$

Initially, pH = 8.0 and $C(IV)_T \approx 5\times10^{-4}$ M. At pH ≈ 8, $\alpha_2^{C(IV)} \approx 4.6\times10^{-3}$; so $\alpha_2^{C(IV)}C(IV)_T = 2.3\times10^{-6}$ M; by the ${}^*K_H^{Fe(II)}$ values in Table 21.6, $\alpha_0^{Fe(II)} \approx 1$. For $FeCO_{3(s)}$ to become saturated at pH ≈ 8, we would need

$$Fe(II)_T \approx 9\times10^{-6}\ M. \quad (22.19)$$

This is slightly more than the initial (am)$Fe(OH)_{3(s)}$. So, we ask if $\alpha_2^{C(IV)}C(IV)_T$ will increase enough during the summer for $FeCO_{3(s)}$ to form.

The most abundant oxidant is O_2, so the greatest effect on pH due to loss of the oxidants will be due to the production of CO_2 by Eq.(21.9). The increase in $C(IV)_T$ will be $(1.057497\times10^{-3}$ eq/L)/(4 eq/L-M) = 2.64×10^{-4} M, bringing $C(IV)_T$ to 7.56×10^{-4} M. With Alk conserved for Eq.(21.9) at 4.840975×10^{-4} eq/L, the new $C(IV)_T$ gives pH ≈ 6.6, and $\alpha_2^{C(IV)}C(IV)_T = 9\times10^{-8}$ M: the pe-pH track will drop down to the left of the $FeCO_{3(s)}$ region that could be drawn for $Fe_T = 8.95\times10^{-6}$ M and $C_{T,free} \approx 10^{-3}$ M (*cf.* Figure 21.8 which is drawn for similar values of $Fe_{T,sys}$ and $C_{T,free}$.

Will $S_{(s)}$ Form in This System? No. In the pe-pH diagram for $S_{T,sys} = 10^{-3}$ M in Figure 21.12, the $S_{(s)}$ wedge extends only to pH = 4.40. For this chapter, $S_{T,sys}$ is ~10^{-4} M. This lake water will not reach sufficiently acidic pH values that $S_{(s)}$ will be able to form.

Will $FeS_{(s)}$ Form in This System? Yes. Once reduction of Fe(III) is essentially complete, reduction of sulfate will begin. At pH ≈ 6.6 where H_2S predominates over HS^-, the net balanced reaction will be mostly

$$\tfrac{2}{6}C_6H_{12}O_6 + 2H^+ + SO_4^{2-} = 2H_2CO_3^* + H_2S. \quad (22.15)$$

Many metal ions, Fe^{2+} included, form very insoluble metal sulfides. The common mineral form of $FeS_{(s)}$ is mackinawite; at 25 °C/1 atm, K_{s0} is very small at $10^{-18.0}$. For $FeS_{(s)}$ to become supersaturated, we need

$$[Fe^{2+}][S^{2-}] = \alpha_0^{Fe(II)}Fe(II)_T\,\alpha_2^{S(-II)}S(-II)_T > 10^{-18.0}. \quad (22.20)$$

Once the system has become sufficiently reducing for some significant degree of $S(-II)_T$ formation to occur, all of the (am)$Fe(OH)_{3(s)}$ will have been reduced to $Fe(II)_T$ (with only exceedingly trace levels of dissolved $Fe(III)_T$ remaining): $Fe(II)_T = 8.953032\times10^{-6}$ M. Taking pH = 6.6 as an estimate for that point, $\alpha_0^{Fe(II)} \approx 1$, and by the p$K$ values in Table 21.8 then $\alpha_2^{S(-II)} = 1.1\times10^{-8}$. $FeS_{(s)}$ will start to form once

$$S(-II)_T \approx \frac{10^{-18.0}}{(1.1\times10^{-8})(1)(8.953032\times10^{-6})} \approx 10^{-5}\ M. \quad (23.21)$$

This is only ~10% of $S_{T,sys}$, so $FeS_{(s)}$ will form at some point after sulfate reduction begins: it will be necessary to test for the point when saturation with respect to $FeS_{(s)}$ occurs. At that point, Period 3 ends.

22.3.2.6 Reduction of SO_4^{2-} and Disproportionation of $C_6H_{12}O_6$ with $FeS_{(s)}$ Present–Period 4

When Period 4 begins, $FeS_{(s)}$ will being precipitating. The dominant redox reaction will be

$$\tfrac{2}{6}C_6H_{12}O_6 + Fe^{2+} + SO_4^{2-} = 2H_2CO_3^* + FeS_{(s)}. \qquad (22.22)$$

The pH will fall due to the addition of the $2H_2CO_3^*$. Once essentially all of the Fe(II) has precipitated, Eq.(22.15) will again become the dominant reaction, and the pH will start to rise. Once the sulfate is exhausted, disproportionation of $C_6H_{12}O_6$ (Eq.(22.15)) will become the dominant redox reaction, and the pH will fall.

22.4 A REDOX TITRATION MODEL FOR WATER IN THE HYPOLIMNION

22.4.1 The Redox Titration Equation (RTE)

22.4.1.1 General

As OC is added to the hypolimnion, the reducing effects of the OC will accumulate. We can develop a titration equation that tracks how many equivalents of electrons have been added. Including the initial OC present and the amount of OC added during the course of the summer,

$$\begin{aligned}\text{initial OC + added OC (eq/L)} &= 3.330281\times10^{-4}\text{ eq/L} + 1.489626\times10^{-5}\text{(eq/L-day) } t \\ &= \text{total eq/L of all reduced species formed.}\end{aligned} \qquad (22.23)$$

Time t has units of days. Eq.(22.23) assumes that all OC reacts, so that no OC remains, an excellent assumption (at equilibrium).

The reducing equivalents of the OC will accumulate in the various reduced species that are formed:

$$\begin{aligned}\text{initial OC + added OC (eq/L)} &= \text{eq/L of } O_2 \text{ reduced (to water or hydroxide)} \\ &+ \text{eq/L of Fe(II) formed + eq/L of new } N_2 \text{ formed} \\ &+ \text{eq/L of S(−II) formed + eq/L of } CH_4 \text{ formed} \\ &+ \text{eq/L of } H_2 \text{ formed.}\end{aligned} \qquad (22.24)$$

22.4.1.2 The Terms in the RTE

The O_2 will be reduced to one oxidation state, O(−II), as water or hydroxide. This requires four electrons per O_2 molecule. The term "eq/L of O_2 reduced" in Eq.(22.24) is a measure of the OC consumed by reaction with O_2. We can do the bookkeeping in terms of the amount of O_2 that has been reduced, rather than in terms of an increase in the concentration total for H_2O plus OH^-.

The term "eq/L of Fe(II) formed" in Eq.(22.24) includes all Fe(II) produced (both dissolved and solid), where the concentration in eq/L is computed for the one electron reduction of Fe(III) to Fe(II). The term "eq/L of S(−II) formed" tracks dissolved $S(-II)_T$ and S(−II) in $FeS_{(s)}$. The term "eq/L of new N_2 formed", which equals "eq/L of NO_3^- reduced", tracks reduction of NO_3^- once the pe is sufficiently low that the initial NO_3^- can become unstable relative to conversion to N_2. (A term for "eq/L of NH_4^+/NH_3 formed" has not been included: reduction of higher oxidation states of nitrogen to N(−III) can usually be neglected (see Figure 21.3).) The term "eq/L of CH_4 formed" tracks disproportionation of the $C_6H_{12}O_6$. Some formation of H_2 will also begin when the system has become very reducing.

Based on the above, Eq.(22.24) becomes

$$\begin{aligned} \text{eq/L of OC added} &= \text{eq/L of } O_2 \text{ reduced} + \text{Fe(II)}_T + [\text{FeS}_{(s)}] \\ &\quad + \text{eq/L of } NO_3^- \text{ reduced} + 8\,\text{S(–II)}_T \\ &\quad + 8[\text{FeS}_{(s)}] + 8[CH_4] + 2[H_2] \end{aligned} \tag{22.25}$$

where $[\text{FeS}_{(s)}]$ has units of FW/(L of water). $[\text{FeS}_{(s)}]$ appears twice because it is counted for both the Fe(II) and S(–II) it contains. The term "eq/L of NO_3^- reduced (to N_2)" has been substituted for "eq/L of new N_2 formed". Prior to the start of the model (*i.e.*, before the Period 0 calculation is executed), it is assumed that at equilibrium, the RHS of Eq.(22.25) is zero.

For S(–II) that has been formed from S(VI), the factor of 8 accounts for the 8 electrons for the reduction. For CH_4 formation, one molecule of $C_6H_{12}O_6$ (which is C(0)) has 24 electrons available for reduction of other species. For the CH_4 that forms by disproportionation of OC, all of the reducing equivalents of the OC are stored in the CH_4. For the C(0) carbons in $C_6H_{12}O_6$, three are oxidized up to C(IV), and three are reduced down to C(–IV) ($3 \times 8 = 24$). For H(0) in H_2 that is formed from H(I), the factor of 2 accounts for the 2 electrons accepted in the reduction.

22.4.1.3 RTE: A Function of pe and pH

For <u>oxygen</u>, the initial level is $[O_2, \text{eq/L}] = 1.057497 \times 10^{-3}$ eq/L. Thereafter,

$$\text{eq/L of } O_2 \text{ reduced} \equiv \underline{\text{OXR}} = \left(1.057497 \times 10^{-3} \text{ eq/L}\right) - \left(\text{eq/L } O_2 \text{ remaining}\right). \tag{22.26}$$

By Eq.(22.6),

$$\text{eq/L of } O_2 \text{ reduced} = \text{OXR} = \left(1.057497 \times 10^{-3} \text{ eq/L}\right) - \left(10^{4(\text{pe}-21.35+\text{pH})}\right). \tag{22.27}$$

For <u>nitrate</u>,

$$\text{eq/L of } NO_3^- \text{ reduced} \equiv \underline{\text{NO3R}} = \left(7.139440 \times 10^{-5} \text{ eq/L}\right) - 5[NO_3^-]. \tag{22.28}$$

Each N(V) takes 5 electrons to reach N(0). From Table 22.3, for equilibrium between NO_3^- and N_2,

$$[NO_3^-] = \frac{[N_2]^{\frac{1}{2}}}{10^{103.7}10^{-6\text{pH}}10^{-5\text{pe}}}. \tag{22.29}$$

With $N_T = 9.985723 \times 10^{-4}$ *M*, the nitrogen MBE is

$$N_T = 2[N_2] + [NO_3^-]. \tag{22.30}$$

Substituting Eq.(22.29),

$$2[N_2] + \frac{[N_2]^{\frac{1}{2}}}{10^{103.7}10^{-6\text{pH}}10^{-5\text{pe}}} - N_T = 0. \tag{22.31}$$

which is of the form $ax^2 + bx + c = 0$, with $x = [N_2]^{\frac{1}{2}}$ and $a = 2$, $b = \left(10^{103.7}10^{-6\text{pH}}10^{-5\text{pe}}\right)^{-1}$, $c = -N_T$. At given values of pe, and pH, Eq.(22.31) can be solved using the quadratic equation. Here, $b > 0$ for all pe and pH, so to get a positive result, we must take the root $[N_2]^{\frac{1}{2}} = \left(-b + \sqrt{(b^2 - 4ac)}\right)\Big/2a$. With $[N_2]$ determined from Eq.(22.31), NO3R can be calculated by Eqs.(22.29) and (22.28), in

TABLE 22.3
Redox Half-Reactions of Interest in the Hypolimnion with log *K* and pe° Values for 25 °C/1 atm

Reaction	log *K*	pe°
$O_2 + 4H^+ + 4e^- = 2H_2O$	86.0	21.50
$NO_3^- + 6H^+ + 5e^- = \frac{1}{2}N_2 + 3H_2O$	103.7	20.74
$SO_4^{2-} + 9H^+ + 8e^- = HS^- + 4H_2O$	34.0	4.25
$H_2CO_3^* + 8H^+ + 8e^- = CH_4 + 3H_2O$	21.6	2.7
$(am)Fe(OH)_{3(s)} + 3H^+ + e^- = Fe^{2+} + 3H_2O$	16.2	16.2
$H_2CO_3^* + 4H^+ + 4e^- = \frac{1}{6}C_6H_{12}O_6 + 2H_2O$	0.67	0.17
$2H^+ + 2e^- = H_2$ (dissolved)	−3.10	−1.55

Note: O_2, N_2, and H_2 as written are dissolved.

that order. (Use of Eq.(22.30) to calculate $[NO_3^-]$ might lead to roundoff error as $2[N_2]$ approaches N_T.) NO3R ≡ 0 until the pe and pH reach the point at which the initial $[NO_3^-]$ is at thermodynamic equilibrium with the initial $[N_2]$: thereafter, as the pe falls, NO_3^- can act as an oxidant. By the redox half-reaction between NO_3^- and N_2, the initial equilibrium conditions will satisfy

$$10^{-5pe}10^{-6pH} = \frac{(4.921467 \times 10^{-4})^{1/2}}{(1.427888 \times 10^{-5})10^{103.7}} \qquad (22.32)$$

or

$$5pe + 6pH = 100.51. \qquad (22.33)$$

Once Eq.(22.33) is first satisfied, nitrate reduction is turned on, and Period 2 begins.

For carbon,

$$\text{eq/L of } CH_4 \text{ formed} = (8 \text{ eq/mol})[CH_4] \equiv \underline{8MEF}. \qquad (22.34)$$

From Table 22.3, for equilibrium between C(IV) and CH_4,

$$[CH_4] = 10^{21.6}[H_2CO_3^*]10^{-8pH}10^{-8pe}. \qquad (22.35)$$

At any point in time, reaction of $C_6H_{12}O_6$ is assumed complete, so

$$C_T = C(IV)_T + [CH_4] = \frac{[H_2CO_3^*]}{\alpha_0^{C(IV)}} + [CH_4] \qquad (22.36)$$

Substituting Eq.(22.35),

$$C_T = \frac{[CH_4]10^{8pH}10^{8pe}10^{-21.6}}{\alpha_0^{C(IV)}} + [CH_4] \qquad (22.37)$$

or

$$[CH_4] = \frac{C_T}{\dfrac{10^{8pH}10^{8pe}10^{-21.6}}{\alpha_0^{C(IV)}} + 1} \qquad (22.38)$$

with

$$C_T = 5.748410 \times 10^{-4}\ M + (3.746566 \times 10^{-6}\ M/\text{day})t. \tag{22.39}$$

The first term on the RHS of Eq.(22.39) represents the sum of the initial $C(IV)_T$ and the initial OC.

$$8\text{MEF} = (8\,\text{eq/mol}) \times \frac{5.748410 \times 10^{-4}\ M + (3.746566 \times 10^{-6}\ M/\text{day})t}{\dfrac{10^{8\text{pH}}10^{8\text{pe}}10^{-21.6}}{\alpha_0^{\text{C(IV)}}} + 1}. \tag{22.40}$$

For hydrogen, for equilibrium between H^+ and H_2 at 25 °C/1 atm,

$$[H_2] = 10^{-3.10}10^{-2\text{pH}}10^{-2\text{pe}} \tag{22.41}$$

$$\text{eq/L of } H_2 \text{ formed} = (2\,\text{eq/mol}) \times [H_2] = 2 \times 10^{-3.10}10^{-2\text{pH}}10^{-2\text{pe}}. \tag{22.42}$$

For iron, as long as (am)$Fe(OH)_{3(s)}$ is present, from Table 21.3,

$$\text{Fe(II)}_T = 10^{16.2}10^{-3\text{pH}}10^{-\text{pe}}/\alpha_0^{\text{Fe(II)}} \tag{22.43}$$

wherein $[Fe^{2+}] = \alpha_0^{\text{Fe(II)}}\text{Fe(II)}_T$. We assume that (am)$Fe(OH)_{3(s)}$ is sufficiently insoluble at ~neutral pH (see Figure 11.3) that it remains present until $\text{Fe(II)}_T = \text{Fe}_T = 8.953032 \times 10^{-6}\ M$. On that basis, (am) $Fe(OH)_{3(s)}$ remains present until Period 2 ends, as marked by

$$3\text{pH} + \text{pe} + \log \alpha_0^{\text{Fe(II)}} = 21.25. \tag{22.44}$$

For sulfur, from Table 21.3,

$$\text{pe} = 4.25 + \tfrac{1}{8}\log \frac{\alpha_2^{\text{S(VI)}}\text{S(VI)}_T[H^+]^9}{\alpha_1^{\text{S(-II)}}\text{S(-II)}_T}. \tag{22.45}$$

Before any $FeS_{(s)}$ precipitates, the sulfur MBE is

$$\text{S(VI)}_T + \text{S(-II)}_T = \text{S}_T. \tag{22.46}$$

At ~neutral pH, $\alpha_2^{\text{S(VI)}} \approx 1$. By Eq.(22.4539)

$$\text{S(VI)}_T = \frac{10^{8(\text{pe}-4.25)}}{10^{-9\text{pH}}}\alpha_1^{\text{S(-II)}}\text{S(-II)}_T. \tag{22.47}$$

Eq.(22.47) with Eq.(22.46) yield

$$\text{S(-II)}_T = \underline{\text{S1}} \equiv \frac{\text{S}_T}{1 + \dfrac{10^{8(\text{pe}-4.25)}}{10^{-9\text{pH}}}\alpha_1^{\text{S(-II)}}}. \tag{22.48}$$

Up to the point that $FeS_{(s)}$ just becomes saturated, S(-II)_T as given by Eq.(22.48) may be used in Eq.(22.25) as the total measure of all S(–II) produced.

Since $K_{s0} = 10^{-18.0}$ for $FeS_{(s)}$, that solid will just become saturated so that Period 4 begins when

$$[Fe^{2+}][S^{2-}] = \alpha_0^{\text{Fe(II)}}\,\text{Fe(II)}_T\,\alpha_2^{\text{S(-II)}}\,\text{S(-II)}_T = 10^{-18.0}. \tag{22.49}$$

All (am)$Fe(OH)_{3(s)}$ will be gone by the time that $FeS_{(s)}$ starts to precipitate. At that point $Fe(II)_T = Fe_{T,sys}$ (dissolved Fe(III) is completely negligible). Eq.(22.49) then becomes

$$\alpha_2^{S(-II)}\, S(-II)_T\, \alpha_0^{Fe(II)}\, 8.95 \times 10^{-6} = 10^{-18.0}. \qquad (22.50)$$

Eq.(22.48) can be used to provide $S(-II)_T$ when using Eq.(22.50) to test for arrival of $FeS_{(s)}$ saturation.
When expressed in units of (FW of solid)/(L of system water), once it starts to form,

$$[FeS_{(s)}] \equiv \underline{FES} = S_T - S(-II)_T - S(VI)_T. \qquad (22.51)$$

Eq.(22.47) and Eq.(22.51) yield

$$[FeS_{(s)}] = S_T - S(-II)_T - \frac{10^{8(pe-4.25)}}{10^{-9pH}}\alpha_1^{S(-II)}\, S(-II)_T. \qquad (22.52)$$

When $FeS_{(s)}$ is present, the MBEs for Fe and S are coupled. With no (am)$Fe(OH)_{3(s)}$ present and only trace levels of dissolved Fe(III) in solution, $Fe(II)_T = Fe_T - [FeS_{(s)}]$. By Eq.(22.52),

$$\alpha_0^{Fe(II)}\left(Fe_T - S_T + S(-II)_T + \frac{10^{8(pe-4.25)}}{10^{-9pH}}\alpha_1^{S(-II)}S(-II)_T\right)\alpha_2^{S(-II)}S(-II)_T = 10^{-18.0} \qquad (22.53)$$

$$\alpha_0^{Fe(II)}\left(1 + \frac{10^{8(pe-4.25)}}{10^{-9pH}}\alpha_1^{S(-II)}\right)\alpha_2^{S(-II)}\, S(-II)_T^2 + \alpha_0^{Fe(II)}(Fe_T - S_T)\alpha_2^{S(-II)}\, S(-II)_T - 10^{-18.0} = 0 \qquad (22.54)$$

which is of the form $ax^2+bx+c = 0$, with $x = S(-II)_T$. At given values of pe, and pH, Eq.(22.54) can be solved using the quadratic equation. Here, $b>0$ for all pe and pH, so to obtain a positive result, we must take the root

$$S(-II)_T = (-b + \sqrt{(b^2 - 4ac)})/2a = \underline{S4} \qquad (22.55)$$

$FeS_{(s)}$ will be forming during Period 4, so we refer to this root as S4. With $S(-II)_T$ and pH, $Fe(II)_T$ can be calculated by Eq.(22.49), and $[FeS_{(s)}]$ = FES can be calculated by Eq.(22.52). We now have expressions for all of the RHS terms of Eq.(22.25) for all four periods. For each value of time t, there are two unknowns, pe and pH. Another equation is needed: the ENE.

22.4.2 The ENE

At the very beginning when pH = 8.00), the ENE is

$$[H^+]_o + [Na^+] = (\alpha_1^{C(IV)}\, C(IV)_T)_o + 2(\alpha_2^{C(IV)}\, C(IV)_T)_o + [NO_3^-]_o + 2[SO_4^{2-}]_o + [OH^-]_o \qquad (22.56)$$

which incorporates all the initial concentrations and assumes that: 1) initial levels of $Fe(II)_T$ and $S(-II)_T$ are negligibly low; 2) $[HSO_4^{2-}]$ is unimportant; 3) the solubility of Fe(III) is negligible; and 4) some ion like Na^+ (or combination of ions) makes up all of the net positive charge needed to obtain initial electroneutrality. Therefore,

$$10^{-8} + [Na^+] = (0.973649)(4.915840 \times 10^{-4}) + 2(4.554097 \times 10^{-3})(4.915840 \times 10^{-4}) + 1.427888 \times 10^{-5} + 2(9.357455 \times 10^{-5}) + 10^{-6}. \qquad (22.57)$$

Na^+ is neither produced nor removed from solution during the summer: during the entire summer,

$$[Na^+] = 6.855255 \times 10^{-4}\, M. \qquad (22.58)$$

Once some oxidation of OC begins, various reduced species begin to form, and the ENE will be

$$[H^+] + [Na^+] + 2[Fe^{2+}] + [FeOH^+] = \alpha_1^{C(IV)}C(IV)_T + 2\alpha_2^{C(IV)}C(IV)_T + [NO_3^-] + 2[SO_4^{2-}] + \alpha_1^{S(-II)}S(-II)_T + [OH^-] \quad (22.59)$$

with

$$2[Fe^{2+}] + [FeOH^+] = (2\alpha_0^{Fe(II)} + \alpha_1^{Fe(II)})Fe(II)_T. \quad (22.60)$$

The concentrations of S^{2-}, $Fe(OH)_3^-$, and all dissolved Fe(III) species are assumed negligible.

With the exception of $[Na^+]$ which is a constant, all of the terms in Eq.(22.59) can be expressed as functions of pe, pH, and time. By Eqs.(22.36) and (22.38),

$$C(IV)_T = C_T - MEF = C_T - \frac{C_T}{\dfrac{10^{8pH}10^{8pe}10^{-21.6}}{\alpha_0^{C(IV)}} + 1}$$

$$= \left(5.748410 \times 10^{-4}\,M + (3.746566 \times 10^{-6}\,M/\text{day})\,t\right) \times \left(1 - \frac{1}{10^{8pH}10^{8pe}10^{-21.6}/\alpha_0^{C(IV)} + 1}\right). \quad (22.61)$$

Eqs.(22.25) and (22.59) can now be used to solve for the corresponding values of pe and pH as functions of "eq/L of OC added" *i.e.*, as functions of time. The solution can be obtained by the Gauss–Seidel method. With that method here, for each time t and corresponding value of (eq/L of OC), a guess for pH is made, and Eq.(22.25) is then solved numerically for the pe that gives LHS – RHS = 0. The resulting pe is used to solve Eq.(22.59) numerically for the pH that gives its LHS – RHS = 0. That pH is then used to solve Eq.(22.25) again for pe, and so on, back and forth until convergence for both pe and pH is attained for the time (*i.e.*, eq/L of OC) of interest. For each time t, this can be done in one step using Solver in Excel by listing both individual LHS – RHS constraints, and having both pe and pH varied to find the solution satisfying both constraints. Table 22.4 provides a summary of the chemical conditions that exist during the four different chemical periods, and affect the different versions of the RTE and ENE (Tables 22.5 and 22.6).

22.4.3 Results

Table 22.7 shows how the pe, the pH, and concentrations change with time. Figure 22.4 plots pH *vs.* time, and Figure 22.5 plots pe *vs.* time. Figure 22.6 plots the pe *vs.* pH track. Figure 22.7 shows the time-dependence of the concentrations of each of the ten redox-active species.

The pe goes up during the first 35 days, because for the O_2/H_2O redox couple, the functional dependence in Eq.(22.4) of pe on pH is stronger than on $[O_2]$: as long as O_2 remains at some non-negligible concentration, lowering the pH due to production of CO_2 (Eq.(22.9)) causes the pe to rise. Only after $[O_2]$ is driven to very low levels and the lowering of the pH slows does the pe finally go down, allowing reduction of NO_3^- to begin. Once the NO_3^- is mostly gone, (am)$Fe(OH)_{3(s)}$ begins to be reduced in significant amounts, and the $Fe(II)_T$ concentration rises rapidly. While some reduction of the SO_4^{2-} to sulfide has been proceeding all along, this reduction starts to take place in significant amounts only after the (am)$Fe(OH)_{3(s)}$ is gone. The $Fe(II)_T$ concentration holds steady until $FeS_{(s)}$ becomes saturated, then $Fe(II)_T$ starts to go down. Significant reduction of C(IV) (*i.e.*, disproportionation of the OC) to CH_4 begins once the SO_4^{2-} is nearly exhausted. At the same time, the reduction of H^+ to H_2 increases, though the concentration of H_2 does not rise very high

TABLE 22.4
Summary of the Chemical Conditions during the Four Periods as They Affect the Versions of the RTE and ENE Used

Period	NO_3^-/N_2 redox active?	(am) $Fe(OH)_{3(s)}$ present?	$FeS_{(s)}$ present?	RTE Version	ENE Version
1	No	Yes	No	RTE-1	ENE-1
2	Yes	Yes	No	RTE-2	ENE-2
3	Yes	No	No	RTE-3	ENE-3
4	Yes	No	Yes	RTE-4	ENE-4

TABLE 22.5
Versions of the Redox Titration Equation (RTE) for Four Different Periods during the Summer

Initial Plus added OC (eq/L)	=	O_2 reduced (eq/L)	+	$Fe(II)_T$ (eq/L)	+	NO_3^- reduced (eq/L)	+	8 S(-II) (dissolved) (eq/L)	+	9[$FeS_{(s)}$] (eq solid/L of water)	+	8MEF (eq/L)	+	2[H_2] (eq/L)

Period 1. RTE-1. NO_3^-/N_2 frozen, $Fe(OH)_{3(s)}$ present, $FeS_{(s)}$ not yet present.

$$(3.330281\times 10^{-4} + 1.498626\times 10^{-5})t = \text{OXR} + \frac{10^{16.2}10^{-3\text{pH}}10^{-\text{pe}}}{\alpha_{0,\text{Fe(II)}}} + 0 + 8\text{S1} + 0$$

$$+ 8\text{MEF} + 2\times 10^{-3.10}10^{-2\text{pe}}10^{-2\text{pH}}$$

Period 2. RTE-2. NO_3^- begins to be reduced, $Fe(OH)_{3(s)}$ present, $FeS_{(s)}$ not yet present.

$$(3.330281\times 10^{-4} + 1.498626\times 10^{-5})t = \text{OXR} + \frac{10^{16.2}10^{-3\text{pH}}10^{-\text{pe}}}{\alpha_{0,\text{Fe(II)}}} + \text{NO3R} + 8\text{S1} + 0$$

$$+ 8\text{MEF} + 2\times 10^{-3.10}10^{-2\text{pe}}10^{-2\text{pH}}$$

Period 3. RTE-3. $Fe(OH)_{3(s)}$ no longer present, $FeS_{(s)}$ not yet present.

$$(3.330281\times 10^{-4} + 1.498626\times 10^{-5})t = \text{OXR} + 8.953032\times 10^{-6} + \text{NO3R} + 8\text{S1} + 0$$

$$+ 8\text{MEF} + 2\times 10^{-3.10}10^{-2\text{pe}}10^{-2\text{pH}}$$

Period 4. RTE-4 $FeS_{(s)}$ present.

$$(3.330281\times 10^{-4} + 1.498626\times 10^{-5})t = \text{OXR} + \frac{10^{-18.0}}{\alpha_0^{\text{Fe(II)}}\alpha_2^{\text{S(-II)}}\text{S4}} + \text{NO3R} + 8\text{S4} + 9\text{FES}$$

$$+ 8\text{MEF} + 2\times 10^{-3.10}10^{-2\text{pe}}10^{-2\text{pH}}$$

(*Continued*)

TABLE 22.5 (CONTINUED)
Versions of the Redox Titration Equation (RTE) for Four Different Periods during the Summer

Note:

$$\mathrm{OXR\ (eq/L)} = (1.057497 \times 10^{-4} - 10^{4(\mathrm{pe}-21.35+\mathrm{pH})})$$

$$\mathrm{8MEF\ (eq/L)} = 8\frac{5.74810 \times 10^{-4} + (3.746566 \times 10^{-6})t}{10^{8\mathrm{pH}}10^{8\mathrm{pe}}10^{-21.6}/\alpha_0^{\mathrm{C(IV)}} + 1}$$

$$\mathrm{S1}(M) = \frac{\mathrm{S_T}}{1 + \dfrac{10^{8(\mathrm{pe}-4.25)}}{10^{-9\mathrm{pH}}}\alpha_1^{S(-\mathrm{II})}}$$

$$\mathrm{S4}\,(M) = \frac{-\mathrm{BSUL} + \sqrt{\mathrm{BSUL}^2 - 4\,\mathrm{ASUL} \times \mathrm{CSUL}}}{2\ \mathrm{ASUL}}$$

$$\mathrm{FES\ (FW/L)} = \mathrm{S_T} - \mathrm{S4} - \frac{10^{8(\mathrm{pe}-4.25)}}{10^{-9\mathrm{pH}}}\alpha_1^{S(-\mathrm{II})}\mathrm{S4}$$

$$\mathrm{NO3R\ (eq/L)} = 7.139440 \times 10^{-5} - 5\left[\frac{-\mathrm{BNIT} + \sqrt{\mathrm{BNIT}^2 - 4\mathrm{ANIT} \times \mathrm{CNIT}}}{2\mathrm{ANIT}}\right]\mathrm{BNIT}$$

$\mathrm{ANIT} = 2$

$\mathrm{BNIT} = (10^{103.7}10^{-6\mathrm{pH}}10^{-5\mathrm{pe}})^{-1}$

$\mathrm{CNIT} = -\mathrm{N_T}$

$$\mathrm{ASUL} = \alpha_0^{\mathrm{Fe(II)}}\left[1 + \frac{10^{8(\mathrm{pe}-4.25)}}{10^{-9\mathrm{pH}}}\alpha_1^{\mathrm{S(-II)}}\right]\alpha_2^{\mathrm{S(-II)}}$$

$\mathrm{BSUL} = \alpha_0^{\mathrm{Fe(II)}}(\mathrm{Fe_T} - \mathrm{S_T})\alpha_2^{\mathrm{S(-II)}}$

$\mathrm{CSUL} = -10^{-18}$

because the drop in pe with time becomes essentially stalled. This is because OC is only a weak reducing agent; it cannot effect much reduction of H(I) to H(0), so disproportionation of the OC is the only redox reaction remaining that can take place in significant measure.

22.5 KINETICS AND LABILE *VS.* NON-LABILE ORGANIC CARBON

It has been assumed that 100% of the OC in the hypolimnion can react immediately, allowing rapid changes in pe near the redox equivalence points. In real systems, OC will not react immediately, and in fact, there is a whole range of reactivities. Some OC is rather labile (as with actual glucose), and some OC is rather stable (as with humic material). For modeling purposes, one could subdivide the system OC into a labile fraction and a non-labile fraction. The non-labile fraction could be assumed to be essentially inert over the timeframe of interest. This type of approach could help incorporate real-world kinetic limitations into an equilibrium-based model. However, obtaining input data to characterize OC lability would be difficult, and so system dependent that it would probably not be worth the effort. Better then to just be satisfied that an equilibrium model will provide a good understanding of the redox sequence according to which things will change.

TABLE 22.6
Versions of the ENE for Four Different Periods during the Summer

$$[H^+] + [Na^+] + (2[Fe^{2+}] + [FeOH^+]) = [HCO_3^-] + 2[CO_3^{2-}] + [NO_3^-] + 2[SO_4^{2-}] + [HS^-] + [OH^-]$$

Period 1. ENE-1. NO_3^-/N_2 frozen, $Fe(OH)_{3(s)}$ present, $FeS_{(s)}$ not yet present.

$$10^{-pH} + 6.855255\times10^{-4} + (1+\alpha_0^{Fe(II)})\frac{10^{16.2}10^{-3pH}10^{-pe}}{\alpha_0^{Fe(II)}} = \alpha_1^{C(IV)}C(IV)_T + 2\alpha_2^{C(IV)}C(IV)_T + 1.427888\times10^{-5}$$

$$+ 2(S_T - S1) + \alpha_1^{S(-II)}S1 + K_w10^{pH}$$

Period 2. ENE-2. NO_3^- begins to reduced, $Fe(OH)_{3(s)}$ present, $FeS_{(s)}$ not yet present.

$$10^{-pH} + 6.855255\times10^{-4} + (1+\alpha_0^{Fe(II)})\frac{10^{16.2}10^{-3pH}10^{-pe}}{\alpha_0^{Fe(II)}} = \alpha_1^{C(IV)}C(IV)_T + 2\alpha_2^{C(IV)}C(IV)_T + NO3$$

$$+ 2(S_T - S1) + \alpha_1^{S(-II)}S1 + K_w10^{pH}$$

Period 3. ENE-3. $Fe(OH)_{3(s)}$ no longer present, $FeS_{(s)}$ not yet present.

$$10^{-pH} + 6.855255\times10^{-4} + (1+\alpha_0^{Fe(II)})(8.95\times10^{-6}) = \alpha_1C(IV)_T + 2\alpha_2C(IV)_T + NO3$$

$$+ 2(S_T - S1) + \alpha_1^{S(-II)}S1 + K_w10^{pH}$$

Period 4. ENE-4. $FeS_{(s)}$ present.

$$10^{-pH} + 6.855255\times10^{-4} + (1+\alpha_0^{Fe(II)})\frac{10^{-18.0}}{\alpha_0^{Fe(II)}\alpha_{2,S(-II)}S4} = \alpha_2^{C(IV)}C(IV)_T + 2\alpha_2^{C(IV)}C(IV)_T + NO3$$

$$+ 2(S_T - S4 - FES) + \alpha_1^{S(-II)}S4 + K_w10^{pH}$$

Note:

$$S1\,(M) = \frac{S_T}{1+\dfrac{10^{8(pe-4.25)}}{10^{-9pH}}\alpha_1^{S(-II)}}$$

$$S4\left(M\right) = \frac{-BSUL+\sqrt{BSUL^2 - 4ASUL\times CSUL}}{2ASUL}$$

$$FES\,(FW/L) = S_T - S4 - \frac{10^{8(pe-4.25)}}{10^{-9pH}}\alpha_1^{S(-II)}S4$$

$$C(IV)_T(M) = 5.748410\times10^{-4} + \left(3.746566\times10^{-6}\right)t - MEF$$

$$NO3\,(M) = \left(\frac{-BNIT + \sqrt{BNIT^2 - 4ANIT\times CNIT}}{2ANIT}\right)BNIT$$

$$ASUL = \alpha_0^{Fe(II)}\left(1 + \frac{10^{8(pe-4.25)}}{10^{-9pH}}\alpha_1^{S(-II)}\right)\alpha_2^{S(-II)}$$

$$BSUL = \alpha_0^{Fe(II)}(Fe_T - S_T)\alpha_2^{S(-II)}$$

$$CSUL = -10^{-18}$$

$$ANIT = 2$$

$$BNIT = (10^{103.7}10^{-6pH}10^{-5pe})^{-1}$$

$$CNIT = -N_T$$

TABLE 22.7
pe, pH, and Concentrations of Various Species of Interest during the Course of the Summer

Part 1.	Time (days)	OC oxidized (eq/L)	pH	pe	TEST	log O_2 (*M*)	log O_2 (eq/L)	log NO_3^- (*M*)	log NO_3^- (eq/L)	log $Fe(OH)_{3(s)}$ (FW/L, eq/L)	log $Fe(II)_T$ (*M*, eq/L)	$\alpha_0^{Fe(II)}$
Period 0. Redox equilibrium not yet established.												
	0	0.00E+00	8.00000	12.60560	111.0280	−3.58	−2.97	−4.85	−4.15	−5.05	–	0.9693
Period 1. NO_3^-/N_2 couple is considered frozen, (am)$Fe(OH)_{3(s)}$ is still present, $FeS_{(s)}$ not yet present. The parameter TEST examines whether the initial NO_3^- can be reduced to N_2: the critical value is 100.51.												
mostly O_2 reduction	0	3.33 E−04	7.075153	13.489852	109.90	−3.74	−3.14	−4.85	−4.15	−5.05	−18.51	0.9963
	5	4.08 E−04	6.994340	13.558811	109.76	−3.79	−3.19	−4.85	−4.15	−5.05	−18.34	0.9969
	10	4.83 E−04	6.926171	13.613671	109.63	−3.84	−3.24	−4.85	−4.15	−5.05	−18.19	0.9973
	15	5.58 E−04	6.867235	13.657437	109.49	−3.90	−3.30	−4.85	−4.15	−5.05	−18.05	0.9977
	20	6.33 E−04	6.815331	13.691700	109.35	−3.97	−3.37	−4.85	−4.15	−5.05	−17.93	0.9979
	25	7.08 E−04	6.768965	13.716994	109.20	−4.06	−3.46	−4.85	−4.15	−5.05	−17.82	0.9981
	30	7.83 E−04	6.727069	13.732718	109.03	−4.16	−3.56	−4.85	−4.15	−5.05	−17.71	0.9983
	35	8.58 E−04	6.688858	13.736372	108.82	−4.30	−3.70	−4.85	−4.15	−5.05	−17.60	0.9985
	40	9.32 E−04	6.653737	13.720507	108.52	−4.51	−3.90	−4.85	−4.15	−5.05	−17.48	0.9986
	45	1.01 E−03	6.621244	13.653687	108.00	−4.90	−4.30	−4.85	−4.15	−5.05	−17.31	0.9987
	46	1.02 E−03	6.615028	13.621302	107.80	−5.06	−4.45	−4.85	−4.15	−5.05	−17.26	0.9987
	47	1.04 E−03	6.608899	13.566980	107.49	−5.30	−4.70	−4.85	−4.15	−5.05	−17.19	0.9987
	48	1.05 E−03	6.602855	13.424641	106.74	−5.89	−5.29	−4.85	−4.15	−5.05	−17.03	0.9987
	48.2	1.06 E−03	6.601656	13.330498	106.26	−6.27	−5.67	−4.85	−4.15	−5.05	−16.93	0.9987
H_2O EP →	48.342	1.06 E−03	6.600807	12.622699	102.72	−9.11	−8.51	−4.85	−4.15	−5.05	−16.22	0.9987
Period 2. NO_3^- begins to be reduced, (am)$Fe(OH)_{3(s)}$ is still present, $FeS_{(s)}$ not yet present. The parameter TEST examines whether (am)$Fe(OH)_{3(s)}$ is still present; the critical value is 21.25. Under the assumption that (am)$Fe(OH)_{3(s)}$ is largely insoluble, and when $FeS_{(s)}$ is not yet present, complete dissolution of (am)$Fe(OH)_{3(s)}$ is also observable when log $Fe(II)_T$ = log Fe_T = −5.05.												
mostly NO_3^- reduction	48.343	1.06 E−03	6.600807	12.180746	31.98	−10.88	−10.27	−4.85	−4.15	−5.05	−15.78	0.9987
	49	1.07 E−03	6.601780	12.166600	31.97	−10.93	−10.33	−4.91	−4.21	−5.05	−15.77	0.9987
	50	1.08 E−03	6.603251	12.140461	31.95	−11.03	−10.43	−5.03	−4.33	−5.05	−15.74	0.9987
	51	1.10 E−03	6.604709	12.104835	31.92	−11.16	−10.56	−5.20	−4.50	−5.05	−15.71	0.9987
	52	1.11 E−03	6.606153	12.047037	31.86	−11.39	−10.79	−5.48	−4.78	−5.05	−15.66	0.9987
	53	1.13 E−03	6.607585	11.841645	31.66	−12.21	−11.60	−6.50	−5.80	−5.05	−15.46	0.9987
N_2 EP →	53.106	1.13 E−03	6.607736	11.295826	31.12	−14.39	−13.79	−9.23	−8.53	−5.05	−14.91	0.9987

(Continued)

TABLE 22.7 (CONTINUED)
pe, pH, and Concentrations of Various Species of Interest during the Course of the Summer

Part 1.	Time (days)	OC oxidized (eq/L)	pH	pe	TEST	log O_2 (*M*)	log O_2 (eq/L)	log NO_3^- (*M*)	log NO_3^- (eq/L)	log $Fe(OH)_{3(s)}$ (FW/L, eq/L)	log $Fe(II)_T$ (*M*, eq/L)	$\alpha_0^{Fe(II)}$
mostly (am)Fe $(OH)_{3(s)}$	53.2	1.13 E−03	6.614057	2.210492	22.05	−50.70	−50.10	−54.61	−53.92	−5.12	−5.852	0.9987
reduction	53.4	1.13 E−03	6.627604	1.674024	21.56	−52.80	−52.19	−57.22	−56.52	−5.34	−5.356	0.9987
	53.65	1.14 E−03	6.644684	1.355421	21.29	−54.00	−53.40	−58.71	−58.01	−6.10	−5.088	0.9986
	53.7	1.14 E−03	6.648123	1.306908	21.25	−54.18	−53.58	−58.93	−58.23	−7.27	−5.050	0.9986
Fe(II) EP →	53.703	1.14 E−03	6.648329	1.304100	21.25	−54.19	−53.59	−58.94	−58.24	−8.04	−5.048	0.9986
Period 3. $Fe(OH)_{3(s)}$ is no longer present, $FeS_{(s)}$ not yet present. The parameter TEST examines the value of $[Fe^{2+}][S^{2-}]$ where the critical value for saturation of $FeS_{(s)}$ is $K_{s0} = 1.00\times10^{-18}$.												
mostly SO_4^{2-} reduction	53.704	1.14 E−03	6.648372	−2.526310	8.83 E−23	−69.51	−68.91	−78.09	−77.39	–	−5.048	0.9986
	54	1.14 E−03	6.648873	−2.888073	6.82 E−20	−70.96	−70.36	−79.90	−79.20	–	−5.048	0.9986
	55	1.16 E−03	6.650551	−2.971320	3.00 E−19	−71.29	−70.68	−80.30	−79.61	–	−5.048	0.9986
	56	1.17 E−03	6.652208	−3.005493	5.35 E−19	−71.42	−70.81	−80.47	−79.77	–	−5.048	0.9986
	57	1.19 E−03	6.653843	−3.028250	7.73 E−19	−71.50	−70.90	−80.57	−79.87	–	−5.048	0.9986
$FeS_{(s)}$ saturation →	57.943	1.20 E−03	6.655367	−3.044861	1.00 E−18	−71.56	−70.96	−80.64	−79.94	–	−5.048	0.9986
Period 4. $FeS_{(s)}$ is present.												
mostly SO_4^{2-} reduction	57.944	1.20 E−03	6.655362	−3.044867	–	−71.56	−70.96	−80.64	−79.94	–	−5.048	0.9986
	58	1.20 E−03	6.655232	−3.045128	–	−71.56	−70.96	−80.65	−79.95	–	−5.050	0.9986
	60	1.23 E−03	6.651459	−3.055334	–	−71.62	−71.02	−80.72	−80.02	–	−5.143	0.9986
	65	1.31 E−03	6.648240	−3.085389	–	−71.75	−71.15	−80.89	−80.19	–	−5.356	0.9986
	70	1.38 E−03	6.650365	−3.116650	–	−71.87	−71.27	−81.03	−80.33	–	−5.529	0.9986
	75	1.46 E−03	6.654759	−3.146960	–	−71.97	−71.37	−81.16	−80.46	–	−5.666	0.9986
	80	1.53 E−03	6.660035	−3.176657	–	−72.07	−71.47	−81.27	−80.57	–	−5.778	0.9986
	85	1.61 E−03	6.665605	−3.206883	–	−72.17	−71.57	−81.39	−80.69	–	−5.873	0.9985
	90	1.68 E−03	6.671209	−3.239547	–	−72.28	−71.67	−81.52	−80.82	–	−5.953	0.9985
	95	1.76 E−03	6.676722	−3.278691	–	−72.41	−71.81	−81.68	−80.98	–	−6.025	0.9985
	100	1.83 E−03	6.682083	−3.338741	–	−72.63	−72.03	−81.95	−81.25	–	−6.088	0.9985
S(−II) EP →	103.665	1.89 E−03	6.685662	−3.563268	–	−73.51	−72.91	−83.05	−82.35	–	−6.129	0.9985

(Continued)

TABLE 22.7 (CONTINUED)
pe, pH, and Concentrations of Various Species of Interest during the Course of the Summer

Part 1.	Time (days)	OC oxidized (eq/L)	pH	pe	TEST	log O_2 (*M*)	log O_2 (eq/L)	log NO_3^- (*M*)	log NO_3^- (eq/L)	log $Fe(OH)_{3(s)}$ (FW/L, eq/L)	log $Fe(II)_T$ (*M*, eq/L)	$\alpha_0^{Fe(II)}$
mostly C(IV) reduction	110	1.98 E−03	6.670750	−3.792801	–	−74.49	−73.89	−84.29	−83.59	–	−6.105	0.9985
(OC) disproportionation	115	2.06 E−03	6.659162	−3.811264	–	−74.61	−74.01	−84.45	−83.75	–	−6.086	0.9986
	120	2.13 E−03	6.647864	−3.818331	–	−74.68	−74.08	−84.56	−83.86	–	−6.067	0.9986
	150	2.58 E−03	6.585456	−3.804453	–	−74.88	−74.28	−84.86	−84.16	–	−5.961	0.9988
	200	3.33 E−03	6.497683	−3.745094	–	−74.99	−74.39	−85.09	−84.39	–	−5.811	0.9990

TABLE 22.7 (CONTINUED)
pe, pH, and Concentrations of Various Species of Interest during the Course of the Summer

Part 2.	Time (days)	log $S(-II)_T$ (*M*)	log $S(-II)_T$ (eq/L)	$\alpha_1^{S(-II)}$	$\alpha_2^{S(-II)}$	log $[FeS_{(s)}]$ (FW/L)	log $FeS_{(s)}$ (eq/L)	log SO_4^{2-} (*M*)	log CH_4 (*M*)	log CH_4 (eq/L)	log $C(IV)_T$ (*M*)	$\alpha_0^{C(IV)}$	log H_2 (*M*)
Period 0. Redox equilibrium not yet established.													
	0	–	–	0.9091	9.09 E–07	–	–	–4.03	–	–	–3.31	0.0218	–
Period 1. NO_3^-/N_2 couple is considered frozen, $(am)Fe(OH)_{3(s)}$ is still present, $FeS_{(s)}$ not yet present. The parameter TEST examines whether the initial NO_3^- can be reduced to N_2C the critical value is 100.51.													
mostly O_2 reduction	0	–141.36	–140.46	0.5431	6.46 E–08	–	–	–4.03	–146.96	–146.06	–3.24	0.1583	–44.23
	5	–141.14	–140.24	0.4967	4.90 E–08	–	–	–4.03	–146.79	–145.88	–3.23	0.1848	–44.21
	10	–140.93	–140.03	0.4576	3.86 E–08	–	–	–4.03	–146.61	–145.71	–3.21	0.2096	–44.18
	15	–140.72	–139.82	0.4241	3.12 E–08	–	–	–4.03	–146.43	–145.53	–3.20	0.2330	–44.15
	20	–140.50	–139.59	0.3952	2.58 E–08	–	–	–4.03	–146.24	–145.33	–3.19	0.2550	–44.11
	25	–140.25	–139.35	0.3700	2.17 E–08	–	–	–4.03	–146.02	–145.12	–3.17	0.2758	–44.07
	30	–139.98	–139.07	0.3478	1.86 E–08	–	–	–4.03	–145.77	–144.87	–3.16	0.2955	–44.02
	35	–139.64	–138.73	0.3281	1.60 E–08	–	–	–4.03	–145.46	–144.55	–3.15	0.3142	–43.95
	40	–139.17	–138.27	0.3106	1.40 E–08	–	–	–4.03	–145.01	–144.11	–3.14	0.3319	–43.85
	45	–138.32	–137.42	0.2948	1.23 E–08	–	–	–4.03	–144.19	–143.28	–3.13	0.3487	–43.65
	46	–138.00	–137.10	0.2918	1.20 E–08	–	–	–4.03	–143.87	–142.97	–3.13	0.3519	–43.57
	47	–137.51	–136.60	0.2889	1.17 E–08	–	–	–4.03	–143.38	–142.48	–3.12	0.3551	–43.45
	48	–136.31	–135.41	0.2860	1.15 E–08	–	–	–4.03	–142.19	–141.28	–3.12	0.3583	–43.15
	48.2	–135.54	–134.64	0.2855	1.14 E–08	–	–	–4 03	–141.42	–140.52	–3.12	0.3590	–42.96
H_2O EP →	48.342	–129.87	–128.97	0.2851	1.14 E–08	–	–	–4.03	–135.75	–134.85	–3.12	0.3594	–41.55

(Continued)

TABLE 22.7 (CONTINUED)
pe, pH, and Concentrations of Various Species of Interest during the Course of the Summer

Part 2.	Time (days)	$\log S(-II)_T$ (*M*)	$\log S(-II)_T$ (eq/L)	$\alpha_1^{S(-II)}$	$\alpha_2^{S(-II)}$	log $[FeS_{(s)}]$ (FW/L)	log $FeS_{(s)}$ (eq/L)	log SO_4^{2-} (*M*)	log CH_4 (*M*)	log CH_4 (eq/L)	log $C(IV)_T$ (*M*)	$\alpha_0^{C(IV)}$	log H_2 (*M*)
Period 2. NO_3^- begins to be reduced, (am)$Fe(OH)_{3(s)}$ is still present, $FeS_{(s)}$ not yet present. The parameter TEST examines whether (am)$Fe(OH)_{3(s)}$ is still present; the critical value is 21.25. Under the assumption that (am)$Fe(OH)_{3(s)}$ is largely insoluble, and when $FeS_{(s)}$ is not yet present, complete dissolution of (am)$Fe(OH)_{3(s)}$ is also observable when $\log Fe(II)_T = \log Fe_T = -5.05$.													
mostly NO_3^- reduction	48.343	−126.34	−125.43	0.2851	1.14 E−08	–	–	−4.03	−132.22	−131.32	−3.12	0.3594	−40.66
	49	−126.23	−125.33	0.2855	1.14 E−08	–	–	−4.03	−132.11	−131.21	−3.12	0.3589	−40.64
	50	−126.04	−125.14	0.2862	1.15 E−08	–	–	−4.03	−131.91	−131.01	−3.12	0.3581	−40.59
	51	−125.77	−124.86	0.2869	1.15 E−08	–	–	−4.03	−131.64	−130.74	−3.12	0.3573	−40.52
	52	−125.32	−124.42	0.2876	1.16 E−08	–	–	−4.03	−131.19	−130.28	−3.11	0.3566	−40.41
	53	−123.69	−122.79	0.2883	1.17 E−08	–	–	−4.03	−129.55	−128.65	−3.11	0.3558	−40.00
N_2 EP →	53.106	−119.33	−118.42	0.2883	1.17 E−08	–	–	−4.03	−125.19	−124.29	−3.11	0.3557	−38.91
mostly (am) $Fe(OH)_{3(s)}$ reduction	53.2	−46.70	−45.80	0.2913	1.20 E−08	–	–	−4.03	−52.56	−51.66	−3.11	0.3524	−20.75
	53.4	−42.54	−41.64	0.2978	1.26 E−08	–	–	−4.03	−48.39	−47.48	−3.11	0.3453	−19.70
	53.65	−40.16	−39.26	0.3061	1.35 E−08	–	–	−4.03	−45.98	−45.08	−3.11	0.3365	−19.10
	53.7	−39.81	−38.90	0.3078	1.37 E−08	–	–	−4.03	−45.63	−44.72	−3.11	0.3347	−19.01
Fe(II) EP →	53.703	−39.79	−38.88	0.3079	1.37 E−08	–	–	−4.03	−45.60	−44.70	−3.11	0.3346	−19.00
Period 3. $Fe(OH)_{3(s)}$ is no longer present, $FeS_{(s)}$ not yet present. The parameter TEST examines the value of $[Fe^{2+}][S^{2-}]$ where the critical value for saturation of $FeS_{(s)}$ is $K_{s0} = 1.00 \times 10^{-18}$.													
mostly SO_4^{2-} reduction	53.704	−9.14	−8.24	0.3079	1.37 E−08	–	–	−4.03	−14.96	−14.06	−3.11	0.3346	−11.34
	54	−6.26	−5.35	0.3082	1.37 E−08	–	–	−4.03	−12.07	−11.17	−3.11	0.3343	−10.62
	55	−5.61	−4.71	0.3090	1.38 E−08	–	–	−4.04	−11.42	−10.52	−3.11	0.3335	−10.46
	56	−5.37	−4.46	0.3098	1.39 E−08	–	–	−4.05	−11.16	−10.25	−3.11	0.3326	−10.39
	57	−5.21	−4.31	0.3106	1.40 E−08	–	–	−4.06	−10.99	−10.08	−3.10	0.3318	−10.35
$FeS_{(s)}$ saturation →	57.943	−5.10	−4.20	0.3114	1.41 E−08	–	–	−4.07	−10.87	−9.96	−3.10	0.3310	−10.32

(Continued)

TABLE 22.7 (CONTINUED)
pe, pH, and Concentrations of Various Species of Interest during the Course of the Summer

Part 2.	Time (days)	log $S(-II)_T$ (*M*)	log $S(-II)_T$ (eq/L)	$\alpha_1^{S(-II)}$	$\alpha_2^{S(-II)}$	log $[FeS_{(s)}]$ (FW/L)	log $FeS_{(s)}$ (eq/L)	log SO_4^{2-} (*M*)	log CH_4 (*M*)	log CH_4 (eq/L)	log $C(IV)_T$ (*M*)	$\alpha_0^{C(IV)}$	log H_2 (*M*)
Period 4. $FeS_{(s)}$ is present.													
mostly SO_4^{2-} reduction	57.944	−5.10	−4.20	0.3114	1.41 E−08	−9.41	−8.45	−4.07	−10.87	−9.96	−3.10	0.3310	−10.32
	58	−5.10	−4.19	0.3113	1.41 E−08	−7.27	−6.32	−4.07	−10.86	−9.96	−3.10	0.3311	−10.32
	60	−5.00	−4.10	0.3094	1.39 E−08	−5.75	−4.80	−4.09	−10.74	−9.84	−3.10	0.3330	−10.29
	65	−4.78	−3.88	0.3078	1.37 E−08	−5.34	−4.39	−4.14	−10.47	−9.56	−3.09	0.3347	−10.23
	70	−4.61	−3.71	0.3089	1.38 E−08	−5.22	−4.27	−4.20	−10.22	−9.32	−3.08	0.3336	−10.17
	75	−4.48	−3.58	0.3111	1.40 E−08	−5.17	−4.21	−4.27	−10.01	−9.11	−3.07	0.3313	−10.12
	80	−4.38	−3.47	0.3137	1.43 E−08	−5.14	−4.18	−4.35	−9.81	−8.91	−3.06	0.3286	−10.07
	85	−4.29	−3.39	0.3164	1.47 E−08	−5.12	−4.16	−4.46	−9.61	−8.70	−3.05	0.3258	−10.02
	90	−4.22	−3.32	0.3192	1.50 E−08	−5.11	−4.15	−4.59	−9.38	−8.48	−3.04	0.3230	−9.96
	95	−4.16	−3.26	0.3220	1.53 E−08	−5.10	−4.14	−4.79	−9.11	−8.21	−3.03	0.3202	−9.90
	100	−4.10	−3.20	0.3247	1.56 E−08	−5.09	−4.14	−5.16	−8.67	−7.76	−3.02	0.3175	−9.79
S(II) EP →	103.665	−4.07	−3.17	0.3265	1.58 E−08	−5.09	−4.13	−6.89	−6.90	−5.99	−3.02	0.3158	−9.34
mostly C(IV) reduction (OC)	110	−4.07	−3.17	0.3190	1.49 E−08	−5.09	−4.13	−8.85	−4.92	−4.02	−3.01	0.3232	−8.86
disproportionation)	115	−4.07	−3.17	0.3132	1.43 E−08	−5.09	−4.14	−9.10	−4.67	−3.77	−3.01	0.3291	−8.80
	120	−4.07	−3.17	0.3077	1.37 E−08	−5.09	−4.14	−9.25	−4.51	−3.61	−3.00	0.3349	−8.76
	150	−4.07	−3.16	0.2779	1.07 E−08	−5.10	−4.15	−9.67	−4.06	−3.16	−2.98	0.3676	−8.66
	200	−4.06	−3.16	0.2392	7.53 E−09	−5.13	−4.18	−9.93	−3.74	−2.84	−2.94	0.4157	−8.61

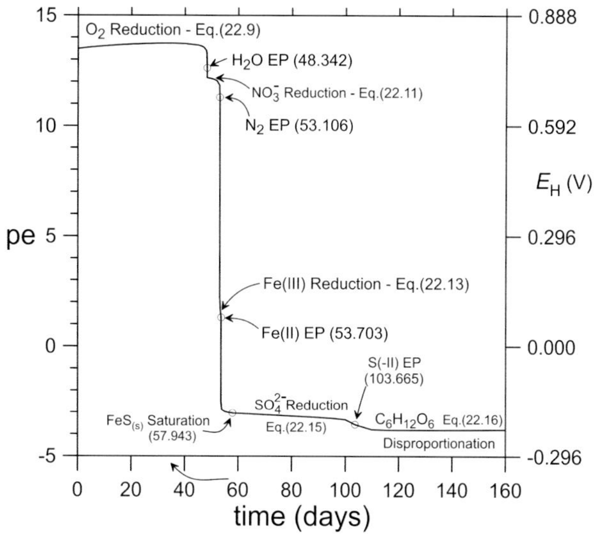

FIGURE 22.4 pe *vs.* time in the hypolimnion. (The initial organic carbon (OC) is considered oxidized at $t = 0$ days.)

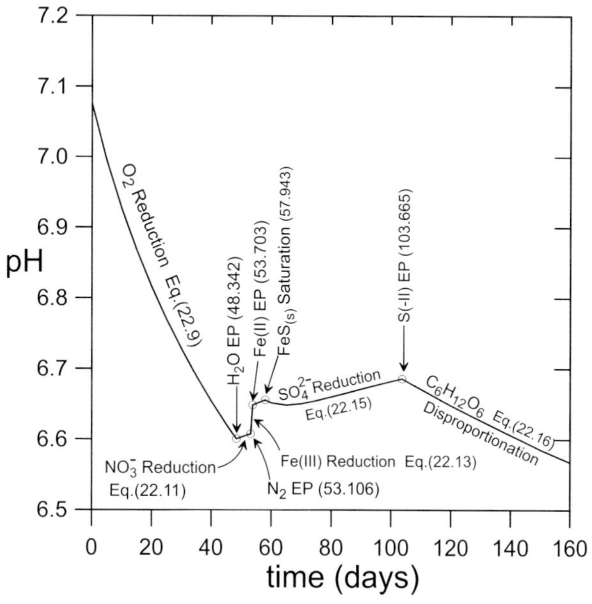

FIGURE 22.5 pH *vs.* time in the hypolimnion. (The initial organic carbon (OC) is considered oxidized at $t = 0$ days.)

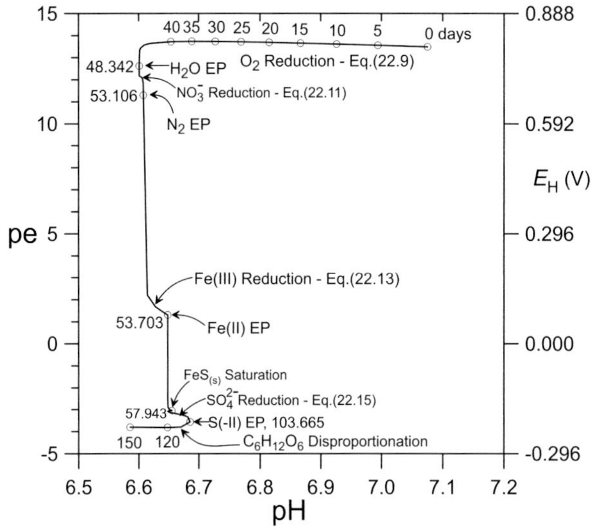

FIGURE 22.6 pe *vs.* pH track in the hypolimnion. (The initial organic carbon (OC) is considered oxidized at $t = 0$ days.)

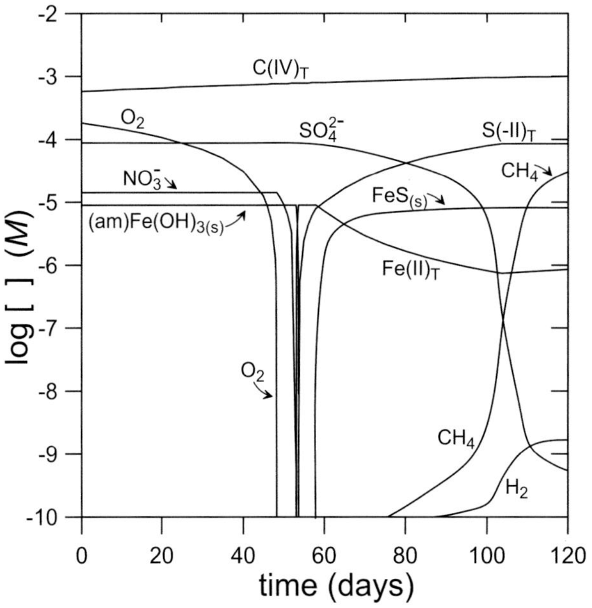

FIGURE 22.7 Log concentration (*M*, or FW/L for solids) *vs.* time in the hypolimnion for various redox species. (The initial organic carbon (OC) is considered oxidized at $t = 0$ days.)

REFERENCES

Fair G M, Geyer JC, Okun DA (1968) *Water and Wastewater Engineering*, Vol 2. John Wiley and Sons, New York.

Imboden DM, Lerman A (1978) Chemical models of lakes. In: Lerman, A (Ed), *Lakes: Chemistry, Geology, and Physics*, Springer, New York, p.363.

Part VI

Effects of Electrical Charges on Solution Chemistry

23 The Debye–Hückel Equation and Its Descendent Expressions for Activity Coefficients of Aqueous Ions

23.1 INTRODUCTION

The development of the Debye–Hückel Law (Debye and Hückel, 1923) is one of the most important 20th-century accomplishments in the field of the physical chemistry of electrolyte solutions. Given its importance and widespread use, here we provide the detailed derivation and describe how the Extended Debye–Hückel Equation and the Davies Equation (see Chapter 2) are developed from it. Other treatments of this subject may be found in Bockris and Reddy (1970), Moore (1972), and Robinson and Stokes (1959); the presentation here follows from that provided by Bockris and Reddy (1970).

As discussed in Chapter 2, on the infinite dilution scale, the activity coefficient γ of a dissolved species equals 1.0 when the solution is in the (hypothetical) infinitely dilute reference state. As the solution becomes more concentrated, the growing interactions among the various dissolved species lead to γ values that tend to deviate from unity. For ionic species, by far the most important interactions are coulombic in nature (*i.e.*, electrostatic). Coulomb's Law states that

$$F = \frac{Kq_1q_2}{\varepsilon d^2} \tag{23.1}$$

where F is the electrostatic force (Newtons, N) between two charges q_1 and q_2 (coulombs, C), K is a constant (= 8.99×10^9 N–m^2/C^2), ε is the dielectric constant of the medium (dimensionless), and d (m) is the distance separating the charges. Applying Eq.(23.1) to individual ions in solution implicitly makes the assumption that ε measured for a bulk medium applies even at such a small scale.

Debye and Hückel (1923) realized that if one could understand how the effects of coulombic interactions depend on the nature and concentrations of the dissolved ions in solution, then it would be possible to predict γ values for ions as a function of the ionic composition of a solution. In particular, by charge balance, an ion of a particular sign will on average be surrounded by a cloud with more charge of the opposite sign than by charge of the same sign. Thus, there will be more attractive forces acting on the ion than repulsive forces. This will tend to lower the activity of the ion, and that lowering will be manifested as $\gamma < 1$. Second, coulombic forces depend inversely on distance. Thus, as an ion becomes closer (on average) to other ions (*i.e.*, more concentrated), the greater will be the strength of the coulombic forces, and therefore the greater the deviation that $\gamma < 1$.

The chemical potential of species i depends on its activity:

$$\mu_i = \mu_i^{\circ} + RT\ln a_i \tag{23.2}$$

$$a_i = m_i\gamma_i \tag{23.3}$$

where μ_i^o is the standard state chemical potential, a_i is the activity, m_i is the molality in the *bulk solution* (essentially the same as M for the solutions of interest here), and γ_i is the activity coefficient. Substituting Eq.(23.3) into (23.2) gives

$$\mu_i = \mu_i^o + RT\ln m_i + RT\ln \gamma_i. \tag{23.4}$$

Assuming that ion-ion interactions are the only types of interactions that affect the activity coefficient of species i, then the activity coefficient term in Eq.(23.4) provides a direct measure of the intensity of those interactions for i. That is, $RT\ln\gamma_i$ represents the portion of μ_i that is due to ion-ion interactions. Since the units of R are energy/mol-K, and since $\ln\gamma_i$ is dimensionless, the units of $RT\ln\gamma_i$ are energy/mol. These ion-ion interactions include those between species i and other types of ions, as well as species i-species i interactions. For example, for the activity coefficient γ_{H^+} of H^+ in a solution of HCl, the Cl^- ions (and the minor OH^- ions) and the H^+ ions themselves play a role.

23.2 THE DEBYE–HÜCKEL LAW

23.2.1 The Poisson Equation and the Local Charge Density

We need a relationship between $RT\ln\gamma_i$ and the solution composition. To obtain this, we examine the magnitude of the free energy change that results when the ions in solution are initially in a hypothetical, uncharged state, and then are slowly brought up at constant T and P to their full levels of charge.

In a constant T and P system, the change in free energy for a given change in state will equal the amount of work that is done in a reversible manner to cause that change in state. Therefore, we are interested in the work required to carry out the process depicted in Figure 23.1. That work is composed of two parts. The first involves the energy needed simply to charge each of the core ions. The second involves the ion-ion *interaction* energy, *i.e.*, the part responsible for the term $RT\ln\gamma_i$. Whether or not it is considered that a core ion has some finite radius depends on the complexity with which the problem is addressed. If a finite radius is acknowledged, then because water is very polar and so attracted to open charge, then it follows that effective radius of the core ion charge will probably be greater than a value found for the radius of the bare ion – more on this in Section 23.3 wherein the Extended Debye–Hückel Law is considered.

The concept of interest now is the electrical potential ψ (volts, V). For a given point, ψ equals the amount of work required to bring a unit positive charge from infinity (where ψ is the baseline value

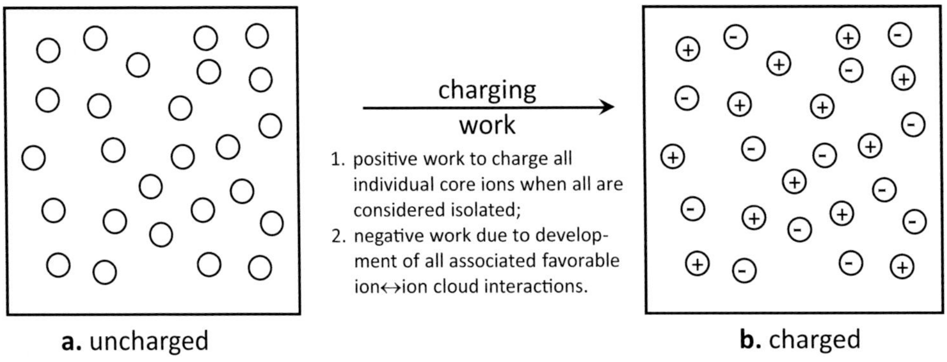

FIGURE 23.1 Work is done when ions in a solution are brought from a hypothetical uncharged state to the fully charged and interacting state. Only a portion of the ions being charged are of type i. There are two portions of the work: 1) positive work due just to placing charge on each of the ions; and 2) negative work due to the attractive interactions that develop during charging between each ion and its surrounding ion cloud.

for the system and can be taken to be zero) to that point. If $\psi = +1$ V at some point, and 1C worth of positive charge was brought to that point without altering ψ, the amount of work done would then be +1 C–V = 1 Joule (J). The charge distribution around a given ion will, on the average, be spherically symmetrical, so ψ will also on average be spherically symmetrical, and only a function of distance r from the center of the ion (see Figure 23.2). In the derivation of the Debye–Hückel Law, all ions are assumed to be point charges with radius zero: any value of $r > 0$ is outside of the core ion. This assumption cannot ever be exactly true, and this is the reason why the Debye–Hückel Law begins to fail as solution concentrations rise to the point that the distance among ions are no longer very large relative the actual non-zero radii of the constituent core ions. (We would then need to use the Extended Debye–Hückel Law, the Davies Equation, or (as in seawater) possibly some empirical algorithm for γ_i.)

An expression for $\psi(r)$ is needed to compute the work done in charging up a given ion surrounded by other ions. The Poisson Equation relates ψ to the local charge density $q(r)$ (C/volume). In spherical coordinates,

$$\frac{1}{r^2}\frac{d}{dr}\left(r^2\frac{d\psi}{dr}\right) = -\frac{1}{\varepsilon\varepsilon_o}q(r) \qquad \text{Poission Equation for Spherical Symmetry} \tag{23.5}$$

where ε_o is the "permittivity of free space" and represents the inherent transmissibility of electrical forces through a perfect vacuum (in Coulomb's Law, $K = 1/(4\pi\varepsilon_o)$).

The sign of $q(r)$ can be positive or negative, and depends on whether there is a net excess of positive or negative charge, respectively, at point r. In a solution if r is measured from the center of a particular positive ion, then: 1) $q(r)$ will be negative on average (ions are bouncing about) for all values of r, and will asymptotically approach zero as $r \to \infty$; and 2) $\psi(r)$ will be positive on average for all r, and will asymptotically approach zero as $r \to \infty$. Thus, to balance the charge of any positive core ion, there must be a negative cloud of charge surrounding that core ion. For a negative core ion, then: 1) $q(r)$ will be positive on average for all r and approach zero as $r \to \infty$; and 2) $\psi(r)$ will

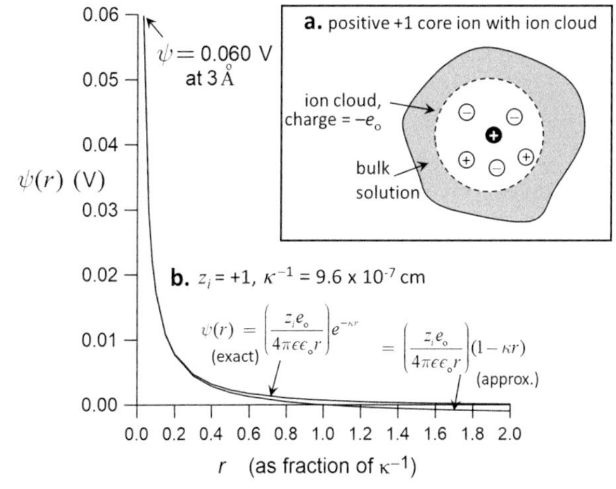

FIGURE 23.2 **a.** A core ion with $z_i = +1$ and charge $+e_o$ as surrounded by an ion cloud with charge $-e_o$. **b.** $\psi(r)$ *vs.* r for $z_i = +1$ where r is viewed as a fraction of κ^{-1} at $I = 0.001$ M ($\kappa^{-1} = 9.6 \times 10^{-7}$ cm). Both the exact function (Eq.(23.39)) and the linearized approximation function (Eq.(23.40)) are plotted. For very small r, the approximation performs well for use in evaluating the ψ_{cloud} component of ψ close to the center of the core ion when evaluating γ_i. At $r \geq \kappa^{-1}$, the approximation clearly fails, since then $\psi(r)$ goes identically to zero at $r = \kappa^{-1}$, then takes on a sign opposite to that near the core ion (in this case, negative) when $r = \kappa^{-1}$.

be negative on average for all r and approach zero as $r \rightarrow \infty$; there is a positive cloud of charge surrounding any negative core ion.

As noted, with r measured from the center of a core ion, theoretically one must go to infinity to reach $q(r) = 0$ and $d\psi(r)/dr = 0$. The reason is that the core ion occupies one point in space, and the entire remainder of the solution is theoretically available for the cloud of opposite charge to provide overall electroneutrality for the system. For all intents and purposes, however, both $q(r)$ and $d\psi(r)/dr$ approach zero very rapidly, usually within a few hundred angstroms (a few tens of nm). As discussed below, the rate of decay is strongly affected by the ionic strength I.

Once $q(r)$ diminishes to essentially zero, the solution is electrically neutral (on average) and one is in the bulk solution, as viewed from a particular core ion (Figure 23.2). Although that medium contains the myriad other core ions and their ion clouds, each of those ion/(ion cloud) pairs is electrically neutral when viewed from the distance of the initial core ion under consideration. The charge on each ion cloud is equal and opposite to the charge on its core ion. For Na^+ which has a charge of $e_o = +1.60 \times 10^{-19}$ C), the charge on the cloud is exactly $-e_o$. For SO_4^{2-}, the charge on the cloud is $+2e_o$.

The charge present in the cloud around a given core ion is made up of portions of charge from many different, surrounding neighboring ions. Continual Brownian motion makes those ions flit in and out of the cloud. On average, when one such ion enters, another leaves, and vice versa. Each of the ions in the cloud contributes only a portion of the charge of the cloud since each of those ions (both positive and negative) is also surrounded by their own ionic cloud, and in turn the core in Figure 23.2 inhabits the ion clouds of its neighbor ions. Throughout this discussion, then, what we are considering is the average nature of a given ion cloud.

23.2.2 $\psi_{ION}(R)$ AND $\psi_{CLOUD}(R)$

Two things contribute to the electrical potential $\psi(r)$: 1) the charge on an ion; and 2) the charge distribution in the surrounding ion cloud. That is,

$$\psi(r) = \psi_{\text{ion}}(r) + \psi_{\text{cloud}}(r). \tag{23.6}$$

The interaction that is causing $\gamma \neq 1$ is that between the ion and its ion cloud.

We first consider what happens when we charge all of the solution ions individually and in isolation from one another. Charging each of those ions involves adding small amounts of either positive or negative charge while building up each $\psi_{\text{ion}}(r)$ until each full charge is reached. For every ion, whether charged positively or negatively, positive work will be required to do this since more and more charge of the same type is being piled onto each core ion, and according to Coulomb's Law, charges of the same type repel one another. For a given ion, none of this work is related to its $RT\ln\gamma_i$ value since all of this work is required for the very existence of the ions in the aqueous medium. This remains true no matter what else is in the solution, or how concentrated the solution is.

Charging up all of the ions in a solution simultaneously builds up all the ion clouds and a $\psi_{\text{cloud}}(r)$ electrical potential distribution around each ion. The overall charging work for a given ion is thereby reduced from that of charging the ion if it were isolated because each ion cloud has a charge that is opposite to that of its core ion. Thus, the sign of $\psi_{\text{cloud}}(r)$ is opposite to that of $\psi_{\text{ion}}(r)$ for all values of r, and it becomes easier to bring in charge from infinity to charge the core ion as $\psi_{\text{cloud}}(r)$ is built up. For each differential portion of added charge, the magnitude of the effect is determined by $\psi_{\text{cloud}}(r)$ evaluated at $r = 0$, that is, by $\psi_{\text{cloud}}(0)$.

To summarize, the work required to charge a core ion is composed of two parts:

1. *Positive work* done in incrementally charging the core ion to build up $\psi_{\text{ion}}(0)$.
2. *Negative work* done in placing the increments of core ion charge at the center of the oppositely charged ion cloud, that is, at $\psi_{\text{cloud}}(0)$.

It is the second component of the work which causes $\gamma \neq 1$. Since that work is negative, then $\gamma < 1$ because Eq.(23.4) tells us that negative work on a chemical species leads to a reduction in its chemical potential.

We need to solve the Poisson Equation to obtain $\psi_{\text{cloud}}(0)$ as a function of the charging. We can then mathematically carry out the charging, and determine the second component of the work as described above. This negative work will be equated with the interaction between the ion and its ion cloud. That is, the work will be equated to $(RT/N_A)\ln\gamma_i$ where we divide by Avogadro's Number N_A to get the work on a *per ion* basis.

In order to solve the Poisson Equation, we need $q(r)$, the local charge density in the cloud. Since $q(r) \neq 0$ in the cloud, the ENE is not satisfied on the very small scale of ions and their clouds. In the aggregate, however, since all ions and their surrounding clouds are of equal and opposite charge, the overall solution satisfies the ENE. The magnitude of $q(r)$ at any given radius r will be determined by the extent to which the local ENE is not satisfied. For the equations and calculations that will ensue, the values of the relevant constants are given in Box 23.1.

23.2.3 The Linearized Poisson–Boltzmann Equation

When a solution satisfies the electroneutrality equation (ENE), then

$$\left(z_{\text{cat},1}n_{\text{cat},1} + z_{\text{cat},2}n_{\text{cat},2} + \ldots\right) + \left(z_{\text{an},1}n_{\text{an},1} + z_{\text{an},2}n_{\text{an},2} + \ldots\right) = 0 \quad (23.7)$$

where $z_{\text{cat},1}$, $z_{\text{cat},2}$, etc. are the nominal charges (with sign) of cations of type 1, 2, etc., and $z_{\text{an},1}$, $z_{\text{an},2}$, etc. are the nominal charges (with sign) of anions of type 1, 2, etc. For example, z_{cat} for Na^+ is +1 and z_{an} for SO_4^{2-} is –2. The n (ions/cm^3) are the corresponding local concentrations. For dilute solutions at 25 °C/atm wherein concentrations in units of molality and molarity are very similar, the bulk solution concentration of i is given by

$$n_i^{\text{o}}(\text{ions/cm}^3) \approx \frac{m_i N_{\text{A}}}{1000\ \text{cm}^3/\text{kg}} \quad (23.8)$$

where N_{A} is Avogadro's number. Species i may be a cation or an anion.

When the two parenthetical sums in Eq.(23.7) combine to equal zero, then, as noted, the ENE is satisfied: the local value of q equals 0, and the solution is electrically neutral. However, inside the ion cloud, Eq.(23.7) is not satisfied: there is an excess of charge with sign opposite to that of the core ion. If the core ion carries a positive charge, then $\left|(z_{\text{an},k}n_{\text{an},k} + z_{\text{an},k}n_{\text{an},k} + \ldots)\right| > (z_{\text{cat},j}n_{\text{cat},j} + z_{\text{cat},j}n_{\text{cat},j} + \ldots)$

BOX 23.1 Values of Physical Constants Relevant for Electrostatic Calculations

$e_{\text{o}} = 1.60 \times 10^{-19}\ \text{C}$

$N_{\text{A}} = 6.02 \times 10^{23}/\text{mol}$

$\varepsilon_{\text{o}} = 8.854 \times 10^{-14}\ \text{C/V-cm}$

$k = 1.381 \times 10^{-23}\ \text{C-V}/K$

$\varepsilon(\text{water})$:

5°C, 85.8; 15°C, 82.0;

20°C, 80.1; 25°C, 78.3;

30°C, 76.5; 37°C, 74.1.

so that each point in the surrounding cloud has a negative charge density. The amount of charge per unit volume at distance r from the center of the core ion is given by

$$q(r) = e_o\left(z_{cat,1}n_{cat,1}(r) + z_{cat,2}n_{cat,2}(r) + \ldots\right) + e_o\left(z_{an,1}n_{an,1}(r) + z_{an,2}n_{an,2}(r) + \ldots\right). \qquad (23.9)$$

$q(r)$ has units of C/cm^3. In Eq.(23.9), each of the two sums is multiplied by e_o, the elementary unit of charge ($+1.60\times 10^{-19}$ C).

To determine the extent to which the ENE is satisfied, we need the concentrations of the various species as a function of r. The average concentration of any ion in the ion cloud around an ion of type i can be determined as a function of r using a form of the Boltzmann Equation:

$$n_j(r) = n_j^o \exp\left(-\frac{U(r)}{kT}\right) \quad \text{Boltzmann Equation} \qquad (23.10)$$

where $U(r)$ = potential energy of an individual ion of type j when located at a distance r from the center of core ion i and n_j^o is the concentration of j in the bulk solution, *i.e.*, far from the center of the core ion. $U(\Delta) = 0$ because the distance from the center of the core ion is so great that the core ion and its cloud have no measurable effects on the concentration: at ∞, $\psi = 0$.

For any given ionic species j of charge z_j, $U(r)$ depends on $\psi(r)$. With z_j being positive for cations and negative for anions,

$$U(r) = z_j e_o \psi(r). \qquad (23.11)$$

Eq.(23.11) is directly related to Eq.(17.7), which states for a redox reaction that $\Delta G = -nFE_{cell}$ wherein n is the number of mols of electrons transferred between the two half cells, and E is the potential difference in volts between the two half cells. For Eq.(17.7), ΔG is the *change* in free energy going from one potential value to another, and not the free energy for some single state, hence the Δ. When $E_{cell} > 0$, the final state has a lower energy than the initial state, hence the negative sign. For Eq.(23.11), $z_j e_o \psi(r)$ represents the energy level of ion j at r, relative to being out in the bulk solution; this equals the amount of work required to bring a charge of $z_j e_o$ from ∞ to distance r.

Depending on the signs of z_j and $\psi(r)$, $z_j e_o \psi(r)$ may be either positive or negative. $z_j e_o \psi(r)$ will be positive when the core ion and ion j have the same sign. In that case, the concentration of ion j will tend on average to be *depleted* in the cloud relative to the bulk solution. Conversely, $z_j e_o \psi(r)$ will be negative when the core ion and ion j have the opposite signs. In that case, the concentration of ion j will tend on average to be *enriched* in the cloud relative to the bulk solution.

Combining Eq.(23.10) and Eq.(23.11) yields

$$n_j(r) = n_j^o \exp\left(\frac{-z_j e_o \psi(r)}{kT}\right). \qquad (23.12)$$

Eq.(23.12) will be obeyed for all of the ions in the solution (cations and anions), including those that are of the type of the core ion (in a solution of NaCl, for a given Na^+ core ion, the ions in the cloud around that Na^+ ion will be H^+, Cl^-, OH^-, as well as other Na^+ ions). Thus,

$$q(r) = \sum_j z_j e_o n_j^o \exp\left(\frac{-z_j e_o \psi(r)}{kT}\right). \qquad (23.13)$$

When the magnitude of the electrical effects as given by $z_j e_o \psi(r)$ is smaller than the thermal effects of the constantly moving water molecules (given by kT), that is, when the absolute value

$$\left|\frac{z_j e_o \psi(r)}{kT}\right| \ll 1 \qquad (23.14)$$

then the arguments of the exponentials in Eq.(23.13) will be small and the exponentials can be linearized. A Taylor series expansion of the function e^y yields

$$e^y = \exp(y) = 1 + y + \frac{y^2}{2!} + \frac{y^3}{3!} + \ldots \tag{23.15}$$

When y is small, only the first two terms of the expansion are required to obtain a good approximation of e^y. In other words, when y is small, $y^2/2!$ and all of the higher order terms are much smaller in magnitude than y. Therefore, for small y, $e^y \approx 1 + y$. (Try it on your calculator for $y = 0.01$.) Under these conditions,

$$\exp\left(\frac{-z_j e_o \psi(r)}{kT}\right) \approx 1 - \frac{z_j e_o \psi(r)}{kT}. \tag{23.16}$$

So, the exponential can be linearized for all r when $\psi(r)$ near the core ion is small.

Substituting Eq.(23.16) into Eq.(23.13) gives

$$q(r) = \sum_j \left(z_j e_o n_j^o \left(1 - \frac{z_j e_o \psi(r)}{kT} \right) \right) \tag{23.17}$$

or

$$q(r) = \sum_j z_j e_o n_j^o - \sum_j \frac{z_j^2 e_o^2 n_j^o \psi(r)}{kT}. \tag{23.18}$$

Since the bulk solution will be electrically neutral, the first term in Eq.(23.18) is zero. Therefore, combining Eq.(23.18) with Eq.(23.5) yields the so-called "Linearized Poisson–Boltzmann Equation" (in radial coordinates)

$$\frac{1}{r^2}\frac{d}{dr}\left(r^2 \frac{d\psi(r)}{dr}\right) = \frac{1}{\varepsilon\varepsilon_o} \sum_j \frac{z_j^2 e_o^2 n_{j,o} \psi(r)}{kT} \qquad \text{Linearized Poisson-Boltzmann Equation} \tag{23.19}$$

$$= \frac{e_o^2 \psi(r)}{\varepsilon\varepsilon_o kT} \sum_j z_j^2 n_{j,o}. \tag{23.20}$$

23.2.4 Integrating the Linearized Poisson–Boltzmann Equation

Since we seek $\psi(r)$ for examination relative to Eq.(23.6), we need to integrate the linearized Poisson–Boltzmann equation. Let

$$\kappa^2 = \frac{e_o^2}{\varepsilon\varepsilon_o kT} \sum_j z_j^2 n_{j,o}. \tag{23.21}$$

The Eq.(23.20) version of the linearized Poisson–Boltzmann equation then becomes

$$\frac{d}{dr}\left(r^2 \frac{d\psi(r)}{dr}\right) = \kappa^2 r^2 \psi(r). \tag{23.22}$$

The parameter κ is most often discussed in terms of its inverse, κ^{-1}, which is called the "Debye length." It provides a measure related to the *thickness of the ion cloud* surrounding an ion.

To solve the differential equation that is Eq.(23.22), we use a change of variable. Let

$$u(r) = r\psi(r), \quad \text{i.e.,} \quad \psi(r) = u(r)/r. \tag{23.23}$$

Since $d(u/v) = (vdu - udv)/v^2$,

$$\frac{d\psi(r)}{dr} = \frac{r\frac{du(r)}{dr} - u(r)}{r^2}. \tag{23.24}$$

Combining Eqs.(23.22) and (23.24) yields

$$\frac{d}{dr}\left(r\frac{du(r)}{dr} - u(r)\right) = \kappa^2 r\, u(r) \tag{23.25}$$

$$\frac{du(r)}{dr} + r\frac{d^2u(r)}{dr^2} - \frac{du(r)}{dr} = \kappa^2 r\, u(r). \tag{23.26}$$

The first and third terms on the LHS of Eq.(23.26) cancel, leaving

$$\frac{d^2u(r)}{dr^2} = \kappa^2 u(r). \tag{23.27}$$

Differential equations like Eq.(23.27) have solutions that are exponential in nature. In particular, $u(r)$ could equal some combination of terms of the forms $Ae^{-\kappa r}$ and $Be^{+\kappa r}$, where A and B are constants to be determined from the boundary conditions of the problem. Therefore, let

$$u(r) = Ae^{-\kappa r} + Be^{+\kappa r}. \tag{23.28}$$

For the first boundary condition, $\psi(r) \to 0$ as $r \to \infty$. When $r = \infty$, since κ is positive, $Ae^{-\kappa r} = 0$ no matter what the value of A so we cannot rule out that $u(r)$ has a contribution from a term of the form $Ae^{-\kappa r}$. However, when $r = \infty$, the value of $Be^{+\kappa r}$, is unbounded for any finite value of B. For example, the value of κ will never be zero since even in pure water the concentrations of H^+ and OH^- are nonzero. This all requires that $B = 0$.

We can now substitute

$$\psi(r) = \frac{u(r)}{r} = \frac{Ae^{-\kappa r}}{r} \tag{23.29}$$

into the basic form of the Poisson Equation

$$\frac{1}{r^2}\frac{d}{dr}\left(r^2\frac{d\psi(r)}{dr}\right) = -\frac{q(r)}{\varepsilon\varepsilon_o}. \tag{23.5}$$

Note that

$$\frac{d\psi(r)}{dr} = -\frac{A}{r^2}e^{-\kappa r} - \frac{A\kappa}{r}e^{-\kappa r} \tag{23.30}$$

$$r^2\frac{d\psi(r)}{dr} = -Ae^{-\kappa r} - rA\kappa e^{-\kappa r}. \tag{23.31}$$

Substituting Eqs.(23.30) and (23.31) into Eq.(23.5),

$$\frac{d}{dr}\left(r^2\frac{d\psi(r)}{dr}\right) = A\kappa e^{-\kappa r} - A\kappa e^{-\kappa r} + rA\kappa^2 e^{-\kappa r} \tag{23.32}$$

and so

$$\frac{1}{r^2}\frac{d}{dr}\left(r^2\frac{d\psi(r)}{dr}\right) = \frac{A\kappa^2}{r}e^{-\kappa r} = -\frac{q(r)}{\varepsilon\varepsilon_o} \tag{23.33}$$

or

$$q(r) = -\frac{A\kappa^2\varepsilon\varepsilon_o}{r}e^{-\kappa r}. \tag{23.34}$$

As discussed in Section 23.2.1, the charge on the ion cloud is equal in magnitude, and opposite in sign to the charge on the ion itself. Thus, if the charge (C) on the ion is $z_i e_o$, then the charge on the cloud is $-z_j e_o$. If $q(r)$ is integrated from $r = 0$ to ∞ (see Figure 23.3), then the integral must equal $-z_i e_o$. We start the integration at $r = 0$ since the core ion itself is assumed (for the moment) to be a *point charge*:

$$4\pi\int_0^\infty r^2 q(r)dr = -z_i e_o. \tag{23.35}$$

By Eq.(23.34), then

$$4\pi A\kappa^2\varepsilon\varepsilon_o\int_0^\infty re^{-\kappa r}dr = z_i e_o. \tag{23.36}$$

With the help of integral tables,

$$\int_0^\infty re^{-\kappa r}dr = \frac{e^{-\kappa r}}{\kappa^2}(\kappa r - 1)\Bigg|_0^\infty = \frac{1}{\kappa^2}. \tag{23.37}$$

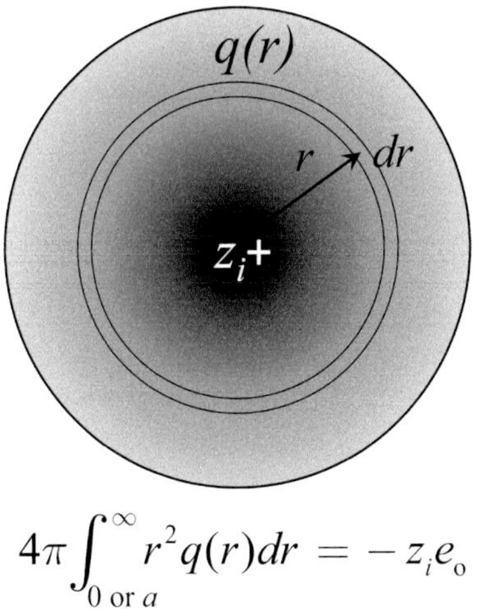

FIGURE 23.3 Integration in spherical coordinates of the charge density $q(r)$ in an ion cloud surrounding a core ion with $z_i = +1$. The charge contained in each successive shell of volume $4\pi r^2 dr$ is equal to $4\pi r^2 q(r)dr$. Integration starting at $r = 0$ leads to the Debye–Hückel Equation. Integration starting at $r = a$ leads to the Extended Debye–Hückel Equation.

Combining Eqs.(23.36) and (23.37), then

$$A = \frac{z_i e_o}{4\pi\varepsilon\varepsilon_o}. \tag{23.38}$$

So, for point charge ions,

$$\boxed{\psi(r) = \left(\frac{z_i e_o}{4\pi\varepsilon\varepsilon_o r}\right)e^{-\kappa r}.} \tag{23.39}$$

The integral in Eq.(23.27) begins at 0 because point charge ions are being assumed. That assumption leads to $\psi(r) = \infty$ at $r = 0$. Obviously the core ion cannot actually be a point charge, so r can never be zero, though the effective radius of the core ion (denoted a) can still be small with respect to κ^{-1} (when ionic strength is relatively low), which is all we need for the Debye–Hückel approach to be valid to a good approximation. This caveat presages the fact when a is not small, then the Debye–Hückel approach will need to be extended to take that fact into consideration; see Section 23.3 and Eq.(23.57). That section carries out the integral from $r = a$ to $r = \infty$; the result approaches $1/\kappa^2$ (as given by Eq.(23.37)) when a is small relative to κ^{-1}.

23.2.5 Determining $\psi_{\text{cloud}}(r)$

Eq (23.39) was developed by considering just the charge density in the cloud as a function of r. Recall now that

$$\psi(r) = \psi_{\text{ion}}(r) + \psi_{\text{cloud}}(r). \tag{23.6}$$

So, the total electrical potential function (including the contribution from the core ion) can be deduced from the charge density in the cloud alone. This is a result of the facts that: 1) the cloud has been assumed to be continuous all the way up to the surface of the ion; and 2) both $\psi_{\text{cloud}}(r)$ *and* $\psi_{\text{ion}}(r)$ affect $q(r)$ in the cloud: evaluating the properties of the cloud reveal information about $\psi_{\text{cloud}}(r)$ as well as about $\psi_{\text{ion}}(r)$.

Near the "surface" of the ion, r will be very small, even if a cannot be zero. Therefore, we can linearize the exponential in Eq.(23.39), with the result that <u>for small r very near the core ion,</u>

$$\psi(r) = \frac{z_i e_o}{4\pi\varepsilon\varepsilon_o r}(1 - \kappa r) \qquad \text{(for small } r\text{)} \tag{23.40}$$

or

$$\psi(r) = \frac{z_i e_o}{4\pi\varepsilon\varepsilon_o r} - \frac{z_i e_o \kappa}{4\pi\varepsilon\varepsilon_o} \qquad \text{(for small } r\text{).} \tag{23.41}$$

From basic electrostatic theory, the first term on the RHS of Eq.(24.41) (which is also the pre-exponential term in Eq.(23.39) is exactly the potential due to a point charge (a.k.a. "the point charge potential") as a function of r for charge $z_i e_o$ in a medium with dielectric constant ε. That is, the first term on the RHS is $\psi_{\text{ion}}(r)$. Therefore,

$$\boxed{\psi_{\text{ion}}(r) = \frac{z_i e_o}{4\pi\varepsilon\varepsilon_o r}, \qquad \psi_{\text{cloud}} = \frac{-z_i e_o \kappa}{4\pi\varepsilon\varepsilon_o} \quad \leftarrow \quad \begin{array}{c}\psi_{\text{cloud}} \text{ at small } r;\\ \text{used to determine } \gamma_i\end{array}.} \tag{23.42}$$

The expression for $\psi_{\text{ion}}(r)$ can be used with varying r, but the expression for ψ_{cloud} refers in particular to the contribution of the cloud to ψ essentially at the center of the cloud where the core ion is located.

Example 23.1 Calculating $\Psi_{\text{ion}}(r)$

Using the values in Box 23.1 and Eq.(23.42), for $z_i = +3$, $+2$, and $+1$, calculate the values of $\psi_{\text{ion}}(r)$ in water at 25 °C when r corresponds to 5 Å.

Solution

The parameter r has units of cm. At 5 Å, then $r = 0.5 \times 10^{-7}$ cm.

z_i	ψ_{ion}
+3	0.1103 V (=110 mV)
+2	0.0735 V (=74 mV)
+1	0.0368 V (=37 mV)

23.2.6 Use of $\psi_{\text{CLOUD}}(r)$ Near the Core Ion to Derive the Debye–Hückel Law

While we know that the parameter z_i is in practice a whole number, there is nothing in the derivation of Eq.(23.42) that requires it to be a whole number. That is good because it will vary during the charging integration for the ion-cloud work. The portion of this change in free energy (*i.e.*, the work done) while building charge on a *single* core ion i that is due to the ψ_{cloud} portion of the potential is given by the integral

$$\Delta G_{\text{ion-cloud}} = \int_0^{z_i e_o} \psi_{\text{cloud}} d(z_i e_o). \tag{23.43}$$

The parentheses around the variable of integration $(z_i e_o)$ are used to set it apart from the final, actual value for the charge on the core ion, which is $z_i e_o$. Only ψ_{cloud} is included in Eq.(23.43) for the charging, since (as discussed in Section 23.2.2) what we are interested in here are the effects that concern γ, and the ψ_{ion} portion of the potential does not affect γ. Substituting Eq.(23.42) into Eq.(23.43) gives

$$\Delta G_{\text{ion-cloud}} = \frac{-\kappa}{4\pi\varepsilon\varepsilon_o} \int_0^{z_i e_o} (z_i e_o) d(z_i e_o). \tag{23.44}$$

Integration yields

$$\Delta G_{\text{ion-cloud}} = \frac{-\kappa}{8\pi\varepsilon\varepsilon_o} z_i^2 e_o^2. \tag{23.45}$$

As expected, regardless of the sign of z_i, the sign of $\Delta G_{\text{ion-cloud}}$ is negative; the ion cloud surrounding each ion lowers the free energy of the ion, or equivalently, it lowers the chemical potential μ_i (free energy per mol of ion i). Recall that for species i

$$\mu_i = \mu_i^o + RT \ln m_i + RT \ln \gamma_i \tag{23.4}$$

where c_i $(\simeq 1000 n_i^o / N_a)$ is the molal concentration of ion i in the *bulk solution*. Since there have been no changes in the concentration of an ion in the bulk solution due to any ion-cloud interactions, if μ_i is to be lowered, this means that the cloud around each such ion must cause $\gamma_i < 1.0$.

Dividing each of the terms on the RHS of Eq.(23.4) by Avogadro's Number (N_A) yields the components of the free energy on a per ion basis. The per-ion free energy due to activity coefficient effects is given by

$$\text{per-ion } G \text{ due to the activity coefficient} = \frac{RT}{N_A} \ln \gamma_i = kT \ln \gamma_i. \tag{23.46}$$

Combining Eqs.(23.45) and (23.46) gives

$$kT \ln \gamma_i = \frac{-\kappa}{8\pi\varepsilon\varepsilon_o} z_i^2 e_o^2 \tag{23.47}$$

$$\ln \gamma_i = \frac{-\kappa}{8\pi\varepsilon\varepsilon_o kT} z_i^2 e_o^2. \tag{23.48}$$

It would be convenient if Eq.(23.48) could be placed into a more usable form. From above,

$$\kappa^2 = \frac{e_o^2}{\varepsilon\varepsilon_o kT} \sum_j n_j^o z_j^2. \tag{23.21}$$

The sum in Eq.(23.21) is directly related to the ionic strength $I(M) = \frac{1}{2}\sum_j c_j z_j^2$ (where we assume equivalency between m and M). The units for all n_j^o are ions/cm³. Thus,

$$I = \frac{1}{2}\sum_j \left(n_j^o \frac{1000 \text{ cm}^3/\text{L}}{N_A} \right) z_j^2 \tag{23.49}$$

$$\sum_j n_j^o z_j^2 = \frac{2N_A}{1000 \text{ cm}^3/\text{L}} I. \tag{23.50}$$

Therefore,

$$\kappa = \left(\frac{2e_o^2 N_A}{\varepsilon\varepsilon_o kT\, 1000 \text{ cm}^3/\text{L}} \right)^{1/2} I^{1/2} \tag{23.51}$$

Using the values of the physical constants in Box 23.1, at $T = 298.15$ K (25 °C),

$$\kappa = 3.29 \times 10^7 I^{1/2} (\text{cm}^{-1}) \tag{23.52}$$

$$\boxed{\kappa^{-1} = 3.04 \times 10^{-8} I^{-1/2} (\text{cm}).} \tag{23.53}$$

And, from Eqs.(23.48) and (23.51),

$$\log_{10} \gamma_i = \frac{-e_o^3 (2N_A)^{1/2}}{2.303\, 8\pi (\varepsilon\varepsilon_o kT)^{3/2} (1000 \text{ cm}^3)^{1/2}} z_i^2 I^{1/2}. \tag{23.54}$$

The dimensions of $I^{1/2}$ are mols$^{1/2}$-L$^{-1/2}$ (assuming equivalency between molality and molarity in water). The expression multiplying $z_i^2 I^{1/2}$ in Eq.(23.54) is usually collected into a single constant $\boldsymbol{A}$ (which is different from the non-boldface A in Eq.(23.28)). Using the values in Box 23.1, at $T = 298.15$ K (25 °C),

$$\boldsymbol{A} = \frac{e_o^3 (2N_A)^{1/2}}{2.303\, 8\pi (\varepsilon\varepsilon_o kT)^{3/2} (1000 \text{ cm}^3)^{1/2}} \tag{23.55}$$

$$\boxed{\log_{10} \gamma_i = -\boldsymbol{A} z_i^2 I^{1/2} \qquad \text{Debye-Hückel Equation.}} \tag{23.56}$$

As required, for $I = 0$, $\gamma_i = 1.0$. At $T = 298.15$ K (25 °C), $A = 0.51$. At temperatures other than 25 °C, neglecting the effects of temperature on the density of water,

$$A = (1.82 \times 10^6)(\varepsilon T)^{-3/2}. \tag{23.57}$$

Example 23.2 Calculating κ, $\Psi_{ion}(r)$, $\Psi_{cloud}(r)$, and $\Psi(r)$

a. The ionic strength I values for the aqueous NaCl solutions at 10^{-4} F, 10^{-3} F, and 10^{-2} F are $I = 10^{-4}$, 10^{-3}, and 0.5×10^{-2} M, respectively. Using the values in Box 23.1 with Eq.(23.53), calculate κ^{-1} at $T = 298.15$ K (25 °C). **c.** For $z_i = +1$, at a distance of 5 Å, obtain the values of ψ_{ion} (see Example 23.1). Assuming 5 Å is small enough for Eqs.(23.40–23.42) to be valid, calculate ψ_{cloud} and the total $\psi = \psi_{ion} + \psi_{cloud}$ for the four values of I. **d.** Comment on the effect of increasing I on the degree of extension of ψ into the solution.

Solution

$z_i = +1$. The parameter r has units of cm. At 5 Å, then $r = 0.5\times10^{-7}$ cm.

I (M)	a. κ^{-1}	b. ψ_{ion} at 5 Å	c. ψ_{cloud} (at small r)†	c. ψ at 5 Å
10^{-4}	3.0×10^{-6} cm (300 Å)	+0.0368 V	−0.0006 V	+0.0361 V
10^{-3}	9.6×10^{-7} cm (96 Å)	+0.0368 V	−0.0019 V	+0.0348 V
10^{-2}	3.0×10^{-7} cm (30 Å)	+0.0368 V	−0.0060 V	+0.0307 V
5×10^{-2}	1.4×10^{-7} cm (14 Å)	+0.0368 V	−0.0135 V	+0.0232 V

† 5 Å is sufficiently small.

d. Increasing I causes the ion cloud to shrink closer around the center of the ion: the volume of surrounding ion cloud is being reduced.

Example 23.3 Ion Depletion and Ion Enrichment in an Ion Cloud

For the NaCl solution with $I = 10^{-3}$, Example 23.2 gives $\kappa^{-1} = 9.6\times10^{-7}$ cm (96 Å). In such a solution, consider an Na^+ core ion so $z_i = +1$. **a.** Use Eq.(23.39) to calculate $\psi(r = \frac{1}{4}\kappa^{-1})$. **b.** Using that value of ψ, use Eq.(23.39) to compute: 1) the factor by which the concentrations of Na^+ and H^+ will be different at $r = \frac{1}{4}\kappa^{-1}$ as compared to the bulk solution; and 2) the factor by which the concentrations of Cl^- and OH^- will be different at $r = \frac{1}{4}\kappa^{-1}$ as compared to the bulk solution.

Solution

a. $\psi(r = \frac{1}{4}\kappa^{-1}) = 0.0060$ V. **b.** For $j = +1$ such as for Na^+ and H^+, the exponential term in Eq.(23.39) equals 0.079: around a +1 ion, at $r = \frac{1}{4}\kappa^{-1}$ there is a 21% depletion of +1 ions relative to the bulk solution. For $j = -1$ such as for Cl^- and OH^-, the exponential term in Eq.(23.39) equals 1.26: around a +1 ion, at $r = \frac{1}{4}\kappa^{-1}$, there is a 26% enrichment of −1 ions relative to the bulk solution.

23.3 DERIVATION OF THE EXTENDED DEBYE–HÜCKEL LAW

The non-boldface pre-exponential factor A in Eq.(23.28) was evaluated using the fact that the charge on the ion cloud is equal in magnitude and opposite in sign to the charge on the core ion. In particular, $q(r)$ was integrated from $r = 0$ to ∞, and the result was equated to $-z_i e_o$. In the derivation of the Extended Debye–Hückel Law, the assumption that the core ion itself is a point charge is not made, and it is acknowledged that the core ion has a finite radius equal to a' (cm) or a (Å). Values of a for various ions of interest as obtained by best fits of data to the Extended Debye–Hückel Law (as obtained below) are collected in Table 2.4. A few examples are compared with the crystal ionic radii for the actual bare ions in Table 23.1: the values for a are clearly much larger than the corresponding actual *bare ion* values (especially for H^+), making clear that for the Extended Debye–Hückel Law the parameter a provides a measure of the distance of closest approach for the *solvated ion*.

For finite a' (cm), Eq.(23.36) becomes

$$4\pi A\kappa^2 \varepsilon\varepsilon_o \int_{a'}^{\infty} re^{-\kappa r}dr = z_i e_o. \tag{23.58}$$

With the help of integral tables,

$$\int_{a'}^{\infty} re^{-\kappa r}dr = \frac{e^{-\kappa r}}{\kappa^2}(\kappa r - 1)\Bigg|_{a'}^{\infty} = \frac{e^{-a'\kappa}(1 + a'\kappa)}{\kappa^2}. \tag{23.59}$$

Combining Eqs.(23.58) and (23.59), we obtain

$$A = \frac{z_i e_o e^{a\kappa}}{4\pi\varepsilon\varepsilon_o(1 + a'\kappa)} \tag{23.60}$$

and so for non-point charge ions,

$$\psi(r) = \frac{z_i e_o e^{a'\kappa}}{4\pi\varepsilon\varepsilon_o(1 + a'\kappa)r} e^{-\kappa r}. \tag{23.61}$$

As also noted above, as a' becomes increasingly small, Eq.(23.59) reduces as expected to Eq.(23.39).

At the surface of the ion, $r = a'$, and so

TABLE 23.1
Selected Ion Size Parameters *a* for the Extended Debye–Hückel Equation (from Kielland 1937) with Corresponding Crystal Ionic Radii

Ion	*a* (Å)	Crystal ionic radii (bare ion) (Å)
Al^{3+}	9	0.68
Ca^{2+}	6	1.14
Cl^-	3	1.67
H^+	9	0.0000085
Mg^{2+}	8	0.86
Na^+	4	1.16

$$\psi(a') = \frac{z_i e_o}{4\pi\varepsilon\varepsilon_o(1 + a'\kappa)a'}. \tag{23.62}$$

Since

$$\psi_{\text{ion}}(a') = \frac{z_i e_o}{4\pi\varepsilon\varepsilon_o a'} \tag{23.63}$$

and since $\psi_{\text{cloud}}(r) = \psi(r) - \psi_{\text{ion}}(r)$, combining Eqs.(23.62) and (23.63) yields

$$\psi_{\text{cloud}}(a') = \frac{z_i e_o}{4\pi\varepsilon\varepsilon_o a'}\left(\frac{1}{(1 + a'\kappa)} - 1\right) = \frac{-\kappa z_i e_o}{4\pi\varepsilon\varepsilon_o(1 + \kappa a')}. \tag{23.64}$$

Substituting Eq.(23.64) into Eq.(23.43) gives

$$\Delta G_{\text{ion-cloud}} = \frac{-\kappa}{4\pi\varepsilon\varepsilon_o(1 + \kappa a')}\int_0^{z_i e_o}(z_i e_o)d(z_i e_o). \tag{23.65}$$

Integrating,

$$\Delta G_{\text{ion-cloud}} = \frac{-\kappa}{8\pi\varepsilon\varepsilon_o(1 + \kappa a')} z_i^2 e_o^2. \tag{23.66}$$

The $(1 + \kappa a')$ term is only difference between Eq.(23.66) and Eq.(23.45). As the parameter a' becomes increasingly small, Eq.(23.66) reduces to Eq.(23.45). The $(1 + \kappa a')$ term propagates through all of the subsequent mathematics. Consequently, the Extended Debye–Hückel analog of Eq.(23.48) is

$$\ln \gamma_i = \frac{-\kappa}{8\pi\varepsilon\varepsilon_o kT(1 + \kappa a')} z_i^2 e_o^2 \tag{23.67}$$

and so we obtain

$$\log_{10}\gamma_i = \frac{-\boldsymbol{A} z_i^2 I^{1/2}}{(1 + \kappa a')} \qquad \text{Extended Debye-Hückel Equation (preliminary form)} \tag{23.68}$$

where $\boldsymbol{A}$ is the same as given in Eq.(23.55). As required, for the limit of $I = 0$, we have $\gamma_i = 1.0$. Eq.(23.68) reduces to the simple Debye–Hückel Law (Eq.(23.55)) for very small a', or very small κ (*i.e.*, large κ^{-1}).

In all portions of the above derivation, the ion size parameter a' has units of cm. If we switch to using a, which has the more relevant scale of angstroms (1 Å = 10^{-8} cm), from Eq.(23.51), $a = a'(10^{-8}\ \text{cm/Å})$, so

$$\kappa a' = \left(\frac{2e_o^2 N_A}{\varepsilon\varepsilon_o kT\, 1000\ \text{cm}^3/\text{L}}\right)^{1/2} I^{1/2}\left(a \times \left(\frac{10^{-8}\text{cm}}{\text{Å}}\right)\right) \tag{23.69}$$

$$= \left(\frac{2e_o^2 N_A}{\varepsilon\varepsilon_o kT\, 1000\ \text{cm}^3/\text{L}}\right)^{1/2} \times \left(\frac{10^{-8}\text{cm}}{\text{Å}}\right) a\, I^{1/2} \tag{23.70}$$

$$\equiv \boldsymbol{B} a I^{1/2} \tag{23.71}$$

So,

$$\log_{10}\gamma_i = \frac{-Az_i^2 I^{1/2}}{(1 + \boldsymbol{B}aI^{1/2})} \quad \text{Extended Debye-Hückel Equation (final form)} \tag{23.72}$$

$$\boldsymbol{B}(\text{Å}^{-1}) = \left(\frac{2e_o^2 N_A}{\varepsilon\varepsilon_o kT\,1000\ \text{cm}^3/\text{L}}\right)^{1/2} \times \left(\frac{10^{-8}\text{cm}}{\text{Å}}\right). \tag{23.73}$$

At 298.15 K (25 °C), $\boldsymbol{B}$ = 0.33. For temperatures other than 298.15 K, neglecting the effects of temperature on the density of water,

$$\boldsymbol{B} = 50.2(\varepsilon T)^{-1/2}(\text{Å}^{-1}). \tag{23.75}$$

23.4 SUMMARY COMMENTS ON THE FOUR ACTIVITY COEFFICIENT EQUATIONS

Debye–Hückel Equation. The four equations used to predict activity coefficients for ionic species in aqueous solutions are summarized in Table 2.3. As noted above, the assumptions used to develop the Debye–Hückel Equation are: 1) the *bulk* dielectric constant ε is applicable even at the very small scale of individual ions; 2) the exponential in Eq.(23.39) can be linearized to obtain Eq.(23.40); and 3) $a = 0$ (ions are point charges). The equation performs well for $I < 0.01$ *M*, and predicts that γ_i for an ion becomes ever smaller towards zero as I increases. However, as I increases, the ions in solution become, on average, closer and closer to one another, and so each individual ion acts less and less like a point charge. This is due to the fact that the Debye length (κ^{-1}) (which provides a measure related to the physical extent of a given ion cloud) decreases as I increases. The volume of the core ion can thus begin to take on a significant fraction of the volume inside the ion cloud, and so the integration of the ionic cloud charge must begin at $r = a$ rather than $r = 0$ (see Eq.(23.58)).

Extended Debye–Hückel Equation. Changing the limits of integration for the ion cloud charge from $r = 0$ to $r = a$ leads to the Extended Debye–Hückel Equation, for which the rate of decrease of γ_i with increasing I is slowed relative to that of the Debye–Hückel Law. The slowing is due to the fact that giving some dimension to the ion reduces the magnitude of $\psi(r)$ very near the ion from what it would be for a point charge. This reduces the interaction between the charge on the ion and the charge in the corresponding ion cloud. The Extended Debye–Hückel Equation performs well for $I < 0.1$, but is inconvenient since each ion of interest has its own ion size parameter a.

Güntelberg Equation. The Güntelberg Equation is an average version of the Extended Debye–Hückel Equation, taking $a \approx 3$ Å for all ions. The Debye–Hückel, the Extended Debye–Hückel, and the Güntelberg Equation all predict that ion activity coefficients tend to forever decrease as I increases. This is not what is observed in real solutions. In fact, the decline in γ_i values weakens beyond what predicted for the Extended Debye–Hückel/Güntelberg Equations, and in fact γ_i values tend to start to move back towards unity at $I \approx 0.3$ (cf. Figure 2.3); at very large I values, $\gamma_i > 1$ are possible. This indicates that some other kind of interaction is becoming operative beyond the simple electrostatic attraction of an ion with its surrounding ion cloud.

Davies Equation. The Davies Equation, with its empirically obtained $-0.3I$ term (which becomes $+0.3I$ after multiplication by the preceding negative sign), provides a means to accomplish a bottoming-out and turn-up in γ_i at large I: ions are starting to become starved for water molecules for their solvation shells, and this starvation causes them become more exposed to one another, and thus more active (see Section 2.8.1). Examples 2.15,

5.8, 5.9, and 11.1 provide some examples of the calculation and use of activity coefficients obtained by the Davies Equation in solving chemical problems. For seawater, a high I matrix, Example 23.4 illustrates the improved (but still imperfect) performance of the Davies Equation, as compared to the Güntelberg Equation.

Example 23.4 The Davies Equation *vs.* the Güntelberg Equation

For seawater, $I \approx 0.7\ M$. In general, amongst different ions with $z_i = \pm 1$ charge in seawater, $\gamma_{\pm 1} \approx 0.7$; amongst different ions with $z_i = \pm 2$, $\gamma_i \approx 0.2$. Compare these values with what is obtained using the Güntelberg Equation and the Davies Equation.

Solution

	Typical Value	Güntelberg	Davies
$\gamma_{\pm 1}$	0.7	0.59	0.75
$\gamma_{\pm 2}$	0.2	0.12	0.32

While the Davies Equation outperforms the Güntelberg Equation for seawater, exact values require empirical equations of the type proposed by Pitzer (see Plummer and Parkhurst, 1990).

REFERENCES

Bockris J O'M, Reddy AKN (1970) *Modern Electrochemistry*, Volume 1, Chapter 3, Plenum Press, New York.
Debye P, Hückel E (1923) Zur Theorie der Elektrolyte. I. Gefrierpunktserniedrigung und verwandte Erscheinungen. *Zeitschrift Physik*, **24**, 185–206.
Kielland J (1937) Individual activity coefficients of ions in aqueous solutions. *Journal of the American Chemical Society*, **59**, 1675–1678.
Moore WJ (1972) *Physical Chemistry*, 4th edn, Prentice-Hall, Englewood Cliffs, NJ.
Plummer LN, Parkhurst DL (1990) Application of the Pitzer equations to the PHREEQE geochemical model. In: Melchior DC, Bassett RL (Eds.), *Chemical Modeling of Aqueous Systems II*, 128–137. American Chemical Society, Washington, D.C.
Robinson RA, Stokes RH (1959) *Electrolyte Solutions*, Butterworths, London.

24 Electrical Double Layers in Aqueous Systems

24.1 INTRODUCTION

When a particle of a mineral solid, a living cell, some organic detrital matter, or even a purely organic phase like mineral oil is equilibrated with liquid water, electrical charges almost always develop at the interface with the aqueous solution. Given the ubiquitous presence of suspended particles and mineral solids, charged interfaces will therefore be found in the natural environment wherever water is found, that is, in aerosol particles/droplets in the atmosphere, in rain, in fresh surface waters of all types, in the oceans, in sediment pore waters of all types, in soil pore waters and in groundwater systems.

There are several reasons why the presence of charges at particle/water interfaces is important. First, for particles suspended in water, the nature of the charged interface and the properties of the surrounding aqueous medium will determine whether or not the particles will tend to coagulate (*i.e.*, agglomerate) into groups that may be removed by sedimentation (density > the water), or flotation (density < the water). Coagulation is of interest because it is a removal mechanism for: 1) natural sediment materials suspended in the waters of lakes, rivers, estuaries, and the oceans; 2) organic and inorganic material in raw drinking water; and 3) organic matter in domestic sewage (about 50% of the organic carbon in the effluent from secondary sewage treatment plants is in particulate form). Removing organic material from raw drinking water prior to disinfection by chlorination is important because of the reactions between organic material and chlorine-containing chemicals that lead to "disinfection by-products" like chloroform, other trihalomethanes, and a host of other halogenated compounds.

Secondly, since the charges on the surfaces of particles affect the fate of suspended particles, they necessarily also affect the fate of any inorganic or organic contaminants that are sorbed to and carried along with those particles. Hydrophobic organic compounds like PCBs, PAHs, and certain pesticides like DDT and dieldrin will be strongly sorbed to particles high in natural organic matter and many metals (*e.g.*, Pb, Cd, etc.) that are of interest from an environmental contamination point of view can be strongly adsorbed to particle surfaces.

Thirdly, since the presence of charges affects ion behavior at a particle/water interface, the very tendency of ions like Pb^{2+} to sorb to an interface will be affected by the nature of the surface charge. Thus, sorption of Pb^{2+} onto a negatively-charged (am)$Fe(OH)_{3(s)}$ surface is stronger than sorption onto a positively-charged (am)$Fe(OH)_{3(s)}$ surface.

24.2 THE ORIGINS OF CHARGE AT SOLID/WATER INTERFACES

24.2.1 Surface Charge Resulting from the Effects of a Potential-Determining Ion (pdi)

24.2.1.1 A Constituent Ion of the Solid Is the pdi

When solids are equilibrated with initially pure water, charges at the solid/water interface will usually arise because the species composing the solid exhibit different tendencies to dissolve into the aqueous phase. For a solid $AB_{(s)}$ composed of A^+ and B^-, assume that A^+ has a greater affinity for the aqueous phase than does B^-. If $AB_{(s)}$ is equilibrated with initially pure water, then the surface of the solid will become negatively charged, since more A^+ will enter the aqueous phase than will B^-. Net charge movement from the solid to the solution leads to an electrical potential at the surface

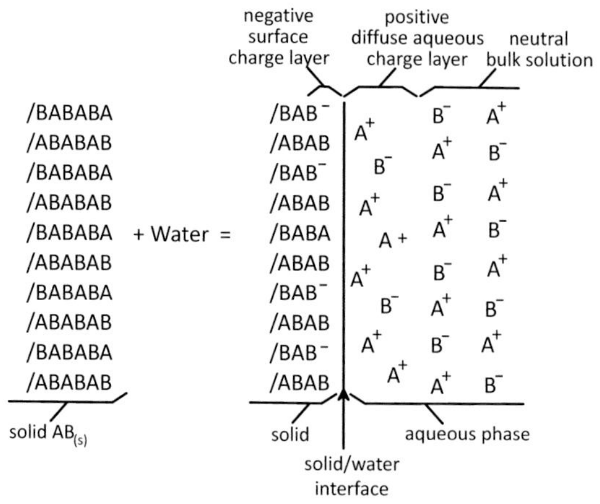

FIGURE 24.1 Dissolution of $AB_{(s)}$ into initially pure water (cross-sectional view). As with $AgI_{(s)}$, it is assumed that A^+ has a greater inherent affinity for the aqueous phase than does B^-: more A^+ than B^- enters the water. Near the surface, $[A^+] > [B^-]$. Out in the bulk solution, $[A^+] = [B^-]$. The aqueous layer near the surface is positively charged; the surface is negatively charged. The behaviors of H^+ and OH^- are not addressed.

and in the nearby water (Figure 24.1). The negative charge at the interface attracts the dissolved A^+ and repels the dissolved B^-. With H^+ and OH^- from the dissociation of water also in the solution, the negative surface charge also attracts the H^+ and repels the OH^-. The net positive charge in the solution is distributed diffusely in the solution, mostly near the surface. Because there are two layers of charge near the interface, one in the solid and one in the water, the two layers are referred to as an "electrical double layer". Like the ion/(ion cloud) system considered in Chapter 23, the integrated diffuse aqueous layer charge (coulombs/area, C/area) is equal in magnitude and opposite in sign to the surface layer charge (C/area).

In the bulk solution, that is, far from the surface, the charge effects of the surface are no longer felt. Thus, when $AB_{(s)}$ dissolves into initially pure water, in the bulk solution, $[A^+] = [B^-]$. Although we do not have $[A^+] = [B^-]$ in the aqueous charge layer near the surface, that charge layer is very thin. Thus, $[A^+] = [B^-]$ as taken over the whole aqueous volume will be sufficiently true for solving for the properties of the aqueous solution, the only exception being very high particle number density systems where the particles are very close to one another and there is no reaching a true bulk solution.

$AgCl_{(s)}$, $AgBr_{(s)}$, and $AgI_{(s)}$ are examples of solids that exhibit the type of behavior depicted in Figure 24.1. In addition to the case of dissolution into initially pure water, double layers will also be present when these solids are equilibrated with solutions containing initial amounts of Ag^+ and/or Cl^-, Ag^+ and/or Br^-, and Ag^+ and/or I^-, respectively. For the general solid $AB_{(s)}$, when $\{A^+\}$ is high and $\{B^-\}$ is low, A^+ can be driven onto the surface (onto B^- surface sites), and a positive surface charge will result. This will remain true regardless of which of the two ions (A^+ or B^-) is inherently more attracted to the aqueous phase. The magnitude of the positive charge density on the surface, as well as the magnitude of the electrical potential at the surface, will be determined by the solution activity $\{A^+\}$ in the bulk solution in conjunction with the relative leaving tendencies of A^+ and B^-. For the solid $AB_{(s)}$, A^+ is referred to as a potential-determining ion (pdi).

A pdi controls both the magnitude and the sign of the surface charge. If the pdi carries positive charge, then increasing the activity of the pdi will make the surface charge more positive (or at least, less negative). If the pdi carries negative charge, then increasing the activity of the pdi will make the surface charge more negative (or at least, less positive). When $\{B^-\}$ is high and $\{A^+\}$ is low, B^- will be driven onto the surface (onto A^+ surface sites), and negative charge will develop there. The degree of the negative charge density as well as the electrical potential at the surface will be determined by the equilibrium value of $\{B^-\}$ in the bulk solution in conjunction with the relative leaving tendencies of A^+ and B^-. Thus, either A^+ and B^- can be viewed as the pdi for an $AB_{(s)}$ surface. Since the bulk solution values of $\{A^+\}$ and $\{B^-\}$ are related through the K_{s0} of the solid, one can parameterize the sign of the surface charge as well as the sign and magnitude of the electrical potential at the surface using the activity of either species.

24.2.1.2 H^+ and OH^- as Potential-Determining Ions

Besides solids of the $AB_{(s)}$ type, solids that can be hydrated, and/or protonated, or deprotonated, can also develop surface charges. For example, with solid silica ($SiO_{2(s)}$), a surface –SiO unit can be protonated directly to form an $-SiOH^+$ group. And a surface –SiO unit can be hydrated by H_2O to form an $-Si<^{OH}_{OH}$ surface group. (The –Si denotes connection of the Si atom to the interior of the solid.) The group $-Si<^{OH}_{OH}$ is acidic, and so can yield an $-Si<^{O^-}_{OH}$ group and a negative contribution to the surface charge. Formation of an $-Si<^{O^-}_{OH}$ group may alternatively be viewed as the direct adsorption of OH^- to a surface –SiO unit. Overall, either H^+ or OH^- (linked by the K_w rather than a K_{s0}) may be considered to be the pdi for the hydrated $SiO_{2(s)}$ surface. It is the pdi role that H^+ plays in creating a double layer on the hydrated $SiO_{2(s)}$ surface of a glass electrode that is used by a pH meter to measure pH (*e.g.*, see Morrison, 1990). Most environmentally-important oxides, hydroxides, and oxyhydroxides (*e.g.*, $Fe_2O_{3(s)}$, $MnO_{2(s)}$, $Fe(OH)_{3(s)}$, $FeOOH_{(s)}$, etc.) follow the pattern of behavior exhibited by the oxide $SiO_{2(s)}$. For obvious reasons, H^+ is usually used as the pdi in double-layer computations involving oxides, hydroxides, and oxyhydroxides. Another class of surfaces for which H^+ is the pdi includes organic biomembranes and organic detrital matter. With these, $\{H^+\}$ controls the magnitude of the negative surface charge by regulating the fraction of surface carboxyl acid groups that are ionized, as with carboxyl, organophosphate, phenol, and protonated amine groups.

24.2.2 Particles with Fixed Surface Charge – Clays

For true clays, the surface charge is not controlled by a pdi. "True clay" refers to small particles that have a crystal structure involving two-dimensional arrays of silicon-oxygen tetrahedra and two-dimensional arrays of aluminum- or magnesium-oxygen-hydroxyl octahedra (van Olphen 1977). These sheets can be layered one on top of one another in different ways to achieve different overall classes of clay crystal structures. For clarity, note that in soil science, "clay" is also used less precisely to refer to all mineral particles with diameters less than about 2 μm. This size fraction is often dominated by true clay particles, but may also include significant amounts of small particles of quartz and other non-clay minerals.

In many true clay minerals, an atom of lower oxidation state replaces an atom of a higher oxidation state in the crystal lattice. On average, the Si^{4+} in a tetrahedral sheet has its charge balanced by two O^{2-} (as is also true for $SiO_{2(s)}$.) Because of the size similarity between Al^{3+} and Si^{4+}, an Al^{3+} can replace an Si^{4+}. The substituting Al^{3+} ion completely fills the lattice site of the Si^{4+} ion. This is "isomorphous substitution". The lower charge of the Al^{3+} relative to the Si^{4+} causes a deficit of positive charge, that is, a negative charge on the lattice. Negative lattice charge can also result from isomorphous substitution of Mg^{2+}, Fe^{2+}, or Cr^{3+} in the octahedral sheet. Given the many different ways in which the sheets can be layered and in which isomorphous substitution can occur, clay chemistry/mineralogy is a complicated and richly diverse field.

When isomorphous substitution leads to net negative charge on the lattice, overall charge balance on a dry clay is achieved through the presence of other, usually +1 cations, adsorbed on the surfaces of the layers and along the edges. These ions might be Na^+, K^+, NH_4^+, or even an organic cation. When placed in water, some clays like the montmorillonites take in significant water and swell (expand) sufficiently that virtually all of the interlayer +1 cations are free to exchange with cations in the aqueous solution. This ability to exchange cations is quantified in terms of a **cation exchange capacity** (CEC) for the clay. The CEC is usually expressed in units of milliequivalents of positive charge per 100 g of air-dried clay. Because they can swell, the CEC for some montmorillonite-type clays can be as high as 100 meq/100 g. (Here, analogously with the derivation of the expression for Alk, meq refers to meq of charge: 100 meq of Na^+ charge corresponds to 0.1 mols of Na^+.)

Important non-swelling clays include the illites, kaolinites, and the chlorites. In non-swelling clays, the layers are held together tightly, and water cannot penetrate and expand the layers. Thus, the interlayer cations balancing the negative-lattice charge caused by isomorphous substitution are unable to exchange with cations in the aqueous solution. Only those balancing cations on the exteriors and edges of the layers will contribute to the CEC values of such clays. The CEC ranges for illites, kaolinites, and chlorites are given in Table 24.1. For some kaolinites, the CEC can be as low as 1 meq/100 g. Since soils are mixtures of different types of materials, the CEC of a soil is a mass-weighted average.

When a true clay particle is equilibrated with liquid water, some of the exchangeable charge will leave the particle and enter the solution. As seen in Figure 24.2, all of the charge that departs from a given clay particle will leave the particle with a corresponding amount of negative charge. The magnitude of the charge will vary from particle to particle in a given sample of clay, and will depend on

TABLE 24.1
Cation Exchange Capacities of Different Clay and Soil Types

True Clay Type	CEC (meq/100 g)
Montmorillonites	60 to 100
Illites	10 to 40
Kaolinites	1 to 10
Chlorites	<10
Soil Types[a]	
Sands	1 to 5
Fine sandy loams	5 to 10
Loams and silt loams	5 to 15
"Clay" loams	15 to 30
"Clays"	1 to 150

[a] Each soil class is a mixture of sands, clays, and organic humic material.

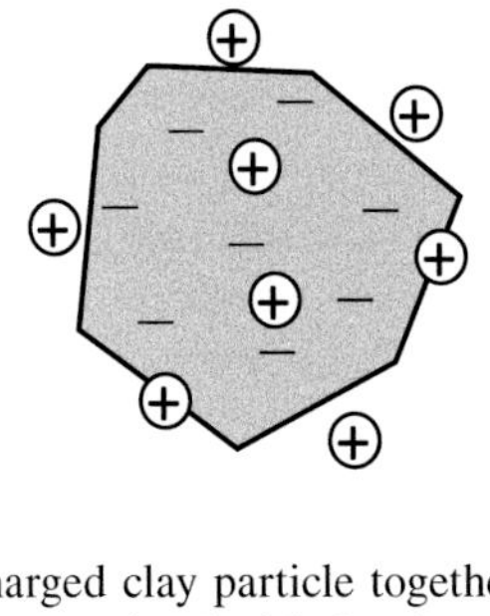

FIGURE 24.2 View of a negatively charged clay particle together with its surrounding, positive aqueous charge layer. Some of the negative charge associated with the clay particle may reside within the clay lattice, and some negative charge may be associated with lattice sites on the surface.

the particle size. The negative charge may be viewed as being distributed over the surface, with an average charge density (C/cm^2). The amount of charge that is sufficiently mobile that it can leave the lattice is usually fixed within the context of time scales of many years or longer.

24.3 SIMILARITIES BETWEEN DOUBLE-LAYERS AND ION/(ION CLOUD) SYSTEMS

The case of the aqueous double layer is a close analog of the ion-ion cloud system of Chapter 23 (though an aqueous double layer does not have the spherical symmetry of the ion-ion cloud system). The charge that develops on the surface of the solid is analogous to the charge on a core ion. The diffuse layer of counterbalancing charge in the aqueous solution near a charged solid surface (Figure 24.3) is analogous to the diffuse sphere of charge in the ion cloud surrounding a core ion. And, just as the charge on a core ion is exactly balanced by the charge in its surrounding spherical ion cloud, so too is the charge density on the solid surface exactly balanced by the charge density in its adjacent planar layer of aqueous charge. In the case of clay particles, the magnitude of the surface charge density (in C/cm^2) is fixed as with the charge (C) on a core ion. When the surface charge is controlled by a pdi, the magnitude and sign will change in response to changes in the solution activity of the pdi.

24.4 THE ELECTROCHEMICAL POTENTIAL

24.4.1 THE ELECTROCHEMICAL POTENTIAL AND EQUILIBRIUM IN THE PRESENCE OF AN ELECTRICAL FIELD

Some aspects of the theoretical discussion provided here will follow the approach of de Bruyn and Agar (1962), though a different specific example will be used to illustrate the basic principles under consideration. Consider now the Figure 24.1 case of an $AB_{(s)}$ surface in equilibrium with initially-pure water. We again assume that A^+ has a greater inherent affinity for the aqueous phase than does B^-, and so at equilibrium the surface will be negatively charged, and the aqueous charge layer will be positively charged. If the movement of A^+ between the solid and the solution were influenced solely by chemical forces, then the dissolution of A^+ would continue until μ_{A^+} was equal in both phases. However, the electrical potential that is present near a charged surface introduces electrical forces. These forces stop the dissolution of A^+ (*i.e.*, and equilibrium is reached) before μ_{A^+} is equal in both phases.

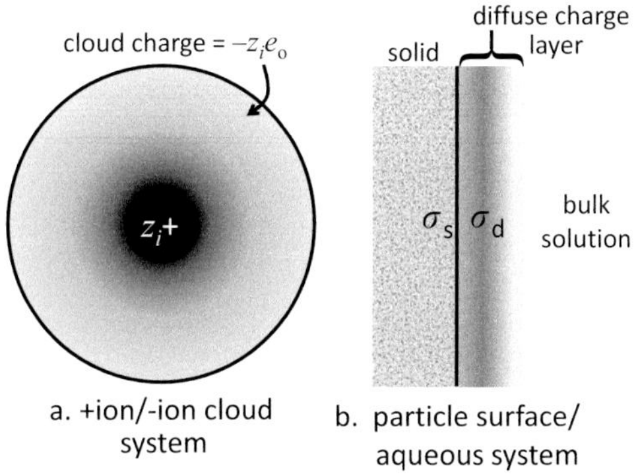

FIGURE 24.3 Comparison of two systems involving charge adjacent to a diffuse aqueous layer of balancing, opposite charge. **a.** A positive ion surrounded by its diffuse, negative ion cloud. **b.** The surface of a particle with surface charge density σ_s (C/cm^2) and its adjacent diffuse, aqueous charge layer with opposite charge density σ_d (C/cm^2); $\sigma_s = -\sigma_d$.

Equilibrium will be attained when an exact counterbalancing is achieved between: 1) the residual chemical force tending to drive A^+ from the solid into the bulk solution, and 2) the attractive electrical force involving negative surface charge that is tending to hold A^+ in the solid phase. At equilibrium

$$\mu^{\beta}_{A+}(0) - \mu^{\alpha}_{A+}(\infty) = z_+ e_o N_A(\psi^{\alpha}(\infty) - \psi^{\beta}(0)) \tag{24.1}$$

$$\text{Chemical Imbalance} = \text{Electrical Imbalance}$$

where:

$\mu^{\beta}_{A+}(0)$ = chemical potential of A^+ at the surface ($x = 0$) in the solid (β) phase
$\mu^{\alpha}_{A+}(\infty)$ = chemical potential of A^+ in the bulk ($x = \infty$) aqueous (α) phase
z_+ = nominal charge on a positive potential-determining ion, +1 for A^+
e_o = elementary charge = 1.60×10^{-19} C = F/N_A
F = Faraday Constant = 96,485 C/mol
N_A = Avogadro's Number = 6.02×10^{23} mol^{-1}
$\psi^{\alpha}(\infty)$ = electrical potential (V) in the bulk ($x = \infty$) aqueous (α) phase
$\psi^{\beta}(0)$ = electrical potential (V) at the surface ($x = 0$) in the solid (β) phase

The terms in Eq.(24.1) have units of energy per mol. The values of some constants relevant for this chapter are collected in Box 24.1. Eq.(24.1) is closely related to the basic thermodynamic equation $\Delta G = -nFE$ which gives the change in free energy when n mols of elementary electrical charge

BOX 24.1 Values of Physical Constants Relevant for Electrostatic Calculations

e_o = elementary charge = 1.60×10^{-19} C=F/N_A

F = Faraday Constant = 96,485 C/mol

N_A = Avogadro's Number = 6.02×10^{23} mol^{-1}

$\varepsilon_o = 8.854 \times 10^{-14}$ C/V-cm

$k = 1.381 \times 10^{-23}$ C-V/K

R = 0.0083145 kJ/mol-K

= 8.3145 V-C/mol-K

$F/RT = 38.92$ V^{-1} at 298.15 K

$\varepsilon_o = 8.85 \times 10^{-14}$ C-V/cm

ε(water):

5 °C, 85.8; 15°C, 82.0;

20 °C, 80.1; 25°C, 78.3;

30 °C, 76.5; 37°C, 74.1.

$2\varepsilon\varepsilon_o RT/F = 3.56 \times 10^{-13}$ C/cm at 298.15 K

move across a potential difference of E volts. Here, the analog of $(\mu^{\beta}_{A+}(0) - \mu^{\alpha}_{A+}(\infty))$ (kJ/mol) is ΔG (kJ), the analog of z_+ (mols of charge/mol of A^+) is n (mols of charge), $e_o N_A$ = Faraday constant F, and $(\psi^{\alpha}(\infty) - \psi^{\beta}(0)) = E$. From Chapter 17, E can be positive or negative.

For the case in Figure 24.1, both sides of Eq.(24.1) are positive. However, Eq.(24.1) is equally valid when a high $\{A^+\}$ in the solution drives positive charge onto the solid surface. In that case, the phase transfer of positive charge again stops before the chemical potential of A^+ can become equal in both phases. Now, both sides of Eq.(24.1) are negative, and the equilibrium position that is attained is the one that is established when an exact counterbalancing is achieved between: 1) the residual chemical force tending to drive A^+ onto the surface; and 2) the repulsive electrical force between the positive surface charge and the positive A^+ ions.

Eq.(24.1) gives

$$\mu^{\beta}_{A+}(0) + z_+ F\psi^{\beta}(0) = \mu^{\alpha}_{A+}(\infty) + z_+ F\psi^{\alpha}(\infty). \tag{24.2}$$

The terms on the LHS of Eq.(24.2) pertain to the solid (β) phase at the surface, and the terms on the RHS pertain to aqueous (α) phase in the bulk solution. Thus, in the presence of electrical forces $\mu_i(x) + z_i F\psi(x) = \bar{\mu}_i$ is the pertinent, equilibrium-controlling potential for species i. This electrochemical potential contains both electrical and chemical components. The criterion for equilibrium across a charged interface is then

$$\bar{\mu}_i^{\beta} = \bar{\mu}_i^{\alpha}. \tag{24.3}$$

Equality of Electrochemical Potentials

When ψ in both phases is either zero or equal, the criterion for equilibrium reduces to $\mu_i^{\alpha} = \mu_i^{\beta}$.

With a negatively charged surface, the electrical forces attract positive ions towards the surface; their concentration is higher near the surface than in the bulk solution (Figure 24.1). So the positive ions tend to diffuse away from the surface, but at the same time are pulled back towards the surface by the electrical forces. The dynamic equilibrium that is established for each distance x from the surface will satisfy Eq.(24.3). In the case of a negatively charged surface and negative ions, their concentration is lower near the surface than in the bulk solution. Diffusion of the negative ions towards the surface will at the same time be counterbalanced by the repulsive electrical forces associated with moving towards the surface. Again, the dynamic equilibrium that is established for each distance x from the surface will satisfy Eq.(24.3).

24.4.2 The Vertical Distribution of Gases in the Atmosphere and Its Relation to the Distant-Dependent Concentration of Aqueous Ions in a Double Layer

The vertical distribution of gas molecules ion the atmosphere provides an example that relates to how a charged surface creates a concentration gradient of an oppositely charged ion in solution. Gas molecules in the atmosphere of the Earth are not uniformly distributed with height. They are most concentrated at the surface of the earth where the total pressure is about 1 atm. As one moves upwards, the pressure decreases exponentially upward according to

$$p(z) = p^{\mathrm{o}} \exp[-M_a z g/(RT)] \tag{24.4}$$

where z is the elevation at sea level, p° is the pressure at sea level, M_a is the average molecular weight of air, and g is the gravitation constant (*i.e.*, the strength of the gravitational field of the Earth).

Gravitation attracts gases towards the surface, and thereby creates a gradient in the concentration. This concentration gradient creates a gradient in the chemical potential: as in the case of an aqueous double layer, equality of the chemical potential is not the criterion for chemical equilibrium. Rather, the criterion for equilibrium may be thought of as being a gravochemical potential. Gases tending to diffuse upward by virtue of the nature of the concentration gradient are at the same time pulled downwards by gravitation. The dynamic equilibrium that is reached is represented by Eq.(24.4).

24.5 DOUBLE-LAYER PROPERTIES AS A FUNCTION OF THE ACTIVITY OF A POTENTIAL-DETERMINING ION

24.5.1 ψ° AS A FUNCTION OF THE ACTIVITY OF A POSITIVE POTENTIAL-DETERMINING ION

Often, when surface charges are present on particles suspended in aqueous systems, the particles in question may be viewed more or less as spheres or long cylinders. When bodies of these geometries bear charges that are uniformly distributed over the surface, electrostatic theory states that the electrical potential ψ inside such a body is constant and equal to the value at the surface. (The same is true for spherical and cylindrical clay particles, if unbalanced charge sites are in the interior are uniformly distributed in the interior of the particle.) Thus, for $AB_{(s)}$ throughout the solid

$$\psi^{\beta} = \psi^{\beta}(0). \tag{24.5}$$

Since the composition of the solid changes very little as a result of the creation of a small amount of charge (*i.e.*, excess A^+ or excess B^-) on the surface, $\mu^{\beta}_{A^+}(0)$ essentially equals the chemical potential of A^+ in pure $AB_{(s)}$, denoted as $\mu_{A^+}^{\beta,\mathrm{pureAB(s)}}$. By substituting the usual equation $\mu_i = \mu_i^{\circ} + RT\ln a_i$ for the chemical potential of A^+ in the bulk aqueous (α) phase in Eq.(24.1), and by using 24.5,

$$\mu_{A+}^{\beta,\mathrm{pureAB(s)}} - (\mu_{A+}^{o,\alpha} + RT\ln a_{A+}^{\alpha}) = z_+F(\psi^{\alpha}(\infty) - \psi^{\beta} \tag{24.6}$$

where a_{A+}^{α} is the activity of A^+ in the bulk aqueous solution.

Equation (24.6) may be rearranged to obtain

$$(\psi^{\beta} - \psi^{\alpha}(\infty)) = \frac{\mu_{A+}^{o,\alpha} - \mu_{A+}^{\beta,\mathrm{pureAB(s)}}}{z_+F} + \frac{RT}{z_+F}\ln a_{A+}^{\alpha}. \tag{24.7}$$

The only activities of interest henceforth are activities in the aqueous phase, so the α superscript in a_{A+}^{α} is no longer needed. The chemical potentials $\mu_{A^+}^{o,\alpha}$ and $\mu_{A^+}^{\beta,\mathrm{pureAB(s)}}$ are constants. Taking the derivative of Eq.(24.7) yields

$$d(\psi^{\beta} - \psi^{\alpha}(\infty)) = \frac{RT}{z_+F}d\ln a_{A+} = \frac{2.303RT}{z_+F}d\log a_{A+}. \tag{24.8}$$

The term $(\psi^{\beta} - \psi^{\alpha}(\infty))$ is abbreviated as $\Delta\psi$. Eq.(24.8) then becomes

$$d(\Delta\psi) = \frac{2.303RT}{z_+F}d\log a_{A+}. \tag{24.9}$$

$2.303RT/F = 0.05916$ V = 59.16 mV.

Equation (24.9) states that the difference in electrical potential between the solid and the aqueous phases is determined by $\log a_{A+}$. Thus, A^+ is a pdi for $AB_{(s)}$. (Note again, clays do not have a pdi.)

Re-integrating Eq.(24.9),

$$\Delta\psi = \frac{2.303RT}{z_+F}\log a_{A+} + C \tag{24.10}$$

where C is a constant of integration. The value of C depends on the properties of water and the solid.

As discussed above, when a_{A+} is very low, then the surface is negatively charged: $\Delta\psi < 0$. When a_{A+} is high, the surface is positively charged: $\Delta\psi > 0$. There must then be a value of a_{A+} for which there is *no net charge* on the surface. This is the **zero point of charge** (zpc) activity of the pdi A^+, denoted a^o_{A+}.

The value of a^o_{A+} is a constant for a given solid, so without any loss of generality let

$$C = \chi - \frac{2.303RT}{z_+F}\log a^o_{A+} \tag{24.11}$$

where χ is another constant. Substituting Eq.(24.11) into Eq.(24.10) gives

$$\Delta\psi = \frac{2.303RT}{z_+F}\log\frac{a_{A+}}{a^o_{A+}} + \chi. \tag{24.12}$$

$\Delta\psi$ is plotted as a function of $\log a_{A+}$ for a generic case in Figure 24.4. When $a_{A+} = a^o_{A+}$, the pdi is at its zpc value, and there is no net charge on the surface. However, $\Delta\psi$ is not zero at that point since χ is not in general zero. As de Bruyn and Agar (1962) point out, usually some polarity in the surface causes a net alignment at the solid/water interface of the oxygen in the water molecules either towards or away from the surface. Since water is a dipole, this organized net charge orientation leads to some amount of potential difference, which equals χ. For our purposes, we do not need to know the value of χ.

The first term in Eq.(24.12) is denoted ψ^o,

$$\Delta\psi = \psi^o + \chi \tag{24.13}$$

$$\psi^o = \frac{2.303RT}{z_+F}\log\frac{a_{A+}}{a^o_{A+}}. \tag{24.14}$$

A plot of ψ^o *vs.* $\log a_{Ag+}$ is provided for $AgI_{(s)}$ in Figure 24.5. How ψ varies as a function of distance into the solution is shown generically in Figure 24.6 for a case when $\Delta\psi > 0$. The manners in which $\bar{\mu}$ and its component terms vary as a function of distance are depicted in Figure 24.7 for a case when $\mu^{\beta,\text{pureAB(s)}}_{A+} < \mu^{\alpha}_{A+}$.

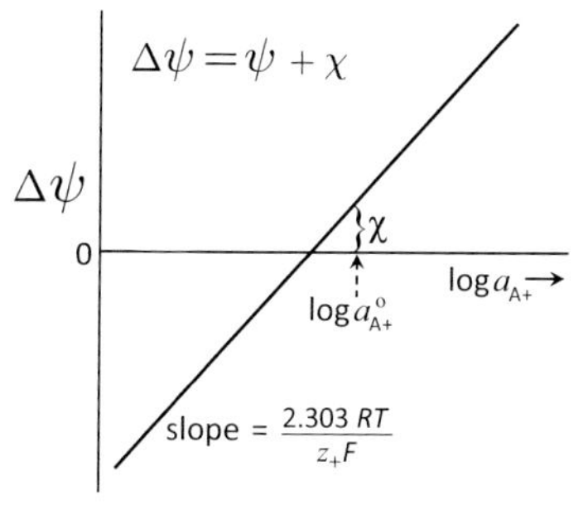

FIGURE 24.4 $\Delta\psi$ *vs.* $\log a_{A+}$ for a case when χ is positive. The slope is positive because the potential determining ion is positive. $\Delta\psi \neq 0$ when $\log a_{A+} = \log a^o_{A+}$ because $\chi \neq 0$.

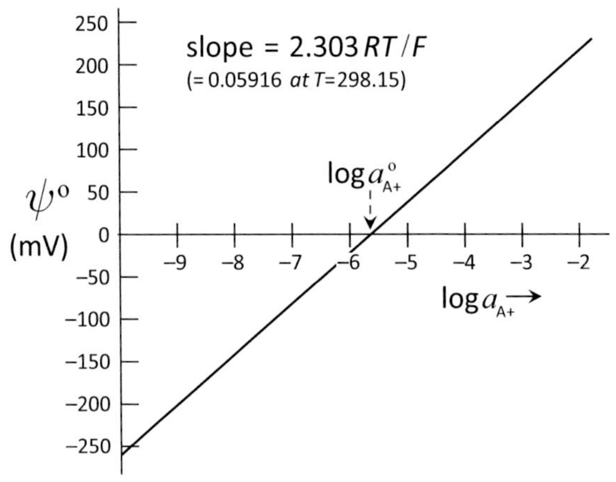

FIGURE 24.5 ψ^{o} *vs.* $\log a_{Ag+}$ for $AgI_{(s)}$ at 25 °C/1 atm. $K_{s0} = 10^{-16.1}$; $a^{o}_{Ag+} = 10^{-5.6}$ and $a^{o}_{I-} = 10^{-10.5}$. $z+ = 1$. When $AgI_{(s)}$ is equilibrated with initially pure water, the AgI surface will be strongly negative because $\log a_{Ag+} = \log a_{I-} = 10^{-8.05}$ in the bulk solution in that system.

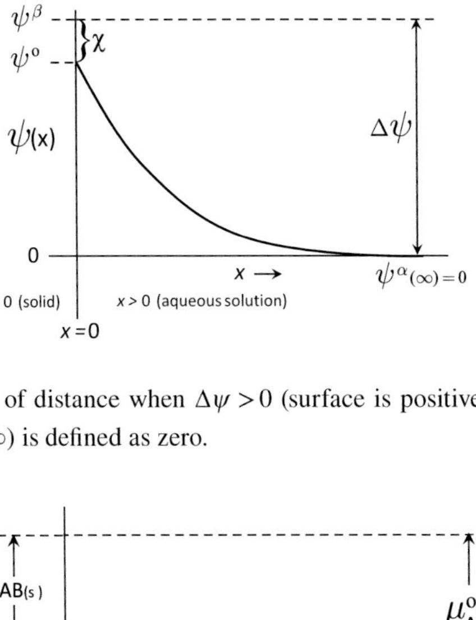

FIGURE 24.6 ψ as a function of distance when $\Delta\psi > 0$ (surface is positively charged), and $\chi > 0$. At the solid/water interface, $x = 0$. $\psi^{\alpha}(\infty)$ is defined as zero.

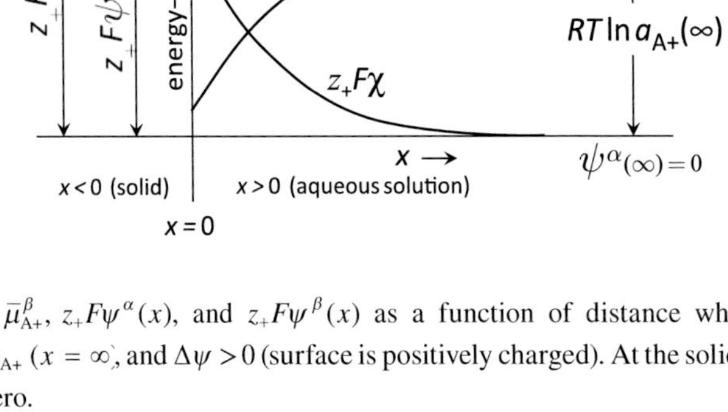

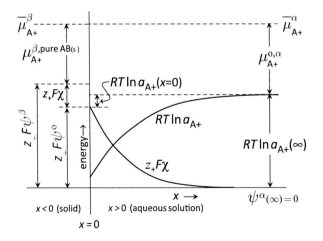

FIGURE 24.7 $\bar{\mu}^{\alpha}_{A+}$, $\bar{\mu}^{\beta}_{A+}$, $z_{+}F\psi^{\alpha}(x)$, and $z_{+}F\psi^{\beta}(x)$ as a function of distance when $\mu^{\beta}_{A+} = \mu^{\beta,\text{pureAB(s)}}_{A+} < \mu^{\alpha}_{A+} = \mu^{o,\alpha}_{A+} + RT \ln a_{A+}$ ($x = \infty$), and $\Delta\psi > 0$ (surface is positively charged). At the solid/water interface, $x = 0$. $\psi^{\alpha}(\infty)$ is defined as zero.

24.5.2 ψ° as a Function of the Activity of the a Negative Potential-Determining Ion

The activity of A^+ was used to derive Eq.(24.14). This could have been done based on the activity of B^-. In particular, when there is equilibrium between the solid and the bulk aqueous solution, the K_{s0} for the solid is satisfied in the bulk solution:

$$a_{A^+}a_{B^-} = K_{s0}. \quad (24.15)$$

Since there can be only one zpc condition, then when $a_{A^+} = a^{o}_{A^+}$, then $a_{B^-} = a^{o}_{B^-}$. As a special case of Eq.(24.15), then

$$a^{o}_{A^+}a^{o}_{B^-} = K_{s0}. \quad (24.16)$$

Combining Eqs.(24.14)–(24.16),

$$\psi^{o} = \frac{2.303RT}{z_+F}\log\frac{K_{s0}/a_{B^-}}{K_{s0}/a^{o}_{B^-}} \quad (24.17)$$

$$= \frac{2.303RT}{z_+F}\log\frac{a^{o}_{B^-}}{a_{B^-}}. \quad (24.18)$$

Since $z_+ = +1 = -z_- = -(-1)$, substituting for z_+ in terms of z_- in Eq.(24.18) yields

$$\psi^{o} = \frac{2.303RT}{z_-F}\log\frac{a_{B^-}}{a^{o}_{B^-}} \quad (24.19)$$

which is analogous to Eq.(24.14).

Figure 24.8 illustrates how taking the pdi to be A^+ is equivalent to taking the pdi to be B^-. In general, for any potential determining ion i of charge z_i (positive or negative),

$$\psi^{o} = \frac{2.303RT}{z_iF}\log\frac{a_i}{a^{o}_i}. \quad (24.20)$$

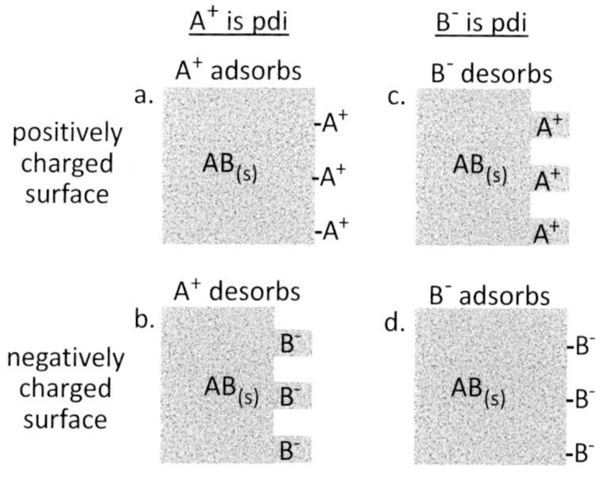

FIGURE 24.8 For the $AB_{(s)}$/water interface, one can view either A^+ or B^- to be the pdi. When A^+ is taken as the pdi, a positively charged surface is the result of adsorption of A^+ on the $AB_{(s)}$ surface (**a**), and negatively charged surface is the result of desorption of A^+ from the $AB_{(s)}$ surface (**b**). When B^- is taken as the pdi, a positively charged surface is the result of desorption of B^- from the $AB_{(s)}$ surface (**c**), and a negatively charged surface is the result of the adsorption of B^- on the $AB_{(s)}$ surface (**d**).

TABLE 24.2
Zero Point of Charge (zpc) Concentrations for Three Silver Halides at 25 °C/l atm

Silver Halide $AgX_{(s)}$	K_{s0}	$a^o_{Ag^+}$	$a^o_{X^-}$	$a^o_{Ag^+}/a^o_{X^-}$
$AgCl_{(s)}$	$10^{-9.7}$	$10^{-4.0}$	$10^{-5.7}$	50
$AgBr_{(s)}$	$10^{-12.3}$	$10^{-5.4}$	$10^{-6.9}$	32
$AgI_{(s)}$	$10^{-16.1}$	$10^{-5.6}$	$10^{-10.5}$	79,000

The ratio $a^o_{Ag^+}/a^o_{X^-}$ provides a measure of the symmetry in the zpc activity values.

24.5.3 Characteristics That Determine the Zero Point of Charge (zpc) Activity of a Potential-Determining Ion

For a solid like $AB_{(s)}$, it is the relative characteristics of the constituent ions that determine their zpc activities (de Bruyn and Agar, 1962). For $AgI_{(s)}$, at 25 °C/1 atm, $K_{s0} = 10^{-16.1}$. Neglecting activity corrections, if both Ag^+ and I^- were equally well attracted to the aqueous phase, $a^o_{Ag^+} = a^o_{I^-} = 10^{-8.05}$ *M*. However, $a^o_{Ag^+} = 10^{-5.6}$ *M* and $a^o_{I^-} = 10^{-10.5}$ *M*; the Ag^+ ion is much more strongly attracted to the aqueous phase than is the I^- ion. When equilibrated with initially-pure water, $[Ag^+] = [Br^-]$ so $a_{Ag^+} \approx a_{Br^-} \approx 10^{-8.05}$ *M*. By Eq.(24.20) $AgI_{(s)}$ in initially pure water takes on a strongly negative charge.

Compared to Ag^+, the I^- ion is bulky, polarizable, and capable of interacting with positive ions in the solid. Thus, I^- prefers to remain in the solid phase. Ag^+, on the other hand, is relatively small in size, and exhibits a strongly negative ΔH of hydration. Significantly more energy is released when a mol of Ag^+ ions becomes solvated with water molecules than is released when a mol of I^- ions is solvated. Overall, $a^o_{Ag^+} = 10^{-5.6} \gg a^o_{I^-} = 10^{-10.5}$. Br^- and Cl^- are less polarizable than is I^-, so the zpc activity values for $AgBr_{(s)}$ and $AgCl_{(s)}$ are more symmetrical (Table 24.2).

24.6 CONCENTRATIONS AS A FUNCTION OF DISTANCE IN THE AQUEOUS CHARGE LAYER

24.6.1 σ_+ AND σ_-

As noted above, the charge on the surface is exactly counterbalanced by the charge in the diffuse aqueous charge layer:

$$\sigma_s + \sigma_d = 0. \tag{24.21}$$

σ_s is the surface charge density (C/cm²), and σ_d is the diffuse aqueous charge density (C/cm²) as integrated with distance from the surface.

Consider a case when the charge on the surface is governed by a pdi, and the pdi has a charge of –1. In Figure 24.9.a, the initial, pre-charged state is an aqueous solution containing the ions of interest at concentrations equal to their bulk phase concentrations all the way up to the solid/water interface. In Figure 24.9.b the solid/water system is taken to be at equilibrium under the assumption that the surface bears a negative charge because $a_i > a^o_i$. The positive ions in the aqueous solution are attracted to the negative surface, and the negative ions are repelled. In the aqueous charge layer, electroneutrality is not obeyed. Out in the bulk solution, electroneutrality applies. Equation (24.21) states that if the surface plus the solution is taken as a whole, then overall electroneutrality is maintained.

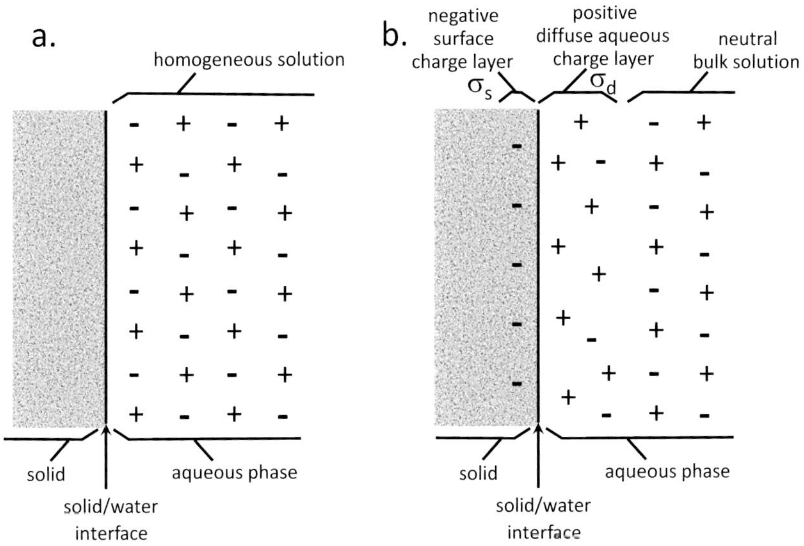

FIGURE 24.9 A surface and aqueous solution (cross-section). **a.** Before some negative charge moves from the solution onto the surface; on average, cations and anions are balanced throughout the solution. **b.** After some negative charge moves onto the surface. $\sigma_s = -\sigma_d$. Near the surface, relative to the bulk solution there is an excess of positive charge and a deficiency of negative charge. In the bulk solution, cations and anions are balanced.

Regardless of the sign of the charge on the surface, the aqueous charge layer is composed of two parts. One part is borne by the cations denoted σ_+ and the second part is borne by the anions denoted σ_-. The result is

$$\sigma_\mathrm{d} = \sigma_+ + \sigma_-. \tag{24.22}$$

Both σ_+ and σ_- always bear the same sign, regardless of the sign of σ_s. For example, for $\sigma_\mathrm{s} < 0$ so that $\sigma_\mathrm{d} > 0$, both $\sigma_+ > 0$ and $\sigma_- > 0$: σ_+ is positive because of enrichment of positive charge in the aqueous charge layer, and σ_- is positive because of depletion of negative charge. To develop expressions for σ_+ and σ_-, the aqueous concentrations of the cations and anions as a function of distance from the surface must be known.

24.6.2 The Boltzmann Distribution in the Aqueous Charge Layer

The electrochemical potential of a given ion i is constant throughout the aqueous solution:

$$\bar{\mu}_i\,(x) = \bar{\mu}_i\,(\infty). \tag{24.23}$$

electrochemical potential at point x = electrochemical potential in bulk solution

At the surface, $x = 0$. The α (aqueous phase) superscript has been dropped since henceforth we are only be concerned with μ and ψ values in the aqueous phase. Eq.(24.23) is expanded to yield

$$\mu_i(x) + z_i F\psi(x) = \mu_i(\infty) + z_i F\psi(\infty). \tag{24.24}$$

The μ_i terms may be expanded using $\mu_i = \mu_i^\mathrm{o} + RT\ln a_i(x)$, giving

$$RT\ln a_i(x) + z_i F\psi(x) = RT\ln a_i(\infty) + z_i F\psi(\infty). \tag{24.25}$$

First, activity corrections in the aqueous phase are now neglected. Second, while molality is the concentration scale preferred by thermodynamicists, concentrations expressed in molality and molarity are in general proportional to one another. Third, the molarity of i is proportional to n_i, the concentration in units of ions/cm^3. Eq.(24.25) then gives

$$\ln\frac{n_i(x)}{n_i(\infty)} = \frac{-z_i F(\psi(x) - \psi(\infty))}{RT} \tag{24.26}$$

or

$$n_i(x) = n_i(\infty)\exp\left[\frac{-z_i F(\psi(x) - \psi(\infty))}{RT}\right]. \tag{24.27}$$

Equation (24.27) is a Boltzmann distribution. If $(\psi(x) - \psi(\infty))$ is negative (*i.e.*, surface is negatively charged), then for a negatively charged ion ($z_i < 0$) we have $\ln\,[n_i(x)/n_i(\infty)] < 0$ for all finite values of x. As $x \to 0$, the value of $(\psi(x) - \psi(\infty))$ becomes increasingly negative, as does $\ln[n_i(x)/n_i\,(\infty)]$. Relative to the bulk solution, the concentration of a negatively charged ion becomes increasingly depleted as $x \to 0$. A positively charged ion becomes increasingly enriched as $x \to 0$. Theoretically, the aqueous charge layer extends infinitely far into the solution (Eq.24.27): $n_i(x) = n_i(\infty)$ occurs only at $x = \infty$. In most cases, however, $n_i(x) \approx n_i(\infty)$ once x reaches a few hundred angstroms.

24.6.3 Reference Value for $\psi(\infty)$

The value of $(\psi(x) - \psi(\infty))$ does not depend on the absolute electrical potential at the point x, only on the difference between $\psi(x)$ and $\psi(\infty)$. Therefore, to obtain $(\psi(x) - \psi(\infty))$, both quantities can be measured relative to some same reference potential at an arbitrary distance. There is no reason why the reference distance for $\psi(x)$ is cannot be ∞, with $\psi(\infty)$ assigned a value of 0. Then $\psi(x) - \psi(\infty) = \psi(x)$:

$$n_i(x) = n_i(\infty)\exp\left[\frac{-z_i F\psi(x)}{RT}\right]. \tag{24.28}$$

Obtaining an equation for $\psi(x)$ will allow calculation of n_i as a function of x for any ion i.

A relation that is useful in conjunction with Eq.(24.28) is

$$\frac{F}{R} = 11{,}604.5\ \text{K/V}. \tag{24.29}$$

At 298.15 K (25 °C),

$$\frac{F}{RT} = 38.92\ \text{V}^{-1}. \tag{24.30}$$

24.6.4 σ_+ and σ_- for $AB_{(S)}$ in Equilibrium with Initially Pure Water

After equilibration of AB(s) with initially pure water, assume for the moment that only A^+ and B^- are present in the solution. The expressions for σ_+ and σ_- each only involve one ion with

$$\begin{aligned}\sigma_+ &= z_+ e_o \int_0^\infty (n_+(x) - n_+(\infty))dx \\ &= \text{charge (C per cm}^2\text{) borne by positive ions}\end{aligned} \tag{24.31}$$

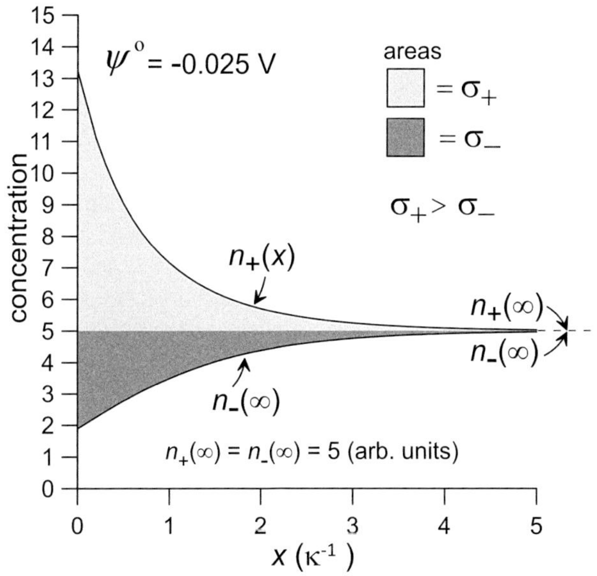

FIGURE 24.10 Concentration as a function of distance x in units of κ^{-1} for $\psi^{o} = -0.025$ V and assuming $z_+ = -z_- = 1$. For negative ψ^{o}, the area under the $n_+(x)$ curve giving σ_+ is always greater than the area for the $n_-(x)$ curve giving σ_-. (*i.e.*, $\sigma_+ > \sigma_-$). The quantity $n_+(x)$ is the concentration sum over all $z_+ = 1$ cations. The quantity $n_-(x)$ is the concentration sum over all $z_- = -1$ anions. $n_+(\infty) = n_-(\infty) = 5$ in some arbitrary units.

$$\sigma_- = z_- e_o \int_0^{\infty} (n_-(x) - n_-(\infty))dx \tag{24.32}$$

$$= \text{charge (C per cm}^2\text{) borne by negative ions.}$$

Figure 24.10 illustrates the nature of the integrals for a negatively charged surface.

24.7 THE DIFFERENTIAL EQUATION GOVERNING THE AQUEOUS CHARGE LAYER

If the electrolyte in the aqueous bulk solution and in the aqueous charge layer is symmetrical, then

$$z_+ = -z_- = z. \tag{24.33}$$

Salts that obey Eq.(24.33) include AgCl, KCl, $CaSO_4$, etc. A salt that does not is $MgCl_2$.

From Eqs.(24.27) and (24.33), for a symmetrical electrolyte solution then

$$n_+(x) = n_+(\infty)\exp\left[\frac{-zF\psi(x)}{RT}\right] \tag{24.34}$$

$$n_-(x) = n_-(\infty)\exp\left[\frac{zF\psi(x)}{RT}\right]. \tag{24.35}$$

When the aqueous solution satisfies the electroneutrality equation at some point x,

$$z_+ n_+(x) + z_- n_-(x) = 0. \tag{24.36}$$

The net amount of charge per unit volume at any distance x from the surface is given by $q(x)$:

$$q(x) = z_{+}e_{o}n_{+}(x) + z_{-}e_{o}n_{-}(x). \tag{24.37}$$

$$q(x) = ze_{o}(n_{+}(x)) - n_{-}(x)). \tag{24.38}$$

Eq.(24.36) is satisfied at $x = \infty$: in the bulk solution $q(\infty) = 0$. The bulk solution is electrically neutral. In the aqueous charge layer, Eq.(24.36) is not satisfied, and is less and less satisfied as one approaches the surface. (Eq.(24.37) is similar to Eq.(23.9) except that x is the coordinate of interest instead of r as used for the spherically symmetrical charge density around a core ion.) As in Eq.(23.9), the concentrations in Eqs.(24.34)–(24.38) have units of ions per cm^3. The factor of e_o (the elementary charge) is included in Eqs.(24.37) and (24.38) so as to obtain the required units of C/cm^3.

The electrical potential ψ at a given point is related to the charge density through the Poisson Equation. Since the dimensions of the particles upon which double layers form are generally very large relative to the thicknesses of their corresponding aqueous charge layers, one can usually assume that the aqueous charge layer is essentially a flat plate of charge. That is, that the properties of the aqueous charge layer vary only with one dimension, x. As a one-dimensional problem in *Cartesian* coordinates, the Poisson Equation is

$$\nabla^2\psi(x) = \frac{d^2\psi(x)}{dx^2} = \frac{-q(x)}{\varepsilon\varepsilon_o} \tag{24.39}$$

where ε is the dielectric constant of water and ε_o is the permittivity of free space ($= 8.85\times10^{-14}$ C–V/cm). Substituting Eq.(24.38) into Eq.(24.39),

$$\frac{d^2\psi(x)}{dx^2} = \frac{-ze_o}{\varepsilon\varepsilon_o}(n_{+}(x) - n_{-}(x)). \tag{24.40}$$

Substituting Eqs.(24.34) and (24.35) into Eq.(24.40),

$$\frac{d^2\psi(x)}{dx^2} = \frac{-ze_o}{\varepsilon\varepsilon_o}\left(n_{+}(\infty)\exp\left[\frac{-zF\psi(x)}{RT}\right] - n_{-}(\infty)\exp\left[\frac{zF\psi(x)}{RT}\right]\right). \tag{24.41}$$

For a symmetrical electrolyte, since $z_{+} = -z_{-}$, by Eq.(24.36),

$$n_{+}(\infty) = n_{-}(\infty) = n(\infty) \tag{24.42}$$

and both bulk phase concentrations may be referred to as $n(\infty)$. The hyperbolic sine function (sinh u) is given by

$$\sinh u = \frac{e^{u} - e^{-u}}{2}. \tag{24.43}$$

The sinh function, like the trigonometric sine function, is dimensionless. The sinh and sine functions bear numerous similarities. One important example is that for small arguments, sinh $u \approx u$ and sin $u \approx u$.

By Eqs.(24.41)–(24.43),

$$\frac{d^2\psi(x)}{dx^2} = \frac{ze_o n(\infty)}{\varepsilon\varepsilon_o}2\sinh\frac{zF\psi(x)}{RT}. \tag{24.44}$$

As with $n_+(x)$ and $n_-(x)$, the dimensions for $n(\infty)$ are ions/cm^3; z has dimensions of units of charge per ion. The units of Eq.(24.44) are then V/cm^2. Multiplying and dividing the RHS of Eq.(24.44) by zF/RT gives

$$\frac{d^2\psi(x)}{dx^2} = \left(\frac{2ze_0 n(\infty)zF}{\varepsilon\varepsilon_0 RT}\right)\left(\frac{\sinh\dfrac{zF\psi(x)}{RT}}{zF/(RT)}\right). \tag{24.45}$$

We have $2z^2n(\infty) = \sum_j z_j^2 n_j(\infty)$ and $F/R = (F/N_A)/(R/N_A) = e_0/k$. Therefore, the first of the two parenthetical factors on the RHS is κ^2 as given by

$$\kappa^2 = \frac{e_0^2}{\varepsilon\varepsilon_0 kT}\sum_j z_j^2 n_j(\infty). \tag{24.46}$$

Eq.(24.45) then simplifies to

$$\frac{d^2\psi(x)}{dx^2} = \kappa^2\frac{\sinh\dfrac{zF\psi(x)}{RT}}{zF/(RT)} \tag{24.47}$$

which is the differential equation that must be solved to obtain $\psi(x)$. For water at 298.15 K (25 °C), from Eqs.(23.52) and (23.53),

$$\kappa = 3.29 \times 10^7 I^{1/2}\ \text{cm}^{-1} \tag{24.48}$$

$$\boxed{\kappa^{-1} = 3.04 \times 10^{-8} I^{-1/2}\ \text{cm}.} \tag{24.49}$$

where I is the ionic strength.

24.8 REAL AQUEOUS SOLUTIONS CONTAIN MULTIPLE CATIONS AND MULTIPLE ANIONS

The inference to this point in this chapter has been that there is only one cation/anion pair in the aqueous solution. For example, in the case of $AB_{(s)}$ dissolving into initially pure water, H^+ and OH^- were not discussed. Multiple cations and multiple anions will, however, generally be present in aqueous solutions, so a discussion of their behavior is needed.

First, note that κ was brought into Eq.(24.45) based on the assumption that only one cation/anion pair was present in the aqueous phase, and that pair moreover satisfies Eq.(24.33). However, even for the very simple case of $AgI_{(s)}$ equilibrated with initially pure water, we will at least have Ag^+, I^-, H^+ and OH^- all present in the solution. The solution to this complication is to note that all of these anions and cations are singly charged, so perhaps $n_+(x)$ as used can refer to the concentration sum for all of the cations at point x, and $n_-(x)$ can refer to the concentration sum for all of the anions at point x. The value of κ can then be computed as usual using the concentrations of all of the ions in the bulk solution. This approach would also apply to solutions containing other salts of the same z value, as might occur in a system containing some $AgI_{(s)}$, some initially pure water, and some dissolved $AgNO_3$ and NaCl.

It is this sums approach that underlies the various double layer equations developed here. However, strictly speaking, the only way that this approach can work is if all of the symmetrically charged ions in solution are simple point charges, and not chemically distinguishable, at least in terms of charge characteristics. Consider the system involving some $AgI_{(s)}$ in equilibrium with some initially-pure water. In the bulk solution, at 25 °C/1 atm, $[H^+] = [OH^-] = 10^{-7}$ M, and $[Ag^+] = [I^-] = 10^{-8.05}$ M. Approaching the surface from out in the bulk solution, $\psi(x)$ becomes more negative, and so according to Eq.(24.27), there is an increase in $[Ag^+]$ from $10^{-8.05}$ M, and a decrease in $[I^-]$ from $10^{-8.05}$ M. The area under the $[Ag^+]$ *vs.* x curve gives the amount of excess Ag^+ that dissolved relative to the line $n_{Ag^+}(x) = n_{Ag^+}(\infty)$. Similarly, the area between the x-axis and the $[I^-]$ curve gives the amount of I^- that stayed on the surface relative to the line $n_{I^-}(x) = n_{I^-}(\infty)$. The sum of these will give σ_d. Near the surface, there will be an enrichment in H^+ relative to the bulk solution, and a corresponding depletion of OH^-. However, unlike Ag^+ and I^-, the integrals from zero to ∞ for H^+ and OH^- relative to the bulk solution values are both zero as neither H^+ nor OH^- is sorbed on the surface or released to the solution from the surface. Thus, the enrichment of H^+ at small x is compensated for by a depletion in H^+ relative to 10^{-7} M at larger x; the depletion of OH^- at small x is compensated for by an enrichment in OH^- relative to 10^{-7} M at larger x. The migration of some H^+ towards the surface allows some Ag^+ to move out into the zone of depletion for H^+. Similarly, the repulsion of OH^- from the surface allows some I^- at moderate distances to move in closer to the surface. Overall, it is still true that $(+1)e_o\int_0^\infty (n_{Ag+}(x) - n_{Ag+}(\infty))\,dx = \sigma_+$ and $(-1)e_o\int_0^\infty (n_{I^-}(x) - n_{I^-}(\infty))\,dx = \sigma_-$, and that $\int_0^\infty (n_{H^+}(x) - n_{H^+}(\infty))\,dx = 0$, and $\int_0^\infty (n_{OH^-}(x) - n_{OH^-}(\infty))\,dx = 0$. When considering all ions to be simple point charges that are distinguished only by the values of their charges (*i.e.*, taking the sums approach), then as discussed above we obtain in this case that

$$\begin{aligned}\sigma_+ &= (+1)e_o\int_0^\infty (n_{Ag^+}(x) - n_{Ag^+}(\infty))\,dx + (+1)e_o\int_0^\infty (n_{H^+}(x) - n_{H^+}(\infty))\,dx \\ &= (+1)e_o\int_0^\infty (n_+(x) - n_+(\infty))\,dx\end{aligned} \tag{24.50}$$

$$\begin{aligned}\sigma_- &= (-1)e_o\int_0^\infty (n_{I^-}(x) - n_{I^-}(\infty)\,dx + (-1)e_o\int_0^\infty (n_{OH}(x) - n_{OH^-}(\infty)\,dx \\ &= (-1)e_o\int_0^\infty (n_-(x) - n_-(\infty))\,dx\end{aligned} \tag{24.51}$$

where $n_+(x)$ is the sum of concentrations at x of all of the cations of charge z_+, and $n_-(x)$ is the sum of concentrations at x of all of the cations of charge z_-, with $z_+ = -z = z$. Most natural waters (including seawater) satisfy the criterion that $z_+ = -z_- = z = 1$ for nearly all ions.

24.9 INTEGRATING THE DIFFERENTIAL EQUATION GOVERNING THE AQUEOUS CHARGE LAYER SO AS TO OBTAIN $\psi(x)$

24.9.1 $z|\psi^o|$ IS SMALL (*I.E.*, SUFFICIENTLY CLOSE TO ZERO)

While for most natural waters z can be taken to be 1, the presence of z will be retained in the presentation. When the absolute value $|zF\psi^o(x)/(RT)| < 1$, the exponential functions comprising the sinh function in Eq.(24.47) can be simplified so that integration of the differential equation is straightforward. A Taylor series expansion of the function e^u yields

$$e^u = 1 + u + u^2/2! + u^3/3! + \cdots. \tag{24.52}$$

For $u = 1$, $e^{1.0} = 2.718\ldots$, so $e^u \approx 1 + u + u^2/2! = 2.50$ is an adequate approximation. The value of $k = 1.381 \times 10^{-23}$ V–C/K and $e_o = 1.602 \times 10^{-19}$ C, so

$$\left|\frac{zF\psi^o}{RT}\right| = \left|\frac{ze_o\psi^o}{kT}\right| = 1 \quad \text{when } z\left|\psi^o\right| = 0.0257 \text{ V at } T = 298.15 \text{ K.} \tag{24.53}$$

Thus, in the context of the above discussion of the approximation $e^u \approx 1 + u + u^2/2!$, a useful "smallness" criterion is that $z\left|\psi^o\right| < 0.025$ V.

For small $z\left|\psi^o\right|$, taking each of the exponentials in the sinh function in Eq.(24.47) as $e^u \approx 1 + u + u^2/2!$ yields

$$\frac{d^2\psi(x)}{dx^2} = \frac{\kappa^2}{zF/(RT)} \frac{\left[\left(1 + \frac{zF\psi(x)}{RT} + \frac{1}{2}\left(\frac{zF\psi(x)}{RT}\right)^2\right) - \left(1 - \frac{zF\psi(x)}{RT} + \frac{1}{2}\left(\frac{zF\psi(x)}{RT}\right)^2\right)\right]}{2}. \tag{24.54}$$

The odd terms of the approximations cancel. Eq.(24.54) reduces to

$$\frac{d^2\psi(x)}{dx^2} = \frac{\kappa^2}{zF/(RT)} \frac{2zF\psi(x)/(RT)}{2} \tag{24.55}$$

$$= \kappa^2\psi(x). \tag{24.56}$$

which is the linearized Poisson-Boltzmann Equation in one-dimensional Cartesian coordinates. The solution to this differential equation is needed. In a manner virtually identical to the approach used with Eq.(23.28) in the development of the Debye–Hückel Equation, we try a solution of the type

$$\psi(x) = Ae^{-\kappa x} + Be^{+\kappa x}. \tag{24.57}$$

Taking the first and second derivatives,

$$d\psi(x)/dx = -A\kappa e^{-\kappa x} + B\kappa e^{+\kappa x} \tag{24.58}$$

$$d^2\psi(x)/dx^2 = A\kappa^2 e^{-\kappa x} + B\kappa^2 e^{+\kappa x} = \kappa^2\psi(x). \tag{24.59}$$

Since κ is always finite and positive, the boundary condition that $d\psi(x)/dx = 0$ at $x = \infty$ can be used together with Eq.(24.58) to deduce that $B = 0$. A second boundary condition is $\psi(0) = \psi^o$. By Eq.(24.57), then $A = \psi^o$. Thus, for small $z\left|\psi^o\right|$,

$$\psi(x) = \psi^o e^{-\kappa x}. \tag{24.60}$$

Eq.(24.60) is consistent with the convention chosen earlier that $\psi(\infty) = 0$.

24.9.2 $z\left|\psi^o\right|$ Is Not Necessarily Small

Eq.(24.47) is considerably more difficult to integrate when it is not true that $z\left|\psi^o\right| < 1$. To carry out the integration, here the approach of Sparnaay (1972) is followed, with added detail. Note first that for any function $y = y(x)$ and any constant C,

$$\frac{d}{dx}\left[\left(\frac{dy}{dx}\right)^2 + C\right] = 2\left(\frac{dy}{dx}\right)\left(\frac{d^2y}{dx^2}\right). \tag{24.61}$$

$$d\left[\left(\frac{dy}{dx}\right)^2 + C\right] = 2\left(\frac{d^2y}{dx^2}\right)dy. \tag{24.62}$$

Integrating,

$$\int d\left[\left(\frac{dy}{dx}\right)^2 + C\right] = 2\int\left(\frac{d^2y}{dx^2}\right)dy \tag{24.63}$$

$$\left(\frac{dy}{dx}\right)^2 + C = 2\int\left(\frac{d^2y}{dx^2}\right)dy. \tag{24.64}$$

If $y(x) = zF\psi(x)/(RT)$, then

$$\frac{RT}{zF}\frac{d^2y}{dx^2} = \frac{d^2\psi(x)}{dx^2}. \tag{24.65}$$

Substituting Eq.(24.65) into Eq.(24.47),

$$\frac{RT}{zF}\frac{d^2y}{dx^2} = \kappa^2\frac{\left[\sinh\frac{zF\psi(x)}{RT}\right]}{zF\,/\,(RT)} \tag{24.66}$$

or, simply

$$\frac{d^2y}{dx^2} = \kappa^2[\sinh y]. \tag{24.67}$$

Combining Eq.(24.67) with Eq.(24.64),

$$\left(\frac{dy}{dx}\right)^2 + C = 2\kappa^2\int\sinh y\,dy. \tag{24.68}$$

Note now that $\int\sinh y\,dy = \cosh y$ where $\cosh y$ is the hyperbolic cosine of y (*i.e.*, $\cosh y = (e^y + e^{-y})/2$). Therefore,

$$\left(\frac{dy}{dx}\right)^2 + C' = 2\kappa^2\cosh y \tag{24.69}$$

where C' is another constant of integration that equals C plus the additional constant of integration that results in going from Eq.(24.68) to Eq.(24.69).

The hyperbolic trigonometric functions have “half-angle” relations just like the regular trigonometric functions. One such relation is

$$\cosh y = 1 + 2\sinh^2(y/2). \tag{24.70}$$

Substituting Eq.(24.70) into Eq.(24.69),

$$\left(\frac{dy}{dx}\right)^2 + C'' = 4\kappa^2\sinh^2(y/2) \tag{24.71}$$

where C'' is another constant ($C'' = C' - 2\kappa^2$). Some boundary conditions are now applied. At $x = \infty$, $d\psi(x)/dx = 0$ so $dy/dx = 0$. Second, at $x = \infty$, $\psi(x) = 0$ so $y(\infty) = zF\psi(\infty)/RT = 0$. Since $\sinh(0) = 0$, at $x = \infty$ $\sinh(y/2) = 0$ at ∞, so $C'' = 0$. Thus,

$$\left(\frac{dy}{dx}\right) = \pm 2\kappa \sinh(y/2). \tag{24.72}$$

The sign properties of the sinh function can be examined to determine whether the positive or negative root is needed. Sinh is an "odd function". That is, $\sinh(-y/2) = -\sinh(y/2)$. (For an even function $f(u)$, $f(-u) = f(u)$.) Thus, as seen in Figure 24.11, for $\psi^o > 0$, $\sinh(y/2) > 0$ for all x; for $\psi^o < 0$, $\sinh(y/2) < 0$ for all x. Now consider the sign behavior of dy/dx. As depicted in Figure 24.12, for $\psi^o > 0$, $dy/dx < 0$ for all values of x. For $\psi^o < 0$, $dy/dx > 0$ for all values of x. These results are summarized in Table 24.3. The only way that Eq.(24.72) can satisfy the conditions in that table is for the applicable root to be the negative one:

$$\left(\frac{dy}{dx}\right) = -2\kappa \sinh(y/2) \tag{24.73}$$

$$\int \frac{dy}{\sinh(y/2)} = -2\kappa \int dx. \tag{24.74}$$

Standard integral tables (*e.g.*, Gradshteyn and Ryzhik, 1962) give

$$\int \frac{du}{\sinh u} = \ln \tanh(u/2) \tag{24.75}$$

where the hyperbolic tangent $\tanh u = (\sinh u)/(\cosh u)$. (Note: since $(\sinh u)^{-1} = \operatorname{csch} u$ (hyperbolic cosecant of u), Eq.(24.75) may be listed in some integral tables as $\int \operatorname{csch} u \, du = \ln \tanh(u/2)$.)

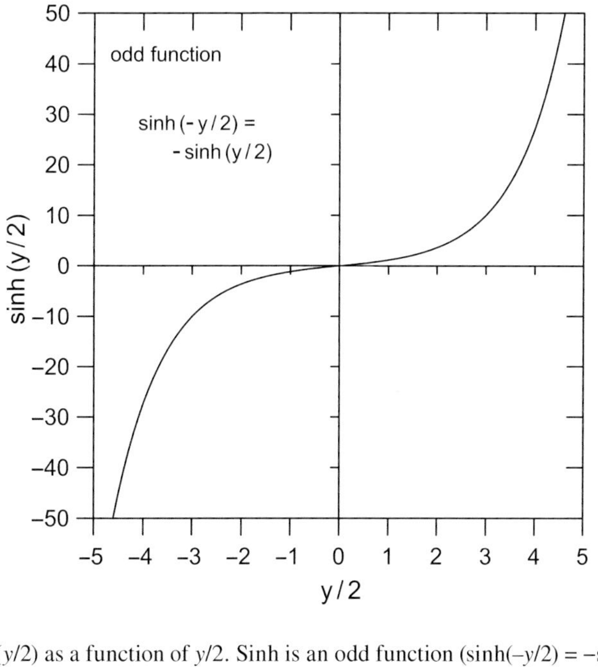

FIGURE 24.11 Sinh(y/2) as a function of y/2. Sinh is an odd function ($\sinh(-y/2) = -\sinh(y/2)$).

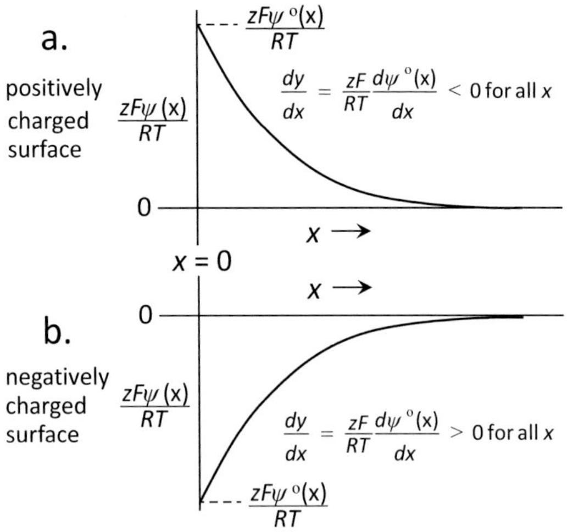

FIGURE 24.12 a. $y = zF\psi(x)/(RT)$ for $\psi^{o} > 0$; $dy/dx < 0$ for all x. b. $y = zF\psi(x)/(RT)$ for $\psi^{o} < 0$; $dy/dx > 0$ for all x.

TABLE 24.3
Sign Properties Governing Eq.(24.72) for All Values of *x*

Positively-Charged Surface	Negatively-Charged Surface
$\psi^{o} > 0$	$\psi^{o} < 0$
$dy/dx < 0$	$dy/dx > 0$
$\sinh(y/2) > 0$	$\sinh(y/2) < 0$

Note: $y = zF\psi(x)/(RT)$; $dy/dx = (zF/(RT))d\psi(x)/dx$.

To use Eq.(24.75) to integrate Eq.(24.74), let $u = y/2$. Since $du = dy/2$, substitute $dy = 2du$ into Eq.(24.74). By Eq.(24.75),

$$\ln \tanh(y/4) + C''' = -\kappa x \tag{24.76}$$

where C''' is some constant.

A boundary condition can be applied to evaluate C'''. For $x = 0$, $y = zF\psi^{o}/(RT)$. Abbreviating $zF\psi^{o}/(RT)$ as $\bar{z}$, then

$$C''' = -\ln \tanh(\bar{z}\ /\ 4) \tag{24.77}$$

$$\kappa x = \ln \tanh(\bar{z}/4) - \ln \tanh(y/4) \tag{24.78}$$

$$\ln \tanh(y/4) = -\kappa x + \ln \tanh(\bar{z}/4). \tag{24.79}$$

Exponentiating both sides,

$$\tanh(y/4) = e^{-\kappa x}\tanh(\bar{z}/4). \tag{24.80}$$

The hyperbolic trigonometric functions have corresponding arc functions just as do the regular trigonometric functions. Thus,

$$y/4 = \tanh^{-1}(e^{-\kappa x}\tanh(\bar{z}/4)). \tag{24.81}$$

One of the properties of $\tanh^{-1} u$ (= arctanh u) is that

$$\tanh^{-1} u = \tfrac{1}{2}\ln\frac{1+u}{1-u}. \tag{24.82}$$

Combining Eqs.(24.81) and (24.82) gives

$$\frac{y}{2} = \ln\frac{1+e^{-\kappa x}\tanh(\bar{z}/4)}{1-e^{-\kappa x}\tanh(\bar{z}/4)} \tag{24.83}$$

with, as defined above, $y = zF\psi(x)/RT$ and $\bar{z} = zF\psi^{o}/RT$. Eq.(24.83) allows calculation of $\psi(x)$ for all ψ^{o}.

24.10 $\psi(x)$ FOR ALL ψ^{o}

Eq.(24.83) can be used to compute $\psi(x)$ for known values of ψ^{o}, κ, and x for all ψ^{o}. It can be placed into a form that involves only exponentials. Note first that $\tanh u = (\sinh u)/(\cosh u)$ so that

$$\tanh(\bar{z}/4) = \frac{e^{\bar{z}/4} - e^{-\bar{z}/4}}{e^{\bar{z}/4} + e^{-\bar{z}/4}}. \tag{24.84}$$

The following operations are now carried out on Eq.(24.83): 1) substitution using Eq.(24.84); and 2) multiplication of top and the bottom of the argument of the ln term by $e^{\bar{z}/4}(e^{\bar{z}/4} + e^{-\bar{z}/4})$; and 3) exponentiate both sides. The result is

$$e^{y/2} = \frac{e^{\bar{z}/2} + 1 + (e^{\bar{z}/2} - 1)e - \kappa x}{e^{\bar{z}/2} + 1 - (e^{\bar{z}/2} - 1)e - \kappa x}. \tag{24.85}$$

Eq.(24.85) is the form of the double layer equation found in many texts on soil colloid chemistry.

24.11 σ_s AND σ_d

Regardless of whether or not $z|\psi^{o}|$ is small, by combining Eqs.(24.22), (24.31), (24.32), and (24.37),

$$-\sigma_s = \sigma_d = \int_0^{\infty} q(x)\,dx. \tag{24.86}$$

Using Poisson's Equation (Eq.(24.40)) to substitute for $q(x)$, Eq.(24.86) becomes

$$-\sigma_s = \sigma_d = -\varepsilon\varepsilon_0\int_0^{\infty} (d^2\psi(x)/dx^2)\,dx. \tag{24.87}$$

Integrating,

$$-\sigma_s = \sigma_d = -\varepsilon\varepsilon_0 \left.\frac{d\psi(x)}{dx}\right|_{x=0}^{x=\infty}. \tag{24.88}$$

At $x = \infty$, $d\psi(x)/dx = 0$. Therefore,

$$-\sigma_s = \sigma_d = \varepsilon\varepsilon_o \left.\frac{d\psi(x)}{dx}\right|_{x=0}. \tag{24.89}$$

For a positively charged surface, $d\psi(x)/dx\big|_{x=0} < 0$, and so Eq.(24.90) gives $\sigma_s > 0$ and $\sigma_d < 0$. For a negatively charged surface, $d\psi(x)/dx\big|_{x=0} > 0$, and $\sigma_s < 0$ and $\sigma_d > 0$.

More quantitatively, from Eq.(24.74),

$$\frac{zF}{RT}\left(\frac{d\psi(x)}{dx}\right) = -2\kappa \sinh\frac{zF\psi(x)}{2RT}. \tag{24.90}$$

Combining Eqs.(24.89) and (24.90),

$$-\sigma_s = \sigma_d = -2\varepsilon\varepsilon_o \frac{RT}{zF}\kappa \sinh\frac{zF\psi^o}{2RT}. \tag{24.91}$$

At 298.15 K (25 °C), $2\varepsilon\varepsilon_o RT/F = 3.56 \times 10^{-13}$ C/cm.

Eq.(24.91) is valid for both large and small $z|\psi^o|$. For small $z|\psi^o|$, by the approximation that $\sinh u \approx u$ for small u, Eq.(24.91) reduces to

$$-\sigma_s = \sigma_d = -\varepsilon\varepsilon_o \kappa \psi^o. \tag{24.92}$$

24.12 THE PHYSICAL SIGNIFICANCE OF κ^{-1}

Chapter 23 discussed the fact that the electrostatic interaction that exists between a core ion and its surrounding ion cloud would be the same if the charge on the ion cloud was concentrated in a shell of radius κ^{-1} (the Debye length) from the core ion. In the case of the aqueous charge layer of a double layer, the Debye length is often referred to as the "thickness of the double layer". From Eq.(24.85), even for large $z\psi^o$, when $x \approx \kappa^{-1}$ then $\psi(x) \approx \psi^o/e = \psi^o/2.718$. Thus, κ^{-1} is the distance at which $\psi(x)$ has dropped to $1/e$ of ψ^o.

Also, when $z|\psi^o|$ is small, then the plane at $x = \kappa^{-1}$ is the x-coordinate for the center of mass of the aqueous charge layer. For a one-dimensional rod of length L, the center of physical mass is the distance x from the end of the rod at which the rod will be perfectly balanced if placed on a fulcrum. The density of the rod might vary along its length, just as the charge density in the aqueous charge layer varies with x. The mathematical definition of the center of mass (com) for the rod is

$$x_{\text{com}} = \frac{1}{M}\int_0^L x\,\rho(x)\,dx \tag{24.93}$$

where M is the total mass of the rod, and $\rho(x)$ is the mass density of the rod as a function of x.

The equation for the center of mass of the aqueous charge layer is

$$x_{\text{com}} = \frac{1}{\sigma_d}\int_0^\infty x\,q(x)\,dx. \tag{24.94}$$

Using the Poisson Equation and Eq.(24.60) to substitute for $q(x)$,

$$x_{\text{com}} = \frac{-\varepsilon\varepsilon_o\kappa^2\psi^o}{\sigma_d}\int_0^\infty x e^{-\kappa x}dx. \tag{24.95}$$

Standard integral tables give

$$\int xe^{-\kappa x}dx = \frac{e^{-\kappa x}}{\kappa^2}(-\kappa x - 1). \tag{24.96}$$

By evaluation at its upper and lower limits, the integral in Eq.(24.95) therefore equals $1/\kappa^2$. Then, with σ_d from Eq.(24.92) into Eq.(24.95),

$$x_{\text{com}} = \kappa^{-1} \qquad \text{small } z\psi^{\text{o}} \tag{24.97}$$

When $z|\psi^{\text{o}}|$ is not small so that Eq.(24.60) cannot be used, the center of mass of the aqueous space charge is closer to the surface than κ^{-1} (Verwey and Overbeek, 1948).

24.13 EFFECTS OF κ^{-1} ON THE PROPERTIES OF THE AQUEOUS CHARGE LAYER

Using Eq.(24.48), κ^{-1} can be computed as a function of *I*. A plot for a range of values at 25 °C/1 atm is given Figure 24.13. For any ψ^{o}, the function $\psi(x) \rightarrow 0$ as *I* increases. In addition to providing a measure of the thickness of the aqueous charge layer, κ^{-1} has implications for the values of ψ^{o} and σ_s. For simplicity, consider the case of a surface for which $z|\psi^{\text{o}}|$ is small enough that Eq.(24.92) applies. For a surface for which ψ^{o} is controlled by a potential-determining ion (pdi), then regardless of the value of *I* (*i.e.*, regardless of the value of κ^{-1}), ψ^{o} will remain unchanged so long as the activity of the pdi remains unchanged. If *I* then increases, κ^{-1} decreases, then by Eq.(24.92) σ_s must increase: as the surface charge is increasingly screened, more charge is needed on the surface to maintain a constant ψ^{o}, so more of the pdi must sorb on the surface (Figure 24.14). In particular, as *I* is increasing, more aqueous ions with a charge opposite to the surface charge contribute to *I* crowd near the solid/water interface. If the surface is positively charged, many anions are near the interface, and *vice versa*.

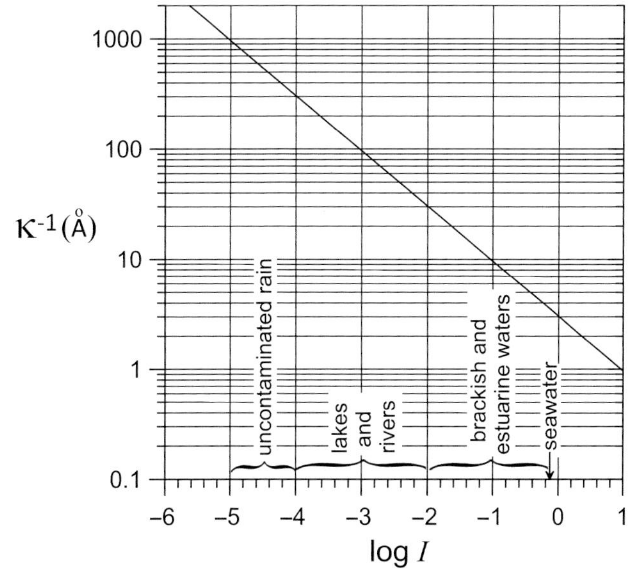

FIGURE 24.13 κ^{-1} (Angstroms, Å) *vs*. log *I* at 25 °C/1 atm. In dilute solutions such as uncontaminated rain, κ^{-1} is large (300 to 1000 Å). In seawater, $\kappa^{-1} \approx 4$ Å.

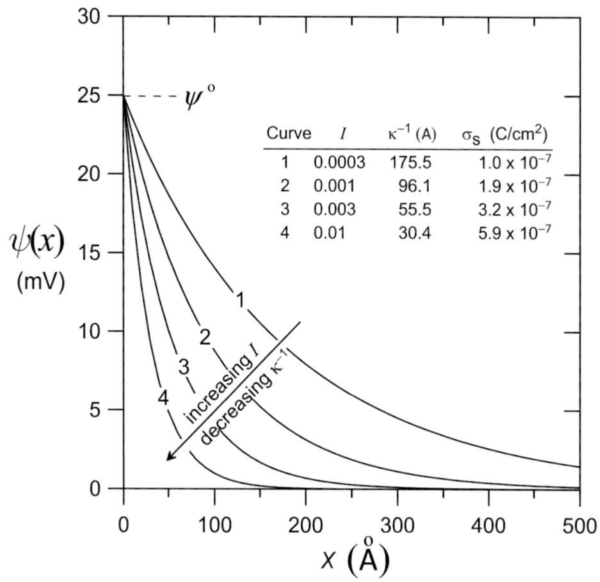

FIGURE 24.14 $\psi^{\text{o}}(x)$ *vs.* distance at different ionic strength (I) values. For all curves, ψ^{o} = 25 mV as controlled by the activity of a potential determining ion (pdi), and z for the solution electrolyte is assumed to be 1. As I increases and κ^{-1} decreases, increasing electrical shielding in the aqueous solution requires σ_s to increase to maintain constant ψ^{o}.

For a clay particle, σ_s will be fixed. Based on Eq.(24.92), increasing I will decrease the magnitude of ψ^{o}. Since there is no pdi to add surface as I increases, the screening of charge at the solid/water interface decreases the magnitude of ψ^{o} (Figure 24.15).

24.14 STRATEGIES FOR USING THE VARIOUS DOUBLE-LAYER EQUATIONS IN SOLVING PROBLEMS

As discussed, double-layer charges can result with or without a pdi being present. When a pdi is present, the steps are:

1. Calculate ψ^{o} based on the activity of the pdi.
2. Calculate κ^{-1} based on the ionic strength I.
3. For $z\left|\psi^{\text{o}}\right| \leq 25\text{ mV}$, calculate $\psi(x)$ based on Eq.(24.60), and for $z\left|\psi^{\text{o}}\right| > 25\text{ mV}$, calculate $\psi(x)$ from Eq.(24.83) (z can be taken to be unity in natural waters).

When σ_{s} is a constant (as with a clay particle), the steps are:

1. Calculate σ_{s} based on a knowledge of the amount of charge present per unit surface area of the solid.
2. Calculate κ^{-1} based on the ionic strength I.
3. Calculate ψ^{o} based on Eq.(24.91).
4. For $z\left|\psi^{\text{o}}\right| \leq 25\text{ mV}$, calculate $\psi(x)$ based on Eq.(24.60), and for $z\left|\psi^{\text{o}}\right| > 25\text{ mV}$, calculate $\psi(x)$ from Eq.(24.83) (z can be taken to be unity in natural waters).

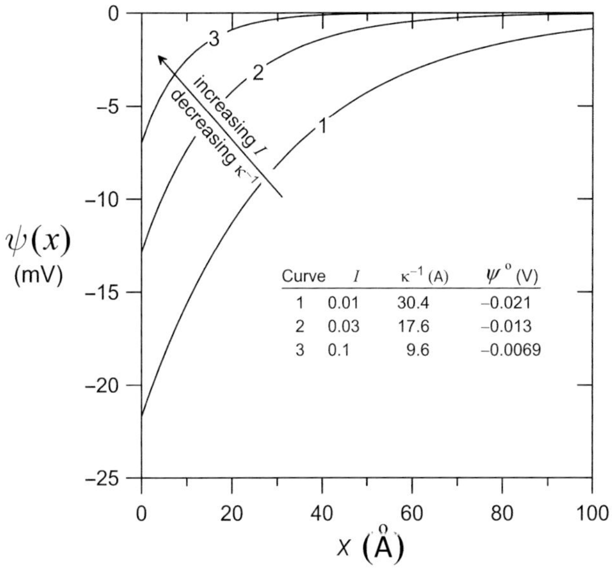

FIGURE 24.15 $\psi(x)$ *vs.* distance at different ionic strength (I) values. For all curves, $\sigma_s = -5\times10^{-7}$ C/cm^2 as might be the case for a clay with a low cation exchange capacity (CEC), and z for the solution electrolyte is assumed to be 1. As I increases and κ^{-1} decreases, the increasing electrical shielding in the aqueous solution causes the magnitude of ψ^o to decrease.

REFERENCES

DeBruyn PL and Agar GE (1962) Surface chemistry of froth flotation. In: Fuerstenau DW (Ed.) *Froth Flotation – 50th Anniversary Volume*, American Institute of Mining, Metallurgical, and Petroleum Engineers, New York, pp.91–138.

Gradshteyn IS and Ryzhik IM (1962) *Tables of Integrals, Series, and Products*, 4th edn, prepared by YuV Geronimus and MYu Tseytlin, Academic Press, New York.

Morrison SR (1990) *The Chemical Physics of Surfaces*, Plenum Press, New York.

Sparnaay MJ (1972) *The Electrical Double Layer*, Volume 4 of Topic 14: The Properties of Interfaces, In: DH Everett (Ed.), *The Encyclopedia of Physical Chemistry and Chemical Physics*, Pergamon Press, Oxford (see in particular pp.57–62).

Van Olphen H (1977) *An Introduction to Clay Colloid Chemistry*, Wiley-Interscience, New York/Amsterdam.

Verwey EJW, Overbeek JTG (1948) *Theory of the Stability of Lyophobic Colloids*, Elsevier Publishing Co, New York/Amsterdam.

25 Colloid Stability and Particle Double Layers

25.1 INTRODUCTION

In 1861, Thomas Graham applied the word *colloid* to refer to any suspension of material in a liquid where the suspension does not separate into two distinct phases, even after a long period of time. Colloidal suspensions in water thus contain small particles or droplets that are stably dispersed within the aqueous phase. An emulsion is a colloidal suspension of liquid droplets.

The colloidal particle size range is generally defined as 30–10,000 Å (3–1000 nm, 0.003–1 μm). Particles smaller than 30 Å are of the same size as some large molecules, and so may be thought of as being essentially "dissolved". It is the constant thermal (*i.e.*, Brownian) bombardment by the surrounding water molecules that keeps colloidal particles/droplets in suspension, and frustrating either gravitational settling (density > aqueous-only density), or flotation (density < aqueous-only density). Once the particles reach a certain size, bombardment by water molecules is not sufficient to prevent net settling or flotation. The latter occurs during the formation of a cream layer on the top of a bottle of whole milk.

Colloidal particles are environmentally important for numerous reasons. First, a significant amount of the suspended particulate matter in lakes and rivers can be colloidal. This is important for sediment transport and accumulation. And, the high surface area per unit volume of colloidal matter allows colloidal particles to sorb and transport organic carbon, nutrients, and organic and inorganic contaminants. Second, removal of colloidal matter from raw drinking water removes viruses and bacteria, improves the aesthetic appeal of the water, and, when chlorination is used for disinfection, reduces the levels of the precursor molecules of disinfection by-products. Third, many anthropogenic wastewaters are suspensions of particles; about 80% of the biological oxygen demand (BOD) of untreated domestic sewage and 50% of the organic carbon in secondary sewage treatment effluent is in suspended particulate form.

Colloids are classified as being either *lyophobic* or *lyophilic*. Lyophobic colloids involve particles that are solvophobic relative to the solvent in which they are dispersed; in water, such particles are thus hydrophobic (clays, metal halides like $AgI_{(s)}$ and $AgCl_{(s)}$, and emulsions of fat (as in whole milk)). Lyophilic colloids involve particles that are solvophilic; in water, such particles are hydrophilic (silica gel (hydrated silica), proteins, and "humic" materials).

25.2 COAGULATION

25.2.1 General

When some of the particles in a suspension stick together when they collide, the suspension is said to be undergoing coagulation. First, pairs of particles form, then triplets, and so on. A continued accretion of more and more particles to each particle agglomerate allows growth to a size such that random Brownian bombardment by the water molecules is no longer able to prevent the agglomerate from either sinking downwards or floating upwards.

Direct hydrogen bonding between lyophilic particles and water can make colloidal suspensions of such materials permanently stable, with essentially zero net coagulation. Colloidal suspensions of lyophobic particles are, on the other hand inherently unstable, and the rate of coagulation will always be greater than zero. If brought sufficiently close together by a mechanism such as

Brownian motion (Figure 25.1), lyophobic particles will stick together and squeeze out much of the intervening aqueous phase. To coagulate lyophilic particles, it is first necessary to convert them to lyophobic particles. In the case of "humic" materials in water, this can be accomplished by adding some strong acid, or a di- or trivalent metal ion; H^+ and metal ions can neutralize the negatively charged carboxylate and phenolate groups that provide "humic" material with its hydrophilicity and stability. This mechanism is involved when metal salt solutions of ferric chloride (with Fe^{3+}) and alum (with Al^{3+}) are added to raw drinking water during the initial steps of raw drinking water treatment.

While the rate of coagulation of a lyophobic colloid is always greater than zero, such suspensions can appear to be stable for months and longer. This is almost always the result of a stabilizing double layer on each particle. Similar particles suspended in the same aqueous medium have surface potential (ψ_o) values of the same sign, so that the particles are surrounded by electric fields that are mutually repulsive. Particles with $\psi_o < 0$ are surrounded by negative electric fields, and particles with $\psi_o > 0$ are surrounded by positive electric fields. For a given pair of particles, the mutually repulsive effects of the two double layers will tend to prevent the two particles from closely approaching one another.

The net potential ψ_{net} between two charged particles separated by distance d is determined by the interaction of the ion clouds of the two double-layers. When the two particles are not too close together, then at the midpoint of the interaction ($d/2$), it can be assumed that the individual $\psi(x)$ values behave approximately independently:

$$\psi_{net}\left(\frac{d}{2}\right) \approx \psi_1\left(\frac{d}{2}\right) + \psi_2\left(\frac{d}{2}\right). \tag{25.1}$$

If $\psi_1(x)$ and $\psi_2(x)$ do behave independently, they can be calculated using Eq.(24.60) or Eq.(24.83), as appropriate.

25.2.2 van der Waals Forces

When suspended particles are highly charged and have thick double layers (κ^{-1} is large), only a low percentage of the collisions will be of high enough energy to overcome the repulsive electrical fields on the particles and bring about direct particle: particle contact. But if such contact does occur, short range *van der Waals* forces can hold the particles together despite the electrostatic repulsion. Three types of interaction are typically included in the van der Waals class: dipole-dipole, dipole-induced dipole, and London dispersion (Figure 25.2).

When a molecule is composed of more than one type of atom, the different atoms will always exhibit at least slightly different affinities for the bonding electrons. For a given bond between two different atoms then, the more electron-attracting ("electronegative") end will be somewhat negative, and the less electronegative end will be somewhat positive. This permanent charge separation

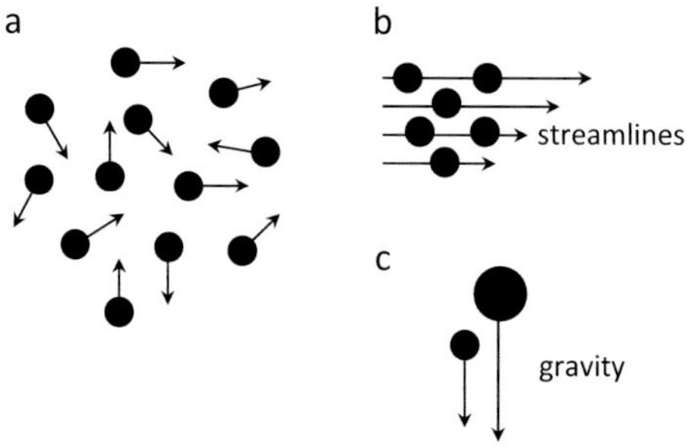

FIGURE 25.1 Collision mechanisms for colloidal particles. a. Brownian motion; b. velocity gradients; c. differential sedimentation.

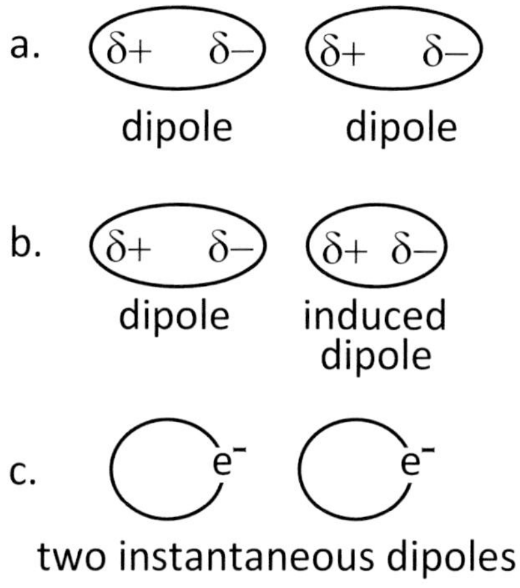

FIGURE 25.2 Types of attractive van der Waals interactions. a. Dipole-dipole; b. Dipole-induced dipole; c. London dispersion (synchronized momentary dipoles).

leads to a permanent dipole and a corresponding "dipole moment." Relative to many other molecules, water has a large molecular dipole moment. Permanent local dipoles can also exist at various places within a large molecule.

Two permanent dipoles will attract one another when they line up in a manner that has the negative end of one interacting with the positive end of the other. This is a *dipole-dipole* interaction. Hydrogen bonding is an extreme type of dipole-dipole interaction. A *dipole-induced dipole* interaction will occur whenever a permanent dipole from one molecule approaches a non-dipole molecule or atom. In this case, the electrical field associated with the permanent dipole distorts the electron distribution in the non-dipole, and induces a dipole with the same orientation. When neither of two species of interest has a permanent dipole, the species can still interact because the local electron circulations in the orbitals in the two different species create instantaneous oscillating dipoles that can get in sync, and then interact as two adjacent dipoles. For materials like hydrocarbon oils that are only weakly polar or completely non-polar (and so have little or no dipole moment), these so-called "London dispersion" interactions are by far the most important source of intermolecular attraction.

25.2.3 Coagulation Mechanisms

First, as noted, particles can collide with one another by Brownian diffusion. Second, if there are velocity gradients in the fluid, then particles are moving at different speeds along adjacent streamlines (Figure 25.1); particles in the faster streamlines will tend to collide with particles in adjacent, slower streamlines. This might seem to suggest that increasing mixing intensity (and therefore the mean velocity gradient) will monotonically increase coagulation rate. Actually, increasing mixing also increases the shear forces operating across the gradients. Thus, because van der Waals attractions are of limited strength, increasing shear forces will eventually tear apart large particle agglomerates, and so limit agglomerate size; during coagulation in water treatment plants, gentle mixing is used.

Third, during quiescent settling, if some of the particles or particle agglomerates are sufficiently large that they are undergoing gravitational settling (or flotation), they will catch up with and collide with other particles that are smaller moving at lower velocities. This is "differential sedimentation". It is effective when large particle agglomerates ("flocs") are settling out of the solution as in drinking water treatment with precipitation of a metal hydroxide material such as (am)$Fe(OH)_{3(s)}$ or $Al(OH)_{3(s)}$ followed by a quiescent settling stage.

25.3 ELECTROSTATIC REPULSION *VS.* VAN DER WAALS ATTRACTION

25.3.1 DLVO Theory

To examine the net energetics of two interacting electrically charged particles, in DLVO theory (Derjaguin, Landau, Verwey, and Overbeek), expressions for the electrostatic (double-layer) repulsion energy and the van der Waals attraction energy are examined. The repulsive contribution for particles separated by distance d (cm) can be approximated as (Berg, 2010)

$$\Phi_{\mathrm{R}}(d) \approx \frac{64\pi a n(\infty)kT}{\kappa^2}\Upsilon_{\mathrm{o}}^2 \exp\left[-\kappa d\right] \tag{25.2}$$

where $\Phi_{\mathrm{R}}(d)$ has units of J, a (cm) is the particle radius, $n(\infty)$ is the bulk solution summed concentration of the symmetrical electrolyte(s) making up the ionic strength (I), and κ is defined below. The parameter $n(\infty)$ gives the number of +/– pairs of ions per cm^3. For a $10^{-3}F$ solution of NaCl, $n(\infty) = 6.023 \times 10^{+20}$ pairs per cm^3. The constant Υ_{o} is given by

$$\Upsilon_{\mathrm{o}} = \frac{\exp\left(\dfrac{zF\psi_{\mathrm{o}}}{2RT}\right) - 1}{\exp\left(\dfrac{zF\psi_{\mathrm{o}}}{2RT}\right) + 1}. \tag{25.3}$$

From Chapter 23, for water at 298.15 K (25 °C),

$$\kappa = 3.29 \times 10^7 I^{1/2}\ \mathrm{cm}^{-1} \tag{23.52}$$

$$\kappa^{-1} = 3.04 \times 10^{-8} I^{-1/2}\ \mathrm{cm} \tag{23.53}$$

where I is the ionic strength.

The attractive van der Waals contribution to the potential energy of interaction between two particles is taken to be (Berg, 2010)

$$\Phi_{\mathrm{A}}(d) \approx -\frac{Aa}{12d} \tag{25.4}$$

where $\Phi_{\mathrm{A}}(d)$ has units of J, and A is the Hamaker constant. The value of A depends on the identity of the material in the interacting particles; a typical range is 10^{-20} to 10^{-19} J. $\Phi_{\mathrm{A}}(d)$ is negative because van der Waals interactions lower the potential energy of the system. Figures 25.3 and 25.4 show how $\Phi_{\mathrm{R}}(d)$ varies as a function of d for $|\psi_{\mathrm{o}}| = 75$ mV and a range of I values, and as a function of d for constant I and a range of $|\psi_{\mathrm{o}}|$ values.

The total, net potential energy of interaction $\Phi_{\mathrm{net}}(d)$ is given by the repulsive contribution plus the attractive contribution:

$$\Phi_{\mathrm{net}}(d) \approx \frac{64\pi a n(\infty)kT}{\kappa^2}\Upsilon_{\mathrm{o}}^2 e^{-\kappa d} - \frac{Aa}{12d}. \tag{25.5}$$

When $d = \infty$ and the particles are infinitely far apart, $\Phi_{\mathrm{net}}(d) = 0$. $\Phi_{\mathrm{net}}(d) > 0$ when the interacting particles are at an energy level that is higher than when $d = \infty$; $\Phi_{\mathrm{net}}(d) < 0$ when the interacting particles are at an energy level that is lower than when $d = \infty$. Figure 25.5 illustrates how $\Phi_{\mathrm{R}}(d)$, $\Phi_{\mathrm{A}}(d)$, and $\Phi_{\mathrm{net}}(d)$ vary with d when $|\psi_{\mathrm{o}}| = 75$ mV, $A = 10^{-19}$ J, $z = 1$, and $I = 0.001$ (by Eq.(23.68), $\kappa = 1.03 \times 10^6$ cm^{-1} at 25 °C/1 atm). The sign of ψ_{o} does not affect the magnitude of $\Phi_{\mathrm{R}}(d)$, so Figure 25.5 applies equally well to both $\psi_{\mathrm{o}} = +75$ mV and $\psi_{\mathrm{o}} = -75$ mV. As I increases and κ^{-1}

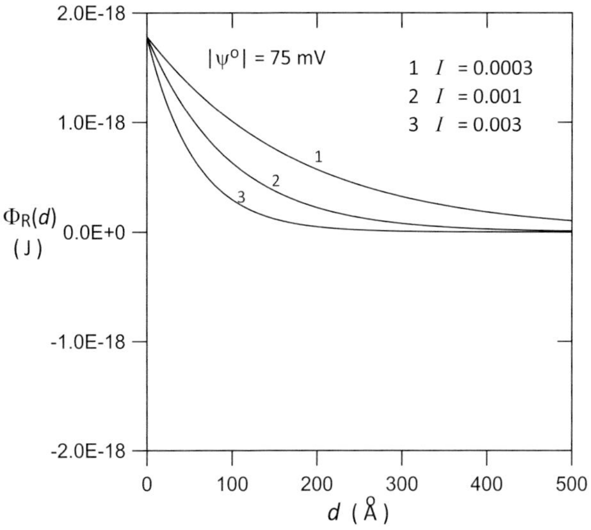

FIGURE 25.3 $\Phi_R(d)$ according to Eq.(25.2) for $|\psi_o| = 75$ mV and a range of I (κ) values (Electrolyte $z = 1$).

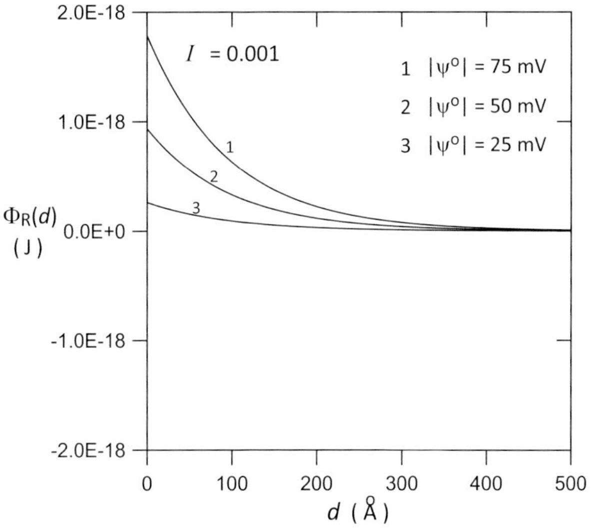

FIGURE 25.4 $\Phi_R(d)$ as a function of d for constant I (κ) and a range of $|\psi_o|$ values (Electrolyte $z = 1$).

decreases, the dependence of $\Phi_A(d)$ on distance remains constant. However, the double layer is compressed, and so $\Phi_R(d)$ as given by Eq.(25.2) decreases at all $d > 0$. Thus, increasing I decreases $\Phi_{net}(d)$ (including at any maximum), and so favors the coagulation process.

A basic equation from chemical kinetics states that the rate of a reaction is proportional to $\exp[-E_a/RT]$ where E_a (J/mol) is the height of the energy barrier separating the initial and final states. That is,

$$\text{reaction rate} \propto \exp[-E_a/RT]. \tag{25.6}$$

For the case at hand, the initial and final states correspond to d values of ∞ and 0, respectively. Thus, E_a here is the maximum value of $\Phi_{net}(d)$. The effects of I on $\Phi_{net}(d)$ are summarized in Figure 25.6 when $|\psi_o| = 75$ mV, $A = 10^{-19}$ J, and $z = 1$. Since increasing I decreases the maximum

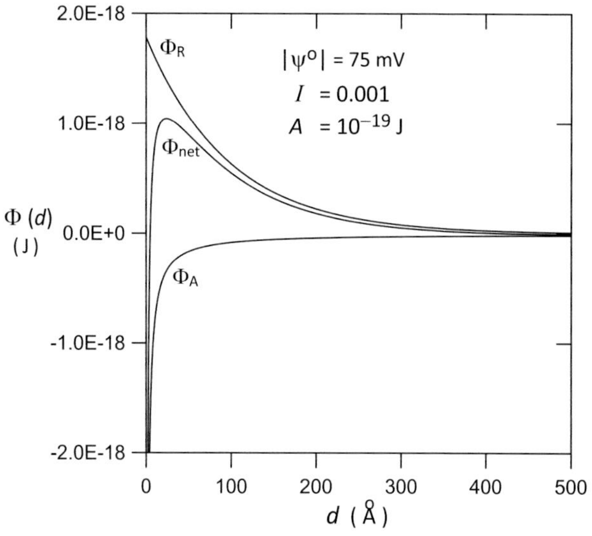

FIGURE 25.5 $\Phi_R(d)$, $\Phi_A(d)$, and $\Phi_{net}(d)$ as a function of d when $|\psi_o| = 75$ mV, $A = 10^{-19}$ J, and ionic strength $I = 0.001$ ($\kappa = 1.03 \times 10^6$ cm^{-1}) (Electrolyte $z = 1$).

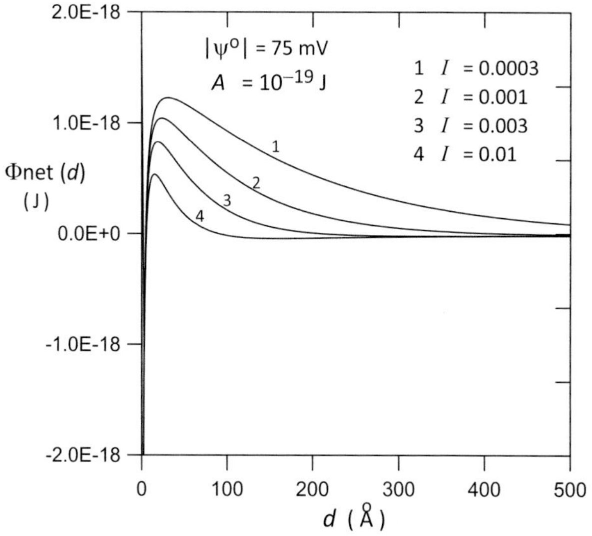

FIGURE 25.6 Effects of ionic strength I on $\Phi_{net}(d)$ when $|\psi_o| = 75$ mV and $A = 10^{-19}$ J (Electrolyte $z = 1$). Increasing I decreases the maximum value of $\Phi_{net}(d)$. Thus, increasing I decreases the height of the activation energy for coagulation.

value of $\Phi_{net}(d)$, and so increases the fraction of the particle collisions that can overcome the repulsion barrier and come into direct contact with $d = 0$. Since the collision efficiency fraction (a.k.a. "sticking factor") cannot be made larger than 1.0, the coagulation rate cannot be increased beyond that observed with $E_a = 0$. At that point, the rate becomes transport controlled (*e.g.*, diffusion controlled).

For given values of ψ_o, A, and z, as I is increased, the coagulation rate should achieve its maximum possible value when E_a becomes zero. This will happen when $\Phi_{net}(d)$ and $d(\Phi_{net}(d)/dd$ are simultaneously zero. Applying these two criteria to Eq.(25.5),

$$\frac{64\pi a n(\infty)kT}{\kappa^2}\Upsilon_o^2 e^{-\kappa d} = \frac{Aa}{12d} \tag{25.7}$$

and

$$\frac{64\pi a n(\infty)kT}{\kappa}\Upsilon_o^2 e^{-\kappa d} = \frac{Aa}{12d^2}. \tag{25.8}$$

Using Eq.(25.7) to substitute for $64n(\infty)kT\Upsilon_o^2 e^{-\kappa d}/\kappa$ into Eq.(25.8), the energy barrier will first be zero and the rate of coagulation will first be maximized for when for the $\Phi_{net}(d)$ it is true that

$$d = \kappa^{-1}. \tag{25.9}$$

Combining Eq.(25.9) with Eq.(25.7) gives

$$\frac{64\pi \mathrm{ccc}(\infty)kT}{\kappa^2}\Upsilon_o^2 e^{-1} = \frac{A}{12\kappa^{-1}} \tag{25.10}$$

where ccc(∞) is the "critical coagulation concentration" value of $n(\infty)$ for the symmetrical electrolyte that determines κ. Rearranging,

$$\kappa^3 = \frac{768\pi\ \mathrm{ccc}(\infty)kT}{A}\Upsilon_o^2 e^{-1}. \tag{25.11}$$

Recall now from Chapter 24 that

$$\kappa^2 = \frac{e_o^2 \sum z^2 n_i(\infty)}{\varepsilon\varepsilon_o kT}. \tag{25.12}$$

For a symmetrical electrolyte, the summation in Eq.(25.12) equals $2z^2n(\infty)$. Therefore when $n(\infty) = \mathrm{ccc}(\infty)$,

$$\kappa^3 = \frac{e_o^3 2^{\frac{3}{2}} z^3 (\mathrm{ccc}(\infty))^{\frac{3}{2}}}{(\varepsilon\varepsilon_o kT)^{3/2}}. \tag{25.13}$$

Substituting into Eq.(25.11) gives

$$\frac{e_o^3 2^{\frac{3}{2}} z^3 (\mathrm{ccc}(\infty))^{\frac{3}{2}}}{(\varepsilon\varepsilon_o kT)^{3/2}} = \frac{768\pi\,\mathrm{ccc}(\infty)kT}{A}\Upsilon_o^2 e^{-1}. \tag{25.14}$$

Rearranging,

$$\mathrm{ccc}(\infty) = \frac{(768\pi kT)^2(\varepsilon\varepsilon_o kT)^3}{A^2 e_o^6 2^3 z^6}\Upsilon_o^4 e^{-2}. \tag{25.15}$$

25.3.2 The Schulze–Hardy Rule

According to Eq.(25.15), for particles, the concentration of background electrolyte required to achieve the maximum rate of coagulation is proportional to z^{-6}. That is,

$$\mathrm{ccc}(\infty) \propto \frac{1}{z^6}. \tag{25.16}$$

For z = 1, 2, and 3 salts, ccc(∞) thus varies as (1/1) to (1/64) to (1/729). Coagulation driven by compression of particle double layers is thus predicted to be much more effective when salts of high z are used. That is, the ccc(∞) for a solution of $MgSO_4$ should be 64 times lower than the ccc(∞) for a solution of NaCl.

Prior to the development of the DLVO theory, an equation similar to Eq.(25.16) had been discovered empirically by Schulze (1882, 1883) and Hardy (1900a,b) for lyophobic colloids. The "Shulze–Hardy Rule" states that

$$\text{ccc}(\infty) \propto \frac{1}{z_{ci}^6} \tag{25.17}$$

where z_{ci} is the charge on the ion of the I-dominating electrolyte whose charge is counter (*i.e.*, opposite) to that of the lyophobic surface itself: when coagulating a negatively charged colloid using a solution of $MgCl_2$ to compress the double layer, $z_{ci} = 2$. According to the Schulze–Hardy Rule, the ccc(∞) values in coagulating a given negatively charged colloid would be predicted to be the same for $MgSO_4$ and $MgCl_2$.

Given the empirical reliability of Eq.(25.17), it was viewed as a success that DLVO theory should lead to Eq.(25.16). However, the charge magnitudes of both the co- and counter-ion affect κ, so the agreement between Eq.(25.16) and Eq.(25.17) is not perfect. The fact that the counter-ion plays a vastly more important role than the co-ion in coagulating lyophobic colloids has been rationalized by saying something like "it is the counter-ion that is pulled into the diffuse, aqueous portion of the double-layer, and therefore it is the counter-ion that neutralizes the surface charge layer". However, Chapter 24 has shown that the charge density q in the aqueous portion of the double-layer is due not only to a relative excess of counter-ions, but also to a relative deficiency of co-ions, though it was also pointed out there that that the excess of counter-ions can be significantly greater than the deficiency of co-ions because of the asymmetry of the exponential function involved (see Figure 24.10). In any case, in view of some of the questionable assumptions involved in developing Eq.(25.16), it is not worthwhile spending a lot of time discussing all the possibilities for how Eq.(25.16) and Eq.(25.17) might be brought into alignment. Indeed, DLVO theory neglects the possible effects of counter-ions adsorbed directly on the surface in a so-called "Stern layer". Second, it was assumed that the net potential at the midpoint between the particles is a simple sum of the individual, non-interacting potentials.

As far as the Schulze–Hardy rule itself is concerned, a significant number of the empirical observations supporting Eq.(25.17) have been based on situations in which a highly charged counterion is in a solution whose pH is such that only an exceedingly small percentage is uncomplexed with OH^-. In contrast, from Chapter 11, we know that adding $FeCl_3$ and $A1_2(SO_4)_2$ to water at pH 6 to 8 to coagulate colloidal matter in drinking water is not going to lead to the presence of much actual Fe^{3+} or Al^{3+}. In such cases, it is more important that a particle-enmeshing floe of (am)$Fe(OH)_{3(s)}$ or $Al(OH)_{3(s)}$ can form, as those very low concentrations of ions like Fe^{3+} and Al^{3+} can compress the double-layer and thereby destabilize the colloid.

REFERENCES

Berg JC (2010) *Introduction to Interfaces and Colloids – The Bridge to Nanoscience.* World Scientific Publishing, Singapore.

Hardy WB (1900a) A preliminary investigation of the conditions which determine the stability of irreversible hydrosols. *Proceedings of the Royal Society (London)*, **66**, 110–125.

Hardy WB (1900b) Colloid solutions, properties of gum Arabic and egg albumin solution. *Zeitshcrift für Physikalische Chemie*, **33**, 385–400.

Schulze H (1882) Schwefelarsen in wässriger Lösung. *Journal für praktische Chemie*, **25**, 431–452.

Schulze H (1883) Antimontrisulfid in wässriger Lösung. *Journal für praktische Chemie*, **27**, 320–332.

Index

B

C

T

U

V

W

Y

Z